REINFORCEMENT LEARNING AND APPROXIMATE DYNAMIC PROGRAMMING FOR FEEDBACK CONTROL

IEEE Press
445 Hoes Lane
Piscataway, NJ 08854

REINFORCEMENT LEARNING AND APPROXIMATE DYNAMIC PROGRAMMING FOR FEEDBACK CONTROL

Edited by

Frank L. Lewis
UTA Automation and Robotics Research Institute
Fort Worth, TX

Derong Liu
University of Illinois
Chicago, IL

IEEE
IEEE PRESS

IEEE Press Series on Computational Intelligence

WILEY
A JOHN WILEY & SONS, INC., PUBLICATION

Cover Illustration: Courtesy of Frank L. Lewis and Derong Liu
Cover Design: John Wiley & Sons, Inc.

Library of Congress Cataloging-in-Publication Data:

Reinforcement learning and approximate dynamic programming for feedback control / edited by Frank L. Lewis, Derong Liu.
p. cm.
ISBN 978-1-118-10420-0 (hardback)
1. Reinforcement learning. 2. Feedback control systems. I. Lewis, Frank L.
II. Liu, Derong, 1963-
Q325.6.R464 2012
003′.5—dc23

2012019014

CONTENTS

PART III FOUNDATIONS IN MDP AND RL

PREFACE

Modern day society relies on the operation of complex systems including aircraft, automobiles, electric power systems, economic entities, business organizations, banking and finance systems, computer networks, manufacturing systems, and industrial processes. Decision and control are responsible for ensuring that these systems perform properly and meet prescribed performance objectives. The safe, reliable, and efficient control of these systems is essential for our society. Therefore, automatic decision and control systems are ubiquitous in human engineered systems and have had an enormous impact on our lives. As modern systems become more complex and performance requirements more stringent, improved methods of decision and control are required that deliver guaranteed performance and the satisfaction of prescribed goals.

Feedback control works on the principle of observing the actual outputs of a system, comparing them to desired trajectories, and computing a control signal based on that error, which is used to modify the performance of the system to make the actual output follow the desired trajectory. The optimization of sequential decisions or controls that are repeated over time arises in many fields, including artificial intelligence, automatic control systems, power systems, economics, medicine, operations research, resource allocation, collaboration and coalitions, business and finance, and games including chess and backgammon. Optimal control theory provides methods for computing feedback control systems that deliver optimal performance. Optimal controllers optimize user-prescribed performance functions and are normally designed offline by solving Hamilton–Jacobi–Bellman (HJB) design equations. This requires knowledge of the full system dynamics model. However, it is often difficult to determine an accurate dynamical model of practical systems. Moreover, determining optimal control policies for nonlinear systems requires the offline solution of nonlinear HJB equations, which are often difficult or impossible to solve. Dynamic programming (DP) is a sequential algorithmic method for finding optimal solutions in sequential decision problems. DP was developed beginning in the 1960s with the work of Bellman and Pontryagin. DP is fundamentally a backwards-in-time procedure that does not offer methods for solving optimal decision problems in a forward manner in real time.

The real-time adaptive learning of optimal controllers for complex unknown systems has been solved in nature. Every agent or system is concerned with acting on its environment in such a way as to achieve its goals. Agents seek to learn how to collaborate to improve their chances of survival and increase. The idea that there is a cause and effect relation between actions and rewards is inherent in animal learning. Most organisms in nature act in an optimal fashion to conserve resources while achieving their goals. It is possible to study natural methods of learning and use them to develop computerized machine learning methods that solve sequential decision problems.

Reinforcement learning (RL) describes a family of machine learning systems that operate based on principles used in animals, social groups, and naturally occurring systems. RL methods were used by Ivan Pavlov in the 1860s to train his dogs. RL refers to an actor or agent that interacts with its environment and modifies its actions, or control policies, based on stimuli received in response to its actions. RL computational methods have been developed by the Computational Intelligence Community that solve optimal decision problems in real time and do not require the availability of analytical system models. The RL algorithms are constructed on the idea that successful control decisions should be remembered, by means of a reinforcement signal, such that they become more likely to be used another time. Successful collaborating groups should be reinforced. Although the idea originates from experimental animal learning, it has also been observed that RL has strong support from neurobiology, where it has been noted that the dopamine neurotransmitter in the basal ganglia acts as a reinforcement informational signal, which favors learning at the level of the neurons in the brain. RL techniques were first developed for Markov decision processes having finite state spaces. They have been extended for the control of dynamical systems with infinite state spaces.

One class of RL methods is based on the actor–critic structure, where an actor component applies an action or a control policy to the environment, whereas a critic component assesses the value of that action. Actor–critic structures are particularly well adapted for solving optimal decision problems in real time through reinforcement learning techniques. Approximate dynamic programing (ADP) refers to a family of practical actor–critic methods for finding optimal solutions in real time. These techniques use computational enhancements such as function approximation to develop practical algorithms for complex systems with disturbances and uncertain dynamics. Now, the ADP approach has become a key direction for future research in understanding brain intelligence and building intelligent systems.

The purpose of this book is to give an exposition of recently developed RL and ADP techniques for decision and control in human engineered systems. Included are both single-player decision and control and multiplayer games. RL is strongly connected from a theoretical point of view with both adaptive learning control and optimal control methods. There has been a great deal of interest in RL and recent work has shown that ideas based on ADP can be used to design a family of adaptive learning algorithms that converge in real-time to optimal control solutions by measuring data along the system trajectories. The study of RL and ADP requires methods from many fields, including computational intelligence, automatic control systems, Markov decision processes, stochastic games, psychology, operations research, cybernetics, neural networks, and neurobiology. Therefore, this book is interested in bringing together ideas from many communities.

This book has three parts. Part I develops methods for feedback control of systems based on RL and ADP. Part II treats learning and control in multiagent games. Part III presents some ideas of fundamental importance in understanding and implementing decision algorithm in Markov processes.

F.L. Lewis
Derong Liu

Fort Worth, TX
Chicago, IL

CONTRIBUTORS

Eduardo Alonso, School of Informatics, City University, London, UK

Charles W. Anderson, Department of Computer Science, Colorado State University, Fort Collins, CO, USA

Titus Appel, MARHES Lab, Department of Electrical & Computer Engineering, University of New Mexico, Albuquerque, NM, USA

Khalid Aziz, Department of Energy Resources Engineering, Stanford University, Stanford, CA, USA

Robert Babuska, Delft Center for Systems and Control, Delft University of Technology, Delft, The Netherlands

S.N. Balakrishnan, Department of Mechanical and Aerospace Engineering, Missouri University of Science and Technology, Rolla, MO, USA

Tamer Başar, Coordinated Science Laboratory, University of Illinois at Urbana-Champaign, Urbana, IL, USA

Dimitri Bertsekas, Department of Electrical Engineering and Computer Science, Massachusetts Institute of Technology, Cambridge, MA, USA

Shubhendu Bhasin, Department of Electrical Engineering, Indian Institute of Technology, Delhi, India

Shalabh Bhatnagar, Department of Computer Science and Automation, Indian Institute of Science, Bangalore, India

V.S. Borkar, Department of Electrical Engineering, Indian Institute of Technology, Powai, Mumbai, India

Lucian Busoniu, Université de Lorraine, CRAN, UMR 7039 and CNRS, CRAN, UMR 7039, Vandœuvre-lès-Nancy, France

Xi-Ren Cao, Shanghai Jiaotong University, Shanghai, China

W. Chen, Coordinated Science Laboratory, Department of Electrical and Computer Engineering, University of Illinois at Urbana-Champaign, Urbana, IL, USA

Vijay Desai, Industrial Engineering and Operations Research, Columbia University, New York, NY, USA

Gianluca Di Muro, Department of Mechanical Engineering and Materials Science, Duke University, Durham, NC, USA

Jie Ding, Department of Mechanical and Aerospace Engineering, Missouri University of Science and Technology, Rolla, MO, USA

Warren E. Dixon, Department of Mechanical and Aerospace Engineering, University of Florida, FL, USA

Louis J. Durlofsky, Department of Energy Resources Engineering, Stanford University, Stanford, CA, USA

Krishnamurthy Dvijotham, Computer Science and Engineering, University of Washington, Seattle, WA, USA

Michael Fairbank, School of Informatics, City University, London, UK

Vivek Farias, Sloan School of Management, Massachusetts Institute of Technology, Cambridge, MA, USA

Silvia Ferrari, Laboratory for Intelligent Systems and Control (LISC), Department of Mechanical Engineering and Materials Science, Duke University, Durham, NC, USA

Rafael Fierro, MARHES Lab, Department of Electrical & Computer Engineering, University of New Mexico, Albuquerque, NM, USA

Haibo He, Department of Electrical, Computer and Biomedical Engineering, University of Rhode Island, Kingston, RI, USA

Ali Heydari, Department of Mechanical and Aerospace Engineering, Missouri University of Science and Technology, Rolla, MO, USA

Dayu Huang, Coordinated Science Laboratory, Department of Electrical and Computer Engineering, University of Illinois at Urbana-Champaign, Urbana, IL, USA

S. Jagannathan, Electrical and Computer Engineering Department, Missouri University of Science and Technology, Rolla, MI, USA

Qing-Shan Jia, Department of Automation, Tsinghua University, Beijing, China

Yu Jiang, Department of Electrical and Computer Engineering, Polytechnic Institute of New York University, Brooklyn, NY, USA

Marcus Johnson, Department of Mechanical and Aerospace Engineering, University of Florida, FL, USA

Zhong-Ping Jiang, Department of Electrical and Computer Engineering, Polytechnic Institute of New York University, Brooklyn, NY, USA

Rushikesh Kamalapurkar, Department of Mechanical and Aerospace Engineering, University of Florida, FL, USA

Kenton Kirkpatrick, Department of Aerospace Engineering, Texas A&M University, College Station, TX, USA

J. Nate Knight, Numerica Corporation, Loveland, CO, USA

F.L. Lewis, UTA Research Institute, University of Texas, Arlington, TX, USA

Derong Liu, State Key Laboratory of Management and Control for Complex Systems, Institute of Automation, Chinese Academy of Sciences, Beijing, P.R. China

Chao Lu, Department of Electrical Engineering, Tsinghua University, Beijing, P. R. China

Ron Lumia, Department of Mechanical Engineering, University of New Mexico, Albuquerque, NM, USA

P. Mehta, Coordinated Science Laboratory, Department of Electrical and Computer Engineering, University of Illinois at Urbana-Champaign, Urbana, IL, USA

Sean Meyn, Department of Electrical and Computer Engineering, University of Florida, Gainesville, FL, USA

Ciamac Moallemi, Graduate School of Business, Columbia University, New York, NY, USA

Remi Munos, SequeL team, INRIA Lille – Nord Europe, France

Zhen Ni, Department of Electrical, Computer and Biomedical Engineering, University of Rhode Island, Kingston, RI, USA

Warren B. Powell, Department of Operations Research and Financial Engineering, Princeton University, Princeton, NJ, USA

L.A. Prashanth, Department of Computer Science and Automation, Indian Institute of Science, Bangalore, India

Danil Prokhorov, Toyota Research Institute North America, Toyota Technical Center, Ann Arbor, MI, USA

Armando A. Rodriguez, School of Electrical, Computer and Energy Engineering, Arizona State University, Tempe, AZ, USA

Brandon Rohrer, Sandia National Laboratories, Albuquerque, NM, USA

Keith Rudd, Laboratory for Intelligent Systems and Control (LISC), Department of Mechanical Engineering and Materials Science, Duke University, Durham, NC, USA

I.O. Ryzhov, Department of Decision, Operations and Information Technologies, Robert H. Smith School of Business, University of Maryland, College Park, MD, USA

John Seiffertt, Department of Electrical and Computer Engineering, Missouri University of Science & Technology, Rolla, MO, USA

Jennie Si, School of Electrical, Computer and Energy Engineering, Arizona State University, Tempe, AZ, USA

A. Surana, United Technologies Research Center, East Hartford, CT, USA

Hamidou Tembine, Telecommunication Department, Supelec, Gif sur Yvette, France

Emanuel Todorov, Applied Mathematics, Computer Science and Engineering, University of Washington, Seattle, WA, USA

Kostas S. Tsakalis, School of Electrical, Computer and Energy Engineering, Arizona State University, Tempe, AZ, USA

John Valasek, Department of Aerospace Engineering, Texas A&M University, College Station, TX, USA

K. Vamvoudaki, Center for Control, Dynamical-Systems and Computation, University of California, Santa Barbara, CA, USA

Benjamin Van Roy, Department of Management Science and Engineering and Department of Electrical Engineering, Stanford University, Stanford, CA, USA

Draguna Vrabie, United Technologies Research Center, East Hartford, CT, USA

Ding Wang, State Key Laboratory of Management and Control for Complex Systems, Institute of Automation, Chinese Academy of Sciences, Beijing, P.R. China

Zheng Wen, Department of Electrical Engineering, Stanford University, Stanford, CA, USA

Paul Werbos, National Science Foundation, Arlington, VA, USA

John Wood, Department of Mechanical Engineering, University of New Mexico, Albuquerque, NM, USA

Don Wunsch, Department of Electrical and Computer Engineering, Missouri University of Science & Technology, Rolla, MO, USA

Lei Yang, College of Information and Control Science and Engineering, Zhejiang University, Hangzhou, China

Qinmin Yang, State Key Laboratory of Industrial Control Technology, Department of Control Science and Engineering, Zhejiang University, Hangzhou, Zhejiang, China

Hassan Zargarzadeh, Embedded Systems and Networking Laboratory, Electrical and Computer Engineering Department, Missouri University of Science and Technology, Rolla, MI, USA

Dongbin Zhao, State Key Laboratory of Management and Control for Complex Systems, Institute of Automation, Chinese Academy of Sciences, Beijing, China

Qianchuan Zhao, Department of Automation, Tsinghua University, Beijing, China

Yanjia Zhao, Department of Automation, Tsinghua University, Beijing, China

Quanyan Zhu, Coordinated Science Laboratory, University of Illinois at Urbana-Champaign, Urbana, IL, USA

PART I

FEEDBACK CONTROL USING RL AND ADP

CHAPTER 1

Reinforcement Learning and Approximate Dynamic Programming (RLADP)—Foundations, Common Misconceptions, and the Challenges Ahead

PAUL J. WERBOS
National Science Foundation (NSF), Arlington, VA, USA

ABSTRACT

Many new formulations of reinforcement learning and approximate dynamic programming (RLADP) have appeared in recent years, as it has grown in control applications, control theory, operations research, computer science, robotics, and efforts to understand brain intelligence. The chapter reviews the foundations and challenges common to all these areas, in a unified way but with reference to their variations. It highlights cases where experience in one area sheds light on obstacles or common misconceptions in another. Many common beliefs about the limits of RLADP are based on such obstacles and misconceptions, for which solutions already exist. Above all, this chapter pinpoints key opportunities for future research important to the field as a whole and to the larger benefits it offers.

1.1 INTRODUCTION

The field of reinforcement learning and approximate dynamic programming (RLADP) has undergone enormous expansion since about 1988 [1], the year of the first NSF workshop on Neural Networks for Control, which evaluated RLADP as one of several important new tools for intelligent control, with or without neural networks. Since

Reinforcement Learning and Approximate Dynamic Programming for Feedback Control, First Edition. Edited by Frank L. Lewis and Derong Liu.

then, RLADP has grown enormously in many disciplines of engineering, computer science, and cognitive science, especially in neural networks, control engineering, operations research, robotics, machine learning, and efforts to reverse engineer the higher intelligence of the brain. In 1988, when I began funding this area, many people viewed the area as a small and curious niche within a small niche, but by the year 2006, when the Directorate of Engineering at NSF was reorganized, many program directors said "we all do ADP now."

Many new tools, serious applications, and stability theorems have appeared, and are still appearing, in ever great numbers. But at the same time, a wide variety of misconceptions about RLADP have appeared, even within the field itself. The sheer variety of methods and approaches has made it ever more difficult for people to appreciate the underlying unity of the field and of the mathematics, and to take advantage of the best tools and concepts from all parts of the field. At NSF, I have often seen cases where the most advanced and accomplished researchers in the field have become stuck because of fundamental questions or assumptions that were taken care of 30 years before, in a different part of the field. The goal of this chapter is to provide a kind of unified view of the past, present, and future of this field, to address those challenges. I will review many points that, though basic, continue to be obstacles to progress. I will also focus on the larger, long-term research goal of building real-time learning systems which can cope effectively with the degree of system complexity, nonlinearity, random disturbance, computer hardware complexity, and partial observability which even a mouse brain somehow seems to be able to handle [2]. I will also try to clarify issues of notation that have become more and more of a problem as the field grows more diverse. I will try to make this chapter accessible to people across multiple disciplines, but will often make side comments for specialists in different disciplines—as in the next paragraph.

Optimal control, robust control, and adaptive control are often seen as the three main pillars of modern control theory. ADP may be seen as part of optimal control, the part that seeks computationally feasible general methods for the nonlinear stochastic case. It may be seen as a computational tool to find the most accurate possible solutions, subject to computational constraints, to the HJB equation, as required by general nonlinear robust control. It may be formulated as an extension of adaptive control which, because of the implicit "look ahead," achieves stability under much weaker conditions than the well-known forms of direct and indirect adaptive control. The most impressive practical applications so far have involved highly nonlinear challenges, such as missile interception [3] and continuous production of carbon–carbon thermoplastic parts [4].

1.2 WHAT IS RLADP?

1.2.1 Definition of RLADP and the Task it Addresses

The term "RLADP" is a broad and an inclusive term, attempting to unite several overlapping strands of research and technology, such as adaptive critics, adaptive dynamic programming (ADP), approximate dynamic programming (ADP), and reinforcement learning (RL).

Because the history through 2005 was very complex [3, 4], it is easier to focus first on one of the core tasks that ADP attempts to solve. Suppose that we are given a stochastic system defined by:

$$X(t+1) = F(X(t), u(t), e_1(t)), \tag{1.1}$$

$$Y(t) = H(X(t), e_2(t)), \tag{1.2}$$

and our goal at every time t is to pick $u(t)$ so as to maximize:

$$J = \left\langle \sum_{\tau=t}^{T} U(Y(\tau), u(\tau))/(1+r)^{\tau-t} \right\rangle, \tag{1.3}$$

where r is a discount rate or interest rate, which may be zero or greater than zero, T is a terminal time, which may be finite or may be infinity, $X(t)$ represents the actual state of the system ("the objective real world") at time t, $Y(t)$ represents what we directly observe about the system at time t, $u(t)$ represents the actions or control we get to decide on at each time t, U represents our utility function, following the definitions of Von Neumann and Morgenstern [5], $e_1(t)$ and $e_2(t)$ are vectors or collections of random numbers, and <>is notation from physics for expectation value.

This task is called a Partially Observed Markov Decision Problem (POMDP), because any system of $X(t)$ governed by Equation (1.1) is a Markov process.

We are asked to develop methods which are general in that they work for any reasonable nonlinear or linear functions F and H, which may also be functions of unknown weights or parameters W. For a true intelligent system, we want to be able to maximize performance for the case where all our knowledge of F and G comes from experience, from the database $\{Y(\tau), u(\tau), \tau = 1 \text{ to } t\}$, and from an "uninformative" prior probability distribution $\Pr(F, H)$ for what they might be [8].

Modern ADP includes any efforts to use, analyze, or develop general-purpose methods to find good approximate answers to this optimization problem, using learning or approximation methods to cope with complexity. Of course, it also includes efforts aimed at the continuous time version of the problem, and hybrid versions with multiple time scales. It also includes efforts to develop general-purpose methods aimed at major special cases of this problem (such as the deterministic case, where there are no vectors e_1 or e_2), or the fully observed case, where $Y = X$), so long as they are useful steps toward the general case, developing the kinds of methods needed for the general case as well, as discussed in Section 1.2.2.

Reinforcement learning (RL) is much older than ADP. As a result, the term RL means different things to different people. RL includes early work by the psychologist Skinner and his followers, such as Harry Klopf, developing models of how animals learn to change their behavior in response to reward (r) and punishment. Some of the recent work in RL still follows that tradition, using "*r*" instead of "*U*," even when the system is intended to solve an optimization problem. Many computer scientists use the term RL to include systems that try to maximize a function $U(u)$ without considering the impact of present actions on future times. A more modern formulation of RL [1] is essentially the same as ADP, except that we are trying to design a system which observes $U(t)$ at each time t, without knowing the function $U(Y, u)$ which underlies it. This is logically just a special case of ADP, since we can add $U(t)$ itself into the list of observed variables included in Y.

Before 1968, research in RL and research related to dynamic programming were two entirely separate areas. Modern ADP dates back, at the earliest, to the 1968 paper [9] in which I first proposed that we can build reinforcement learning systems through adaptive approximation to the Bellman equation, as will be discussed in Section 1.2.2.

In a recent conference on modernizing the electric power grid [10], I heard a key researcher say "We really need new general methods to solve these complex multistage stochastic optimization problems, but ADP does not work so well. We need to develop better methods for this purpose." Logically this does not make sense, because we have *defined* this field to include any such "better methods." The researcher was actually thinking of one particular set of ADP tools, which do not represent the full capabilities of the field as it exists now, let alone in the future.

Equations (1.1)–(1.3) do not yet give a complete problem specification, but first I need to give some more explanations.

The utility function U is our statement of what *we*—the users, system engineers, or policy makers—want this computer system to do for us. It is a statement of our basic value system, our bottom line, and no computer system can tell us what it should be. There has been a huge amount of literature developed on how to choose U [11–13], which is essential to proper use of such tools. Many system engineers have observed that bad outcomes in large engineering projects result from bad choices of U, or failure to have some kind of U in mind, just as often as they do from failure to maximize U effectively over time. If we pick U just to make the optimization problem easy to solve, we very often will end up with a policy that does a poor job of accomplishing what we really care about.

In many practical applications, people say that they want to minimize something, like cost, instead of maximizing something. Of course, we could just set U equal to minus cost, or reverse the signs of the entire discussion with no real change in the mathematics. Here for simplicity I will stick with the positive formulation.

Many computer scientists have simplified the appearance of Equation (1.3) by defining:

$$\gamma = \frac{1}{1+r}, \tag{1.4}$$

where they call γ a "discount factor." Simplifying algebra often has its uses, but it is extremely important to remember what the real starting point here is [7]. The choice of r is part of our statement as users or policy makers of *what we want our system to do*. It is a key part of our overall utility function [12]. It is crucial in maintaining the connections between economics and the other domains where ADP is relevant.

In general, it is usually much easier to solve a myopic decision problem, where r is large (or T is finite), than to solve a problem with a commitment to the future, where r is zero. Yet the risks of myopia can be seen at many levels, from the instabilities it can cause in traditional adaptive control to the risks of extinction it poses for the human species as a whole in the face of very complex decision problems. In many situations, the best approach is to start by solving the problem for large r, and then ratcheting r down as close to zero as possible, step by step, by using the policies, weights and parameters of the previous step as initial guesses for the next step. This is one example of the general strategy which Barto has called "shaping," [4] and is now often called "transfer learning." This strategy has led to great results in many practical applications (like some of the earlier work of Jay Farrell), but it can also be used in more automated learning systems. In the limit, when r is zero, the basic theorems of dynamic programming need to be modified; that is why much of my earlier work on ADP [13–15] referred to the seminal work of Ron Howard [16] on that general case, rather than the work of Richard Bellman which it was built on.

Furthermore, in Equations (1.1)–(1.3), I have allowed for the possibility that the system state X and the observables Y may be complex structures, made up of continuous variables, discrete variables, or variables defined over a variable structure graph. The problem specification is also incomplete, insofar as I have said nothing about the possibilities for the functions F and H.

There is an important special case of Equation (1.1), in which: (1) X and Y are simply fixed vectors, $\underline{x}$ and $\underline{y}$, for any given learning system, each made up of a fixed number of continuous and (binary) discrete variables; (2) we implicitly assume that F and H are sampled from some kind of "uninformative prior" distribution, favoring smooth functions and so on, which is natural for such vectors, and does not favor strange higher-order symmetry relations between components of the vector. I call this "vector intelligence [2]," and say more about the crucial concept of uninformative priors in a recent talk for the Erdos Lectures series [8]. One of the two great challenges for basic research in RLADP in coming years is to prove theorems showing that certain families of RLADP design are "optimal" in some sense, in making full use of data from limited experience, in addressing the problem of vector intelligence. Of course, we also need to make such general-purpose tools widely available to the larger community, both for conventional and megacore computer hardware.

As recently as 1990 [4], I hoped that the higher intelligence of simple mammal brains could be matched by such an optimal vector intelligence; however, by 1998 [17], I realized there are fundamental general principles at work in those brains, which provide additional capabilities in handling spatial complexity, complex time structure, and a new level of stochastic creativity. This leads to a roadmap for more advanced ADP systems [2, 8], involving, in order: (1) more powerful systems for approximating complicated nonlinear functions, to better address complexity in X; (2) new extensions of the Bellman equation and methods for approximating *these extensions* efficiently, to address multiple time intervals; and (3) at the highest level, tight new coupling of the stochastic capabilities of the prediction system in the brain to the ADP circuits proper, supporting a higher level of creativity. Exciting as the new Bellman equations are, the issue of spatial complexity is currently the "other main fundamental challenge" for fundamental RLADP in the coming decades, and will itself require an enormous amount of new effort. Of course, these grand challenges also entail many important research opportunities to get us closer to the larger goals. Equally important are the grand challenges of using ADP in "reverse engineering the brain," and in using these methods to maximum benefit in the three crucial areas of achieving sustainability on earth (e.g., via new energy technologies), achieving economically sustainable human settlement of space (e.g., by solving crucial design and control problems in low-cost access to space), and by better supporting "inner space," realizing the full potential of human intelligence [13].

A major part of the work on RLADP and of dynamic programming deals with the special case where our decision-making system or control system can "see everything" in the plant to be controlled, such that $Y = X$. In that case, Equations (1.1)–(1.3) reduce to:

$$X(t+1) = F(X(t), u(t), e(t)), \tag{1.5}$$

$$J = \left\langle \sum_{\tau=t}^{T} U(X(\tau), u(\tau))/(1+r)^{\tau-t} \right\rangle. \tag{1.6}$$

This is called a Markhov Decision Process (MDP). Some of the theoretical literature on POMDP and MDP assumes that X is just a finite integer, between 1 and N, where N is the number of possible states of the system X. That special case might even be called "lookup table intelligence." It yields important theoretical insights but also some pitfalls, similar to the insights and pitfalls which come in physics when people assume, for simplicity, that the Hamiltonian operator H is a finite matrix [18]. Most of the work on POMDP and MDP in engineering (especially the work on practical applications) now assumes that X is actually a vector $\underline{x}$ in a vector space R^n, where n is the number of *state variables*. That work is well represented in this book, and in its two predecessors [4, 6]. (Unfortunately, some practical engineers refer to n at times as the number of "states.") Systems where X is a combination of a vector $\underline{x}$ and a set of discrete or binary variables are usually called "hybrid systems," or, more precisely, hybrid discrete-continuous systems.

Much of the new work on ADP in operations research (e.g., [19–22]) addresses the case where X is a combination of discrete and integer variables, subject to some combination of equality and inequality constraints. In the special case where $T = 1$, this is called a one-stage decision problem or "stochastic program." The deterministic case of that is called a mixed integer program. As of 2012, decisions about who generates electricity, from day to day or from 5 min interval to 5 min interval, to serve the large-scale electric power market, are made by Independent System Operators (ISO) such as PJM (see www.pjm.org), based on new mixed integer linear programming systems, which have proven that they can handle many thousands of variables quickly enough for practical use in real time; however, because power flows are highly nonlinear, new nonlinear algorithms for alternating current optimal power flow (such as those of Marija Ilic or James Momoh) have demonstrated great improvements in performance. Unfortunately, the power of all these methods in coping with many thousands of variables has depended on the development of general heuristic tricks, developed by insightful intuitive trial and error, which are mostly proprietary and held very tightly as secrets. The more open literature on stochastic programming, using open software systems like COIN-OR, has some important relations to ADP; there is an emerging community in "stochastic optimization" in OR which tries to bring both together, and to explore new stochastic methods for deterministic problems as well.

The current smart grid policy statement from the White House [23] states: "NSF is supporting research to develop a 'fourth generation intelligent grid' that would use intelligent system-wide optimization to better allow renewable sources and pluggable electric vehicles without compromising reliability or affordability [8]." The paper which it refers to [10] describes substantial opportunities for new applications of ADP, at all level of the electric power system, of great importance as part of the larger effort to make a transition to a sustainable global energy system.

1.2.2 Basic Tools—Bellman Equation, and Value and Policy Functions

Dynamic programming and ADP were originally developed for the MDP case, Equations (1.5) and (1.6).

Before we can build systems that learn to solve MDPs, we first need to define more precisely what we mean by "picking $u(t)$ to maximize J." Looking at Equations (1.1–3), you can see that J depends on future choices of u; therefore, we must make some kind of assumption about future choices of u, to state the problem more precisely. Intuitively, we want to pick $u(t)$ *at all times* to maximize J. We want to pick the value of $u(t)$ at time t, so as to maximize the best we can do in future times to keep on maximizing it.

To translate these intuitive concepts into mathematics, we must rely on the concept of a "policy." A policy π is simply a rule for saying what we will do under all circumstances, now and in the future:

$$u(t) = \pi(X(t)). \tag{1.7}$$

In earlier work [9], we sometimes called this a "strategy." In some RLADP systems, we do rely on explicit policies or "controllers" or "action networks," and in some we do not, but the *mathematical concept* of "policy" underlies all of ADP. In all of modern RLADP, we are trying to converge as closely as possible to the performance of the optimal policy, the policy which maximizes J as defined in Equation (1.6).

Following the notation of Bryson and Ho [24], we may define the function J^*:

$$J^* = \max_{\pi} \left\langle \sum_{\tau=t}^{T} U(X(\tau), \pi(X(\tau)))/(1+r)^{\tau-t} \right\rangle. \tag{1.8}$$

This leads directly to the key equation derived by Richard Bellman (in updated notation):

$$J^*(X(t)) = \max_{u(t)} \left\langle U(X(t), u(t)) + J^*(X(t+1))/(1+r) \right\rangle . \tag{1.9}$$

In dynamic programming (DP) proper, the user must specify a function U, the interest rate r, the function F shown in Equation (1.5), and the set of allowed values which $u(t)$ may be taken from. (When there are constraints on u, the Bellman Equation (1.9) takes care of them automatically; for example, it is still valid in the example where u is taken from the subspace of R^{n} defined by any number of constraints.) With that information, it is possible to solve for the function J^* which satisfies this equation. The original theorems of DP tell us that J^* exists, and that maximizing $U + J^*/(1+r)$ as shown in the Bellman equation gives us an optimal policy. Note, however, that this depends heavily on the assumption that we can observe the entire state vector (or graph) X; *when people use Equation (1.9) directly on partially observed or nonMarkhovian systems, they often end up with seriously inferior performance.* General-purpose ADP software needs to include tools to address this problem.

In control theory, Equation (1.9) is often called the Hamilton–Jacobi–Bellman (HJB) equation, though this can be misleading. Bellman was the first to solve the stochastic problem in Equations (1.5) and (1.6). Hamilton and Jacobi, in physics, derived an equation similar to Equation (1.9), for the deterministic case, where there is no random noise e and no reference to expectation values and no concept of cardinal utility U. Many of the most important results in nonlinear robust control [25] require that we "solve" (or approximate) the full stochastic Bellman's equation, and not just the deterministic special case.

The function $J^*(X)$ is often called the *value function*, and denoted as $V(X)$.

In theory, exact DP should be able to outperform all other methods for addressing Equations (1.5) and (1.6), including all problems in nonlinear robust control, except for just one difficulty—computational cost. The "curse of dimensionality" for exact DP, and with the simpler forms of RL, is well known. In my Harvard Ph.D. proposal of 1972, and in a journal paper published in 1977 [15], I proposed a general solution to this problem: why not *approximate* the function $J^*(X)$ with an approximation function or model $J^\wedge(X, W)$, with tunable weights W, as in the models we use to make predictions in statistics?

I also provided a general algorithm for training $J^\wedge$, which I called heuristic dynamic programming (HDP), which is essentially the same as what Richard Sutton called TD in his well-known work of 1990 [1].

Because HDP allows for any tunable choice of $J^\wedge$, it is possible to write computer code or pseudocode [4] for HDP which gives the user a wide range of choices, including options such as user-specified models (as in statistics packages), elastic fuzzy logic [26], or universal nonlinear function approximators such as Taylor series or neural networks [27, 28].

In the 1980s, I defined the new term "adaptive critic" to refer to any approximator of J^* or of something like J^*, which contains tunable weights or parameters W and for which we have a general method to train, adapt or tune those weights. More generally, any RLADP system is an adaptive critic system, if it contains such a system to approximate the value function "or something like the value function."

What else would we want to approximate, other than J^* itself?

When X is actually a vector $\underline{x}$ in R^n and F is differentiable, we usually get better results by approximating:

$$\underline{\lambda}(\underline{x}) = \nabla J^*(\underline{x}). \tag{1.10}$$

The $\underline{\lambda}$ vector is fundamental across many disciplines, and is essential to understanding how decisions and control fit together across different fields. For example, in control theory, the components of $\underline{\lambda}$ are often called the "costate variables." In the deterministic case, they may be found by solving the Pontryagin equation, which is closely related to the original Hamilton–Jacobi equation. In Chapter 13 of [4], I showed how to derive an equation for the stochastic case (a stochastic Pontryagin equation), simply by differentiating the Bellman equation I also specified an algorithm, Dual Heuristic Programming (DHP), for training a critic to approximate $\underline{\lambda}$, and showed that it converges to the right answer at least in the usual multivariate linear/stochastic case.

In economics, the "value" of a commodity x_i is its "marginal utility," which is essentially just λ_i; thus the output of a DHP critic is essentially just a kind of price signal. In applications like electric power, the $\underline{\lambda}$ vector is simply a price vector. It fits Dynamic Stochastic General Equilibrium economics better than the conventional "locational marginal cost" now used in pricing electricity, because it accounts for important effects like the impact of present decisions on future scarcity and congestion. For Freudian psychology, λ_i would represent the emotional value or affect attached to a variable or object, which Freud called "cathexis" or "psychic energy."

Early simulations studies verified that DHP has substantial benefits in performance over HDP [29]. Using a DHP critic, Balakrishnan reduced errors in hit-to-kill missile interception by more than an order of magnitude, compared to all previous methods, in work that has reached many applications. Ferrari and Stengel have also demonstrated its power in applications like reconfigurable flight control.

All of this is what one would expect, based on a simple analysis of learning rates, feedback, and the requirements of local control [4].

Nevertheless, HDP does have the advantage of ensuring that the approximation is globally consistent, and of being able to handle state variables that are not continuous. In order to combine the best advantages of DHP and HDP together, I proposed a different way to approximate J^* in 1987 [30], in which we keep updating the weights W so as to reduce the error measure:

$$\begin{aligned} E = {} & (J^\wedge(X(t+1))/(1+r) - (U(t) + J^\wedge(X(t), W))^2 \\ & + \sum_{i=1}^{n} \alpha_i \left(\frac{\partial J^\wedge}{\partial x_i}(X(t+1))/(1+r) - \left(\frac{\partial U(t)}{\partial x_i} + \frac{\partial J^\wedge}{\partial x_j}(X(t), W) \right) \right)^2, \end{aligned} \tag{1.11}$$

where n is the number of continuous variables in the state description X. I called this globalized DHP, or GDHP.

Note that I do not include W in the left-hand sides of these terms, the sides representing $J^*(t+1)$. In 1998 [5], I analyzed the stability and convergence properties of all these methods in the linear-quadratic case, with and without noise. When W is included on "both sides," in variations of the methods that I called "Galerkinized," the weights do not converge to the correct values in the stochastic case, though convergence is guaranteed robustly in the deterministic case. In Section 9 of [5], I described new variations of the methods which should possess both strong robust stability and converge to the right answer in the stochastic case; however, it is not clear whether the additional complexity is worthwhile in practical applications, or whether the brain itself possesses that kind of robust stability. For now, it is often best to train a controller or a policy using the original methods, and then verify convergence and stability for the outcome [3].

In the general case, GDHP requires the use of second-order backpropagation to compute all the derivatives [4, 30]. However, Wunsch et al. [31] have proposed a way to train an additional critic to approximate λ, intended to approximate GDHP without the need for second order derivatives. Liu et al. [32] have recently reported new stability results and simulations for GDHP. In applications like operations research, when we restrict our attention to special forms of value function approximator, GDHP reduces to a very convenient and simple form [22].

Besides approximating $J^*(X)$ or $\underline{\lambda}(X)$, it is also possible to approximate:

$$J'(X(t), u(t)) = Q(X(t), u(t)) = U(X(t), u(t)) + \underset{u(t+1)}{\text{Max}} \left\langle J^*(X(t+1))/(1+r)\right\rangle . \tag{1.12}$$

Note that J' and Q are the same thing. In 1989, Watkins used the term "Q" in his seminal Ph.D. thesis [33], addressing the case where X is an integer (a lookup table), in a process called "Q learning." In the same year [34], independently, I proposed the use of universal approximators to approximate J', in action-dependent HDP. Action-dependent HDP was the method used by White and Sofge [4] in their breakthrough control for the continuous production of thermoplastic carbon–carbon parts, a technology which is now of enormous importance to the aircraft industry, as in the recent breakthrough commercial airplane, the Boeing 787. This is also the approach taken in the recent work by Si, with some variation in how the training is done. Action-dependent versions of DHP [4] and GDHP also exist.

In addition to approximating the function J^* (or $\underline{\lambda}$ or Q), we often need to approximate the optimal policy by using an action function, action network or "actor":

$$u(t) = A(X(t), W, e). \tag{1.13}$$

In other words, if we cannot realistically explore the space of all possible policies π, we can use an approximation function or model A, and explore the space of those policies defined by tuning the weights W in that model. As in standard statistics or advanced neural networks, we can also add and delete terms in A automatically, based on what fits in the data. In most applications today, we do not actually include a random term (e) in the action network, but stochastic exploration of the physical world is an important part of animal learning, and may become more important in challenging future applications.

These fundamental methods are described in great detail in Handbook of Intelligent Control [4]. Many applications and variations and special cases have appeared since, in [6] and in this book, for example. But there is still a basic choice between approximating J^* (as in HDP), approximating $\underline{\lambda}$ (as in DHP) and approximating J^* while accounting for gradient error (as in GDHP), with or without a dependence on the actions $u(t)$ (as in the action-dependent variations).

A more complete review of the early history through 1998, as well as extensive robust stability results extending to the stochastic case, may be found in [35].

1.2.3 Optimization Over Time Without Value Functions

The value function of dynamic programming, $J^*(t)$, essentially represents the value of a whole complex range of possible future trajectories for later times, which cannot be tabulated explicitly because there are so many of them, because of the uncertainty.

But what about the case where there is no uncertainty (no random disturbance e), or where the uncertainty is so simple that we do not need to account for more than a few possible trajectories?

In those kinds of situations, we do not need to use value functions or ADP. We can try to solve for a fixed schedule of actions, $\{u(1), \ldots, u(T)\}$, by calculating the fixed trajectory they lead to, and calculating J explicitly, and minimizing it by use of classical methods. Bryson and Ho [24] give several methods for doing this. Recent work in receding horizon control and model predictive control (MPC) takes the same approach.

In those situations, we can also use the same kind of direct method to calculate the optimal weights W of an action network like Equation (1.13). This may not give as good performance, in theory, as finding the optimal schedule of action, but the resulting action network may carry over better to future decisions when the time horizon T moves further into the future.

Even when this kind of direct method works, the sheer cost of running forward, say, a hundred time points into the future, and optimizing over choices of trajectory, can be a major limiting factor. So long as F is any differentiable function, one can use backpropagation through time to calculate the gradient of J exactly at low cost, and complementary methods to make better use of those gradients (see Chapter 10 of [4]). This is especially easy when F is a neural network model of the plant; this may be called neural model predictive control (NMPC). Widrow has shown that a neural network action network trained by backpropagation through time [1] can learn amazing performance in the task of backing up a truck, both in simulation and on a physical testbed. This is a highly nonlinear task, and he proved that the system trained on a limited set of states could generalize to perform well across the entire range of possible starting states. Suykens et al. [36] have proven that this method offers far stronger robust stability guarantees than traditional neural adaptive control, which offers guarantees similar to traditional linear adaptive control [37]. NMPC has been extremely successful in many applications in the automobile industry, such as idle speed control at Ford and 15% improvement of mpg in the Prius hybrid [38]. At Neurodimensions (www.nd.com), Curt Lefebvre developed general software for NMPC, working with Jose Principe, and later reported that his intelligent control system is used in 25% by US coal-fired generators [10].

Can ADP work better in some automotive applications than NMPC? The answer is not clear. Using ADP, Sarangapani has shown mpg improvements more like 5–6% compared to the best conventional controllers for conventional and diesel car engines, and a 98% reduction in NO_x emissions. This is simply a different application.

These direct methods are not part of ADP, since they do not approximate the Bellman equation or any of its relatives. But at the same time, they provide a relatively simple, high quality comparison with ADP. Thus, they should be part of any general software package for useful ADP. Many computer scientists would call such methods "direct policy reinforcement learning," and include them in RLADP. There has been impressive success in dextrous robotics and in understanding human motor control using both approaches; Schaal [39] and Atkeson have mainly used the direct methods, while Todorov [40] has used ADP and hybrids of NMPC and ADP.

One of the most important methods in this class is "differential dynamic programming" (DDP) by Jacobson and Mayne [41]. DDP is not really a form of DP or ADP, but it does address the case where stochastic disturbances e exist. Their method for handling e is very rigorous, and very straightforward; it underlies all my own work on the nonlinear stochastic case. In essence, we simply treat the random variables as additional arguments to F and to other functions in the system. Most methods of handling random noise in reinforcement learning are like what statisticians call "unpaired comparisons" [42]; this method is like "paired comparisons," and far more efficient in using limited data and computational resources. More precisely, paired comparisons tend to reduce error by a factor of sqrt(N), where N is the number of simulated cases.

Just as neural networks and backpropagation through time can improve the performance of conventional MPC, they can also be used in Neural DDP—NDDP? DDP itself already includes a propagation of information backwards through time, based on a global Jacobian, but true backpropagation [43] does better by exploiting the structure of nonlinear dynamical systems.

For relatively simple problems, NMPC and DDP can sometimes outperform adaptive critic methods, but they cannot explain or replicate the kind of complexity we see in intelligent systems like the mammal brain. Unlike ADP, they do not offer a true brain-like real-time learning option. Even in using NMPC in receding horizon control, one can often improve performance by training a critic network to evaluate the final state $X(T)$.

1.3 SOME BASIC CHALLENGES IN IMPLEMENTING ADP

Among the crucial choices in using ADP are

- discrete time versus continuous time,
- how to account for the effect of unseen variables,

- offline controller design versus real-time learning,
- "model-based methods" like HDP and DHP versus "model free methods" like ADHDP and Q learning,
- how to approximate the value function effectively,
- how to pick $u(t)$ at each time t even knowing the value function,
- how to use RLADP to build effective cooperative multiagent systems?

Equations (1.1) through (1.11) all formulate the optimization problem in discrete time—from t to $t+1$, and so on, up to some final time T. This chapter will not discuss the continuous-time versions, in part because Frank Lewis will address that in his sections. In my own work, I have been motivated most of all by the ultimate goal of understanding and replicating the optimization and prediction capabilities of the mammal brain [2, 13]. The higher levels of the brain like the cerebral cortex and the limbic system are tightly controlled by regular "clock signals" broadcast from the nonspecific thalamus, enforcing basic rhythms of about 8 Hz ("alpha") and 4 Hz ("theta"). However, the brain also includes a faster lower-level motor control system, based on the cerebellum, running more like 200 Hz, acting as a kind of responsive slave to the higher system. Like some of Frank Lewis's designs, it does use a 4 Hz feedback/control signal from higher up, even though its actual operation is much faster. But perhaps some version of discrete-time ADHDP would be almost equivalent to Lewis's continuous time method here. Since I do not know, I will focus instead on the three other challenges here.

1.3.1 Accounting for Unseen Variables

Most engineering applications of ADP do not really fit Equations (1.5) and (1.6). Designs which assume that all the important state variables X are observed directly often perform poorly in the real world. In linear/quadratic optimal control, methods to cope with unobserved variables play a central role in practical systems [24]. In neural network control, variations between different engines and robots and generators play a central role [44]. In the brain itself, reconstruction of reality by the cerebral cortex plays a central role [2]. This is why we need to build general systems that address the general case given in Equations (1.1) through (1.3).

What happens when we apply a simple model-free form of ADP, like ADHDP or Q learning, directly to a system governed by Equations (1.1) and (1.2)? If we pick actions u so as to maximize $Q^\wedge(Y, u)$, our critic $Q^\wedge$ simply does not have the information we need to make the best decisions. The true Q function is a function of X and u, in effect.

The obvious way to do better is to create some kind of updated estimate of X, which may be called $\underline{x}^\sim$ (following [24]) or $X^\sim$. For example, the real-world successes of White and Sofge in using ADHDP [4] depended on the fact that they used Extended Kalman Filtering (EKF) to create that kind of estimate. They used ADHDP to train a critic which approximated Q (J') as a function of the $X^\sim$ and of u.

Notice that we do not really need to estimate the "true" value of X. Engineers sometimes worry that we could simply measure X in different units, and end up with a different function F, which still fits our observations of Y just as well as the original model. This is called an "identifiability" problem. But we do not really need to know the true "units" in which X should be measured. All we need is the *information* in X. More precisely, if we can develop any updated estimate of $g(X)$, where g is any invertible function, and use that as an input to our Critic, then we should be able to approximate J^* or Q as well as we could if we had an estimate of X itself.

There are four standard ways to develop a state estimate R which can be used as the main input to a critic network or action network, in the general nonlinear case:

- extended Kalman Filter (EKF),
- particle Filter,
- training a time-lagged recurrent network (TLRN) to predict X from Y in simulated data [45, 46],
- extracting the output of the recurrent nodes of a neural network used to model the plant (the system to be controlled) [4, 8].

EKF and particle filters are large subjects, beyond the scope of this chapter—but the work of Feldkamp and Prokhorov at Ford [46] strongly suggests that we do not really need to use them here. In simulation studies of automotive engines, they found that TLRNs and particle filters both performed much better than EKF in state estimation, but that TLRNs had much smaller computational cost. They also fit well on the new embedded control chips used in Ford cars.

But estimating X is still not the best way, in the general case.

In linear/quadratic control, it is well known [24] that we can get optimal control by using "dual control," in which $\underline{x}^{\sim}$ estimated by a Kalman Filter is treated as if it were the true state vector. But in the general nonlinear case, it is not so simple. For example, at the ADP conference in Cocoyoc, Mexico, in 2005, I faced a nonlinear optimization problem, in trying to find two small boys who had run off in the area. In theory, I could estimate the center of gravity of their possible locations, and walk straight there... but I was already close to that center of gravity (where they started from). I had to consider the probabilities of places where they might be, and plan to "buy information" on where they might be. In general, in studies of POMDP, it is well known that the optimal action $u(t)$ at any time depends on the entire "belief state," $\Pr(X)$, rather than just the most likely state.

Fortunately, the problem of optimal state estimation also depends on the entire belief state. Even in linear Kalman filtering [24], it is necessary to update matrices like "P" which represent the uncertainties in state estimation. Thus, the successful results in [46] tell us that the *information* in the belief state is encoded somehow into the recurrent nodes of the TLRN. That is necessary, to get to minimum square error in predicting the state variables, which training to least square error enforces. James Lo [47] has proven more general and formal mathematical results supporting this conclusion.

In summary, at the end of the day, it is good enough to use a TLRN or similar predictor [8] to model the plant, and feed its recurrent nodes into the critic and action networks, as described in [4].

1.3.2 Offline Controller Design Versus Real-Time Learning

Many control engineers start from a detailed model of the function F. Their goal is simply to derive a controller, like the action network of Equation (1.13), which optimizes performance or achieves nonlinear robust control [25]. Often, the most practical approach [10] is to maximize a utility function that combines the two objectives, by adding a term which represents value added by the plant to a term, which represents undesired breakdowns. The action network or controller could be anything from a linear controller, to a soft switching controller of settings for PID controllers, to elastic fuzzy logic [26], to a domain-dependent algorithm in need of calibration, to a neural network. The general algorithms for adapting the parameters W of $A(X, W)$ are the same, across all these choices of A. All the basic methods of ADP can be used in this way. In 1986 [48], I described one way to build a general ADP command within statistical software packages like Troll or SAS to make these kinds of capabilities more widely available.

On the other hand, the mammal brain is based entirely on real-time learning, at some level. At each time t, we observe $Y(t)$ and decide on $u(t)$, and adapt the networks in our brain, and move on to the next time period. Traditional adaptive control [36] takes a similar approach. To fully understand and replicate the intelligence of the mammal brain, we need to develop ADP systems that can operate successfully in this mode.

The tradeoffs between offline learning and real-time learning are not so simple as they appear at first. For example, in 1990, Miller [1] showed how he could train a neural network by traditional adaptive control to push an unstable cart forwards around a figure 8 track. When he doubled the weight on the cart, the network learned to rebalance and recover after only two laps around the track. This seemed very impressive, but we can do much better by training a recurrent network offline [44] to perform this kind of task. Given a database of cart behavior across a wide range of variations in weights, the recurrent network can learn to detect and respond immediately to changes in weight, even though it does not observe the weight of the cart directly. This trick, of "learning offline to be adaptive online," underlies the great successes of Ford in many applications. Presumably the brain itself includes a *combination* of recurrent neurons to provide this kind of quick adaptive response (like the "working memory" studied by Goldmann–Rakic and other neuroscientists) along with real-time learning to handle more novel kinds of changes in the world. It also includes some ability to learn from reliving past memories [8], which could also be used to enhance the prediction systems we use in engineering and social science.

Full, stochastic ADP also makes it possible to develop a controller for a new airplane, for example, which does not require the usual lengthy and expensive period of tweaking the controller when real-time data come in from flight tests. One can build a kind of "metamodel" which allows random variables to change coupling constants and other parameters which are uncertain, and then train the controller to perform well across the entire range of possibilities [49]. This is different, however, from the type of "metamodeling" which is growing in importance in operations research [22].

Strictly speaking, one might ask: "If brains use TLRNs, how can they adapt them in real time?" For engineering today, backpropagation through time (BTT) is the practical method to use, but for strict real-time operation one may use an "error critic" (Chapter 13 of [4]) to approximate it.

1.3.3 "Model-Based" Versus "Model Free" Designs

DHP requires some kind of model of F, in order to train the λ Critic. Even with HDP, a model of F is needed to find the actions $u(t)$ or train the action network A to maximize $U(t) + \langle J(t+1) \rangle)/(1+r)$.

On the other hand, ADHDP and Q learning do not require such a model. This is an important distinction, but its implications are often misunderstood.

Some researchers have even said that the "pure trial and error" character of ADHDP and Q learning make them more plausible as models of how brains work. This ignores the huge literature in animal learning and neuroscience, starting from Pavlov, showing that brains include neural networks which learn to predict or model their environments, and to perform some kind of state estimation as discussed in Section 1.3.1. Because we need state estimation anyway in the general case, it does not cost us much to exploit the information which results from it.

Intuitively, in ADHDP, we choose actions $u(t)$ which are similar to actions which have worked well in the past. In HDP, we pick actions $u(t)$ which are expected to lead to better outcomes at time $t + 1$, based on our understanding of what causes what (our model of F). The optimal approach, in principle, is to *combine* both kinds of information. This does require some kind of model of F, but also requires some way to be robust with respect to the uncertainties in that model. For brain-like real-time learning when we cannot use multistreaming [49], this calls for some kind of new hybrid of DHP (or GDHP) and ADHDP.

That will be an important area for research, especially when tools for DHP and HDP proper become more widely available and user-friendly.

Given a straight choice between DHP and ADHDP, the best information we have now [4, 29] suggests that DHP develops more and more advantage as the number of state variables grows. Balakrishnan and Lendaris have done simulation studies showing that the performance of DHP is not so dependent in practice to the details of the model of F. Nevertheless, more research would be useful in providing more systematic and analytical information about these tradeoffs.

In the stochastic case, when we build a model of F by training a neural network or some other universal approximator, we usually train the weights so as to minimize some measure of the error in predicting $Y(t)$ from past data [8]. We assume that the random disturbances added to each variable Y_i are distributed like the errors we see when trying to predict Y_i. When Y_i is a continuous variable, we usually assume that the disturbances follow a Gaussian a distribution. When Y_i is a binary variable, we use a logistical distribution and error function [42]. This usually works well enough when the sampling time (the difference between t and $t + 1$) is not so large. However, when Y is something complicated, this does not tell us how random disturbances affecting one component of Y correlate with random disturbances in other components. A more general method to model F as a truly stochastic system, accounting for such correlations, is the Stochastic Encoder–Decoder Predictor (SEDP), described in Chapter 13 of [4]. In essence, SEDP is the nonlinear generalization of a method in statistics called maximum likelihood factor analysis. It may also be viewed as a more rigorous and general version of the encoder–decoder "bottleneck" architectures which have shown great success in pattern recognition recently [50–53]; those networks may be seen as the nonlinear generalization of principal component analysis (PCA). I would claim that capabilities like those of SEDP will be necessary in explaining how the mammal brain builds up its model of the world, and that the giant pyramid cells of the cerebral cortex have a unique architecture well-designed to implement that kind of architecture [2].

SEDP and other traditional stochastic modeling tools assume that causality always moves forwards in time. This underlying assumption is implemented by assuming that the random disturbance terms may correlate with later values of the state variables, but not with earlier values [54, 55]. However, a re-examination of the foundations and empirical evidence of quantum mechanics suggests that quantum effects do not fit that assumption [56]. A serious literature now exists on mixed forwards–backwards stochastic differential equations [57]. Human foresight can cause what appear to be backwards causal effects in economic systems, leading to what George Soros calls "reflexive" situations and to multiple solutions in general equilibrium models such as the long-term energy analysis program whose evaluation I once led at the Department of Energy [58]. Hans Georg Zimmermann of Siemens has built neural network modeling systems which allow for time-symmetry in causation, which have been successful in real-world economic and trading applications. However, it would not be easy to build general ADP systems to fully account for time symmetry effects in quantum mechanics, because of the great challenges in building and modeling computing hardware that makes such capabilities available to the systems designer [59]. Hameroff has argued that the intelligence of mammal brains (and earlier brains) may be based on some kind of quantum computing effects, but, like most neuroscientists, I find it hard to believe that such capabilities exist at the systems level in what we can see in mammal brains.

1.3.4 How to Approximate the Value Function Better

Before the time of modern ADP, many people would try to approximate the value function J^* by using simple lookup tables or decision trees. This led to the famous curse of dimensionality. For example, if X consists of 10 continuous variables, and if we consider just 20 possible levels for each of these variables, the lookup table would contain 20^{10} numbers. It would be an enormous, unrealistic computational task to fill in that lookup table, but 20 possible levels might still be too coarse for useful performance.

For a general-purpose ADP system, one would want to be able to approximate J^* or $\underline{\lambda}$ with a universal nonlinear function approximator. Many such universal approximators have been proven to exist, for a large class of possible functions. For example, multivariable Taylor series have often been seen as the most respectable of universal approximators, because of their long history. However, Andrew Barron [27] has proven that all such "linear basis function approximators" show the same basic problem as lookup tables: they require an exponential growth in complexity (number of weights) as the number of state variables grows, for a given quality of approximation of a smooth function. Furthermore, because many polynomial terms correlate heavily with each other, Taylor series approximations usually have severe problems with numerical conditioning [60].

For many applications, then, the best choice for a Critic is the Multilayer Perceptron (MLP), the traditional, simple workhorse of neural network engineering. Barron has proven [27, 28] that the required level of complexity of an MLP grows only as a low power of the number of state variables in approximating a smooth function. This has worked very well in difficult nonlinear applications requiring, say, 10–20 variables. One can use simpler approximations like Gaussians or Radial Basis Functions or differentiable CMAC [4] or kernel methods or adaptive resonance theory in cases where there are fewer state variables or where the system seems to be restricted in practice to a few clusters within the larger state space. All of these neural network approximations also provide an effective way to make full use of the emerging most powerful "megacore" chips. Hewlett–Packard has even developed a software platform which allows use of powerful chips available today, with seamless automatic upgrade to the much more massive capabilities expected in just a few years [61, 62]. There is a reason to suspect that Elastic fuzzy logic [26] might offer performance similar to MLPs in function approximation, since it looks like an exponentiated version of MLP, but this has yet to be proved or disproved.

In 1996, Bertsekas and Tsitsiklis [63] proposed another way to approximate the value function, using user-specified basis functions ϕ_i:

$$J^\wedge(X, W) = \sum_{i=1}^{n} W_i \phi_i(X). \tag{1.14}$$

This approximation is still governed, in principle, by Barron's results on linear basis function approximators, but if the users supply basis functions suited to his particular problem, it might allow better performance in practice. This is analogous to those statistical software packages that allow users to estimate models which are "linear in parameters but nonlinear in variables." It is also analogous to the use of user-defined "features," like HOG or SIFT features, in traditional image processing. It also opens the door to many special cases of general-purpose ADP methods, and new special-purpose methods for the linear case, such as the use of linear programming to estimate the weights W [64].

Common sense and Barron's theorems suggest that much better performance could be achieved if the basis functions ϕ_i could be learned, or adaptively improved, instead of being held fixed—but that brings us back, in principle, to the problem of how to adapt a general nonlinear Critic $J^\wedge(X, W)$ or $\underline{\lambda}^\wedge(X, W)$, which existing methods already address. In image processing, "deep learning" of neural networks has already led to major breakthroughs, outperforming old feature-based methods developed over decades of intense domain-specific work, and has sometimes reduced error by a factor of two or more compared with traditional methods [50–53]. One would expect deep learning to yield similar benefits here, especially if there is more research in this area.

On the other hand, neurodynamic programming (NDP) as in Equation (1.14) could become an alternative general-purpose method for ADP, if powerful enough methods were found to adapt the basis functions ϕ_i here or in linearized DHP (LDHP), defined by:

$$\lambda_i^\wedge(X) = \sum_{j=1}^{n} W_{ij}\phi_j. \tag{1.15}$$

In discussions at an NSF workshop on ADP in Mexico, Van Roy suggested that we could solve this problem by using nonlinear programming somehow. James Momoh suggested that his new implementation of interior point methods for nonlinear programming might make this practical, but there has been no follow-up on this possibility. It leads to technical challenges to be discussed in Section 1.4. Other approaches to this problem have not worked out very well, so far as I know. Of course, it would be easy enough to implement an NDP/HDP hybrid:

- pick the functions ϕ_i themselves to be tunable functions $\phi_i(X, W^{[i]})$,
- adapt the outer weights W_i by NDP,
- treat the outer weights as fixed, adapt the inner weights $\{W^{[i]}\}$ by HDP.

The same could be done with NDP/DHP. However, so far as I know, no one has explored this kind of hybrid.

In presentations at the INFORMS conference years ago, Warren Powell reported that he could get much better results than NDP or deterministic methods, in solving a large-scale stochastic optimization problem from logistics, by using a different simple value function approximator [15]. MLPs are not suitable for tasks like this, in logistics or electric power at the grid level [10], where the number of state variables number in the thousands. They are not a plausible model of Critic networks in the brain, either, because the brain itself can also handle this kind of spatial complexity. In fact, they are not able to handle the level of complexity we see in raw images, in pattern recognition.

To address the challenge of spatial complexity, we now have a whole "ladder" of ever more powerful new neural network designs, starting with "Convolutional Neural Networks" [50–53], going up to Cellular Simultaneous Recurrent Networks (CSRN) [65, 66] and feedforward ObjectNets [67, 68], up to true recurrent object nets [65]. Convolutional Neural Networks are a special case of CSRN and of feedforward ObjectNets, while CSRN are a special case of ObjectNets. The Convolutional Networks, embodying a few simple design tricks, are the ones that have recently beaten many records in image recognition, phoneme recognition, video, and text analysis. (They should not be confused with the Cellular Neural Networks of Chua and Roska, which are essentially a type of chip architecture.) CSRNs have been shown to do an accurate job of value function approximation for the *generalized* maze navigation problem, a task in which convolutional neural networks failed badly. A machine using a feedforward object net as a Critic was the first system which actually learned master class performance in chess on its own, without a human telling it rules or guidelines for how to play the game, and without a supercomputer [68]. There is every reason to believe that proper use of these new types of neural network could have great power in general-purpose complex applications of ADP. All these network designs are also highly compatible with fast Graphical Processing Unit chips, with Cellular Neural Networks, and with new emerging chips based on memristors [61, 62].

ObjectNets themselves are essentially just a practical approximation of a more complex model of the way in which spatial complexity is handled in the mammal brain and many other vertebrate brains [2, 8].

Grossberg [69] has described models of how the required kind of multiplexing and resetting may be performed, after learning, in the cerebral cortex.

1.3.5 How to Choose $u(t)$ Based on a Value Function

When you have a working estimate of one of the basic value functions, J^* or $\underline{\lambda}$ or Q, the choice of $u(t)$ at time t may be trivial or extremely challenging, depending on the specific application. Of course, as your policy for choosing $u(t)$ changes, your estimate of the value function should change along with it. In my earlier work, I proposed concurrent adaptation of the critic network and the action network (and the network which models the world), as in the brain.

In some applications, even when F represents a serious and challenging engineering plant, the choice of actions $u(t)$ may be just a short list of discrete possibilities. For example, Liu et al. [70] describe a problem in optimal battery management where the task at each time t is to choose between three possible actions:

- charge the battery,
- discharge the battery,
- do neither.

This decision may be very difficult, because it may depend on what we expect the price of electricity to do, and it may depend on things like changing weather or driving plans if the battery is located in a home or a car. It may require a sophisticated stochastic model of price fluctuations and a model of battery lifetime. But despite that complexity, we may still face a simple choice in the end, at each time. In that case, the Q function would be very complex, but we can still just compute $Q^\wedge(X, u)$ for each of the three choices for u, and pick whichever gives the highest value of $Q^\wedge$. If we use a J^* critic, trained by HDP or GDHP, we can still use our model of the system to predict $U(t) + \langle J^*(t+1)/(1+r)\rangle$ for each of the three options, and pick the option which scores highest. The recent work by Powell and by Liu on the battery problem suggests that the quality of the control algorithm may often be crucial in deciding whether a new battery is money-maker or a money-loser for the people who buy it. Of course, Liu et al. used ADHDP rather than Q learning, because the state vector included continuous variables.

The choice of u is also relatively simple in cases where the sample time is relatively brief, such that the state X is not likely to change dramatically from time t to $t+1$, for feasible control actions $u(t)$. In such situations, Equation (1.5) may be represented accurately enough by an important special case:

$$\underline{x}(t+1) = F(\underline{x}(t)) + G(\underline{x}(t))\underline{u}(t), \tag{1.16}$$

with:

$$U(\underline{x}(t), \underline{u}(t)) = U_1(\underline{x}(t)) - c\underline{u}^T(t)R\underline{u}(t). \tag{1.17}$$

This is a variation of the "affine control problem" well-known in control theory. Intuitively, the penalty term $\mathrm{u^TRu}$ prevents us from using really large control vectors which would change $\underline{x}$ dramatically over one sample time. In many physical plants, like airplanes or cars, our controls are limited in any case, and it costs some energy to step on the gas too hard. If we use a $\underline{\lambda}$ Critic, as in DHP, the critic already tells us which way we want to move in state space. The optimal policy is simply to move in that direction, as fast as we reasonably can, limited by the penalty term. Using our estimate of $\underline{\lambda}$ from DHP, we can simply solve directly for the optimal value of $\underline{u}(t)$, by algebraically minimizing the quadratic function of $\underline{\lambda}$, G and R implied here. This trick is the basis of Balakrishnan's Single Network Adaptive Critic (SNAC) system [71], which has been studied further by Sarangapani [72]. Many of the recent stability theorems for real-time ADP have also focused on this case of affine control. Note that Balakrishnan's recent results with SNAC, which substantially outperform all other methods in the much-tested application of hit-to-kill missile interception, still use DHP for the training of the Critic.

For the more general case shown in Equation (1.5), we can simply train an action network $A(X, W)$ using the methods given in [4] and in many papers by researchers using that approach. Intuitively, the idea is to train the parameters W so that the resulting $u(t)$ performs the maximization shown in the Bellman Equation (1.9). Regardless of whether A is a neural network or some other differentiable system, backpropagation can be used to calculate the gradient of the function to be maximized with respect to all of the weights, in an efficient and accurate closed form calculation [42]. So long as this maximization problem is convex, a wide variety of gradient-based methods should converge to the correct solution.

When the optimization problem for $u(t)$ in Equation (1.9) is very complicated and nonconvex, more complicated methods may be needed. This often happens when the time interval between t and $t+1$ is large. For example, in large logistic problems, $u(t)$ may represent a plan of action for day t, and T may be chosen to be $t+7$, a week-ahead planning problem. There are several possible approaches here, some available today and some calling for future research.

In electric power, Marija Ilic has recently shown that large savings are possible if large-scale system operators calculate optimal power, voltage and frequency instructions for all generators across the system, every 5 min or so, using a nonlinear one-stage optimization system called "AC optimal power flow" (ACOPF). James Momoh also developed an ACOPF system for the Electric Power Research Institute many years ago [6]. These ACOPF systems already provide good practical approximate solutions to the highly complex, nonlinear problem of maximizing some measure of utility in this domain. One can apply ADP here simply by training a Critic (like an ObjectNet trained by DHP or GDHP), and using the existing ACOPF package to find the optimal $u(t)$. In effect, this adds a kind of foresight capability to the ACOPF decisions. Momoh and I have called this "dynamic stochastic OPF" (DSOPF) [6].

Venayagamoorthy has recently developed a parallel chip implementation of his ObjectNet ADP system [66], which he claims is fast enough to let us unify the new ACOPF capabilities of decisions made every 15 min. with grid regulation decisions which must be made every 2 s, to stabilize frequency and phase and voltage. If this is possible, it would allow local value signals (like the λ outputs of DHP, which value specific outputs form specific units, like price signals in economics) to manage the power electronics of renewable energy systems, so that they shift from being a major cost to the grid to a major benefit; if so, this would have major implications for the economics of renewable energy.

Nonconvexity becomes especially important for biological brains that address the problem of optimization over multiple time intervals [2, 17]. These are similar to robots which need to decide on when to invoke "behaviors" or "motor primitives"; such decisions have a direct impact which goes all the way from time t to the completion of the action, not just to $t + 1$. Use of Action Networks, without additional systems, can lead to suboptimal decisions for such behaviors. In practical terms, this simply implies that our controller has lack of creativity in finding new, out of the box (out of the current basin of attraction) options for higher-level decisions. In [2], I proposed that the modern mouse is more creative than the dinosaur, because of a new system for cognitive mapping of the space of possible decisions, exploiting certain features in its six-layer cerebral cortex that do not exist in the reptile. This is an important area for research, but several steps beyond the useful systems we can build today. It is not even clear how much creativity we would want our robots to have. For the time being, ideas such as stochastic optimization, metamodelling [48], brain-like stochastic search and Powell's work on the exploration gradient present more tangible opportunities.

Nonconvexity can be especially nasty when we are developing multistage versions of problems which are currently treated as single-stage linear problems using Mixed Integer Linear Programming (MILP). Unlike ACOPF, the MILP packages do not have the ability to maximize a general *nonlinear* utility function. Thus for many applications today, the easiest starting point for overcoming myopia and using ADP is to use the existing packages as the Action Network, and train a Critic network to define the actual objective function which the single-stage optimizer is asked to maximize [19, 22].

It is difficult to document and prove the performance tradeoffs in this case, because the leading MILP system (Gurobi)—like the ACOPF systems of Ilic and Momoh—is highly proprietary. Within linear programming, two major methods have competed through the years—the classical simplex method and the the interior point method pioneered by Karmarkar and Shanno [73]. The simplex method is very specific to the linear case. There are parallelized versions that make effective use of computers with 1–14 processors or so. The interior point method is more compatible with massively parallel processors and with nonlinearity. Shanno even suggested in 1988 [1] that some varieties or adaptations of interior point methods might work better than anything known today to speed up offline learning in neural networks. At that time, interior point was also beginning to perform better on large-scale linear programming problems. But in recent years, Gurobi has developed a variety of highly proprietary heuristics for using simplex for MILP, which outperform what they have for interior point.

Breakthrough performance in this area may well require a new emphasis on interior point methods, using open-source packages like COIN-OR on new more massively parallel computing platforms [62, 63]. Because major users of MILP worry a lot more about physical clock time than they do about the number of processors, massively parallel interior point methods should be able to overtake the classical methods now in use based on simplex. The nonlinear versions of these methods should interface well with nonlinear critics and with neural networks, both in physical interface and at the mathematical level.

1.3.6 How to Build Cooperative Multiagent Systems with RLADP

Multiagent systems (MAS) have grown more popular and more important in recent years. In practice, there are two general types of multiagent systems:

- systems which use distributed control to maximize some kind of global performance of the system as a whole,
- systems truly intended to balance the goals of different humans, with different goals.

For distributed control, it is extremely important to remember that model networks, action networks, and critics can all be *networks*. In other words, global ADP mathematics automatically gives us a recipe for how to manage and tune a global system made up of widely distributed parts.

In some applications, like electric power [10], it has become fashionable to assign a complex decision problem to a large number of independent agents, commonly agents trained by reinforcement learning. It is common to show by simulation or mathematics how these kinds of systems may converge to a Nash "solution." Unfortunately, many researchers do not understand that a Nash equilibrium is not really an optimal outcome or solution in the general case. It is well-known in game theory that Nash equilibria are commonly far inferior to Pareto optima, which is what we really want in multiplayer systems. In the design of games or of markets, we do often work hard to get to those special cases where a Nash equilibrium would be close to a Pareto optimum. Of course, economics has a lot to tell us about that kind of challenge.

In artificial systems, one can avoid these difficulties by designing a distributed system to be one large ADP system, mathematically, even though the components function independently. With large markets involving people, like power grids, one can use a DHP critic at the systems operator level [10] (or gradients of a GDHP critic) to determine price signals, which can then be sent to independent ADP agents of homeowners that respond to those prices, to organize a market.

In the special case where there, only two actors, in total opposition to each other, one arrives at a Hamilton Jacobi Isaacs equation (HJI). That case is important to higher-order robust control and to many types of military system. Frank Lewis has recently published results generalizing ADP to that case, and begun exploring ADP to seek Pareto optima in multiplayer games of mixed cooperation and conflict.

There has been substantial progress in the electric power sector recently [10] to avoid pathological gaming and Nash equilibrium effects in the power sector. In a global economy which is currently in a Nash equilibrium, far inferior to the sustainable growth which should be possible, it is interesting to consider how optimization approaches might help in getting us closer to a Pareto optimum. More generally, the large complex networks that move electricity, communications, freight, money, water are among the areas where multiagent extensions of ADP have potential to help.

DISCLAIMER

The views herein represent no one's official views but the chapter was written on U.S. government time.

REFERENCES

1. W.T. Miller, R. Sutton, and P. Werbos, editors. *Neural Networks for Control*, MIT Press, Cambridge, MA, 1990.
2. P. Werbos, Intelligence in the brain: a theory of how it works and how to build it. *Neural Networks*, 22(3):200–212, 2009. Related material is posted at www.werbos.com/Mind.htm.
3. S.N. Balakrishnan, J. Ding, and F.L. Lewis. Issues on stability of ADP feedback controllers for dynamical systems. *IEEE Transactions on Systems, Man, and Cybernetics, Part B: Cybernetics*, 38(4):913–917, 2008. URL: http://ieeexplore.ieee.org/stamp/stamp.jsp?tp = &arnumber = 4554210&isnumber = 4567535.
4. D. White and D. Sofge, editors. *Handbook of Intelligent Control*, Van Nostrand, 1992. The prefaces and three chapters in the government public domain are posted at www.werbos.com/Mind.htm#mouse.

5. P. Werbos. *Stable Adaptive Control Using New Critic Designs.* xxx.lanl.gov: adap-org/9810001 (October). 1998
6. J. Si, A.G. Barto, W.B. Powell, and D. Wunsch, editors. *Handbook of Learning and Approximate Dynamic Programming* (IEEE Press Series on Computational Intelligence), Wiley-IEEE Press, 2004.
7. J.V. Neumann and O. Morgenstern. *The Theory of Games and Economic Behavior*, Princeton University Press, Princeton, NJ, 1953.
8. P. Werbos. *Mathematical foundations of prediction under complexity*, Erdos Lecture series, 2011. Available at http://www.werbos.com/Neural/Erdos_talk_Werbos_final.pdf
9. P. Werbos. The elements of intelligence. Cybernetica (Namur), No. 3, 1968.
10. P.J. Werbos. Computational Intelligence for the smart grid-history, challenges, and opportunities, *Computational Intelligence Magazine, IEEE*, 6(3):14–21, 2011. URL: http://ieeexplore.ieee.org/stamp/stamp.jsp?tp = arnumber = 5952103&isnumber = 5952082.
11. H. Raiffa. *Decision Analysis*, Addison-Wesley, Reading, MA, 1968.
12. P. Werbos. Rational approaches to identifying policy objectives. *Energy: The International Journal*, 15(3/4):171–185, 1990.
13. P. Werbos. Neural networks and the experience and cultivation of mind, *Neural Networks*, 32:86–85, 2012.
14. P. Werbos. Changes in global policy analysis procedures suggested by new methods of optimization, *Policy Analysis and Information Systems*, 3(1): 1979.
15. P. Werbos. Advanced forecasting for global crisis warning and models of intelligence, *General Systems Yearbook*, 1977, p. 37.
16. R. Howard. *Dynamic Programming and Markhov Processes*, MIT Press, Cambridge, MA, 1960.
17. P. Werbos. A brain-like design to learn optimal decision strategies in complex environments. In M. Karny, K. Warwick, and V. Kurkova, editors. *Dealing with Complexity: A Neural Networks Approach.* Springer, London, 1998. Also in S. Amari and N. Kasabov, *Brain-Like Computing and Intelligent Information Systems.* Springer, 1998. See also international patent application #WO 97/46929, filed June 1997, published December 11.
18. T. Kato. *Perturbation Theory for Linear Operators*, Springer, Berlin, 1995.
19. W.B. Powell. *Approximate Dynamic Programming: Solving the Curses of Dimensionality*, 2nd edition, Wiley Series in Probability and Statistics, 2011.
20. B. Palmintier, M. Webster, J. Morris, N. Santen, and B. Ustun. 2010. *ADP Toolbox—A Modular System for Rapid Experimentation with Dynamic and Approximate Dynamic Programming*, Massachusetts Institute of Technology, Cambridge, MA.
21. C. Cervellera, A. Wen and V.C.P. Chen. Neural network and regression splinevalue function approximations for stochastic dynamic programming. *Computers and Operations Research*, 34, 70–90, 2007.
22. L. Werbos, R. Kozma, R. Silva-Lugo, G.E. Pazienza, and P. Werbos. Metamodeling and critic-based approach to multi-level optimization, *Neural Networks*, 32:179–185, 2012.
23. A POLICY FRAMEWORK FOR THE 21st CENTURY GRID: *Enabling Our Secure Energy Future* http://www.whitehouse.gov/sites/default/files/microsites/ostp/nstc-smart-grid-june2011.pdf
24. A. Bryson and Y.C. Ho. *Applied Optimal Control*, Ginn, 1969.

25. J.S. Baras and N.S. Patel. Information state for robust control of set-valued discrete time systems, *IEEE, Proceedings of 34th Conference Decision and Control (CDC)*, 1995. p. 2302.
26. P. Werbos. Elastic fuzzy logic: a better fit to neurocontrol and true intelligence. *Journal of Intelligent and Fuzzy Systems*, 1:365–377, 1993. Reprinted and updated in M. Gupta, ed, *Intelligent Control*, IEEE Press, New York, 1995.
27. A.R. Barron. Universal approximation bounds for superpositions of a sigmoidal function, *IEEE Transactions on Information Theory*, IT-39:930–944, 1993.
28. A.R. Barron. Approximation and estimation bounds for artificial neural networks, *Machine Learning*, 14(1):113–143, 1994.
29. D.V. Prokhorov, R.A. Santiago, and D.C. Wunsch II. Adaptive critic designs: a case study for neurocontrol, *Neural Networks*, 8(9):1367–1372, 1995.
30. P. Werbos. Building and understanding adaptive systems: a statistical/numerical approach to factory automation and brain research, *IEEE Transactions of SMC*, 17(1): 1987.
31. D.V. Prokhorov and D.C. II Wunsch. Adaptive critic designs. *Neural Networks, IEEE Transactions on*, 8(5):997–1007, 1997. URL: http://ieeexplore.ieee.org/stamp/stamp.jsp?tp = &arnumber = 623201&isnumber = 13541.
32. D. Liu, D. Wang, and D. Zhao. Adaptive dynamic programming for optimal control of unknown nonlinear discrete-time systems, *2011 IEEE Symposium on, Adaptive Dynamic Programming And Reinforcement Learning (ADPRL)*, vol., no., pp. 242–249, 2011 11–15 April doi: 10.1109/ADPRL.2011.5967357 URL: http://ieeexplore.ieee.org/stamp/stamp.jsp?tp=& arnumber=5967357& isnumber=5967347.
33. C.J.C.H., Watkins. Learning from delayed rewards. Ph.D. thesis, Cambridge University, 1989.
34. P. Werbos. Neural networks for control and system identification. In: IEEE Conference on Decision and Control (Florida), IEEE, New York. 1989.
35. P. Werbos. *Stable Adaptive Control Using New Critic Designs.* xxx.lanl.gov: adap-org/9810001(October 1998).
36. J.A. Suykens, B. DeMoor, and J. Vandewalle. NLQ theory: a neural control framework with global asymptotic stability criteria *Neural Networks*, 10(4):615–637, 1997.
37. K. Narendra and A. Annaswamy. *Stable Adaptive Systems*, Prentice-Hall, Englewood, NJ, 1989.
38. D. Prokhorov. Prius HEV neurocontrol and diagnostics. *Neural Networks*, 21(2–3):458–465, 2008.
39. S. Schaal. Learning Motor Skills in Humans and Humanoids, plenary talk presented at International Joint Conference on Neural Networks 2011 (IJCNN2011). Video forthcoming from the IEEE CIS Multimedia tutorials center, currently http://ewh.ieee.org/cmte/cis/mtsc/ieeecis/video_tutorials.htm.
40. E. Todorov. Efficient computation of optimal actions. *PNAS* 106:11478–11483, 2009.
41. D. Jacobson and D. Mayne. *Differential Dynamic Programming*, American Elsevier, 1970.
42. T.H. Wonnacott and R.J. Wonnacott. *Introductory Statistics for Business and Economics*, 4th edition, Wiley, 1990.

43. P. Werbos. Backwards differentiation in AD and neural nets: Past links and new opportunities. In H.M. Bucker, G. Corliss, P. Hovland, U. Naumann, and Boyana Norris, editors. *Automatic Differentiation: Applications, Theory and Implementations*, Springer, New York, 2005.

44. P. Werbos. Neurocontrollers. In J. Webster, editor. *Encyclopedia of Electrical and Electronics Engineering*, Wiley, 1999.

45. Y.H. Kim and F.L. Lewis. *High-Level Feedback Control with Neural Networks,* World Scientific Series in Robotiocs and Intelligent Systems, Vol. 21, 1998.

46. L.A. Feldkamp and D.V. Prokhorov. Recurrent neural networks for state estimation, in. Proceedings of the Workshop on adaptive and learning systems, Yale University (Narendra ed.), 2003. Posted with authors' permission at http://www.werbos.com/FeldkampProkhorov2003.pdf. Also see http://home.comcast.net/~dvp/.

47. J.T.-H. Lo. Synthetic approach to optimal filtering. *IEEE Transactions on Neural Networks*, 5(5):803–811, 1994. See also the relaxation of required assumptions in James Ting-Ho Lo and Lei Yu, Recursive Neural Filters and Dynamical Range Transformers, Invited paper, *Proceedings of The IEEE,* 92(3):514–535, March 2004.

48. P. Werbos. Generalized information requirements of intelligent decision-making systems, *SUGI 11 Proceedings*, Cary, NC: SAS Institute, 1986.

49. L. Feldkamp, D. Prokhorov, C. Eagen, and F. Yuan. *Enhanced Multi-Stream Kalman Filter Training for Recurrent Networks*. In J. Suykens and J. Vandewalle, editors. *Nonlinear Modeling: Advanced Black-Box Techniques*. Kluwer Academic, 1998, pp. 29–53. URL: http://home.comcast.net/~dvp/bpaper.pdf. See also L.A. Feldkamp, G.V. Puskorius, and P.C. Moore, Adaptive behavior from fixed weight networks. *Information Sciences*, 98(1–4):217–235, 1997.

50. K. Kavukcuoglu, P. Sermanet, Y-Lan Boureau, K. Gregor, M. Mathieu, and Y. LeCun. Learning convolutional feature hierachies for visual recognition, *Advances in Neural Information Processing Systems (NIPS 2010)*, 2010.

51. Y. LeCun, K. Kavukvuoglu, and C. Farabet. Convolutional Networks and Applications in Vision, *IEEE Proceedings of International Symposium on Circuits and Systems (ISCAS'10)*, 2010.

52. J. Schmidhuber, Neural network ReNNaissance, plenary talk presented at International Joint Conference on Neural Networks 2011 (IJCNN2011). Video forthcoming from the IEEE CIS Multimedia tutorials center currently at http://ewh.ieee.org/cmte/cis/mtsc/ieeecis/video_tutorials.htm.

53. A. Ng. Deep Learning and Unsupervised Feature Learning, plenary talk presented at International Joint Conference on Neural Networks 2011 (IJCNN2011). Video forthcoming from the IEEE CIS Multimedia tutorials center, currently http://ewh.ieee.org/cmte/cis/mtsc/ieeecis/video_tutorials.htm.

54. G.E.P. Box and G.M. Jenkins. *Time-Series Analysis: Forecasting and Control*, Holden-Day, San Francisco, 1970.

55. D.F. Walls and G.F. Milburn. *Quantum Optics*, Springer, New York, 1994.

56. P. Werbos. Bell's theorem, many worlds and backwards-time physics: not just a matter of interpretation. *International Journal of Theoretical Physics*, 47(11):2862–2874, 2008. URL: http://arxiv.org/abs/0801.1234.

57. N. El-Karoui and L. Mazliak. *Backward Stochastic Differential Equations.* Addison-Wesley Longman, 1997.

58. C.R. Weisbin, R.W. Peelle, and R.G. Alsmiller, Jr. An assessment of The Long-Term Energy Analysis Program used for the EIA 1978 report to Congress, Energy Volume 7, Issue 2, February 1982, pp. 155–170.

59. P. Werbos. Circuit design methods for quantum separator (QS) and systems to use its output. URL: http://arxiv.org/abs/1007.0146.

60. T. Sauer. *Numerical Analysis*, 2nd edition, Addison-Wesley, 2011.

61. G. Snider, R. Amerson, D. Carter, H. Abdalla, M.S. Qureshi, J. Léveillé, M. Versace, H. Ames, S. Patrick, B. Chandler, A. Gorchetchnikov, and E. Mingolla. From synapses to circuitry: using memristive memory to explore the electronic brain, *Computer*, 44(2):21–28, 2011. URL: http://ieeexplore.ieee.org/stamp/stamp.jsp?tp = &arnumber = 5713299&isnumber = 713288.

62. R. Kozma, R. Pino, and G. Pazienza. *Advances in Neuromorphic Memristor Science and Applications*, Springer, 2012.

63. D.P. Bertsekas and J.N. Tsisiklis. *Neuro-Dynamic Programming*. Athena Scientific, Belmont, MA, 1996.

64. D.P. De Farias and B.V. Roy. The linear programming approach to approximate dynamic programming, *Operations Research* 51(6):850–865, 2003. Article Stable URL: http://www.jstor.org/stable/4132447.

65. R. Ilin, R. Kozma, and P.J. Werbos. Beyond backpropagation and feedforward models: a practical training tool for more efficient universal approximator. *IEEE Transactions of Neural Networks* 19(3):929–937, 2008.

66. K.Y. Ren, K.M. Iftekharuddin, and E. White. Large-scale pose invariant face recognition using cellular simultaneous recurrent network, *Applied Optics, Special Issue in Convergence in Optical and Digital: Ettern Recognition*, 49:B92, 2010.

67. S. Mohagheghi, G.K. Venayagamoorthy, and R.G. Harley. Optimal wide area controller and state predictor for a power system. *IEEE Transactions on Power Systems*, 22(2):693–705, 2007.

68. D.B. Fogel, T.J. Hays, S.L. Han, and J. Quon. A self-learning evolutionary chess program. *Proceedings of IEEE*, 92(12):1947–1954, 2004.

69. Y. Cao, S. Grossberg, and J. Markowitz. How does the brain rapidly learn and reorganize view- and positionally-invariant object representations in inferior temporal cortex? *Neural Networks*, 24:1050–1061, 2011.

70. T. H. D. Liu, Residential energy system control and management using adaptive dynamic programming, *IEEE Proceedings of the International Joint Conference on Neural Networks IJCNN2011*, 2011.

71. S. Chen, Y. Yang, S.N. Balakrishnan, N.T. Nguyen, K. Krishnakumar, In: IEEE Proceedings of the International Joint Conference on Neural Networks 2009 (IJCNN2009), 2009.

72. S. Mehraeen and S. Jagannathan. Decentralized near optimal control of a class of interconnected nonlinear discrete-time systems by using online Hamilton–Bellman–Jacobi formulation, *IEEE Transactions on Neural Networks*, 22(11):1709–1722, 2011.

73. J.N.S. Wright. *Numerical Optimization*, Springer, 2006.

CHAPTER 2

Stable Adaptive Neural Control of Partially Observable Dynamic Systems

J. NATE KNIGHT and CHARLES W. ANDERSON
Department of Computer Science, Colorado State University, Ft Collins, CO, USA

ABSTRACT

The control of a nonlinear, uncertain, partially observable, and multiple spring–mass–damper system is considered in this chapter. The system is a simple instance of a larger class of models representing physical systems such as flexible manipulators and active suspension systems. A recurrent neural controller for the system is presented which guarantees the stability of the system during adaptation of the controller. A reinforcement learning algorithm is used to update the recurrent neural network weights to optimize the performance of the control system. A stability analysis based on integral quadratic constraint (IQC) models is used to reject weight updates that cannot be guaranteed to result in stable behavior. The basic stable learning algorithm suffers from performance problems when the controller parameters are near the boundary of the provable stable part of the parameter space. A derivative of the closed loop gain is obtained from the IQC computations and is shown to bias the parameter trajectory away from this boundary. In the example control problem this bias is shown to improve the performance of the algorithm.

2.1 INTRODUCTION

The robust control of physical systems requires an accurate accounting of uncertainty. Uncertainty enters the control problem in at least three ways: unmeasured states, unknown dynamics, and uncertain parameters. Modern robust control theory is based on explicit mathematical models of uncertainty [1]. If it is possible to describe what is unknown about a system, stronger assurances can be made about its stability and

Reinforcement Learning and Approximate Dynamic Programming for Feedback Control, First Edition. Edited by Frank L. Lewis and Derong Liu.

performance. The automation of robust controller design relies on the tractable representation of the uncertainty in a system. Some types of uncertainty can be described by integral quadratic constraints (IQCs) and lead to representations of uncertainty as convex sets of linear operators. Linear systems are a particularly tractable type of model, and the design of feedback controllers for linear systems is a well-understood problem. Most physical systems, however, exhibit some nonlinear dynamics and linear models are generally insufficient for accurately describing them. Unmodeled nonlinear dynamics can often be treated as uncertainty. Because robust controllers must be insensitive to inaccuracies and uncertainties in system models, performance is often suboptimal on the actual system to which the controller is applied. Additional loss in performance is often introduced by restricting controllers to be linear and of low order. This restriction is usually made because linear controllers can be easily analyzed and understood. Control performance can often be improved in these situations by the use of nonlinear and adaptive control techniques. Guaranteeing stability and performance, however, is more difficult in this environment. In this chapter, we examine the use of adaptive, recurrent neural networks in control systems where stability must be assured.

Recurrent neural networks are, in some respects, ideal for applications in control. The nonlinearity in neural networks allows for the compensation of nonlinearities in system dynamics that is not generally possible with low order, linear controllers. The dynamics of recurrent neural networks allow internal models of unmeasured states to be produced and used for control. The difficulty in applying recurrent neural networks in control systems, however, is in the analysis and prediction of the system's behavior for the purpose of stability analysis.

The control of a nonlinear, uncertain, partially observable, multiple spring–mass–damper system is considered in this chapter. The system is a simple instance of a larger class of models representing physical systems such as flexible manipulators and active suspension systems. We present a recurrent neural controller for the system with guaranteed stability during operation and adaptation. A reinforcement learning algorithm is used to update the recurrent neural network's weights to optimize performance and a stability analysis based on IQC models is used to reject weight updates that cannot be guaranteed to result in stable behavior [2, 3]. In addition, we developed a stability bias that is applied to the network's weight trajectory to improve the performance of the algorithm.

2.2 BACKGROUND

The following simple approach is one way to guarantee the stability of an adaptive control system as it changes through time. Compute a set of bounds on the variation of the control parameters within which stability is guaranteed, and then filter out any parameter updates that put the parameters outside of the computed bounds. Such an approach is at one end of a trade-off between computational cost and conservativeness. Only a single stability analysis is required, and the cost of rejecting weight updates that cannot be proved stable is trivial. On the other hand, the initial weights may not be close to the optimal weights and the bounds may limit optimization of the problem objective. In Ref. [4], this approach was applied to train an recurrent neural network (RNN) to model a chaotic system. Given a *good* initial weight matrix, the learning algorithm was able to improve the model within the specified stability bounds. In general, however, it cannot be expected that the optimal weight matrix, $\bar{W}$, for a problem will be reachable from an initial weight matrix, $W(0)$, while still respecting the initial stability constraints. The relative inexpensiveness of this approach has as its price a reduction in the achievable performance. At the other end of the spectrum is an algorithm that recomputes the bounds on weight variations at every update to the weights. The algorithm does not ensure that every update is accepted, but it does, in theory, result in the acceptance of many more updates than the simple approach. It also allows, again, in theory, better performance to be achieved. The computational cost of the algorithm is, however, prohibitively expensive because of the large number of stability analysis computations required.

In previous work [2, 3, 5, 6], we describe an algorithm that falls somewhere between these two extremes allowing better optimization of the objective than the first approach with less computational expense than the second. The algorithm assumes that changes to the control parameters are proposed by an external agent, such as the reinforcement learning agent described in a later section, at discrete time steps indexed by the variable k. The algorithm ensures the stability of an adaptive control system by filtering parameter updates that cannot be guaranteed to result in stable behavior. A constraint set, $\mathcal{C}_j(W)$ is a set of bounds, $\{\underline{\Delta}, \overline{\Delta}\}$, on the variation in the parameters centered on the fixed parameter vector W. Variations in $W(k)$ that stay within these bounds can be assured not to result in instability. The constraint set, $\mathcal{C}_j(W)$, is a volume centered on W with each element of $W(k)$ constrained by

$$W_i - \underline{\Delta}_i \leq W_i(k) \leq W_i + \overline{\Delta}_i.$$

When an update causes $W(k)$ to lie outside of the constraint set, a new set of constraints is computed if updates to $W(k)$ have occurred since the last set of constraints was constructed. Otherwise, the update is rejected. Given this new set of constraints centered on the most recently seen stable W, the current update $W(k-1) + \Delta W$ is again checked for validity. If the update fails to satisfy the new constraints it is then rejected. Rather than rejecting the update outright, the procedure described here makes better use of the available parameter update suggestions from the adaptation algorithm. An illustration of the algorithm is given in Figure 2.1.

Recurrent neural networks (RNNs) are a large class of both continuous and discrete time dynamical systems. RNN formulations range from simple ordinary differential equation (ODE) models to elaborate distributed and stochastic system models. The main focus of this work is on the application of RNNs to control problems. In this context, RNNs can be seen as input–output maps for modeling data or acting as controllers. For these types of tasks, it will be sufficient to restrict attention to continuous

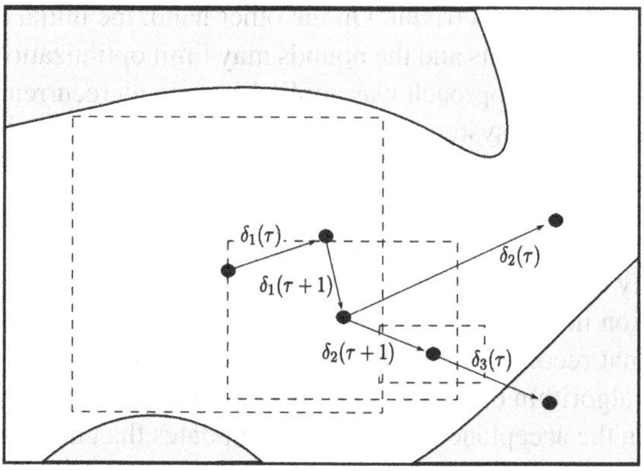

FIGURE 2.1 An illustration of the stable control algorithm. Updates outside of the provably stable update region are rejected.

time RNN formulations, primarily of the form

$$\begin{aligned} \dot{x} &= -Cx + W\Phi(x) + u, \\ y &= x. \end{aligned} \tag{2.1}$$

Here, x is the state of the RNN, u a time-varying input, y the output of the network, C a diagonal matrix of positive time constants, W the RNN's *weight matrix*, and Φ a nonlinear function of the form

$$\Phi(x) = [\phi(x_1)\ \phi(x_2)\ \ldots\ \phi(x_n)]^T.$$

The function $\phi(x)$ is a continuous one dimensional map, and generally a sigmoid-like function, such as $\tanh(x)$. Since the RNN will be applied as an input–output map, the output, denoted by y, is defined to be the state x. More general models allow the selection of certain states as outputs or an additional mapping to be applied at the output layer. These modifications do not affect the stability analysis of the RNNs dynamics, but need to be considered when the network is used in a control system.

Many algorithms exist for adapting the weights of RNNs. A survey of gradient-based approaches can be found in [7]. To solve the example learning problem in this chapter, the real-time recurrent learning (RTRL) algorithm is applied. RTRL is a simple stochastic gradient algorithm for minimizing an error function over the parameters of a dynamic system. It is used here because it is an online algorithm applicable in adaptive control systems. Computing the gradient of an error function with respect to the parameters of a dynamic system is difficult because the system dynamics introduce temporal dependencies between the parameters and the error function. Computation of the error gradient requires explicitly accounting for these dependencies, and RTRL provides one way of doing this.

Letting $F(x, u; C, W) = -Cx + W\Phi(x) + u$ in Equation (2.1), the gradient of the error function $E(||x(t) - \bar{x}(t)||_2^2)$ with respect to the weight matrix, W, is given in

RTRL by the equations

$$\frac{\partial E}{\partial W} = \int_{t_0}^{t_1} s \frac{\partial E}{\partial y} dt,$$

$$\frac{\partial s}{\partial t} = \frac{\partial F(x, u; C, W)}{\partial W} + \frac{\partial F(x, u; C, W)}{\partial x} s,$$

where $s(t_0) = 0$. The variable s is a rank three tensor with elements s^l_{ij} corresponding to the sensitivity of x_l to changes in the weight W_{ij}. RTRL requires simulation of the s variables forward in time along with the dynamics of the RNN. The gradient, however, need not necessarily be integrated. The weights can instead be updated by

$$W \leftarrow W - \eta s(t) \frac{\partial E(t)}{\partial y(t)},$$

for the update time t. The parameter η is a learning rate that determines how fast the weights change over time. The stochastic gradient algorithm requires that this parameter decrease to zero over time, but often it is simply fixed to a small value. The algorithm has an asymptotic cost of $\Theta(n^4)$ for each update to s when all the RNN weights are adapted. For large networks the cost is impractical, but improvements have been given. For example, an exact update with complexity $\Theta(n^3)$ is given in [8] and an approximate update with an $\Theta(n^2)$ cost is given in [9]. The basic algorithm is practical for small networks and is sufficient for illustrating the properties of the proposed stable learning algorithm.

2.3 STABILITY BIAS

The algorithm used to adjust the weights of the RNN will not, in general, have any information about the stability constraints on the system under control. Because of this, the algorithms will often push the system near the boundary of the region in which the system can be proved stable. Near this boundary the stable learning algorithm described above becomes increasingly inefficient because vary little variation can be safely allowed in the RNN weights. To improve this situation, we introduce in this section a bias that can be applied to the weight trajectory in order to force it away from this boundary. To bias the learning trajectories away from the boundary, a measure of closeness to this boundary is needed. Fortunately, the IQC analysis used to perform the stability analysis of the RNN and control system provides an estimate of the gain of the control system. This gain increases rapidly near the boundary of the region that can be proven stable. While the magnitude of this gain is not immediately useful as a bias, the gradient of this value with respect to the RNN weights carries information about closeness to the boundary and of its direction from the current weights. The computation of this derivative is now summarized.

Consider the general, nonlinear, semidefinite program (SDP) given in [10]

$$\begin{aligned} p^* = \min b^T x \quad \text{s.t.} \quad & x \in \mathbb{R}^n, \\ & \mathcal{B}(x) \preceq 0, \\ & c(x) \le 0, \\ & d(x) = 0. \end{aligned} \tag{2.2}$$

The Lagrangian of this problem $\mathcal{L} : \mathbb{R}^n \times \mathbb{S}^m \times \mathbb{R}^p \times \mathbb{R}^q \to \mathbb{R}$, is defined by [10]

$$\mathcal{L}(x, Y, u, v) = b^T x + \mathcal{B}(x) \bullet Y + u^T c(x) + v^T d(x), \tag{2.3}$$

where $Y \in \mathbb{S}^m$, $u \in \mathbb{R}^p$, and $v \in \mathbb{R}^q$ are the Lagrange multiplier variables. The Lagrangian dual function, defined as, [11]

$$g(Y, u, v) = \inf_x \mathcal{L}(x, Y, u, v), \tag{2.4}$$

is a lower bound on the optimal value of Equation (2.2) for all values of the multiplier variables. When the best lower bound given by Equation (2.4), that is,

$$d^* = \max_{Y,u,v} g(Y, u, v) \quad \text{s.t. } Y \succeq 0, u \succeq 0, \tag{2.5}$$

is equal to, p^*, the optimal value of Equation (2.2), the problem is said to satisfy a *strong duality* condition. For convex optimization problems a sufficient condition, known as Slater's condition, for strong duality is the existence of a strictly feasible point. So, if $\mathcal{B}(x)$, $c(x)$, and $d(x)$ are convex functions of x, and there exists an x satisfying, $\mathcal{B}(x) \preceq 0$ and $c(x) < 0$ then $d^* = p^*$.

Often, rather than considering a single SDP a set of related SDPs parameterized by some data θ is of interest. For example, the linear matrix inequality (LMI) stability condition for RNNs with time invariant weights forms a set of SDPs parameterized by W. For parameterized SDPs, the Lagrangian, p^*, and d^* are functions of the problem data and are written $\mathcal{L}(x, Y, u, v; \theta)$, $p^*(\theta)$, and $d^*(\theta)$. Of specific interest is the set of θ for which the SDP satisfies the strong duality condition. Over this set, the affect of perturbing the data, θ, on the optimal solution, $p^*(\theta)$, can be estimated with a Taylor series expansion using the gradient defined by

$$\nabla_\theta p^*(\theta) = \left[\frac{\partial}{\partial \theta_i} p^*(\theta)\right] = \left[\frac{\partial}{\partial \theta_i} d^*(\theta)\right] = \left[\frac{\partial}{\partial \theta_i} \mathcal{L}(\bar{x}, \bar{Y}, \bar{u}, \bar{v}; \theta)\right].$$

This gradient is well defined when the Lagrangian is differentiable with respect to the data which is always the case when it is linear in the parameters of interest. The gradient is a first order approximation to the function $p^*(\theta)$ and gives the direction in the parameter space in which $p^*(\theta)$ increases the most in this approximation.

To specialize this gradient for the stability of an RNN, consider the following optimization problem associated with proving the stability of a time invariant RNN [2]:

$$\bar{\gamma} = \inf_{\gamma, T, P} \gamma \quad \text{s.t.} \tag{2.6}$$

$$\begin{bmatrix} -CP - PC + I & P & PW + T \\ P & -\gamma I & 0 \\ W^T P + T & 0 & -2T \end{bmatrix} < 0, \quad P = P^T, \quad T \in \mathcal{D}_+. \tag{2.7}$$

Here, the decision variables are $x = (\gamma, T, P)$. An upper bound on the gain of the RNN with weight matrices C and W is given by $\sqrt{\bar{\gamma}}$. The LMI constraint in the problem has an associated Lagrange multiplier denoted Y. Take the Lagrangian to be a function of the weight matrix, W, and write $\mathcal{L}(x, Y; W)$. The problem is convex and by the definition of $\mathcal{W}_{ss}^n$ satisfies Slater's condition for all W in $\mathcal{W}_{ss}^n$. The Lagrangian takes the value $\bar{\gamma}$ at the solution to Equation (2.6). Since the Lagrangian is linear in W, and thus differentiable, the gradient of $\bar{\gamma}$ with respect to W can be computed by the formula

$$\nabla_W \bar{\gamma} = \left[\frac{\partial}{\partial W_{ij}} \mathcal{L}\left(\bar{x}, \bar{Y}; W\right) \right]. \tag{2.8}$$

For conciseness $\nabla_W \bar{\gamma}$ will be denoted by ∇_s throughout the chapter. The Lagrangian in Equation (2.3) specializes to

$$\mathcal{L}(x, Y; W) = \gamma + \mathcal{B}(x) \bullet Y,$$

in this case. The function $\mathcal{B}(x)$ corresponds to the left-hand side of the LMI constraint in Equation (2.7). It is a function of the problem data W given by

$$\mathcal{B}(x; W) = \sum_i x_i B^{(i)}(W),$$

where the $B^{(i)}$s are linear functions taking W into $\mathbb{S}^m$. Since this is the only point at which the problem data enters the Lagrangian, its gradient is easily computed as

$$\left[\frac{\partial}{\partial W_{ij}} \mathcal{L}\left(\bar{x}, \bar{Y}; W\right) \right] = \left(\sum_{i=1}^{n} \bar{x}_i \frac{\partial}{\partial W_{ij}} B^{(i)}(W) \right) \bullet \bar{Y}. \tag{2.9}$$

Since $\mathcal{B}(x; W)$ is linear in W, the partial derivatives $\frac{\partial}{\partial W_{ij}} B^{(i)}(W)$ are constant and can be computed without the knowledge of a specific W. Evaluation of the gradient is a simple operation but requires the optimal values $\bar{x}$ and $\bar{Y}$ associated with a given W. This in turn requires that the optimization problem (2.6) be solved for each evaluation of the stability bias. Solving the optimization problem (2.6) dominates the computational cost of computing ∇_s. This computation has a run time on the order of $\Theta(n^3)$ to $\Theta(n^6)$ depending on the IQCs used and the choice of SDP solver [2].

Ideally, the stability bias would affect only the path the parameters take to an optimal solution and not the final point to which they converge. To achieve this ideal behavior the effect of the stability bias on the weight trajectories should be small relative to the effect of the updates given by the learning algorithm and should diminish over time. For example, the update rule might scale the magnitude of ∇_s with respect to the magnitude of the learning update, $\eta\nabla_l$, where, for example, $\nabla_l = s(t)\frac{\partial E(t)}{\partial y(t)}$ in the case of RTRL. Such an update would look like

$$W \leftarrow W - \left(\eta\nabla_l + \eta_2\frac{\|\eta\nabla_l\|_2}{\|\nabla_s\|_2}\nabla_s\right), \tag{2.10}$$

with $\eta_2 < 1$ and decreasing with time. A potential problem with this approach is that the effect of the stability bias is small regardless of closeness to the boundary of $\mathcal{W}_{ss}^n$. An alternate updating scheme varies the scale of the stability bias with the magnitude of the $\mathcal{L}_2$-gain bound, γ. The update is given by

$$\begin{aligned} \eta_3 &\leftarrow \eta_2\frac{3}{4}(\tanh(\gamma - \hat{\gamma}) + 1), \\ W &\leftarrow W - \left(\eta\nabla_l + \eta_3\frac{\|\eta\nabla_l\|_2}{\|\nabla_s\|_2}\nabla_s\right). \end{aligned} \tag{2.11}$$

where $\eta_2 \leq 1$ and decreases with time as before. The scaling η_3 is a monotonically increasing function of γ with a transition from low to high centered at $\hat{\gamma}$. This approach has the benefit of not affecting the weight trajectory when it is far from the boundary of $\mathcal{W}_{ss}^n$, but requires the transition point $\hat{\gamma}$ to be chosen. The scaling η_2 must decrease to zero with time to ensure that asymptotically the stability bias has no effect. If a certain maximum gain value is desired for the system, allowing η_3 to remain positive and setting $\hat{\gamma}$ to the intended gain bound helps to bias the system toward satisfying this condition. Methods for choosing a decay schedule for the parameter η_2 are rather adhoc, but the learning rate η for the RTRL updates has the same requirements and associated problems.

2.4 EXAMPLE APPLICATION

The example application described here illustrates the ability of the simplified stability algorithm to improve control performance by removing instability from the control loop. The stability bias proposed is integral to the success of the approach.

2.4.1 The Simulated System

The two degree of freedom spring–mass–damper system is adapted from the work in [12]. Certain features of the simulated model are considered unknown and not used

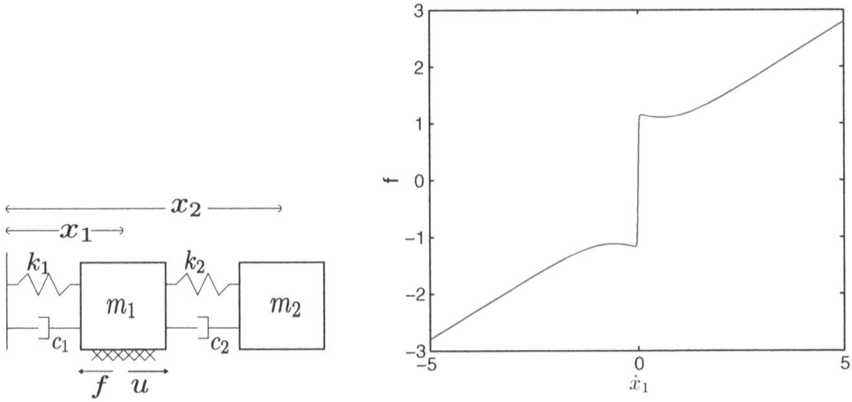

(a) (b)

FIGURE 2.2 (a) A multiple spring–mass–damper. (b) A continuously differentiable friction model.

explicitly in the adaptation of controllers or stability analysis. The nonlinear model and simulation details are provided in this section. Additionally, an uncertain model of the plant is developed, and an IQC analysis of the closed loop control system is described.

A diagram of the simulated plant is shown in Figure 2.2a. Two masses are connected by nonlinear springs and linear dampers. The first mass is attached via a spring and damper to a stationary point. A control force is be applied to the first mass which is also acted upon by a nonlinear, static, and friction force. A position sensor is attached to the second mass. The goal of the control problem is for the second mass to track a time-varying reference signal given by an external agent.

The plant dynamics are governed by the ordinary differential equations

$$\begin{aligned} m_1\ddot{x}_1 + c_1\dot{x}_1 + c_2(\dot{x}_1 - \dot{x}_2) + k_1 x_1 \\ + k_2(x_1 - x_2) + h_1 x_1^3 + h_2(x_1 - x_2)^3 = u - f(\dot{x}_1), \\ m_2\ddot{x}_2 + c_2(\dot{x}_2 - \dot{x}_1) + k_2(x_2 - x_1) + h_2(x_2 - x_1)^3 = 0 \end{aligned} \tag{2.12}$$

where u is the control force applied to the first mass. Actuator dynamics are ignored for the purpose of these experiments, and the control enters the system linearly. The parameters c_1 and c_2 govern the damping force of the two dampers. The spring dynamics are governed by the spring constants k_1 and k_2 and the spring hardening constants $h_1 = h^2 k_1$ and $h_2 = h^2 k_2$ with $h \geq 0$. The spring hardening constant was set to $h = 0.1$ for all simulations of the system. The friction force is modeled by the equation

$$f(\dot{x}_1) = g_7(g_1(\tanh(g_2\dot{x}_1) - \tanh(g_3\dot{x}_1)) + g_4\tanh(g_5\dot{x}_1) + g_6\dot{x}_1),$$

with

$$(g_1, g_2, g_3, g_4, g_5, g_6, g_7) = (12.5, 50, 1, 11, 50, 9, 0.05).$$

The magnitude of the friction force is shown in Figure 2.2b for $\dot{x}_1 \in [-5, 5]$. The friction equation and parameters are taken from Ref. [13] and model both stiction and Coulomb-type friction.

All simulations of the system were performed using a variable step size, Dormand–Prince algorithm with an absolute error tolerance of 10^{-5}. Changes in the control signal occur at a rate of 10 Hz, and observations are sampled at the same rate. For the purposes of simulation, the parameters were set to $m_1 = 2$, $m_2 = 2.01$, $c_1 = 1.05$, $c_2 = 0.97$, $k_1 = 5.3$, and $k_2 = 4.8$.

2.4.2 An Uncertain Linear Plant Model

For the purposes of controller design and stability analysis an uncertain linear plant model is developed as follows. The parameters of the system are measured as $m_1 = m_2 = 2$, $c_1 = c_2 = 1$, and $k_1 = k_2 = 5$ with an uncertainty of 2%, 10%, and 10%, respectively. The parameters are thus assumed to lie in the ranges $m_i \in [1.96, 2.04]$, $c_i \in [.91.1]$, and $k_i \in [4.5, 5.5]$. The simulated plant's parameters are within the assumed measurement errors, and uncertainty models based on these error estimates will be valid.

A linear model of the plant is easily constructed for use in control design and analysis. Ignoring the hardened spring and friction effects in Equation (2.12) yields the linear model

$$\begin{bmatrix} \dot{y}_1 \\ \dot{y}_2 \\ \dot{y}_3 \\ \dot{y}_4 \end{bmatrix} = \begin{bmatrix} 1 & & & \\ & 1 & & \\ & & m_1 & \\ & & & m_2 \end{bmatrix}^{-1} \begin{bmatrix} 0 & 0 & 1 & 0 \\ 0 & 0 & 0 & 1 \\ -(k_1 + k_2) & k_2 & -(c_1 + c_2) & c_2 \\ k_2 & -k_2 & c_2 & -c_2 \end{bmatrix} \begin{bmatrix} y_1 \\ y_2 \\ y_3 \\ y_4 \end{bmatrix}.$$

The set of linear models derived by taking the parameters in the given uncertainty ranges does not completely capture all of the possible dynamics of the system that is to be controlled. The uncertainty model should also account for the nonlinearity in the spring response and the friction force. It is also possible that the system exhibits other unmodeled dynamics. To account for the effect of the unmodeled dynamics, a multiplicative input uncertainty is added to the model. If $G(s)$ is the transfer function of the nominal linear plant model, then the uncertainty model looks like

$$G_{\text{unc}}(s) = G(s)(1 + 0.2\Delta_{\text{lti}}),$$

where Δ_{lti} is an unknown, linear, and time-invariant system with $||\Delta_{\text{lti}}||_\infty < 1$. The unknown linear time-invariant (LTI) system can model up to a 20% deviation in the plant input.

An uncertain real scalar, such as any of the parameters of the linear plant model, can be modeled with the representation $p = (p_n + \delta_p m_p)$, where p_n is the nominal value of the parameter, δ_p unknown constant satisfying $|\delta_p| \leq 1$ and m_p scaling factor on the uncertainty [14]. To represent the uncertain parameter, $k_1 \in [4.5, 5.5]$, for example, consider $p_n = 5$ and $m_p = 0.5$. The uncertain parameters in the model are assumed to be constant. The representation of time-varying parameters is essentially the same, but δ_p is allowed to vary with time. Time-varying parameters can have a more varied impact on a system's behavior than constant uncertain parameters. The stability analysis presented in the following section takes advantage of the fact that the parameters are constant to reduce the conservativeness of the analysis.

To perform robustness analysis and control design for the uncertain plant, the uncertainties must be factored out of the plant to form a feedback loop between the known LTI plant and the uncertainties: Δ_{lti} and the δ_p's for the uncertain parameters. The details of this procedure can be found in standard robust control references such as [1, 14]. Standard robustness analysis shows that the linear plant model is stable for all possible values of the uncertain parameters and all possible input perturbations allowed by the model [14]. An IQC model can be developed and used to show the same result [2]. A bound on the $\mathcal{L}_2$-gain from $u \to y$, computed using the IQC stability theorem, has a value of $\gamma = 1.4755$. The finite gain is a proof of stability for all plants covered by the uncertainty model.

2.4.3 The Closed Loop Control System

The closed loop control system under investigation is depicted in Figure 2.3. A reference signal enters the system from an external source and specifies the desired value of the plant output. In this case, the position of the second mass in the spring–mass–damper system is to be controlled. For the experiments that follow, the reference signal takes values in the range $[-2, 2]$ and can change every 50 s. The reference signal might represent, for example, the desired position of a read head in a hard drive or the desired location of the end point of a flexible manipulator.

Standard robust control designs for reference tracking problems generally use the error signal, $e = r - y$, as the input to the controller. This particular representation is invariant with respect to the position of the observed mass in the multiple spring–mass–damper system. It does not allow the controller to adequately compensate for nonlinearity in the spring, since the nonlinearity in Equation (2.12) is a function of x_1 and $x_1 - x_2$. Even though the position of the first mass is unobserved, the trajectory of x_2 contains information about the true state of the system and thus x_1. It is possible for a recurrent neural network to use this inferred information to improve control

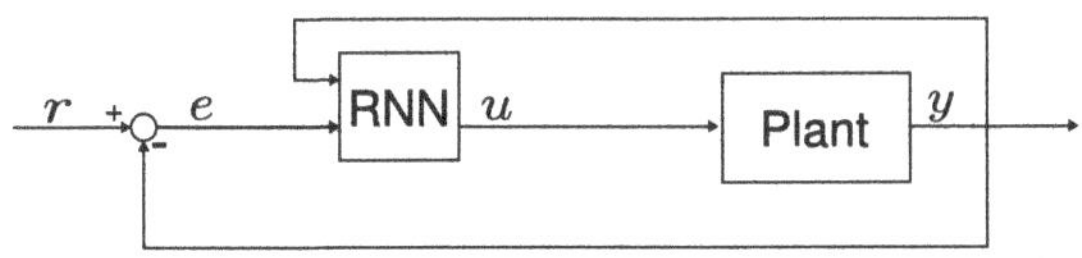

FIGURE 2.3 The closed loop control system.

performance. For this reason, y is also included as an input to the RNN controller. The output of the RNN is the control action, u. The control signal is fed directly into the plant since actuator dynamics are being ignored.

The basic RNN equations must be adapted to fit the desired control structure. The simple RNN model considered in earlier chapters had the same number of inputs, outputs, and states. In the desired control configuration the network has two inputs, a single output, and a number of states set to determine the learning complexity. Input and output weights are added to the RNN in the following manner

$$\begin{aligned} \dot{x}_r &= -Cx_r + W\Phi(x_r) + W^{\mathrm{i}} \begin{bmatrix} e & y \end{bmatrix}^T, \\ u &= W^{\mathrm{o}} x_r. \end{aligned} \tag{2.13}$$

Different configurations of W^{i} and W^{o} affect the behavior of the network. A common configuration, for instance [15], feeds each input into a different node and reads the output from another node. This leads to

$$W^{\mathrm{i}} = \begin{bmatrix} 1 & 0 \\ 0 & 1 \\ 0 & 0 \\ & \vdots \\ 0 & 0 \end{bmatrix} \text{ and } W^{\mathrm{o}} = \begin{bmatrix} 0\,0\,1\,0\ldots 0 \end{bmatrix}.$$

RNN stability analysis can easily be adapted to include the input and output weights [2]. These weights do not affect analysis of the internal stability of the RNN, but they do affect the gain measured from the input to output signals. For instance, if the output weights are all zero then the gain from input to output is zero regardless, even, of the internal stability of the network. It is important, then, that these weights be included in the analysis since the measured gain affects the stability analysis of the closed loop control system.

The analysis of robust control systems focuses on two properties of an uncertain system model: robust stability and robust performance. A robustly stable system is stable for all systems in the given uncertainty set. A system with robust performance, on the other hand, is guaranteed to meet certain performance requirements for all systems in the uncertainty set. Robust stability is necessary for robust performance, but it is not sufficient. Robust stability of a set of uncertain systems implies stability of an actual physical system, *if* the physical system is adequately described by the uncertain model. Increasing the size of the uncertainty set—that is, increasing the range of system dynamics it covers—can allow better assurances to be made about the stability of an actual system. At the same time, increasing the size of the uncertainty set can make it more difficult to prove robust performance.

Performance of a robust control system is measured by the $\mathcal{H}_\infty$ norm of the system. In other words, performance is measured in terms of the $\mathcal{L}_2$-gain from the system's input to its output. Performance objectives in robust control systems are specified by augmenting a given control system with weighted outputs that specify the desired behavior. For example, the weighted outputs might be designed to penalize large actions or low frequency deviations of the plant output from the reference signal. Robust performance of a system is given if the gain from the input to the weighted outputs is less than one. Robust controller synthesis methods, such as the D–K iteration [1], are computational approaches to designing linear controllers that have robust performance for a given uncertain system model.

The synthesis of neural controllers with robust performance guarantees is a difficult problem. The main difficulty arises from treating the RNN nonlinearity as uncertainty in the analysis. The uncertainty descriptions of recurrent neural networks in terms of IQCs have the serious drawback that they do not explicitly model the *boundedness* of the nonlinearities. The resulting uncertainty model and analysis can not distinguish between say, $\phi(x) = x$ and $\phi(x) = \tanh(x)$. Obviously, the resulting dynamics of the two RNNs will be very different. Recent work in [16] on generalized sector conditions may provide a way to incorporate the boundedness of $\phi(x)$ into the analysis, but at present this work is in its infancy. Until such improvements are made, it is necessary to restrict attention to the analysis of robust stability in neural control systems. The inaccuracy in the analysis is slightly less troublesome here because the question being addressed—stability of the uncertain, closed loop system—is less specific. The analysis requires only that the gain from reference input to plant output is finite and not that it meet a prescribed bound.

While a robust synthesis method for neural controllers would be useful in generating the initial neural control design for a system, adaptation is necessary to specialize the controller to a given plant. In the context of adaptive control, robust performance is forgone in the name of performance on a specific plant. This is convenient in the case of neural control because of the problems discussed earlier with analyzing the robust performance of recurrent neural network controllers. Robust stability, however, is still desirable. Given that the plant is not completely known, and that even models adapted to the plant online cannot be completely accurate, assuring stability with respect to a set of plant models makes sense. The robust stability of the closed control loop can be addressed using the IQC descriptions of the plant and recurrent neural network.

A linear fractional representation (LFR) of the closed loop control system is needed to apply the IQC analysis results. Representations of the control loop containing an RNN with time-invariant weights and an RNN with time-varying weights are both needed. The details of constructing LFRs can be found in [14]. The uncertain, nonlinear operator in the resulting model has the structure

$$\Delta(s) = \mathrm{diag}\{\Delta_{\mathrm{lti}}(s), \delta_{m_1}, \delta_{m_2}, \delta_{k_1}, \delta_{k_2}, \delta_{c_1}, \delta_{c_2}, \Phi(\cdot)\},$$

when the RNN weights are static. The operator is augmented with the time-varying coefficients for the full, time-varying control loop model,

$$\Delta(s) = \mathrm{diag}\{\Delta_{\mathrm{lti}}(s), \delta_{m_1}, \delta_{m_2}, \delta_{k_1}, \delta_{k_2}, \delta_{c_1}, \delta_{c_2}, \Phi(\cdot), \overline{\delta}_{ij}, \underline{\delta}_{ij}\}.$$

Using IQCs for the different uncertainties and nonlinearities that have been previously described, an LMI problem can be constructed to assess the stability of the closed loop system and compute a bound on its gain. If the resulting LMI is feasible, the control loop is proved stable for all plants in the uncertainty set and additionally, for all controllers satisfying the IQC description of the RNN.

As an example, consider the RNN with $n = 3$, $C = I$, and

$$W = \begin{bmatrix} -1.4992 & 0.5848 & 0.5417 \\ 0.3425 & 0.4623 & 0.4551 \\ 0.7165 & 0.0323 & -0.2045 \end{bmatrix}.$$

The gain across the RNN is bounded from above by $\gamma_{\mathrm{r}} = 0.9751$. The bound was computed using the Popov IQC and taking $T \in \mathcal{M}_{dd}$, the doubly dominant nonlinearity IQC. Since the gain of the uncertain plant model is bounded by $\gamma_{\mathrm{p}} = 1.4755$ and $\gamma_{\mathrm{r}} > 1/\gamma_{\mathrm{p}} \approx 0.6777$, the small gain theorem fails to prove stability. The full IQC analysis, on the other hand provides a gain bound of $\gamma_{\mathrm{cl}} = 1.7938$ proving the stability of the closed loop for all systems in the uncertainty set. A sense of the effect the plant uncertainty has on the analysis of the closed loop can be gained by measuring the gain of the RNN in a loop with just the nominal plant. The gain of this system is bounded above by $\gamma_{\mathrm{n}} = 1.4830$. In this particular case the plant uncertainty has only a mild effect on the estimated loop gain, but this is not always the case.

Bounds on the allowable variation in the RNN parameters can be computed using the full closed loop system model in the analysis. For the weight matrix given above, variation bounds are computed as

$$\overline{\Delta} = \begin{bmatrix} 0.1000 & 0.1000 & 0.1000 \\ 0.1000 & 0.1000 & 0.1000 \\ 0.1000 & 0.1000 & 0.1000 \end{bmatrix} \quad \text{and} \quad \underline{\Delta} = \begin{bmatrix} 0.3409 & 0.1210 & 0.2290 \\ 0.8891 & 0.4266 & 0.3296 \\ 1.8523 & 0.2090 & 0.4622 \end{bmatrix},$$

using the LMI approximation approach and restricting the bounds to be greater than or equal to 0.1 [2]. The uncertainty in the plant has a large effect on this particular computation. Computing the variation bounds using just the nominal plant model leads to an increase in the sum of the bounds by 1.8220 and for some weights doubles the amount of variation that can be tolerated. Inaccuracies in the uncertain plant model can thus negatively impact the performance of the stable learning algorithm presented in the previous chapter by allowing less variation in the weights than can be safely tolerated by the actual system.

2.4.4 Determining RNN Weight Updates by Reinforcement Learning

The RNN weights are adapted using a reinforcement learning approach of the reference tracking problem following [17]. Since the algorithms are generally applied in discrete time to sampled data, the following presentation uses a discrete time representation of certain parts of the problem. Given a deterministic, dynamical system

$$\dot{x} = f(x, u),$$

where $x \in \mathbb{R}^n$ is the system state and $u \in \mathbb{R}^m$ control input, find a control law, or *policy*, $\mu(x) : \mathbb{R}^n \to \mathbb{R}^m$, mapping x to u that minimizes

$$Q^{\mu}(x(t), u(t)) = \int_t^{\infty} e^{-\frac{s-t}{\tau}} c(x(s), u(s))\, ds, \quad u(s) = \begin{cases} \mu(x(s)) & s > t, \\ u(t) & s = t. \end{cases} \tag{2.14}$$

The function $c(x, u) : \mathbb{R}^n \times \mathbb{R}^m \to \mathbb{R}$ describes the cost of being in a particular system state and taking a particular control action. For example, in the reference tracking problem, a simple cost function is $c(x(t), u(t)) = \|r(t) - x(t)\|$, where $r(t)$ is the current reference signal. The design of the cost function is one of the most important parts of the problem specification since it determines the behavior of the optimal solution [18]. The parameter, τ, is a discount factor that determines the time horizon over which the algorithm attempts to minimize the cost.

To use a reinforcement learning approach within the context of the proposed stable learning algorithm, an explicit representation of the control policy is needed. Thus, we use an actor–critic approach to generate an explicit control policy that is modeled with an RNN. Two of the basic assumptions of the reinforcement learning approach are that the state of the environment is fully observable and that the dynamics satisfy the Markov property. The Markov property requires that the observed effect of a control action at time t is a function of only the state at the current time and does not have any dependence on the past state trajectory. The two assumptions of state observability and Markov dynamics are closely related. In the control problem under consideration, the state of the plant is not fully observable. Because of this, the observed dynamics do not satisfy the Markov property. One approach to solving this type of problem requires modeling the control problem as a type of partially observable Markov decision process or POMDP. Another approach requires that a representation of the system that satisfies the Markov property be developed. Recurrent neural networks are useful for this purpose because of their ability to model temporal sequences and hidden dynamics [19]. To simplify the computations of this example application and as the structure of the critic does not affect the stability analysis [5], a standard feedforward neural network with complete state input and 50 hidden units is trained to model $Q^{\mu}(x(t), u(t))$ in the following experiments.

The RNN used for the actor has three states represented by three units, thus $W \in \mathbb{R}^{3\times 3}$. Updates to the actor weights are made using stochastic gradient descent on the gradient of the Q-function with respect to the control inputs,

$$W \leftarrow W - \eta_a \frac{\partial Q(x, u)}{\partial u} \frac{\partial u}{\partial W}.$$

The update can be computed by using $-\frac{\partial Q(x,u)}{\partial u}$ as the error function in the RTRL algorithm. The learning rate was varied for some experiments, but as a default $\eta_a = 0.001$. The updates to the actor weights proposed by this algorithm are exactly the updates that must be monitored by the stable learning algorithm to ensure stability.

2.4.5 Results

The following experiments are designed to illustrate the properties of the stable learning algorithm more than to simulate the actual application of these techniques in practice. For instance, in practice, a robust controller would be designed for the plant and the actor–critic model would simply modify the output of the robust controller in some way [5]. The combination of robust controller and actor–critic system allows a certain amount of performance to be guaranteed at the deployment of the system. In the experiments that follow, less attention is payed to the quality of the learned control law than to the stability properties of the adaptation.

The actor and critic models were initialized by simulating them for 5000 s on the nominal, linear, plant model described in Section 2.4.2. The simulation was done without regard for stability since the actor and critic adaptation was performed on a model and not the real system. Initializing the actor and critic in this way allowed them to be hooked into the real system with some *a priori* knowledge. All of the experiments that follow begin with these initialized actor and critic parameters. Three sets of results are reported for actor–critic learning without stability analysis, with step-wise stability analysis, and with step-wise stability plus stability bias analysis. Before summarizing the results, details of how the stability analysis is applied in this example are presented.

During the training of the actor and critic the stable learning algorithm is used to filter the actor updates. The stability bias is used with two slight modifications. First, to make better use of the available information, updates that are rejected by the original algorithm are instead rescaled such that they satisfy the current constraint set. This allows progress to be made at every step. Second, rather than compute the stability bias at every step, the bias computation is turned on and off based on whether or not the previous steps were accepted without modification. Initially, the stability bias is not used, but after a sequence of three updates in a row have been rescaled due to violation of the constraints the stability bias computation is turned on. The bias computation remains on until a sequence of five steps in a row are accepted without scaling.

The initial actor weights, generated by training on the linear plant, can not be proved stable. To initialize the stable learning algorithm the weights are scaled down to satisfy the stability constraints. The resulting initial network weights produced a gain of $\gamma_{cl} = 0.600$.

The system was simulated for 1100 s and the following results were observed. Unlike the previous example, no instability was observed in the controlled system during the adaptation. Out of the 11,000 updates generated by the actor–critic algorithm, 2541, roughly one in four, required rescaling to satisfy the stability constraints. The stability bias was computed 4200 times, or on almost half of the steps. The stability

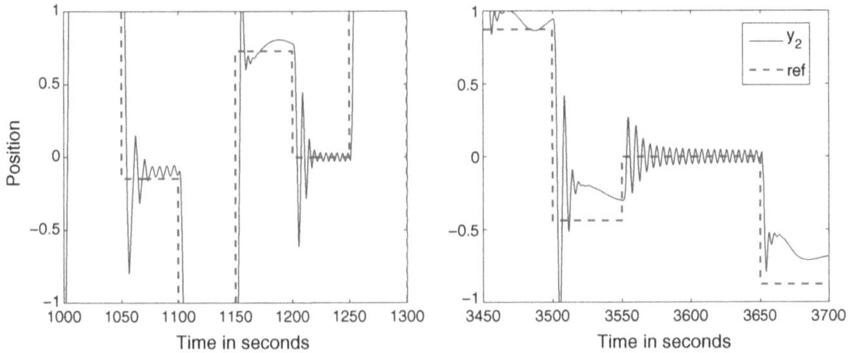

FIGURE 2.4 An example of unstable behavior exhibited during the training of the actor–critic system.

bounds were recomputed 6540 times. The mean variation allowed per weight over these constraint sets was 0.028. Because of repeated failures to find bounds that satisfied the minimum constraint, the lower bound was decreased repeatedly. The amount of variation allowed in the weights is, on average, very less. This causes the number of constraint computations to rise and increases the cost of the algorithm drastically. The gains computed in the stability bias computations remained relatively small, $\gamma < 5$, during the simulation. This suggests that the small amounts of allowable variation are not due to closeness of the controller to the boundary of the stable weight set, $\mathcal{W}_{ss}^n$. It appears that conservativeness in the analysis of the time-varying RNN is hindering the application of the stable learning algorithm. The algorithm succeeds in keeping the control system stable, but has an excessive cost.

2.4.5.1 Actor–Critic Learning without Stability Analysis Beginning with the actor and critic trained on the nominal, linear plant model, the control system was simulated for 1000 s on the actual plant without any constraints on the stability of the system. At a sampling rate of 10 Hz, the actor and critic weights were updated 10000 times. In Figure 2.4 a portion of the recorded system behavior is shown. Clearly, the system exhibits instability for reference signals near zero.

The closed loop system can not be proved stable by the IQC analysis.

2.4.5.2 Step-wise Stable Actor–Critic Learning Because the conservativeness in the analysis of the time-varying RNN limits the amount of variation that can be tolerated under the stability constraints, it seems worthwhile to consider what benefit might be had from accepting weaker stability guarantees. Rather than constraining the variation in the weights using the full static and dynamic stability analysis, the weights of the actor are simply constrained to remain in $\mathcal{W}_{ss}^n$ at all times; weight updates that push the weights out of $\mathcal{W}_{ss}^n$ are rejected. This guarantees that stopping the adaptation at any point will always result in a stable controller. This type of stability guarantee has been called step-wise stability [18]. The weaker stability algorithm does

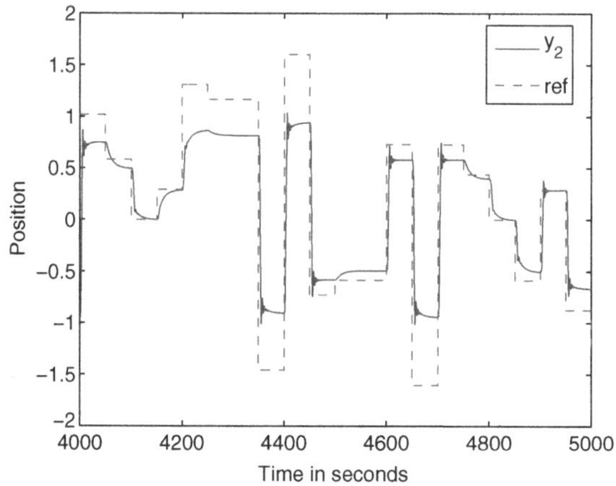

FIGURE 2.5 Example of the behavior of the step-wise stable adaptation.

not protect against the type of dynamic instability due to switching problems. Such problems might occur due to cycles in the weight settings cause by a nondecaying step size in the actor updates. This type of problem was never encountered during simulation, however.

Starting from the initialized actor used in the previous section, this new algorithm was simulated for 5000 s on the actual plant. Of the 50,000 updates generated, only 10,364 were accepted. The updates were rejected when the weights approached the boundary between provably stable and possibly unstable weight matrices. The behavior during the last 1000 s is shown in Figure 2.5. No instability was seen over the entire 5000- s history.

2.4.5.3 Actor–Critic Learning with a Stability Bias

Only one in four updates to the actor was accepted using the simple step-wise stability algorithm. To test whether the addition of the stability bias term to the weight updates increases the number of accepted updates, the previous experiment is repeated with the stability bias term added to the actor RNN's weight updates. The weight update with stability bias in (2.11) was applied over two 5000- s adaptation trials. The weight update is parameterized by $\hat{\gamma}$ which determines at what $\mathcal{L}_2$-gain the bias begins to have a major effect on the updates. The first trial used $\hat{\gamma} = 5$ and the second used $\hat{\gamma} = 50$. The weighting parameters, η_2 followed the decay schedule $\eta_2 = \frac{8000-t}{8000}$. Figure 2.6a shows the last 1000 s of the episode and the weight trajectories for $\hat{\gamma} = 5$. Figure 2.6b shows the same data, but from the trial with $\hat{\gamma} = 50$. Several observations can be made about the data. Compared to the previous experiment where no stability bias was used, both trials of this experiment perform better at tracking the reference signal. Also, the weights grow larger in these two trials. The lack of stability bias in the previous experiment caused the actor to suffer from a lack of progress due to discarded updates.

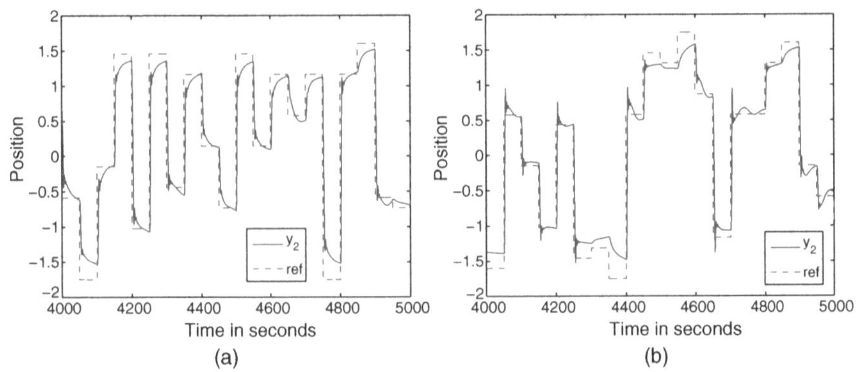

FIGURE 2.6 Examples of the behavior of the step-wise stable adaptation and the stability bias. (a) $\hat{\gamma} = 5$. (b) $\hat{\gamma} = 50$.

In the case of $\hat{\gamma} = 5$ all 50 000 updates were accepted. When $\hat{\gamma} = 50$ nearly 82% of the updates were accepted. The larger value of $\hat{\gamma}$ in this trial allowed the network weights to approach the stability boundary more closely. This increases the likelihood that weight updates will be rejected. The dynamics of the plant output in the low $\hat{\gamma}$ experiment are smoother than those of the high $\hat{\gamma}$ trial. This is due, again, to the soft bound on the network gain that results from the update functions dependence on γ and $\hat{\gamma}$.

The performance of the two actor–critic systems was compared by running the systems for an additional 5000 s and comparing the mean squared tracking error and mean squared control output over the 5000- s window. The trial with $\hat{\gamma} = 5$ had a mean squared tracking error of 0.1858 and a mean squared control output of 23.1557. For the $\hat{\gamma} = 50$ case these values were 0.2050 and 26.1874, respectively. Neither of the controllers perform extremely well, but the system with the lower $\hat{\gamma}$ value performs slightly better. This may be due to the fact that the smaller gain constraint results in smoother dynamics which may in turn help to regularize the learning dynamics.

To complete the analysis, the controllers learned after the 5000 step trials of this and the previous section were compared on a fixed set of reference changes over 1000 steps. No adaptation was performed during these test trials. The mean-squared tracking error and mean-squared control action were recorded. These results are reported in Table 2.1 along with the gain of the controllers. The results reported for the stability bias trial are for $\hat{\gamma} = 5.0$. The results show that the step-wise stability approach with the stability bias has the best overall mean-squared tracking error and a lower mean-squared control output than the unconstrained case. The gain of the controller in the step-wise stability constrained trial where no stability bias was used is very large, providing further evidence of the controller parameters getting limited near the stability boundary. The controller gain when stability bias is introduced is the lowest, showing that this controller is most robust to parameter variations within the uncertainty ranges.

TABLE 2.1 Performance on a Fixed Sequence of Reference Changes After 5000 s of Training

	Mean Squared Tracking Error	Mean Squared Control Output	γ
Initial actor–critic after learning:	0.35	6.31	13.62
Without stability analysis	0.11	24.94	∞
With step-wise analysis	0.25	8.92	643.93
With step-wise and gain analysis	0.10	17.13	4.00

2.4.6 Conclusions

The application of a stable learning framework to an uncertain, partially observable, spring–mass–damper system was illustrated. A reinforcement learning algorithm governs the learning of a control policy in an RNN. Updates to the RNN weights are constrained to guarantee stability of the closed loop system represented with uncertain parameters and bounded nonlinearities. The introduction of a stability bias to direct RNN weight updates away from regions of instability improved the ability of the actor–critic system to learn under the stability constraints. Instability was observed in an actor–critic control system when no stability constraints were explicitly enforced. Application of even the relaxed step-wise stability constraint kept the actor from instability.

REFERENCES

1. G.E. Dullerud and F. Paganini. *A Course in Robust Control Theory*. Springer, 2000.
2. J.N. Knight. *Stability Analysis of Recurrent Neural Networks with Applications*. Ph.D. thesis, Department of Computer Science, Colorado State University, 2008.
3. J.N. Knight and C.W. Anderson. Stable reinforcement learning with recurrent neural networks. *Journal of Control Theory and Applications*, 9(3):410–420, 2011.
4. J. Steil. *Input–Output Stability of Recurrent Neural Networks*. Ph.D. Thesis, Der Technischen Fakultät der Universität Bielefeld, 1999.
5. C.W. Anderson, P.M. Young, M. Buehner, J.N. Knight, K.A. Bush, and D.C. Hittle. Robust reinforcement learning control using integral quadratic constraints for recurrent neural networks. *IEEE Transactions on Neural Networks*, 18(4):993–1002, 2006.
6. R.M. Kretchmar. *A Synthesis of Reinforcement Learning and Robust Control Theory*. Ph.D. thesis, Computer Science Department, Colorado State University, 2000.
7. B.A. Pealmutter. Gradient calculations for dynamic recurrent neural networks: a survey. *IEEE Transactions on Neural Networks*, 6(5):1212–1228, 1995.
8. G.-Z. Sun, H.-H. Chen, and Y.C. Lee. Green's function method for fast online learning algorithm of recurrent neural networks. In J.E. Moody, S.J. Hanson, and R.P. Lippmann, editors. *Advances in Neural Information Processing Systems 4*, MIT Press, 1991, pp. 333–340.

9. A.F. Atiya and A.G. Parlos. New results on recurrent network training: unifying the algorithms and accelerating convergence. *IEEE Transactions on Neural Networks*, 11(3):697–709, 2000.
10. R.W. Freund and F. Jarre. A sensitivity analysis and a convergence result for a sequential semidefinite programming method. Technical Report Numerical Analysis Manuscript 03-4-08, Bell Laboratories, 2003.
11. S. Boyd and L. Vandenberghe. *Convex Optimization*. Cambridge University Press, 2004.
12. J.-J. Kim and T. Singh. Desensitized control of vibratory systems with friction: linear programming approach. *Optimal Control Applications and Methods*, 25:165–180, 2004.
13. C. Makkar, W.E. Dixon, W.G. Sawyer, and G. Hu. A new continuously differentiable friction model for control systems design. In *Proceedings of the 2005 IEEE/ASME International Conference on Advanced Intelligent Mechatronics*, 2005, pp. 600–605.
14. U. Mackenroth. *Robust Control Systems: Theory and Case Studies*, Springer-Verlag, 2004.
15. U.D. Schiller. *Analysis and Comparison of Algorithms for Training Recurrent Nerual Networks*. Ph.D. Thesis, University of Bielefeld, 2003.
16. T. Hu, B. Huang, and Z. Lin. Absolute stability with a generalized sector condition. *IEEE Transactions on Automatic Control*, 49(4):535–548, 2004.
17. K. Doya. Reinforcement learning in continuous time and space. *Neural Computation*, 12(1):219–245, 2000.
18. G.G. Lendaris and J.C. Neidhoefer. Guidance in the use of adaptive critics for control. In J. Si, A.G. Barto, W.B. Powell, and D. Wunsch, editors, *Handbook of Learning and Approximate Dynamic Programming*, Wiley-IEEE Press, 2004, pp. 97–124.
19. K. Bush. *An Echo State Model of Non-Markovian Reinforcement Learning*. Ph.D. thesis, Colorado State University, 2008.

CHAPTER 3

Optimal Control of Unknown Nonlinear Discrete-Time Systems Using the Iterative Globalized Dual Heuristic Programming Algorithm

DERONG LIU and DING WANG

State Key Laboratory of Management and Control for Complex Systems, Institute of Automation, Chinese Academy of Sciences, Beijing, PR, China

ABSTRACT

In this chapter, a neuro-optimal control scheme for unknown nonlinear discrete-time systems with discount factor in the cost function is developed. The iterative adaptive dynamic programming (ADP) algorithm using globalized dual heuristic programming (GDHP) technique is introduced to obtain the optimal controller with convergence analysis in terms of cost function and control law. It is proved that the cost function and control law sequences of the iterative ADP algorithm can converge to the optimal ones. A neural network is constructed first to identify the unknown nonlinear system. Then, based on the learned system model, two other neural networks are used as parametric structures to facilitate the implementation of the iterative algorithm, which aims at approximating at each iteration the cost function and the control law, respectively. Finally, two simulation examples are provided to verify the effectiveness of the present optimal control approach. Simulation results show that the iterative ADP algorithm using GDHP technique can obtain better control performance than basic ADP schemes, for example, heuristic dynamic programming.

Reinforcement Learning and Approximate Dynamic Programming for Feedback Control, First Edition. Edited by Frank L. Lewis and Derong Liu.

3.1 BACKGROUND MATERIAL

The optimal control of nonlinear systems often requires to solve the nonlinear Hamilton–Jacobi–Bellman (HJB) equation, which has been a challenging research topic for many years. For example, the discrete-time HJB (DTHJB) equation is more difficult to deal with than the Riccati equation because it involves solving nonlinear partial difference equations. Hence, the adaptive/approximate dynamic programming (ADP) algorithm, which combines dynamic programming (DP) [1] with artificial neural networks (ANN or NN) [2], was proposed in [3–6] to solve the problem forward-in-time. There are several synonyms used for ADP, including "adaptive critic designs" [7, 8], "adaptive dynamic programming" [9–13], "approximate dynamic programming" [4, 14–16], "neural dynamic programming" [17–19], "neuro-dynamic programming" [20], and "reinforcement learning" (RL) [10, 14]. According to [4, 7], ADP approach was classified into several main schemes: heuristic dynamic programming (HDP), action-dependent HDP (ADHDP), also known as Q-learning, dual heuristic dynamic programming (DHP), ADDHP, globalized DHP (GDHP), and ADGDHP. Good surveys about ADP were given in Wang et al. [9], Lewis and Vrabie [10], Lendaris [12], and Balakrishnan et al. [13].

As an emerging and promising intelligent control method, in recent years, ADP and related research have gained much attention from researchers [21–53]. Abu-Khalaf and Lewis [25] designed the nearly optimal control laws for nonlinear systems with saturating actuators using an NN HJB approach. Al-Tamimi et al. [21] proposed a discrete-time nonlinear HJB solution using ADP and also gave the related convergence analysis. Vrabie and Lewis [51] presented an online approach to direct adaptive optimal control with infinite horizon cost for nonlinear systems. Vamvoudakis and Lewis [36] discussed an online algorithm based on policy iteration for learning the continuous-time optimal control solution for nonlinear systems with known dynamics. Liu et al. [22, 28, 31, 32] and Zhang et al. [23, 30] studied the optimal control of discrete-time nonlinear systems by making use of ADP-based techniques. Zhang et al. [33] introduced an iterative ADP method for solving a class of nonlinear zero-sum differential games. Han and Balakrishnan [40] provided a state-constrained agile missile control scheme with adaptive–critic-based NNs. Dierks et al. [26] tackled the optimal control of unknown affine nonlinear discrete-time systems using offline-trained NNs with proof of convergence. Jagannathan and He [42] derived an NN-based state feedback control approach for nonlinear discrete-time systems in nonstrict feedback form. Venayagamoorthy et al. [48] made a comparison between HDP and DHP for neurocontrol of a turbogenerator. Adhyaru et al. [50] dealt with the bounded robust control of nonlinear systems using NN-based HJB solution.

In this chapter, we solve the optimal regulator design problem for unknown nonlinear discrete-time systems with discount factor in the cost function using the GDHP technique. Consider the nonlinear systems described by

$$x_{k+1} = f(x_k) + g(x_k)u(x_k), \tag{3.1}$$

where $x_k \in \mathbb{R}^n$ is the state vector and $u(x_k) \in \mathbb{R}^m$ the control vector, and $f(\cdot)$ and $g(\cdot)$ are differentiable in their arguments with $f(0) = 0$. Assume that $f + gu$ is Lipschitz continuous on a set Ω in $\mathbb{R}^n$ containing the origin, and that the system (3.1) is controllable in the sense that there exists a continuous control on Ω that asymptotically drives the system to the origin. In this chapter, $u(x_k)$ is also denoted by u_k for simplicity.

Let x_0 be an initial state and define $\underline{u}_0^{N-1} = (u_0, u_1, \ldots, u_{N-1})$ be a control sequence with which the system (3.1) gives a trajectory starting from x_0: $x_1 = f(x_0) + g(x_0)u_0$, $x_2 = f(x_1) + g(x_1)u_1$, $\ldots$, $x_N = f(x_{N-1}) + g(x_{N-1})u_{N-1}$. We call the number of elements in the control sequence $\underline{u}_0^{N-1}$ the length of $\underline{u}_0^{N-1}$ and denote it as $\left|\underline{u}_0^{N-1}\right|$. Then, $\left|\underline{u}_0^{N-1}\right| = N$. The final state under the control sequence $\underline{u}_0^{N-1}$ can be denoted as $x^{(f)}\left(x_0, \underline{u}_0^{N-1}\right) = x_N$. When the control sequence starting from u_0 has infinite length, we denote it as $\underline{u}_0^{\infty} = (u_0, u_1, \ldots)$ and then the corresponding final state can be written as $x^{(f)}(x_0, \underline{u}_0^{\infty}) = \lim_{k\to\infty} x_k$.

Let $\underline{u}_k^{\infty} = (u_k, u_{k+1}, \ldots)$ be the control sequence starting at k. It is desired to find the control sequence $\underline{u}_k^{\infty}$ which minimizes the infinite horizon cost function given by

$$J(x_k, \underline{u}_k^{\infty}) = \sum_{i=k}^{\infty} \gamma^{i-k} U(x_i, u_i), \tag{3.2}$$

where U is the utility function, $U(0, 0) = 0$, $U(x_i, u_i) \geq 0$ for $\forall\, x_i, u_i$, and γ is the discount factor with $0 < \gamma \leq 1$. In this chapter, the utility function is chosen as the quadratic form $U(x_i, u_i) = x_i^T Q x_i + u_i^T R u_i$, where Q and R are positive definite matrices.

Note that the designed feedback control in optimal control problem must be admissible. Now, we give the following definition.

Definition 3.1 *A control sequence $\underline{u}_k^{\infty}$ is said to be admissible for a state $x_k \in \mathbb{R}^n$ with respect to (3.2) on Ω if u is continuous on a compact set $\Omega_u \in \mathbb{R}^m$, $u(0) = 0$, $x^{(f)}(x_k, \underline{u}_k^{\infty}) = 0$, and $J(x_k, \underline{u}_k^{\infty})$ is finite.*

Let $\mathfrak{A}_{x_k} = \left\{\underline{u}_k^{\infty} \colon x^{(f)}(x_k, \underline{u}_k^{\infty}) = 0\right\}$ be the set of all infinite horizon admissible control sequences of x_k. Define the optimal cost function as

$$J^*(x_k) = \inf_{\underline{u}_k^{\infty}} \left\{ J(x_k, \underline{u}_k^{\infty}) \colon \underline{u}_k^{\infty} \in \mathfrak{A}_{x_k} \right\}. \tag{3.3}$$

According to Bellman's optimality principle, $J^*(x_k)$ satisfies the DTHJB equation

$$J^*(x_k) = \min_{u_k} \left\{ x_k^T Q x_k + u_k^T R u_k + \gamma J^*(x_{k+1}) \right\}, \tag{3.4}$$

where $x_{k+1} = f(x_k) + g(x_k)u_k$. In addition, the optimal control u^* satisfies the first-order necessary condition, which is given by the gradient of the right-hand side of (3.4) with respect to u_k, that is,

$$u^*(x_k) = -\frac{\gamma}{2} R^{-1} g^T(x_k) \frac{\partial J^*(x_{k+1})}{\partial x_{k+1}}. \tag{3.5}$$

Consequently, the DTHJB Equation (3.4) becomes

$$J^*(x_k) = x_k^T Q x_k + \frac{\gamma^2}{4}\left(\frac{\partial J^*(x_{k+1})}{\partial x_{k+1}}\right)^T g(x_k) R^{-1} \times g^T(x_k) \frac{\partial J^*(x_{k+1})}{\partial x_{k+1}} + \gamma J^*(x_{k+1}).$$

When dealing with the linear quadratic regulator (LQR) problem, the DTHJB equation reduces to the Riccati equation which can be efficiently solved. In the general nonlinear case, however, it cannot be solved exactly. Therefore, we will develop a novel algorithm to approximate the cost function iteratively in the following section.

Remark 3.1 *It is important to note that a control law sequence* $\{\eta_i\} = (\eta_N, \ldots, \eta_1, \eta_0)$, $N \to \infty$, *is called admissible if the resultant control sequence* $(u_0, u_1, \ldots, u_N)$ *is admissible. It means that each control action obeys a different control law, that is, the control action* u_i *is produced by the control law* η_{N-i} *or* $u_i = \eta_{N-i}(x_i)$, *for* $i = 0, 1, \ldots, N$, $N \to \infty$.

3.2 NEURO-OPTIMAL CONTROL SCHEME BASED ON THE ITERATIVE ADP ALGORITHM

3.2.1 Identification of the Unknown Nonlinear System

We choose the three-layer feedforward NN as the function approximation structure. Let the number of hidden layer neurons be denoted by l, the ideal weight matrix between the input layer and hidden layer be denoted by ν_m^*, and the ideal weight matrix between the hidden layer and output layer be denoted by ω_m^*. According to the universal approximation property of NN, the system dynamics (3.1) has an NN representation on a compact set S, which can be written as

$$x_{k+1} = \omega_m^{*T} \sigma\left(\nu_m^{*T} z_k\right) + \theta_k = \omega_m^{*T} \sigma\left(\bar{z}_k\right) + \theta_k, \tag{3.6}$$

where $\bar{z}_k = \nu_m^{*T} z_k$, $\bar{z}_k \in \mathbb{R}^l$. In Equation (3.6), $z_k = [x_k^T \; u_k^T]^T$ is the NN input, θ_k is the bounded NN functional approximation error according to the universal approximation property, and the activation functions $\sigma(\cdot)$ are defined by $[\sigma(\xi)]_i = (e^{\xi_i} - e^{-\xi_i})/(e^{\xi_i} + e^{-\xi_i})$, $i = 1, 2, \ldots, l$, for $\xi \in \mathbb{R}^l$. Additionally, the NN activation functions are bounded such that $\|\sigma(\bar{z}_k)\| \le \sigma_M$ for a constant σ_M.

In the system identification process, we keep the weight matrix between the input layer and the hidden layer as constant while only tune the weight matrix between the hidden layer and the output layer. Hence, we define the NN system identification scheme as

$$\hat{x}_{k+1} = \omega_m^T(k) \sigma(\bar{z}_k), \tag{3.7}$$

where $\hat{x}_k$ is the estimated system state vector, and $\omega_m(k)$ is the estimation of the constant ideal weight matrix ω_m^*.

Denote $\tilde{x}_k = \hat{x}_k - x_k$ as the state estimation error. Then, combining Equations (3.6) and (3.7), we can obtain the state estimation error dynamics as

$$\begin{aligned}\tilde{x}_{k+1} &= \hat{x}_{k+1} - x_{k+1} \\ &= \tilde{\omega}_m^T(k)\sigma\left(v_m^{*T}\left[(\hat{x}_k - \tilde{x}_k)^T \ \ u_k^T\right]^T\right) - \theta_k,\end{aligned} \tag{3.8}$$

where $\tilde{\omega}_m(k) = \omega_m(k) - \omega_m^*$. The system (3.8) is in fact

$$\tilde{x}_{k+1} = \tilde{\omega}_m^T(k)\sigma(\bar{z}_k) - \theta_k. \tag{3.9}$$

Let $\psi_k = \tilde{\omega}_m^T(k)\sigma(\bar{z}_k)$. Then, Equation (3.9) can be rewritten as

$$\tilde{x}_{k+1} = \psi_k - \theta_k. \tag{3.10}$$

The weights in the system identification process are updated to minimize the performance measure $E_{k+1} = \tilde{x}_{k+1}^T\tilde{x}_{k+1}/2$. Using the gradient-based adaptation rule, the weights can be updated as

$$\begin{aligned}\omega_m(k+1) &= \omega_m(k) - \alpha_m\left[\frac{\partial E_{k+1}}{\partial \omega_m(k)}\right] \\ &= \omega_m(k) - \alpha_m\sigma(\bar{z}_k)\tilde{x}_{k+1}^T,\end{aligned} \tag{3.11}$$

where $\alpha_m > 0$ is the NN learning rate. Accordingly, we obtain the weight estimation error dynamics as

$$\begin{aligned}\tilde{\omega}_m(k+1) &= \omega_m(k+1) - \omega_m^* \\ &= \tilde{\omega}_m(k) - \alpha_m\sigma(\bar{z}_k)\left(\sigma^T(\bar{z}_k)\tilde{\omega}_m(k) - \theta_k^T\right).\end{aligned} \tag{3.12}$$

The following assumption is necessary before we present the stability analysis of error dynamics (3.8) and (3.12) using the Lyapunov theory.

Assumption 3.1 *The NN approximation error term θ_k is assumed to be upper bounded by a function of the state estimation error $\tilde{x}_k$, that is,*

$$\theta_k^T\theta_k \leq \delta_M\tilde{x}_k^T\tilde{x}_k,$$

where δ_M is a constant value.

Assumption 3.1 has been used by many researchers [26, 27] and is considered mild in comparison with the approximation error being bounded by a known constant.

Theorem 3.1 *Let the identification scheme (3.7) be used to identify the nonlinear system (3.1), and let the weight update law given in Equation (3.11) be used for tuning the NN weights. Then, the state estimation error dynamics (3.8) is asymptotically stable while the weight estimation error dynamics (3.12) is stable in the sense of Lyapunov.*

Proof: Consider the following positive definite Lyapunov function candidate

$$L_k = L_{1k} + L_{2k}, \tag{3.13}$$

where $L_{1k} = \tilde{x}_k^T \tilde{x}_k$ and $L_{2k} = \mathrm{tr}\{\tilde{\omega}_m^T(k)\tilde{\omega}_m(k)\}/\alpha_m$. Taking the first difference of (3.13) and substituting (3.10) and (3.11) reveal that

$$\begin{aligned}
\Delta L_{1k} &= \tilde{x}_{k+1}^T \tilde{x}_{k+1} - \tilde{x}_k^T \tilde{x}_k \\
&= \psi_k^T \psi_k - 2\psi_k^T \theta_k + \theta_k^T \theta_k - \tilde{x}_k^T \tilde{x}_k, \\
\Delta L_{2k} &= \frac{1}{\alpha_m}\mathrm{tr}\{\tilde{\omega}_m^T(k+1)\tilde{\omega}_m(k+1) - \tilde{\omega}_m^T(k)\tilde{\omega}_m(k)\} \\
&= \frac{1}{\alpha_m}\mathrm{tr}\{-2\alpha_m \psi_k \tilde{x}_{k+1}^T + \alpha_m^2 \tilde{x}_{k+1}\sigma^T(\bar{z}_k)\sigma(\bar{z}_k)\tilde{x}_{k+1}^T\} \\
&= -2\psi_k^T \tilde{x}_{k+1} + \alpha_m \sigma^T(\bar{z}_k)\sigma(\bar{z}_k)\tilde{x}_{k+1}^T \tilde{x}_{k+1}.
\end{aligned}$$

After applying the Cauchy–Schwarz inequality $\big((a_1 + a_2 + \cdots + a_n)^T(a_1 + a_2 + \cdots + a_n) \le n(a_1^T a_1 + a_2^T a_2 + \cdots + a_n^T a_n)\big)$ to ΔL_{2k}, we have

$$\Delta L_{2k} \le -2\psi_k^T(\psi_k - \theta_k) + 2\alpha_m \sigma^T(\bar{z}_k)\sigma(\bar{z}_k)\big(\psi_k^T \psi_k + \theta_k^T \theta_k\big).$$

Therefore, we can find that

$$\Delta L_k \le -\psi_k^T \psi_k + \theta_k^T \theta_k - \tilde{x}_k^T \tilde{x}_k + 2\alpha_m \sigma^T(\bar{z}_k)\sigma(\bar{z}_k)\big(\psi_k^T \psi_k + \theta_k^T \theta_k\big).$$

Considering $\|\sigma(\bar{z}_k)\| \le \sigma_M$ and Assumption 3.1, we can obtain

$$\Delta L_k \le -\big(1 - 2\alpha_m \sigma_M^2\big)\|\psi_k\|^2 - \big(1 - \delta_M - 2\alpha_m \delta_M \sigma_M^2\big)\|\tilde{x}_k\|^2. \tag{3.14}$$

Define $\alpha_m \le \rho^2/(2\sigma_M^2)$, then (3.14) becomes

$$\begin{aligned}
\Delta L_k &\le -\big(1 - \rho^2\big)\|\psi_k\|^2 - \big(1 - \delta_M - \delta_M \rho^2\big)\|\tilde{x}_k\|^2 \\
&= -\big(1 - \rho^2\big)\big\|\tilde{\omega}_m^T(k)\sigma(\bar{z}_k)\big\|^2 - \big(1 - \delta_M - \delta_M \rho^2\big)\|\tilde{x}_k\|^2. \qquad (3.15)
\end{aligned}$$

From (3.15), we can conclude that $\Delta L_k \leq 0$ if $0 < \delta_M < 1$ and if

$$\max\left\{-\sqrt{\frac{1-\delta_M}{\delta_M}}, -1\right\} \leq \rho \leq \min\left\{\sqrt{\frac{1-\delta_M}{\delta_M}}, 1\right\},$$

where $\rho \neq 0$. As long as the parameters are selected as discussed earlier, $\Delta L_k \leq 0$ in (3.15) which shows stability of systems (3.8) and (3.12) in the sense of Lyapunov. Therefore, $\tilde{x}_k$ and $\tilde{\omega}_m(k)$ are bounded, provided $\tilde{x}_0$ and $\tilde{\omega}_m(0)$ are bounded in the compact set S. Furthermore, by summing both sides of (3.15) to infinity and taking account of $\Delta L_k \leq 0$, we have

$$\left|\sum_{k=0}^{\infty} \Delta L_k\right| \geq \sum_{k=0}^{\infty}\left\{\left(1-\rho^2\right)\left\|\tilde{\omega}_m^T(k)\sigma(\bar{z}_k)\right\|^2 + \left(1-\delta_M-\delta_M\rho^2\right)\|\tilde{x}_k\|^2\right\}$$

and

$$\left|\sum_{k=0}^{\infty} \Delta L_k\right| = \left|\lim_{k\to\infty} L_k - L_0\right| < \infty.$$

Accordingly,

$$\sum_{k=0}^{\infty}\left\{\left(1-\rho^2\right)\left\|\tilde{\omega}_m^T(k)\sigma(\bar{z}_k)\right\|^2 + \left(1-\delta_M-\delta_M\rho^2\right)\|\tilde{x}_k\|^2\right\} < \infty.$$

It can be concluded that the state estimation error approaches zero, that is, $\|\tilde{x}_k\| \to 0$ as $k \to \infty$. ■

According to Theorem 3.1, after a sufficient learning session, the state estimation error converges to zero, that is, we have

$$f(x_k) + \hat{g}(x_k)u_k = \omega_m^T(k)\sigma(\bar{z}_k), \tag{3.16}$$

where $\hat{g}(x_k)$ denotes the estimated value of the control coefficient matrix $g(x_k)$. Taking the partial derivative of both sides of (3.16) with respect to u_k, $\hat{g}(x_k)$ can be obtained as follows:

$$\begin{aligned}\hat{g}(x_k) &= \frac{\partial\left(\omega_m^T(k)\sigma(\bar{z}_k)\right)}{\partial u_k} \\ &= \omega_m^T(k)\frac{\partial\sigma(\bar{z}_k)}{\partial\bar{z}_k}v_m^{*T}\begin{bmatrix}0_{n\times m}\\ \hdashline I_m\end{bmatrix},\end{aligned} \tag{3.17}$$

where $0_{n\times m}$ is an $n \times m$ zero matrix and I_m is an $m \times m$ identity matrix.

3.2.2 Derivation of the Iterative ADP Algorithm

Now, we present the iterative ADP algorithm. First, we start with the initial cost function

$$V_0(\cdot) = 0, \tag{3.18}$$

and solve the initial control law by

$$\begin{aligned} v_0(x_k) &= \arg\min_{u_k} \left\{x_k^T Q x_k + u_k^T R u_k + \gamma V_0(x_{k+1})\right\} \\ &= \arg\min_{u_k} \left\{x_k^T Q x_k + u_k^T R u_k + \gamma V_0(f(x_k) + g(x_k)u_k)\right\}. \end{aligned} \tag{3.19}$$

Then, we update the cost function by

$$\begin{aligned} V_1(x_k) &= \min_{u_k} \left\{x_k^T Q x_k + u_k^T R u_k + \gamma V_0(x_{k+1})\right\} \\ &= x_k^T Q x_k + v_0^T(x_k) R v_0(x_k) + \gamma V_0(f(x_k) + g(x_k)v_0(x_k)). \end{aligned} \tag{3.20}$$

For $i = 1, 2, \ldots,$ the iteration can be implemented between

$$\begin{aligned} v_i(x_k) &= \arg\min_{u_k} \left\{x_k^T Q x_k + u_k^T R u_k + \gamma V_i(x_{k+1})\right\} \\ &= \arg\min_{u_k} \left\{x_k^T Q x_k + u_k^T R u_k + \gamma V_i(f(x_k) + g(x_k)u_k)\right\}, \end{aligned} \tag{3.21}$$

and

$$\begin{aligned} V_{i+1}(x_k) &= \min_{u_k} \left\{x_k^T Q x_k + u_k^T R u_k + \gamma V_i(x_{k+1})\right\} \\ &= x_k^T Q x_k + v_i^T(x_k) R v_i(x_k) + \gamma V_i(f(x_k) + g(x_k)v_i(x_k)). \end{aligned} \tag{3.22}$$

In the earlier iteration process, i is the iteration index of the control law and cost function, while k is the time index of the system's control and state trajectories. The cost function and control law are updated until they converge to the optimal ones. In the following part, we will present the convergence analysis of the iteration process (3.18)–(3.22) to show that the cost function $V_i \rightarrow J^*$ and the control law $v_i \rightarrow u^*$ as $i \rightarrow \infty$.

3.2.3 Convergence Analysis of the Iterative ADP Algorithm

Lemma 3.1 (cf. [21]) *Let $\{\mu_i\}$ be an arbitrary sequence of control laws and $\{v_i\}$ be the control law sequence described in (3.21). Define V_i as in (3.22) and Λ_i as*

$$\Lambda_{i+1}(x_k) = x_k^T Q x_k + \mu_i^T(x_k) R \mu_i(x_k) + \gamma \Lambda_i(x_{k+1}),$$

where $x_{k+1} = f(x_k) + g(x_k)\mu_i(x_k)$. *If* $V_0(\cdot) = \Lambda_0(\cdot) = 0$, *then* $V_{i+1}(x) \leq \Lambda_{i+1}(x)$, $\forall i$.

Lemma 3.2 *Let the cost function sequence* $\{V_i\}$ *be defined as in (3.22). If the system is controllable, then there is an upper bound* Y *such that* $0 \leq V_i(x_k) \leq Y$, $\forall i$.

Proof: Let $\{\eta_i(x)\}$ be a sequence of admissible control laws, and $V_0(\cdot) = Z_0(\cdot) = 0$, where V_i is updated as in (3.22) and Z_i is updated by

$$Z_{i+1}(x_k) = x_k^T Q x_k + \eta_i^T(x_k) R \eta_i(x_k) + \gamma Z_i(x_{k+1}),$$

where $x_{k+1} = f(x_k) + g(x_k)\eta_i(x_k)$. It is clear that

$$Z_i(x_{k+1}) = x_{k+1}^T Q x_{k+1} + \eta_{i-1}^T(x_{k+1}) R \eta_{i-1}(x_{k+1}) + \gamma Z_{i-1}(x_{k+2}),$$

where $x_{k+2} = f(x_{k+1}) + g(x_{k+1})\eta_{i-1}(x_{k+1})$. Noticing that $U(x_k, \eta_i(x_k)) = x_k^T Q x_k + \eta_i^T(x_k) R \eta_i(x_k)$, we can further obtain

$$\begin{aligned}
Z_{i+1}(x_k) &= U(x_k, \eta_i(x_k)) + \gamma U(x_{k+1}, \eta_{i-1}(x_{k+1})) + \gamma^2 Z_{i-1}(x_{k+2}) \\
&= U(x_k, \eta_i(x_k)) + \gamma U(x_{k+1}, \eta_{i-1}(x_{k+1})) \\
&\quad + \gamma^2 U(x_{k+2}, \eta_{i-2}(x_{k+2})) + \gamma^3 Z_{i-2}(x_{k+3}) \\
&\ \ \vdots \\
&= U(x_k, \eta_i(x_k)) + \gamma U(x_{k+1}, \eta_{i-1}(x_{k+1})) + \gamma^2 U(x_{k+2}, \eta_{i-2}(x_{k+2})) \\
&\quad + \cdots + \gamma^i U(x_{k+i}, \eta_0(x_{k+i})) + \gamma^{i+1} Z_0(x_{k+i+1}),
\end{aligned} \tag{3.23}$$

where $Z_0(x_{k+i+1}) = 0$. Then, (3.23) can be written as

$$\begin{aligned}
Z_{i+1}(x_k) &= \sum_{j=0}^{i} \gamma^j U(x_{k+j}, \eta_{i-j}(x_{k+j})) \\
&= \sum_{j=0}^{i} \gamma^j \left(x_{k+j}^T Q x_{k+j} + \eta_{i-j}^T(x_{k+j}) R \eta_{i-j}(x_{k+j}) \right) \\
&\leq \lim_{i \to \infty} \sum_{j=0}^{i} \gamma^j \left(x_{k+j}^T Q x_{k+j} + \eta_{i-j}^T(x_{k+j}) R \eta_{i-j}(x_{k+j}) \right).
\end{aligned}$$

Since $\{\eta_i(x)\}$ is an admissible control law sequence, from Definition 3.1, we have $x_k \to 0$ as $k \to \infty$, and there exists an upper bound Y such that

$$Z_{i+1}(x_k) \leq \lim_{i \to \infty} \sum_{j=0}^{i} \gamma^j U(x_{k+j}, \eta_{i-j}(x_{k+j})) \leq Y, \ \forall i.$$

By Lemma 3.1, we can obtain

$$V_{i+1}(x_k) \leq Z_{i+1}(x_k) \leq Y, \ \forall i.$$

Thus, we complete the proof. ■

Theorem 3.2 *Define the cost function sequence $\{V_i\}$ as in (3.22) with $V_0(\cdot) = 0$, and the control law sequence $\{v_i\}$ as in (3.21). Then, $\{V_i\}$ is a monotonically nondecreasing sequence satisfying $V_i \leq V_{i+1}$, $\forall i$.*

Proof: Define a new sequence

$$\Phi_{i+1}(x_k) = x_k^T Q x_k + v_{i+1}^T(x_k) R v_{i+1}(x_k) + \gamma \Phi_i(x_{k+1}),$$

with $\Phi_0(\cdot) = V_0(\cdot) = 0$, where $x_{k+1} = f(x_k) + g(x_k)v_{i+1}(x_k)$. Let the control law sequence $\{v_i\}$ and the cost function sequence $\{V_i\}$ be updated as in (3.21) and (3.22), respectively.

Now, we prove $\Phi_i(x_k) \leq V_{i+1}(x_k)$ by mathematical induction. First, we prove that it holds for $i = 0$. Considering

$$V_1(x_k) - \Phi_0(x_k) = x_k^T Q x_k + v_0^T(x_k) R v_0(x_k) \geq 0,$$

for $i = 0$, we get

$$\Phi_0(x_k) \leq V_1(x_k).$$

Second, we assume that it holds for $i - 1$, that is, $\Phi_{i-1}(x_k) \leq V_i(x_k)$, $\forall x_k$. Then, for i, noticing that

$$V_{i+1}(x_k) = x_k^T Q x_k + v_i^T(x_k) R v_i(x_k) + \gamma V_i(x_{k+1}),$$

and

$$\Phi_i(x_k) = x_k^T Q x_k + v_i^T(x_k) R v_i(x_k) + \gamma \Phi_{i-1}(x_{k+1}),$$

we can get

$$V_{i+1}(x_k) - \Phi_i(x_k) = \gamma(V_i(x_{k+1}) - \Phi_{i-1}(x_{k+1})) \geq 0,$$

that is,

$$\Phi_i(x_k) \leq V_{i+1}(x_k).$$

Thus, we complete the proof through mathematical induction.

Furthermore, from Lemma 3.1, we know that $V_i(x_k) \leq \Phi_i(x_k)$. Therefore, we have

$$V_i(x_k) \leq \Phi_i(x_k) \leq V_{i+1}(x_k),$$

and complete the proof. ■

We have derived the conclusion that the cost function sequence $\{V_i\}$ is a monotonically nondecreasing sequence with an upper bound, and therefore, its limit exists. Define $\lim_{i\to\infty} V_i(x_k) = V_\infty(x_k)$ as the limit of the cost function sequence $\{V_i\}$. Now, we prove the following theorem.

Theorem 3.3 *Define the cost function sequence $\{V_i\}$ as in (3.22) with $V_0(\cdot) = 0$. If the system state x_k is controllable, then J^* is the limit of the cost function sequence $\{V_i\}$, that is, $\lim_{i\to\infty} V_i(x_k) = J^*(x_k)$, where J^* is defined in (3.3).*

Proof: Let $\{\eta_i^{(l)}\}$ be the lth admissible control law sequence. We construct the associated sequence $\{P_i^{(l)}(x)\}$ as follows:

$$P_{i+1}^{(l)}(x_k) = x_k^T Q x_k + \left(\eta_i^{(l)}(x_k)\right)^T R \eta_i^{(l)}(x_k) + \gamma P_i^{(l)}(x_{k+1}),$$

with $P_0^{(l)}(\cdot) = 0$, where $x_{k+1} = f(x_k) + g(x_k)\eta_i^{(l)}(x_k)$. Similar to the derivation of (3.23), we get

$$P_{i+1}^{(l)}(x_k) = \sum_{j=0}^{i} \gamma^j U\left(x_{k+j}, \eta_{i-j}^{(l)}(x_{k+j})\right). \tag{3.24}$$

Using Lemmas 3.1 and 3.2, we have

$$V_{i+1}(x_k) \leq P_{i+1}^{(l)}(x_k) \leq Y_l, \ \forall l, i,$$

where Y_l is the upper bound associated with the sequence $\{P_i^{(l)}(x_k)\}$. Denote $\lim_{i\to\infty} P_i^{(l)}(x_k) = P_\infty^{(l)}(x_k)$. Then, we can obtain

$$V_\infty(x_k) \leq P_\infty^{(l)}(x_k) \leq Y_l, \ \forall l. \tag{3.25}$$

Let the corresponding control sequence associated with (3.24) be

$$\begin{aligned} {}^{(l)}\hat{\underline{u}}_k^{k+i} &= \left({}^{(l)}\hat{u}_k, {}^{(l)}\hat{u}_{k+1}, \ldots, {}^{(l)}\hat{u}_{k+i}\right) \\ &= \left(\eta_i^{(l)}(x_k), \eta_{i-1}^{(l)}(x_{k+1}), \ldots, \eta_0^{(l)}(x_{k+i})\right). \end{aligned}$$

Then, we have

$$J\big(x_k, {}^{(l)}\underline{\hat{u}}_k^{k+i}\big) = \sum_{j=0}^{i} \gamma^j U\big(x_{k+j}, \eta_{i-j}^{(l)}(x_{k+j})\big) = P_{i+1}^{(l)}(x_k).$$

Letting $i \to \infty$, and denoting the admissible control sequence related to $P_{\infty}^{(l)}(x_k)$ with length ∞ as ${}^{(l)}\underline{\hat{u}}_k^{\infty}$, we get

$$J\big(x_k, {}^{(l)}\underline{\hat{u}}_k^{\infty}\big) = \sum_{j=0}^{\infty} \gamma^j U\big(x_{k+j}, {}^{(l)}\hat{u}_{k+j}\big) = P_{\infty}^{(l)}(x_k). \tag{3.26}$$

Then, according to the definition of $J^*(x_k)$ in (3.3), for any $\varepsilon > 0$, there exists a sequence of admissible control laws $\{\eta_i^{(M)}\}$ such that the associated cost function

$$J\big(x_k, {}^{(M)}\underline{\hat{u}}_k^{\infty}\big) = \sum_{j=0}^{\infty} \gamma^j U\big(x_{k+j}, {}^{(M)}\hat{u}_{k+j}\big) = P_{\infty}^{(M)}(x_k),$$

satisfies $J\big(x_k, {}^{(M)}\underline{\hat{u}}_k^{\infty}\big) \leq J^*(x_k) + \varepsilon$. Combining with (3.25), we have

$$V_{\infty}(x_k) \leq P_{\infty}^{(M)}(x_k) \leq J^*(x_k) + \varepsilon.$$

Since ε is chosen arbitrarily in the above expression, we get

$$V_{\infty}(x_k) \leq J^*(x_k). \tag{3.27}$$

On the other hand, as $V_{i+1}(x_k) \leq P_{i+1}^{(l)}(x_k) \leq Y_l, \forall l, i$, we can get $V_{\infty}(x_k) \leq \inf_l\{Y_l\}$. According to the definition of admissible control law sequence, the control law sequence associated with the cost function $V_{\infty}(x_k)$ must be an admissible control law sequence. Then, we can obtain that there exists a sequence of admissible control laws $\{\eta_i^{(N)}\}$ such that $V_{\infty}(x_k) = P_{\infty}^{(N)}(x_k)$. Combining with (3.26), we get $V_{\infty}(x_k) = J\big(x_k, {}^{(N)}\underline{\hat{u}}_k^{\infty}\big)$. Since $J^*(x_k)$ is the infimum of cost functions derived from all admissible control sequences starting at k with length ∞, we can obtain

$$J^*(x_k) \leq V_{\infty}(x_k). \tag{3.28}$$

Based on (3.27) and (3.28), we can conclude that J^* is the limit of the cost function sequence $\{V_i\}$, that is, $V_{\infty}(x_k) = J^*(x_k)$. ■

From Theorems 3.2 and 3.3, we can obtain that the cost function sequence $\{V_i(x_k)\}$ converges to the optimal cost function $J^*(x_k)$ of the DTHJB equation, that is, $V_i \to J^*$ as $i \to \infty$. Then, according to (3.5) and (3.21), we can conclude the convergence of the corresponding control law sequence. Now, we present the following corollary.

Corollary 3.1 *Define the cost function sequence $\{V_i\}$ as in (3.22) with $V_0(\cdot) = 0$, and the control law sequence $\{v_i\}$ as in (3.21). If the system state x_k is controllable, then the sequence $\{v_i\}$ converges to the optimal control law u^* as $i \to \infty$, i.e.,* $\lim_{i\to\infty} v_i(x_k) = u^*(x_k)$.

3.2.4 Design Procedure of the Iterative ADP Algorithm

The detailed design procedure for optimal control of nonlinear systems using the iterative ADP algorithm is as follows.

Step 1: Specify the maximum number of iterations $i_{\max}$, the prespecified accuracy ε, the initial state x_0 and the weight matrices Q and R.

Step 2: Set $i = 0$, let $V_0(x_k) = 0$ and solve the initial control law $v_0(x_k)$ by (3.19).

Step 3: Update the cost function $V_1(x_k)$ by (3.20).

Step 4: Set $i = i + 1$.

Step 5: Compute $v_i(x_k)$ by (3.21) and update the corresponding cost function $V_{i+1}(x_k)$ by (3.22).

Step 6: If $|V_i(x_k) - V_{i+1}(x_k)| \leq \varepsilon$, then go to Step 8; otherwise, go to Step 7.

Step 7: If $i > i_{\max}$, then go to Step 8; otherwise, go to Step 4.

Step 8: Stop.

Remark 3.2 *Similar to (3.23), when expanding (3.22), we can obtain a control law sequence $(v_i, v_{i-1}, \ldots, v_0)$ and then derive the resultant control sequence $(v_i(x_0), v_{i-1}(x_1), \ldots, v_0(x_i))$. Then, using the corresponding control sequence, we can obtain a state trajectory. However, it is not derived from a single control law. With the iteration number increasing to ∞, the derived control law sequence has the length of ∞. For infinite horizon optimal control problem, what we should get is a unique optimal control law under which we can obtain the optimal state trajectory. Therefore, we only use the optimal control law u^* obtained in Corollary 3.1 to produce a control sequence when we apply the algorithm to practical systems.*

3.2.5 NN Implementation of the Iterative ADP Algorithm Using GDHP Technique

In the iterative GDHP algorithm, there are three networks, which are model network, critic network, and action network. All the networks are chosen as three-layer feedforward NNs. The diagram of the whole structure is shown in Figure 3.1, where

$$\text{DER} = \left(\frac{\partial \hat{x}_{k+1}}{\partial x_k} + \frac{\partial \hat{x}_{k+1}}{\partial \hat{v}_{i-1}(x_k)} \frac{\partial \hat{v}_{i-1}(x_k)}{\partial x_k} \right)^T.$$

It can be seen that the critic network of the iterative GDHP algorithm outputs both the cost function and its derivatives, which is different from HDP and DHP algorithms.

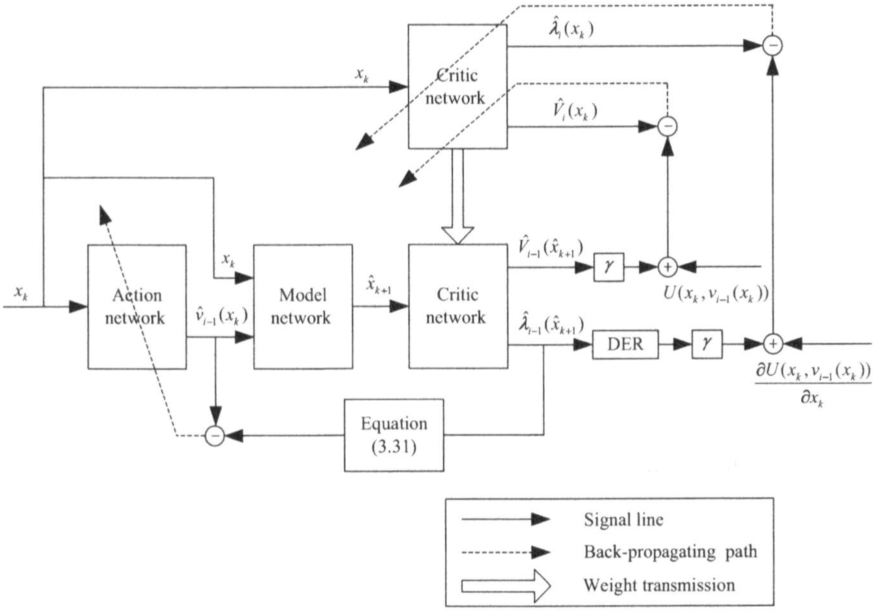

FIGURE 3.1 The structure diagram of the iterative GDHP algorithm.

The training of the model network is completed after the system identification process and its weights are kept unchanged. Then, according to Theorem 3.1, when given x_k and $v_i(x_k)$, we can compute $\hat{x}_{k+1}$ by (3.7). Besides, the $\hat{g}(x_k)$ obtained in (3.17) can be applied to express the control law. As a result, we avoid the requirement of knowing $f(x_k)$ and $g(x_k)$ during the implementation of the iterative GDHP algorithm. Next, the learned NN system model will be used in the process of training the critic network and the action network.

The critic network is used to approximate both $V_i(x_k)$ and its derivative $\partial V_i(x_k)/\partial x_k$, which is named costate function and denoted as $\lambda_i(x_k)$. The output of the critic network is formulated as

$$\begin{bmatrix} \hat{V}_i(x_k) \\ \hat{\lambda}_i(x_k) \end{bmatrix} = \begin{bmatrix} \omega_{ci}^{1T} \\ \omega_{ci}^{2T} \end{bmatrix} \sigma\left(\nu_{ci}^T x_k\right) = \omega_{ci}^T \sigma\left(\nu_{ci}^T x_k\right),$$

where $\omega_{ci} = \left[\omega_{ci}^1 \;\; \omega_{ci}^2\right]$. Then, we have

$$\hat{V}_i(x_k) = \omega_{ci}^{1T} \sigma\left(\nu_{ci}^T x_k\right),$$

and

$$\hat{\lambda}_i(x_k) = \omega_{ci}^{2T} \sigma\left(\nu_{ci}^T x_k\right).$$

The target functions can be written as

$$V_i(x_k) = x_k^T Qx_k + v_{i-1}^T(x_k)Rv_{i-1}(x_k) + \gamma \hat{V}_{i-1}(\hat{x}_{k+1}),$$

and

$$\begin{aligned}\lambda_i(x_k) &= \frac{\partial\big(x_k^T Qx_k + v_{i-1}^T(x_k)Rv_{i-1}(x_k)\big)}{\partial x_k} + \gamma\frac{\partial \hat{V}_{i-1}(\hat{x}_{k+1})}{\partial x_k}\\ &= 2Qx_k + 2\left(\frac{\partial v_{i-1}(x_k)}{\partial x_k}\right)^T Rv_{i-1}(x_k)\\ &\quad + \gamma\left(\frac{\partial \hat{x}_{k+1}}{\partial x_k} + \frac{\partial \hat{x}_{k+1}}{\partial \hat{v}_{i-1}(x_k)}\frac{\partial \hat{v}_{i-1}(x_k)}{\partial x_k}\right)^T \hat{\lambda}_{i-1}(\hat{x}_{k+1}).\end{aligned}$$

Then, we define the error functions for training the critic network as

$$e_{cik}^1 = \hat{V}_i(x_k) - V_i(x_k),$$

and

$$e_{cik}^2 = \hat{\lambda}_i(x_k) - \lambda_i(x_k).$$

The objective function to be minimized in the training process of critic network is $E_{cik} = (1-\beta)E_{cik}^1 + \beta E_{cik}^2$, where $E_{cik}^1 = e_{cik}^{1T}e_{cik}^1/2$ and $E_{cik}^2 = e_{cik}^{2T}e_{cik}^2/2$. The weight updating rule for training the critic network is also gradient-based adaptation given by

$$\omega_{ci}(j+1) = \omega_{ci}(j) - \alpha_c\left[(1-\beta)\frac{\partial E_{cik}^1}{\partial \omega_{ci}(j)} + \beta\frac{\partial E_{cik}^2}{\partial \omega_{ci}(j)}\right], \tag{3.29}$$

$$\nu_{ci}(j+1) = \nu_{ci}(j) - \alpha_c\left[(1-\beta)\frac{\partial E_{cik}^1}{\partial \nu_{ci}(j)} + \beta\frac{\partial E_{cik}^2}{\partial \nu_{ci}(j)}\right], \tag{3.30}$$

where $\alpha_c > 0$ is the learning rate of the critic network, j inner-loop iterative step for updating the weight parameters, and $0 \le \beta \le 1$ parameter that adjusts how HDP and DHP are combined in GDHP. For $\beta = 0$, the training of the critic network reduces to a pure HDP, while $\beta = 1$ does the same for DHP.

Remark 3.3 *According to Theorem 3.3, $V_i(x_k) \to J^*(x_k)$ as $i \to \infty$. Since $\lambda_i(x_k) = \partial V_i(x_k)/\partial x_k$, we can conclude that the costate function sequence $\{\lambda_i(x_k)\}$ is also convergent with $\lambda_i(x_k) \to \lambda^*(x_k)$ as $i \to \infty$.*

Remark 3.4 *From Figure 3.1, we can see that the outputs of critic network of the iterative GDHP algorithm contain not only the cost function but also its derivatives. This is important because the information associated with the cost function is as useful* *as the knowledge of its derivatives. Besides, as is shown in (3.29) and (3.30), training the critic network of the iterative GDHP algorithm utilizes an error measure which is a combination of the error measures of HDP and DHP. Though more complicated to do this, the resulting behavior is expected to be superior to simple ADP schemes.*

The action network is used to approximate the control law and its output can be expressed as

$$\hat{v}_{i-1}(x_k) = \omega_{a(i-1)}^T \sigma\left(\nu_{a(i-1)}^T x_k\right).$$

According to (3.21), the target control input is given by

$$v_{i-1}(x_k) = -\frac{\gamma}{2} R^{-1} \hat{g}^T(x_k) \hat{\lambda}_{i-1}(\hat{x}_{k+1}). \tag{3.31}$$

The error function of the action network is defined as

$$e_{a(i-1)k} = \hat{v}_{i-1}(x_k) - v_{i-1}(x_k).$$

The weights of the action network are updated to minimize the error measure $E_{a(i-1)k} = e_{a(i-1)k}^T e_{a(i-1)k}/2$. Similarly, the weight updating algorithm is

$$\omega_{a(i-1)}(j+1) = \omega_{a(i-1)}(j) - \alpha_a \left[\frac{\partial E_{a(i-1)k}}{\partial \omega_{a(i-1)}(j)}\right],$$
$$\nu_{a(i-1)}(j+1) = \nu_{a(i-1)}(j) - \alpha_a \left[\frac{\partial E_{a(i-1)k}}{\partial \nu_{a(i-1)}(j)}\right],$$

where $\alpha_a > 0$ is the learning rate of the action network, and j is the inner-loop iterative step for updating the weight parameters.

3.3 GENERALIZATION

Now, we consider the nonlinear discrete-time systems given by

$$x_{k+1} = F(x_k, u_k), \; k = 0, 1, 2, \ldots, \tag{3.32}$$

where $F(x_k, u_k)$ is continuous for $\forall\, x_k, u_k$ and $F(0, 0) = 0$. Note that $x = 0$ is an equilibrium state of system (3.32) under the control $u = 0$. In this case, the iteration between (3.21) and (3.22) becomes

$$\begin{aligned} v_i(x_k) &= \arg\min_{u_k} \left\{x_k^T Q x_k + u_k^T R u_k + \gamma V_i(x_{k+1})\right\} \\ &= \arg\min_{u_k} \left\{x_k^T Q x_k + u_k^T R u_k + \gamma V_i(F(x_k, u_k))\right\}, \end{aligned} \tag{3.33}$$

and

$$\begin{aligned} V_{i+1}(x_k) &= \min_{u_k} \left\{ x_k^T Q x_k + u_k^T R u_k + \gamma V_i(x_{k+1}) \right\} \\ &= x_k^T Q x_k + v_i^T(x_k) R v_i(x_k) + \gamma V_i(F(x_k, v_i(x_k))). \end{aligned} \quad (3.34)$$

Using the iteration process between (3.33) and (3.34), we can design the optimal controller for system (3.32). The NN implementation of the iterative algorithm is shown in Section 3.2.5.

3.4 SIMULATION STUDIES

Example 3.1 Consider the following nonlinear system derived from [26]:

$$x_{k+1} = \begin{bmatrix} -\sin(0.5x_{2k}) \\ -\cos(1.4x_{2k})\sin(0.9x_{1k}) \end{bmatrix} + \begin{bmatrix} 0 \\ 1 \end{bmatrix} u_k, \quad (3.35)$$

where $x_k = [x_{1k}\ x_{2k}]^T \in \mathbb{R}^2$ and $u_k \in \mathbb{R}$ are the state and control variables, respectively. The cost function is chosen as $U(x_k, u_k) = x_k^T x_k + u_k^T u_k$.

We choose three-layer feedforward NNs as model network, critic network and action network with the structures of 3–8–2, 2–8–3, 2–8–1, respectively. In the system identification process, the initial weights between the input layer and the hidden layer, and between the hidden layer and the output layer are chosen randomly in $[-0.5, 0.5]$ and $[-0.1, 0.1]$, respectively. We collected 500 data samples for the NN identification scheme with learning rate $\alpha_m = 0.05$. The model network validation result is shown in Figure 3.2 for 100 time steps. It is clearly observed that the NN identifier successfully learns the unknown nonlinear system. Then, we finish the training of the model network and keep its weights unchanged.

The initial weights of the critic network and action network are all set to be random in $[-0.1, 0.1]$. Then, let the discount factor $\gamma = 1$ and the adjusting parameter $\beta = 0.5$. We apply the iterative ADP algorithm for 10 iterations (i.e., for $i = 1, 2, \ldots, 10$) with 2000 training epochs for each iteration to make sure the prespecified accuracy of 10^{-6} is reached for critic network and action network training. In the training process, the learning rates are chosen as $\alpha_c = \alpha_a = 0.05$. The convergence process of the cost function and its derivatives of the iterative GDHP algorithm at time instant $k = 0$ are shown in Figure 3.3. We can see that the iterative cost function sequence does converge to the optimal cost function quite rapidly, which also indicates the effectiveness of the iterative GDHP algorithm.

Moveover, to make comparison with the iterative ADP algorithm using HDP and DHP techniques, we also present the controllers designed by iterative HDP algorithm and iterative DHP algorithm, respectively. Then, for the given initial state $x_{10} = 0.5$ and $x_{20} = 0.5$, we apply the optimal control laws designed by iterative GDHP, HDP,

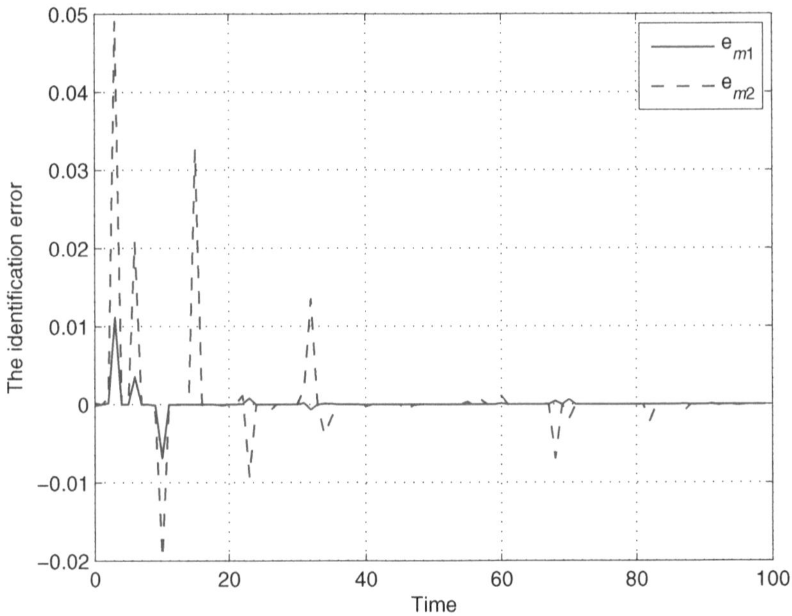

FIGURE 3.2 The system identification error ($\tilde{x}_{k1}$ and $\tilde{x}_{k2}$ are denoted as e_{m1} and e_{m2}, respectively).

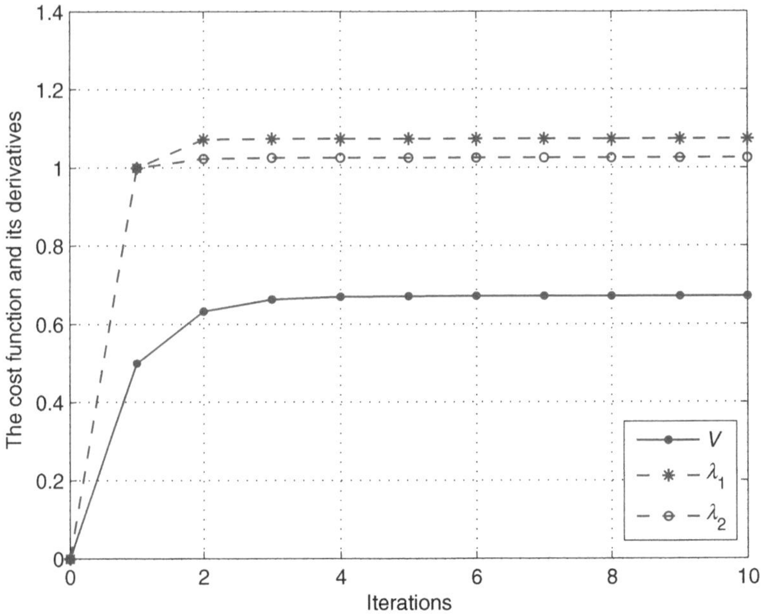

FIGURE 3.3 The convergence process of the cost function and its derivatives of the iterative GDHP algorithm.

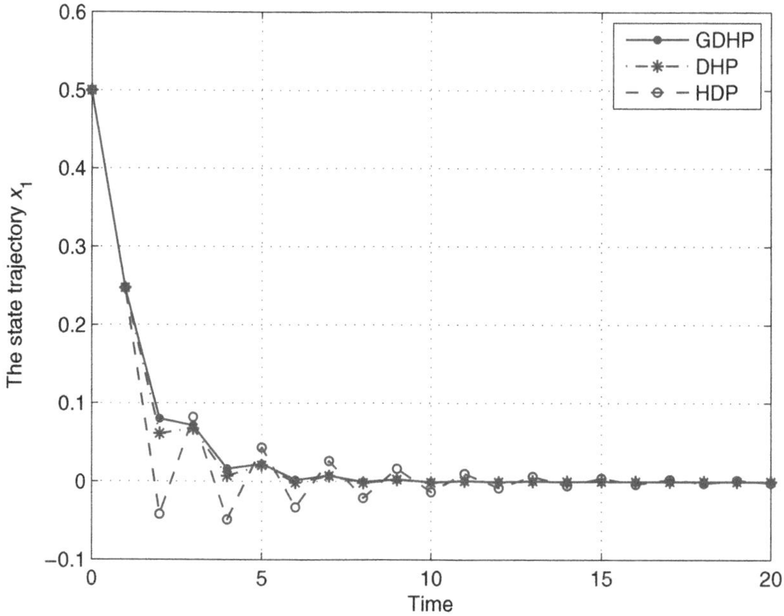

FIGURE 3.4 The state trajectory x_1.

and DHP algorithms to the controlled system (3.35) for 20 time steps, respectively, and obtain the state and control curves are shown in Figures 3.4–3.6. It can be seen from the simulation results that the controller designed by the iterative GDHP algorithm has better performance than the iterative HDP algorithm and iterative DHP algorithm.

Example 3.2 Consider the nonlinear discrete-time system

$$x_{k+1} = x_k + \sin(x_k + u_k), \tag{3.36}$$

where $x_k \in \mathbb{R}$, $u_k \in \mathbb{R}$, $k = 1, 2, \ldots$. The utility function is also chosen as $U(x_k, u_k) = x_k^T x_k + u_k^T u_k$. It can be seen that $x_k = 0$ is an equilibrium state of system (3.36). However, the system is unstable at this equilibrium, since $(\partial x_{k+1}/\partial x_k)|_{(0,0)} = 2 > 1$.

Three NNs are chosen with the structures of 2–8–1, 1–8–2, 1–8–1, respectively, and their initial weights are all set to be random in $[-1, 1]$. First, we train the model network using 500 data samples under the learning rate $\alpha_m = 0.1$. Then, let the discount factor $\gamma = 1$ and the adjusting parameter $\beta = 0.5$. We apply the iterative ADP algorithm for 120 iterations (i.e., for $i = 1, 2, \ldots, 120$) with 2000 training epochs for each iteration to make sure the prespecified accuracy of 10^{-6} is reached for critic

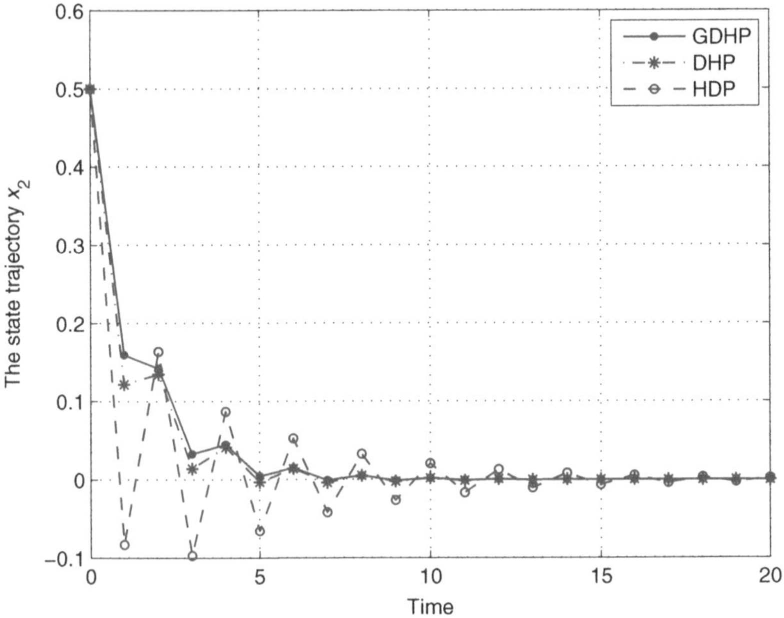

FIGURE 3.5 The state trajectory x_2.

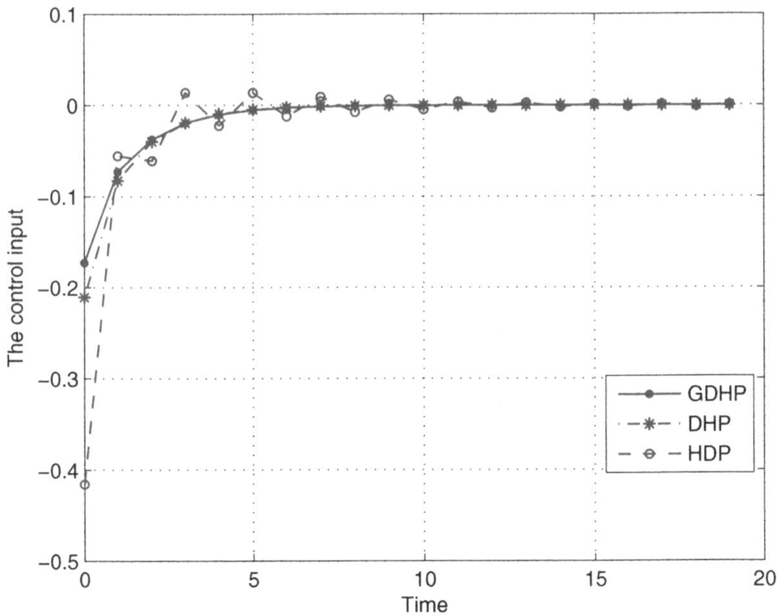

FIGURE 3.6 The control input u.

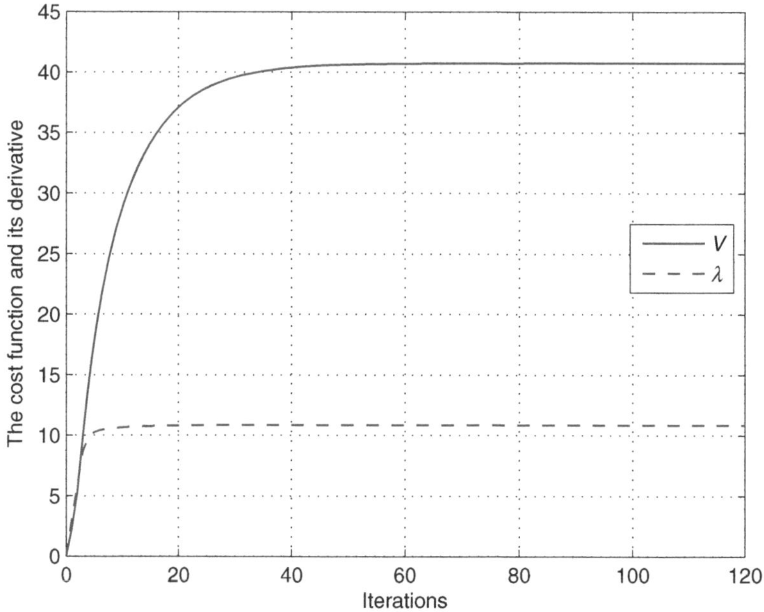

FIGURE 3.7 The convergence process of the cost function and its derivatives of the iterative GDHP algorithm.

network and action network training. The learning rates are chosen as $\alpha_c = \alpha_a = 0.05$. The convergence process of the cost function and its derivatives of the iterative GDHP algorithm are shown in Figure 3.7, for $k = 0$ and $x_0 = 1.5$. For the same problem, the iterative GDHP algorithm takes about 16 seconds while HDP takes about 117 s before satisfactory results are obtained.

To show the impact of discount factor on our iterative algorithm, we choose the discount factor $\gamma = 0.9$ and set other parameters the same as above. Then, we apply the iterative ADP algorithm for 80 iterations to reach the prespecified accuracy of 10^{-6} for critic network and action network training, which demonstrates that smaller discount factor can ensure quicker convergence of the cost function sequence. Next, we show the discrepancy of the state and control curves under different iterations to prove the usefulness of the iterative algorithm. For the same initial state $x_0 = 1.5$, we apply different control laws to the controlled plant (3.36) for 15 time steps and obtain the simulation results as follows. The state curves are shown in Figure 3.8, and the corresponding control inputs are shown in Figure 3.9. From the simulation results, we can see that the closed-loop system is divergent when using the control law obtained in the first iteration. However, the system responses become better and better as the iteration numbers increasing from 3 to 80. Besides, the responses basically remain unchanged when the iteration number is larger than 5, which verifies the effectiveness of the proposed iterative GDHP algorithm.

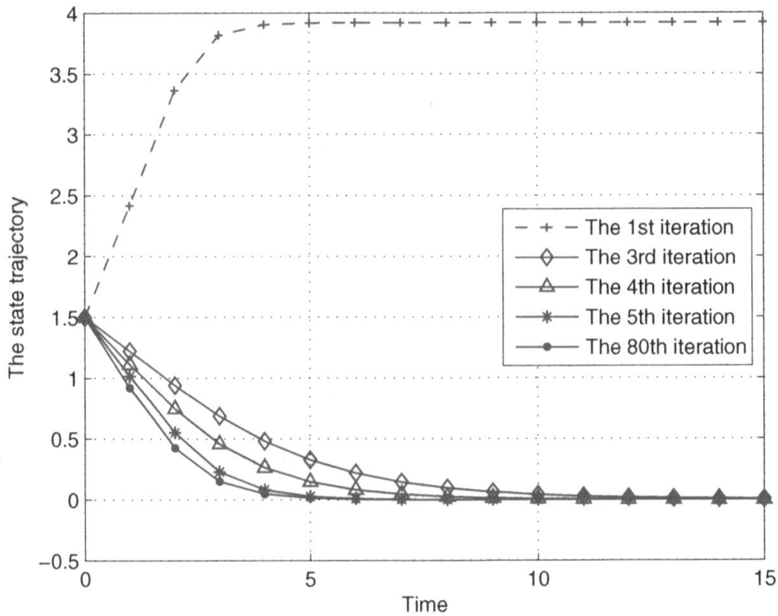

FIGURE 3.8 The state trajectory x.

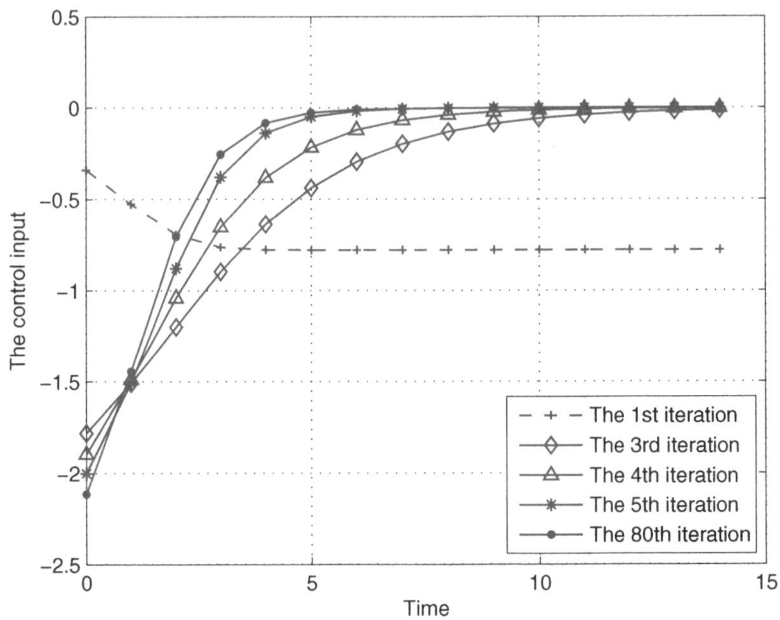

FIGURE 3.9 The control input u.

3.5 SUMMARY

Since it is computationally expensive to use DP to solve the nonlinear optimal control problems, the ADP method, which combines RL, NN, and DP, has become an important methodology in the area of computational intelligence. Werbos [6] indicated that ADP is one of the core technologies to potentially achieve the brain-like general-purpose intelligence and bridge the gap between theoretic study and challenging real-world engineering applications.

In this chapter, an effective iterative algorithm is developed for the design of a near optimal controller for unknown nonlinear discrete-time systems with discount factor in the cost function. The iterative GDHP algorithm is introduced to solve the cost function of the DTHJB equation with convergence analysis. Three NNs are used as parametric structures to approximate at each iteration the cost function, the control law and the unknown nonlinear system, respectively. The simulation studies demonstrated the validity of the control approach and showed that the iterative GDHP algorithm outperforms simple ADP schemes to a large extent.

REFERENCES

1. R.E. Bellman. *Dynamic Programming*. Princeton University Press, Princeton, NJ, 1957.
2. S. Haykin. *Neural Networks: A Comprehensive Foundation*. Prentice-Hall, Upper Saddle River, NJ, 1999.
3. P.J. Werbos. Advanced forecasting methods for global crisis warning and models of intelligence. *General Systems Yearbook*, 22:25–38, 1977.
4. P.J. Werbos. Approximate dynamic programming for real-time control and neural modeling. In *Handbook of Intelligent Control: Neural, Fuzzy, and Adaptive Approaches*. D.A. White and D.A. Sofge, editors. Chapter 13, Van Nostrand Reinhold, New York, 1992.
5. P.J. Werbos. ADP: the key direction for future research in intelligent control and understanding brain intelligence. *IEEE Transactions on Systems, Man, and Cybernetics–Part B: Cybernetics*, 38(4):898–900, 2008.
6. P.J. Werbos. Intelligence in the brain: a theory of how it works and how to build it. *Neural Networks*, 22(3):200–212, 2009.
7. D.V. Prokhorov and D.C. Wunsch. Adaptive critic designs. *IEEE Transactions on Neural Networks*, 8(5):997–1007, 1997.
8. Z. Sun, X. Chen, and Z. He. Adaptive critic designs for energy minimization of portable video communication devices. *IEEE Transactions on Circuits and Systems for Video Technology*, 20(1):27–37, 2010.
9. F.Y. Wang, H. Zhang, and D. Liu. Adaptive dynamic programming: an introduction. *IEEE Computational Intelligence Magazine*, 4(2):39–47, 2009.
10. F.L. Lewis and D. Vrabie. Reinforcement learning and adaptive dynamic programming for feedback control. *IEEE Circuits and Systems Magazine*, 9(3):32–50, 2009.
11. F.Y. Wang, N. Jin, D. Liu, and Q. Wei. Adaptive dynamic programming for finite-horizon optimal control of discrete-time nonlinear systems with ε-error bound. *IEEE Transactions on Neural Networks*, 22(1):24–36, 2011.

12. G.G. Lendaris. A retrospective on adaptive dynamic programming for control. In *Proceedings of International Joint Conference on Neural Networks*, Atlanta, GA, pp. 1750–1757, 2009.
13. S.N. Balakrishnan, J. Ding, and F.L. Lewis. Issues on stability of ADP feedback controllers for dynamic systems. *IEEE Transactions on Systems, Man, and Cybernetics–Part B: Cybernetics*, 38(4):913–917, 2008.
14. J. Si, A.G. Barto, W.B. Powell, and D.C. Wunsch, editors. *Handbook of Learning and Approximate Dynamic Programming*, IEEE Press/Wiley, New York, 2004.
15. W.B. Powell. *Approximate Dynamic Programming: Solving the Curses of Dimensionality*, Wiley, Hoboken, NJ, 2007.
16. J. Seiffertt, S. Sanyal, and D.C. Wunsch. Hamilton–Jacobi–Bellman equations and approximate dynamic programming on time scales. *IEEE Transactions on Systems, Man, and Cybernetics–Part B: Cybernetics*, 38(4):918–923, 2008.
17. D. Liu and H. Zhang. A neural dynamic programming approach for learning control of failure avoidance problems. *International Journal of Intelligent Control and Systems*, 10(1):21–32, 2005.
18. J. Si and Y.T. Wang. On-line learning control by association and reinforcement. *IEEE Transactions on Neural Networks*, 12(2):264–276, 2001.
19. R. Enns and J. Si. Helicopter trimming and tracking control using direct neural dynamic programming. *IEEE Transactions on Neural Networks*, 14(4):929–939, 2003.
20. D.P. Bertsekas and J.N. Tsitsiklis. *Neuro-Dynamic Programming*, Athena Scientific, Belmont, MA, 1996.
21. A. Al-Tamimi, F.L. Lewis, and M. Abu-Khalaf. Discrete-time nonlinear HJB solution using approximate dynamic programming: convergence proof. *IEEE Transactions on Systems, Man, and Cybernetics–Part B: Cybernetics*, 38(4):943–949, 2008.
22. D. Liu, Y. Zhang, and H. Zhang. A self-learning call admission control scheme for CDMA cellular networks. *IEEE Transactions on Neural Networks*, 16(5):1219–1228, 2005.
23. H. Zhang, Y. Luo, and D. Liu. Neural-network-based near-optimal control for a class of discrete-time affine nonlinear systems with control constraints. *IEEE Transactions on Neural Networks*, 20(9):1490–1503, 2009.
24. D. Wang and D. Liu. Optimal control for a class of unknown nonlinear systems via the iterative GDHP algorithm. In *Proceedings of 8th International Symposium on Neural Networks*, Guilin, China, pp. 630–639, 2011. (See Advances in Neural Networks – ISNN 2011, Lecture Notes in Computer Science, Vol. 6676, D. Liu, H. Zhang, M. Polycarpou, C. Alippi, and H. He, editors. Springer-Verlag, Berlin, 2011.)
25. M. Abu-Khalaf and F.L. Lewis. Nearly optimal control laws for nonlinear systems with saturating actuators using a neural network HJB approach, *Automatica*, 41(5):779–791, 2005.
26. T. Dierks, B.T. Thumati, and S. Jagannathan. Optimal control of unknown affine nonlinear discrete-time systems using offline-trained neural networks with proof of convergence. *Neural Networks*, 22(5–6):851–860, 2009.
27. S. Jagannathan. *Neural Network Control of Nonlinear Discrete-Time Systems*, CRC Press, Boca Raton, FL, 2006.
28. D. Liu, H. Javaherian, O. Kovalenko, and T. Huang. Adaptive critic learning techniques for engine torque and air–fuel ratio control. *IEEE Transactions on Systems, Man, and Cybernetics–Part B: Cybernetics*, 38(4):988–993, 2008.

29. D. Liu. Approximate dynamic programming for self-learning control. *Acta Automatica Sinica*, 31(1):13–18, 2005.

30. H. Zhang, Q. Wei, and Y. Luo. A novel infinite-time optimal tracking control scheme for a class of discrete-time nonlinear systems via the greedy HDP iteration algorithm. *IEEE Transactions on Systems, Man, and Cybernetics–Part B: Cybernetics*, 38(4):937–942, 2008.

31. D. Liu, D. Wang, and D. Zhao. Adaptive dynamic programming for optimal control of unknown nonlinear discrete-time systems. In *Proceedings of IEEE Symposium on Adaptive Dynamic Programming and Reinforcement Learning*, Paris, France, pp. 242–249, 2011.

32. D. Liu, D. Wang, and D. Zhao. Neural-network-based optimal control for a class of nonlinear discrete-time systems with control constraints using the iterative GDHP algorithm. In *Proceedings of International Joint Conference on Neural Networks*, San Jose, CA, pp. 53–60, 2011.

33. H. Zhang, Q. Wei, and D. Liu. An iterative adaptive dynamic programming method for solving a class of nonlinear zero-sum differential games. *Automatica*, 47(1):207–214, 2011.

34. F.L. Lewis and K.G. Vamvoudakis. Reinforcement learning for partially observable dynamic processes: adaptive dynamic programming using measured output data. *IEEE Transactions on Systems, Man, and Cybernetics–Part B: Cybernetics*, 41(1):14–25, 2011.

35. K.G. Vamvoudakis and F.L. Lewis, Multi-player non-zero-sum games: online adaptive learning solution of coupled Hamilton–Jacobi equations. *Automatica*, 47(8):1556–1569, 2011.

36. K.G. Vamvoudakis and F.L. Lewis. Online actor-critic algorithm to solve the continuous-time infinite horizon optimal control problem. *Automatica*, 46(5):878–888, 2010.

37. J.H. Kim and F.L. Lewis. Model-free H_∞ control design for unknown linear discrete-time systems via Q-learning with LMI. *Automatica*, 46(8):1320–1326, 2010.

38. V. Yadav, R. Padhi, and S.N. Balakrishnan. Robust/optimal temperature profile control of a high-speed aerospace vehicle using neural networks. *IEEE Transactions on Neural Networks*, 18(4):1115–1128, 2007.

39. R. Padhi and S.N. Balakrishnan. Optimal management of beaver population using a reduced-order distributed parameter model and single network adaptive critics. *IEEE Transactions on Control Systems Technology*, 14(4):628–640, 2006.

40. D. Han and S.N. Balakrishnan. State-constrained agile missile control with adaptive-critic-based neural networks. *IEEE Transactions on Control Systems Technology*, 10(4):481–489, 2002.

41. R. Padhi, S.N. Balakrishnan, and T. Randolph. Adaptive–critic based optimal neuro control synthesis for distributed parameter systems. *Automatica*, 37(8):1223–1234, 2001.

42. S. Jagannathan and P. He. Neural-network-based state feedback control of a nonlinear discrete-time system in nonstrict feedback form. *IEEE Transactions on Neural Networks*, 19(12):2073–2087, 2008.

43. Q. Yang, J.B. Vance, and S. Jagannathan. Control of nonaffine nonlinear discrete-time systems using reinforcement-learning-based linearly parameterized neural networks. *IEEE Transactions on Systems, Man, and Cybernetics–Part B: Cybernetics*, 38(4):994–1001, 2008.

44. Z. Chen and S. Jagannathan. Generalized Hamilton–Jacobi–Bellman formulation based neural network control of affine nonlinear discrete-time systems. *IEEE Transactions on Neural Networks*, 19(1):90–106, 2008.

45. P. Shih, B.C. Kaul, S. Jagannathan, and J.A. Drallmeier. Reinforcement-learning-based dual-control methodology for complex nonlinear discrete-time systems with application to spark engine EGR operation. *IEEE Transactions on Neural Networks*, 19(8):1369–1388, 2008.

46. P. He and S. Jagannathan. Reinforcement learning neural-network-based controller for nonlinear discrete-time systems with input constraints. *IEEE Transactions on Systems, Man, and Cybernetics–Part B: Cybernetics*, 37(2):425–436, 2007.

47. G.K. Venayagamoorthy, R.G. Harley, and D.C. Wunsch. Dual heuristic programming excitation neurocontrol for generators in a multimachine power system. *IEEE Transactions on Industry Applications*, 39(2):382–394, 2003.

48. G.K. Venayagamoorthy, R.G. Harley, and D.C. Wunsch. Comparison of heuristic dynamic programming and dual heuristic programming adaptive critics for neurocontrol of a turbogenerator. *IEEE Transactions on Neural Networks*, 13(3):764–773, 2002.

49. J. Fu, H. He, and X. Zhou. Adaptive learning and control for MIMO system based on adaptive dynamic programming. *IEEE Transactions on Neural Networks*, 22(7):1133–1148, 2011.

50. D.M. Adhyaru, I.N. Kar, and M. Gopal. Bounded robust control of nonlinear systems using neural network-based HJB solution. *Neural Computing and Applications*, 20(1):91–103, 2011.

51. D. Vrabie and F.L. Lewis. Neural network approach to continuous-time direct adaptive optimal control for partially unknown nonlinear systems. *Neural Networks*, 22(3):237–246, 2009.

52. D. Vrabie, O. Pastravanu, M. Abu-Khalaf, and F.L. Lewis. Adaptive optimal control for continuous-time linear systems based on policy iteration. *Automatica*, 45(2):477–484, 2009.

53. T. Cheng, F.L. Lewis, and M. Abu-Khalaf. A neural network solution for fixed-final time optimal control of nonlinear systems. *Automatica*, 43(3):482–490, 2007.

CHAPTER 4

Learning and Optimization in Hierarchical Adaptive Critic Design

HAIBO HE,[1] ZHEN NI,[1] and DONGBIN ZHAO[2]

[1]University of Rhode Island, Kingston, RI, USA
[2]Chinese Academy of Sciences, Beijing, China

ABSTRACT

This chapter introduces a novel hierarchical adaptive critic design to improve learning and optimization over time. Specifically, we propose to integrate a hierarchical *goal generator network* to provide the learning system a more informative and detailed goal representation to guide its decision making. The motivations for this idea is twofold. First, instead of using a typical binary reinforcement signal (e.g., 0 or 1) to represent "success" or "failure" of the system, we propose a more informative reinforcement signal representation for the intelligent system to make better choice of actions. Second, in order to mimic certain levels of brain-like intelligence, we consider it is important to introduce a multilevel goal representation into the adaptive critic design to guide the system's decision-making to accomplish the long-term goal over time. We present the detailed system architecture, learning, and adaptation procedure, and a case study of the ball-and-beam system to demonstrate the learning and control capability of this approach.

4.1 INTRODUCTION

Understanding of brain intelligence and developing self-adaptive systems to potentially mimic such a level of intelligence is still one of the greatest unsolved scientific challenges [1, 2]. With the recent developments of brain research and modern technologies, scientists, and engineers will hopefully find efficient ways to build complex systems that are highly adaptive, robust, and fault tolerant to uncertain and

Reinforcement Learning and Approximate Dynamic Programming for Feedback Control, First Edition. Edited by Frank L. Lewis and Derong Liu.

unstructured environment. However, although many important fundamental research as well as critical engineering applications have been successfully developed, there is still a long way to go to achieve truly brain-like general-purpose intelligent machines. One of the key fundamental challenges is how to design intelligent systems to be able to "learn to optimize" and "learn to predict" overtime to achieve goals. In this chapter, we present our latest research on a hierarchical adaptive critic design to tackle this problem.

Over the past decades, we have witnessed great efforts and progresses in machine intelligence research from different disciplines in engineering and science. Among many promising methodologies, adaptive dynamic programming (ADP) has been widely recognized as *"the only general-purpose way to learn to approximate the optimal strategy of action in the general case"* [1], and also demonstrated great potential and capability to many critical complex system applications to bring such a level of intelligence closer to reality. Briefly speaking, the key idea of ADP is built on the Bellman equation [3] to achieve optimization over time. Specifically, given a system with a performance cost function, the objective of dynamic programming is to choose control sequence $u(t)$ so that the cost function is minimized:

$$J^*(X(t)) = \min_{u(t)}\{U(X(t), u(t)) + \alpha J^*(X(t+1))\}. \tag{4.1}$$

where $X(t)$ is the state vector of the system, $u(t)$ the control action, U the utility function, and α a discount factor. For implementations, ADP can use a function approaching structure to approximate the cost function. For instance, universal approximators like neural networks with the backpropagation method have been widely adopted in the community, and extensive theoretical studies and their applications to different domains such as industrial control [4], engine air–fuel ratio control [5], helicopter control [6], power system control [7], and many others [8–13].

Generally speaking, the existing approaches of ADP design can be categorized into three major groups: heuristic dynamic programming (HDP), dual heuristic dynamic programming (DHP), and globalized dual heuristic dynamic programming (GDHP) [14, 15]. Various variations of these major categories of approaches, such as the action dependent (AD) version of these methods by taking action value as an additional input to the critic network, have also been developed in the community. Recently, important discussions on ADP development include the "two brains in one model" in the second generation of ADP design, and the "mouse level intelligence" in the third and fourth generation of ADP design [1]. Interested readers can refer to [1] for a detailed discussion on this.

In this chapter, we study a hierarchial structure in the ADP design for improved learning and optimization [16, 17]. In the traditional ADP and reinforcement learning (RL) design, the reinforcement signal is normally set as a binary signal, such us using either a "+1" or a "−1" to represent "success" or "failure" of the system, or as a "reward" or a "punishment" to guide the system's decision-making process. To provide more informative reinforcement signals in many complex problems, different techniques have been proposed to use nonbinary reinforcement signals to improve the learning and optimization performance [6, 14, 18]. Another motivation for using more informative reinforcement signal is based on the multistage goal representations in biological systems. For instance, a biological system may have different goals at different stages, or very likely to have multiple goals simultaneous at the same time. For example, for an animal to be able to survive in a hostile environment, it needs to consider the trade-off between different levels of goals, such as getting some food and/or escaping from a hunter (see the interesting discussions of "mouse–fox–cheese" example in [1]). Motivated by such observations, *our hypothesis in this chapter is that using a multiple-level reinforcement signal representation will be able to facilitate the development of hierarchical internal goal representations to build a value system, which will trigger learning and association in both top–down and bottom–up pathways to support intelligent decision-making process.* This hypothesis provides the major motivation and foundation of the proposed work in this chapter.

4.2 HIERARCHICAL ADP ARCHITECTURE WITH MULTIPLE-GOAL REPRESENTATION

4.2.1 System Level Structure

Figure 4.1a shows the conceptual diagram of the proposed actor–critic architecture with a hierarchical goal representation network to build different level of goals. Compared to the traditional actor–critic design as shown in Figure 4.1b, the key idea is to introduce another level of design to represent multiple level goals through the learning process. In this way, different stages of objectives can be built internally within the hierarchical structure to guide the system to accomplish the final goal. Each higher

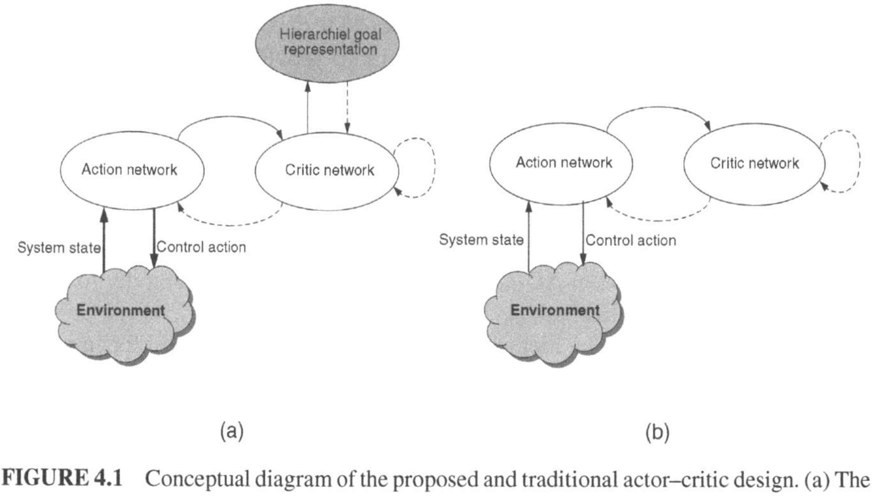

FIGURE 4.1 Conceptual diagram of the proposed and traditional actor–critic design. (a) The proposed actor–critic architecture with a hierarchical goal representation; (b) The traditional actor–critic design with action network and critic network.

hierarchical level will provide suggestions (i.e., a top–down approach) for the lower hierarchical level learning, which will be combined with the interactions with the critical network through evaluating the current control action from the action network (i.e., a bottom–up approach). In this way, this structure combines both top–down and bottom–up approaches together to accomplish the final goal.

From Figure 4.1a we can see, the essential part of this architecture is how to develop such a hierarchical goal representation to guide the system's behavior. That is to say, we need to address two important questions. First, how to build such a hierarchical goal representation in a principled and systematic way. Second, how learning and optimization can be achieved through the interactions of the hierarchical goal representation and the critic network and action network, either directly or indirectly. The first question is related to architecture design and implementation, and the second question concerns about specific learning procedures in such an architecture. We now proceed to discuss these two important questions.

4.2.2 Architecture Design and Implementation

Generally speaking, the proposed hierarchical structure can be designed in different types of ADP structures, such as HDP, DHP, or GDHP. In this chapter, we use the action-dependent heuristic dynamic programming (ADHDP) as an example to demonstrate how to implement this idea in reality. Specifically, Figure 4.2 shows the detailed system architecture.

In this design, the lower part of the system (i.e., the action network and critic network) is similar to the neural dynamic programming design in [18, 19]. Specifically, this part of the design is closely related to ADHDP, however, one major difference is that this structure does not have a system model to predict the future system state value and consequently the cost-to-go for the next time step [18]. Instead, one can simply record the previous J value. Once the current J value is available, one can then obtain the temporal difference (TD) and use such information for training. In this way, this approach can resolve the dilemma of either needing an additional model network to predict future system state (i.e., increased computational cost) or ignoring the derivative calculation term of $J(t+1)$ over the critic network weights (i.e., sacrificing learning accuracy). For more detailed discussions on this, interested readers can refer to [18] for further details.

The key part of our design is the upper part in Figure 4.2. This part includes a series of hierarchically organized *goal generator networks* to represent different levels of goals to facilitate learning, which corresponds to the highlighted "hierarchical goal representation" box in Figure 4.1a. In this design, a total of L goal generator networks are shown. Each level of goal generator network will observe the current system state $X(t)$, the current control action $u(t)$, and the goal signal output $s(t)$ from a higher hierarchical level. At the highest hierarchical level, the primary reinforcement signal $r(t)$ is presented to the top-level goal generator network, which indicates to the system about the status of the final objective, such as "success" or "failure" of the system. Based on such a primary reinforcement signal, each lower goal generator network will output a goal signal $s(t)$, to build an internal value system to tell the system how

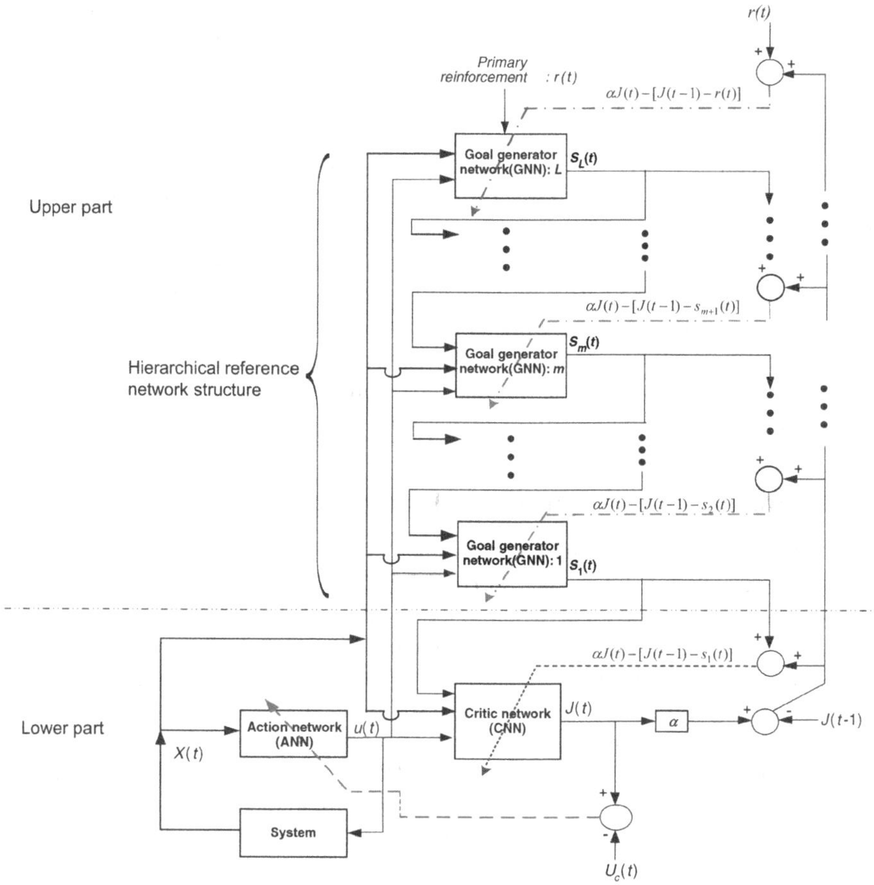

FIGURE 4.2 Architecture design of the proposed hierarchical ADP approach based on online ADHDP.

"good" or "bad" the current control action is. Therefore, one might also consider the internal goal signals as different levels of secondary reinforcement signals. Such internal goal representation will provide a more informative learning objectives of the entire system at different levels, therefore hopefully guide the system to improve its learning performance.

To see how such goal generator networks can interact with the critic network (directly) and action network (indirect), we now analyze how to formulate the error function for optimization. By observing Figure 4.2 we can see, the error function $E_{fm}(t)$ used to update the parameters in the goal generator network at hierarchical level m ($m = 1, \dots, L-1$) can be defined as:

$$e_{fm}(t) = \alpha J(t) - [J(t-1) - s_{m+1}(t)]; \qquad E_{fm}(t) = \frac{1}{2}e_{fm}^2(t). \tag{4.2}$$

At the top level L, the $s_{m+1}(t)$ in Equation (4.2) will be replaced with the primary reinforcement signal $r(t)$. Once the goal generator network at level m outputs the $s_m(t)$ signal, it will be used as an input to the one-level lower critic network (i.e., level $m-1$), and also used to define the error function to adjust the parameters of that level goal generator network. At the lowest hierarchical level, the output $s_1(t)$ will then be used to define the error function for the critic network as in the follows:

$$e_c(t) = \alpha J(t) - [J(t-1) - s_1(t)]; \qquad E_c(t) = \frac{1}{2}e_c^2(t). \tag{4.3}$$

4.2.3 Learning and Adaptation in Hierarchical ADP

In this architecture, chain backpropagation rule [20, 21] is the key to the learning and adaptation of the parameters of all networks (action network, critic network, and goal generator networks). To focus on the learning principles, we adopt multilayer perceptron (MLP) neural network with one-layer hidden neuron structure in all three types of networks. Figure 4.3a–c shows the neural network structure for action network, m-level goal generator network, critic network, respectively. Here we assume the system state X has n variables, therefore, the number of input signals are n, $(n+2)$, and $(n+2)$ for the action network, goal generator network, and critic network, respectively. One should also note that for the top hierarchical goal generator network design, it will only include $(n+1)$ inputs since there is no additional higher level goal signal input. We would also like to note that the learning principles discussed here can also be generalized to any arbitrary function approximators by properly applying backpropagation rule.

Since the learning in the action network is similar to that of [18], in this chapter we focus on the discussion of goal generator network and critic network to show how this proposed architecture can learn through interactions between them.

To calculate the backpropagation, we first need to define the critic network output and goal generator network output. For the critic network, the output $J(t)$ can be calculated in the following way.

$$J(t) = \sum_{i=1}^{N_h} w_{c_i}^{(2)}(t) p_i(t), \tag{4.4}$$

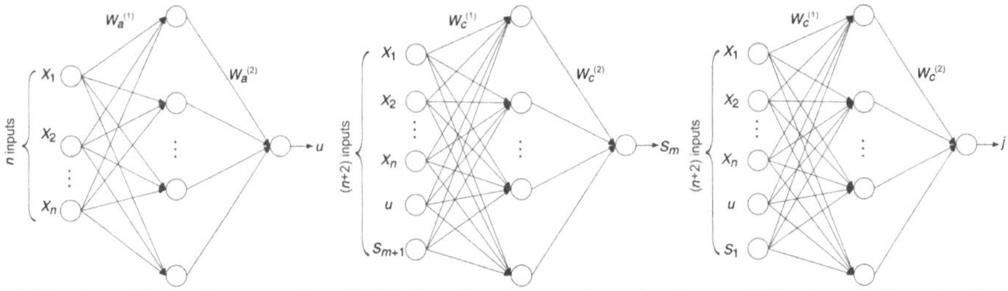

(a) Design of the action network (b) Design of goal generator network (at m hierarchical level) (c) Design of critic network

FIGURE 4.3 Design of action network, goal generator network, and critic network.

$$p_i(t) = \frac{1 - \exp^{-q_i(t)}}{1 + \exp^{-q_i(t)}}, \quad i = 1, \ldots, N_h, \tag{4.5}$$

$$q_i(t) = \sum_{j=1}^{n+2} w_{c_{i,j}}^{(1)}(t) x_j(t), \quad i = 1, \ldots, N_h. \tag{4.6}$$

Where q_i and p_i are the input and output of the ith hidden node of the critic network, respectively, and $(n + 2)$ is the total number of inputs to the critic network including the action value $u(t)$ from the action network and the goal signal $s_1(t)$ from the lowest hierarchical goal generator network.

For the goal generator network at level m, $m = 1, \ldots, L - 1$, the output is defined as follows:

$$s_m(t) = \frac{1 - \exp^{-k_m(t)}}{1 + \exp^{-k_m(t)}}. \tag{4.7}$$

$$k_m(t) = \sum_{i=1}^{N_h} w_{f_{mi}}^{(2)}(t) y_{mi}(t), \tag{4.8}$$

$$y_{mi}(t) = \frac{1 - \exp^{-z_{mi}(t)}}{1 + \exp^{-z_{mi}(t)}}, \quad i = 1, \ldots, N_h, \tag{4.9}$$

$$z_{mi}(t) = \sum_{j=1}^{n+2} w_{f_{m,(i,j)}}^{(1)}(t) x_j(t), \quad i = 1, \ldots, N_h \tag{4.10}$$

where subscript m means the variables refer to the m-level goal generator network, z_i is the ith hidden node input of the goal generator network and y_i is the corresponding output of the hidden node, k is the input to the output node of the goal generator network before the sigmoid function, N_h is the number of hidden neurons of the goal generator network, and $(n + 2)$ is the total number of inputs to the goal generator network including the action value $u(t)$ from the action network and the s_{m+1} from the one-level up in the hierarchical goal generator network structure. One should note that at the highest hierarchical level (i.e., $m = L$), the goal generator network will only have $(n + 1)$ inputs and one can easily modify the above equations accordingly for that case.

Since the output $s_m(t)$ at m level is an input to the one-level lower goal generator network in the hierarchical structure (i.e., level $(m - 1)$), this builds a chain of connection until it reaches the critic network. In this way, backpropagation can be applied here through the chain rule to adapt the parameters of the goal generator network w_{fm}. This procedure is illustrated as follows:

(1) $\Delta w_{fm}^{(2)}$: m-level goal generator network weight adjustment for the hidden to the output layer.

$$\Delta w_{f_{m,i}}^{(2)} = \eta_f(t)\left[-\frac{\partial E_{fm}(t)}{\partial w_{f_{m,i}}^{(2)}(t)}\right], \tag{4.11}$$

$$\begin{aligned}\frac{\partial \mathrm{E_{fm}}(t)}{\partial w_{f_{m,i}}^{(2)}(t)} &= \frac{\partial E_{fm}(t)}{\partial J(t)}\frac{\partial J(t)}{\partial s_1(t)}\frac{\partial s_1(t)}{\partial s_2(t)}\cdots\frac{\partial s_{m-1}(t)}{\partial s_m(t)}\frac{\partial s_m(t)}{\partial k_m(t)}\frac{\partial k_m(t)}{\partial w_{f_{m,i}}^{(2)}(t)}\\ &= \frac{\partial E_{fm}(t)}{\partial J(t)}\frac{\partial J(t)}{\partial s_1(t)}\left(\prod_{j=1}^{m-1}\frac{\partial s_j(t)}{\partial s_{j+1}(t)}\right)\frac{\partial s_m(t)}{\partial k_m(t)}\frac{\partial k_m(t)}{\partial w_{f_{m,i}}^{(2)}(t)},\end{aligned} \tag{4.12}$$

where

$$\frac{\partial E_{fm}(t)}{\partial J(t)} = \alpha \cdot e_{fm}(t), \tag{4.13}$$

$$\frac{\partial J(t)}{\partial s_1(t)} = \frac{\partial J(t)}{\partial p_i(t)}\frac{\partial p_i(t)}{\partial q_i(t)}\frac{\partial q_i(t)}{\partial s_1(t)} = \sum_{i=1}^{N_h}\left[w_{ci}^{(2)}(t)\frac{1}{2}(1-p_i^2(t))w_{ci,(n+2)}^{(1)}(t)\right], \tag{4.14}$$

$$\begin{aligned}\prod_{j=1}^{m-1}\frac{\partial s_j(t)}{\partial s_{j+1}(t)} &= \prod_{j=1}^{m-1}\left(\frac{\partial s_j(t)}{\partial k_j(t)}\frac{\partial k_j(t)}{\partial y_j(t)}\frac{\partial y_j(t)}{\partial s_{j+1}(t)}\right)\\ &= \prod_{j=1}^{m-1}\left(\sum_{i=1}^{N_h}\left[\frac{1}{2}\left(1-s_j^2(t)\right)w_{fj,i}^{(2)}(t)\frac{1}{2}\left(1-y_{j,i}^2(t)\right)w_{fj,(i,n+2)}^{(1)}(t)\right]\right),\end{aligned} \tag{4.15}$$

$$\frac{\partial s_m(t)}{\partial k_m(t)} = \frac{1}{2}\left(1-s_m^2(t)\right), \tag{4.16}$$

$$\frac{\partial k_m(t)}{\partial w_{f_{m,i}}^{(2)}(t)} = y_{mi,(n+2)}(t). \tag{4.17}$$

(2) $\Delta w_f^{(1)}$: m-level goal generator network weight adjustments for the input to the hidden layer.

$$\Delta w_{fm,(i,j)}^{(1)} = \eta_f(t)\left[-\frac{\partial E_{fm}(t)}{\partial w_{fm,(i,j)}^{(1)}(t)}\right], \tag{4.18}$$

$$\frac{\partial E_{fm}(t)}{\partial w_{fm,(i,j)}^{(1)}(t)} = \frac{\partial E_{fm}(t)}{\partial J(t)}\frac{\partial J(t)}{\partial s_1(t)}\left(\prod_{j=1}^{m-1}\frac{\partial s_j(t)}{\partial s_{j+1}(t)}\right)\frac{\partial s_m(t)}{\partial k_m(t)}\frac{\partial k_m(t)}{\partial y_{mi}(t)}$$
$$\times\frac{\partial y_{mi}(t)}{\partial z_{mi}(t)}\frac{\partial z_{mi}(t)}{\partial w_{fm,(i,j)}^{(1)}(t)}. \tag{4.19}$$

The first four terms of Equation (4.19) have already been given in Equations (4.13–4.16), respectively. The remaining terms are defined as follows:

$$\frac{\partial k_m(t)}{\partial y_{mi}(t)} = w_{fmi}^{(2)}(t), \tag{4.20}$$

$$\frac{\partial y_{mi}(t)}{\partial z_{mi}(t)} = \frac{1}{2}(1 - y_{mi}^2(t)), \tag{4.21}$$

$$\frac{\partial z_{mi}(t)}{\partial w_{fm,(i,j)}^{(1)}(t)} = x_j(t). \tag{4.22}$$

One should note that for the lowest hierarchical level of goal generator network (i.e., $m = 1$), the calculations of the backpropagation will need to be modified accordingly since it involves the interaction with the weights of the critic network. Similar modifications are also needed for the highest hierarchical level ($m = L$) since it only include $(n + 1)$ inputs. Since such modifications are very straightforward following the above procedure, we refrain from giving the detailed equations for these two cases for space considerations.

For the critic network, we can refer to Figure 4.2 and update the weights according to backpropagation rule. This is summarized as follows:

(1) $\Delta w_c^{(2)}$: Critic network weight adjustments for the hidden to the output layer.

$$\Delta w_{c_i}^{(2)} = \eta_c(t)\left[-\frac{\partial E_c(t)}{\partial w_{c_i}^{(2)}(t)}\right], \tag{4.23}$$

$$\frac{\partial E_c(t)}{\partial w_{c_i}^{(2)}(t)} = \frac{\partial E_c(t)}{\partial J(t)}\frac{\partial J(t)}{\partial w_{c_i}^{(2)}(t)} = \alpha e_c(t)\cdot p_i(t), \tag{4.24}$$

(2) $\Delta w_c^{(1)}$: Critic network weight adjustments for the input to the hidden layer.

$$\Delta w_{c_{i,j}}^{(1)} = \eta_c(t)\left[-\frac{\partial E_c(t)}{\partial w_{c_{i,j}}^{(1)}(t)}\right], \tag{4.25}$$

$$\frac{\partial E_c(t)}{\partial w_{c_{i,j}}^{(1)}(t)} = \frac{\partial E_c(t)}{\partial J(t)}\frac{\partial J(t)}{\partial p_i(t)}\frac{\partial p_i(t)}{\partial q_i(t)}\frac{\partial q_i(t)}{\partial w_{c_{i,j}}^{(1)}(t)} = \alpha e_c(t)\cdot w_{c_i}^{(2)}(t)\cdot\frac{1}{2}(1-p_i^2(t))\cdot x_j(t), \tag{4.26}$$

In summary, in this section we have presented the complete learning and adaptation process for the goal generator networks and critic network in the proposed hierarchical actor–critic design. From this discussion we can see, the key idea of this approach is to develop a series of interconnected goal generator networks to build the internal goal-representation to facilitate learning. The $s(t)$ signal between two adjacent hierarchical levels provides the connection to allow backpropagation to be calculated in this structure for weights updating. We now proceed to demonstrate the application of this architecture on a popular benchmark, the ball and beam system, to show its performance.

4.3 CASE STUDY: THE BALL-AND-BEAM SYSTEM

The ball-and-beam system is a popular laboratory model for testing different control approaches [22, 23]. Normally, this system consists of a long beam which can be tilted by a servo or electric motor together with a ball rolling back and forth on top of the beam. There are several versions of the ball and beam system, in this chapter we investigate the system as shown in Figure 4.4 to demonstrate the learning and control

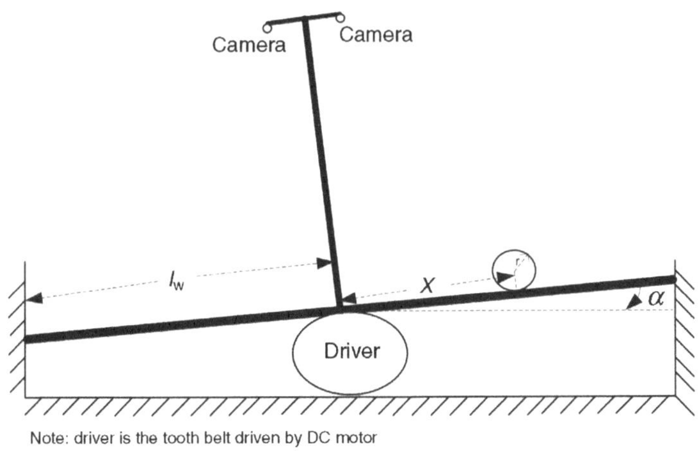

FIGURE 4.4 A schematic diagram of ball-and-beam system.

performance of our hierarchial adaptive critic design. In this system, the driver is located in the center of the beam. The angle of the beam to the horizonal axis is measured by an incremental encoder, and the position of the ball can be obtained with the cameras mounted on top of the system.

4.3.1 Problem Formulation

From [22], we can obtain the motion equations from Lagrange equation as following.

$$\left(m + \frac{I_b}{r^2}\right)\ddot{x}' + (mr^2 + I_b)\frac{1}{r}\ddot{\alpha} - mx'\dot{\alpha}^2 = mg(\sin\alpha), \tag{4.27}$$

$$\begin{aligned}[m(x')^2 + I_b + I_\omega]\ddot{\alpha} + (2m\dot{x}'x' + bl^2)\dot{\alpha} + Kl^2\alpha \\ + (mr^2 + I_b)\frac{1}{r}\ddot{x}' - mgx'(\cos\alpha) = ul(\cos\alpha),\end{aligned} \tag{4.28}$$

where m, $0.0162kg$, the mass of the ball; r, $0.02m$, the roll radius of the ball; I_b, $4.32 \times 10^{-5}kg \cdot m^2$, the inertia moment of the ball; b, $1N_s/m$, the friction coefficient of the drive mechanics; l, $0.48m$, the radius of force application; l_ω, $0.5m$, the radius of beam; K, $0.001N/m$, the stiffness of the drive mechanics; g, $9.8N/kg$, the gravity; I_ω, $0.14025kg \cdot m^2$, the inertia moment of the beam; u, the force of the drive mechanics.

To simplify the system model function, we define that $x_1 = x'$ represents the position of the ball, $x_2 = \dot{x}'$ represents the velocity the ball, $x_3 = \alpha$ is the angle of the beam with respect to the horizontal axis, and $x_4 = \dot{\alpha}$ is the angular velocity of the beam. In this way, the system function (4.27) and (4.28) can be transformed into the following forms:

$$\left(m + \frac{I_b}{r^2}\right)\dot{x}_2 + (mr^2 + I_b)\frac{1}{r}\dot{x}_4 = mx_1x_4^2 + mg(\sin x_3), \tag{4.29}$$

$$\begin{aligned}(mr^2 + I_b)\frac{1}{r}\dot{x}_2 + [mx_1^2 + I_b + I_\omega]\dot{x}_4 = (ul + mgx_1)\cos x_3 \\ - (2mx_2x_1 + bl^2)x_4 - Kl^2x_3,\end{aligned} \tag{4.30}$$

To be more clear, we re-write the (4.29) and (4.30) into one matrix formula as following.

$$\begin{bmatrix} A & B \\ C & D \end{bmatrix} \cdot \begin{bmatrix} \dot{x}_2 \\ \dot{x}_4 \end{bmatrix} = \begin{bmatrix} P \\ Q \end{bmatrix}, \tag{4.31}$$

where,

$$P = m x_1 x_4^2 + mg(\sin x_3), \tag{4.32}$$

$$Q = (ul + mgx_1)\cos x_3 - (2mx_2x_1 + bl^2)x_4 - Kl^2 x_3, \tag{4.33}$$

$$A = m + \frac{I_b}{r^2}, \tag{4.34}$$

$$B = (mr^2 + I_b)\frac{1}{r}, \tag{4.35}$$

$$C = (mr^2 + I_b)\frac{1}{r}, \tag{4.36}$$

$$D = mx_1^2 + I_b + I_\omega. \tag{4.37}$$

Therefore, we can obtain the general form of this problem as described in (4.38)

$$\begin{bmatrix} \dot{x}_2 \\ \dot{x}_4 \end{bmatrix} = \begin{bmatrix} A & B \\ C & D \end{bmatrix}^{-1} \begin{bmatrix} P \\ Q \end{bmatrix}, \tag{4.38}$$

and the other two terms in the state vector can be expressed as $\dot{x}_1 = x_2$ and $\dot{x}_3 = x_4$. Note that here the state vector is $X = \begin{bmatrix} x_1 & x_2 & x_3 & x_4 \end{bmatrix}$.

4.3.2 Experiment Configuration and Parameters Setup

For comparative study, we demonstrate the performance of our proposed adaptive critic design in Figure 4.2 and the architecture in [18]. For structure and parameters setting, we use neural network with multi-layer perceptron in both approaches. Since the state vector has four elements, in the method of [18], the action network adopts a 4-6-1 structure (i.e., four input neurons, six hidden layer neurons, and one output neuron) and the critic network adopts a 5-6-1 structure. For our proposed approach, the action network, critic network, and goal generator network adopt a structure of 4-6-1, 6-6-1, 5-6-1, respectively. All of the parameters we used are summarized in Table 4.1 and the notations are defined as following:

α : discount factor;
$l_c(0)$: initial learning rate of the critic network;
$l_a(0)$: initial learning rate of the action network;
$l_r(0)$: initial learning rate of the goal generator network;
$l_c(t)$: learning rate of the critic network at time t, which is decreased by 0.05 every 5 time steps until it reach $l_c(f)$ and stay thereafter;
$l_a(t)$: learning rate of the action network at time t, which is decreased by 0.05 every 5 time steps until it reach $l_a(f)$ and stay thereafter;
$l_r(t)$: learning rate of the goal generator network at time t, which is decreased by 0.05 every 5 times step until it reach $l_r(f)$ and stay thereafter;

TABLE 4.1 Summary of the Parameters Used in Ball and Beam Balancing Task

Para.	$l_c(0)$	$l_a(0)$	$l_r(0)$	$l_c(f)$	$l_a(f)$	$l_r(f)$	*
value	0.3	0.3	0.3	0.005	0.005	0.005	*
Para.	N_c	N_a	N_r	T_c	T_a	T_r	α
value	80	100	50	0.05	0.005	0.05	0.95

N_c : internal cycle of the critic network;
N_a : internal cycle of the action network;
N_r : internal cycle of the goal generator network;
T_c : internal training error threshold for the critic network;
T_a : internal training error threshold for the action network;
T_r : internal training error threshold for the goal generator network.

All simulation results presented in this chapter are based on 100 random runs with random initial neural network weights. The initial conditions of the the ball-and-beam system are set up as follows: The ball position (x_1) and the angle of the beam to the horizonal axis (x_3) are uniformly distributed in the range of $[-0.1m, 0.1m]$ and $[-0.1rad, 0.1rad]$, respectively, and the ball velocity (x_2) and the angular velocity (x_4) are set to be zero. For fair comparison, in each run, we also set the neural network initial weights and initial conditions of the beam and ball system to be the same for both methods (i.e., both methods will start from the same conditions in each run). The objective of the task is to keep the balance of the ball on the beam (i.e., the ball remain on the beam and the maximum angle of the beam to the horizontal axis is less or equal than $0.24rad$) for a certain period of time. Specifically, in the current study, each run consists of a maximum of 1000 trials, and a run will be considered successful if the last trial (i.e., trial number less or equal than 1000) can maintain balance of the ball for 10,000 time steps with a step size of $0.02s$. In addition to noise-free situation, we also tested the performance of both methods subject to different amounts of noises at the actuators and sensors of the positions and angles.

4.3.3 Simulation Results and Analysis

Table 4.2 demonstrates the successful rate, required average number of trials to balance the task and its associated standard deviation for both approaches under the 100 random runs. In this table, the 2nd column presents the statistical results for our hierarchical adaptive critic design and the 3rd column represents the results based on the method in [18]. From these results, one can see that under noise-free condition, our proposed approach can improve the performance in terms of both successful rate and required number of average trials. Under different noise conditions, it seems our approach is also able to improve the learning and control performance in general. One should note that when the sensor of position is subjected to the uniform 5% noise, the required average number of trials for our approach is larger than that of existing method (28.9 vs 22.4) though our method can improve the successful rate

TABLE 4.2 Simulation Results on Ball-and-Beam Balancing Task. The *2nd* Column is with our Proposed Hierarchical Adaptive Critic Design, While the *3rd* Column is with the Approach in [18]

Noise Type	Success Rate (%)	♯ of trial (%)	σ	Success Rate (%)	♯ of trial (%)	σ
Noise free	100	17.3	20.8	99	25.7	104.2
Uniform 5% a.[1]	100	13.1	15.0	99	47.3	94.6
Uniform 10% a.	100	17.0	22.4	98	33.5	59.1
Uniform 5% x.[2]	97	28.9	191.4	95	22.4	218.9
Uniform 10% x.	99	14.9	100.4	98	35.9	171.1
Uniform 5% α.[3]	100	33.7	96.7	95	41.0	234.3
Uniform 10% α.	98	24.7	150.6	99	34.9	121.6

σ : standard deviation,

[1]a. : actuators are subject to the noise,

[2]x. : sensors of positions are subject to the noise,

[3]α. : sensors of angles are subject to the noise.

The number of trials and standard deviation are calculated based on the successful runs.

(97% vs 95%). When the sensor of angle is subject to the uniform 10% noise, our proposed approach has lower successful rate (98% versus 99%) while it requires less number of average trials to balance the task (24.7 versus 34.9). In all other cases, our approach can improve the performance in terms of both successful rate and required number of trials to achieve the control objective. Another important observation from this table is that from the standard deviation information one can see, our approach has much less standard deviations compared to the existing approach except the last row. This may suggest our approach is much more robust under different conditions. Computational complexity is another interesting point to analyze. From the architecture design point of view, it can be observed that our approach has more goal generator networks needed to learn and tune. On the other hand, since our approach can provide more informative goal representation to guide the system to learn and optimize, it might require less iterations to decide the optimal action at each time instance. Our empirical study on the computational time analysis shows that on an average our approach requires 0.706 ms to decide each action, while the regular ADP approach requires 0.708 ms (simulations are conducted based on Sun server with 16 GB memory, Intel Xeon CPU 3.60 GHz and Matlab R2011b). These results suggest that the computational costs of both methods are comparable, while our approach seems to be able to improve the learning and control performance.

To have a closer observation of the learning and control process, we demonstrate several snapshots of the state vector of a typical successful run. Figure 4.5a shows the ball position (solid line) and ball velocity (dash line) in a successful run under noise free condition. From this figure one can see that ball starts at a random position and rolls forth and back at the early stage. As our system continues to learn to control the ball, it converges to the balance point after about 3500 time steps. From this figure, one can also clearly see the correlation between the ball velocity and position. For

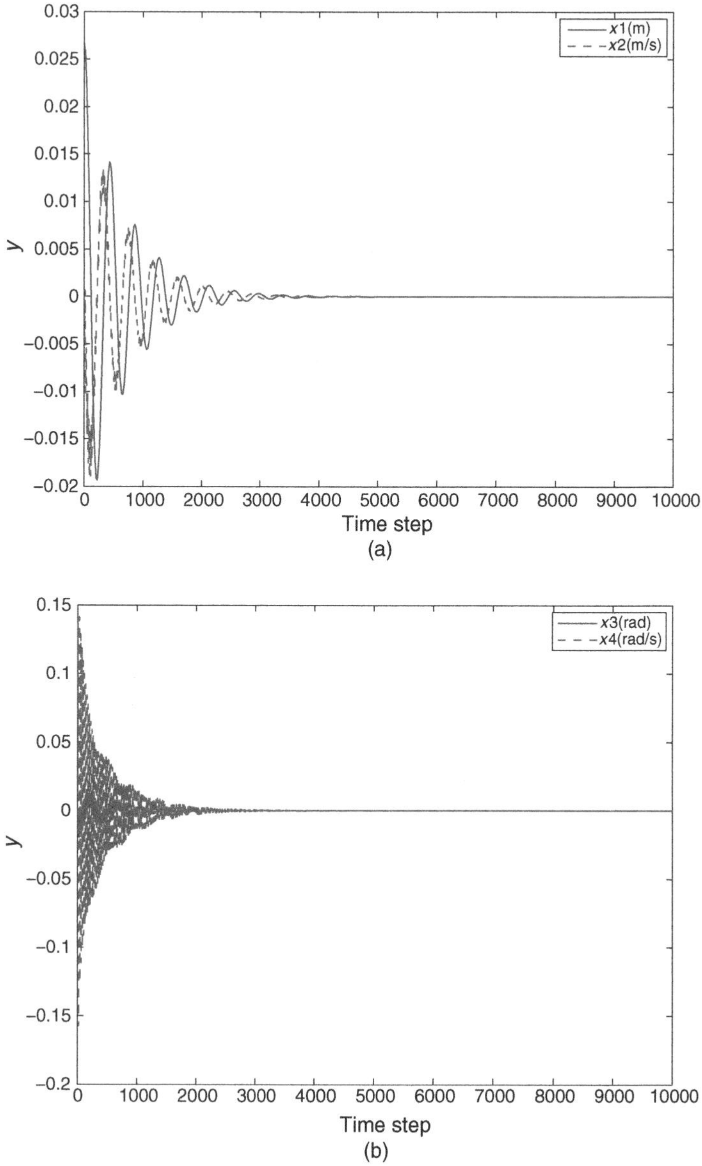

FIGURE 4.5 The typical trajectory of the state vectors of the system under the condition of noise free. (a) The typical trajectory of x_1 and x_2, (b) The typical trajectory of x_3 and x_4.

instance, when the ball velocity is at the peak point, the ball is rolling with the highest speed to cross the zero point. On the other hand, when the ball velocity is zero, the ball is in the maximum position which it can reach. Figure 4.5b shows the angle of the beam with respect to the horizontal axis (solid line) and the beam angular velocity

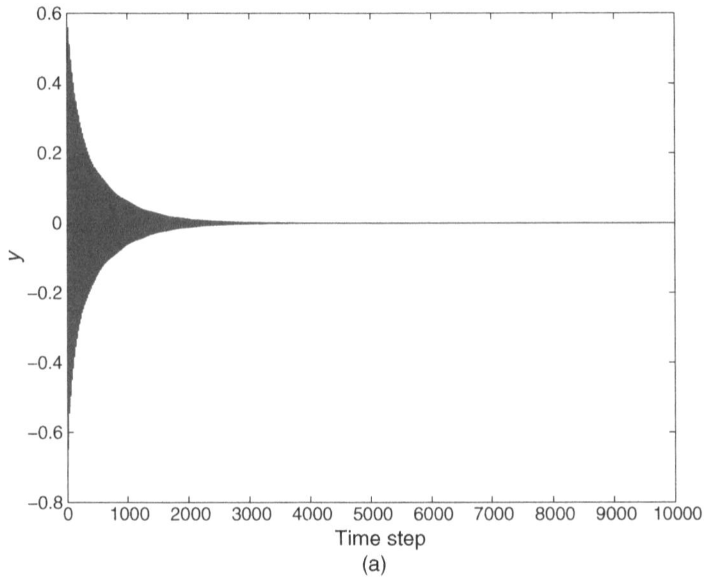

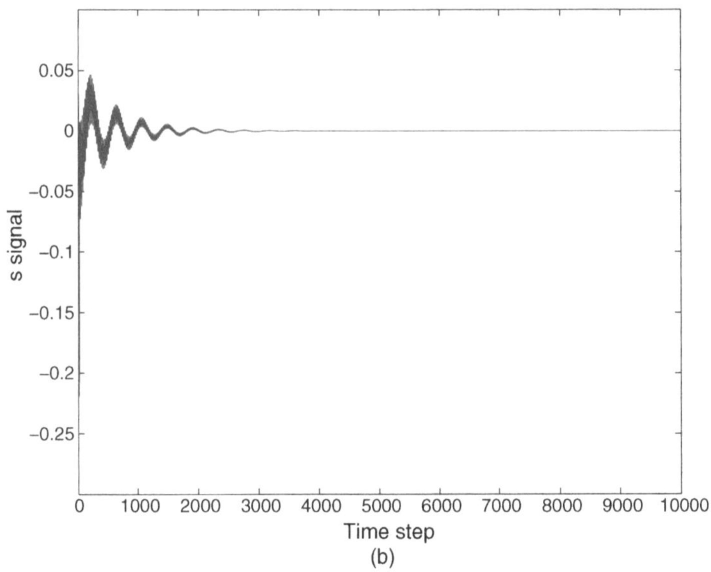

FIGURE 4.6 The typical trajectory of control action and the internal goal signal s. (a) The typical trajectory of control action under the condition of noise free, (b) The typical value of the internal goal signal s under the condition of noise free.

(dash line), which also clearly shows in this case the system approaches balance at around 3500 time steps. To have an in-depth observation of the proposed approach, Figure 4.6a shows the control action of a typical run under noise free condition, which indicates how the system learns to appropriately adjust the force to balance the task.

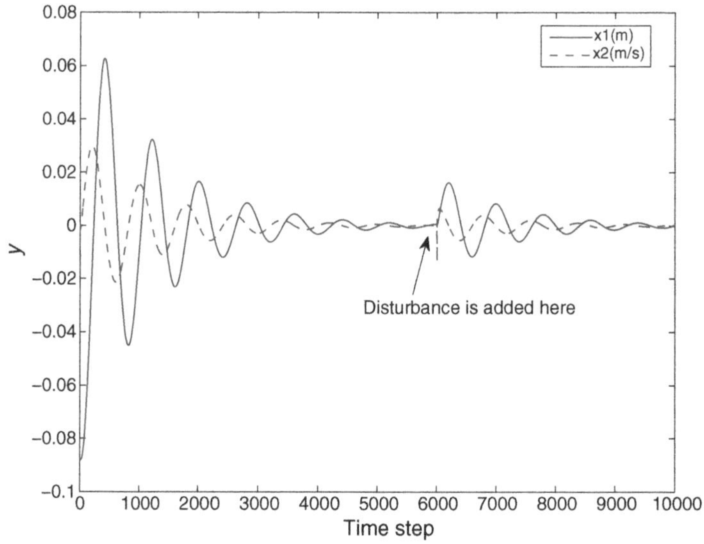

FIGURE 4.7 A typical trajectory of x_1 and x_2 with disturbance in 6000 time step.

The internal goal signal $s(t)$ provided by the goal generator network is shown in Figure 4.6b, which provides informative goal representations to guide the system to learn to control the task. Compared to the binary-value reinforcement signals in many of the existing ADP/RL approaches, our research in this paper demonstrates the improved learning and control capability by providing the intelligent system such continuous internal goal representations.

To show the robustness and adaptability of our proposed method, here we conduct an interesting experiment to add a large disturbance after the system has learned to balance the task. The goal is to see whether the system can adaptively learn again on-the-fly with some kind of disturbances after it reaches a balancing point. To do this, we apply a fix and relatively large force $u = 10$ when it is at the 6000 time step. Figure 4.7 shows our approach can adaptively learn to control it in this case. From the state vector x_1 and x_2 one can see, under such a large disturbance, our approach can effectively learn to control it and balance the ball again. This figure demonstrates good robustness of our learning and control approach presented in this chapter.

4.4 CONCLUSIONS AND FUTURE WORK

In this chapter, we propose a hierarchical adaptive critic design for improved learning and optimization. The key idea of this approach is to develop a hierarchical goal representation structure to interact with the critic network and action network, either directly or indirectly, to provide a multi-stage goal representations to facilitate learning. In our current design, we use a series of interconnected goal generator networks to represent the hierarchical goal representation. In this way, the top hierarchical goal generator network will receive the primary reinforcement signal from the external environment to represent the final objective of the learning system, while the internal goal generator networks in the hierarchy will provide an informative goal representation about how "good" or "bad" the current action is. We present the detailed system architecture of this design and its learning and adaptation procedure. Simulation results on a popular benchmark, the ball-and-beam system, demonstrate the effectiveness of our approach when it is compared to the existing method.

As hierarchical learning is a critical part to understand brain-like intelligence, there are numerous interesting future research directions along this topic. For instance, our work in this chapter mainly focuses on architecture design, implementation, learning process, and simulation studies. It would be of critical importance to study the theoretical aspects such as convergence and stability of the proposed adaptive critic structure. Such analytical analysis will provide deep understanding of the foundation of this approach. Furthermore, our currently implementation in this work is closely related to the ADHDP design. As there are several major groups of ADP design methods, it would be interesting to see how to extend and integrate this structure into other ADP design methods such as DHP and GDHP. Furthermore, our case study and simulation analysis in this chapter is based on the ball and beam system. It would be useful to test and demonstrate its performance under other benchmarks and real complex control problems to hopefully bring this technique closer to reality. We are currently investigating all these issues and will report the results in the near future. Motivated by our research results in this work, we hope the proposed hierarchical adaptive critic design will not only provide useful suggestions and insights about the fundamental ADP research for machine intelligence, but it will also provide new techniques and tools for complex engineering applications.

ACKNOWLEDGMENTS

We gratefully acknowledge the support from National Science Foundation (NSF) under CAREER grant ECCS 1053717, National Natural Science Foundation of China (NSFC) under Grant Nos. 51228701, 61273136, 60874043, 60921061, and 61034002, and the Toyota Research Institute North America. The authors are also gratefully to Danil V. Prokhorov for his constructive suggestions and comments for the development of this chapter.

REFERENCES

1. P.J. Werbos. Intelligence in the brain: a theory of how it works and how to build it, *Neural Networks*. 22(3):200–212, 2009.
2. P.J. Werbos. Using ADP to understand and replicate brain intelligence: the next level design, in *IEEE International Symposium on Approximate Dynamic Programming and Reinforcement Learning*. 2007, pp. 209–216.

3. R.E. Bellman. *Dynamic Programming*. Princeton University Press, Princeton, NJ, 1957.
4. J. Fu, H. He, and X. Zhou. Adaptive learning and control for mimo system based on adaptive dynamic programming, *IEEE Transactions on Neural Networks*. 22(7):1133–1148, 2011.
5. D. Liu, H. Javaherian, O. Kovalenko, and T. Huang. Adaptive critic learning techniques for engine torque and air–fuel ratio control, *IEEE Transactions on System, Man and Cybernetics, Part B*. 38(4):988–993, 2008.
6. R. Enns and J. Si. Helicopter flight control using direct neural dynamic programming, *Handbook of Learning and Approximate Dynamic Programming*. IEEE Press, 2004, pp. 535–559.
7. W. Qiao, G. Venayagamoorthy, and R. Harley. DHP-based wide-area coordinating control of a power system with a large wind farm and multiple FACTS devices, in *Proceeding of IEEE International Conference on Neural Networks*. 2007, pp. 2093–2098.
8. D. Liu, Y. Zhang, and H.G. Zhang. A self-learning call admission control scheme for cdma cellular networks, *IEEE Transactions on Neural Networks*. 16(5):1219–1228, 2005.
9. F.Y. Wang, N. Jin, D. Liu, and Q. Wei. Adaptive dynamic programming for finite-horizon optimal control of discrete-time nonlinear systems with ε-error bound, *IEEE Transactions on Neural Networks*. 22(1):24–36, 2011.
10. H.G. Zhang, Y.H. Luo, and D. Liu. Neural-network-based near-optimal control for a class of discrete-time affine nonlinear systems with control constraints, *IEEE Transactions on Neural Networks*. 20(9):1490–1503, 2009.
11. H.G. Zhang, Q.L. Wei, and D. Liu. An iterative approximate dynamic programming method to solve for a class of nonlinear zero-sum differential games, *Automatica*. 47(1):207–214, 2011.
12. Q.L. Wei, H.G. Zhang, D. Liu, and Y. Zhao. An optimal control scheme for a class of discrete-time nonlinear systems with time delays using adaptive dynamic programming, *ACTA Automatica Sinica*. 36(1):121–129, 2010.
13. H.G. Zhang, Q.L. Wei, and Y.H. Luo. A novel infinite-time optimal tracking control scheme for a class of discrete-time nonlinear systems via the greedy HDP iteration algorithm, *IEEE Transactions on System, Man and Cybernetics, Part B*. 38(4):937–942, 2008.
14. D.V. Prokhorov and D.C. Wunsch. Adaptive critic designs, *IEEE Transactions on Neural Networks*. 8(5):997–1007, 1997.
15. P.J. Werbos. Neuralcontrol and supervised learning: an overview and evaluation, *Handbook of Intelligent Control*. Van Nostrand, New York, 1992.
16. H. He, Z. Ni, and J. Fu. A three-network architecture for on-line learning and optimization based on adaptive dynamic programming, *Neurocomputing*, 78(1):3–13, 2012.
17. H. He and B. Liu. A hierarchical learning architecture with multiple-goal representations based on adaptive dynamic programming, In *Proceedings IEEE International Conference on Networking, Sensing, and Control (ICNSC'10)*. 2010.
18. J. Si and Y.T. Wang. On-line learning control by association and reinforcement, *IEEE Transactions on Neural Networks*. 12(2):264–276, 2001.
19. J. Si and D. Liu. Direct neural dynamic programming, *Handbook of Learning and Approximate Dynamic Programming*. IEEE Press, pp. 125–151, 2004.
20. P.J. Werbos. Backpropagation through time: What it does and how to do it, in *Proceedings IEEE*. 78:1550–1560, 1990.

21. P.J. Werbos. *The Roots of Backpropagation: From Ordered Derivatives to Neural Networks and Political Forecasting*. Wiley-Interscience, 1994.
22. T.L. Chien, C.C. Chen, Y.C. Huang, and W.-J. Lin. Stability and almost disturbance decoupling analysis of nonlinear system subject to feedback linearization and feedforward neural network controller, *IEEE Transactions Neural Networks*. 19(7):1220–1230, 2008.
23. P. H. Eaton, and D. V. Prokhorov, and D. C. Wunsch II., Neurocontroller alternatives for fuzzy ball-and-beam systems with nonuniform nonlinear friction, *IEEE Transactions on Neural Networks*. 11(2):423–435, 2000.

CHAPTER 5

Single Network Adaptive Critics Networks—Development, Analysis, and Applications

JIE DING, ALI HEYDARI, and S.N. BALAKRISHNAN

Department of Mechanical and Aerospace Engineering, Missouri University of Science and Technology, Rolla, MO, USA

ABSTRACT

Solving infinite time optimal control problems in an approximate dynamic programing framework with two-network structure has become popular in recent years. In this chapter, an alternative to the two-network structure is provided. We develop single network adaptive critics (SNAC) which eliminate the need to have a separate action network to output control. Two versions of SNAC are presented. The first version, called SNAC outputs the costates and the second version called J-SNAC outputs the cost function values. A special structure called Finite-SNAC to efficiently solve finite-time problems is also presented. Illustrative infinite time and finite time problems are considered; numerical results clearly demonstrate the potential of the single network structures to solve optimal control problems.

5.1 INTRODUCTION

There are very few open loop control based systems in practice. Feedback or closed loop control is desired as a hedge against noise that systems encounter in their operations and modeling uncertainties. Optimal control based formulations have been shown to yield desirable stability margins in linear system. For linear or nonlinear systems, dynamic programing formulations offer the most comprehensive solutions

Reinforcement Learning and Approximate Dynamic Programming for Feedback Control, First Edition. Edited by Frank L. Lewis and Derong Liu.

for obtaining feedback control [1, 2]. For nonlinear systems, solving the associated Hamilton–Jacobi–Bellman (HJB) equation is well-nigh impossible due to the associated number of computations and storage requirements. Werbos introduced the concept of adaptive critics (AC) [3], a dual neural network structure to solve an "approximate dynamic programing" (ADP) formulation which uses the discrete version of the HJB equation.

Many researchers have embraced and researched the enormous potential of AC based formulations over the last two decades [4–10]. Most of the model based adaptive critic controllers [11–13] and others [4, 5] were targeted for systems described by ordinary differential equations. Ref. [14] extended the realm of applicability of the AC design to chemical systems characterized by distributed parameter systems. Authors of [15] formulated a global adaptive critic controller for a business jet. Authors of [16] applied an adaptive critic based controller in an atomic force microscope based force controller to push nano-particles on the substrates. Ref. [17] showed in simulations that the ADP could prevent cars from skidding when driving over unexpected patches of ice.

There are many variants of AC designs [12] of which the most popular ones are the heuristic dynamic programing (HDP) and the dual heuristic programing (DHP) architectures. In the HDP formulation, one neural network, called the "critic" network maps the input states to output the optimal cost and another network called the "action" network outputs the optimal control with states of the system as its inputs [12]. In the DHP formulation, while the action network remains the same as with the HDP, the critic network outputs the optimal costate vector with the current states as inputs [11, 15]. Note that the AC designs are formulated in a discrete framework. Ref. [18] considered the use of continuous time adaptive critics. More recently, some authors have pursued continuous-time controls [19–22]. In [19, 20] the suggested algorithms for the online optimal control of continuous systems are based on sequential updates of the actor and critic networks.

Many nice theoretical developments have taken place in the last few years, primarily due to the groups led by Lewis and Sarangapani which have made the reinforcement learning based AC controllers acceptable for mainstream evaluation. However, there has not been much work in alleviating the computational load with the AC paradigms from a practical standpoint. Balakrishnan's group has been working from this perspective and formulated the single network adaptive critic designs. Authors of [23–25] have proposed a single network adaptive critic (SNAC) related to DHP designs. SNAC eliminates the usage of one neural network (namely the action network) that is a part of a typical dual network AC setup. As a consequence, the SNAC architecture offers three advantages: a simpler architecture for implementation, less computational load for training and recall process, and elimination of the approximation error associated with the eliminated network. In Ref. [25], it is shown through comparison studies that the SNAC costs about half the computation time as that of a typical dual network AC structure for the same problem.

SNAC is applicable to a wide class of nonlinear systems where the optimal control equation can be explicitly expressed in terms of the state and costate vectors. Most of the problems in aerospace, automobile, robotics, and other engineering disciplines can be characterized by *nonlinear control–affine equations* that yield such a relation. SNAC based controllers have yielded excellent tracking results in micro-electromechanical systems and chemical reactor applications, [14, 25]. Authors of [25] have proved that for linear systems the SNAC converges to discrete Riccati equation solution.

Motivated by the simplicity of the training and implementation phase resulted from SNAC, Ref. [26] developed an HDP based SNAC scheme. In their cost function based single network adaptive critic, or J-SNAC, the critic network outputs the cost instead of the costate vector. There are some advantages with the J-SNAC formulation: first, a smaller neural network in needed for J-SNAC because the output is only a scalar, while in SNAC, the output is the costate vector which has as many elements as the order of the system. The smaller network size results in less training and recall effort in J-SNAC. Second, in J-SNAC the output is the optimal cost that has some physical meaning for the designer, as opposed to the costate vector in SNAC.

Note that these developments introduced above have mainly addressed only the infinite horizon problems. On the other hand, finite-horizon optimal control is an important branch of optimal control theory. Solutions to the resulting time varying HJB equation are usually very difficult to obtain. In the finite-horizon case, the optimal cost-to-go is not only a function of the current states, but also a function of how much time is left (time-to-go) to accomplish the goal. There is hardly any work in the neural network literature to solve this class of problems [28, 29].

To deal with finite horizon problems, a single neural network based solution, called Finite-SNAC, was developed in [27] which embeds solutions to the time-varying HJB equation. Finite-SNAC is a DHP based NN controller for finite-horizon optimal control of discrete-time input-affine nonlinear systems and is developed based on the idea of using a single set of weights for the network. The Finite-SNAC has the current time *as well as* the states as inputs to the network. This feature results in much less required storage memory compared to the other published methods [28, 29]. *Note that once trained, Finite-SNAC provides optimal feedback solutions to any different final time as long as it is less than the final time for which the network is synthesized.*

Rest of this chapter is organized as follows: the approximate dynamic programing formulation is discussed in Section 5.2 followed by presentation of the SNAC architecture for infinite-horizon problems in Section 5.3; J-SNAC architecture and simulation results are presented in Section 5.4; and Finite-SNAC architecture for finite-time optimal control problems and its simulation results are presented in Section 5.5.

5.2 APPROXIMATE DYNAMIC PROGRAMING

The dynamic programing formulation offers a comprehensive optimal control solution in a state feedback form, however, it is handicapped by computational and storage requirements. The approximate dynamic programing (ADP) formulation implemented with an adaptive critic (AC) neural network structure has evolved as a powerful alternative technique that obviates the need for excessive computations and storage requirements in solving optimal control problems.

In this section, the principles of ADP are described. An interested reader can find more details about the derivations in [3, 11]. Note that a prime requirement to apply the ADP is to formulate the problem in discrete-time. The control designer has the freedom to use any appropriate discretization scheme. For example, one can use the Euler approximation for the state equation and trapezoidal approximation for the cost function [30].

In a discrete-time formulation, the objective is to find an admissible control u_k, which causes the system given by

$$x_{k+1} = F(x_k, u_k), \tag{5.1}$$

to follow an admissible trajectory from an initial point x_0 to a final desired point x_N while minimizing a desired cost function J given by

$$J = \sum_{k=0}^{N-1} \Psi_k(x_k, u_k), \tag{5.2}$$

where $x_k \in \mathbb{R}^n$ and $u_k \in \mathbb{R}^l$ denote the state vector and the control at time step k, respectively. n is the order of the system and l is the system's number of inputs. The functions $F_k : \mathbb{R}^n \times \mathbb{R}^l \to \mathbb{R}^n$ and $\Psi_k : \mathbb{R}^n \times \mathbb{R}^l \to \mathbb{R}$ are assumed to be differentiable with respect to both x_k and u_k. Moreover, Ψ_k is assumed to be convex (e.g., a quadratic function in x_k and u_k). One can notice that when $N \to \infty$, this cost function leads to a regulator (infinite time) problem.

The steps in obtaining optimal control are now described. First, the cost function (5.2) is rewritten to start from time step k as

$$J_k = \sum_{\tilde{k}=k}^{N-1} \Psi_{\tilde{k}}(x_{\tilde{k}}, u_{\tilde{k}}). \tag{5.3}$$

The cost, J_k, can be split into

$$J_k = \Psi_k + J_{k+1}, \tag{5.4}$$

where Ψ_k and $J_{k+1} = \sum_{\tilde{k}=k+1}^{N-1} \Psi_{\tilde{k}}$ represent the "utility function" at time step k and the cost-to-go from time step $k+1$ to N, respectively.

The $n \times 1$ costate vector at time step k is defined as

$$\lambda_k \equiv \frac{\partial J_k}{\partial x_k}, \tag{5.5}$$

The necessary condition for optimality is given by

$$\frac{\partial J_k}{\partial u_k} = 0. \tag{5.6}$$

Equation (5.6) can be further expanded as

$$\frac{\partial J_k}{\partial u_k} = \frac{\partial \Psi_k}{\partial u_k} + \frac{\partial J_{k+1}}{\partial u_k} = \frac{\partial \Psi_k}{\partial u_k} + \left(\frac{\partial x_{k+1}}{\partial u_k}\right)^{\mathrm{T}} \frac{\partial J_{k+1}}{\partial x_{k+1}}. \tag{5.7}$$

The optimal control equation can, therefore, be written as

$$\frac{\partial \Psi_k}{\partial u_k} + \left(\frac{\partial x_{k+1}}{\partial u_k}\right)^{\mathrm{T}} \frac{\partial J_{k+1}}{\partial x_{k+1}} = 0. \tag{5.8}$$

The costate equation is derived in the following way

$$\lambda_k = \frac{\partial \Psi_k}{\partial x_k} + \frac{\partial J_{k+1}}{\partial x_k} = \frac{\partial \Psi_k}{\partial x_k} + \left(\frac{\partial x_{k+1}}{\partial x_k}\right)^{\mathrm{T}} \frac{\partial J_{k+1}}{\partial x_{k+1}} = \frac{\partial \Psi_k}{\partial x_k} + \left(\frac{\partial x_{k+1}}{\partial x_k}\right)^{\mathrm{T}} \lambda_{k+1}. \tag{5.9}$$

To synthesize an optimal controller, (5.1), (5.8), and (5.9) have to be solved simultaneously, along with appropriate boundary conditions. For regulator problems, the boundary conditions usually take the form: x_0 is fixed and $\lambda_N \to 0$ as $N \to \infty$. If the state equation and cost function are such that one can obtain an explicit relationship for the control variable in terms of the state and the cost variables from Equation (5.8), the ADP formulation can be solved through SNAC. Note that control affine nonlinear systems (of the form $x_{k+1} = f(x_k) + g(x_k)u_k$) with a quadratic cost function (of the form $J = \frac{1}{2}\sum_{k=1}^{\infty}(x_k^T Q x_k + u_k^T R u_k)$) fall under this class.

5.3 SNAC

In this section, an improvement and modification to the AC architecture, called the "single network adaptive critic (SNAC)" related to the DHP designs is presented. In the SNAC design, the critic network captures the functional relationship between the state x_k and the optimal costate λ_{k+1}. Denoting the neural network functional mapping with $NN(.)$ one has

$$\lambda_{k+1} = NN(x_k). \tag{5.10}$$

For the input-affine system and the quadratic cost function described in the following equation, once λ_{k+1} is calculated, one can generate the optimal control through Equation (5.13).

$$x_{k+1} = f(x_k) + g(x_k)u_k, \tag{5.11}$$

$$J = \frac{1}{2}\sum_{k=1}^{\infty}\left(x_k^{\mathrm{T}} Q x_k + u_k^{\mathrm{T}} R u_k\right), \tag{5.12}$$

$$u_k = -R^{-1} g(x_k)^{\mathrm{T}} \lambda_{k+1}. \tag{5.13}$$

Note that, for this case, Equation (5.9) reads

$$\lambda_{k+1} = Q x_{k+1} + \left(\frac{\partial\left(f\left(x_{k+1}\right) + g\left(x_{k+1}\right) u_{k+1}\right)}{\partial x_{k+1}}\right)^{\mathrm{T}} \lambda_{k+2}. \tag{5.14}$$

5.3.1 State Generation for Neural Network Training

State generation is an important part of the training process for the SNAC outlined in the next subsection. For this purpose, define $S_i = \{x_k : x_k \in \Omega\}$, where Ω denotes the domain in which the system operates. This domain is so chosen that its elements cover a large number of points of the state space in which the state trajectories are expected to lie. For a systematic training scheme, a "telescopic method" is arrived at as follows:

For $i = 1, 2, \ldots,$ define the set $S_i = \{x_k : \|x_k\|_\infty \le c_i\}$, where c_i is a positive constant. At the beginning, a small c_1 is fixed and the network is trained with the states generated in S_1. After the network converges (the convergence condition will be discussed in the next subsection), c_2 is chosen such that $c_2 > c_1$. Then the network is trained again for states within S_2 and so on. The network training is continued until $i = I$, where S_I covers the domain of interest Ω.

5.3.2 Neural Network Training

The steps for training the SNAC network (see Figure 5.1) are as follows:

(1) generate S_i. For each element x_k of S_i, follow the steps
  (a) input x_k to the critic network to obtain $\lambda_{k+1} = \lambda_{k+1}^a$,
  (b) calculate u_k from the optimal control equation with x_k and λ_{k+1},

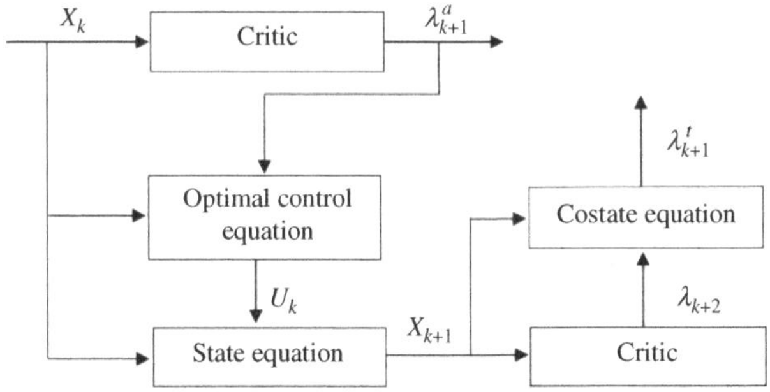

FIGURE 5.1 SNAC training scheme.

   (c) get x_{k+1} from the state equation using x_k and u_k,
   (d) input x_{k+1} to the critic network to get λ_{k+2},
   (e) using x_{k+1} and λ_{k+2}, calculate λ_{k+1}^t from costate Equation (5.14)

(2) train the critic network for all x_k in S_i; the target being λ_{k+1}^t

(3) check the convergence of the network. If it is converged, revert to step 1 with $i = i + 1$ until $i = I$. Otherwise, repeat steps 1–2.

5.3.3 Convergence Condition

To check the convergence of the critic network, a set of new states, S_i^c and target outputs are generated as described in Section 5.3.2. Let the target outputs be λ_{k+1}^t and the outputs from the trained networks (using the same inputs from the set S_i^c) be λ_{k+1}^a. A tolerance value, tol_c is used as the convergence criterion for the critic network. By defining the relative error $e_{c_k} \equiv \left(\lambda_{k+1}^t - \lambda_{k+1}^a / \lambda_{k+1}^t\right)$ and $e_c \equiv \left\{e_{c_k}\right\}$, $k = 1, \ldots, |S_I|$, the training process is stopped when $\|e_c\| < tol_c$.

5.4 J-SNAC

In this section, the cost function based single network adaptive critic, called J-SNAC is presented. In J-SNAC the critic network outputs cost instead of the costate as in SNAC. This approach is applicable to the class of nonlinear systems of $x_{k+1} = f(x_k) + gu_k$ with a constant g matrix. As mentioned in the introduction section, the J-SNAC technique retains all the powerful features of the AC methodology while eliminating the action network completely. In the J-SNAC design, the critic network captures the functional relationship between the state x_k and the optimal cost J_k.

Denoting the functional mapping of J-SNAC with $NN(.)$, one has

$$J_k = NN(x_k). \tag{5.15}$$

One can calculate λ_k through $\lambda_k = \frac{\partial J(x_k)}{\partial x_k}$, and rewrite the costate Equation (5.14), for the quadratic cost function (5.12), in the following form

$$\lambda_{k+1} = \left(\frac{\partial f(x_k)}{\partial x_k}\right)^{-\mathrm{T}} (\lambda_k - Qx_k). \tag{5.16}$$

Optimal control can now be calculated as

$$u_k = -R^{-1} g^{\mathrm{T}} \lambda_{k+1} = -R^{-1} g^{\mathrm{T}} \left(\frac{\partial f(x_k)}{\partial x_k}\right)^{-\mathrm{T}} \left(\frac{\partial NN(x_k)}{\partial x_k} - Qx_k\right). \tag{5.17}$$

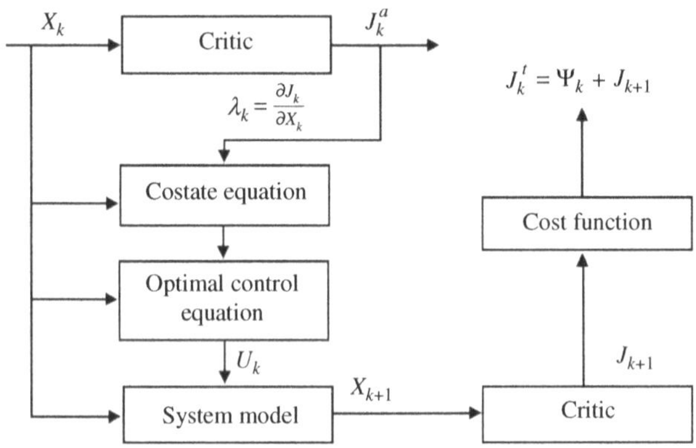

FIGURE 5.2 J-SNAC training scheme.

5.4.1 Neural Network Training

Using a similar state generation and convergence check as discussed in the SNAC training procedure, the steps for training the J-SNAC network are as follows (Figure 5.2):

(1) generate S_i. For each element x_k of S_i, follow the steps:
  (a) input x_k to the critic network to obtain $J_k = J_k^a$,
  (b) calculate $\lambda_k = \partial J_k / \partial x_k$ and then λ_{k+1} using Equation (5.16),
  (c) calculate u_k from Equation (5.17),
  (d) Use x_k and u_k to get x_{k+1} from the state equation,
  (e) Input x_{k+1} to the critic network to get J_{k+1},
  (f) Use x_k, u_k and J_{k+1}, to calculate J_k^t using Equation (5.4),
(2) train the critic network for all x_k in S_i with the output being J_k^t.
(3) check the convergence of the critic network. If yes, revert to step 1 with $i = i + 1$ until $i = I$. Otherwise, repeat steps 1–2.

5.4.2 Numerical Analysis

For illustrative purposes, a satellite attitude control problem is selected.

5.4.2.1 Modeling the Attitude Control Problem Consider a rigid spacecraft controlled by reaction wheels. It is assumed that the control torques are applied through a set of reaction wheels along three orthogonal axes. The spacecraft rotational motion is described by [31]

$$I\dot{\omega} = p \times \omega + \tau, \tag{5.18}$$

where I is the matrix of moment of inertia, ω is the angular velocity of the body frame, with respect to the inertial frame, p is the total spacecraft angular momentum expressed in the spacecraft body frame, and τ is the torque applied to the spacecraft by the reaction wheels.

Using the Euler angles ϕ, θ, and ψ, the kinematics equation describing the attitude of the spacecraft may be written as

$$\frac{\mathrm{d}}{\mathrm{d}t}[\phi, \theta, \psi]^{\mathrm{T}} = J(\phi, \theta, \psi)\,\omega, \tag{5.19}$$

where

$$J(\phi, \theta, \psi) = \begin{bmatrix} 1 & \sin(\phi)\tan(\theta) & \cos(\phi)\tan(\theta) \\ 0 & \cos(\phi) & -\sin(\phi) \\ 0 & \sin(\phi)\sec(\theta) & \cos(\phi)\sec(\theta) \end{bmatrix}, \tag{5.20}$$

The total spacecraft angular momentum p is written as

$$p = R(x)\,p^{I}, \tag{5.21}$$

where $p^{I} = [1, -1, 0]^{T}$ is the (constant) inertial angular momentum and

$$R(x) = \begin{bmatrix} \cos(\psi)\cos(\theta) & \cos(\psi)\sin(\theta)\sin(\phi) - \sin(\psi)\cos(\phi) & \cos(\psi)\sin(\theta)\sin(\phi) + \sin(\psi)\sin(\phi) \\ \sin(\psi)\cos(\theta) & \sin(\psi)\sin(\theta)\sin(\phi) + \cos(\psi)\cos(\phi) & \sin(\psi)\sin(\theta)\cos(\phi) - \cos(\psi)\sin(\phi) \\ -\sin(\psi) & \cos(\psi)\sin(\theta) & \cos(\psi)\cos(\theta) \end{bmatrix}, \tag{5.22}$$

Choosing $x = \left[\phi, \theta, \psi, \omega_x, \omega_y, \omega_z\right]^{\mathrm{T}}$ as the states and $u = [\tau_1, \tau_2, \tau_3]^{\mathrm{T}}$ as the controls, the dynamics of the system, that is, (5.18) and (5.19) can be rewritten as

$$\begin{bmatrix} \dot{\phi} \\ \dot{\theta} \\ \dot{\psi} \\ \dot{\omega}_x \\ \dot{\omega}_y \\ \dot{\omega}_z \end{bmatrix} = \begin{bmatrix} 0_{3\times3} & J(\phi, \theta, \psi) \\ 0_{3\times3} & I^{-1}p\times \end{bmatrix} \begin{bmatrix} \phi \\ \theta \\ \psi \\ \omega_x \\ \omega_y \\ \omega_z \end{bmatrix} + \begin{bmatrix} 0_{3\times3} \\ I^{-1} \end{bmatrix} \begin{bmatrix} \tau_1 \\ \tau_2 \\ \tau_3 \end{bmatrix}, \tag{5.23}$$

The control objective is to drive all the states to zero as $t \to \infty$. A quadratic cost function, J_c is selected as

$$J_c = \int_0^{\infty} \left[\frac{1}{2}x^{\mathrm{T}}Qx + \frac{1}{2}u^{\mathrm{T}}Ru\right] \mathrm{d}t, \tag{5.24}$$

where $Q \in \mathbb{R}^{6\times6}$ is a positive semi-definite and $R \in \mathbb{R}^{3\times3}$ is a positive definite matrix for penalizing the states and controls, respectively.

Denoting the time step by Δt, the state equation is discretized as

$$x_{k+1} = x_k + \Delta t \left[f(x_k) + g u_k \right], \tag{5.25}$$

where $f(x_k)$ and g are given in the state Equation (5.23). The quadratic cost function (5.24) is also discretized as

$$J = \sum_{k=0}^{N\to\infty} \left(\frac{1}{2} x_k^{\mathrm{T}} Q x_k + \frac{1}{2} u_k^{\mathrm{T}} R u_k \right) \Delta t. \tag{5.26}$$

The optimality condition leads to the control equation

$$u_k = -(\Delta t R)^{-1} \Delta t g^{\mathrm{T}} \left(\left(\frac{\partial x_{k+1}}{\partial x_k} \right)^{\mathrm{T}} \right)^{-1} \left(\frac{\partial J_k}{\partial x_k} - \Delta t Q_W x_k \right). \tag{5.27}$$

5.4.2.2 *Simulation* In numerical simulations, $\Delta t = 0.005$, $Q = diag(20, 20, 20, 0, 0, 0)$, and $R = diag(10^{-3}, 10^{-3}, 10^{-3})$ are selected. Note that regulating the Euler angles will automatically regulate the angular rates as well, hence, penalizing the first three element of the state vector through the selected Q, will guarantee the regulation of the whole state vector. The inertia matrix is selected as an identity matrix.

A single layer neural network of the form $J_k = W^T \phi(x_k)$ is selected, where W denotes the neural network weights and $\phi(.)$ the basis function. The network weights are initialized to zero and the basis functions $\phi(x)$ are selected as

$$\phi(x) = \left[x_1, x_2, \ldots, x_6, x_1^2, x_2^2, \ldots, x_6^2, x_1x_2, x_1x_3, \ldots, x_5x_6 \right], \tag{5.28}$$

where x_i, $i = 1, 2, \ldots, 6$ denotes the ith element of state vector x. The initial condition is selected as $x_0 = [45°, -45°, 35°, 0, 0, 0]^{\mathrm{T}}$.

Histories of the Euler angles and rotation rates with time are shown in Figure 5.3. It can be seen that all the states go to zero within 5 s. Moreover, as seen

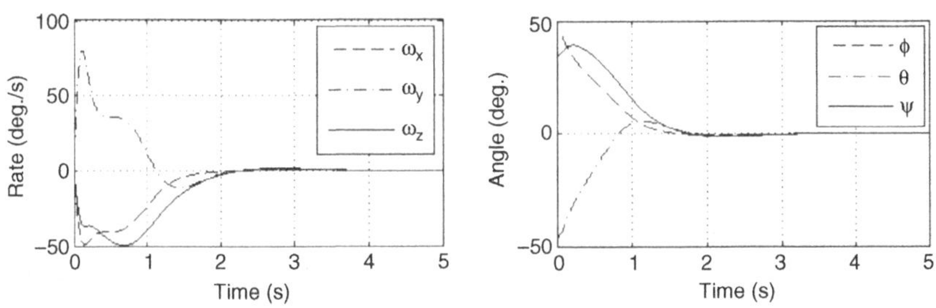

FIGURE 5.3 Histories of angles and body rates.

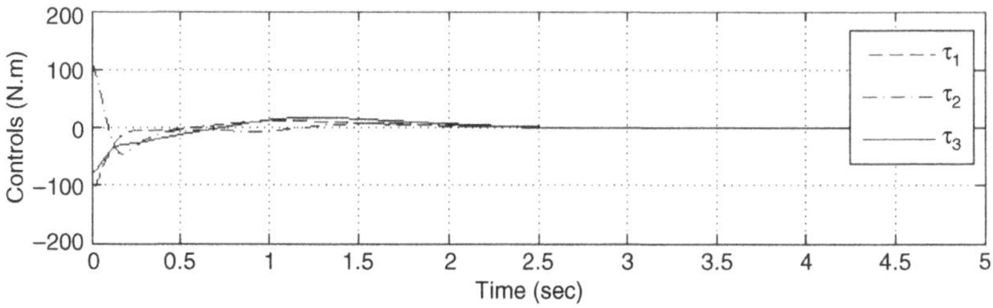

FIGURE 5.4 Histories of applied controls.

in Figure 5.4 which shows the history of the applied controls, the goal is achieved through applying bounded controls about the three axes of the spacecraft.

5.5 FINITE-SNAC

Finite-SNAC solves *finite horizon optimal control* of nonlinear input-affine systems based on a DHP formulation. The discrete-time nonlinear input-affine system and a quadratic cost function to be minimized are given in (5.29) and (5.30).

$$x_{k+1} = f(x_k) + g(x_k)u_k, \tag{5.29}$$

$$J = \frac{1}{2}x_N^T Q_f x_N + \sum_{i=0}^{N-1} \frac{1}{2}(x_i^T Q x_i + u_i^T R u_i), \tag{5.30}$$

where $x_k \in \mathbb{R}^n$ and $u_k \in \mathbb{R}^l$ denote the state vector and the control at time step k, respectively. n is the order of the system and l is the number of inputs. $f(.) \in \mathbb{R}^n$ and $g(.) \in \mathbb{R}^{n\times l}$ are the system dynamics and $Q_f \in \mathbb{R}^{n\times n}$, $Q \in \mathbb{R}^{n\times n}$ and $R \in \mathbb{R}^{l\times l}$ are weighting matrices for the final states, states, and control vectors, respectively. Q_f and Q are positive semi-definite or positive definite matrices and R is a positive definite matrix. Finally, N is the total (fixed) number of time steps and superscript T denotes transposition.

Denoting the neural network mapping by $NN(.)$, a single neural network called Finite-SNAC is developed to output the desired costate vector based on the state vector and the time-to-go as inputs.

$$\lambda_{k+1} = NN(x_k, N-k, W), \; 0 \le k < N-1, \tag{5.31}$$

where $\lambda_{k+1} \in \mathbb{R}^n$ is the system costate vector at time step $k+1$ and W denotes the network weight matrix.

The neural network in this section is selected in the form of linear in the weights.

$$NN(x, N-k, W) = W^T \phi(x, N-k), \tag{5.32}$$

where $\phi(.) \in \mathbb{R}^m$ is composed of m linearly independent scalar basis functions and $W \in \mathbb{R}^{m\times n}$ where m is the number of neurons.

Following the ADP framework discussed in Section 5.2, the costate equation will be

$$\lambda_{k+1} = Qx_{k+1} + \left(\frac{\partial (f(x_{k+1}) + g(x_{k+1})u_{k+1})}{\partial x_{k+1}}\right)^{\mathrm{T}} \lambda_{k+2}, \ 0 \leq k < N-1. \quad (5.33)$$

Note that since

$$J_N = J(x_N, N) = \frac{1}{2} x_N^{\mathrm{T}} Q_f x_N, \quad (5.34)$$

the final condition on the costate vector is

$$\lambda_N = Q_f x_N. \quad (5.35)$$

The network-training target, denoted by λ^t, can be calculated using following two equations.

$$\lambda_{k+1}^t = Qx_{k+1} + \left(\frac{\partial (f(x_{k+1}) + g(x_{k+1})u_{k+1})}{\partial x_{k+1}}\right)^{\mathrm{T}} \lambda_{k+2}, \quad 0 \leq k < N-1, \quad (5.36)$$

$$\lambda_N^t = Q_f x_N, \quad (5.37)$$

In the training process, λ_{k+2} in the right-hand side of (5.36) will be substituted by $NN(x_{k+1}, N-(k+1), W)$ as described earlier in the SNAC training process.

Once the network is trained, it can be used for optimal feedback control in the sense that in the online implementation, the states and the time-to-go will be fed into the network to generate the optimal costate vector and the optimal control will be calculated through (5.38).

$$u_k = -R^{-1} g(x_k)^T \lambda_{k+1}, \quad 0 \leq k < N-1. \quad (5.38)$$

5.5.1 Neural Network Training

The Finite-SNAC training should be done in such a way that along with learning the target given in (5.36) for every state x_k and time k, the final condition (5.37) is also satisfied. The authors suggest augmenting the training input-target pairs in such a way that the final condition is forced to be met in each learning iteration. To do so, one can define following augmented parameters:

$$\bar{\lambda} \equiv [\lambda_{k+1} \quad \lambda_N], \quad (5.39)$$

$$\bar{\phi} \equiv [\phi(x_k, N-k) \quad \phi(x_{N-1}, 1)]. \quad (5.40)$$

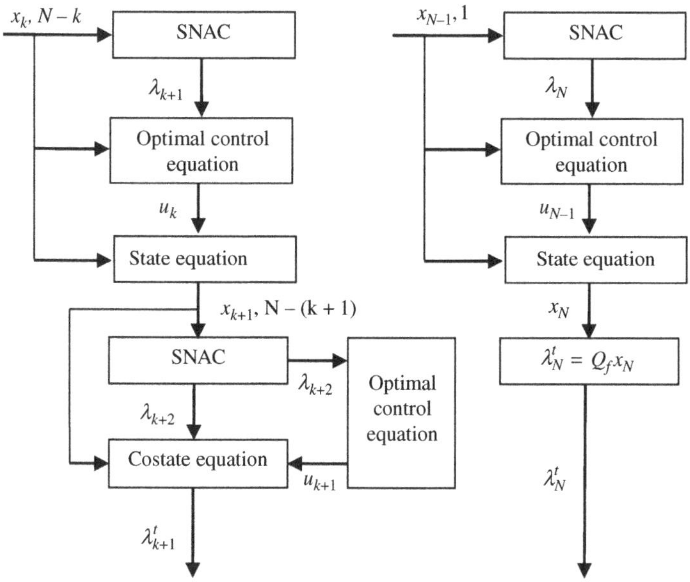

FIGURE 5.5 Finite-SNAC training scheme.

Now, the network output and the target to be learned are

$$\bar{\lambda} = W^{\mathrm{T}}\bar{\phi}, \tag{5.41}$$

$$\bar{\lambda}^t \equiv [\lambda^t_{k+1} \ \lambda^t_N]. \tag{5.42}$$

The training error is given by

$$e \equiv \bar{\lambda} - \bar{\lambda}^t = W^{\mathrm{T}}\bar{\phi} - \bar{\lambda}^t. \tag{5.43}$$

Now, in each iteration along with selecting a random state x_k, a random time k, k, $0 \le k < N-1$, will also be selected. Feeding x_k and $N-k$ into (5.31) results in a costate vector which will be used for calculating u_k through (5.38). Having x_k and u_k one can propagate x_k to x_{k+1} using (5.29). Having x_{k+1} fed into (5.31) once more, λ_{k+2} will be resulted which is the desired unknown needed for target λ^t_{k+1} calculation through (5.36). Then, to calculate λ^t_N through (5.37), a randomly selected state will be considered as x_{N-1} and propagated to x_N using the similar process discussed earlier for propagating x_k to x_{k+1}, and fed to (5.37). Finally $\bar{\lambda}^t$ will be formed using (5.42). This process is depicted graphically in Figure 5.5. In this diagram, the left column follows (5.36) and the right column follows (5.37) for the target calculation.

Having the input-target pair $\left\{\left[(x_k, N-k)\,(x_{N-1}, 1)\right],\ \left[\lambda^t_{k+1}\ \lambda^t_N\right]\right\}$ calculated, the network can be trained using any training method. The selected training law in this study is least squares.

5.5.2 Convergence Theorems

The proposed algorithm for the Finite-SNAC training is based on the DHP scheme, in which starting with an initial value for the costate vector one iterates to converge to the optimal costate. Denoting the iteration index by a superscript and the time index by a subscript, the learning algorithm for finite horizon optimal control starts with an initial value assignment to λ_k^0 for all k's, for example, $\lambda_k^0 = 0\,\forall k$, and repeating below three calculations for different i's from zero to infinity.

$$u_k^i = -R^{-1} g(x_k)^{\mathrm{T}} \lambda_{k+1}^i, \tag{5.44}$$

$$\lambda_k^{i+1} = Q x_k + A\left(x_k, u_k^i\right)^{\mathrm{T}} \lambda_{k+1}^i, \tag{5.45}$$

$$\lambda_N^{i+1} = Q_f x_N. \tag{5.46}$$

The last equation is actually the final condition of the optimal control problem. Note that,

$$A\left(x_k, u_k^i\right) \equiv \frac{\partial\left(f(x_k) + g\left(x_k\right) u_k^i\right)}{\partial x_k}, \tag{5.47}$$

$$\lambda_{k+1}^i \equiv \lambda^i\left(x_{k+1}\right) = \lambda^i\left(f\left(x_k\right) + g\left(x_k\right) u_k^i\right), \tag{5.48}$$

The problem is to prove that this iterative procedure results in the optimal values for the costate λ and control u. The convergence proofs presented in [27] is based on the convergence of the HDP, in which the parameter subject to evolution is the cost function J whose behavior is much simpler to discuss as compared to that of the costate vector λ.

In HDP, the cost function J needs to be initialized, e.g. $J^0\left(x_k, k\right) = 0\,\forall k$, and is iteratively updated through the following steps.

$$J^{i+1}\left(x_k, k\right) = \frac{1}{2}\left(x_k^T Q x_k + u_k^{i^T} R u_k^i\right) + J^i(x_{k+1}, k+1), \tag{5.49}$$

$$u_k^i = \operatorname{argmin}_{\mathrm{u}}\left(J^{i+1}\left(x_k, k\right)\right) = -R^{-1} g(x_k)^T \frac{\partial J_{k+1}^i}{\partial x_{k+1}}. \tag{5.50}$$

For the finite horizon case, the final condition given in (5.51) is satisfied at every iteration.

$$J^{i+1}\left(x_N, N\right) = \frac{1}{2} x_N^T Q_f x_N. \tag{5.51}$$

Note that $J_k \equiv J(x_k, k)$ and

$$J_{k+1}^i \equiv J^i\left(f\left(x_k\right) + g\left(x_k\right) u_k^i, k+1\right). \tag{5.52}$$

The convergence of the aforementioned reinforcement learning schemes are given below and their proofs, given in [27], are skipped because of the page constraint. Note that the proofs of the finite-time cases are inspired by the proof for the HDP convergence for infinite-horizon case in [6].

Theorem 5.1 HDP Convergence *The sequence of J^i iterated through (5.49) to (5.51), in case of $J^0(x_k) = 0$ converges to the finite-horizon optimal solution.*

Theorem 5.2 DHP Convergence *The sequences λ_k^i and u_k^i defined by Equations (5.44)–(5.46), in case of $\lambda_k^0 = 0\,\forall k$, converges to the finite-horizon optimal solution for the given nonlinear control–affine system.*

5.5.3 Numerical Analysis

An orbital maneuver problem is provided to motivate the readers on the Finite-SNAC usage. A spacecraft is supposed to move from a certain orbit to another orbit in a fixed final time. This problem is known as the *Rendezvous* problem in the orbital mechanics literature.

5.5.3.1 Modeling the Orbital Maneuver Problem Denoting the displacement vector of the center of mass of a rigid spacecraft from the center of the orbital frame in the destination orbit by $\delta\boldsymbol{r} \equiv [X\ Y\ Z]^{\mathrm{T}}$, where X, Y, and Z are the components of the vector $\delta\boldsymbol{r}$ in the orbital frame, the nondimentionalized equation of motion of a spacecraft in the gravity field is given as [32],

$$\begin{aligned} &\ddot{X} - 2\dot{Y} + (1+X)\left(\frac{1}{r^3} - 1\right) = u_x \\ &\ddot{Y} + 2\dot{X} + Y\left(\frac{1}{r^3} - 1\right) = u_y \\ &\ddot{Z} + \frac{1}{r^3}Z = u_z, \end{aligned} \tag{5.53}$$

where the dots denote the time-derivatives; u_x, u_y, and u_z respectively denote the three components of the nondimensionalized total force exerted on the spacecraft in x, y, and z axes and $r \equiv \sqrt{(1+X)^2 + Y^2 + Z^2}$.

Selecting the state vector as $x = \begin{bmatrix} X & Y & Z & \dot{X} & \dot{Y} & \dot{Z} \end{bmatrix}^{\mathrm{T}}$ and the control vector as $u = \begin{bmatrix} u_x & u_y & u_z \end{bmatrix}^{\mathrm{T}}$, the state equation of the orbital maneuver problem reads:

$$\frac{d}{dt}\begin{bmatrix} x_1 \\ x_2 \\ x_3 \\ x_4 \\ x_5 \\ x_6 \end{bmatrix} = \begin{bmatrix} x_4 \\ x_5 \\ x_6 \\ 2x_5 + (1+x_1)\left(1/r^3 - 1\right) \\ -2x_4 - x_2(1/r^3 - 1) \\ -x_3(1/r^3) \end{bmatrix} + \begin{bmatrix} 0 \\ 0 \\ 0 \\ u_1 \\ u_2 \\ u_3 \end{bmatrix}, \tag{5.54}$$

TABLE 5.1 The Spec of the Destination and Source Orbit

The Characteristic	Source Orbit	Destination Orbit
Orbit semimajor axes	10,000 km	10,000 km
Right ascension of the ascending node	15°	0°
Inclination	85°	90°
Eccentricity	0	0

where the components of the state vector are denoted by x_i, $1 \leq i \leq 6$, and the control vector are denoted by u_i, $1 \leq i \leq 3$. Note that this state equation is in a nonlinear input-affine form, suitable for the Finite-SNAC application.

Now, the problem is to apply optimal control to force the states x_i, $1 \leq i \leq 6$, to go to zero in a predetermined and fixed final time. Convergence of the states to zero is equivalent to performing the orbital maneuver and locating at the destination point in the dictated time.

5.5.3.2 Simulation The nondimentionalized state equation can cover any circular orbit as the rendezvous position. The mean anomaly of both of the maneuvering and the destination spacecraft are zero at the start of the maneuver. The specifications of the two orbits are given in Table 5.1.

For the given characteristics, the initial condition will be

$$x_0 = [0.0\ 0.2588\ 0.0341\ -0.0038\ 0.0842\ 0.0226]^T. \tag{5.55}$$

The fixed final time is selected as 3 time units. The time unit is $t = 1.58 \times 10^3$ s and the reference length is $\bar{R} = 10{,}000$ km, [32]. In this condition, each control unit will be $\bar{R}/t^2 = 0.004\ \text{m/s}^2$. Note that, this implementation is based on having continuous thrust for actuation. The weight matrices are selected through trial and error as $Q_f = diag\,(2000\ 2000\ 2000\ 200\ 200\ 200)$, $Q = diag\,(10\ 10\ 10\ 1\ 1\ 1)$, and $R = diag(1\ 1\ 1)$. The values of the elements of Q_f are selected much higher than those of Q to force the error of the states at the final time to be small.

The basis functions selected for the neural network are combinations of the following polynomials: x_i, x_i^2, x_i^3 for $i = 1, 2, \ldots, 7$ and $x_i x_j$, $x_i x_j x_7$, $x_i x_7^2$, $x_i x_7^3$, $x_i e^{-x_7}$, $x_i^2 e^{-x_7}$, $x_i x_j e^{-x_7}$ for $i, j = 1, 2, \ldots, 6$ and $i \neq j$. Note that x_i is the ith input of the neural network and x_7 is the time-to-go, normalized through dividing it by the total time, and its contribution in the basis functions is selected through some trial and error such that the network error is as small as possible.

The total number of basis functions is equal to 96. The network is trained for 200 epochs with the number of states selected at each iteration equal to 300 for creating a mesh over the region of interest as explained in [27]. For this simulation, the states are selected in such a way that each element belongs to the interval of $(-0.3,\ 0.3)$. Performing the weight update, the weights converged as shown in Figure 5.6.

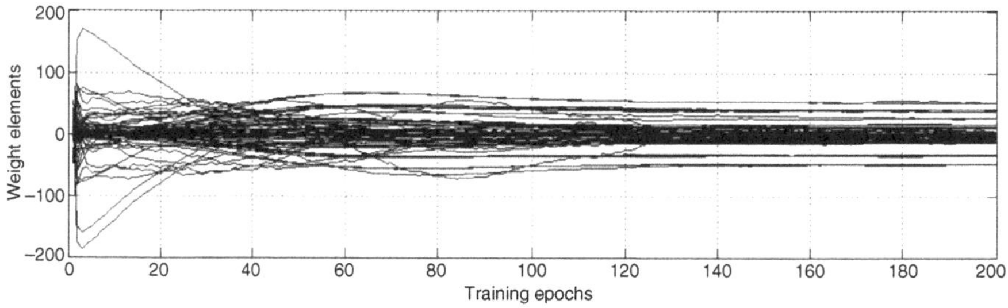

FIGURE 5.6 Histories of some of the elements of the weights matrix during the training iterations.

Histories of the position elements of the state vector and the control are given in Figure 5.7. To evaluate the performance of the controller, a numerical solution to this problem was calculated through an iterative process for the given initial condition and the results are depicted using dash plots in Figure 5.7, while solid plots denote those of the Finite-SNAC. As seen through the plots, the proposed controller has been able to force the states to converge to the origin in the fixed final time of 3 units and the Finite-SNAC trajectories are close to those of the optimal open loop numerical solution. The cost-to-go of the Finite-SNAC turned out to be only 3% more than the open loop optimal solution. While the open loop solution is only applicable to the preselected initial condition and the fixed time-to-go, the Finite-SNAC solutions are more versatile as will be shown in the discussions below.

To demonstrate the versatility of the HJB based finite-horizon controller, *the same network without retraining* is used for another maneuver with the same initial condition but with a shorter time-to-go, that is 2 time units. The trajectories of the position elements of the state vector and the control histories for the shorter horizon are superimposed on the result of the previous simulation of the Finite-SNAC controller and shown in Figure 5.8. In the two figures, the solid plots denote the results of the maneuver with time-to-go of 3, and the dash plots denote those of the maneuver with time-to-go of 2 time units.

As could be seen in Figure 5.8, the controller has applied a different control history on the spacecraft to accomplish the same maneuver in a shorter time. This shows that

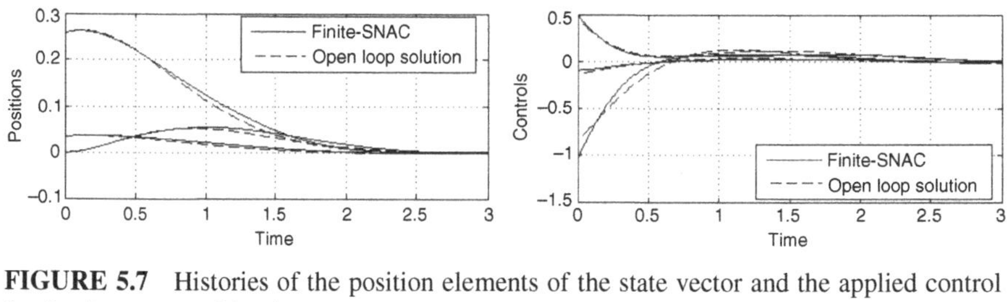

FIGURE 5.7 Histories of the position elements of the state vector and the applied control for the time-to-go of 3 units.

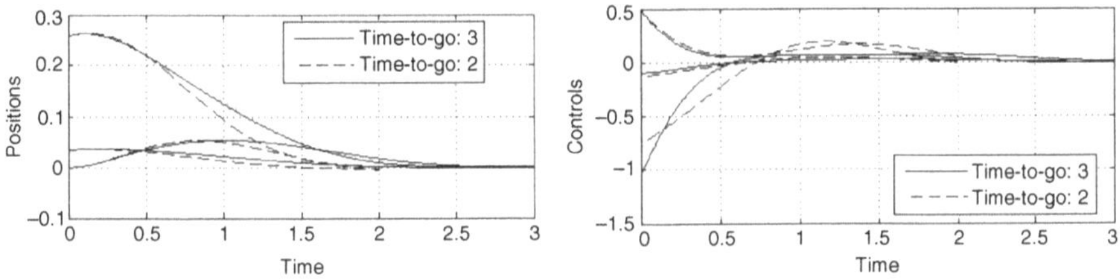

FIGURE 5.8 Histories of the position elements of the state vector and the applied control; Finite-SNAC with different final times.

the network has learned the *time-dependency* of the optimal control in finite-horizon problems.

The developed controller is able to perform well under different initial conditions as long as the resulting state trajectory belongs to the domain on which the network is trained. To evaluate the performance of the controller with different initial conditions, another source orbit is selected with a of semimajor axis of 11,000 km, a right ascension of $-15°$ and an inclination of $95°$ which leads to a new set of initial conditions given below.

$$x_0 = [0.0 \;\; -0.2847 \;\; -0.0625 \;\; -0.0502 \;\; -0.0803 \; 0.0215]^{\mathrm{T}}. \tag{5.56}$$

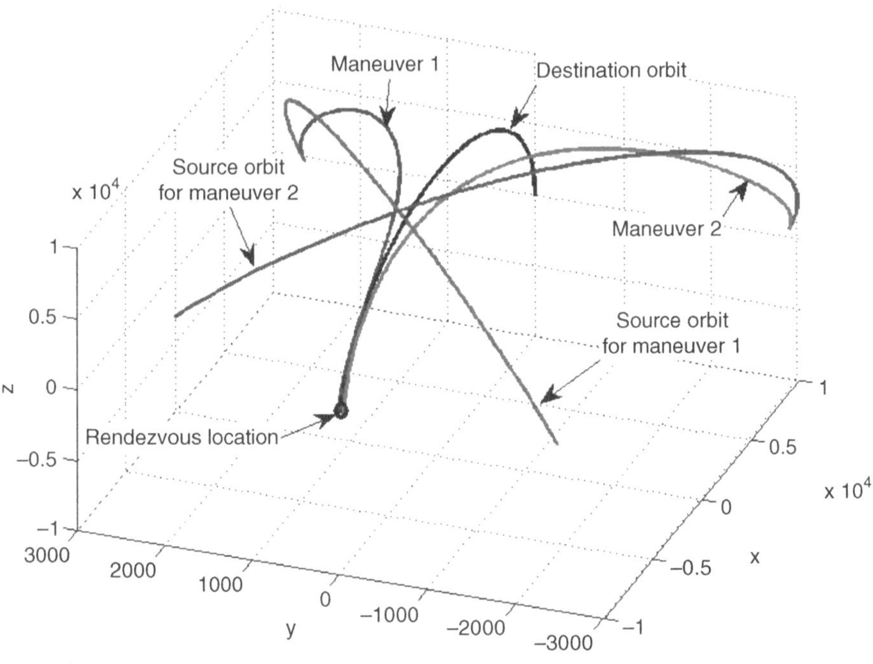

FIGURE 5.9 Three-dimensional trajectories of the maneuvers with different source orbits, that is, different initial conditions.

The simulation is performed using the same trained network for the new initial condition and the resulting position trajectory, called maneuver 2, is plotted with the results from the previous initial conditions, called maneuver 1 and shown in a 3D plot in Figure 5.9. As seen, the new maneuver is accomplished in the fixed set time, as well, confirming the applicability of the controller for different initial conditions.

5.6 CONCLUSIONS

Efficient neural network structures to solve approximate dynamic programing based control problems were presented in this chapter. Since they eliminate a network from the typical dual-network adaptive critic formulations, computing time is significantly reduced. These structures called single network adaptive critics (SNAC, J-SNAC) are applicable to a wide variety of engineering problems. Numerical results from limited applications and the underlying simpler structures seem to portend that they have a great potential for implementation.

ACKNOWLEDGMENTS

Support for this study from NASA under Grant No.: ARMD NRA NNH07ZEA001N-IRAC1 and the National Science Foundation are gratefully acknowledged. The views of the authors do not necessarily represent the views of the NSF or NASA.

REFERENCES

1. F. Lewis. *Applied Optimal Control and Estimation*. Prentice-Hall, 1992.
2. A.E. Bryson and Y.C. Ho. *Applied Optimal Control*. Taylor and Francis, 1975.
3. P.J. Werbos. Approximate dynamic programming for real-time control and neural modeling. In *Handbook of Intelligence Control*. Multiscience Press, 1992.
4. D.A. White and D.A. Sofge. Applied learning-optimal control for manufacturing. In *Handbook of Intelligent Control*. Van Nostrand Reinhold, New York, p. 9, 1992.
5. J. Si, A.G. Barto, W.B. Powel, and D. Wunsch. Handbook of learning and approximate dynamic programming, *IEEE Press Series on Computational Intelligence*. Wiley-IEEE Press, New York, 2004.
6. A. Al-Tamimi, F.L. Lewis, and M. Abu-Khalaf. Discrete-time nonlinear HJB solution using approximate dynamic programming: convergence proof. *IEEE Transactions on Systems, Man, and Cybernetics B*, 38(4):943–949, 2008.
7. S.N. Balakrishnan, J. Ding, and F.L. Lewis. Issues on stability of ADP feedback controllers for dynamical systems, *IEEE Transactions on Systems, Man, and Cybernetics B*, 38(4):913–917, 2008.
8. T. Dierks, B. Thumati, and S. Jagannathan. Optimal control of unknown affine nonlinear discrete-time systems using offline-trained neural networks with proof of convergence. *Neural Networks*, 22:851–860, 2009.

9. B. Li and J. Si. Robust dynamic programming for discounted infinite-horizon markov decision processes with uncertain stationary transition matrices, *IEEE International Symposium on Approximate Dynamic Programming and Reinforcement Learning*. Honolulu, HI, 2007, pp. 96–102.
10. P.J. Werbos. Using ADP to understand and replicate brain intelligence: the next level design, *IEEE International Symposium on Approximate. Dynamic Programming and Reinforcement Learning*. Honolulu, HI, 2007, pp. 209–216.
11. S.N. Balakrishnan and V. Biega. Adaptive–critic based neural networks for aircraft optimal control. *Journal of Guidance, Control and Dynamics*, 19:893–898, 1996.
12. D. Prokhorov and D. Wunsch. Adaptive critic designs. *IEEE Transactions on Neural Networks*, 8:997–1007, 1997.
13. G. Venayagamoorthy, R. Harley, and D. Wunsch. Dual heuristic programming excitation neurocontrol for generators in a multimachine power system. *IEEE Transactions on Industry Applications*, 39:382–384, 2003.
14. R. Padhi and S.N. Balakrishnana. Proper orthogonal decomposition based optimal neurocontrol synthesis of a chemical reactor process using approximate dynamic programming. *Neural Networks*. 16(5–6):719–28, 2003.
15. S. Ferrari and R. Stengel. Online adaptive critic flight control. *Journal of Guidance, Control and Dynamics*, 27(5):777–786, 2004.
16. Q. Yang and S. Jagannathan. Adaptive critic neural network force controller for atomic force microscope-based nanomanipulation. In *Proceedings of IEEE International Symposium on Intelligence Control*, pp. 464–469, 2006.
17. G. Lendaris, L. Schultz, and T. Shannon. Adaptive critic design for intelligent steering and speed control of a 2-axle vehicle. In *Proceedings International Joint Conference on Neural Networks*. Como, Italy, 2000.
18. T. Hanselmann, L. Noakes, and A. Zaknich. Continuous time adaptive critics. *IEEE Transcations on Neural Network*, 18(3):631–647, 2007.
19. D. Vrabie and F. Lewis. Adaptive optimal control algorithm for continuous-time nonlinear systems based on policy iteration. In *Proceedings of IEEE Conference on Decision and Control*. Cancun, 2008, pp. 73–79.
20. D. Vrabie, O. Pastravanu, F. Lewis, and M. Abu-Khalaf. Adaptive optimal control for continuous-time linear systems based on policy iteration. *Automatica*, 45(2):477–484, 2009.
21. K. Vamvoudakis and F. Lewis. Online actor–critic algorithm to solve the continuous-time infinite horizon optimal control problem. *Automatica*, 46:878–888, 2010.
22. T. Dierks and S. Jagannathan. Optimal control of affine nonlinear continuous-time systems. In *Proceedings of the American Control Conference*. Marriott Waterfront, Baltimore, 2010, pp. 1568–1573.
23. R. Padhi and S.N. Balakrishnan. Optimal beaver population management using reduced order distributed parameter model and single network adaptive critics, In *Proceedings of American Control Conference*. Boston, MA, 2004, pp. 1598–1603.
24. V. Yadav, R. Padhi, and S.N. Balakrishnan. Robust/optimal temperature profile control using neural networks. In *Proceedings of IEEE International Conference on Control Applications*. Munich, Germany, 2006, pp. 3169–3174.

25. R. Padhi, N. Unnikrishnan, X. Wang, and S.N. Balakrishnan. A Single network adaptive critic (SNAC) architecture for optimal control synthesis for a class of nonlinear systems. *Neural Networks*, 19:1648–1660, 2006.

26. J. Ding, S.N. Balakrishnan, and F.L. Lewis. A cost function based single network adaptive critic architecture for optimal control synthesis for a class of nonlinear systems. In *Proceedings of IJCNN*. Barcelona, Spain, 2010.

27. A. Heydari and S.N. Balakrishnan. Finite-horizon input-constrained nonlinear optimal control using single network adaptive critics. In *Proceedings of the American Control Conference*, 2011.

28. D. Han and S.N. Balakrishnan. State-constrained agile missile control with adaptive–critic-based neural networks. *IEEE Transactions on Control Systems Technology*, 10(4):481–489, 2002.

29. T. Cheng, F.L. Lewis, and M. Abu-Khalaf. A neural network solution for fixed-final time optimal control of nonlinear systems. *Automatica*, 43:482–490, 2007.

30. S.K. Gupta. *Numerical Methods for Engineers*. Wiley Eastern Ltd and New Age International Ltd, 1995.

31. J.-J.E. Slotine and W. Li. *Applied Nonlinear Control*. Prentice-Hall, 1991.

32. C. Park, V. Guibout, and D. Scheeres. Solving optimal continuous thrust rendezvous problems with generating functions. *Journal of Guidance, Control, and Dynamics*, 29(2):321–331, 2006.

CHAPTER 6

Linearly Solvable Optimal Control

K. DVIJOTHAM[1] and E. TODOROV[2]

[1]Department of Computer Science and Engineering, University of Washington, Seattle, WA, USA
[2]Department of Computer Science and Engineering and Applied Mathematics, University of Washington, Seattle, WA, USA

ABSTRACT

We summarize the recently developed framework of linearly solvable stochastic optimal control. Using an exponential transformation, the (Hamilton–Jacobi) Bellman equation for such problems can be made linear, giving rise to efficient numerical methods. Extensions to game theory are also possible and lead to linear Isaacs equations. The key restriction that makes a stochastic optimal control problem linearly solvable is that the noise and the controls must act in the same subspace. Apart from being linearly solvable, problems in this class have a number of unique properties including: path-integral interpretation of the exponentiated value function, compositionality of optimal control laws, duality with Bayesian inference, and trajectory-based maximum principle for stochastic control. Development of a general class of more easily solvable problems tends to accelerate progress—as linear systems theory has done. The new framework may have similar impact in fields where stochastic optimal control is relevant.

6.1 INTRODUCTION

Optimal control is of interest in many fields of science and engineering [1, 2] and is arguably at the core of robust-yet-efficient animal behavior [3, 4]. Apart from the fact that "optimal" tends to be good even when it is not exactly optimal, this

Reinforcement Learning and Approximate Dynamic Programming for Feedback Control, First Edition. Edited by Frank L. Lewis and Derong Liu.

approach to control engineering is appealing because one can, in principle, define a high-level cost function specifying the task goal, and leave the hard work of synthesizing a controller to numerical optimization software. This leads to better automation, especially when compared to the manual designs often used in engineering practice. Yet, optimizing controllers for real-world tasks is very challenging even numerically, and this book explores the state-of-the-art approaches to overcome this challenge.

One of the most productive lines of attack when it comes to solving hard problems is to identify restricted problem formulations that can be solved efficiently, and use these restricted formulations to approximate (perhaps iteratively) the harder problem. An example is the field of numerical optimization, where the only multivariate function we know how to optimize analytically is the quadratic—and so we model every other function as being locally quadratic. This is the key idea behind all second-order methods. The situation is similar in optimal control and control theory in general, where the only systems we truly understand are linear—and so we often approximate many other systems as being linear, either locally or globally. An example of an optimal control method relying on iterative linearizations of the dynamics (and quadratizations of the cost) is the iterative LQG method [5].

This general approach to solving hard problems relies on having restricted problem formulations that are computationally tractable. For too long, linear systems theory has remained pretty much the only item on the menu. Recently, we and others have developed a restricted class of stochastic optimal control problems that are linearly solvable [6, 7]. The dynamics in such problems can be nonlinear (and even nonsmooth), the costs can be nonquadratic, and the noise can be non-Gaussian. Yet, the problem reduces to solving a linear equation, which is a minimized and exponentially transformed Bellman equation. This is not nearly as tractable as an LQG problem, because the linear equation is question is a functional equation characterizing a scalar function (the exponent of the value function) over a high-dimensional continuous state space. Nevertheless, solving such problems is much easier computationally than solving generic optimal control problems.

The key restriction that makes a stochastic optimal control problem linearly solvable is that the noise and the controls are interchangeable, that is, anything that the control law can accomplish could also happen by chance (however, small the probability may be) and vice versa. The control cost associated with a given outcome is inversely related to the probability of the same outcome under the passive/uncontrolled dynamics. The form of this control cost is fixed, while the state cost can be arbitrary.

Apart from being linearly solvable, problems in this class have unique properties that enable specialized numerical algorithms. These can be summarized as follows:

- The solution can be expressed as an expectation/path-integral, which enables sampling approximations. This yields a model-free reinforcement learning method which only estimates the value function, as opposed to the much larger Q-function estimated in Q-learning.

- The most likely trajectory under the optimally controlled stochastic dynamics coincides with the optimal trajectory in a related deterministic problem, giving rise to the first trajectory-based maximum principle for stochastic control.
- The state density under the optimal controls coincides with the Bayesian posterior in a related inference problem, giving rise to a general duality between Bayesian inference and stochastic optimal control.
- The optimal solutions to first-exit and finite-horizon problems with identical dynamics and running cost, but different final costs, can be used as control primitives: they can be combined analytically so as to yield provably optimal solutions to new problems.
- Bellman residual minimization reduces to a linear algebraic equation.
- Natural policy gradient for linearly parameterized policies is possible by estimating only the value function, as opposed to the Q-function.
- Inverse optimal control, that is the problem of inferring the cost from state space trajectories of the optimally controlled system, reduces to an unconstrained convex optimization problem and does not require solving the forward problem.
- Extensions to risk sensitive and game theoretic control yield linear Isaacs equations.

6.1.1 Notation

Before we proceed, we summarize notational conventions that will be used throughout this chapter. Let S be a set, $\mathcal{P}[S]$ the set of probability distributions over S, and $S^{\mathbf{R}^+}$ the set of positive real-valued functions on S. For any $p \in \mathcal{P}[S]$, let $H\left[p\right] = \mathrm{E}_p\left[-\log(p)\right]$ denote the entropy. If f is a real-valued function on S, the expectation of f under p is denoted $\mathrm{E}_p\left[f\right] = \sum_s p(s)f(s)$. Define the function

$$\Psi_p^{\alpha}\left[f\right] = \alpha^{-1}\log\left(\mathop{\mathrm{E}}_{p}\left[\exp(\alpha f)\right]\right), \Psi_p\left[f\right] = \Psi_p^{1}\left[f\right].$$

One can prove that in the limit $\alpha \to 0$ this is just the expectation, so we define $\Psi_\pi^0\left[f\right] = \mathrm{E}_\pi\left[f\right]$. Given two positive functions $p, q \in S^{\mathbf{R}^+}$, define the distribution

$$(p \otimes q)(s) = (q \otimes p)(s) = \frac{p(s)q(s)}{\sum_{s \in S} p(s)q(s)}.$$

We will use the shorthand notation Pol for policy, Dyn for dynamics, Co for cost, and OP for optimal policy. In general, we will use boldface for vectors or discrete symbols, and italics for scalar valued functions.

6.1.2 Markov Decision Processes

Markov Decision Processes (MDPs) are a widely used framework for specifying and solving optimal control problems. MDPs are formally defined by specifying:

- A state space $\mathcal{X}$. We use $\mathbf{x}$ to denote states, $\mathbf{x} \in \mathcal{X}$. This could be continuous (subset of $\Re^n$), discrete (set of nodes in a graph) or a mixture of both.
- An action space $\mathcal{U}(\mathbf{x})$ for each state. Actions are denoted by $\mathbf{u}$. We denote policies by the same letter $\mathbf{u}(\mathbf{x}) \in \mathcal{U}(\mathbf{x})$.
- A stochastic dynamics $\mathbb{P}(\mathbf{x}, \mathbf{u})$, which is the probability distribution over the next state given the current state $\mathbf{x}$ and action $\mathbf{u} \in \mathcal{U}(\mathbf{x})$.
- An immediate cost function $\ell_t(\mathbf{x}, \mathbf{u})$.

At any time t, an action $\mathbf{u}$ is chosen depending on the current state and the system transitions into a new state sampled from the stochastic dynamics. The objective of the control is to minimize the expected cost accumulated over time. The precise notion of accumulation can vary, giving rise to different problem formulations as follows. Finite horizon (FH) problems are specified by a horizon T, a running cost $\ell_t(\mathbf{x}, \mathbf{u})$, and a terminal cost $\ell_f(\mathbf{x}, \mathbf{u})$. First exit (FE) problems are specified by a set of terminal states $\mathcal{T}$, a running cost $\ell(\mathbf{x}, \mathbf{u})$, and a terminal cost $\ell_f : \mathcal{T} \rightarrow \Re$. Infinite horizon average cost (IH) problems are specified just by a running cost $\ell(\mathbf{x}, \mathbf{u})$, and infinite horizon discounted cost problems are specified by a running cost $\ell(\mathbf{x}, \mathbf{u})$ and a discount factor γ. Discounted cost problems are very popular in reinforcement learning [3], however, we do not consider them here as they do not lead to linear Bellman equations. All other problem formulations lead to linear Bellman equations.

The optimal cost-to-go function (or optimal value function) $v_t(\mathbf{x})$ is defined as the expected cumulative cost for starting at state $\mathbf{x}$ at time t and acting optimally thereafter. This function is characterized by the Bellman equation (BE):

$$v_t(\mathbf{x}) = \min_{\mathbf{u}} \ell_t(\mathbf{x}, \mathbf{u}) + \mathop{\mathrm{E}}_{\mathbb{P}(\mathbf{x},\mathbf{u})} \left[v_{t+1}\right], \tag{6.1}$$

$$\mathbf{u}^*(\mathbf{x}; t) = \operatorname*{argmin}_{\mathbf{u}} \ell_t(\mathbf{x}, \mathbf{u}) + \mathop{\mathrm{E}}_{\mathbb{P}(\mathbf{x},\mathbf{u})} \left[v_{t+1}\right].$$

$\mathbf{u}^*(\cdot; t)$ is called the *optimal policy*.

For most control problems of practical interest, solving the Bellman equation is computationally intractable. This is because one needs to store the value function at each state $\mathbf{x}$ and the number of states could be very large (infinite if $\mathcal{X}$ is a continuous domain). This has led to a variety of approximation schemes. Many of these rely on solving the BE approximately. However, getting such schemes to work often requires a lot of problem-specific tuning, and even then may not scale to genuinely hard problems. Part of the difficulty is the highly nonlinear nature of the BE which is a result of the $\min_{\mathbf{u}}$ term. A key advantage of linearly solvable MDPs (see below) is that the minimization over actions can be done analytically given the value function. The minimized Bellman equation can then be made linear by exponentiating the value function.

6.2 LINEARLY SOLVABLE OPTIMAL CONTROL PROBLEMS

6.2.1 Probability Shift: An Alternate View of Control

Conventionally, we think of control signals as quantities that modify the system behavior in some prespecified manner. In our framework, it is more convenient to work with a somewhat different notion of control, which is nevertheless largely equivalent to the conventional notion, allowing us to model problems of practical interest. To motivate this alternate view, consider a control-affine diffusion:

$$\mathrm{d}\,\mathbf{x} = (\mathsf{a}(\mathbf{x}) + \mathsf{B}(\mathbf{x})\,\mathbf{u})\,\mathrm{d}t + \mathsf{C}(\mathbf{x})\mathrm{d}\omega.$$

This is a stochastic differential equation specifying the infinitesimal change in the state $\mathbf{x}$, caused by a passive/uncontrolled drift term $\mathsf{a}(\mathbf{x})$, a control input $\mathbf{u}$ scaled by a control gain $\mathsf{B}(\mathbf{x})$, and Brownian motion noise with amplitude $\mathsf{C}(\mathbf{x})$. Subject to this system dynamics, the controller seeks to minimize a cost function of the form

$$\ell(\mathbf{x}) + \frac{1}{2}\mathbf{u}^T\,\mathbf{u}\,.$$

In terms of MDPs, the transition probability may be written as

$$\mathbb{P}\,(\mathbf{x}, \mathbf{u}) = \mathcal{N}(\mathbf{x} + \delta(\mathsf{a}(\mathbf{x}) + \mathsf{B}(\mathbf{x})\,\mathbf{u}),\, \Sigma),$$

where we have discretized time using a time step δ. Thus, one way of thinking of the effect of control is that it changes the distribution of the next state from $\mathcal{N}(\mathbf{x} + \delta a(\mathbf{x}), \Sigma)$ to $\mathcal{N}(\mathbf{x} + \delta(\mathsf{a}(\mathbf{x}) + \mathsf{B}(\mathbf{x})\,\mathbf{u}), \Sigma)$. In other words, the controller shifts probability mass from one region of the state space to another. More generally, we can think of the system as having an uncontrolled dynamics, which gives a distribution p over future states. The controller acts by modifying this distribution by probability shift to get a new distribution: $u \otimes p = \frac{pu}{\mathrm{E}_p[u]}$. This causes the probability mass in p to shift toward areas where u is large (Figure 6.1). The controllers in our framework will act on the

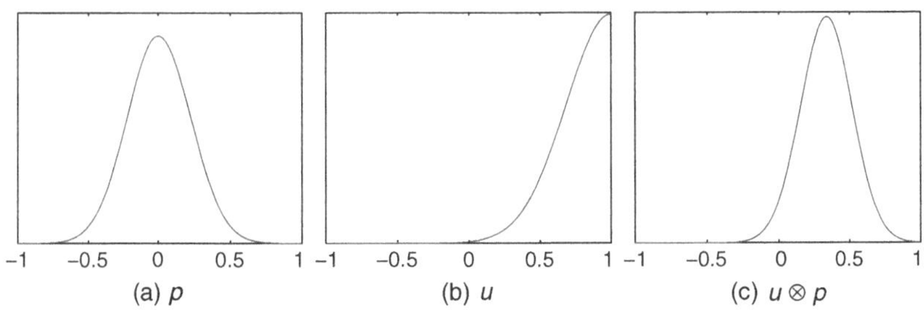

FIGURE 6.1 Probability shift.

system dynamics by performing such probability shifts. The control signals will be positive scalar functions over the state space, rather than vectors or discrete symbols.

6.2.2 Linearly Solvable Markov Decision Processes (LMDPs)

Here, we introduce the framework of linearly solvable optimal control in discrete time. Such problems, called LMDPs, can be viewed in two mathematically equivalent ways. We shall describe both, since they both offer useful perspectives and illustrate the relationship to traditional MDPs in complementary ways.

In traditional MDPs, the controller chooses a control signal or action **u** that determines the distribution of the next state $\mathbf{x}' \sim \mathbb{P}(\mathbf{x}, \mathbf{u})$. In LMDPs, we assume that there is uncontrolled or passive dynamics $\Pi^0(\mathbf{x})$ for each state **x** that gives the distribution of the next state. The controller can change this distribution by picking a probability shift $u \in \mathcal{X}^{\mathbf{R}^+}$. This causes the distribution of the next state to change: $\mathbf{x}' \sim u \otimes \Pi^0(\mathbf{x})$. However, the controller must pay a price for doing so, given by the KL divergence between the controlled distribution $u \otimes \Pi^0(\mathbf{x})$ and the uncontrolled distribution $\Pi^0(\mathbf{x})$, which is a measure of the amount of change in the dynamics due to the controller. The Bellman equation for LMDPs is nonlinear in terms of the value function, but using an exponential transformation $z_t = \exp(-v_t)$ yields a linear equation in z. We call this the desirability function, since it is inversely related to the cost-to-go. The desirability function also gives the optimal shift policy $u^*(\mathbf{x}; t) = z_{t+1}$, so the optimal controller is always trying to shift the uncontrolled dynamics toward more desirable states. The key results and their analogs for traditional MDPs are summarized in the following table:

	MDPs	LMDPs
Pol	$\mathbf{u}: \mathcal{X} \to \mathcal{U}$	$u: \mathcal{X} \to \mathcal{X}^{\mathbf{R}^+}$
Dyn	$\mathbf{x} \xrightarrow{\mathbf{u}} \mathbf{x}' \sim \mathbb{P}(\mathbf{x}, \mathbf{u}(\mathbf{x}))$	$\mathbf{x} \xrightarrow{u} \mathbf{x}' \sim u(\mathbf{x}) \otimes \Pi^0(\mathbf{x})$
Co	$\ell_t(\mathbf{x}, \mathbf{u}(\mathbf{x}))$	$\ell_t(\mathbf{x}) + \mathrm{KL}\left(u(\mathbf{x}) \otimes \Pi^0(\mathbf{x}) \parallel \Pi^0(\mathbf{x})\right)$
BE	$v_t(\mathbf{x}) = \min_{\mathbf{u}} \ell_t(\mathbf{x}, \mathbf{u}) + \mathop{\mathrm{E}}_{\mathbb{P}(\mathbf{x}, \mathbf{u}(\mathbf{x}))}\left[v_{t+1}\right]$	$z_t(\mathbf{x}) = \exp(-\ell_t(\mathbf{x})) \mathop{\mathrm{E}}_{\Pi^0(\mathbf{x})}\left[z_{t+1}\right]$
OP	$\mathbf{u}^*(\mathbf{x}; t) = \mathop{\mathrm{argmin}}_{\mathbf{u}} \ell_t(\mathbf{x}, \mathbf{u}) + \mathop{\mathrm{E}}_{\mathbb{P}(\mathbf{x}, \mathbf{u}(\mathbf{x}))}\left[v_{t+1}\right].$	$u^*(\mathbf{x}; t) = z_{t+1}.$

6.2.3 An Alternate View of LMDPs

In an alternate view, LMDPs are almost the same as traditional MDPs with deterministic dynamics and stochastic policies, except for two differences: we impose an additional cost that encourages policies with high entropy, and we compute the cost based not on the action that happened to be sampled from the stochastic policy, but by taking an expectation over all actions that could have been sampled. In this view, the relation between traditional deterministic MDPs and LMDPs is summarized as:

	Deterministic MDPs with stochastic policies	LMDPs
Pol	$u : \mathcal{X} \rightarrow \mathcal{P}[\mathcal{U}]$	$u : \mathcal{X} \rightarrow \mathcal{P}[\mathcal{U}]$
Dyn	$\mathbf{u} \sim u(\mathbf{x})$ $\mathbf{x}' = \mathrm{f}(\mathbf{x}, \mathbf{u})$	$\mathbf{u} \sim u(\mathbf{x})$ $\mathbf{x}' = \mathrm{f}(\mathbf{x}, \mathbf{u})$
Co	$\ell_t(\mathbf{x}, \mathbf{u})$	$\underset{\mathbf{u} \sim u(\mathbf{x})}{\mathrm{E}} [\ell_t(\mathbf{x}, \mathbf{u})] - H(u(\mathbf{x}))$
BE	$v_t(\mathbf{x}) =$ $\min_{u(\mathbf{x})} \underset{u(\mathbf{x})}{\mathrm{E}} \left[\ell_t(\mathbf{x}, \mathbf{u}) + v_{t+1}(\mathrm{f}(\mathbf{x}, \mathbf{u}))\right]$	$z_t(\mathbf{x}) =$ $\sum_{\mathbf{u}} \exp(-\ell_t(\mathbf{x}, \mathbf{u}))\, z_{t+1}(\mathrm{f}(\mathbf{x}, \mathbf{u}))$
OP	$u^*(\mathbf{x}; t) = \delta(\mathbf{u}^*)$ $\mathbf{u}^* = \underset{\mathbf{u}}{\mathrm{argmin}}\, \ell_t(\mathbf{x}, \mathbf{u}) + v_{t+1}(\mathrm{f}(\mathbf{x}, \mathbf{u}))$	$u^*(\mathbf{x}; t) = z_{t+1}.$

We can rewrite the BE for LMDPs in this interpretation as:

$$v_t(\mathbf{x}) = -\log\left(\sum_{\mathbf{u}} \exp(-\ell_t(\mathbf{x}, \mathbf{u}) - v_{t+1}(\mathrm{f}(\mathbf{x}, \mathbf{u})))\right).$$

The relationships between MDPs and LMDPs is now clear: the hard minimum in the Bellman equation for MDPs is replaced by a soft minimum for LMDPs, namely, $-\log(\sum(\exp(-\ldots)))$. If we replace the cost $\ell_t(\mathbf{x}, \mathbf{u})$ by a scaled version $\gamma\, \ell_t(\mathbf{x}, \mathbf{u})$, as γ increases we move closer and closer to the hard minimum, and in the limit $\gamma \rightarrow \infty$ we recover the Bellman equation for MDPs. Thus, any deterministic MDP can be obtained as a limit of LMDPs.

The relationship between the two interpretations can be understood as follows. Define a passive dynamics with support only on the states immediately reachable from $\mathbf{x}$ under some action $\mathbf{u}$:

$$\Pi^0(\mathrm{f}(\mathbf{x}, \mathbf{u}) | \mathbf{x}) \propto \exp(-\ell_t(\mathbf{x}, \mathbf{u})).$$

For states not immediately reachable from $\mathbf{x}$, the probability under the passive dynamics is 0. Given any control (probability shift) $u \in \mathcal{X}^{\mathbf{R}^+}$, we have:

$$\begin{aligned} \mathrm{KL}\left(u \otimes \Pi^0(\mathbf{x}) \,\|\, \Pi^0(\mathbf{x})\right) &= -H\left[u \otimes \Pi^0(\mathbf{x})\right] + \underset{u \otimes \Pi^0(\mathbf{x})}{\mathrm{E}}\left[-\log\left(\Pi^0(\mathbf{x})\right)\right] \\ &= -H\left[u \otimes \Pi^0(\mathbf{x})\right] + \underset{u \otimes \Pi^0(\mathbf{x})}{\mathrm{E}} [\ell_t(\mathbf{x}, \mathbf{u})] - \ell_t(\mathbf{x}), \end{aligned}$$

where $\ell_t(\mathbf{x}) = -\log\left(\sum_{\mathbf{u}} \exp(-\ell_t(\mathbf{x}, \mathbf{u}))\right)$. Thus, the alternate interpretation is equivalent to the original interpretation with passive dynamics proportional to $\exp(-\ell_t(\mathbf{x}, \mathbf{u}))$ and cost function $-\log\left(\sum_{\mathbf{u}} \exp(-\ell_t(\mathbf{x}, \mathbf{u}))\right)$.

6.2.4 Other Problem Formulations

Thus far, we focused on the FH problem formulation. We can obtain linearly solvable problems with other problem formulations as well. The corresponding BEs are

$$\text{FE} \quad z(\mathbf{x}) = \exp(-\ell(\mathbf{x})) \underset{\Pi^0(\mathbf{x})}{\mathrm{E}}[z] \text{ if } \mathbf{x} \notin \mathcal{T}$$
$$z(\mathbf{x}) = \exp(-\ell_f(\mathbf{x})) \text{ if } \mathbf{x} \in \mathcal{T}.$$

$$\text{IH} \quad z(\mathbf{x}) = \exp(c - \ell(\mathbf{x})) \underset{\Pi^0(\mathbf{x})}{\mathrm{E}}[z],\ c \text{ is the optimal average cost.}$$

In the IH case, the linear BE becomes an eigenvalue problem, with eigenvalue $\exp(-c)$ where c is the average cost. It can be shown that the solution to the optimal control problem corresponds to the principal eigenpair.

6.2.5 Applications

We now give some examples of how commonly occurring control problems can be modeled as LMDPs.

6.2.5.1 Shortest Paths Consider the shortest path problem defined on a graph. We can view this as an MDP with nodes corresponding to states and edges corresponding to actions. A stochastic version of this problem is one where the action does not take you directly to where you intend, but possibly to the end of one of the other outgoing edges from that node. We can define an LMDP with passive dynamics at a node to be the uniform distribution over all nodes reachable in one step. The cost is a constant cost per unit time and the problem is a FE problem with the goal state as the state to which the shortest path is being computed. By scaling up the constant cost by ρ, in the limit as $\rho \to \infty$ we recover the traditional deterministic shortest paths problem. This yields an efficient approximation algorithm for the shortest paths problem, by solving an LMDPs with sufficiently large ρ, see [8].

6.2.5.2 Discretizing Continuous Problems We can construct efficient solutions to problems with continuous state spaces and continuous time, provided the state space can be discretized to a reasonable size (LMDPs can easily handle problems with millions of discrete states). We consider a simple problem that has been a standard benchmark in the reinforcement learning literature, the mountain–car problem. In this problem, the task is to get a car to drive down from a hill into a valley and park on another hill on the other side of the valley. The control variable is the acceleration of the car, and the state consists of the position and velocity of the car. We impose limits on all these quantities and discretize the state space to within those limits. The dynamics is completely determined by gravity and the shape of the hill. We plot results in Figure 6.2 comparing the LMDP discretization and an iterative solution of the LMDP to a standard MDP discretization and using policy/value iteration to solve that. It can be seen that the LMDP solution converges faster to the optimal policy. See [8].

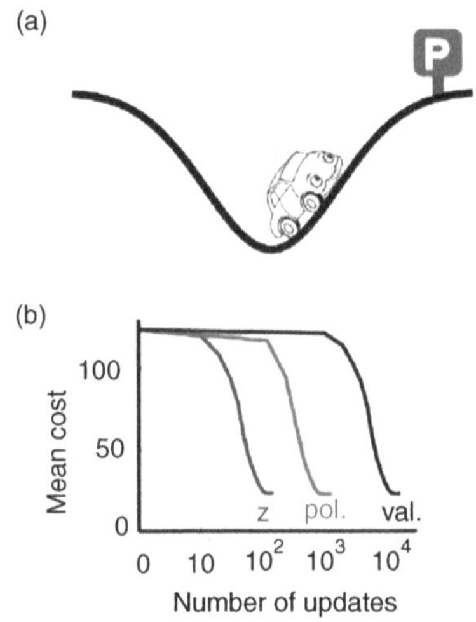

FIGURE 6.2 Continuous problems. Comparison of our MDP approximation and a traditional MDP approximation on a continuous car-on-a-hill problem. (a) Terrain, (b) Z iteration (ZI) (z), policy iteration (PI) (Pol), and value iteration (VI) (val) converge to control laws with identical performance; ZI is 10 times faster than PI and 100 times faster than VI. Horizontal axis is on log scale. Figure taken from Ref. [8].

6.2.6 Linearly Solvable Controlled Diffusions (LDs)

Although the focus of this chapter is on discrete-time problems (i.e., LMDPs), here we summarize related results in continuous time. The linearly solvable optimal control problems in continuous time are control-affine diffusions with dynamics

$$d\mathbf{x} = \mathsf{a}(\mathbf{x})dt + \mathsf{B}(\mathbf{x})\,\mathbf{u}\,dt + \sigma\,\mathsf{B}(\mathbf{x})d\omega$$

and cost rate

$$\ell_t(\mathbf{x}) + \frac{1}{2\sigma^2}\|\mathbf{u}\|^2.$$

The unusual aspects of this problem are that (i) the noise and the control act in the same subspace spanned by the columns of $\mathsf{B}(\mathbf{x})$ and (ii) the control cost is scaled by σ^{-2}, thus increasing the noise in the dynamics makes the controls cheaper.

For problems in this class, one can show that the optimal control law is

$$\mathbf{u}^*(\mathbf{x};t) = \frac{\sigma^2}{z_t(\mathbf{x})}\mathsf{B}(\mathbf{x})^T\frac{\partial z_t(\mathbf{x})}{\partial \mathbf{x}}$$

and the Hamilton–Jacobi–Bellman (HJB) equation expressed in terms of z becomes linear and is given by

$$\frac{\partial z_t(\mathbf{x})}{\partial t} = \ell_t(\mathbf{x})\, z_t(\mathbf{x}) - \mathcal{L}[z_t](\mathbf{x}). \tag{6.2}$$

Here $\mathcal{L}$ is a second-order linear differential operator known as the generator of the passive dynamics:

$$\mathcal{L}\left[f\right](\mathbf{x}) = a(\mathbf{x})^T \frac{\partial f(\mathbf{x})}{\partial \mathbf{x}} + \frac{\sigma^2}{2}\mathrm{tr}\left(\frac{\partial^2 f(\mathbf{x})}{\partial \mathbf{x} \partial \mathbf{x}^T}\,\mathsf{B}(\mathbf{x})\mathsf{B}(\mathbf{x})^T\right). \tag{6.3}$$

This operator computes expected directional derivatives of functions along trajectories of the passive dynamics. We call problems of this kind linearly solvable controlled diffusions (LDs).

6.2.7 Relationship Between Discrete and Continuous-Time Problems

If we take the first view of LMDPs that uses the notion of a stochastic passive dynamics, we can interpret the above linearly solvable diffusion as a continuous-time limit of LMDPs. This can be done by discretizing the time axis of the diffusion process with time step h using the Euler approximation:

$$\mathbf{x}(t+h) = \mathbf{x}(t) + h\,\mathsf{a}(\mathbf{x}) + h\,\mathsf{B}(\mathbf{x})\,\mathbf{u} + \epsilon,$$

where $\epsilon \sim \mathcal{N}\left(0, h\sigma^2\,\mathsf{B}(\mathbf{x})\mathsf{B}(\mathbf{x})^T\right)$. The covariance is scaled by h since for Brownian noise the standard deviation grows as the square root of time. The discrete-time cost becomes $h\,\ell_t(\mathbf{x}) + h\frac{1}{2\sigma^2}\mathbf{u}^T\,\mathbf{u}$. We will now construct an LMDP that resembles this time-discretized LD. To do this, we define the passive dynamics at state $\mathbf{x}$ to be the Euler approximation of the distribution of $\mathbf{x}(t+h)$ given $\mathbf{x}(t) = \mathbf{x}$:

$$\Pi^0(\mathbf{x}) = \mathcal{N}\left(\mathbf{x} + h\,\mathsf{a}(\mathbf{x}), h\sigma^2\,\mathsf{B}(\mathbf{x})\mathsf{B}(\mathbf{x})^T\right).$$

This converges to the continuous time LD dynamics with $\mathbf{u} = 0$ as $h \to 0$. Now, consider a family of probability shifts $u^{\mathbf{u}}$ parameterized by $\mathbf{u}$ such that

$$u^{\mathbf{u}} \otimes \Pi^0(\mathbf{x}) = \mathcal{N}\left(\mathbf{x} + h\,\mathsf{a}(\mathbf{x}) + h\,\mathsf{B}(\mathbf{x})\,\mathbf{u}, h\sigma^2\,\mathsf{B}(\mathbf{x})\mathsf{B}(\mathbf{x})^T\right).$$

This distribution is the Euler discretization of the LD dynamics under control $\mathbf{u}$. It can be shown that $\mathrm{KL}\left(u^{\mathbf{u}} \otimes \Pi^0(\mathbf{x}) \parallel \Pi^0(\mathbf{x})\right) = h\frac{1}{2\sigma^2}\mathbf{u}^T\,\mathbf{u}$. Thus, for every $\mathbf{u}$, there is a probability shift $u^{\mathbf{u}}$ that matches the Euler approximation of the LD dynamics under control $\mathbf{u}$ and also matches the time-discretized control cost. We define the state cost to be $h\,\ell_t(\mathbf{x})$. This LMDP is very close to the MDP corresponding to the time discretized LD, the only difference being that we allow probability shifts that are not equal to $u^{\mathbf{u}}$ for any $\mathbf{u}$. However, it turns out that this extra freedom does not change the optimal control law, at least in the limit $h \to 0$. The BE corresponding to this LMDP is:

$$z_t(\mathbf{x}) = \exp(-h\, \ell_t(\mathbf{x})) \underset{\mathcal{N}\left(\mathbf{x}+h\, \mathsf{a}(\mathbf{x}), h\sigma^2\, \mathsf{B}(\mathbf{x})\mathsf{B}(\mathbf{x})^T\right)}{\mathrm{E}} \left[z_{t+h}\right].$$

It can be shown that after some algebra and taking the limit $h \to 0$, we recover the linear HJB Equation (6.2).

6.2.8 Historical Perspective

Linearly solvable optimal control is a rich mathematical framework that has recently received a lot of attention, following Kappen's work on control-affine diffusions in continuous time [6], and our work on Markov decision processes in discrete time [7]. Both groups have since then obtained many additional results: see [9–12] and [8, 13–21], respectively. Other groups have also started to use and further develop this framework [22–25].

The initial studies [6, 7] were done independently, yet they both built upon the same earlier results which we discuss here. For more than 30 years, these earlier results had remained a curious mathematical fact that was never actually used to solve control problems, which, unfortunately, is not uncommon in control theory.

In continuous time, the trick that makes the HJB equation linear is

$$v_{xx} - v_x v_x^T = -\frac{z_{xx}}{z}, \quad \text{where } z = \exp(-v).$$

Applying this exponential (or logarithmic) transformation to second-order PDEs has a long history in Physics [26, 27]. Its first application to control was due to Fleming and Mitter, who showed that nonlinear filtering corresponds to a stochastic optimal control problem whose HJB equation can be made linear [28]. Kappen generalized this idea, and noted that the solution to the resulting linear PDE is also a path integral, which yields sampling approximations to the optimal value function [6].

Our work [7] was motivated by the same earlier results but in a more abstract way: we asked, are there classes of linearly solvable optimal control problems involving arbitrary dynamics? This led to the LMDP framework summarized here. In discrete time, the trick that makes the Bellman equation linear is

$$\min_q \{\mathrm{KL}(q \parallel p) + \underset{q}{\mathrm{E}}[v]\} = -\log \underset{p}{\mathrm{E}}\left[\exp(-v)\right],$$

where the minimum is achieved at $q^* = \exp(-v) \otimes p$. We introduced this trick in [7], although it turned out to have been used earlier to derive a variational characterization of the Bayesian posterior [29]. Indeed, if p is a prior and v is a negative log-likelihood, then the above q^* is a Bayesian posterior.

6.3 EXTENSION TO RISK-SENSITIVE CONTROL AND GAME THEORY

6.3.1 Game Theoretic Control: Competitive Games

Here, we briefly introduce the notion of game theoretic control or robust control [30]. In this setting, the system can be influenced by another agent (adversary) in addition to the controller. The controller needs to design a strategy that achieves the control objective in spite of the adversarial disturbances. We shall focus on the simplest case of two-player zero-sum dynamic games, where the adversary is trying to maximize the same cost that the controller is trying to minimize. The game proceeds as follows: (1) the adversary and controller pick actions $\mathbf{u}_a$, $\mathbf{u}_c$, respectively, (2) the controller pays cost $\ell_t(\mathbf{x}, \mathbf{u}_c, \mathbf{u}_a)$ and adversary pays $-\ \ell_t(\mathbf{x}, \mathbf{u}_c, \mathbf{u}_a)$, and (3) the system transitions to state $\mathbf{x}' \sim \mathbb{P}(\mathbf{x}'|\mathbf{x}, \mathbf{u}_c, \mathbf{u}_a)$. The solution to such a game can be formulated using the Bellman–Isaacs equations:

$$v_t(\mathbf{x}) = \max_{\mathbf{u}_a \in \mathcal{U}_a(\mathbf{x}, u_c)} \min_{\mathbf{u}_c \in \mathcal{U}(\mathbf{x})} \ell_t(\mathbf{x}, \mathbf{u}_c, \mathbf{u}_a) + \mathop{\mathrm{E}}_{\mathbb{P}(\mathbf{x}, \mathbf{u}_c, \mathbf{u}_a)} \left[v_{t+1}\right].$$

We call such problems Markov Games or MGs. If the min, max can be interchanged without changing the optimal policies for either the controller or the adversary, we say that the game has a *saddle-point equilibrium.* If not, then it matters which player plays first and we have corresponding *upper* and *lower* value functions.

We have recently discovered a class of linearly solvable Markov games (LMGs), where the Bellman–Isaacs equation can be made linear as explained below. But first, we need to introduce a class of divergence measures between probability distributions that will play a key role in LMGs.

6.3.2 Rényi Divergence

Rényi divergences are a generalization of the KL divergence. For distributions $p, q \in \mathcal{P}[\mathcal{X}]$, the Rényi divergence of order α is defined as

$$\mathbb{D}_\alpha(p \parallel q) = \frac{\operatorname{sign}(\alpha)}{\alpha - 1} \log\left(\mathop{\mathrm{E}}_{p}\left[\left(\frac{q}{p}\right)^{1-\alpha}\right]\right).$$

For any fixed p, q, it is known that $\mathbb{D}_\alpha$ is always non-negative, decreasing for $\alpha < 0$, and increasing for $\alpha > 0$. It is also known that $\lim_{\alpha \to 1} \mathbb{D}_\alpha(p \parallel q) = \mathrm{KL}(p \parallel q)$.

6.3.3 Linearly Solvable Markov Games

An LMG proceeds as follows:

- The system in state $\mathbf{x}$ at time t.
- The adversary picks controls $u_a \in \mathcal{X}^{\mathbf{R}^+}$.

- The controller picks controls $u_c \in \mathcal{X}^{\mathbf{R}^+}$.
- The system transitions into a state $\mathbf{x}' \sim u_c \otimes u_a \otimes \Pi^0(\mathbf{x})$.

The cost function is

$$\begin{aligned} \ell_t(\mathbf{x}, u_c, u_a) &= \ell_t(\mathbf{x}). \\ &+ \mathrm{KL}\left(u_c \otimes u_a \otimes \Pi^0(\mathbf{x}) \parallel u_a \otimes \Pi^0(\mathbf{x})\right) \text{ (control cost)}. \\ &- \mathbb{D}_{\frac{1}{\alpha}}\left(\Pi^0(\mathbf{x}) \parallel u_a \otimes \Pi^0(\mathbf{x})\right) \text{ (control cost for adversary)}. \end{aligned}$$

We focus on competitive games and require that $\alpha > 0, \alpha \neq 1$. Also, the dynamics of the game is such that the adversary plays first, so the controller has a chance to respond to the adversarial disturbance. Thus, it is a maximin problem where we work with the lower value function. Later, we describe the case $\alpha < 0$ which leads to cooperative games.

The differences between standard MGs and LMGs can be summarized as follows:

	MGs	LMGs
Pol	$\mathbf{u}_c : \mathcal{X} \times \mathcal{U}_a \to \mathcal{U}$ $\mathbf{u}_a : \mathcal{X} \to \mathcal{U}_a$	$u_c : \mathcal{X} \times \mathcal{X}^{\mathbf{R}^+} \to \mathcal{X}^{\mathbf{R}^+}$ $u_a : \mathcal{X} \to \mathcal{X}^{\mathbf{R}^+}$
Dyn	$\mathbf{u}_a = \mathbf{u}_a(\mathbf{x}), \mathbf{u}_c = \mathbf{u}_c(\mathbf{x}, \mathbf{u}_a)$ $\mathbf{x} \xrightarrow{u_c, u_a} \mathbf{x}' \sim \mathbb{P}(\mathbf{x}' \mid \mathbf{x}, \mathbf{u}_c, \mathbf{u}_a)$	$u_a = u_a(\mathbf{x}), u_c = u_c(\mathbf{x}, u_a)$ $\mathbf{x} \xrightarrow{u_c, u_a} \mathbf{x}' \sim u_c \otimes u_a \otimes \Pi^0(\mathbf{x})$
Co	$\ell_t(\mathbf{x}, \mathbf{u}_c, \mathbf{u}_a)$	$\ell_t(\mathbf{x}) - \mathbb{D}_{\frac{1}{\alpha}}\left(\Pi^0(\mathbf{x}) \parallel u_a \otimes \Pi^0(\mathbf{x})\right)$ $+ \mathrm{KL}\left(u_c \otimes u_a \otimes \Pi^0(\mathbf{x}) \parallel u_a \otimes \Pi^0(\mathbf{x})\right)$
BE	$v_t(\mathbf{x}) = \max_{\mathbf{u}_a} \min_{\mathbf{u}_c} \ell_t(\mathbf{x}, \mathbf{u}_c, \mathbf{u}_a)$ $+ \mathrm{E}_{\mathbb{P}(\mathbf{x}, \mathbf{u}_c, \mathbf{u}_a)}\left[v_{t+1}\right]$	$z_t(\mathbf{x}) = \mathcal{Q}_t(\mathbf{x}) \underset{\Pi^0(\mathbf{x})}{\mathrm{E}}\left[z_{t+1}\right]$ $z_t(\mathbf{x}) = \exp((\alpha - 1) v_t(\mathbf{x}))$ $\mathcal{Q}_t(\mathbf{x}) = \exp((\alpha - 1)\ell_t(\mathbf{x}))$
OP	$\mathbf{u}_c^*(\mathbf{x}, \mathbf{u}_a; t) = \underset{\mathbf{u}_c}{\mathrm{argmin}}\, \ell_t(\mathbf{x}, \mathbf{u}_c, \mathbf{u}_a)$ $+ \mathrm{E}_{\mathbb{P}(\mathbf{x}, \mathbf{u}_c, \mathbf{u}_a)}\left[v_{t+1}\right].$	$u_c^*(\mathbf{x}, u_a; t) = z_{t+1}^{\frac{1}{1-\alpha}}$

6.3.3.1 LMDPs as a Special Case of LMGs As $\alpha \to 0$, we recover the LMDP Bellman equation. We can explain this by looking at the cost function. It is known that $\lim_{\alpha \to 0} \mathbb{D}_{1/\alpha}(p \parallel q) \to \log\left(\sup_{\mathbf{x}} p(\mathbf{x})/q(\mathbf{x})\right)$. For this cost, the optimal strategy for the adversary is to always leave the passive dynamics unchanged, that is, $u_a^*(\mathbf{x}; t) = 1$. Intuitively, this says that the control cost for the adversary is high enough and the optimal strategy for them is to do nothing. Thus, the problem reduces to the LMDP setting.

6.3.3.2 Effect of α As α increases, the relative control cost of the controller with respect to the adversary increases, so, effectively, the adversary becomes more powerful. This makes the controller more conservative (or risk-averse), since it is fighting a stronger adversary.

6.3.3.3 Cooperative LMGs We have also derived a cooperative LMG where two agents collaborate to accomplish the same control task. The game proceeds similar to a competitive game, however, now both agents pay the same cost and are trying to minimize it in collaboration. The cost function for cooperative LMGs (for both agents) is:

$$\ell_t(\mathbf{x}) + \mathbb{D}_{1/\alpha}\left(u_a \otimes \Pi^0(\mathbf{x}) \parallel \Pi^0(\mathbf{x})\right) + \mathrm{KL}\left(u_c \otimes u_a \otimes \Pi^0(\mathbf{x}) \parallel u_a \otimes \Pi^0(\mathbf{x})\right),$$

where $\alpha < 0$. As $|\alpha|$ gets bigger, the control cost for the helper gets smaller and the helper contributes more toward accomplishing the control task while the controller contributes less. The resulting BE is similar to the competitive case:

$$z_t(\mathbf{x}) = \exp\left((\alpha - 1)\,\ell_t(\mathbf{x})\right) \mathop{\mathrm{E}}_{\Pi^0(\mathbf{x})}\left[z_{t+1}\right]$$

$$z_t(\mathbf{x}) = \exp\left((\alpha - 1)v_t(\mathbf{x})\right).$$

In this case, again we can recover LMDPs by taking $\alpha \to 0$ and making the control cost for the helper effectively large enough that the helper always chooses not to change the passive dynamics.

6.3.3.4 Examples We illustrate the effect of α with a simple control problem that requires one to drive up as high as possible on a hilly terrain. The cost function encourages one to drive up to the highest point, but the highest point is the peak of a steep hill, so that even a small perturbation from the adversary can push one downhill quickly. On the other hand, there is a shorter but less steep hill, where the adversary cannot have as much of an effect. The problem is formulated in the IH setting, so we are looking for a control strategy that achieves low average cost over a very long horizon. The terrain and cost function are plotted in Figure 6.3. The stationary distributions over $\mathcal{X}$ under optimal control for different values of α are plotted in

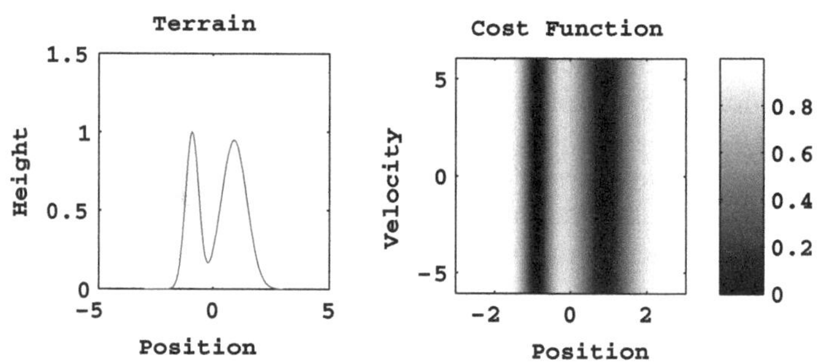

FIGURE 6.3 Terrain and cost function for LMG example

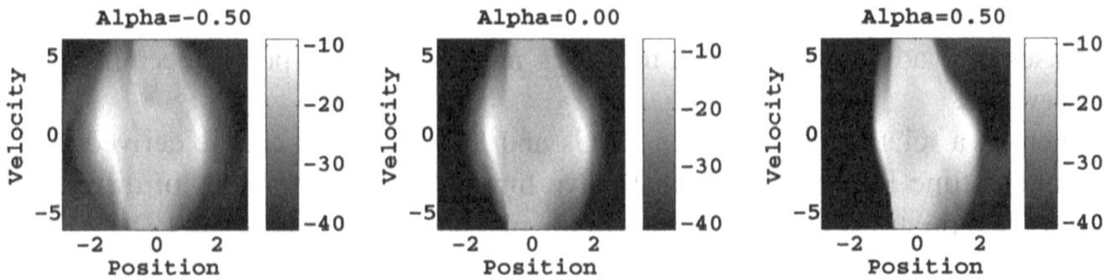

FIGURE 6.4 Logarithm of stationary distribution under optimal control versus α.

Figure 6.4. It can be seen that when $\alpha < 0$ (cooperative case), the controller places more probability on the riskier but more rewarding option (steeper/higher hill) but when $\alpha > 0$, the controller is more conservative and chooses the safer but less rewarding option (shorter/less steep hill). In the LMDP case, the solution splits its probability more or less evenly between the two options.

6.3.4 Linearly Solvable Differential Games

In this section, we consider differential games (DGs) which are continuous-time versions of MGs. A differential game is described by a stochastic differential equation

$$d\mathbf{x} = \left(\mathsf{a}(\mathbf{x}) + \mathsf{B}(\mathbf{x})\,\mathbf{u}_{\mathrm{c}} + \sqrt{\alpha}\,\mathsf{B}(\mathbf{x})\,\mathbf{u}_{\mathrm{a}}\right) dt + \sigma\,\mathsf{B}(\mathbf{x})d\omega.$$

The infinitesimal generator $\mathcal{L}[\cdot]$ for the uncontrolled process ($\mathbf{u}_{\mathrm{c}}, \mathbf{u}_{\mathrm{a}} = 0$) can be defined similarly to (6.3). We also define a cost rate

$$\ell_{\mathrm{t}}(\mathbf{x}, \mathbf{u}_{\mathrm{c}}, \mathbf{u}_{\mathrm{a}}) = \underbrace{\ell_{\mathrm{t}}(\mathbf{x})}_{\text{state cost}} + \underbrace{\frac{1}{2\sigma^2}{\mathbf{u}_{\mathrm{c}}}^T\,\mathbf{u}_{\mathrm{c}}}_{\text{control cost for controller}} - \underbrace{\frac{1}{2\sigma^2}{\mathbf{u}_{\mathrm{a}}}^T\,\mathbf{u}_{\mathrm{a}}}_{\text{control cost for adversary}}.$$

Like LMGs, these are two-player zero-sum games, where the controller is trying to minimize the cost function, whereas the adversary tries to maximize the same cost. It can be shown that the optimal solution to differential games based on diffusion processes is characterized by a nonlinear PDE known as the Isaacs equation [30]. However, for the kinds of differential games we described here, the Isaacs equation expressed in terms of $z_t = \exp\left((\alpha - 1)v_t\right)$ becomes linear and is given by:

$$\frac{\partial z_t(\mathbf{x})}{\partial t} = (1-\alpha)\,\ell_{\mathrm{t}}(\mathbf{x})\,z_t(\mathbf{x}) - \mathcal{L}[z_t](\mathbf{x}),$$

$$\mathbf{u}_{\mathrm{c}}^*(\mathbf{x};t) = \frac{\sigma^2}{(\alpha-1)\,z_t(\mathbf{x})}\mathsf{B}(\mathbf{x})^T\frac{\partial z_t(\mathbf{x})}{\partial \mathbf{x}},$$

$$\mathbf{u}_{\mathrm{a}}^*(\mathbf{x};t) = \frac{-\sqrt{\alpha}\sigma^2}{(\alpha-1)\,z_t(\mathbf{x})}\mathsf{B}(\mathbf{x})^T\frac{\partial z_t(\mathbf{x})}{\partial \mathbf{x}}.$$

When $\alpha = 0$, the adversarial control $\mathbf{u}_{\mathrm{a}}$ has no effect and we recover LDs. As α increases, the adversary's power increases and the control policy becomes more conservative.

There is a relationship between LDGs and LMGs. LDGs can be derived as the continuous time limit of LMGs that solve time-discretized versions of differential games. This relationship is analogous to the one between LMDPs and LDs.

6.3.4.1 Connection to Risk-Sensitive Control Both LMGs and LDGs can be interpreted in an alternate manner, as solving a sequential decision-making problem with an alternate objective: Instead of minimizing expected total cost, we minimize the expectation of the exponential of the total cost:

$$\mathop{\mathrm{E}}_{\mathbf{x}_{t+1} \sim u_{\mathrm{c}}(\mathbf{x}_t) \otimes \Pi^0(\mathbf{x}_t)} \left[\exp \left(\sum_{t=0}^{T} \alpha\, \ell_{\mathrm{t}}(\mathbf{x}_t) + \mathbb{D}_{\alpha} \left(u_{\mathrm{c}}(\mathbf{x}_t) \otimes \Pi^0(\mathbf{x}_t) \parallel \Pi^0(\mathbf{x}_t) \right) \right) \right].$$

This kind of objective is used in risk-sensitive control [31] and it has been shown that this problem can also be solved using dynamic programming giving rise to a risk-sensitive Bellman equation. It turns out that for this objective, the Bellman equation is exactly the same as that of an LMG. The relationship between risk-sensitive control and game theoretic or robust control has been studied extensively [30], and it also shows up in the context of linearly solvable control problems.

6.3.5 Relationships Among the Different Formulations

Linearly Solvable Markov Games (LMGs) are the most general class of linearly solvable control problems, to the best of our knowledge. As the adversarial cost increases ($\alpha \to 0$), we recover linearly solvable MDPs (LMDPs) as a special case of LMGs. When we view LMGs as arising from the time-discretization of linearly solvable differential games (LDGs), we recover LDGs as a continuous time limit ($\mathrm{d}t \to 0$). Linearly solvable controlled diffusions (LDs) can be recovered either as the continuous time limit of an LMDP, or as the nonadversarial limit ($\alpha \to 0$) of LDGs. The overall relationships between the various classes of linearly solvable control problems is summarized in the figure below:

$$\begin{array}{ccc} \text{LMGs} & \xrightarrow[\alpha \to 0]{} & \text{LMDPs} \\ \downarrow {\scriptstyle \mathrm{d}t \to 0} & & \downarrow {\scriptstyle \mathrm{d}t \to 0} \\ \text{LDGs} & \xrightarrow{\alpha \to 0} & \text{LDs} \end{array}.$$

6.4 PROPERTIES AND ALGORITHMS

6.4.1 Sampling Approximations and Path-Integral Control

For LMDPs, it can be shown that the FH desirability function equals the expectation

$$z_0(\mathbf{x}_0) = \mathop{\mathrm{E}}_{\mathbf{x}_{t+1} \sim \Pi^0(\mathbf{x}_t)} \left[\exp \left(-\ell_{\mathrm{f}}(\mathbf{x}_T) - \sum_{t=1}^{T-1} \ell_{\mathrm{t}}(\mathbf{x}_t) \right) \right]$$

over trajectories $\mathbf{x}_1 \cdots \mathbf{x}_T$ sampled from the passive dynamics starting at $\mathbf{x}_0$. This is also known as a path-integral. It was first used in the context of linearly solvable controlled diffusions [6] to motivate sampling approximations. This is a model-free method for reinforcement learning [3], however, unlike Q-learning (the classic model-free method) which learns a Q-function over the state-action space, here we only learn a function over the state space. This makes model-free learning in the LMDP setting much more efficient [8].

One could sample directly from the passive dynamics, however, the passive dynamics are very different from the optimally controlled dynamics that we are trying to learn. Faster convergence can be obtained using importance sampling:

$$z_0(\mathbf{x}_0) = \mathop{\mathrm{E}}_{\mathbf{x}_{t+1} \sim \Pi^1(\mathbf{x}_t)} \left[\exp\left(-\ell_{\mathrm{f}}(\mathbf{x}_T) - \sum\nolimits_{t=1}^{T-1} \ell_{\mathrm{t}}(\mathbf{x}_t) \right) \frac{p^0(\mathbf{x}_1 \cdots \mathbf{x}_T \mid \mathbf{x}_0)}{p^1(\mathbf{x}_1 \cdots \mathbf{x}_T \mid \mathbf{x}_0)} \right].$$

Here $\Pi^1(\mathbf{x}_{t+1} \mid \mathbf{x}_t)$ is a proposal distribution and p^0, p^1 denote the trajectory probabilities under Π^0, Π^1. The proposal distribution would ideally be Π^*, the optimally controlled distribution, but since we do not have access to it, we use the approximation based on our latest estimate of the function z. We have observed that importance sampling speeds up convergence substantially [8]. Note however that to evaluate the importance weights p^0/p^1, one needs a model of the passive dynamics.

6.4.2 Residual Minimization via Function Approximation

A general class of methods for approximate dynamic programming is to represent the value function with a function approximator, and tune its parameters by minimizing the Bellman residual. In the LMDP setting, such methods reduce to linear algebraic equations. Consider the function approximator

$$\widehat{z}(\mathbf{x}; w, \theta) = \sum\nolimits_i w_i f_i(\mathbf{x}; \theta), \tag{6.4}$$

where w are linear weights while θ are location and shape parameters of the bases f. The reason for separating the linear and nonlinear parameters is that the former can be computed efficiently by linear solvers. Choose a set of "collocation" states $\{\mathbf{x}_n\}$ where the residual will be evaluated. Defining the matrices F and G with elements

$$\begin{aligned} F_{ni} &= f_i(\mathbf{x}_n) \\ G_{ni} &= \exp(-\ell(\mathbf{x}_n)) \mathop{\mathrm{E}}_{\Pi^0(\mathbf{x}_n)}[f_i] \end{aligned}$$

the linear Bellman equation (in the IH case) reduces to

$$\lambda F(\theta) w = G(\theta) w.$$

One can either fix θ and only optimize λ, w using a linear solver, or alternately implement an outer loop in which θ is also optimized—using a general-purpose method such as Newton's method or conjugate gradient descent. When the bases are localized (e.g., Gaussians), the matrices F, G are sparse and diagonally dominant, which speeds up the computation [14]. This approach can be easily extended to the LMG case.

6.4.3 Natural Policy Gradient

The residual in the Bellman equation is not monotonically related to the performance of the corresponding control law. Thus, many researchers have focused on policy gradient methods that optimize control performance directly [32–34]. The remarkable finding in this literature is that, if the policy is parameterized linearly and the Q-function for the current policy can be approximated, then the gradient of the average cost is easy to compute.

Within the LMDP framework, we have shown [17] that the same gradient can be computed by estimating only the value function. This yields a significant improvement in terms of computational efficiency. The result can be summarized as follows. Let $g(\mathbf{x})$ denote a vector of bases, and define the control law

$$u^{(s)}(\mathbf{x}) = \exp\left(-s^{\mathsf{T}} g(\mathbf{x})\right).$$

This coincides with the optimal control law when $s^{\mathsf{T}} g(\mathbf{x})$ equals the optimal value function $v(\mathbf{x})$. Now let $v^{(s)}(\mathbf{x})$ denote the value function corresponding to control law $u^{(s)}$, and let $v(\mathbf{x}) = r^{\mathsf{T}} g(\mathbf{x})$ be an approximation to $v(\mathbf{x})$, obtained by sampling from the optimally controlled dynamics $u^{(s)} \otimes \Pi^0$ and following a procedure described in [17]. Then it can be shown that the natural gradient [35] of the average cost with respect to the Fisher information metric is simply $s - r$. Note that these results do not extend to the LMG case since the policy-specific Bellman equation is nonlinear in this case.

6.4.4 Compositionality of Optimal Control Laws

One way to solve hard control problems is to use suitable primitives [36, 37]. The only previously known primitives that preserve optimality were Options [36], which provide temporal abstraction. However, what makes optimal control hard is space rather than time, that is, the curse of dimensionality. The LMDP framework for the first time provided a way to construct spatial primitives, and combine them into provably optimal control laws [15, 23]. This result is specific to FE and FH formulations. Consider a set of LMDPs (indexed by k) which have the same dynamics and running cost, and differ only by their final costs ${\ell_{\mathrm{f}}}^{(k)}(\mathbf{x})$. Let the corresponding desirability functions be $z^{(k)}(\mathbf{x})$. These will serve as our primitives. Now define a new (composite) problem whose final cost can be represented as

$$\ell_{\mathrm{f}}(\mathbf{x}) = -\log\left(\sum\nolimits_k w_k \exp\left(-{\ell_{\mathrm{f}}}^{(k)}(\mathbf{x})\right)\right)$$

for some constants w_k. Then, the composite desirability function is

$$z(\mathbf{x}) = \sum\nolimits_k w_k \, z^{(k)}(\mathbf{x})$$

and composite optimal control law is

$$u^*(\mathbf{x}) = \sum\nolimits_k w_k \, u^{*(k)}(\mathbf{x}).$$

One application of these results is to use LQG primitives, which can be constructed very efficiently by solving Riccati equations. The composite problem has linear dynamics, Gaussian noise, and quadratic cost rate, however, the final cost no longer has to be quadratic. Instead, it can be the log of any Gaussian mixture. This represents a substantial extension to the LQG framework. These results can also be applied in infinite-horizon problems where they are no longer guaranteed to yield optimal solutions, but nevertheless may yield good approximations in challenging tasks such as those studied in computer graphics [23]. These results extend to the LMG case as well, by simply defining the final cost as $\ell_f(\mathbf{x}) = \frac{1}{\alpha-1}\log\left(\sum_k w_k \exp\left((\alpha-1)\ell_f{}^{(k)}(\mathbf{x})\right)\right)$.

6.4.5 Stochastic Maximum Principle

Pontryagin's maximum principle is one of the two pillars of optimal control theory (the other being dynamic programming and the Bellman equation). It applies to deterministic problems, and characterizes locally optimal trajectories as solutions to an ODE. In stochastic problems, it seemed impossible to characterize isolated trajectories, because noise makes every trajectory dependent on its neighbors. There exist results called stochastic maximum principles, however, they are PDEs that characterize global solutions, and in our view they are closer to the Bellman equation than the maximum principle.

The LMDP framework provided the first trajectory-based maximum principle for stochastic control. In particular, it can be shown that the probability of a trajectory $\mathbf{x}_1 \cdots \mathbf{x}_T$ starting from $\mathbf{x}_0$ under the optimal control law is

$$p^*(\mathbf{x}_1, \cdots \mathbf{x}_T \mid \mathbf{x}_0) = \frac{\exp(-\ell_f(\mathbf{x}_T))}{z_0(\mathbf{x}_0)} \exp\left(-\sum\nolimits_{t=1}^{T-1} \ell_t(\mathbf{x}_t)\right) p^0(\mathbf{x}_1 \cdots \mathbf{x}_T \mid \mathbf{x}_0).$$

Note that $z_0(\mathbf{x}_0)$ acts as a partition function. Computing z_0 for all $\mathbf{x}_0$ would be equivalent to solving the problem globally. However, in FH formulations where $\mathbf{x}_0$ is known, $z_0(\mathbf{x}_0)$ is merely a normalization constant. Thus, we can characterize the most likely trajectory under the optimal control law, without actually knowing what the optimal control law is. In terms of negative log-probabilities, the most likely trajectory is the minimizer of

$$J(\mathbf{x}_1, \cdots \mathbf{x}_T \mid \mathbf{x}_0) = \ell_f(\mathbf{x}_T) + \sum\nolimits_{t=0}^{T-1} \ell_t(\mathbf{x}_t) - \log \Pi^0(\mathbf{x}_{t+1} \mid \mathbf{x}_t).$$

Interpreting $-\log \Pi^0(\mathbf{x}_{t+1} \mid \mathbf{x}_t)$ as a control cost, J becomes the total cost for a deterministic optimal control problem [18].

Similar results are also obtained in continuous time, where the relation between the stochastic and deterministic problems is particularly simple. Consider a FH problem with dynamics and cost rate

$$\begin{aligned} d\mathbf{x} &= \mathsf{a}(\mathbf{x})\,dt + \mathsf{B}(\mathbf{x})(u\,dt + \sigma d\omega) \\ \ell(\mathbf{x}, u) &= \ell(\mathbf{x}) + \frac{1}{2\sigma^2}\|u\|^2. \end{aligned}$$

It can be shown that the most likely trajectory under the optimally controlled stochastic dynamics coincides with the optimal trajectory for the deterministic problem

$$\begin{aligned} \dot{\mathbf{x}} &= \mathsf{a}(\mathbf{x}) + \mathsf{B}(\mathbf{x})\,u \\ \ell(\mathbf{x}, u) &= \ell(\mathbf{x}) + \frac{1}{2\sigma^2}\|u\|^2 + \frac{1}{2}\operatorname{div}\mathsf{a}(\mathbf{x}). \end{aligned} \tag{6.5}$$

The extra divergence cost pushes the deterministic dynamics away from states where the drift $\mathsf{a}(\mathbf{x})$ is unstable. Note that the latter cost still depends on σ, and so the solution to the deterministic problem reflects the noise amplitude in the stochastic problem [18]. The maximum principle does extend to the LMG case, and it characterizes the mostly likely trajectory of the closed loop system that includes both the controller and the adversary. For the discrete-time problem, the maximum principle reduces to minimizing

$$J^{\alpha}(\mathbf{x}_1, \cdots \mathbf{x}_T \mid \mathbf{x}_0) = (1-\alpha)\ell_f(\mathbf{x}_T) + \sum_{t=0}^{T-1}(1-\alpha)\ell_t(\mathbf{x}_t) - \log \Pi^0(\mathbf{x}_{t+1} \mid \mathbf{x}_t).$$

Thus, when $\alpha < 1$, the most likely trajectory is trying to minimize accumulated state costs, while when $\alpha > 1$, the most likely trajectory is trying to maximize state costs. This gives us the interpretation that the controller "wins" the game for $\alpha < 1$, whereas the adversary "wins" the game for $\alpha > 1$.

6.4.6 Inverse Optimal Control

Consider the problem of getting a robot to perform locomotion or manipulation. Designing optimal controllers for these tasks is a computationally daunting task but biological systems accomplish these tasks with ease. Given this, a promising approach to designing controllers is to learn from biological systems and apply the same principles to robotic systems. There are reasons to believe that biological systems are optimal or near optimal, having been shaped by the processes of evolution and learning [4]. This motivates the problem of inverse optimal control, that is, inferring the control law and cost function given state space trajectories of the optimally controlled system. Traditionally, this has been done [38–40] by guessing a cost function, solving the (forward) optimal control problem, and adjusting the cost function so that the resulting optimal behavior matches the observed behavior. However, this approach defeats one of the main motivations of studying inverse optimal control, which is to leverage observed behavior of biological systems to design controllers without having to solve optimal control problems from scratch. We present an efficient algorithm that circumvents this problem, by using the framework of LMDPs to infer state cost functions given the passive dynamics Π^0 and state trajectories of the optimally-controlled system. Given a set of observed state transitions $\{(\mathbf{x}_n, \mathbf{x'}_n)\}$, the log-likelihood of the data up to a constant offset is

$$\sum_n -v(\mathbf{x'}_n;\theta) - \log\left(\mathop{\mathrm{E}}_{\Pi^0(\mathbf{x}_n)}\left[\exp\left(-v(\cdot;\theta)\right)\right]\right),$$

where $v(\mathbf{x};\theta)$ is a parameterized value function. We choose θ by maximizing the above log-likelihood, yielding an optimal estimate $v(\cdot;\theta^*)$ of the value function within our parametric family. Once we have inferred the value function, we can recover the cost function using $\ell(\mathbf{x}) = v(\mathbf{x};\theta^*) + \log\left(\mathrm{E}_{\Pi^0(\mathbf{x})}\left[\exp\left(-v(\cdot;\theta^*)\right)\right]\right)$. When we use a linear parametrization, $v(\mathbf{x};\theta) = \theta^T f(\mathbf{x})$, the likelihood maximization problem is a convex optimization problem and can be solved efficiently. However, to cope with high dimensional continuous state spaces, one needs to be able to adapt the features $f(\mathbf{x})$ as well, and we describe a nonconvex optimization approach to do this in [16]. Provided we know the risk parameter α, we can extend these results in a straightforward manner to LMGs.

6.5 CONCLUSIONS AND FUTURE WORK

Linearly solvable optimal control is an exciting new development in control theory and has been the subject of many papers over the past few years. In this chapter, we have attempted to provide a unified treatment of the developments in this area. The work so far has been mostly aimed at understanding the framework and its properties. We are now at a stage where the framework is mature and well understood and can lead to the development of algorithms that scale to hard real-world control problems from various application domains. Impressive results in robotics [24] and character animation [23] have recently been obtained. We feel that the surface has barely been scratched in terms of developing more efficient numerical methods for stochastic optimal control.

REFERENCES

1. D. Bertsekas. *Dynamic Programming and Optimal Control*, 2nd edition, Athena Scientific, Bellmont, MA, 2001.
2. R. Stengel. *Optimal Control and Estimation*. Dover, New York, 1994.
3. R.S. Sutton and A.G. Barto. *Reinforcement Learning: An Introduction*. MIT Press, Cambridge MA, 1998.

4. E. Todorov. Optimality principles in sensorimotor control. *Nature Neuroscience*, 7(9):907–915, 2004.
5. E. Todorov and W. Li. A generalized iterative LQG method for locally-optimal feedback control of constrained nonlinear stochastic systems. *American Control Conference*, Portland, USA, pp. 300–306, 2005.
6. H.J. Kappen. Linear theory for control of nonlinear stochastic systems. *Physical Review Letters*, 95(20):200201, 2005.
7. E. Todorov. Linearly-solvable Markov decision problems. *Advances in Neural Information Processing Systems*, 19:1369, 2007.
8. E. Todorov. Efficient computation of optimal actions. *Proceedings of the National Academy of Sciences*, 106(28):11478, 2009.
9. W. Wiegerinck, B. Broek, and H. Kappen. Stochastic optimal control in continuous space-time multi-AgentSystems. *22nd Annual Conference on Uncertainty in Artificial Intelligence*, Cambridge, USA, 2006.
10. T. Mensink, J. Verbeek, and H. Kappen. EP for efficient stochastic control with obstacles. In *European Conference on Artificial Intelligence Proceedings*, IOS Press, Amsterdam, 2010.
11. J. Broek, W. Wiegerinck, and H. Kappen. Risk sensitive path integral control. *Uncertainty in Artificial Intelligence*, AUAI Press, Corvallis, Oregon, 2010.
12. J. Broek, W. Wiegerinck, and H. Kappen. Stochastic optimal control of state constrained systems. *International Journal of Control*, 84(3):597–615, 2011.
13. E. Todorov. General duality between optimal control and estimation. *IEEE Conference on Decision and Control*, 47:4286–4292, 2008.
14. E. Todorov. Eigenfunction approximation methods for linearly-solvable optimal control problems. In *IEEE International Symposium on Adaptive Dynamic Programming and Reinforcemenet Learning*, Nashville, TN, USA, 2009.
15. E. Todorov. Compositionality of optimal control laws. In Y. Bengio, D. Schuurmans, J. Lafferty, C.K.I. Williams, and A. Culotta, editors. *Advances in Neural Information Processing Systems 22*, MIT Press, Cambridge, MA, USA, 2009, pp. 1856–1864.
16. K. Dvijotham and E. Todorov. Inverse optimal control with linearly-solvable MDPs. In J. Fürnkranz and T. Joachims, editors. *International Conference on Machine Learning*, pp. 335–342. Omnipress, 2010.
17. E. Todorov. Policy gradients in linearly-solvable mdps. In J. Lafferty, C.K.I. Williams, J. Shawe-Taylor, R.S. Zemel, and A. Culotta, editors. *Advances in Neural Information Processing Systems* 23, 2010, pp. 2298–2306.
18. E. Todorov. Finding the most likely trajectories of optimally-controlled stochastic systems. In *World Congress of the International Federation of Automatic Control (IFAC)*, Milan, Italy, 2011.
19. K. Dvijotham and E. Todorov. A unifying framework for linearly solvable control. In *Proceedings of the Twenty-Seventh Conference Annual Conference on Uncertainty in Artificial Intelligence (UAI-11)*, pp. 179–186, Corvallis, Oregon, 2011. AUAI Press.
20. M. Zhong and E. Todorov. Aggregation methods for linearly-solvable MDPs. *IFAC World Congress*, Milan, Italy, 2011.
21. M. Zhong and E. Todorov. Moving least-squares approximations for linearly-solvable stochastic optimal control problems. *Journal of Control Theory and Applications*, 9:451–463, 2011.

22. M. Toussaint. Robot trajectory optimization using approximate inference. *International Conference on Machine Learning*, 26:1049–1056, 2009.

23. M. Da Silva, F. Durand, and J. Popović. Linear Bellman combination for control of character animation. *ACM Transactions on Graphics (TOG)*, 28(3):1–10, 2009.

24. E. Theodorou, J. Buchli, and S. Schaal. Reinforcement learning of motor skills in high dimensions: A path integral approach. In *2010 IEEE International Conference on Robotics and Automation (ICRA)*, pp. 2397–2403. IEEE, Anchorage, Alaskta, 2010.

25. E.A. Theodorou. *Iterative Path Integral Stochastic Optimal Control: Theory and Applications to Motor Control*. Ph.D. thesis, University of Southern California, 2011.

26. B. Hopf. The partial differential equation $u_t + uu_x = \mu u_{xx}$. *Communications on Pure and Applied Mathematics*, 3:201–230, 1950.

27. C. Holland. A new energy characterization of the smallest eigenvalue of the Schrödinger equation. *Communications on Pure and Applied Mathematics*, 30:755–765, 1977.

28. W. Fleming and S. Mitter. Optimal control and nonlinear filtering for nondegenerate diffusion processes. *Stochastics*, 8:226–261, 1982.

29. S. Mitter and N. Newton. A variational approach to nonlinear estimation. *SIAM Journal on Control and Optimization*, 42:1813–1833, 2003.

30. T. Başar and P. Bernhard. *H-Infinity Optimal Control and Related Minimax Design Problems: A Dynamic Game Approach*. Birkhauser, 1995.

31. S.I. Marcus, E. Fernández-Gaucherand, D. Hernández-Hernandez, S. Coraluppi, and P. Fard. Risk sensitive Markov decision processes. In C.I. Byrnes, B.N. Datta, C.F. Martin, editors. *Systems and Control in the Twenty-First Century*, Progress in Systems and Control Theory, Vol. 22 (1996), Birkhäuser, Boston, 29, 1997.

32. R. Williams. Simple statistical gradient following algorithms for connectionist reinforcement learning. *Machine Learning*, 8(3):229–256, 1992.

33. R. Sutton, D. Mcallester, S. Singh, and Y. Mansour. Policy gradient methods for reinforcement learning with function approximation. In *Advances in Neural Information Processing Systems*, 2000.

34. J. Peters and S. Schaal. Natural actor-critic. *Neurocomputing*, 71(7-9):1180–1190, 2008.

35. S.I. Amari. Natural gradient works efficiently in learning. *Neural Computation*, 10(2):251–276, 1998.

36. D. Precup, R.S. Sutton, and S. Singh. Multi-time models for temporally abstract planning. In *Advances in Neural Information Processing Systems 11*, 1998.

37. S. Mahadevan and M. Maggioni. Proto-value functions: A laplacian framework for learning representation and control in markov decision processes. *Journal of Machine Learning Research*, 8:2169–2231, 2007.

38. P. Abbeel and A. Ng. Apprenticeship learning via inverse reinforcement learning. *International Conference on Machine Learning*, 2600 Anderson Street Madison, WI 53704, 21, 2004.

39. A.Y. Ng and S. Russell. Algorithms for inverse reinforcement learning. In *Proceedings of the Seventeenth International Conference on Machine Learning*, pp. 663–670. Morgan Kaufmann Publishers Inc., 2000.

40. B.D. Ziebart, A. Maas, J.A. Bagnell, and A.K. Dey. Maximum entropy inverse reinforcement learning. In *Proceedings of AAAI Conference on Artificial Intelligence* (AAAI 2008), Chicago, IL, USA, pp. 1433–1438, 2008.

CHAPTER 7

Approximating Optimal Control with Value Gradient Learning

MICHAEL FAIRBANK,[1] DANIL PROKHOROV,[2] and EDUARDO ALONSO[1]

[1]Department of Computing, School of Informatics, City University London, London, UK
[2]Toyota Research Institute NA, Toyota Technical Center, Ann Arbor, MI, USA

ABSTRACT

In this chapter, we extend the ADP algorithm, dual heuristic programming (DHP), to include a "bootstrapping" parameter λ, analogous to that used in the reinforcement learning algorithm TD(λ). The resulting algorithm, which we call VGL(λ) for value-gradient learning, is proven to produce a weight update that can be equivalent to backpropagation through time (BPTT) applied to a greedy policy on a critic function. This provides a surprising connection between the two alternate methods of BPTT and DHP. Under certain smoothness conditions, VGL($\lambda = 1$) with a greedy policy acquires the strong convergence conditions of BPTT, while using a general function approximator for the critic. We show that this can lead to increased stability in the learning of control problems by a neural network.

7.1 INTRODUCTION

Adaptive dynamic programming (ADP) is the study of how an agent can learn actions that maximize a long-term reward [1]. For example, a typical scenario is an agent wandering around in an environment, such that at time t it has state vector $\vec{x}_t$. At each time t the agent chooses an action $\vec{a}_t$ which takes it to the next state according to the environment's (Possibly stochastic) model function $\vec{x}_{t+1} = f(\vec{x}_t, \vec{a}_t)$, and gives it an immediate scalar reward r_t, given by the reward function $r_t = r(\vec{x}_t, \vec{a}_t)$. The agent keeps moving, forming a trajectory of states $(\vec{x}_0, \vec{x}_1, \ldots)$, which terminates if and

Reinforcement Learning and Approximate Dynamic Programming for Feedback Control, First Edition. Edited by Frank L. Lewis and Derong Liu.

when a designated terminal state is reached. The ADP problem is for the agent to learn how to choose actions so as to maximize the expectation of the total reward received, $\left\langle \sum_t \gamma^t r_t \right\rangle$, where $\gamma \in [0, 1]$ is a constant *discount factor* that specifies the importance of long-term rewards over short-term ones. Specifically, the problem is to find a policy function $\pi(\vec{x}, \vec{z})$, where $\vec{z}$ is the parameter vector of a function approximator that calculates which action $\vec{a} = \pi(\vec{x}, \vec{z})$ to take for any given state $\vec{x}$, such that this total reward sum is maximized.

In this chapter, we concentrate on the subproblem of when the model functions $f(\vec{x}, \vec{a})$ and $r(\vec{x}, \vec{a})$ are comprised of a twice-differentiable deterministic part and optionally some additive noise. We assume that these functions are unchanging and either already known analytically, or already learned beforehand by a system identification process, for example, by using a neural network as described by [2].

One paradigm of ADP is to learn a *value function*, $V(\vec{x})$, from Bellman's optimality principle [3], and then choose a policy function that is "greedy" on that value function. A greedy policy is the one that always chooses actions which lead to states with the highest V value (whilst also taking into account the immediate short-term reward in getting there). These methods use an approximated function (e.g., the output of a neural network) to represent the learned V function, and this approximated function is called the *critic*. We call this paradigm *critic-learning*. These critic-learning methods include heuristic dual programming, dual heuristic programming (DHP), and globalized dual heuristic programming (GDHP) from the ADP literature [1, 4, 5], and TD(λ) and Q-learning from the reinforcement learning (RL) literature [6, 7].

Critic learning methods can be very effective and computationally efficient, but proving their convergence when a general function approximator is used for V is challenging, and even more challenging when combined with a greedy policy.

One reason that convergence analysis is difficult with a greedy policy is that in the Bellman condition, V depends on the policy π, which, being greedy, depends on V; so making progress in learning one of them can undo progress in learning the other. We make an important insight into this difficulty by showing (in Lemma 7.4) that the dependency of a greedy policy on a value function is through what we call the *value-gradient*, which we define to be $\frac{\partial V}{\partial \vec{x}}$. Hence, a value-gradient analysis seems necessary at some level to provide a theoretical gateway to analyzing the convergence properties of any critic weight update that uses a greedy policy in a continuous state space. We refer to algorithms that specifically aim to learn the value-gradient $\frac{\partial V}{\partial \vec{x}}$ as "value-gradient learning" (VGL) algorithms. The existing ADP algorithms DHP and GDHP are examples of VGL algorithms.

We extend the DHP and GDHP algorithms into a new algorithm that we call VGL(λ). This extension algorithm includes a constant parameter $\lambda \in [0, 1]$ analogous to that used in TD(λ), such that VGL(0) is equivalent to DHP. The motivation for doing this extension is similar to the motivation for using TD(λ) in preference to TD(0): By choosing λ carefully, we might get faster and more stable learning.

An alternate paradigm of ADP is to use backpropagation through time (BPTT, [8]), which is gradient ascent on the total long-term reward with respect to the policy's parameter vector $\vec{z}$. Since this is gradient ascent on a function that is bound above, proving convergence is more straightforward.

In this chapter, we demonstrate a theoretical connection between these two paradigms of ADP (i.e., between critic learning and BPTT). We prove that, under certain strict conditions, doing BPTT on a greedy policy function is equivalent to VGL(1) with a carefully chosen learning parameter, "Ω_t," described below. This proof of equivalence is interesting because it provides a way to ensure the same convergence guarantees of BPTT will also apply to this critic learning method, with a greedy policy, and a general function approximator for V. Hence, we provide a convergence proof for this instance of the algorithm in this chapter.

The convergence proof requires a particular choice of a learning parameter matrix, Ω_t. This was defined by Werbos for the algorithm GDHP (e.g., see [9, equation 32]). Previously, there was little guidance on how to set this matrix, but our proof and discussions (in Sections 7.3.4 and 7.4.3) cast some new light on this problem.

Other convergence proofs do exist for critic learning methods, but they do not generally apply to a greedy policy or a critic function implemented by a general function approximator. For example, the RL algorithm TD(λ) [6] has been proven to converge [10] provided the function approximator for V is linear in its parameter vector, and the policy is fixed (i.e., that excludes the greedy policy). This particular algorithm is not proven to converge when a general function approximator is used to represent the critic (e.g., a neural network) or when a greedy policy is used. Divergence examples exist for a nonlinear function approximator [10], and where the function approximator for V is linear in its weight vector but where a greedy policy is used (diverging for both $\lambda = 0$ and $\lambda = 1$; see [11]). A convergence proof for general critic learning was given by [12], although this assumed the critic function being used could always be made to exactly learn any given target values, and so this is not realistic for a general function approximator. In the case of a general function approximator and a greedy policy, popular ADP methods such as DHP and GDHP can also be forced to diverge [11].

The structure of this chapter is as follows: We define the VGL(λ) and BPTT algorithms in Section 7.2. We prove their equivalence in Section 7.3, and discuss the conditions for convergence in Section 7.3.3, and insights into the Ω_t matrix in Sections 7.3.4 and 7.4.3. In Section 7.4, we define a simple computer experiment and demonstrate the improved convergence properties of the VGL algorithm over DHP. We also define an efficient form for the greedy policy and Ω_t matrix in Section 7.4.2. Finally, in Section 7.5, we give conclusions.

7.2 VALUE GRADIENT LEARNING AND BPTT ALGORITHMS

First, we give some preliminary definitions and notation, and then follow with the algorithm definitions.

7.2.1 Preliminary Definitions

***Critic Function*:** We define $\widetilde{V}(\vec{x}, \vec{w})$ to be the scalar output of a smooth function approximator with weight vector $\vec{w}$. This can also be known as the *approximate value function*.

Critic Gradient Function: We define the critic gradient function to be $\widetilde{G}(\vec{x}, \vec{w}) \equiv \frac{\partial \widetilde{V}(\vec{x}, \vec{w})}{\partial \vec{x}}$. The VGL and DHP algorithms attempt to learn this quantity. The critic gradient can also be known as the *approximate value gradient.*

Smooth Policy Function: We define a general smooth policy function $\pi(\vec{x}, \vec{z})$ to be the output of a smooth function approximator with output vector of dimension $\dim(\vec{a})$, and with parameter vector $\vec{z}$. This is also referred to in the literature as an "action network" or "actor."

Approximate Q Function: We define the function $\widetilde{Q}(\vec{x}, \vec{a}, \vec{w})$ to be

$$\widetilde{Q}(\vec{x}, \vec{a}, \vec{w}) = r(\vec{x}, \vec{a}) + \gamma \widetilde{V}(f(\vec{x}, \vec{a}), \vec{w}). \tag{7.1}$$

This function comes up often in this chapter in the greedy policy and its related lemmas. It is related to, but not identical to, the Q-function defined by [7].

Matrix and Vector Notation Used: We make a notational convention that all vectors are columns, and differentiation of a scalar by a vector gives a column vector. So for example $\widetilde{G}$, $\vec{x}$, and $\frac{\partial \widetilde{V}}{\partial \vec{x}}$ are all columns. We define differentiation of a vector function by a vector as follows:[1] $\frac{\partial \widetilde{G}}{\partial \vec{w}}$ is a matrix with element (i, j) equal to $\frac{\partial \widetilde{G}(\vec{x}, \vec{w})^j}{\partial \vec{w}^i}$. Similarly, $\frac{\partial f}{\partial \vec{x}}$ is the matrix with element $\left(\frac{\partial f}{\partial \vec{x}}\right)^{ij} = \frac{\partial f(\vec{x}, \vec{a})^j}{\partial \vec{x}^i}$.

Trajectory Shorthand Notation: All subscripted "t" indices refer to the time-step of a trajectory and provide corresponding arguments $\vec{x}_t$ and $\vec{a}_t$ where appropriate; so that for example $\widetilde{V}_{t+1} \equiv \widetilde{V}(\vec{x}_{t+1}, \vec{w})$, $r_t \equiv r(\vec{x}_t, \vec{a}_t)$, $\widetilde{G}_t \equiv \widetilde{G}(\vec{x}_t, \vec{w})$, and $\left(\frac{\partial \widetilde{Q}}{\partial \vec{a}}\right)_t$ is shorthand for the function $\frac{\partial \widetilde{Q}(\vec{x}, \vec{a}, \vec{w})}{\partial \vec{a}}$ evaluated at $(\vec{x}_t, \vec{a}_t, \vec{w})$. Similarly, $\left(\frac{\partial \widetilde{G}}{\partial \vec{w}}\right)_t \equiv \left.\frac{\partial \widetilde{G}}{\partial \vec{w}}\right|_{(\vec{x}_t, \vec{w})}$.

7.2.2 VGL(λ) Algorithm

The VGL algorithm is an extension of the DHP algorithm [4] to include a constant parameter λ analogous to that used in the algorithm TD(λ). The algorithm requires the derivatives of $f(\vec{x}, \vec{a})$, $r(\vec{x}, \vec{a})$, $\pi(\vec{x}, \vec{z})$, and $\widetilde{V}(\vec{x}, \vec{w})$ all to exist at every time step along a trajectory.

Using the notation conventions of Section 7.2.1, and the implied matrix products, the VGL(λ) algorithm is defined to be a critic weight update of the form:

$$\Delta \vec{w} = \alpha \sum_t \left(\frac{\partial \widetilde{G}}{\partial \vec{w}}\right)_t \Omega_t (G'_t - \widetilde{G}_t), \tag{7.2}$$

[1] This is the transpose of a common convention.

where α is a small positive constant; $\Omega_t \in \Re^{\dim(\vec{x})\times\dim(\vec{x})}$ is an arbitrary positive definite matrix described further below; $\widetilde{G}_t$ is the critic gradient; and G'_t is the "target value gradient" defined recursively by:

$$G'_t = \left(\frac{Dr}{D\vec{x}}\right)_t + \gamma\left(\frac{Df}{D\vec{x}}\right)_t\left(\lambda G'_{t+1} + (1-\lambda)\widetilde{G}_{t+1}\right), \tag{7.3}$$

with $G'_t = \vec{0}$ at any terminal state, and where $\lambda \in [0, 1]$ is a fixed constant; and where $\frac{D}{D\vec{x}}$ is shorthand for

$$\frac{D}{D\vec{x}} \equiv \frac{\partial}{\partial\vec{x}} + \frac{\partial\pi}{\partial\vec{x}}\frac{\partial}{\partial\vec{a}}; \tag{7.4}$$

and where all of these derivatives are assumed to exist. We ensure the recursion in Equation (7.3) converges by requiring that either $\gamma\lambda < 1$, or the environment is such that the agent is guaranteed to reach a terminal state at some finite time (i.e., the environment is "episodic").

Equations (7.2), (7.3), and (7.4) define the VGL(λ) algorithm. These equations can be implemented by unrolling a whole trajectory and then working backward along it applying the recursion of Equation (7.3), as demonstrated in the pseudocode of Algorithm 7.1, which runs in time $O(\dim(\vec{w}))$ per trajectory step. Alternately, it is possible to apply the weight update in an on-line manner as described by [13, Appendix B], but this runs in a longer time of $O(\dim(\vec{w})\dim(\vec{x})^2)$ per trajectory step.

When we choose the parameter $\lambda = 0$, then we get the VGL(0) algorithm which is equivalent to DHP. Like DHP, the objective of the VGL(λ) weight update is to make the values $\widetilde{G}_t$ move toward the target values G'_t.

Ω_t was introduced by [9, Equation 32] for the GDHP algorithm, and is included in our weight update for full generality. This positive definite matrix can be set as required by the experimenter since its presence ensures every component of $\widetilde{G}_t$ will move toward the corresponding component of G'_t (in any basis). However choosing what value to use for Ω_t is not obvious, so it is often just taken to be the identity matrix for all t. We make an insight into how to choose Ω_t by finding an explicit formula for it in Section 7.3.2.

Algorithm 7.1

VGL(λ) algorithm {
  $t \leftarrow 0$, $\Delta\vec{w} \leftarrow \vec{0}$
  while `not terminated(`$\vec{x}_t$`)`
    $\vec{a}_t \leftarrow \pi(\vec{x}_t, \vec{z})$
    $\vec{x}_{t+1} \leftarrow f(\vec{x}_t, \vec{a}_t)$
    $t \leftarrow t+1$
  end while
  $F \leftarrow t$
  $\vec{p} \leftarrow \vec{0}$

for $t = F - 1$ `to` 0 `step` -1

$G'_t \leftarrow \left(\frac{Dr}{D\vec{x}}\right)_t + \gamma \left(\frac{Df}{D\vec{x}}\right)_t \vec{p}$

$\Delta\vec{w} \leftarrow \Delta\vec{w} + \left(\frac{\partial\widetilde{G}}{\partial\vec{w}}\right)_t \Omega_t \left(G'_t - \widetilde{G}_t\right)$

$\vec{p} \leftarrow \lambda G'_t + (1 - \lambda)\widetilde{G}_t$

end for

$\vec{w} \leftarrow \vec{w} + \alpha\Delta\vec{w}$

}.

The policy function $\pi(\vec{x}, \vec{z})$ needs training concurrently with the VGL algorithm. This could be achieved by the same weight update as used by the DHP algorithm's policy function weight update, that is,

$$\Delta\vec{z} = \beta \sum_t \left(\frac{\partial\pi}{\partial\vec{z}}\right)_t \left(\left(\frac{\partial r}{\partial\vec{a}}\right)_t + \gamma \left(\frac{\partial f}{\partial\vec{a}}\right)_t \widetilde{G}_{t+1}\right), \tag{7.5}$$

where β is a separate learning rate for the policy function.

There are a variety of schemes to do this concurrently with the critic training. For example, the policy could be trained to completion in between every critic function weight update (a process known as value iteration), or the vice-versa arrangement could be applied (a process known as policy iteration). Or a simple concurrent scheme of doing alternating iterations of each could be applied. A final option is to use a greedy policy function, as described in Section 7.3.1, which obviates the need for training the policy, and can be viewed as an extreme form of value iteration.

In Algorithm 7.1, the critic function $\widetilde{G}(\vec{x}, \vec{w})$ can be interpreted as either a synonym for $\frac{\partial\widetilde{V}(\vec{x},\vec{w})}{\partial\vec{x}}$, which implies that $\frac{\partial\widetilde{G}}{\partial\vec{w}} \equiv \frac{\partial^2\widetilde{V}}{\partial\vec{w}\partial\vec{x}}$, or alternately $\widetilde{G}(\vec{x}, \vec{w})$ could be implemented more simply as the output of a smooth vector function approximator of dimension $\dim(\vec{x})$. We call the first of these two options a "GDHP-style critic" and the second a "DHP-style critic." A GDHP-style critic is the harder of the two options to implement, since it requires second order backpropagation to find $\frac{\partial\widetilde{G}}{\partial\vec{w}}$, while a DHP-style critic only requires first-order backpropagation for this quantity. In the experiments of Section 7.4.4, we use a DHP-style critic, for simplicity.

7.2.3 BPTT Algorithm

We define $R(\vec{x}_0, \vec{z})$ to be the total discounted reward encountered by an agent starting at state $\vec{x}_0$ and then following a policy $\pi(\vec{x}, \vec{z})$ until termination, so that $R(\vec{x}_0, \vec{z}) = \sum_t \gamma^t r_t$. This function can be witten recursively as:

$$R(\vec{x}, \vec{z}) = r(\vec{x}, \pi(\vec{x}, \vec{z})) + \gamma R(f(\vec{x}, \pi(\vec{x}, \vec{z})), \vec{z}), \tag{7.6}$$

with $R(\vec{x}, \vec{z}) = 0$ at any terminal state.

BPTT is gradient ascent on $R(\vec{x}_0, \vec{z})$ with respect to $\vec{z}$, that is, $\Delta\vec{z} = \beta\left(\frac{\partial R}{\partial \vec{z}}\right)_0$ for some small positive constant β. Expanding the term $\left(\frac{\partial R}{\partial \vec{z}}\right)_t$ gives:

$$\left(\frac{\partial R}{\partial \vec{z}}\right)_t = \left(\frac{\partial}{\partial \vec{z}}(r(\vec{x}, \pi(\vec{x}, \vec{z})) + \gamma R(f(\vec{x}, \pi(\vec{x}, \vec{z})), \vec{z}))\right)_t \quad \text{by Equation (7.6)}$$

$$= \left(\frac{\partial \pi}{\partial \vec{z}}\right)_t \left(\left(\frac{\partial r}{\partial \vec{a}}\right)_t + \gamma \left(\frac{\partial f}{\partial \vec{a}}\right)_t \left(\frac{\partial R}{\partial \vec{x}}\right)_{t+1}\right) + \gamma \left(\frac{\partial R}{\partial \vec{z}}\right)_{t+1},$$

where we used the chain rule, trajectory shorthand notation, $\vec{a}_t = \pi(\vec{x}_t, \vec{z})$ and $\vec{x}_{t+1} = f(\vec{x}_t, \vec{a}_t)$. Expanding this recursion gives:

$$\left(\frac{\partial R}{\partial \vec{z}}\right)_0 = \sum_{t \geq 0} \gamma^t \left(\frac{\partial \pi}{\partial \vec{z}}\right)_t \left(\left(\frac{\partial r}{\partial \vec{a}}\right)_t + \gamma \left(\frac{\partial f}{\partial \vec{a}}\right)_t \left(\frac{\partial R}{\partial \vec{x}}\right)_{t+1}\right).$$

It is common practice to drop the γ^t factor in this equation. Combining this with the gradient ascent equation, $\Delta\vec{z} = \beta\left(\frac{\partial R}{\partial \vec{z}}\right)_0$, gives:

$$\Delta\vec{z} = \beta \sum_{t \geq 0} \left(\frac{\partial \pi}{\partial \vec{z}}\right)_t \left(\left(\frac{\partial r}{\partial \vec{a}}\right)_t + \gamma \left(\frac{\partial f}{\partial \vec{a}}\right)_t \left(\frac{\partial R}{\partial \vec{x}}\right)_{t+1}\right). \tag{7.7}$$

This equation is the BPTT weight update. It refers to the quantity $\frac{\partial R}{\partial \vec{x}}$ which can be found recursively by differentiating Equation (7.6) and using the chain rule, giving

$$\left(\frac{\partial R}{\partial \vec{x}}\right)_t = \left(\frac{Dr}{D\vec{x}}\right)_t + \gamma \left(\frac{Df}{D\vec{x}}\right)_t \left(\frac{\partial R}{\partial \vec{x}}\right)_{t+1}, \tag{7.8}$$

with $\frac{\partial R}{\partial \vec{x}} = \vec{0}$ at any terminal state.

Equation (7.8) can be understood to be backpropagating the quantity $\left(\frac{\partial R}{\partial \vec{x}}\right)_{t+1}$ through the actor network, model and reward functions to obtain $\left(\frac{\partial R}{\partial \vec{x}}\right)_t$, and giving the algorithm its name.

7.3 A CONVERGENCE PROOF FOR VGL(1) FOR CONTROL WITH FUNCTION APPROXIMATION

In this section, we define an equivalence proof between VGL(1) and BPTT applied to a greedy policy. Since BPTT is gradient ascent on a function that is bound above, it has relatively good convergence guarantees, and hence the equivalence proof can be used to make a convergence guarantee for VGL(1).

First, we describe the greedy policy with some useful lemmas, then we prove the equivalence of BPTT to VGL(1), and then discuss the convergence conditions. For the equivalence and convergence to hold, we require a specifically chosen time-dependent Ω_t matrix, which is discussed in Section 7.3.4.

7.3.1 Using a Greedy Policy with a Critic Function

The greedy policy is defined to choose actions as follows:

$$\vec{a} = \arg\max_{\vec{a}\in\Re^n}(\widetilde{Q}(\vec{x}, \vec{a}, \vec{w})), \tag{7.9}$$

where $\widetilde{Q}(\vec{x}, \vec{a}, \vec{w})$ is defined in Equation (7.1).

For VGL, we required all of the constituent functions of Equation (7.1) to be smooth, therefore under this requirement, $\widetilde{Q}$ is a smooth function with respect to all of its parameters.

The greedy policy depends on $\widetilde{V}(\vec{x}, \vec{w})$, that is, uses the weight vector $\vec{w}$ instead of $\vec{z}$. This means the greedy policy is a function $\pi(\vec{x}, \vec{w})$ (instead of the usual policy dependency $\pi(\vec{x}, \vec{z})$). Hence from now on, when considering the greedy policy, it is the *same* weight vector $\vec{w}$ that controls the policy as is used for the critic function, and so we will write $\pi(\vec{x}, \vec{w})$ to always specifically mean the greedy policy. Any change to the critic function will immediately affect the greedy policy and move trajectories, and we need to take this into account when proving convergence.

***Greedy Actions*:** A greedy action is one that satisfies Equation (7.9).

Because a greedy action selects a *maximum* with respect to $\vec{a}$ of the smooth function $\widetilde{Q}(\vec{x}, \vec{a}, \vec{w})$, the following two consequences hold:

Lemma 7.1 *For a greedy action $\vec{a}$, $\frac{\partial \widetilde{Q}}{\partial \vec{a}} = \vec{0}$.*

Lemma 7.2 *For a greedy action $\vec{a}$, $\frac{\partial^2 \widetilde{Q}}{\partial \vec{a} \partial \vec{a}}$ is a negative semi-definite matrix.*

Furthermore, we prove three less obvious lemmas about greedy actions and a greedy policy:

Lemma 7.3 *The greedy policy implies $\left(\frac{\partial r}{\partial \vec{a}}\right)_t = -\gamma \left(\frac{\partial f}{\partial \vec{a}}\right)_t \widetilde{G}_{t+1}$.*

Proof: First, we note that differentiating Equation (7.1) gives

$$\left(\frac{\partial \widetilde{Q}}{\partial \vec{a}}\right)_t = \left(\frac{\partial r}{\partial \vec{a}}\right)_t + \gamma \left(\frac{\partial f}{\partial \vec{a}}\right)_t \widetilde{G}_{t+1}. \tag{7.10}$$

Substituting this into Lemma 7.1 and solving for $\frac{\partial r}{\partial \vec{a}}$ completes the proof. ■

Lemma 7.4 *When* $\left(\frac{\partial \pi}{\partial \vec{w}}\right)_t$ *and* $\left(\frac{\partial^2 \widetilde{Q}}{\partial \vec{a} \partial \vec{a}}\right)_t^{-1}$ *exist for an action* $\vec{a}_t$*, the greedy policy implies*

$$\left(\frac{\partial \pi}{\partial \vec{w}}\right)_t = -\gamma \left(\frac{\partial \widetilde{G}}{\partial \vec{w}}\right)_{t+1} \left(\frac{\partial f}{\partial \vec{a}}\right)_t^T \left(\frac{\partial^2 \widetilde{Q}}{\partial \vec{a} \partial \vec{a}}\right)_t^{-1}.$$

Proof: We use implicit differentiation. The dependency of $\vec{a}_t = \pi(\vec{x}_t, \vec{w})$ on $\vec{w}$ must be such that Lemma 7.1 is always satisfied, since the policy is greedy. This means that $\left(\frac{\partial \widetilde{Q}}{\partial \vec{a}}\right)_t \equiv \vec{0}$, both before and after any infinitesimal change to $\vec{w}$. Therefore, the function $\pi(\vec{x}_t, \vec{w})$ must be such that,

$$\begin{aligned}
\vec{0} &= \frac{\partial}{\partial \vec{w}} \left(\frac{\partial \widetilde{Q}(\vec{x}_t, \pi(\vec{x}_t, \vec{w}), \vec{w})}{\partial \vec{a}_t} \right), \\
&= \frac{\partial}{\partial \vec{w}} \left(\frac{\partial \widetilde{Q}(\vec{x}_t, \vec{a}_t, \vec{w})}{\partial \vec{a}_t} \right) + \left(\frac{\partial \pi}{\partial \vec{w}}\right)_t \frac{\partial}{\partial \vec{a}_t} \left(\frac{\partial \widetilde{Q}(\vec{x}_t, \vec{a}_t, \vec{w})}{\partial \vec{a}_t} \right), \\
&= \frac{\partial}{\partial \vec{w}} \left(\left(\frac{\partial r}{\partial \vec{a}}\right)_t + \gamma \left(\frac{\partial f}{\partial \vec{a}}\right)_t \widetilde{G}_{t+1} \right) + \left(\frac{\partial \pi}{\partial \vec{w}}\right)_t \left(\frac{\partial^2 \widetilde{Q}}{\partial \vec{a} \partial \vec{a}}\right)_t, \\
&= \frac{\partial}{\partial \vec{w}} \left(\left(\frac{\partial r}{\partial \vec{a}}\right)_t + \gamma \sum_i \left(\frac{\partial (f)^i}{\partial \vec{a}}\right)_t (\widetilde{G}_{t+1})^i \right) + \left(\frac{\partial \pi}{\partial \vec{w}}\right)_t \left(\frac{\partial^2 \widetilde{Q}}{\partial \vec{a} \partial \vec{a}}\right)_t, \\
&= \gamma \sum_i \left(\frac{\partial (f)^i}{\partial \vec{a}}\right)_t \frac{\partial (\widetilde{G}_{t+1})^i}{\partial \vec{w}} + \left(\frac{\partial \pi}{\partial \vec{w}}\right)_t \left(\frac{\partial^2 \widetilde{Q}}{\partial \vec{a} \partial \vec{a}}\right)_t, \\
&= \gamma \left(\frac{\partial \widetilde{G}}{\partial \vec{w}}\right)_{t+1} \left(\frac{\partial f}{\partial \vec{a}}\right)_t^T + \left(\frac{\partial \pi}{\partial \vec{w}}\right)_t \left(\frac{\partial^2 \widetilde{Q}}{\partial \vec{a} \partial \vec{a}}\right)_t.
\end{aligned}$$

In the above six lines of algebra, line 2 is by the chain rule and substitution of $\vec{a}_t = \pi(\vec{x}_t, \vec{w})$; line 3 is by Equation (7.10); line 4 just expands an inner product; line 5 follows since $\frac{\partial r}{\partial \vec{a}}$ and $\frac{\partial f}{\partial \vec{a}}$ are not functions of $\vec{w}$; and line 6 just forms an inner product.

Then solving the final line for $\left(\frac{\partial \pi}{\partial \vec{w}}\right)_t$ proves the lemma. ■

Lemma 7.5 *When* $\left(\frac{\partial \pi}{\partial \vec{x}}\right)_t$ *and* $\left(\frac{\partial^2 \widetilde{Q}}{\partial \vec{a} \partial \vec{a}}\right)_t^{-1}$ *exist for an action* $\vec{a}_t$*, the greedy policy implies*

$$\left(\frac{\partial \pi}{\partial \vec{x}}\right)_t = -\gamma \left(\frac{\partial^2 \widetilde{Q}}{\partial \vec{x} \partial \vec{a}}\right)_t \left(\frac{\partial^2 \widetilde{Q}}{\partial \vec{a} \partial \vec{a}}\right)_t^{-1}.$$

This lemma is useful because it provides the quantity $\left(\frac{\partial \pi}{\partial \vec{x}}\right)_t$, which is used in the VGL(λ) algorithm definition, in Equation (7.4). This enables us to use the VGL(λ) algorithm with a greedy policy.

Proof: The proof is virtually the same as that of Lemma 7.4, but with the start point changed to $\vec{0} = \frac{\partial}{\partial \vec{x}}\left(\frac{\partial \widetilde{Q}(\vec{x}_t, \pi(\vec{x}_t, \vec{w}), \vec{w})}{\partial \vec{a}_t}\right)$. ■

7.3.2 The Equivalence of VGL(1) to BPTT

BPTT can be defined on *any* smooth policy function $\pi(\vec{x}, \vec{z})$. A greedy policy, $\pi(\vec{x}, \vec{w})$, can be defined on *any* critic function $\widetilde{G}(\vec{x}, \vec{w})$. In this section, we apply BPTT to the greedy policy function $\pi(\vec{x}, \vec{w})$, and we observe the resulting combined weight update that emerges. Surprisingly, we find that this weight update is identical to the VGL(1) weight update, provided the Ω_t matrix is chosen carefully. This proves that the VGL(λ) weight update of Equation (7.2), with $\lambda = 1$ and a carefully chosen Ω_t matrix, is equivalent to BPTT on a greedy policy.

First, we note that by comparing Equations (7.3) and (7.8), we see that

$$G'_t \equiv \left(\frac{\partial R}{\partial \vec{x}}\right)_t \qquad \text{when } \lambda = 1. \tag{7.11}$$

The BPTT gradient ascent weight update, following on from Equation (7.7), but now using a greedy policy $\pi(\vec{x}, \vec{w})$ instead of a general policy function $\pi(\vec{x}, \vec{z})$, is

$$\begin{aligned}
\Delta\vec{w} &= \beta \sum_{t \geq 0} \left(\frac{\partial \pi}{\partial \vec{w}}\right)_t \left(\left(\frac{\partial r}{\partial \vec{a}}\right)_t + \gamma \left(\frac{\partial f}{\partial \vec{a}}\right)_t \left(\frac{\partial R}{\partial \vec{x}}\right)_{t+1}\right) && \text{(Equation 7.7)} \\
&= \beta \sum_{t \geq 0} \left(\frac{\partial \pi}{\partial \vec{w}}\right)_t \gamma \left(\frac{\partial f}{\partial \vec{a}}\right)_t \left(-\widetilde{G}_{t+1} + \left(\frac{\partial R}{\partial \vec{x}}\right)_{t+1}\right) && \text{by Lemma 7.3} \\
&= \beta \sum_{t \geq 0} \gamma \left(\frac{\partial \pi}{\partial \vec{w}}\right)_t \left(\frac{\partial f}{\partial \vec{a}}\right)_t \left(G'_{t+1} - \widetilde{G}_{t+1}\right) && \text{by Equation (7.11)} \\
&= \beta \sum_{t \geq 0} -\gamma^2 \left(\frac{\partial \widetilde{G}}{\partial \vec{w}}\right)_{t+1} \left(\frac{\partial f}{\partial \vec{a}}\right)_t^T \left(\frac{\partial^2 \widetilde{Q}}{\partial \vec{a} \partial \vec{a}}\right)_t^{-1} \left(\frac{\partial f}{\partial \vec{a}}\right)_t (G'_{t+1} - \widetilde{G}_{t+1}) && \text{by Lemma 7.4} \\
&= \beta \sum_{t \geq 0} \gamma^2 \left(\frac{\partial \widetilde{G}}{\partial \vec{w}}\right)_t \Omega_t (G'_t - \widetilde{G}_t),
\end{aligned} \tag{7.12}$$

where

$$\Omega_t = \begin{cases} -\left(\frac{\partial f}{\partial \vec{a}}\right)_{t-1}^T \left(\frac{\partial^2 \widetilde{Q}}{\partial \vec{a} \partial \vec{a}}\right)_{t-1}^{-1} \left(\frac{\partial f}{\partial \vec{a}}\right)_{t-1} & \text{if } t > 0 \\ 0 & \text{if } t = 0 \end{cases}, \tag{7.13}$$

and is positive semi-definite, by the greedy policy (Lemma 7.2).

Equation (7.12) is identical to a VGL weight update equation (Equation 7.2), with a carefully chosen matrix for Ω_t, and $\lambda = 1$, provided $\left(\frac{\partial \pi}{\partial \vec{w}}\right)_t$ and $\left(\frac{\partial^2 \widetilde{Q}}{\partial \vec{a} \partial \vec{a}}\right)_t^{-1}$ exist for all t. If $\left(\frac{\partial \pi}{\partial \vec{w}}\right)_t$ does not exist, then $\frac{\partial R}{\partial \vec{w}}$ is not defined either.

This completes the demonstration of the equivalence of a critic learning algorithm (VGL(1), with the conditions stated above) to BPTT on a greedy policy $\pi(\vec{x}, \vec{w})$ (where $\vec{w}$ is the weight vector of a critic function $\widetilde{G}(\vec{x}, \vec{w})$ defined for Algorithm 7.1), when $\frac{\partial R}{\partial \vec{w}}$ exists.

7.3.3 Convergence Conditions

Good convergence conditions exist for BPTT since it is gradient ascent on a function that is bound above, and therefore convergence is guaranteed if that surface is smooth and the learning step size is sufficiently small. If the ADP problem is such that $\frac{\partial \pi}{\partial \vec{w}}$ always exists, and we choose Ω_t by Equation (7.13), then the above equivalence proof shows that the good convergence guarantees of BPTT will apply to VGL(1). Significantly, $\frac{\partial \pi}{\partial \vec{w}}$ always does exist in a continuous time setting when a value-gradient policy using the technique of Section 7.4.2 is used.

In addition to smoothness of the policy, we also require smoothness of the functions, r and f, for VGL to be defined; and also for the convergence of BPTT that we have proved equivalence to, the weight vector for the policy must only traverse smooth regions of the surface of $R(\vec{x}, \vec{w})$.

If all of these conditions are satisfied, then this approximated-critic value-iteration scheme will converge.

This has been a nontrivial accomplishment in proving convergence for a smoothly approximated critic function with a greedy policy, even though it is only proven for $\lambda = 1$. Other related algorithms with $\lambda = 1$, such as TD(1) and Sarsa(1), and VGL(1) *without* the specially chosen Ω_t matrix, can all be made to diverge under the same conditions when a greedy policy is used [11]. Algorithms with $\lambda = 0$, such as TD(0), Sarsa(0), DHP and GDHP are also shown to diverge with a greedy policy by [11].

While the smoothness of all functions is required for provable convergence, in practice, a sufficient condition appears to be piece-wise continuity, as BPTT has been applied successfully to systems with friction and dead-zones.

7.3.4 Notes on the Ω_t Matrix

The Ω_t matrix that we derived in Equation (7.13) differs from the previous instances of its use in the literature (e.g., [9, Equation 32]):

- First, our Ω_t matrix is time dependent, whereas previous usages of it have not used a t subscript.
- Second, we have found an exact equation on how to choose it (i.e., equation 7.13). Previous guidance on how to choose it has been only intuitive.

- Third, Equation (7.13) often only produces a positive *indefinite* matrix, which is problematic for the case of $\lambda < 1$. If we have $\dim(\vec{x}) > \dim(\vec{a})$ then the matrix $\frac{\partial f}{\partial \vec{a}}$ will be wider than it is tall, and so the matrix product in Equation (7.13) will yield an Ω_t matrix that is rank deficient (i.e., positive indefinite). It seems that it is not a problem to have a rank-deficient Ω_t matrix when $\lambda = 1$ (as Section 7.3.2 effectively proves), but it is a problem when $\lambda < 1$. A rank-deficient Ω_t matrix will have some zero eigenvalues, and the components of $\widetilde{G}$ corresponding to these missing eigenvalues will not be learned at all by Equation (7.2). However, in the case of $\lambda < 1$, the definition of G'_t in Equation (7.3) depends upon potentially *all* of the components of $\widetilde{G}_{t+1}$ via the multiplication in Equation (7.3) by $\left(\frac{Df}{D\vec{x}}\right)_t$. So if some of the components of $\widetilde{G}_{t+1}$ are missing, then the target gradients G'_t will be wrong, and so the VGL(λ) algorithm will be badly defined.

  This view is necessary for Ω_t to be full rank for $\lambda < 1$ is consistent with the original positive-definite requirement made by Werbos for GDHP, which is a $\lambda = 0$ algorithm [9].

Consequently, our choice of Ω_t matrix is best used for the situation of $\lambda = 1$ and a greedy policy. But, we feel it may provide some guidance in how to choose Ω_t in other situations, especially if working in a problem where $\dim(\vec{x}) \leq \dim(\vec{a})$. And even if the policy is not greedy, then Equation (7.13) might still be a useful *guiding* choice for Ω_t, since it is the objective of the training algorithm for the actor network to always try to make the policy greedy.

The Ω_t matrix definition in Equation (7.13) requires an inverse of the following rather cumbersome looking matrix:

$$\left(\frac{\partial^2 \widetilde{Q}}{\partial \vec{a} \partial \vec{a}}\right)_t = \frac{\partial}{\partial \vec{a}}\left(\left(\frac{\partial r}{\partial \vec{a}}\right)_t + \gamma \left(\frac{\partial f}{\partial \vec{a}}\right)_t \widetilde{G}_{t+1}\right) \quad \text{by Equation (7.10)}$$

$$\Rightarrow \left(\frac{\partial^2 \widetilde{Q}}{\partial \vec{a}^i \partial \vec{a}^j}\right)_t = \left(\frac{\partial^2 r}{\partial \vec{a}^i \partial \vec{a}^j}\right)_t + \gamma \left(\frac{\partial^2 f}{\partial \vec{a}^i \partial \vec{a}^j}\right)_t \widetilde{G}_{t+1} + \gamma \left(\frac{\partial f}{\partial \vec{a}^i}\right)_t \left(\frac{\partial \widetilde{G}}{\partial \vec{x}}\right)_{t+1} \left(\frac{\partial f}{\partial \vec{a}^j}\right)_t^T. \tag{7.14}$$

Hence to evaluate the Ω_t matrix, we could require knowledge of the functions, f and r, so that the first and second order derivatives in Equations (7.13) and (7.14) could be manually computed. Computing Equation (7.13) is no more challenging to implement than computing $\frac{\partial \pi}{\partial \vec{x}}$ by Lemma 7.5, which is a necessary step to implement the VGL(λ) algorithm with a greedy policy. In many cases, such as in Section 7.4.2, both of these computations simplify considerably, for example, if the functions are linear in $\vec{a}$, or in a continuous time situation. Alternately, if a neural network is used to represent the functions f and r, then we would require first and second order backpropagation through the neural network work to find these necessary derivatives.

We make further observations on the role of the Ω_t matrix in Section 7.4.3.

7.4 VERTICAL LANDER EXPERIMENT

We describe a simple computer experiment which shows VGL learning with a greedy policy and demonstrates increased learning stability of VGL(1) compared to DHP (VGL(0)). We also demonstrate the value of using the Ω_t matrix as defined by Equation (7.13), which can make learning progress achieve consistent convergence to local optimality with $\lambda = 1$.

After defining the problem in Section 7.4.1, we derive an efficient formula for the greedy policy and Ω_t matrix (Section 7.4.2), which provides some further insights into the purpose of Ω_t (Section 7.4.3), before giving the experimental results in Section 7.4.4.

7.4.1 Problem Definition

A spacecraft is dropped in a uniform gravitational field, and its objective is to make a fuel-efficient gentle landing. The spacecraft is constrained to move in a vertical line, and a single thruster is available to make upward accelerations. The state vector $\vec{x} = (h, v, u)^T$ has three components: height (h), velocity (v), and fuel remaining (u). The action vector is one-dimensional (so that $\vec{a} \equiv a \in \Re$) producing accelerations $a \in [0, 1]$. The Euler method with time step Δt is used to integrate the motion, giving functions:

$$\begin{aligned} f((h, v, u)^T, a) &= (h + v\Delta t, v + (a - k_g)\Delta t, u - (k_u)a\Delta t)^T \\ r((h, v, u)^T, a) &= -(k_f)a\Delta t + r^c(a)\Delta t. \end{aligned} \tag{7.15}$$

Here, $k_g = 0.2$ is a constant giving the acceleration due to gravity; the spacecraft can produce greater acceleration than that due to gravity. $k_f = 1$ is a constant giving fuel penalty. $k_u = 1$ is a unit conversion constant. Δt was chosen to be 1. $r^c(a)$ is an "action cost" function described further below that ensures the greedy policy function chooses actions satisfying $a \in [0, 1]$.

Trajectories terminate as soon as the spacecraft hits the ground ($h = 0$) or runs out of fuel ($u = 0$). For correct gradient calculations, clipping is needed at the terminal time step, and differentiation of the functions needs to take account of this clipping. Further details of clipping are given by [14, Appendix E.1].

In addition to the reward function $r(\vec{x}, a)$ defined above, a final impulse of reward equal to $-\frac{1}{2}mv^2 - m(k_g)h$ is given as soon as the lander reaches a terminal state, where $m = 2$ is the mass of the spacecraft. The terms in this final reward are cost terms for the kinetic and potential energy, respectively. The first cost term penalises landing too quickly. The second term is a cost term equivalent to the kinetic energy that the spacecraft would acquire by crashing to the ground under freefall.

A sample of 10 optimal trajectories in state space is shown in Figure 7.1.

For the action cost function, we follow the methods of [15] and [16], and choose

$$r^c(a) = -\int g^{-1}(a)da, \tag{7.16}$$

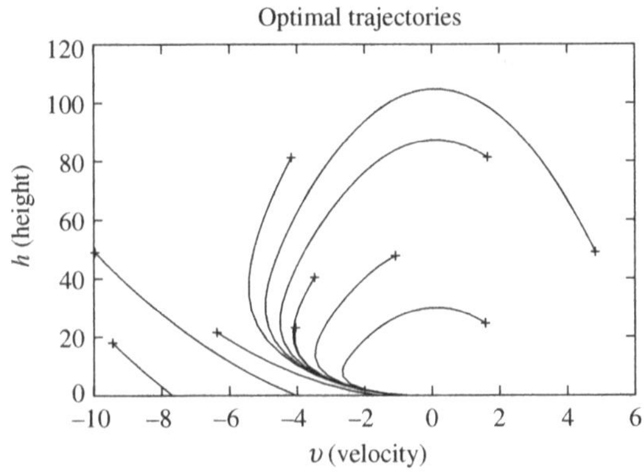

FIGURE 7.1 State space view of a sample of optimal trajectories in the vertical lander problem. Each trajectory starts at the cross symbol, and ends at $h = 0$. The u-dimension (fuel) of state space is not shown.

where $g(x)$ is a chosen sigmoid function, as this will force a to be bound to the range of the chosen function $g(x)$, as illustrated in the following subsection. Hence, to ensure $a \in [0, 1]$, we use

$$g(x) = \frac{1}{2}(\tanh(x/c) + 1), \tag{7.17}$$

where $c = 0.2$ is a sharpness constant, and therefore

$$r^c(a) = c\left(a \operatorname{arctanh}(1 - 2a) - \frac{1}{2}\ln(2 - 2a)\right).$$

7.4.2 Efficient Evaluation of the Greedy Policy

The greedy policy $\pi(\vec{x}, \vec{w})$ is defined to choose the maximum with respect to $\vec{a}$ of the $\tilde{Q}(\vec{x}, \vec{a}, \vec{w})$ function. This function has been defined to be smooth, so a numerical solver could be used to maximize this function, while introducing some inefficiency. A technical difficulty is that there might be multiple local maxima, and this means that as $\vec{w}$ or $\vec{x}$ change, the global maximum could hop from one local maximum to another, meaning the derivatives $\frac{\partial \pi}{\partial \vec{x}}$ and $\frac{\partial \pi}{\partial \vec{w}}$ would not be defined in these instances.

We can get around these problems and derive a closed form solution to the greedy policy by following the method of [15]. This leads to a very efficient and practical solution to using a greedy policy, and it avoids the need to use an actor network altogether. To achieve this though, we do have to transfer to a continuous time analysis, that is, we consider the case in the limit of $\Delta t \to 0$. The most important benefit that this delivers is that it forces the greedy policy function to be always differentiable, and hence for the VGL(λ) algorithm to be always defined.

We make a first order Taylor series expansion of the $\widetilde{Q}(\vec{x}, \vec{a}, \vec{w})$ function (Equation 7.1) about the point $\vec{x}$:

$$\begin{aligned}\widetilde{Q}(\vec{x}, \vec{a}, \vec{w}) &\approx r(\vec{x}, \vec{a}) + \gamma \left(\left(\frac{\partial \widetilde{V}}{\partial \vec{x}} \right)^T (f(\vec{x}, \vec{a}) - \vec{x}) + \widetilde{V}(\vec{x}, \vec{w}) \right) \\ &= r(\vec{x}, \vec{a}) + \gamma \left(\widetilde{G}(\vec{x}, \vec{w}) \right)^T (f(\vec{x}, \vec{a}) - \vec{x}) + \gamma \widetilde{V}(\vec{x}, \vec{w}). \end{aligned} \tag{7.18}$$

This approximation becomes exact in continuous time. We next define a greedy policy that maximizes Equation (7.18). Differentiating Equation (7.18) gives

$$\begin{aligned}\left(\frac{\partial \widetilde{Q}}{\partial a} \right)_t &= \left(\frac{\partial r}{\partial a} \right)_t + \gamma \left(\frac{\partial f}{\partial a} \right)_t \widetilde{G}_t \quad \text{by Equation (7.18)} \\ &= -(k_f)\Delta t + \left(\frac{\partial r^c(a)\Delta t}{\partial a} \right)_t + \gamma \Delta t (0, 1, -1) \widetilde{G}_t \quad \text{by Equation (7.15)} \\ &= \left(-(k_f) - g^{-1}(a_t) + \gamma(0, 1, -1)\widetilde{G}_t \right) \Delta t \quad \text{by Equation (7.16)} \end{aligned} \tag{7.19}$$

For the greedy policy to satisfy Equation (7.9), we must have $\frac{\partial \widetilde{Q}}{\partial a} = 0$. Therefore,

$$\begin{aligned} 0 &= -(k_f) - g^{-1}(a_t) + \gamma(0, 1, -1)\widetilde{G}_t \quad \text{by Equation (7.19)} \\ \Rightarrow a_t &= g \left(-(k_f) + \gamma(0, 1, -1)\widetilde{G}_t \right). \end{aligned} \tag{7.20}$$

This closed form greedy policy is efficient to calculate, bound to [0, 1], and most importantly, always differentiable. This has achieved the objectives we aimed for by moving to continuous time. Furthermore, we get a simplified expression for the Ω_t matrix in continuous time.

Because $\dim(\vec{a}) = 1$, $\frac{\partial^2 \widetilde{Q}}{\partial a \partial a}$ is a scalar:

$$\begin{aligned}\left(\frac{\partial^2 \widetilde{Q}}{\partial a \partial a} \right)_t &= \frac{\partial \left(-(k_f) - g^{-1}(a_t) + \gamma(0, 1, -1)\widetilde{G}_t \right)}{\partial a_t} \Delta t \quad \text{by Equation (7.19)} \\ &= -\frac{\partial g^{-1}(a_t)}{\partial a_t} \Delta t \\ &= -\frac{1}{g'\left(g^{-1}(a_t)\right)} \Delta t \quad \text{differentiating an inverse} \\ &= -\frac{1}{g'\left(-(k_f) + \gamma(0, 1, -1)\widetilde{G}_t \right)} \Delta t \quad \text{by Equation (7.20)} \end{aligned} \tag{7.21}$$

Here, g' is the derivative of the sigmoidal function g given by Equation (7.17). Substituting Equation (7.21) into Equation (7.13) gives,

$$\Omega_t = (0, 1, -1)^T g' \left(-(k_f) + \gamma(0, 1, -1)\widetilde{G}_t \right) (0, 1, -1)\Delta t. \tag{7.22}$$

This is a much simpler version of the Ω_t matrix than that described by Equations (7.13) and (7.14). The simplicity arose because of the linearity with respect to a of the function $f(\vec{x}, a)$ and because of the change to continuous time.

Since we have moved to continuous time for the sake of deriving this efficient and always differentiable greedy policy, there are some consequential minor changes that we should make to the VGL algorithm. First, if we were to re-derive Lemmas 7.3, 7.4, and 7.5 using the $\widetilde{Q}$ function of Equation (7.18), then the references in the lemmas to $\widetilde{G}_{t+1}$ would change to $\widetilde{G}_t$. For example, Lemma 7.4 would change to:

$$\left(\frac{\partial \pi}{\partial \vec{w}}\right)_t = -\gamma \left(\frac{\partial \widetilde{G}}{\partial \vec{w}}\right)_t \left(\frac{\partial f}{\partial \vec{a}}\right)_t^T \left(\frac{\partial^2 \widetilde{Q}}{\partial \vec{a} \partial \vec{a}}\right)_t^{-1}. \tag{7.23}$$

The VGL(λ) weight update would be the same as Equation (7.12), but we would use Ω_t as given by Equation (7.22), and the greedy policy given by Equation (7.20). Also, $\frac{\partial \pi}{\partial x}$ (which is needed in the VGL(λ) algorithm in Equation 7.4) is found most easily by differentiating Equation (7.20), as opposed to using Lemma 7.5.

We note that Equation (7.23), when combined with Equation (7.21), is consistent with what is obtained by differentiating Equation (7.20) directly.

7.4.3 Observations on the Purpose of Ω_t

Now that we have a simple expression for Ω_t, we can make some observations on its purpose. Substituting Ω_t of Equation (7.22) into the VGL weight update (Equation 7.2) gives:

$$\Delta \vec{w} = \alpha \gamma^2 (\Delta t) \sum_{t \geq 0} \left(\frac{\partial \widetilde{G}}{\partial \vec{w}}\right)_t (0, 1, -1)^T g' \left(-(k_f) + \gamma(0, 1, -1)\widetilde{G}_t \right)$$
$$(0, 1, -1)({G'}_t - \widetilde{G}_t). \tag{7.24}$$

This has similarities in form to a weight update for the supervised learning neural network problem. Consider a neural network output function $y = g(s(\vec{x}, \vec{w}))$ with sigmoidal activation function g, summation function $s(\vec{x}, \vec{w})$, input vector $\vec{x}$, and weight vector $\vec{w}$. To make the neural network learn targets t_{p} for input vectors $\vec{x}_{\mathrm{p}}$ (where p is a "pattern" index), the gradient descent weight update would be:

$$\begin{aligned} \Delta \vec{w} &= \alpha \sum_p \frac{\partial y_p}{\partial \vec{w}} (t_{\mathrm{p}} - y_{\mathrm{p}}), \\ &= \alpha \sum_p \frac{\partial s(\vec{x}_p, \vec{w})}{\partial \vec{w}} g' \left(s(\vec{x}_p, \vec{w}) \right) (t_{\mathrm{p}} - y_{\mathrm{p}}). \end{aligned} \tag{7.25}$$

The similarities in Equations (7.24) and (7.25) give hints at the purpose of Ω_t, since it is the Ω_t matrix that introduces the g' term into the VGL weight update equation (Equation 7.24). In neural network training, we would not omit the g' term from a weight update, and we would not treat it as a constant; and likewise we deduce that we should not omit the Ω_t matrix or treat it as fixed in the VGL learning algorithm. In neural network training, some algorithms choose to give the g' term an artificial boost to help escape plateaus of the error surface in weight space (e.g., Fahlman's method [17] which replaces g' by $g' + k$, for a small constant k), but this comes at the expense of the learning algorithm no longer being true gradient descent, and hence it not being as stable. Choosing to set $\Omega_t \equiv I$, the identity matrix, is like doing an extreme version of Fahlman's method on the VGL algorithm. This can help the learning algorithm escape plateaus of the R surface very effectively, but may lead to divergence. Plateaus are a severe problem when c is small in Equation (7.17), since then $g' \approx 0$ which will make learning by Equation (7.24) grind to a halt.

Having made these deductions about the role of Ω_t in VGL, we should make the caveat that these deductions only strictly apply to VGL(1) with a greedy policy, as is the algorithm that Ω_t was derived for.

7.4.4 Experimental Results for Vertical Lander Problem

A DHP-style critic, $\widetilde{G}(\vec{x}, \vec{w})$, was provided by a fully connected multilayer perceptron (MLP) (see [18] for details). The MLP had three inputs, two hidden layers of six units each, and three units in the output layer. Additional shortcut connections were present fully connecting all pairs of layers. The weights were initially randomized uniformly in the range $[-1, 1]$. The activation functions were logistic sigmoid functions in the hidden layers, and a linear function with slope 0.1 in the output layer. The input to the MLP was a rescaling of the state vector, given by $D(h, u, v)^T$, where $D = \text{diag}(0.01, 0.1, 0.02)$, and the output of the MLP gave $\widetilde{G}$ directly. In our implementation, we also defined the function $f(\vec{x}, \vec{a})$ to input and output coordinates rescaled by D, the intention being to ensure that the value gradients would be more appropriately scaled too.

Three algorithms were tested: VGL(0) with $\Omega_t = I$, the identity matrix; VGL(1) with $\Omega_t = I$; and VGL(1) with Ω_t given by Equation (7.13) (denoted by throughout by "VGLΩ(1)"). Each algorithm was set the task of learning a group of 10 trajectories with randomly chosen fixed start points (the 10 start points used in all experiments are those shown in Figure 7.1), and with initial fuel $u = 30$. In each iteration of the learning algorithm, the weight update was first accumulated for all 10 trajectories, and then this aggregate weight update was applied. In some experiments, RPROP was used to accelerate this aggregate weight update at each iteration, with its default parameters defined by [19].

Figure 7.2 shows learning performance of the three algorithms, both with and without RPROP. These graphs show the clear stability and performance advantages of using $\lambda = 1$ and the chosen Ω_t matrix.

The VGLΩ(1) algorithm shows near-to-monotonic progress in the later stages of learning. The large kink in learning performance in the early iterations of RPROP is

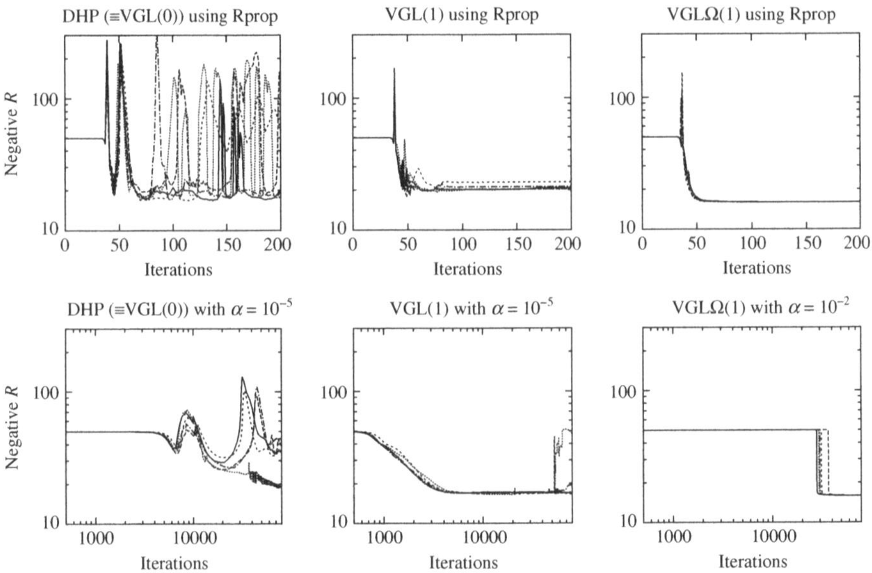

FIGURE 7.2 Results show learning progress for five typical random weight initialisations, for the problem of trying to learn 10 different trajectories. Results show increasing effectiveness (particularly in reduced volatility) for the three learning algorithms being considered, in the order that the graphs appear from left to right. The top row of graphs are all using RPROP to accelerate learning. The bottom row of graphs all use a fixed step-size parameter α.

present because RPROP causes the weight vector to traverse a significant discontinuity in the value function that exists at $h = 0$, $v = 0$.

VGL(0) shows very far-from-monotonic behavior in this problem.

7.5 CONCLUSIONS

We have defined the VGL(λ) algorithm and proven its equivalence under certain conditions to BPTT. VGL(1) with an Ω_t matrix defined in Equation (7.13) is thus a critic learning algorithm that is proven to converge, under conditions stated in Section 7.3.3, for a greedy policy and general smooth approximated critic. Although the proof does not extend to VGL(0), that is, DHP, we hope that it might provide a pointer for research in that direction, particularly with the publication of Lemma 7.4. This convergence proof has also given us insights into how the Ω_t matrix can be chosen and what its purpose is, at least for the case of $\lambda = 1$ with a greedy policy, and we speculate that similar choices could be valid for $\lambda < 1$ or nongreedy policies. In our experiment, we used a simplified Ω_t matrix that was analytically derived and easy to compute; but this may not always be possible, so an approximation to Equation (7.13) may be necessary.

Our experiment has been a simple one with known analytical functions, but it has demonstrated effectively the convergence properties of VGL(1) with the chosen Ω_t matrix, and the relative ease with which it can be accelerated using RPROP. In this experiment, we found the convergence behavior and optimality attained by VGL(1) with the chosen Ω_t matrix to be superior to VGL(1) with $\Omega_t = I$, which in turn has proved superior to VGL(0) (DHP) with $\Omega_t = I$. The given experiment was quite problematic for VGL(0) to learn and produce a stable solution, partly because in this deceptively simple environment the major proportion of the total reward arrives in the final time step, and partly because the low c value chosen for Equation (7.17) makes the function g into approximately a step-function, which implies that the surface $R(\vec{x}, \vec{w})$ will be riddled with flat plateaus separated by steep cliffs.

It was surprising to the authors that the VGL(1) weight update has been proven to be equivalent to *gradient ascent on R* when previous research has always expected DHP (and therefore presumably its variant, VGL(1)) to be *gradient descent on E*, where E is the error function $E = \sum_t \left(G'_t - \widetilde{G}_t\right)^T \Omega_t \left(G'_t - \widetilde{G}_t\right)$.

REFERENCES

1. F.-Y. Wang, H. Zhang, and D. Liu. Adaptive dynamic programming: an introduction. *IEEE Computational Intelligence Magazine*, 4(2):39–47, 2009.
2. P.J. Werbos. Neural networks, system identification, and control in the chemical process industries. In White and Sofge, editors. *Handbook of Intelligent Control*. Van Nostrant Reinhold, New York, 1992, pp. 283–356.
3. R.E. Bellman. *Dynamic Programming*. Princeton University Press, Princeton, NJ, 1957.
4. P.J. Werbos. Approximating dynamic programming for real-time control and neural modeling. In White and Sofge, editors. *Handbook of Intelligent Control*. Van Nostrant Reinhold, New York, 1992, pp. 493–525.
5. D. Prokhorov and D. Wunsch. Adaptive critic designs. *IEEE Transactions on Neural Networks*, 8(5):997–1007, 1997.
6. R.S. Sutton. Learning to predict by the methods of temporal differences. *Machine Learning*, 3:9–44, 1988.
7. C.J.C.H. Watkins. *Learning from Delayed Rewards*. PhD thesis, Cambridge University, 1989.
8. P.J. Werbos. Backpropagation through time: What it does and how to do it. *Proceedings of the IEEE*, 78(10):1550–1560, 1990.
9. P.J. Werbos. Stable adaptive control using new critic designs. *eprint arXiv:adap-org/9810001*, 1998.
10. J.N. Tsitsiklis and B. Van Roy. An analysis of temporal-difference learning with function approximation. Technical Report LIDS-P-2322, 1996.
11. M. Fairbank and E. Alonso. The divergence of reinforcement learning algorithms with value-iteration and function approximation. *eprint arXiv:1107.4606*, 2011.
12. S. Ferrari and R.F. Stengel. Model-based adaptive critic designs. In J. Si, et al., editors. *Handbook of Learning and Approximate Dynamic Programming*. Wiley-IEEE Press, New York, 2004, pp. 65–96.

13. M. Fairbank and E. Alonso. The local optimality of reinforcement learning by value gradients and its relationship to policy gradient learning. *eprint arXiv:1101.0428*, 2011.
14. M. Fairbank. Reinforcement learning by value gradients. *eprint arXiv:0803.3539*, 2008.
15. K. Doya. Reinforcement learning in continuous time and space. *Neural Computation*, 12(1):219–245, 2000.
16. Ali Heydari and S.N. Balakrishnan. Finite-horizon input-constrained nonlinear optimal control using single network adaptive critics. American Control Conference ACC, 2011, pp. 3047–3052.
17. S.E. Fahlman. Faster-learning variations on back-propagation: an empirical study. In *Proceedings of the 1988 Connectionist Summer School*, pp. 38–51, San Mateo, CA, 1988. Morgan Kaufmann.
18. C.M. Bishop. *Neural Networks for Pattern Recognition*. Oxford University Press, 1995.
19. M. Riedmiller and H. Braun. A direct adaptive method for faster backpropagation learning: The RPROP algorithm. In *Proceedings of the IEEE International Conference on Neural Networks*, pp. 586–591, San Francisco, CA, 1993.

CHAPTER 8

A Constrained Backpropagation Approach to Function Approximation and Approximate Dynamic Programming

SILVIA FERRARI, KEITH RUDD, and GIANLUCA DI MURO

Department of Mechanical Engineering, Duke University, Durham, NC, USA

ABSTRACT

The ability of preserving prior knowledge in an artificial neural network (ANN) while incrementally learning new information is important to many fields, including approximate dynamic programming (ADP), feedback control, and function approximation. Although ANNs exhibit excellent performance and generalization abilities when trained in batch mode, when they are trained incrementally with new data they tend to forget previous information due to a phenomenon known as interference. McCloskey and Cohen [1] were the first to suggest that a fundamental limitation of ANNs is that the process of learning a new set of patterns may suddenly and completely erase a network's knowledge of what it had already learned. This phenomenon, known as catastrophic interference or catastrophic forgetting, seriously limits the applicability of ANNs to adaptive feedback control, and incremental function approximation. Natural cognitive systems learn most tasks incrementally and need not relearn prior patterns to retain them in their long-term memory (LTM) during their lifetime. Catastrophic interference in ANNs is caused by their very ability to generalize using a single set of shared connection weights, and a set of interconnected nonlinear basis functions. Therefore, the modular and sparse architectures that have been proposed so far for suppressing interference also limit a neural network's ability to approximate and generalize highly nonlinear functions. This chapter describes how constrained backpropagation (CPROP) can be used to preserve prior knowledge while training

Reinforcement Learning and Approximate Dynamic Programming for Feedback Control, First Edition. Edited by Frank L. Lewis and Derong Liu.

ANNs incrementally through ADP and to solve differential equations, or approximate smooth nonlinear functions online.

8.1 BACKGROUND

Some of the most significant advances on preserving memories in ANNs to date have been made in the field of self-organizing networks and associative memories (e.g., [2, 3]). In these networks, the neurons use competitive learning to recognize groups of similar input vectors and associate them with a particular output by allowing neurons that are physically near each other to respond to similar inputs. Interference has also been suppressed successfully in NN classifiers, using Learn++ algorithms that implement a weighted voting procedure to retain long-term episodic memory [4]. Although these methods are very important to pattern recognition and classification, they are not applicable to preserving functional knowledge in ANNs, as may be required, for example, by feedback control. Although ADP aims to improve existing ANN approximations of the control law and value function, it can greatly benefit from the ability to retain control knowledge in the long term, and generally from the elimination of interference. While associative memories in self-organizing networks resemble declarative memories for recalling episodes or facts, constrained backpropagation (CPROP) aims to establish *procedural memories*, which refer to cognitive and motor skills, such as the ability to ride a bike or fly an airplane [5].

The problem of interference in nonlinear differentiable ANNs has been addressed along two main lines of research. One approach presents some long-term memory (LTM) data, or the information for the unknown function that must be preserved at all times, together with short-term memory (STM) data, or data to be learned. This approach has been proven effective for supervised radial-basis networks with compact support [4]. While useful, this approach is not suited to ANN implementations that require LTM to be preserved reliably (e.g., control systems), nor to implementations that have stringent computational requirements due, for example, to high dimensional input–output spaces, large training sets, or repeated incremental training over time, such as ADP. Another approach consists of partitioning the weights into two subsets, one that is used to preserve LTM by holding the weights' values constant, and one that is updated using the new STM data [6]. Although effective in some applications, this approach cannot guarantee LTM preservation and may not suppress interference in nonlinear neural networks with global support (Section 8.2.3). Similarly in [6], CPROP partitions the weights into STM and LTM subsets. However, both subsets are updated at every epoch of the training algorithm. STM weights are updated to learn new STMs, and LTM weights are updated to preserve LTM.

8.2 CONSTRAINED BACKPROPAGATION (CPROP) APPROACH

Neural network training is typically formulated as an unconstrained optimization problem involving a scalar function $e : \mathbb{R}^N \rightarrow \mathbb{R}$, with respect to the network weights $\mathbf{w} \in \mathbb{R}^N$. This scalar function may consist of the the neural network output error, or of an indirect measure of performance, such as the cost-to-go in an ADP algorithm. By optimizing e, the training algorithm seeks to obtain a neural network representation of an unknown vector function $\mathbf{y} = \mathbf{h}(\mathbf{p})$, with input $\mathbf{p} \in \mathbb{R}^r$, and output $\mathbf{y} \in \mathbb{R}^m$. Assume the LTM knowledge of the function can be embedded into a functional relationship describing the network weights such as,

$$g(\mathbf{w}_L, \mathbf{w}_S) = \mathbf{0}. \tag{8.1}$$

Then, training preserves the LTM expressed by (8.1) provided it is carried out according to the following constrained optimization problem:

$$\begin{aligned} &\text{minimize } e(\mathbf{w}_L, \mathbf{w}_S) \\ &\text{subject to } g(\mathbf{w}_L, \mathbf{w}_S) = \mathbf{0}. \end{aligned} \tag{8.2}$$

The solution of a constrained optimization problem can be provided by the method of Lagrange multipliers or by direct elimination. If (8.1) satisfies the implicit function theorem, then it uniquely implies the function,

$$\mathbf{w}_L = \mathcal{C}(\mathbf{w}_S) \tag{8.3}$$

and the method of direct elimination can be applied by expressing the error function as,

$$E(\mathbf{w}_S) = e(\mathcal{C}(\mathbf{w}_S), \mathbf{w}_S) \tag{8.4}$$

such that the value of $\mathbf{w}_S$ can be determined independently of $\mathbf{w}_L$. In this case, the solution of (8.2) is an extremum of (8.4) that obeys,

$$\nabla_{\mathbf{w}_S} E = \mathbf{0}, \tag{8.5}$$

where the gradient ∇ is defined as a column vector of partial derivatives taken with respect to every element of the subscript $\mathbf{w}_S$. Once the optimal value of $\mathbf{w}_S$ is determined, the optimal value of the weights $\mathbf{w}_L$ can be obtained from $\mathbf{w}_S$ using (8.3).

Hereon, it is assumed that the equality constraint can be written in explicit form (8.3). Furthermore, since (8.3) can be very involved, its substitution in the error function is circumvented by seeking the extremum defined by the *adjoined error gradient*, obtained by the chain rule,

$$\nabla_{\mathbf{w}_S} E := \left\{ \frac{\partial E}{\partial w_{S_i}} \right\} = \left\{ \frac{\partial e}{\partial w_{S_i}} + \frac{\partial e}{\partial \mathcal{C}} \frac{\partial \mathcal{C}}{\partial w_{S_i}} \right\}, \tag{8.6}$$

where w_{S_i} is the ith element of $\mathbf{w}_S$.

The constrained training approach is applicable to incremental training of neural networks for smooth function approximation under the following assumptions:

(1) *a priori* knowledge of the function is available locally in its domain (e.g., a batch training set or a physical model); (2) it can be expressed as an equality constraint on the neural network weights; (3) it is desirable to preserve this prior knowledge during future training sessions; (4) new functional information must be assimilated incrementally through domain exploration; and (5) the new information is consistent with the prior knowledge. Then, equations used in constrained training are given in the following sections.

8.2.1 Neural Network Architecture and Procedural Memories

A feedforward, one-hidden-layer, sigmoidal architecture is chosen because of its universal function approximation ability and its broad applicability. The hidden layer can be represented by an operator with repeated sigmoids $\mathbf{\Phi}(\mathbf{n}) := [\sigma(n_1)\cdots\sigma(n_s)]^T$, where n_i denotes the ith component of the input-to-node vector $\mathbf{n} \in \mathbb{R}^{s\times 1}$. The sigmoidal function $\sigma(n_i) : \mathbb{R} \to \mathbb{R}$ is assumed to be a bounded measurable function on $\mathbb{R}$ for which $\sigma(n_i) \to 1$ as $n_i \to \infty$, and $\sigma(n_i) \to -1$ as $n_i \to -\infty$. In this chapter, the sigmoid of choice is $\sigma(n_i) := (e^{n_i} - 1)/(e^{n_i} + 1)$. Then, the neural network input–output equation,

$$\hat{\mathbf{y}}(\mathbf{p}) = \mathbf{V}\mathbf{\Phi}(\mathbf{W}\mathbf{p} + \mathbf{b}) := \mathbf{V}\mathbf{\Phi}[\boldsymbol{\nu}(\mathbf{p})] \tag{8.7}$$

can be written in terms of linear *input-to-node operator*, $\boldsymbol{\nu} : \mathbb{R}^r \to \mathbb{R}^s$, which maps the input space into node space, where, $\mathbf{b} \in \mathbb{R}^{s\times 1}$, $\mathbf{W} \in \mathbb{R}^{s\times r}$, and $\mathbf{V} \in \mathbb{R}^{m\times s}$, are the adjustable bias, and input and output ANN weights, respectively.

The LTM is defined as the input–output and gradient information for the unknown function $\mathbf{h} : \mathbf{p} \to \mathbf{y}$ that must be preserved at all times during incremental training sessions. The LTM may comprise sampled output and derivative information, or information about the functional form over a bounded subset $\mathcal{D} \subset \mathcal{P}$. The STM is defined as the sequence of skills (e.g., control laws) or information that must be learned through one or more training functions $\{e_k(\mathbf{w})\}_{k=1,2,\ldots}$. CPROP utilizes algebraic neural network training [7], and the adjoined Jacobian described in Section 8.2.2. By this approach, any backpropagation-based algorithm can be modified to retain the ANN's LTM during training.

8.2.2 Derivation of LTM Equality Constraints and Adjoined Error Gradient

Classical backpropagation-based algorithms minimize the scalar function e using an unconstrained optimization approach that is based on the gradient of e with respect to the ANN weights, $\nabla_{\mathbf{w}} e \in \mathbb{R}^N$. Because e often represents the ANN output error, they are commonly referred to as error backpropagation (EBP) algorithms. Examples of unconstrained optimization algorithms that have been utilized for ANN training are steepest descent, conjugate gradient, and Newton's method. In every case, the first- or second-order derivatives of e with respect to $\mathbf{w}$ are utilized to minimize e, and backpropagation refers to a convenient approach for computing these derivatives across the ANN hidden layer(s), for example, Φ in (8.7). Then, the definition of e determines the training style. In supervised training, e is an error function representing the distance between the ANN output and the output data sampled from $\mathbf{h}$, and organized in a training set $\mathcal{T} = \{\mathbf{p}_k, \mathbf{y}_k\}_{k=1,2,\ldots}$, where every sample satisfies the function to be approximated, that is, $\mathbf{y}_k = \mathbf{h}(\mathbf{p}_k)$, for all k. In reinforcement learning (RL) and ADP, e may be an indirect measure of performance, such as the value function, or the improved policy, in which case the weight update includes the temporal difference error. In batch training, the information is presented all at once, by defining e as a sum over all training pairs, whereas in incremental training, the information is presented one sample at-a-time, or one subset at-a-time in batch mode. In every one of these instances, the chosen backpropagation-based algorithm can be constrained to preserve LTM by backpropagating a so-called adjoined gradient (or Jacobian, depending on the algorithm) that can be computed conveniently across the hidden layer using the approach described in this section.

The approach is illustrated for LTM that can be expressed by a training set of input–output samples and derivative information denoted by $\mathcal{T}_L = \{\mathbf{p}_{L_\ell}, \mathbf{y}_{L_\ell}, \chi_{L_\ell}\}_{\ell=1,\ldots,K}$, where $\mathbf{y}_{L_\ell} = \mathbf{h}(\mathbf{p}_{L_\ell})$, $\chi_{L_\ell} = \nabla_{\mathbf{p}}\mathbf{h}(\mathbf{p}_{L_\ell})$, where $\mathbf{p}_{L_\ell} \in \mathcal{D} \;\; \forall \ell = 1, \ldots, K$. If the functional form of $\mathbf{h}$ over $\mathcal{D}$ is known, then it can be sampled to produce $\mathcal{T}_L$. Whenever possible, derivative information should be incorporated to improve ANN generalization and prevent overfitting. In general, the ANN performance may depend on a vector function $\boldsymbol{\epsilon}$, such as the ANN output error, or the gradient of the cost-to-go. Then, the STM scalar function to be minimized during training can be expressed by the quadratic form,

$$e(\mathbf{w}) = \frac{1}{q}\sum_{k=1}^{q} \boldsymbol{\epsilon}_k^T(\mathbf{w})\boldsymbol{\epsilon}_k^T(\mathbf{w}), \tag{8.8}$$

where q is the number of STM samples available during the training session. In this work, the Levenberg–Marquardt (LM) is the EBP algorithm of choice, because of its excellent convergence and stability properties.

In the classical, unconstrained case, the LM algorithm iteratively minimizes e with respect to $\mathbf{w}$, based on the *unconstrained* Jacobian, $\mathbf{J} = \partial\boldsymbol{\epsilon}_k/\partial\mathbf{w}$, which is based on the Jacobian of the ANN and/or one of its derivatives. For a training set $\mathcal{T}$ with q samples, let $\mathbf{P} \in \mathbb{R}^{r\times q}$ and $\mathbf{B} \in \mathbb{R}^{s\times q}$ be defined as

$$\mathbf{P} := [\mathbf{p}_1 \cdots \mathbf{p}_q] \quad \text{and} \quad \mathbf{B} := [\underbrace{\mathbf{b}\cdots\mathbf{b}}_{q}]. \tag{8.9}$$

Then the q sample can be arranged into the matrix equation

$$\mathbf{Y} = \mathbf{V}\Phi(\mathbf{WP} + \mathbf{B}). \tag{8.10}$$

A similar equation can be derived for the matrix representation of derivative information. Consider the general partial derivative

$$\frac{\partial^\gamma \tilde{\mathbf{y}}(\mathbf{p})}{\partial^{m_1} p_1 \ldots \partial^{m_r} p_r}, \tag{8.11}$$

where $\gamma = m_1 + ... + m_r$. Let $\boldsymbol{\omega}_j$ represent a diagonal matrix of the jth column of W, and let

$$\boldsymbol{\Lambda} = \prod_{j=1}^{r} \boldsymbol{\omega}_j^{m_j}. \tag{8.12}$$

Then the matrix equation for the general derivative is given by

$$\chi = \mathbf{V}\boldsymbol{\Lambda}\boldsymbol{\Phi}^{\gamma}(\mathbf{WP} + \mathbf{B}), \tag{8.13}$$

where $\boldsymbol{\Phi}^{\gamma}(\cdot)$ indicates γ derivatives of the transfer function. Note that in the case that $\gamma = 0$, (8.13) reduces to (8.10).

As e is problem-specific, the derivatives of e with respect to the weights cannot be given explicitly. However, these derivatives are composed of derivatives of (8.13). Let $\alpha_{ij} = m_j w_{ij}^{m_j - 1} \prod_{k \neq j}^{r} w_{ik}^{m_k}$, $\boldsymbol{\Phi}_i(\cdot)$ be the ith row of $\boldsymbol{\Phi}(\cdot)$, V_i be the ith column of V, and $\mathbf{D}_j$ be a diagonal matrix with the jth row of $\mathbf{P}$ along the diagonal, such that,

$$\frac{\partial \chi}{\partial w_{ij}} = \mathbf{V}_i \left(\alpha_{ij} \boldsymbol{\Phi}_i^{\gamma}(\mathbf{WP} + \mathbf{B}) + \boldsymbol{\Lambda}_{ii} \boldsymbol{\Phi}_i^{\gamma+1}(\mathbf{WP} + \mathbf{B})\mathbf{D}_j \right), \tag{8.14}$$

$$\frac{\partial \chi}{\partial b_i} = \mathbf{V}_i \boldsymbol{\Lambda}_{ii} \boldsymbol{\Phi}_i^{\gamma+1}(\mathbf{WP} + \mathbf{B}), \tag{8.15}$$

$$\frac{\partial \chi}{\partial v_{ij}} = \boldsymbol{\Lambda}_{jj} \boldsymbol{\Phi}_j^{\gamma}(\mathbf{WP} + \mathbf{B}). \tag{8.16}$$

To preserve the LTM, the equality constraints (8.1) are derived from the training set $\mathcal{T}_L$. According to [7], a training set in the form of $\mathcal{T}_L$ can be matched exactly by an ANN with Equation (8.7) provided there are $s_L = K$ (LTM) nodes in the hidden layer. An additional number of nodes, denoted by s_S, is added to allow the ANN to assimilate additional (STM) information during incremental training. Subsequently, the ANN weights associated with the LTM nodes, denoted by $\mathbf{W}_L$, $\mathbf{b}_L$, and $\mathbf{V}_L$, are used to satisfy the LTM constraints at all times, and the ANN weights associated with the STM nodes, denoted by $\mathbf{W}_S$, $\mathbf{b}_S$, and $\mathbf{V}_S$, are used to acquire the STM, via constrained LM.

The LTM training set is satisfied at all times provided the following set of algebraic equations is:

$$\mathbf{Y}_L^T = \boldsymbol{\Phi}\left(\mathbf{P}_L^T \left[\mathbf{W}_L^T \ \mathbf{W}_S^T\right] + \left[\mathbf{B}_L^T \ \mathbf{B}_S^T\right]\right) \begin{bmatrix} \mathbf{V}_L^T \\ \mathbf{V}_S^T \end{bmatrix}, \tag{8.17}$$

$$\chi_L^T = \boldsymbol{\Phi}^{\gamma}\left(\mathbf{P}_L^T \left[\mathbf{W}_L^T \ \mathbf{W}_S^T\right] + \left[\mathbf{B}_L^T \ \mathbf{B}_S^T\right]\right) \begin{bmatrix} \boldsymbol{\Lambda}_L & \mathbf{0} \\ \mathbf{0} & \boldsymbol{\Lambda}_S \end{bmatrix} \begin{bmatrix} \mathbf{V}_L^T \\ \mathbf{V}_S^T \end{bmatrix}, \tag{8.18}$$

where $\mathbf{P}_L$ is a matrix of LTM input information. The above equality constraint, in the form (8.1), can be transformed into the explicit form (8.3) by rewriting (8.17) and (8.18) in terms of the STM and LTM weights,

$$\begin{aligned}\mathbf{M} &:= \begin{bmatrix} \mathbf{Y}_L^T \\ \chi_L^T \end{bmatrix} \\ &= \begin{bmatrix} \mathbf{\Phi}\left(\mathbf{P}_L^T\mathbf{W}_L^T + \mathbf{B}_L^T\right) \\ \mathbf{\Phi}^\gamma\left(\mathbf{P}_L^T\mathbf{W}_L^T + \mathbf{B}_L^T\right)\mathbf{\Lambda}_L \end{bmatrix}\mathbf{V}_L^T + \begin{bmatrix} \mathbf{\Phi}\left(\mathbf{P}_L^T\mathbf{W}_S^T + \mathbf{B}_S^T\right) \\ \mathbf{\Phi}^\gamma\left(\mathbf{P}_L^T\mathbf{W}_S^T + \mathbf{B}_S^T\right)\mathbf{\Lambda}_S \end{bmatrix}\mathbf{V}_S^T \\ &:= \mathbf{\Psi}\mathbf{V}_L^T + \mathbf{\Omega}\mathbf{V}_S^T. \end{aligned} \tag{8.19}$$

Then, the explicit constraint (8.3) takes the form,

$$\mathbf{V}_L^T = \mathbf{\Psi}^{-1}[\mathbf{M} - \mathbf{\Omega}\mathbf{V}_S^T], \tag{8.20}$$

where $\mathbf{\Psi}$ is a known, constant matrix. If it is nonsquare, then the pseudo-inverse can be computed. This explicit representation can then be used to compute $\frac{\partial \mathcal{C}}{\partial w_{S_i}}$ from (8.6).

$$\frac{\partial \mathbf{V}_L^T}{\partial w_{S_{ij}}} = -\Psi^{-1}\begin{bmatrix} \mathbf{D}_{L_j}\mathbf{\Phi}_i'(\mathbf{W}_S\mathbf{P}_L + \mathbf{B}_S)^T \\ \alpha_{S_{ij}}\mathbf{\Phi}_i^\gamma(\mathbf{W}_S\mathbf{P}_L + \mathbf{B}_S)^T + \mathbf{\Lambda}_{S_{ii}}\mathbf{D}_{L_j}\mathbf{\Phi}_i^{\gamma+1}(\mathbf{W}_S\mathbf{P}_L + \mathbf{B}_S)^T \end{bmatrix}\mathbf{V}_{S_i}^T, \tag{8.21}$$

$$\frac{\partial \mathbf{V_L}^T}{\partial b_{S_i}} = -\Psi^{-1}\begin{bmatrix} \mathbf{\Phi}_i'(\mathbf{W}_S\mathbf{P}_L + \mathbf{B}_S)^T \\ \mathbf{\Lambda}_{S_{ii}}\mathbf{\Phi}_i^{\gamma+1}(\mathbf{W_S}\mathbf{P}_S + \mathbf{B}_S)^T \end{bmatrix}\mathbf{V}_{S_i}^T, \tag{8.22}$$

$$\frac{\partial \mathbf{V_L}^T}{\partial v_{S_{ij}}} = -\Psi^{-1}\begin{bmatrix} \mathbf{\Phi}_j(\mathbf{W}_S\mathbf{P}_L + \mathbf{B}_S)^T \\ \mathbf{\Lambda}_{S_{jj}}\mathbf{\Phi}_j^\gamma(\mathbf{W}_S\mathbf{P}_L + \mathbf{B}_S)^T \end{bmatrix}. \tag{8.23}$$

8.2.3 Example: Incremental Function Approximation

In this section, CPROP is illustrated through a simple example taken from [8], involving a scalar nonlinear function $h : p \to y$, to be approximated over a bounded domain $p \in \mathcal{P}$ by a one-layer sigmoidal neural network (8.7). Consider the nonlinear function plotted by a dashed line in Figure 8.1 over a domain $\mathcal{P} = [0,\ 3\pi] \subset \mathbb{R}$. Suppose the shape of the function over a bounded subset $\mathcal{D} = [0,\ \pi] \subset \mathcal{P}$ is known *a priori* to be a sine function, and the LTM training set $\mathcal{T}_L$ is formed using the LTM samples shown in Figure 8.1. Then, the LTM training set takes the form $\mathcal{T}_L = \{p_{L\ell}, y_{L\ell}, \partial y/\partial p|_{p_{L\ell}}\}_{\ell=1,\ldots,K}$, with $K = 4$, and $p_{L_\ell} \in \mathcal{P}$ for all ℓ. After learning $\mathcal{T}_L$ in batch mode, the ANN is re-trained using new STM data, $\mathcal{T}_S = \{p_{S_k}, y_{S_k}\}_{k=1,\ldots,q}$, where in this case $y_{S_k} \in \{\mathcal{P}/\mathcal{D}\}$, even though it is not a requirement for the algorithm.

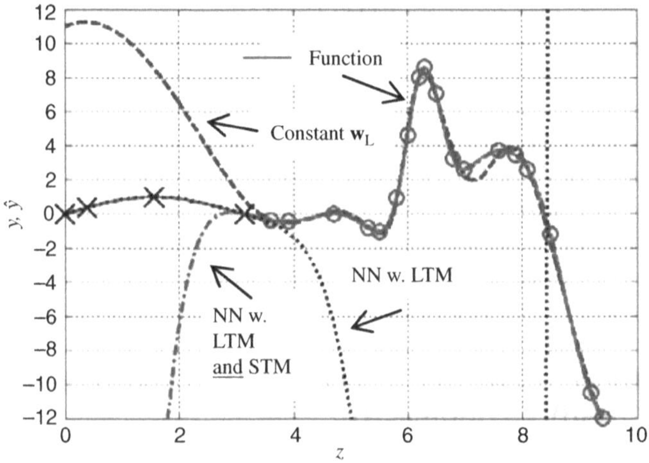

FIGURE 8.1 Function to be approximated incrementally and performance of existing algorithms (taken from [8]). (*See the color version of this figure in color plates section.*)

Thus, in this example, two subsets of sampled data are presented incrementally to the ANN, as could come about when the two subsets are too large to be presented at once, or when $\mathcal{T}_S$ becomes available at a later time, for example, in on-line learning.

As illustrated in Figure 8.1, when trained with existing algorithms, the neural network may forget the LTM while learning the STM data due to interference. Here, a sigmoidal ANN with 15 hidden nodes is trained to approximate $\mathcal{T}_L$ using the LM algorithm available from the MATLAB® Neural Network Toolbox. Later, 18 new STM samples become available (Figure 8.1), the NN is retrained by the same LM algorithm using $\mathcal{T}_S$. If training is conducted sequentially, without re-using the LTM

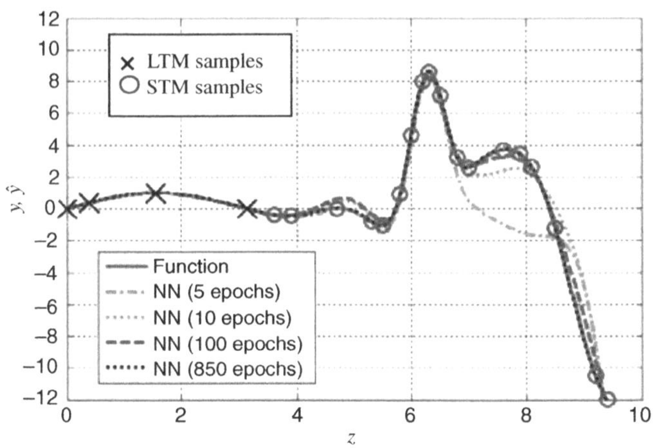

FIGURE 8.2 CPROP preserves LTM accurately at every epoch, until it properly learns the STM at approximately 850 epochs (taken from [8]). (*See the color version of this figure in color plates section.*)

data, the NN starts out with proper LTM (dotted line in Figure 8.1), but then experiences catastrophic interference and, although it learns STM well, it forgets the LTM entirely in the process (dashed–dotted line in Figure 8.1). Figure 8.1 also shows the performance of a NN that is trained with the incremental training method proposed by Mandziuk [6]. That is, LTM weights ($\mathbf{w}_L$) are held constant while learning from $\mathcal{T}_S$. As shown by the dashed line in Figure 8.1, when the NN is trained by this method, it still experiences interference and forgets the LTM. Instead, when the LM algorithm is constrained by implementing the above Jacobian and memory constraint, the LTM is preserved at every epoch, and the NN learns the STM without forgetting the LTM, as shown in Figure 8.2.

8.3 SOLUTION OF PARTIAL DIFFERENTIAL EQUATIONS IN NONSTATIONARY ENVIRONMENTS

Neural networks are often implemented to provide functional representations of numerical partial differential equation (PDE) solutions that are amenable to mathematical analysis and data assimilation algorithm. In many applications, a numerical algorithm is first utilized to obtain a solution in discrete form, then the solution is utilized to train a neural network. Methods have also been proposed to determine the PDE solution in one step, by training an ANN to minimize an error function formulate in terms of the differential operator. Different techniques have been developed to take into account the boundary conditions (BCs). One line of research [9] expresses the solution as the sum of two functions. One function is problem dependent and is designed by the user to satisfy the BCs with no adjustable parameters. The second function is an ANN trained to approximate the differential operator. However, this approach only is applicable to PDEs defined on orthogonal box domains, and cannot be extended to nonstationary environments in which the BCs change over time because part of the solution must be designed by the user off-line. A more recent study [10] overcomes some of these limitations, but adopts radial basis functions to correct the solution on the boundaries, making difficult to implement this technique in nonstationary environments, when the shape of the boundaries may change over time.

Another approach consists of embedding the BCs in the cost function [11, 12]. Although this approach has been shown effective in solving PDEs, to match BCs with high accuracy, which may be crucial in many engineering applications, it requires many points on the integration domain boundaries, thereby increasing the computational cost dramatically. Also, it relies on evolutionary algorithms that are computationally expensive. By incorporating equality constraints of any form systematically in any nonlinear ANN training process, CPROP overcomes these existing limitations.

8.3.1 CPROP Solution of Boundary Value Problems

Consider the nonlinear boundary value problem (BVP),

$$\mathcal{L}\left[u(\mathbf{p})\right] = f(\mathbf{p}), \tag{8.24}$$

where $\mathcal{L}(\cdot)$ is a nonlinear differential operator, $\mathbf{p} \in \mathcal{I} \subset \mathbb{R}^r$, and $\mathcal{I} \in \mathbb{R}^r$ is a compact set with associated BCs,

$$\mathcal{G}\left[u(\mathbf{p})\right] = h(\mathbf{p}). \tag{8.25}$$

$\mathcal{G}(\cdot)$ is a linear operator of order less than $\mathcal{L}(\cdot)$. The functions $\mathbf{p} \in \partial\mathcal{I} \subset \mathbb{R}^r$ and $f, h : \mathbb{R}^r \rightarrow \mathbb{R}$ are assumed to be continuous and known.

The solution $\hat{u}(\mathbf{p})$ is provided by the output of an ANN in the form (8.7), with scalar output ($m = 1$), and adjustable parameters $\mathbf{W}$, $\mathbf{b}$, and $\mathbf{V}$.

$$\hat{u}(\mathbf{p}) = \mathbf{V}\boldsymbol{\Phi}\left[\nu(\mathbf{p})\right], \tag{8.26}$$

where $\mathbf{V} \in \mathbb{R}^{1\times s}$ is a vector of output weights. Substituting (8.26) into (8.24), the differential operator is applied to the ANN,

$$\mathcal{L}\left\{\mathbf{V}\boldsymbol{\Phi}\left[\nu(\mathbf{p})\right]\right\} = f(\mathbf{p}). \tag{8.27}$$

The STM set is formed by input samples $\mathcal{T}_S = \{\mathbf{p}_k \in \mathcal{I}, k = 1, \ldots, q\}$, which are used to define P_S. Replacing $\mathbf{P}_L$ with $\mathbf{P}_S$ in Equations (8.17) and (8.18) defines $\mathbf{Y}_S$ and χ_S, which are used to compute $\mathcal{L}\left[\hat{u}(\mathbf{p})\right]|_{\mathbf{p}\in\mathcal{T}_{\mathrm{STM}}}$. Introducing the vector $\mathbf{f} := f(\mathbf{p})|_{\mathbf{p}\in\mathcal{T}_S} \in \mathbb{R}^q$, the function to be minimized may be defined as,

$$e(\mathbf{w}) = \frac{1}{2}\boldsymbol{\epsilon}^T\boldsymbol{\epsilon}, \tag{8.28}$$

where

$$\boldsymbol{\epsilon} := \mathbf{f} - \mathcal{L}\left[\hat{u}(\mathbf{p})\right]|_{\mathbf{p}\in\mathcal{T}_{\mathrm{STM}}}. \tag{8.29}$$

The LTM is given by the BC (8.25). The training set is given by $\mathcal{T}_L = \{\mathbf{p}_{L\ell}, h(\mathbf{p}_{L\ell})\}$. The training points from $\mathcal{T}_L$ can be substituted directly into (8.17), and in the case of Neumman or Dirichlet BCs, the constraint is given explicitly by (8.20). If the BC is of type Robin, or mixed, then the constraint will be slightly different. However, because $\mathcal{G}$ was defined as a linear operator, $\mathbf{V}_L$ can be factored and the form will remain the same. Now using (8.14)–(8.16) and (8.21)–(8.23) the weights can be trained to satisfy the BVP.

8.3.2 Example: PDE Solution on a Unit Circle

This example illustrates how an ANN trained via CPROP is capable to adapt and approximate a changing PDE solution in a nonstationary environment. The nonlinear

PDE is forced by a known term $f_j(p_1, p_2)$ that is subject to change, and must be solved over a unit circle centered at the origin,

$$\nabla^2 u + u \frac{\partial u}{\partial p_2} = f_j(p_1, p_2), \quad j = 1, 2, 3, \tag{8.30}$$

where, three forcing terms, index by $j = 1, 2, 3$ are considered in this example. The above PDE solution must satisfy the BCs,

$$u(p_1, p_2) = 1 \tag{8.31}$$

on the unit circle defined as $\mathcal{I} := \{p_1, p_2 \mid p_1^2 + p_2^2 = 1\}$; where, in (8.30), the solution $u(p_1, p_2)$ is denoted by u for brevity, and the functions $f_j(p_1, p_2)$, $j = 1, 2, 3$ are assumed to be continuous and known. Then, by defining a two-dimensional grid over $\mathcal{I}$, the STM training set $\mathcal{T}_{S_j} = \{\mathbf{p}_k \in \mathcal{I}, k = 1, \ldots, q\}$ is obtained for every j, and the values of the present (j) forcing function evaluated, and organized in the $q \times 1$ vector $\mathbf{f}_j := [f_j(\mathbf{p}_1) \cdots f_j(\mathbf{p}_q)]^T$. The STM training sets are used to adapt the ANN solution $\hat{u}(\mathbf{p})$ incrementally, by presenting each one as the forcing function changes from f_1 to f_2, and then from f_2 to f_3. These functions are chosen such that the above PDE has the following analytical solutions: $u = p_1^2 + p_2^2$ when $j = 1$, $u = 1.15(p_1^2 + p_2^2) - 0.15$ when $j = 2$, and $u = \exp[-(p_1^2 + p_2^2)] + 1.1(p_1^2 + p_2^2) - e^{-1} - 0.1$ when $j = 3$. Plugging the various pieces into (8.29) gives the error function to this example:

$$\begin{aligned}
\boldsymbol{\epsilon} = \mathbf{f} &- \boldsymbol{\Phi}''\left(\mathbf{P}_S^T \mathbf{W}_L^T + \mathbf{B}_L\right)(\boldsymbol{\omega}_{L_1}^2 + \boldsymbol{\omega}_{L_2}^2)\mathbf{V}_L^T \\
&- \boldsymbol{\Phi}''\left(\mathbf{P}_S^T \mathbf{W}_S^T + \mathbf{B}_S^T\right)(\boldsymbol{\omega}_{S_1}^2 + \boldsymbol{\omega}_{S_2}^2)\mathbf{V}_S^T \\
&- \left(\boldsymbol{\Phi}\left(\mathbf{P}_S^T \left[\mathbf{W}_L^T \ \mathbf{W}_S^T\right] + \left[\mathbf{B}_L^T \ \mathbf{B}_S^T\right]\right) \begin{bmatrix} \mathbf{V}_L^T \\ \mathbf{V}_S^T \end{bmatrix}\right) \\
&\circ \left(\boldsymbol{\Phi}'\left(\mathbf{P}_S^T \mathbf{W}_L^T + \mathbf{B}_L\right)\boldsymbol{\omega}_{L_2}\mathbf{V}_L^T + \boldsymbol{\Phi}'\left(\mathbf{P}_S^T \mathbf{W}_S^T + \mathbf{B}_S^T\right)\boldsymbol{\omega}_{S_2}\mathbf{V}_S^T\right),
\end{aligned} \tag{8.32}$$

where $\circ$ represents the Schur product. The LTM input data is given by points along the boundary: $\mathcal{T}_L = \{\mathbf{p}_{L\ell}\}$ where $p_{L_{1\ell}}^2 + p_{L_{2\ell}}^2 = 1$ for any $1 \leq \ell < K$. The constraint defined in (8.20) for this example becomes

$$\mathbf{V}_L^T = \left[\boldsymbol{\Phi}\left(\mathbf{P}_L^T \mathbf{W}_L^T + \mathbf{B}_L^T\right)\right]^{-1} \left(\begin{bmatrix} 1 \\ \vdots \\ 1 \end{bmatrix} - \boldsymbol{\Phi}(\mathbf{P}_L^T \mathbf{W}_S^T + \mathbf{B}_S^T)\mathbf{V}_S^T\right). \tag{8.33}$$

For the numerical solutions, a 100-points grid was used, with points chosen from 10 concentric circles, equidistant from the center, by dividing each circle into 10 equal sectors. To avoid a homogenous radial distribution of points, which may lead to singularities, a positive, counter-clockwise $\theta = \pi/8$-swirl was imposed between

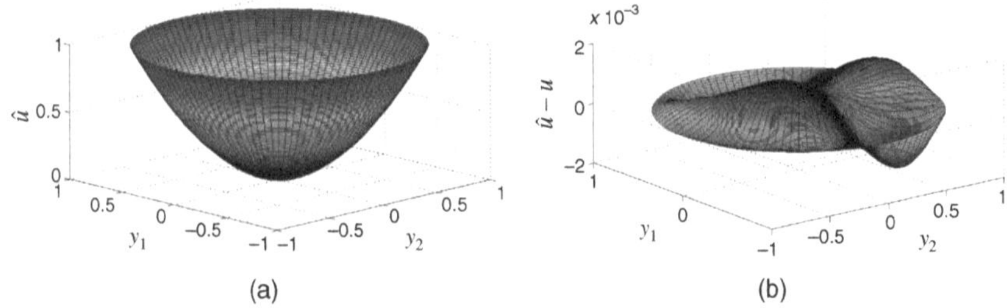

FIGURE 8.3 Neural network solution (a) and error surface (b) when $j = 1$. (*See the color version of this figure in color plates section.*)

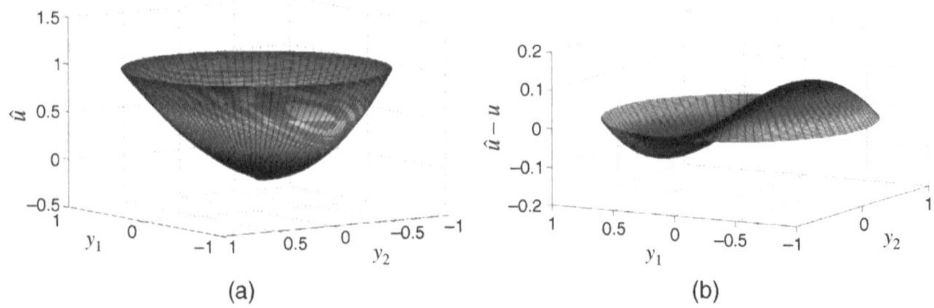

FIGURE 8.4 Neural network solution (a) and error surface (b) when $j = 2$. (*See the color version of this figure in color plates section.*)

circles during sampling. The ANN was trained with $q = 7$ STM samples, and contained $s_L = 25$ LTM nodes to be able to match the BCs on the unit circle (8.31) at all times. For every forcing term f_j, CPROP training is conducted for approximately 50 epochs, after which it is assumed that the environmental conditions have changed, and the ANN-PDE solution is adapted to the new term f_{j+1}. The final training has been run until satisfactory convergence has occurred and no further improvement seemed possible (after 450 epochs). For validation, the ANN solution is tested and plotted over a much denser grid obtained using 90 circles subdivided into 20 sectors. The results plotted in Figures 8.3–8.5 show that the ANN output approximates every one of the three solutions with good accuracy, excellent generalization, and without any overfitting.

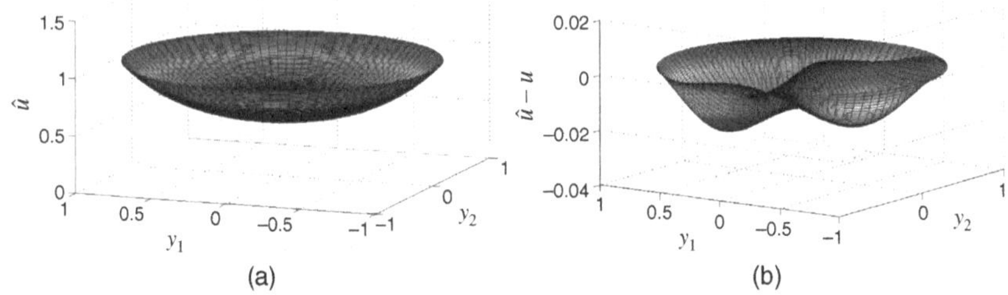

FIGURE 8.5 Neural network solution (a) and error surface (b) when $j = 3$. (*See the color version of this figure in color plates section.*)

8.3.3 CPROP Solution to Parabolic PDEs

Consider the nonlinear initial boundary value problem (IBVP),

$$\mathcal{L}\left[u(\mathbf{p})\right] = f(\mathbf{p}), \tag{8.34}$$

where again $\mathcal{L}$ is a nonlinear differential operator, $\mathbf{p} \in \mathcal{I} = \mathcal{H} \times [t_0, t_f) \subset \mathbb{R}^r$, $\mathcal{H}$ being a compact set. The associated Dirichlet BCs are

$$u(\mathbf{p}) = h(\mathbf{p}) \tag{8.35}$$

for $\mathbf{p} \in \partial\mathcal{H} \times [t_0, t_f)$. IBVPs also have an initial condition associated with one of the variables. Without loss of generality let p_r be that variable, where $t_0 \leq p_r \leq t_f$. Then, the initial condition is given by

$$u(p_1, \ldots, p_{r-1}, t_0) = g(p_1, \ldots, p_{r-1}). \tag{8.36}$$

The form of the approximate solution is different from that of the BVP, and is given by

$$\hat{u}(\mathbf{p}) = \tilde{h}(\mathbf{p}) + q(p_1, \ldots, p_{r-1}) \left(\mathbf{\Phi}(\mathbf{p}^T\mathbf{W}_L^T + \mathbf{b}_L^T)\mathbf{V}_L^T + \mathbf{\Phi}(\mathbf{p}^T\mathbf{W}_S^T + \mathbf{b}_S^T)\mathbf{V}_S^T\right), \tag{8.37}$$

where $\tilde{h}(\mathbf{p})$ is differentiable for all $\mathbf{p}$ in $\mathcal{I}$ and satisfies the BCs, and $q(p_1, \ldots, p_{r-1})$ is differentiable for all $(p_1, \ldots, p_{r-1}) \in \mathcal{H}$ and is zero along $\partial\mathcal{H}$ but nowhere in the interior of $\mathcal{H}$. This structure was explored by [13] for solving BVP. In [13], a general method for constructing $q(\cdot)$ for nonrectangular domains is described. This method, as mentioned previously, has the limitation that $q(\cdot)$ must be designed off-line. However, for IBVP this structure has an added benefit in that the BC is automatically satisfied, leaving the only constraint to be the initial condition. Then, the LTM input data is given by $\mathcal{T}_L = \{\mathbf{p}_{L\ell} = [p_1, \ldots, p_{r-1}, t_0]_{L\ell}^T\}$. Define M from (8.20) to be $\left.\frac{u(\mathbf{p})-\tilde{h}(\mathbf{p})}{q(\mathbf{p})}\right|_{\mathbf{p}=\mathbf{p}_{L\ell}}$, then the constraint takes on the same form as the BVP. Additionally, the problem can be shifted so that $t_0 = 0$. Then, the constrant is *independent* of the rth column of $\mathbf{W}_L$, $\mathbf{W}_{L_r}$. Since $\mathbf{W}_{L_r}$ is no longer present in the constraint, it can be trained with the other STM weights. Training of $\mathbf{W}_S$, $\mathbf{V}_S$, and $\mathbf{b}_S$ is done the same way that it is done for the BVP.

8.4 PRESERVING PRIOR KNOWLEDGE IN EXPLORATORY ADAPTIVE CRITIC DESIGNS

The advantages brought about by using classical control theory in conjunction with nonlinear NNs have long been recognized [5, 14–20]. In particular, using classical controllers to obtain the starting neural control design has been shown to be a key step in the development of highly effective adaptive neural controllers. One reason is that the starting design provides adequate performance while the adaptation compensates for nonlinearities and unmodeled dynamics. Another reason is that many popular adaptation schemes, such as, adaptive critics, improve iteratively upon the existing approximation of the control law. Therefore, the classical controller provides a performance baseline for the adaptive neural controller. Typically, a classical controller is used to pre-train the neural controller in a supervised fashion, and the adaptation is carried out by RL, since the ideal control law for nonlinear and unmodeled dynamics is unknown. However, due to interference, the neural controller may rapidly forget the performance baseline provided by the classical controller during adaptation.

A popular approach for combining these two styles of learning, recently reviewed in [21], consists of computing the control as a weighted sum of a supervisor's control and an exploratory policy. Although the supervisor provides the nominal control signal to the system to be controlled or plant, the exploratory policy or actor is updated by linearly interpolating between the RL weight update and a supervised learning (SL) weight update. Although this approach may not suppress interference, the supervisor overrides bad control signals from the actor and, in this fashion, guarantees a minimum performance baseline [21]. The main drawback of this approach is that the linear superposition of control policies may not lead to any performance improvement when the plant exhibits highly nonlinear dynamics. Similarly, computing the linear interpolation between the RL and SL weight updates may prove ineffective for learning highly nonlinear (and nonconvex) control laws.

8.4.1 Derivation of LTM Constraints for Feedback Control

Consider a plant whose dynamics can be approximated by the nonlinear differential equation and output equation,

$$\dot{\mathbf{x}}(t) = \mathbf{f}[\mathbf{x}(t), \mathbf{p}_m(t), \mathbf{u}(t)], \tag{8.38}$$

$$\mathbf{y}(t) = \mathbf{h}[\mathbf{x}(t), \mathbf{u}(t)], \tag{8.39}$$

where $\mathbf{x} \in \mathcal{X} \subset \mathbb{R}^{n\times 1}$ is the state, and $\mathbf{u} \in \mathcal{U} \subset \mathbb{R}^{m\times 1}$ is the control. The differential equation structure and parameters, $\mathbf{p}_m$, are not always known *a priori*, and are subject to change during the lifetime of the plant. It is assumed that perfect knowledge of the state $\mathbf{x}$ is available at a present time t, based on error-free measurements of the output $\mathbf{y}$. Under restricted operating conditions the plant operates in a subset of the state space $\mathcal{X}_{\mathrm{LPV}} \subset \mathcal{X}$, referred to as linear-parameter-varying or LPV regime, in which plant dynamics can be closely approximated by a linear-time-invariant (LTI) model near every equilibrium or operating point. Then, a finite set of p equilibria in $\mathcal{X}_{\mathrm{LPV}}$ can be selected, and indexed by a corresponding scheduling vector $\boldsymbol{\zeta} \in \mathbf{R}^{z\times 1}$, which contains time-varying physical parameters that significantly influence plant dynamics. Let $\mathcal{Z} = \{\boldsymbol{\zeta}_1, \ldots, \boldsymbol{\zeta}_p\} \in \mathcal{X}_{\mathrm{LPV}}$ denote a finite set of scheduling vectors, such that at every $\boldsymbol{\zeta}_j \in \mathcal{X}_{\mathrm{LPV}}$ the plant can be represented by the transfer function,

$$\mathbf{H}_j(s) : \begin{cases} \Delta\dot{\mathbf{x}} = \mathbf{A}_j\Delta\mathbf{x} + \mathbf{B}_j\Delta\mathbf{u} \\ \Delta\mathbf{y} = \mathbf{C}_j\Delta\mathbf{x} + \mathbf{D}_j\Delta\mathbf{u} \end{cases}. \tag{8.40}$$

Then, the set of transfer functions $\mathcal{H} = \{\mathbf{H}_1(s), \ldots, \mathbf{H}_p(s)\}$ can be used to design a gain-scheduled controller that interpolates between the LTI controllers derived from each transfer function in $\mathcal{H}$.

Several gain-scheduling techniques, including multivariable control, μ-synthesis, and H_∞ have been developed to design state feedback linear control systems that interpolate among p transfer functions of the form,

$$K_j(s) : \begin{cases} \dot{\mathbf{x}}_K = \mathbf{A}_{K_j}\mathbf{x}_K + \mathbf{B}_{K_j}\Delta\mathbf{x} \\ \mathbf{u}_K := \mathbf{C}_{K_j}\mathbf{x}_K + \mathbf{D}_{K_j}\Delta\mathbf{x} \end{cases}, \tag{8.41}$$

to control the plant in the LPV regime, where, $\mathbf{x}_K \in \mathbf{R}^{\kappa\times 1}$ is the controller state. Several techniques, including convex interpolation, can be utilized to obtain a gain-scheduled controller from the set of LTI controllers $\mathcal{K} = \{K_1, \ldots, K_p\}$. In this chapter, the gain-scheduled controller is obtained by means of a recurrent neural network,

$$ANN : \begin{cases} \dot{\mathbf{x}}_K = \mathbf{A}_K(\boldsymbol{\zeta})\mathbf{x}_K + \mathbf{B}_K(\boldsymbol{\zeta})\Delta\mathbf{x} \\ \mathbf{u}_N := \mathbf{v}\ \boldsymbol{\Phi}(\mathbf{W}\mathbf{x}_a) \end{cases}, \tag{8.42}$$

with input $\mathbf{x}_a := [\boldsymbol{\chi}^T \quad \boldsymbol{\zeta}^T]^T$. Where, $\boldsymbol{\chi} := [\Delta\mathbf{x}^T \quad \mathbf{x}_K^T]^T \in \mathbf{R}^{\nu\times 1}$, and it can be seen from (8.40) that $\mathbf{x}_K$ is a function of $\mathbf{u}_N$ because, once the controller is implemented, $\Delta\mathbf{u} = \mathbf{u}_N$. Hereon, the control input is assumed to be scalar to simplify the presentation.

Then, using algebraic training [5], it can be shown that given a set of LTI controllers $\mathcal{K} = \{K_1, \ldots, K_p\}$ there exists an ANN controller (8.42) with $l = p$ sigmoidal nonlinearities that is input–output equivalent to (8.41) at the equilibria $\{\boldsymbol{\zeta}_1, \ldots, \boldsymbol{\zeta}_p\} \in \mathcal{X}_{\mathrm{LPV}}$, that is, satisfies the closed-loop requirements,

$$u_N(\boldsymbol{\zeta}_j) = u_K(\boldsymbol{\zeta}_j), \quad \text{and} \quad \frac{\partial u_N}{\partial \dot{\boldsymbol{\chi}}}(\boldsymbol{\zeta}_j) = \frac{\partial u_K}{\partial \dot{\boldsymbol{\chi}}}(\boldsymbol{\zeta}_j), \quad \text{for} \quad j = 1, \ldots, p \tag{8.43}$$

provided the ANN weights satisfy the algebraic equations,

$$\begin{aligned} \mathbf{N} &= \mathbf{W}_\zeta \mathbf{Z} \\ \mathbf{S}\mathbf{v}^T &= \mathbf{b} \\ \mathbf{D}\mathbf{V}\mathbf{W}_\chi &= \mathbf{M}_2 \end{aligned} \tag{8.44}$$

and provided the matrices $(\mathbf{I} - \mathbf{D}_j\mathbf{D}_{K_j})$ are invertible. The ANN inputs weights are partitioned into $\mathbf{W} = [\mathbf{W}_\chi \quad \mathbf{W}_\zeta]$ based on the corresponding inputs, $\mathbf{V} := \mathrm{diag}(\mathbf{v})$ and $\mathbf{b} := \tilde{b}\,\boldsymbol{\imath}_p$, where, $\boldsymbol{\imath}_p := 1_{p\times 1}$, $\tilde{b}$ is an arbitrary constant bias, and $\boldsymbol{\Phi}'(n) := [\sigma'(n_1)\cdots\sigma'(n_l)]^T$. The remaining matrices are defined in the Appendix. A proof is provided in [5], where the properties of the resulting ANN controller, including closed-loop stability, are also proven.

8.4.2 Constrained Adaptive Critic Design

The advantage of the ANN controller (8.42) over other gain-scheduled controllers is that it is nonlinear and adaptive. Therefore, an approximate dynamic programming (ADP) algorithm can be used to improve its performance online through incremental training. This may be necessary when the plant (8.38) operates outside of the LPV regime, that is, $\mathbf{x} \in \{\mathcal{X}/\mathcal{X}_{\mathrm{LPV}}\}$, and in the presence of parameter variations or unmodeled dynamics. In this case, the observed state $\hat{\mathbf{x}}$ differs from that estimated from the dynamic equation. Also, since the underlying assumptions are violated, the gain-scheduled controller's performance ceases to be optimal. However, if the weights of the ANN controller are updated via ADP using a conventional backpropagation algorithm, the algebraic equations in (8.44) may no longer be satisfied due to interference. Subsequently, when the plant returns to the LPV regime, the neural network controller will no longer meet the original LPV performance and stability guarantees. By viewing the algebraic equations (8.44) as the ANN LTM, it is possible to preserve the same guarantees of closed-loop performance and stability in $\mathcal{X}_{\mathrm{LPV}}$, while the ANN controller is adapted incrementally online via ADP. In this case, the ADP policy improvement routine and value determination operation provide STMs to be learned incrementally over time, subject to the LTM constraints in (8.44).

The state feedback control law is adapted by updating the weights of the recurrent neural network (8.42) through several iterations of the ADP algorithm that are indexed by k and take place over time. Then, letting the adaptive control law be defined as,

$$c_k[\mathbf{x}_a(t)] := \mathbf{v}^{[k]}\mathbf{\Phi}[\mathbf{W}^{[k]}\,\mathbf{x}_a(t)], \quad k = 0, 1, 2, \ldots \tag{8.45}$$

the weights can be determined by a constrained ADP approach that improves performance over time, while preserving the LTM. Where, $\mathbf{v}^{[k]}$ and $\mathbf{W}^{[k]}$ denote the values of $\mathbf{v}$ and $\mathbf{W}$ at the kth iteration of the ADP algorithm. The same control objectives used to specify the LPV performance, such as, H_∞ and H_2 performance, and pole placement, are used to express the desired system performance in $\mathcal{X}$. When the plant dynamics are nonlinear, equivalent control objectives can be formulated by means of an integral cost function,

$$J = \lim_{t_f \to \infty} \left\{ \frac{1}{t_f} \int_{t_0}^{t_f} \sum_{i=1}^{q} \boldsymbol{\xi}_i^T(\tau)\boldsymbol{\xi}_i(\tau)\mathrm{d}\tau \right\}, \tag{8.46}$$

where each control objective defines a quadratic cost index $\boldsymbol{\xi}_i$, which is formulated in terms of the plant state and control vectors. For example, both H_2 performance and pole placement can be expressed by a cost index $\boldsymbol{\xi}_i(t) = \mathbf{M}_i^{1/2}[\mathbf{x}_a^T(t) \quad \mathbf{u}^T(t)]^T$, through a symmetric weighting matrix $\mathbf{M}_i$ of design parameters [22, Section 6.3]. Then, the value function,

$$V_k[\mathbf{x}_a(t), c_k] := \int_t^{t_f} \sum_{i=1}^{q} \left[\mathbf{x}_a^T(\tau)\, c_k^T[\mathbf{x}_a(\tau)]\right] \mathbf{M}_i^{[k]} \left[\mathbf{x}_a^T(\tau)\, c_k^T[\mathbf{x}_a(\tau)]\right]^T \mathrm{d}\tau \tag{8.47}$$

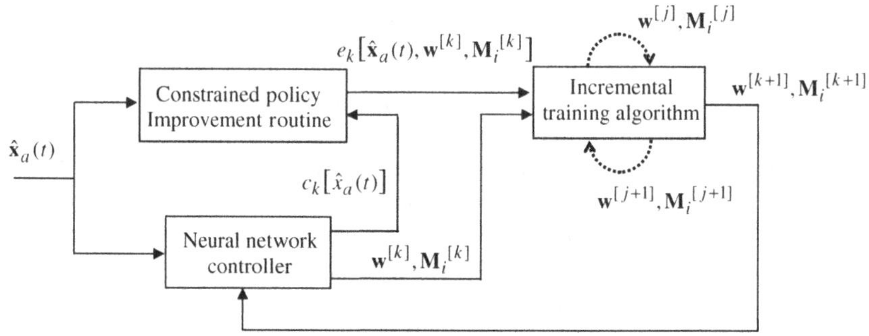

FIGURE 8.6 Algorithmic diagram of constrained policy improvement routine applied to the neural network controller (8.45).

can be updated through the ADP value-determination operation to learn the optimal cost-to-go for the actual system dynamics, where $\mathbf{M}_i^{[k]}$ denotes the value of $\mathbf{M}_i$ at the kth iteration of the ADP algorithm.

In the LPV regime, the values of the design matrices $\mathbf{M}_i$ are obtained using classical control techniques, such as implicit model following and dynamic compensation [22, Section 6.3], and (8.46) is optimized by the same values of $\mathbf{v}$ and $\mathbf{W}$ that satisfy (8.44). Therefore, at $k = 0$ all design parameters are set equal to these initial LPV values, obtained from (8.44), which hereon are denoted by $\mathbf{v}^{[0]}$, $\mathbf{W}^{[0]}$, and $\mathbf{M}_i^{[0]}$. Subsequently, if the plant leaves $\mathcal{X}_{\text{LPV}}$, or experiences parameter variations or unmodeled dynamics, the controller violates the optimality condition

$$\nabla_{\mathbf{u}(t)}\mathcal{L}[\mathbf{x}(t), \mathbf{u}(t)] + \nabla_{\mathbf{u}(t)} V_k[\mathbf{f}[\mathbf{x}(t), \mathbf{p}_m(t), \mathbf{u}(t)], c_k]\} = \mathbf{0}. \tag{8.48}$$

and sets off the ADP adaptation. The ADP adaptation optimizes the neural network control law (8.45) by cycling between the policy improvement routine and the value determination operation repeatedly over time, as illustrated in Figure 8.6.

The unconstrained policy improvement routine can be written as,

$$\begin{aligned} c_{k+1}[\mathbf{x}(t)] &= \arg\min_{\mathbf{u}(t)}\{\mathcal{L}[\mathbf{x}(t), \mathbf{u}(t)] + V_k[\mathbf{f}[\mathbf{x}(t), \mathbf{p}_m(t), \mathbf{u}(t)], c_k]\} && (8.49)\\ &=: \arg\min_{\mathbf{u}(t)}\{e_k(\mathbf{w})\}, \quad k = 0, 1, 2, \ldots. && (8.50)\end{aligned}$$

such that $V_k[\mathbf{x}(t), c_{k+1}] \leq V_k[\mathbf{x}(t), c_k]$, for $\forall \mathbf{x}(t) \in \mathcal{X}$. In order for the neural network controller to preserve its optimal performance in $\mathcal{X}_{\text{LPV}}$, a set of constraints obtained from (8.44) are adjoint leading to the following constrained optimization problem:

$$\begin{aligned} \text{minimize} \quad & E_k(\mathbf{w}) := \sum_{i=1}^{q} \hat{\boldsymbol{\xi}}_i^T(t)\hat{\boldsymbol{\xi}}_i(t) + V_k\left[\mathbf{f}\left[\hat{\mathbf{x}}(t), \mathbf{p}_m(t), c_k[\hat{\mathbf{x}}_a(t)]\right], c_k[\hat{\mathbf{x}}_a(t)]\right] \\ \text{subject to} \quad & \mathbf{g}(\mathbf{w}) := \begin{bmatrix} \mathbf{N} - \mathbf{W}_\zeta \mathbf{Z} \\ \mathbf{S}\mathbf{v}^T - \mathbf{b} \\ \mathbf{DVW}_\chi - \mathbf{M}_2 \end{bmatrix} = \mathbf{0}, \end{aligned} \tag{8.51}$$

where $\hat{\boldsymbol{\xi}}_i(t) := \mathbf{M}_i^{1/2}[\hat{\mathbf{x}}_a^T(t) \quad c_k^T[\hat{\mathbf{x}}_a(t)]]^T$ and $\hat{\mathbf{x}}_a(t)$ is the actual value of the augmented state observed at the present time t. The adjustable parameters $\mathbf{v}$ and $\mathbf{W}$ are rearranged into $\mathbf{w} \in \mathbf{R}^{N\times 1}$, which is the variable of the above constrained minimization problem, where, $\mathbf{Z}$, $\mathbf{b}$, $\mathbf{M}_2$, and $\mathbf{R}_2$ are known constants defined in Section 8.4.1 the Appendix. The matrices $\mathbf{N}$, $\mathbf{S}$, $\mathbf{D}$, and $\mathbf{V}$ are functions of $\mathbf{w}$, as defined in Section 8.4.1. Finally, the parameters $\mathbf{M}_i$, $i = 1, \ldots, q$, in (8.47) are updated according to the value determination operation,

$$V_{k+1}[\mathbf{x}(t), c_{k+1}] = \mathcal{L}[\mathbf{x}(t), \mathbf{u}(t)] + V_k\{\mathbf{f}[\mathbf{x}(t), \mathbf{p}_m(t), \mathbf{u}(t)], c_{k+1}\}. \tag{8.52}$$

Since in this case the equality constraint (8.51) may not satisfy the implicit function theorem, the method of Lagrange multipliers can be used to seek the solution of (8.51)–(8.51). In this method, reviewed in [22, pp. 36-41], the equality constraint (8.51) is adjoined to the function to be minimized (8.51) by defining an augmented error function

$$E_{a_k}(\mathbf{w}) \equiv e_k(\mathbf{w}) - \boldsymbol{\lambda}^T \mathbf{g}(\mathbf{w}). \tag{8.53}$$

The vector of Lagrange multipliers $\boldsymbol{\lambda}$ contains as many unknowns as there are equality constraints. As shown in [22, pp. 36-41], in the vicinity of an extremum of (8.53), $\boldsymbol{\lambda}$ takes the value,

$$\boldsymbol{\lambda}^* = -\left(\frac{\partial \mathbf{g}}{\partial \mathbf{w}}\right)^{-T} (\nabla_{\mathbf{w}} e_k)^T, \tag{8.54}$$

where $-T$ denotes the inverse transpose of a matrix. Then, the optimal values of $\mathbf{w}$ can be determined from,

$$\nabla_{\mathbf{w}} e_k - (\boldsymbol{\lambda}^*)^T \frac{\partial \mathbf{g}}{\partial \mathbf{w}} = \mathbf{0} \tag{8.55}$$

using a Newton–Raphson algorithm.

8.5 SUMMARY

Because of their ability to approximate nonlinear functions, and to generalize and learn from sampled data, adaptive ANNs have been able to solve a broad range of problems in the sciences and engineering, such as, differential equations, system identification, and control. But, as demonstrated by natural cognitive systems, it is only by learning tasks and information incrementally over time that intelligent systems can reach high levels of complexity and performance. The CPROP approach described in this chapter provides a rigorous computational framework for procedural memory formation in incremental training. As shown through a few sample applications, CPROP allows ANNs to solve adapt problem solutions repeatedly, through multiple training sessions, while retaining a desired baseline performance at all times. CPROP offers a unified view of memory formation and retention in ANNs that does not rely on presenting the same information to the ANN repeatedly over time. Thus, similarly to natural cognitive systems, CPROP ANNs are capable of retaining a wide variety of long-term memories, including memory of input–output patterns (function approximation), knowledge of a system's behavior (system ID), BCs (differential equations), and optimal motor and control skills, while learning new skills or short-term memories through new training sessions.

APPENDIX: ALGEBRAIC ANN CONTROL MATRICES

$$\mathbf{M}_{1_j} := (I - D_j D_{K_j})^{-1} \begin{bmatrix} C_j & D_j C_{K_j} \end{bmatrix} \tag{A.1}$$

$$\mathbf{M}_{2_j} := \begin{bmatrix} 0 & C_{K_j} \end{bmatrix} + D_{K_j} \mathbf{M}_{1_j} \tag{A.2}$$

$$\mathbf{M}_2 := \begin{bmatrix} \mathbf{M}_{2_1}^T \cdots \mathbf{M}_{2_p}^T \end{bmatrix}^T \tag{A.3}$$

$$Z := [\zeta_1 \cdots \zeta_p] \tag{A.4}$$

$$N := \begin{bmatrix} n^1 \cdots n^p \end{bmatrix} \tag{A.5}$$

$$S := \begin{bmatrix} \mathbf{\Phi}(n^1) \cdots \mathbf{\Phi}(n^p) \end{bmatrix}^T \tag{A.6}$$

$$D := \begin{bmatrix} \mathbf{\Phi}'(n^1) \cdots \mathbf{\Phi}'(n^p) \end{bmatrix}^T . \tag{A.7}$$

REFERENCES

1. M. McCloskey and N. Cohen. Catastrophic interference in connectionist networks: the sequential learning problem. *The Psychology of Learning and Motivation*, 24(109–164), 1989.
2. G.A. Carpenter. Neural network models for pattern recognition and associative memory. *Neural Networks*, 2(4):243–257, 1989.
3. A. Zhao. Global stability of bidirectional associative memory neural networks with distributed delays. *Physics Letters*, 297(3–4):182–190, 2002.
4. R. Polikar, L. Udpa, S.S. Udpa, and V. Honavar. Learn++: An incremental learning algorithm for supervised neural networks. *IEEE Transactions on Systems, Man, and Cybernetics - Part C*, 31(4):497–508, 2001.
5. S. Ferrari. Multiobjective algebraic synthesis of neural control systems by implicit model following. *IEEE Transactions on Neural Networks*, 20(3):406–419, 2009.
6. J. Mándziuk and L. Shastri. Incremental class learning—an approach to longlife and scalable learning. In *Proceedings of the International Joint Conference on Neural Networks*. Washington DC, 1999.

7. S. Ferrari and R.F. Stengel. Smooth function approximation using neural networks. *IEEE Transactions on Neural Networks*, 16(1):24–38, 2005.

8. G. DiMuro and S. Ferrari. A constrained-optimization approach to training neural networks for smooth function approximation and system identification. In *Proceedings of the IEEE World Congress on Computational Intelligence*, pp. 2354–2360, Hong Kong, China, June 2008.

9. I.E. Lagaris, A. Likas, and D.I. Fotiadis. Artificial neural networks for solving ordinary and partial differential equations. *IEEE Transactions on Neural Networks*, 9(5):987–1000, 1998.

10. I.E. Lagaris, A. Likas, and D. Papageorgio. Neural-network methods for boundary value problems whith irregular boundaries. *IEEE Transactions on Neural Networks*, 11(5):1041–1049, 2000.

11. L. Jianyu, L. Siwei, Q. Yingjian, and H. Yaping. Numerical solution of elliptic partial differential equation using radial basis function neural networks. *Neural Networks*, 16(5–6):729–734, 2003.

12. Y. Shirvany, M. Hayati, and R. Moradian. Numerical solution of the nonlinear schrodinger equation by feedforward neural networks. *Communications in Nonlinear Science and Numerical Simulation*, 13(10):2132–2145, 2008.

13. K.S. McFall and J.R. Mahan. Artificial neural network method for solution of boundary value problems with exact satisfaction of arbitrary boundary conditions. *IEEE Transactions on Neural Networks*, 20(8):1221–1233, 2009.

14. S. Ferrari and M. Jensenius. A constrained optimization approach to preserving prior knowledge during incremental training. *IEEE Transactions on Neural Networks*, 19(6):996–1009, 2008.

15. S. Ferrari and R.F. Stengel. On-line adaptive critic flight control. *Journal of Guidance, Control, and Dynamics*, 27(5):777–786, 2004.

16. S. Ferrari and R.F. Stengel. Model-based adaptive critic designs. In J. Si, A. Barto, and W. Powell, editors. *Learning and Approximate Dynamic Programming*. John Wiley and Sons, 2004.

17. S. Ferrari and R.F. Stengel. Classical/neural synthesis of nonlinear control systems. *Journal of Guidance, Control, and Dynamics*, 25(3):442–448, 2002.

18. F.-L. Lewis, S. Jagannathan, and A. Yesilderek. *Neural Network Control of Robot Manipulators and Nonlinear Systems*. Taylor and Francis, London, England, 1999.

19. S. Limanons and J. Si. Neural-network-based control design: An LMI approach. *IEEE Transactions on Neural Networks*, 9(6):1422–1429, 1998.

20. R. Rysdyk and A.J. Calise. Robust nonlinear adaptive flight control for consistent handling qualities. *IEEE Transactions on Control Systems Technology*, 13(6):896–910, 2005.

21. M.T. Rosenstein and A.G. Barto. Supervised actor-critic reinforcement learning. In J. Si, A. Barto, and W. Powell, editors. *Learning and Approximate Dynamic Programming*. John Wiley and Sons, 2004.

22. R.F. Stengel. *Optimal Control and Estimation*. Dover Publications, Inc., 1986.

CHAPTER 9

Toward Design of Nonlinear ADP Learning Controllers with Performance Assurance

JENNIE SI,[1] LEI YANG,[2] CHAO LU,[3] KOSTAS S. TSAKALIS,[1] and ARMANDO A. RODRIGUEZ[1]

[1]School of Electrical, Computer, and Energy Engineering, Arizona State University, Tempe, AZ, USA
[2]College of Information and Control Science and Engineering, Zhejiang University, Hangzhou, China
[3]Department of Electrical Engineering, Tsinghua University, Beijing, China

ABSTRACT

This chapter discusses nonlinear control system design using approximate/adaptive dynamic programming (ADP). ADP algorithms based on learning and approximation have shown great promise to reduce the curses of dimensionality suffered by dynamic programming (DP). They benefited from the design techinques of artificial neural networks and other function approximators, which have developed principled ways for universal function approximation. Direct heuristic dynamic programming (HDP) was introduced as an on-line learning control scheme inspired by adaptive critique designs, a family of ADP algorithms. Applications of the direct HDP to large and complex problems have demonstrated the feasibility and scalability of the learning controller design. The results, such as Apache helicopter control and coordination of large power networks for damping low-frequency oscillation, are encouraging and promising as proof of concepts toward scalable ADP designs, however, real controllers demand performance assurances, not merely a statistical learning success rate indicating that most of the time the controller works. With this in mind, this chapter discusses some recent developments in this direction.

Reinforcement Learning and Approximate Dynamic Programming for Feedback Control, First Edition. Edited by Frank L. Lewis and Derong Liu.

9.1 INTRODUCTION

Dynamic programming (DP) is an approach to computing optimal control policies over time for systems in the general case of nonlinearity with random disturbances by using the principle of optimality. This basic principle has allowed for breaking down a complex optimization over time problem into subproblems, which can be solved by recursion. The simplicity of the principle and its potential applicablity to complex problems truly is appealing. However, the algorithm becomes elusive in reality since the computational efforts and the data storage requirement increase exponentially with the dimensionality of the system, which are reflected in the three curses from the state space, the observation space, and the action space. Because of this, traditional DP approaches have been limited to solving small-size problems.

Approximate dynamic programming (ADP) [1] was proposed to address these issues by using approximation and/or learning methods in a DP problem centered on solving the Bellman equation [2]. The main idea is to approximate DP solutions by using function approximation structures such as neural networks to approximate the cost function as well as the control policy. Therefore, neural networks or any other universal function approximators are building blocks in approximation and learning-based ADP. The function approxmation objectives are formulated according to the principle of optimality. Therefore, the quality or the optimality of the ADP solution has a direct correlation with that of the function approximator.

Several early results attempted to obtain approximate solutions of the Hamilton–Jacobi–Bellman (HJB) equation [3]. There also exist a large literature on ADP designs with convergence guarantees under a discrete-time finite-state and finite-control Markov decision process (MDP) framework [4]. Reinforcement learning techniques [5] as well as adaptive critics [6, 7] opened new opportunities by introducing approximating functions into the iterative process of updating the value function estimate, and sometimes the control policy. They become popular and have shown great promise to address the curse of dimenationality.

The temporal difference (TD) method by Sutton [5] has introduced a new way of approximating the value function by prediction based on TDs. The method is effective for approximating the value function, however, it does not provide a proposal to address problems with large control spaces. Heuristic dynamic programming (HDP) [6, 7], one of the adaptive critic design techniques, created a whole new way of computing both the value function and the optimal control by approximations. With this new idea, the curse of dimensionality is finally, and in a principled way, possible to be challenged. To actually compute the optimal solutions efficiently and reliably, Werbos proposed additional adaptive critic structures [6, 7]: dual heuristic programming (DHP), and globalized DHP. As the structures become more complex, the design aims at accommodating more complex terms in the computation of optimal control. Applications of these designs to a single control simulated aircraft landing problem also concur that the more complex the adaptive critic techniques, the tighter the control results are [8]. Results of adaptive critic optimal control for a missile guidance problem [9] also illustrated the potential of adaptive critiques for continuous state and continuous control problems. Additional illustrative examples of adaptive critic designs include power system excitation and stability control problems [10]. Given those simulation-based results of adaptive critic designs, especially its great promise to scalable applications including continuous state and continuous control problems, it is desirable to develop learning control algorithms with some performance guarantees.

Direct HDP was introduced in Ref. [11]. It takes advantage of the potential scalability of the adaptive critic designs and the intuitiveness of the actor-critic design in Ref. [12]. It is an on-line learning scheme that simultaneously update the value function approximation and the action network approximation. It has been applied to some realistic and complex control problems that involve continuous states and control. In Refs. [13–15], the direct HDP controller was designed to stabilize, maneuver, and reconfigure after fault an Apache helicopter using an industrial scale Apache simulator. In another report [16], the direct HDP was used to provide coordinated control for damping low-frequency oscillation in a large power system.

Previous case studies are encouraging in the sense that adaptive critic designs and the direct HPD design have shown their potential in addressing large scale, realistic control problems. However, challenges remain for the current ADP algorithms, including adaptive critics and direct HDP, and algorithms of similar nature [17]. First and foremost, how to achieve efficient learning in ADP such that learning controller parameters converge and converge fast to an meaningful equilibrium? How to guarantee to meet the various performance measures in time and frequency domain of the control system at the equilibrium point of the learning controller? Answering these questions are necessary and essential to the success of the ADP algorithms. In the following, we provide some of our preliminary results in these regards for the direct HDP.

9.2 DIRECT HEURISTIC DYNAMIC PROGRAMMING

Central to DP is the solution to the Bellman equation, which is the value function. It describes the optimal cost-to-go for a dynamic process with an associated cost function. The underlying optimality principle decomposes a dynamic optimziation problem into recursive subproblems. However, solving the recursive equation for optimal cost-to-go and consequently the optimal control is not as easy as the equation appears to be.

A key idea in the family of adaptive critique designs [6, 7] is to tune the parameters in the value function approximator by the residual of the Bellman equation. Another feature of this class of ADP designs is co-adaptation of the control output and the value function. Because of this functional approximation-based approach, it is possible for these ADP controls to potentially scale up to solve large-size problems, including continuous state and control problems. The direct HDP is also based on these basic principles. Figure 9.1 is a schematic diagram of the direct HDP, which was first introduced in Ref. [11] as an on-line learning control paradigm. In direct HDP, there are two major approximation blocks implemented using neural networks, the critic network, and the action network. Note however that, other universal approximators

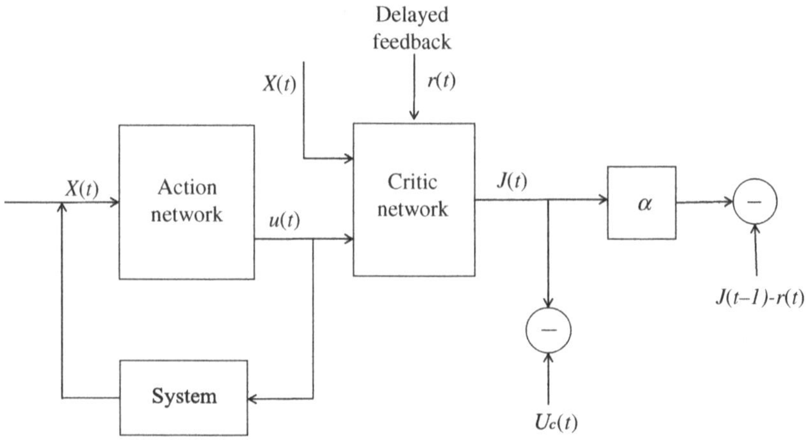

FIGURE 9.1 Schematic diagram of the direct HDP.

can also be used in place of neural networks. The critic network output is an estimate of the optimal-cost-to-go or the value function. The critic network learning is tied into the Bellman optimality principle. The action network is coupled with the critic network by providing a control input to the critic and by constructing a state feedback control law via backpropagated residual of the value function approximation. The reinforcement signal $r(t)$ is provided from the external environment. It can be a less informative form of feedback with only a binary reprentation signifying, for example, "success" or "failure," "on" or "off," and "good" or "bad." As such, $r(t)$ may be considered, for example, a final equilibrium of a controled dynamic process, or a delayed feedback from the system. When a desired final state of the dynamic process is reached, $r(t)$ may be given a certain designated numeric value. Otherwise, $r(t)$ toggles its value state to the opposite. In the development of optimal control theory, $r(t)$ as a performance measure usually is more informative than a binary indicator. For example, it can be directly associated with the energy consumption of both the state variable as well as the control variable, namely, $x^2(t)$ and $u^2(t)$ where $x(t)$ and $u(t)$ are the state and control varialbes at time t, respectively.

The output of the critic element, the J function, approximates the discounted total cost-to-go $R(t)$, that is,

$$R(t) = r(t+1) + \alpha r(t+2) + \cdots, \tag{9.1}$$

where α is a discount factor for the infinite-horizon problem and it is usually that $0 < \alpha < 1$.

The direct HDP can be used for either on-line or off-line learning. During on-line learning, the controller initially is "naive," namely, the action network and the critic network are both randomly initialized in their weights/parameters. Once a system state is observed, an action will be subsequently produced based on the parameters in the action network. A better control value under the specific system state will reduce the residual of the Bellman optimality equation. This set of system operations will be reinforced through memory or association between states and control outputs in the action network. Otherwise, the control value will be adjusted through tuning the weights in the action network.

Training the action and the critic network starts with well-defined objective functions for the two networks, which are formulated to minimize the residual in the Bellman equation as least square problems.

Specifically, the prediction error for the critic element as

$$e_{\mathrm{c}}(t) = \alpha J(t) - [J(t-1) - r(t)], \tag{9.2}$$

and the objective function to be minimized in the critic network is

$$E_{\mathrm{c}}(t) = \frac{1}{2} e_{\mathrm{c}}^2(t). \tag{9.3}$$

The principle in adapting the action network is to indirectly back-propagate the error between the desired ultimate objective, denoted by U_{c}, and the approximate J function from the critic network. For example, if we designate "0" as the reinforcement signal for "success," U_{c} is set to "0" in implementation. In the action network, the state measurements are used as inputs to create a control as the output of the network. In turn, the action network can be implemented by either a linear or a nonlinear network, depending on the complexity of the problem. The objective function for training the action network can be formulated as follows. Let

$$e_{\mathrm{a}}(t) = J(t) - U_{\mathrm{c}}(t). \tag{9.4}$$

The weights in the action network are updated to minimize the following performance error measure,

$$E_{\mathrm{a}}(t) = \frac{1}{2} e_{\mathrm{a}}^2(t). \tag{9.5}$$

In Ref. [11], a gradient descent algorithm was developed for training the two multilayer perceptron neural networks to meet the two objectives as described in (9.3) and (9.5), respectively. However, the approximation structures or the approximators as well as the associated training algorithms are not only limited to those mentioned earlier.

9.3 A CONTROL THEORETIC VIEW ON THE DIRECT HDP

Classical control designs start with meeting performance measures in mind. System performance specifications such as stability, disturbance rejection, rise time and overshoot, gain and phase margin, and bandwidth are commonly used in feedback controller evaluation. Elegant robust control design methodologies with thorough analysis on controller performances are widely available. Modern control system designs aim at controlling complex, nonlinear, multiple input, and multiple output systems. Similar design performance requirements are still desired but however, they are not as easily and as directly represented in controller parameters during design as the case for single input single output system. For example, in linear quadratic regulator (LQR) designs, choosing the two weighing matrices on states x and control u is not straight forward. Under a linear system setting, the two matrices are required to be positive definite to ensure the existence of solutions to the algebraic Riccati equation to formulate the linear state feedback law. The usual performance measure in LQR attempts to strike a balance between the rate of convergence of the control solutions and the cost of the control. Some inappropriate choices of the two matrices can result in poor, for example, oscillatory controllers.

For a learning controller designed to meet the two network training objectives (9.3) for the critic and (9.5) for the action network, respectively, there is no performance guarantee in the traditional sense for rise time, overshoot, bandwith, and other measures to be within desired bounds. To gain insight on the issue to improve learning controller design, in the following, a detailed quantiative performance analysis of direct HDP learning controller is presented where an LQR controller is used in the backdrop as a reference. Specificially, we make use of the cart-pole balancing benchmark as an example to demonstrate practical approaches to assessing the quality of a controller. Actually, these approaches of using control theoretic analysis and its associated performance measures are applicable to other general nonlinear learning control problems beyond the cart-pole example.

9.3.1 Problem Setup

Under a general problem setting, a learning controller can be designed based on a delayed feedback, which is a more general or less constrained performance measure than, for example, that used in LQR. As a result, the instantaneous feedback signal accessible to the control agent is incomplete. Take, for example, the cart-pole problem. The feedback to the control agent is only qualitative indicating the pole is in one of the two possible states: up or down, not explicit as for how far it is from the perfect straight-up position, nor other constraints related to its angular velocity or acceleration.

The learning controller design based on this incomplete feedback does learn to keep the pole in the up state [11]. Without time and frequency domain performance measures expressed in the learning objective, what characteristics would a learning controller exhibit under disturbances? What about its response time and settling time? In the following, we demonstrate how to make use of control theoretic analysis to characterize a learning controller.

We use the same cart-pole problem as the one in Ref. [11] to carry on our discussions. Direct HDP designs were implemented using nonlinear feedforward neural networks with one-layer hidden neurons in both the action and the critic networks. The direct HDP had no prior knowledge about the plant nor about a stabilizing controller. In this sense, the direct HDP was "naïve." A binary reinforcement signal $r(t)$ was used, provided from the external environment, which was either a "0" or a "-1" corresponding to "success" or "failure," respectively. Specificially, $r(x(t)) = -1$ corresponds to either the case of the angle (θ) falling outside the range $[-12°,\ 12°]$

and/or the case of the cart position (z) being beyond the range of $[-2.4,\ 2.4]$ meters in reference to the central position on the track; $r(x(t)) = 0$ corresponds to the case within the designated bounds. In all simulations, the initial state of the pole angle (θ) was set randomly between $[-0.5, 0.5]$ degrees, one run consists of a maximum of 50 consecutive trials, where a trial is a complete process for the pendulum from start to fall. A run was considered successful if the last trial (trial number less than 50) of the run had lasted 600,000 time steps. Otherwise, if the controller was unable to learn to balance the cart-pole within 50 trials (i.e., none of the 50 trials had lasted over 600,000 time steps), then the run was considered unsuccessful.

Given this, the control system performance analysis of the direct HDP controller evolves in two phases, namely the "learning" and "evaluation" phases. During learning, the randomly initialized HDP controller normally encounters several failures before it learns to stabilize the system according to the failure conditions. After each failure, the network weights of the critic and action networks are retained as the initial weights for the next trial. During the evaluation phase, the learned HDP controller is evaluated using the control theoretic sensitivity analysis framework discussed below.

To develop a general understanding of the direct HDP controller performance, different initial (random) weights were selected for both the action and the critic networks. Figure 9.2 depicts a typical direct HDP learning process by showing sample

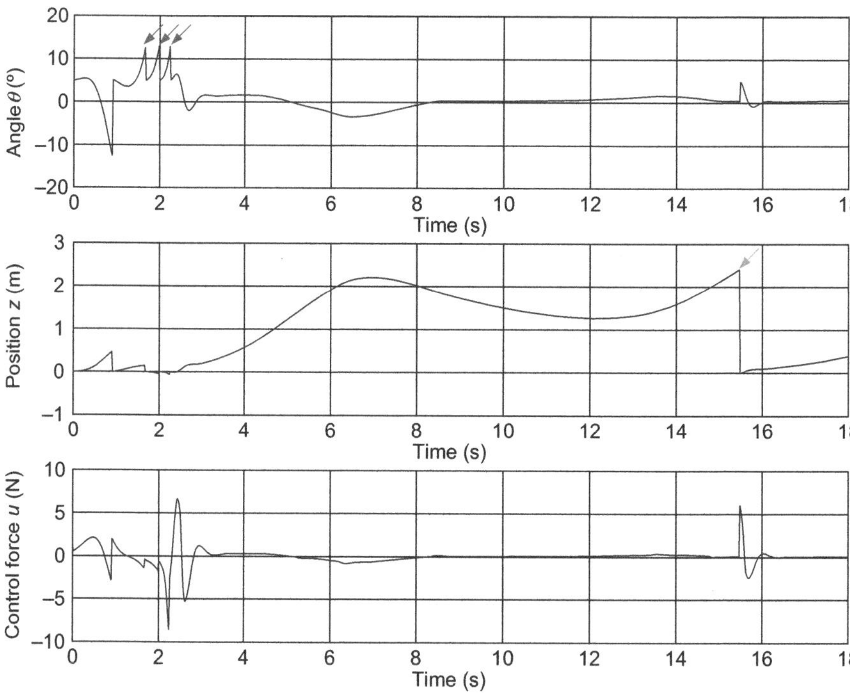

FIGURE 9.2 Trajectories of a typical learning control run of the direct DP to balance the cart-pole system.

trajectories of the pole angle θ, the cart position z, and the corresponding applied control force u. The three arrows (aligned approximately with the time at 2 s) in the plot indicate the positions where failures occurred due to the pole angle being out of range ($|\theta| > 12°$); while the single arrow (aligned approximately with the time below 16 s) points to a failure due to the cart reaching the track boundary ($|z| > 2.4m$). In this specific experiment, the HDP controller learned to balance the cart-pole system from its naïve state after the above four initial failures. Note that different sets of initial weights used in the action and the critic networks, as well as different learning parameters used in the training algorithms, result in different controllers. Simulation results show that, for every run initialized from random weights in the networks, the direct HDP controller had always been able to successfully learn to balance the cart-pole within 50 trials.

9.3.2 Frequency Domain Analysis of Direct HDP

Figure 9.3 illustrates the schematic for the proposed analytical framework, where P represents the plant. The controller block is used to represent the possible control laws to be studied. Also, d_x, d_u represent an exogenous, additive disturbances associated with system states x and the plant input u_p, respectively; y_r denotes an exogenous reference command for output y; and $e = y_r - x$ denotes the command tracking error.

For the direct HDP, the proposed analysis is based on linearizing the closed loop system about the origin (i.e., cart motionless with pole upright). Linearization replaces the action network by a linear map of the following form $u = -K(x - r_x + d_x)$, where $K = [\ k_1\ \ k_2\ \ k_3\ \ k_4\ ] \in \mathcal{R}^{1\times 4}$ is a constant state feedback gain matrix. Given the above, the following closed loop sensitivity maps are examined for the direct HDP:

$$S_u = 1 - K(sI - A + BK)^{-1}B \qquad (d_u \to u_p) \quad (9.6)$$

$$S_z = 1 - [\ 0\ \ 0\ \ 1\ \ 0\ ](sI - A + BK)^{-1}Bk_3 \qquad (x_{rz} \to [x_{rz} - z]) \quad (9.7)$$

$$S_\theta = 1 - [\ 1\ \ 0\ \ 0\ \ 0\ ](sI - A + BK)^{-1}Bk_1 \qquad (d_\theta \to [\theta + d_\theta]). \quad (9.8)$$

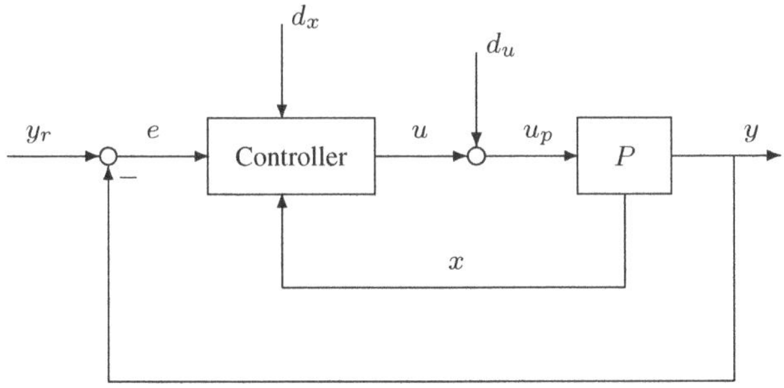

FIGURE 9.3 A control theoretic analysis setup.

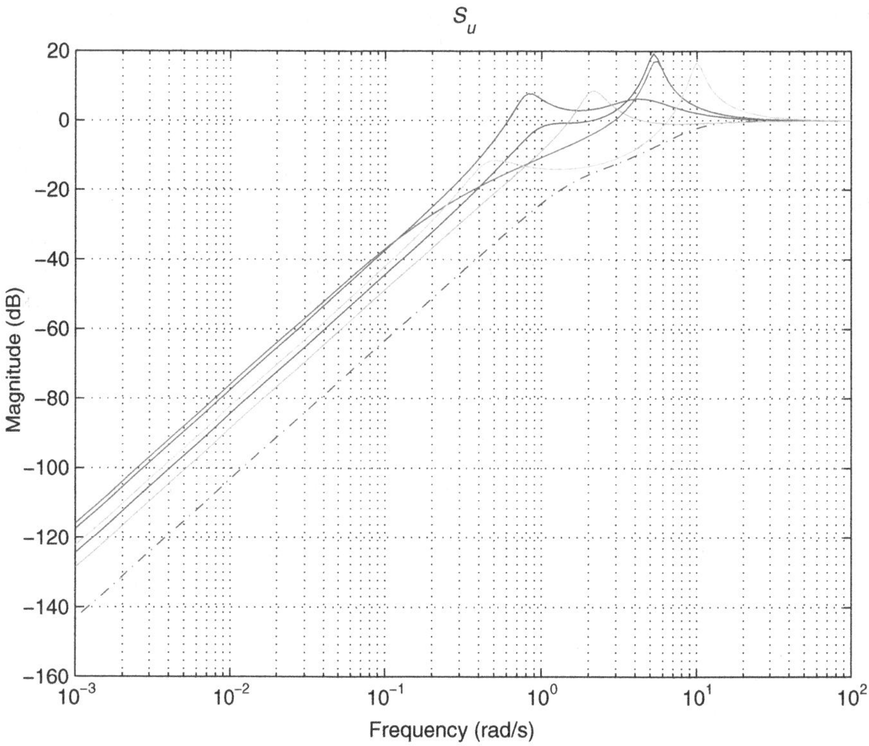

FIGURE 9.4 Sensitivity (S_u) plots for five direct HDP controller realizations (solid) corresponding to five different action network weight initializations.

The corresponding complementary sensitivity maps are given by

$$T_u = 1 - S_u \qquad (d_u \to u) \tag{9.9}$$

$$T_z = 1 - S_z \qquad (x_{rz} \to z) \tag{9.10}$$

$$T_\theta = 1 - S_\theta \qquad (d_\theta \to -\theta). \tag{9.11}$$

To carry out the sensitivity analysis, the closed loop system, consisting of the cart-pole and the direct HDP stabilizing controller, is linearized about the equilibrium $x_e = [\ 0\ \ 0\ \ 0\ \ 0\ ]^{\mathrm{T}}$ (pole upright position). The resulting sensitivities S_u and S_z were plotted in Figures 9.4 and 9.5 using solid lines to facilitate comparison with the LQR design. Note that the multiple solid lines signify different direct HDP designs, each corresponding to one design after the learning phase with randomly initialized network weights of both action and critic networks. The magnitude responses in Figures 9.4 and 9.5 show that other than some peaking and bandwidth differences, direct HDP (solid lines) has resulted in closed loop sensitivity characteristics similar to LQR (dashed line). Peaking in some of these direct HDP designs, however, are undesirable. This is caused by lightly damped closed loop poles resulting from certain sets of

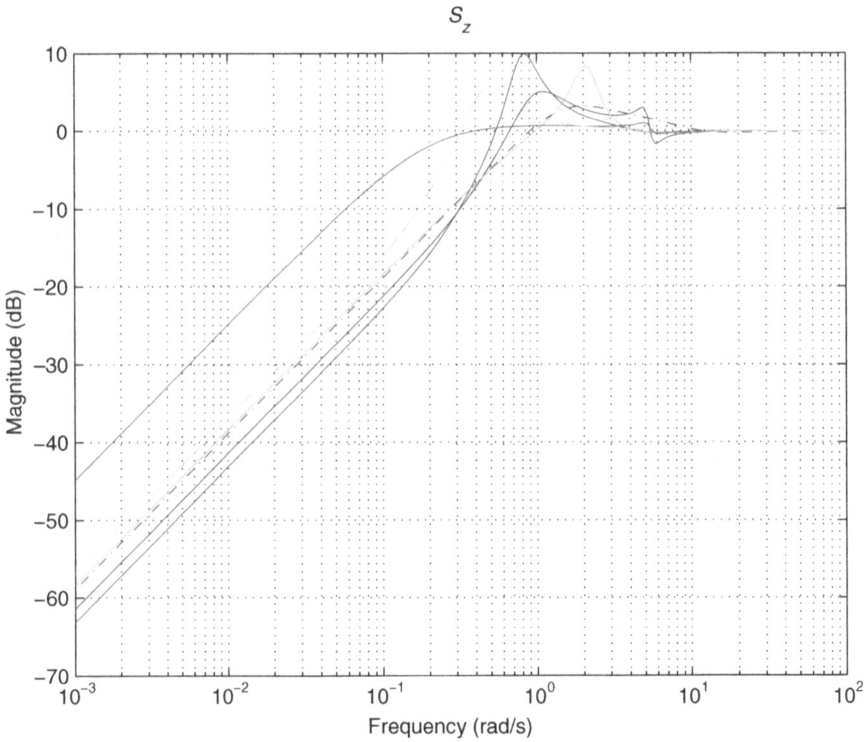

FIGURE 9.5 Sensitivity (S_z) plots for five direct HDP controller realizations (solid) corresponding to five different action network weight initializaations.

initial weights in the network. Also, it is very noticeable that the direct HDP with randomly initialized weights can potentially result in controllers with a large variety of closed loop performance characteristics. This is not surprising since the direct HDP was only based on a very weak performance measure, essentially reflecting a simple stabilization objective. It is therefore desirable for the learning controller to have additional closed loop sensitivity properties. They include small sensitivity magnitude at low frequencies and a reduction in excessive peaking. The former implies good disturbance attenuation, whereas the latter is associated with robustness of stability and low resonance effects in the response. Sensitivity measures of S_{LQ} and T_{LQ}, and correspondingly S_u and T_u, can be designed to possess "nice" properties, such as (1) infinite upward gain margin or robustness with respect to arbitrary upward (simultaneous and independent) gain perturbations in each control channel; (2) at least −6 dB downward gain margin or robustness with respect to arbitrary downward (simultaneous and independent) gain perturbations in each control channel; and (3) at least ±60° phase margin or robustness with respect to arbitrary (simultaneous and independent) phase perturbations in the range (−60°, 60°) within each control channel, which implies good low frequency command following as well as good low frequency disturbance attenuation. It should be noted that, other closed loop maps (e.g., S_z, T_z, S_θ, and T_θ) may not possess these properties. Typically, a designer must trade-off good properties at different loop breaking points. For general multivariable plants, such trade-offs may be exacerbated by poor conditioning of plant singular values.

9.3.3 Insight from Comparing Direct HDP to LQR

During our control theoretic analysis, we used either the LQR gain controller obtained via linear system analysis or the nonlinear direct HDP controller obtained via learning into the closed loop system shown in Figure 9.3 as a feedback controller. The system was then linearized about the origin, which was far away from the control boundary. This analysis is therefore local with the above-mentioned constraints on the control. Additional testing will be provided for nonlinear controllers beyond the linear region.

For comparison purposes, the LQR control design method was applied to the cart-pole system using the following design parameters of Q and R and thus the respective infinite horizon quadratic cost functional is given by

$$J(u) = \int_0^{\infty} \left[4\,\theta^2(t) + 0.25\, z^2(t) + 0.001\, u^2(t) \right] \, \mathrm{d}t. \tag{9.12}$$

The resulting (unique) optimal feedback control gain is given by

$$K = [\ 111.7706 \quad 23.7900 \quad 15.8114 \quad 18.0419\ ]\,. \tag{9.13}$$

Using the data provided above for the LQR design, closed loop sensitivity plots S_u and S_z are obtained and provided in Figures 9.4 and 9.5 (dashed lines), respectively: S_u lies below 0 dB as expected, S_z guarantees −20 dB command following for sinusoidal commands below $\omega \approx 0.09$ rad/s. With slight peaking, S_z is very acceptable.

Conversely, to further analyze the possible relationships between the direct HDP and LQR design, one specific direct HDP design, corresponding to one set of initialization of the weights in action and critic networks, is selected from its class after the learning phase. As shown in Figure 9.4, the input sensitivity plot has a nice closed-loop shape with maximum not exceeding 0 dB (i.e., $\max_{\omega \geq 0} |S_u(j\omega)| \leq 1$). The direct HDP design implemented as nonlinear multilayer network is then linearized near the unstable equilibrium $\tilde{x}_e = \{0, 0, 0, 0\}$, written as $u = -K_{\mathrm{HDP}}x$. The LQR design method is then applied to approximate the linearized direct HDP controller (K_{HDP}) with an LQR controller (K_{LQR}).

The linearization of the direct HDP controller near the system's unstable equilibrium x_e is performed and the following linear feedback gain

$$K^*_{\mathrm{HDP}} = [382.27 \quad 58.32 \quad 42.22 \quad 38.31] \tag{9.14}$$

is obtained.

Next, we examine the existence of an LQR controller by determining appropriate design parameters $\tilde{Q}$ and $\tilde{R}$ that yield an LQR state feedback control gain matrix

$\tilde{K}_{\mathrm{LQR}}$ that minimizes the Euclidean norm:

$$\min_{\tilde{Q},\tilde{R}} ||K^*_{\mathrm{HDP}} - \tilde{K}_{\mathrm{LQR}}(\tilde{Q}, \tilde{R})||_2. \tag{9.15}$$

Such optimization resulted in

$$\tilde{K}_{\mathrm{LQR}} = [382.24 \quad 58.31 \quad 42.21 \quad 38.30]. \tag{9.16}$$

with the corresponding $\tilde{Q}$ and $\tilde{R}$ satisfying the standard LQR detectability, symmetry, and sign-definiteness constraints [18]. In comparing the closed loop sensitivity maps $S_u(K^*_{\mathrm{HDP}})$ and $S_u(\tilde{K}_{\mathrm{LQR}})$, one finds that: $||S_u(K^*_{\mathrm{HDP}}) - S_u(\tilde{K}_{\mathrm{LQR}})||_\infty = 8.59 \times 10^{-7}$.

Results show that even though the direct HDP designs typically take on a more relaxed form of cost objective than the LQR quadratic cost, it is possible that the direct HDP controller can actually be linearly approximated by an LQR design with parameters Q and R satisfying all the necessary design constraints of LQR.

9.4 DIRECT HDP DESIGN WITH IMPROVED PERFORMANCE CASE 1—DESIGN GUIDED BY *A PRIORI* LQR INFORMATION

In Section 9.3.3, it is shown that direct HDP designs based on delayed feedback exhibit a wide spectrum of close loop characteristics due to their diverse converging states of the network weights in the learning controller. This is not surprising since only a simple reward functional was used in the controller performance specification. This, on the other hand, has created an opportunity for ADP algorithms to be applied to a much wider class of dynamic optimization problems without instantaneous feedback or reward information for each decision and control state. Nonetheless, it is highly desirable to obtain learning controllers with more specific closed loop properties. Toward this goal, this section and Section 9.4.1 are dedicated to introduce two design approaches that aim at improving controller performance by integrating a priori information into the learning controller design and show how much it impacts on the overall controller performance.

9.4.1 Direct HDP Design Guided by *a Priori* LQR Information

We first provide a set of approximation results from four direct HDP designs. They are direct HDP with randomly initialized action and critic networks; with LQR-guided action network; with LQR-guided critic network; and with LQR-guided action and critic networks. One approach will be used as *a priori* information to guide the direct HDP learning.

The implementation procedure for the above approaches is as follows: (1) By appropriately choosing the design parameters Q and R based on the system linearization about the equilibrium x_e, an LQR constant gain state feedback controller is selected with desirable closed loop sensitivity properties. (2) The off-line training data, including the measurements of the state and control variables, are collected by applying the LQR controller to the dynamic system under all possible operating conditions. (3) The pre-training of the critic network is performed on a stand-along neural network, starting with the random initial weights and the data collected during the previous step. The inputs of the critic network are the system states $x(t)$ and current control signal $u(t)$ from the off-line data; and the output is the overall objective value $J(t)$. The training objective is to minimize $||J(t) - x(t)^T Sx(t)||_2$, in which S is the unique solution of the algebraic Riccati equation. (4) The pre-training of the action network is implemented by a stand-along neural network, conducted to approximate the LQR control gain matrix K_{LQR} using the data collected in the previous step 2). The inputs of the action network are the system states $x(t)$ from the off-line data; and the output is the corresponding control signal $u(t)$. The training objective is to minimize $||u(t) - (-K_{\text{LQR}}x(t))||_2$. (5) Training of the critic network and the action network concludes when the mean square validation error drops below a specified threshold (e.g., 10^{-6}) at which point the network weights are saved. The resulting weights are then used to initialize the direct HDP on-line learning/design process.

For each of the four HDP implementations, 15 runs were conducted with respective initialization conditions. The control performance analysis was then conducted to evaluate the learning controller.

The above experiments resulted in sensitivity functions S_u and S_z that can give us a collective impression of the four designs. It is comforting to know from the sensitivity plots [17] that S_u is small at low frequencies, and near unity at high frequencies while $T_u = 1 - S_u$ is near unity at low frequencies and small at high frequencies. These imply that low frequency input disturbances d_u should result in a small (steady state) plant input u_p. Similarly, S_z and T_z we conclude that low-frequency position commands x_{rz} should result in a small (steady state) tracking error $x_{rz} - z$.

It is also noted that some direct HDP implementations exhibit more peaking than others and more variations in the resulting system characteristics than others. Specifically, when both the action and critic networks are randomly initialized, the variations in the plots are quite substantial. This is expected as discussed in Section 9.4. When only the critic network is initialized with supervision from LQR controller, the variations in the plots remain as large as the randomly initialized. It appears that knowledge of the optimal cost has not helped much. This may be due to the fact that the ever changing weights in the action network, and hence the action network outputs, have adversely affected the weights in the critic network. Even though the critic network has been previously trained off-line, the on-line updating in the action network has caused the un-learning of the critic network. This argument is also partially supported by the next set of experiments where only the action network was initialized by LQR guidance. The variations in system transfer functions decrease substantially. That is, the plots have become closer to the associated LQR plots than the previous two sets of experiments. The improvement is even more significant when both the action and critic networks are initialized by LQR guidance. The variations in sensitivity become almost negligible. That is, the plots are almost identical to those for the associated LQR design—with little peaking. This suggests that knowledge of a good controller and the optimal cost (i.e., a Lyapunov function) may significantly enhance the learning process.

The direct HDP controller time domain performance was also evaluated. During this phase of the study, the L_∞ measurements for the pole angle θ and cart position z were collected from 15 independent runs. The weights in the action and the critic networks were initiated accordingly as described earlier. For time domain performance analysis, a test initial state condition $x(0) = [\ 5\ \ 0\ \ 0\ \ 0\ ]^{\mathrm{T}}$ was used in the results reported below.

The L_∞ norm of the cart position trajectories for the randomly initialized, LQR-guided action network, and LQR-guided action and critic networks are 0.160 ± 0.050, 0.140 ± 0.030, and 0.130 ± 0.001 m, respectively, comparing to 0.110 m for the corresponding LQR design. The numerical values reported represent the mean and the standard deviation of the cart position on the track. For the pole angle trajectories obtained from the HDP designs with LQR-guided action and critic networks, the L_∞ norm of each trajectory is hardly differentiable from those obtained from the LQR design.

9.4.2 Performance of the Direct HDP Beyond Linearization

In this section, we provide a series of studies to demonstrate the extensibility and flexibility of the direct HDP controllers even though they were initially guided by linear state feedback controller on a linearized system model.

Even for the cart-pole balancing problem that LQR controller can handle well, the LQR-guided direct HDP controller may still outperform the linear gain LQR due to its learning in the nonlinear control region. To show that, analyzing direct HDP as a nonlinear controller, both the LQR gain controller and the nonlinear direct HDP controller are substituted into the closed loop system of the cart-pole; the time domain data for stabilization task are analyzed; and the command following tasks are evaluated. Note that the direct HDP learning is performed for the original nonlinear cart-pole system, and the resulting controller is also nonlinear.

Figure 9.6 shows responses to various step commands in cart position, for the LQR-controlled closed loop system and for an improved direct HDP design. Results show that for small commands, the LQR-guided direct HDP and LQR responses are identical, which is expected. However, the LQR-guided direct HDP command following is better than LQR for large reference commands following. Specifically, for larger reference commands, the LQR-guided direct HDP design results in a smaller rise time, a smaller settling time, and less overshoot. This implies that the LQR-guided direct HDP design has the potential to outperform the original LQR controller.

To evaluate the performance of the proposed LQR-guided direct HDP design, the cart-pole and triple-link pendulum balancing problems were implemented under different operating settings [17]. Two cases are studied through simulations corresponding to the dynamic system operating near the linearization region and away from the region due to system parameter changes, respectively.

It is not surprising that the two controller designs, both the nominal LQR controller and the LQR-guided direct HDP design, achieve their control performance very close

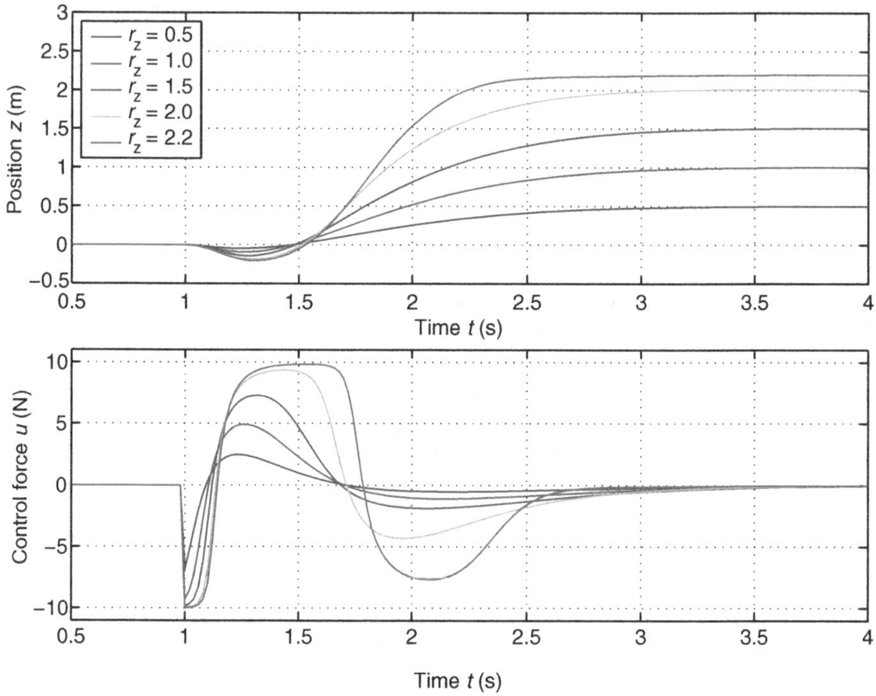

FIGURE 9.6 Responses of step command following for LQR-guided direct HDP designs.

to each other in the first case, where the system operates near the linearization region because the critic network has already been pre-trained by the optimal cost function near that region. In the meanwhile, the action network does not provide additional control as the nominal LQR controller already achieved the optimal control objective. Therefore, not much learning took place while the LQR controller tried to stabilize the system.

However, the situation becomes different when system parameter change occurs. It drives the system away from the region that the nominal LQR controller was designed for. With the same structure of the reinforcement signal provided, the critic network is essentially learning the global quadratic cost function corresponding to the modified system and the design parameter, resulting in the action network outputs that achieve a better performance.

Both the stand along nominal LQR controller and the LQR-guided direct HDP design are applied to the cart-pole system with the system settings identical to those in Section 9.4.1. The direct HDP algorithm implementation and parameters are the same as the one used in previous simulations. To verify the learning and the robustness of the LQR-guided direct HDP design, a large parameter change is introduced to the cart-pole system. The length of the pendulum was changed from 0.5 to 0.7 m. As a result, the dynamic system no longer operates in the linearization region for which it was designed. The results are shown in Figure 9.7.

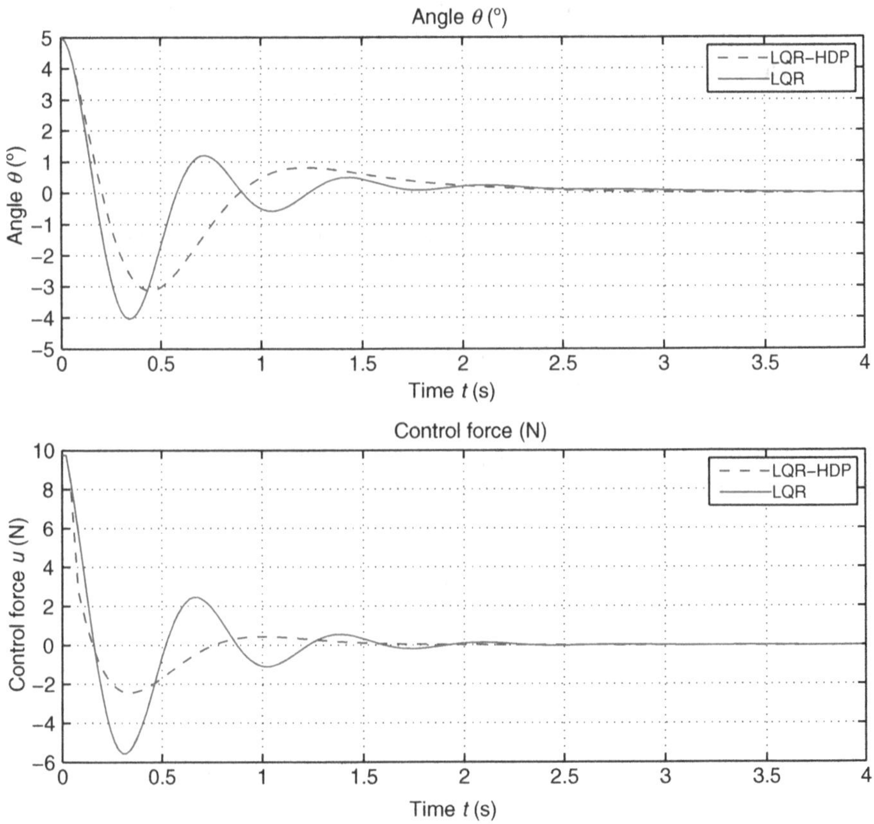

FIGURE 9.7 Comparisons of trajectories of the cart-pole system subject to major system parameter (pendulum length) change for the LQR design and LQR-guided direct HDP design.

The cart was motionless at the beginning and the initial angle between the pole and the vertical axis was selected as $\theta(0) = 5.0°$. The two control designs, the stand along nominal LQR controller and the LQR-guided direct HDP design, are applied to the system. Typical trajectories of the system's control and states for both controller designs show that although both designs are able to stabilize the system with parameter changes, the LQR-guided direct HDP design outperform the nominal LQR controller with less oscillation, a smaller settling time, and less overshoot.

A trajectory following scenario is also implemented for the system. A desired state trajectory is selected, which is the cart position that changes from 0 to 0.9 m in the first second of the simulation time. The typical trajectories of the system's control and states for both controller designs show that although both designs are able to follow the trajectory under system parameter changes, the LQR-guided direct HDP design outperforms the stand along nominal LQR controller with less oscillation, faster tracking time, and less overshoot.

9.5 DIRECT HDP DESIGN WITH IMPROVED PERFORMANCE CASE 2—DIRECT HDP FOR COORINDATED DAMPING CONTROL OF LOW-FREQUENCY OSCILLATION

The China southern power grid (CSG) is an AC/DC hybrid power system and is composed of five provincial grids: Yunnan, Guizhou, Guangxi, Guangdong, and Hainan. It is one of the six power networks in China. The transmission distance from the west to the east is over 1000 km. The large capacity of power transmission takes place through six 500 kV AC lines and two high voltage DC (HVDC) links (Gaozhao and Tianguang) in parallel in 2006. This study is based on the 2006 network data and geometry. Low-frequency oscillations have become a prominent problem and created potential threats to the system stability. Under such operating condition, the power modulation control using HVDC is an attractive alternate given its huge adjustable capacity and its fast responding characteristics. The structure of a conventional modulation controller is usually based on pole placement design and the multiple HVDC modulation controllers are usually designed independently. Coordination of the controllers is a challenging issue, which has not been addressed. It was attempted by using the direct HDP method.

The supplementary control using HVDC is like that through the generator or static var compensator (SVC), an oscillating power control signal ΔP is produced to damp the swing.

Most of the current power system stability controllers are composed of basic elements such as integral, differential, inertial, and lead-lag blocks. The universal approximation capability of neural networks makes it possible to replace the traditional blocks. As an example, consider a lead-lag block of the following form:

$$Y(s) = X(s)(1 + T_1 s)/(1 + T_2 s), \tag{9.17}$$

where $1/T_1$ and $1/T_2$ correspond to the poles of the lead and lag compensators, respectively.

A discrete time realization of the lead-lag compensator is:

$$y(n+1) = u(n+1) + K_1(y(n) - u(n+1)) + K_2(u(n) - u(n+1)), \tag{9.18}$$

where $K_1 = (T_2 - \Delta t/2)/(T_2 + \Delta t/2)$, $K_2 = (\Delta t/2 - T_1)/(T_2 + \Delta t/2)$, and Δt is the discretization step size. Equation (9.18) can, in principle, be implemented using a neural network with K_1 and K_2 as weights in the network. Because this is a dynamic system, it can be implemented using a feedforward network with recurrency. For the ease of discussion, this implementation is referred to as a lead-lag recurrent neural network (LLRNN) in the following.

The direct HDP controller can then be embedded into traditional controllers as shown in Figure 9.8. The parameters K, T_w, T_3, and T_4 other than those in the direct HDP are designed using traditional methods or based on knowledge of the system. The time constants in the lead-lag block must be greater than zero, consequently,

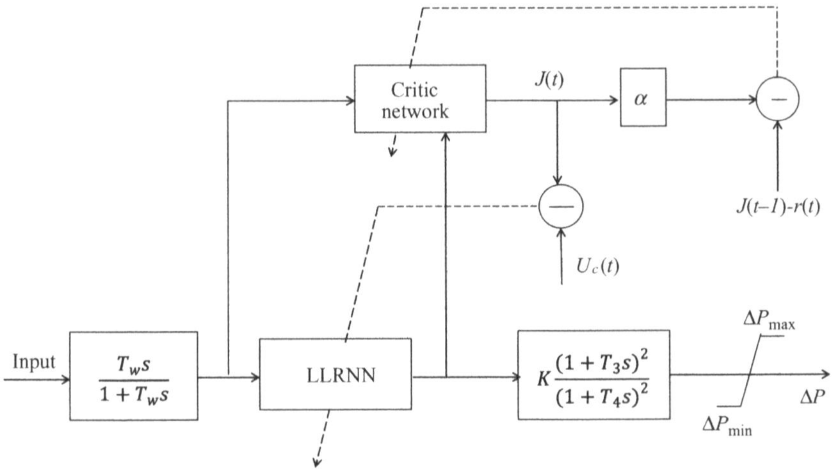

FIGURE 9.8 Coordinated HVDC power control using direct HDP to damp low-frequency oscillation.

K_1 and K_2 in the LLRNN are also constrained. At each time step, T_1 and T_2 are calculated based on K_1 and K_2. If one of the two parameters becomes less than zero, the K values should be retained at the previous level unchanged. The supplementary damping controllers of the Gaozhao (Gaopo-Zhaoqing) and Tianguang (TSQ-Beijiao) HVDC links are implemented using the direct HDP controller as shown in Figure 9.8, where the input signals are chosen as the active power of the two parallel 500kV AC lines based on the analyses of observability and dynamic relative gain array. The reinforcement signal $r(t)$ is defined as

$$r(t) = -(b_1 \Delta\omega^2{}_{\text{inter}-1} + b_2 \Delta\omega^2{}_{\text{inter}-2} + b_3 \Delta\omega^2{}_{\text{local}-1} + b_4 \Delta\omega^2{}_{\text{local}-2}) \quad (9.19)$$

where $\Delta\omega_{\text{inter}-1} = ((\omega_{\text{gui}-1} + \omega_{\text{gui}-2}) - (\omega_{\text{yun}-1} + \omega_{\text{yun}-2}))/2$ represents the oscillation mode between Yunan and Guizhou, $\Delta\omega_{\text{inter}-2} = ((\omega_{\text{gui}-1} + \omega_{\text{yun}-1}) - (\omega_{\text{gua}-1} + \omega_{\text{gua}-2}))/2$ represents the oscillation mode between Yunnan-Guizhou and Guangdong, and, $\Delta\omega_{\text{local}-1} = \omega_{\text{yun}-1} - \omega_{\text{yun}-2}$, $\Delta\omega_{\text{local}-2} = \omega_{\text{gui}-1} - \omega_{\text{gui}-2}$, represent the local modes in Yunan and Guizhou, respectively. Therefore, most of the important modes are simultaneously considered in the cost function.

Because the terminals of the two DC links are not far from each other, and two more HVDC links will be in service in the coming years, the controllers must be coordinated. In previous sections, the learning ability of the direct HDP was demonstrated. In this section, a coordinated design of damping control will be examined. If only one of Gaozhao and Tianguang HVDC power modulation controller is used, the parameters can be tuned to achieve the improved control performance after a disturbance of 40 ms short circuit on the 500 kV Huishui-Hechi AC line. However, if these independently designed controllers are put to action together, the interactions between the two HVDC

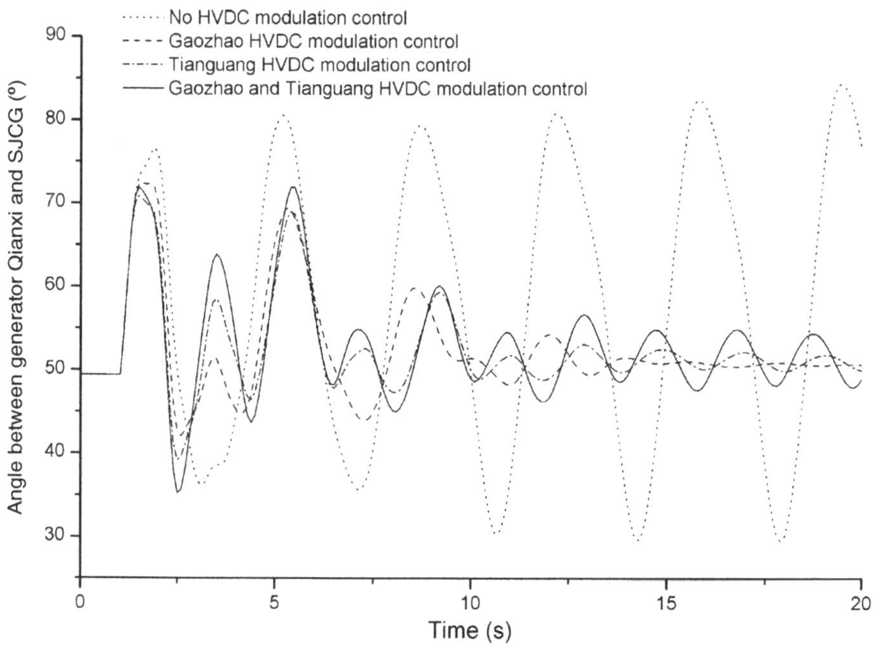

FIGURE 9.9 Rotor angle bewteen generators Qianxi and SJCG.

links can sometimes even facility low-frequency oscillation, contrarary to damping the oscillation. Figure 9.9 is an illustration of reduced damping due to interactions between the HVDC links. In a multi-infeed DC system, the supplementary controllers must be designed on a coordinated basis to avoid any unexpected excitations of a new, lightly damped mode [16].

Because the control law update is instructed by the cost function, which is defined as the weighted sum of the squares of the relative rotor speed differences reflecting multiple interprovince and local oscillation modes. It reflects the stability of the entire system. If only one oscillation mode is suppressed, the cost is not minimal, and the controller parameters need to be adjusted further. Starting from the independently designed Gaozhao and Tianguang DC power modulation controllers, the new direct HDP controller can learn a set of coordinated parameters. Results show that after learning with coordinated DC line control expressed clearly in the learning objective, low-frequency oscillation including multiple modes can be properly damped, with much improved performance over the traditional controllers designed independently. As can be seen from Figure 9.10, coordinated low-frequency damping control using direct HDP has resulted in pushing the closed loop poles further to the left compared to the independently designed damping control, and the respective system damping ratio has improved from 3.8% for the independent design to 14.3% for the direct HDP design.

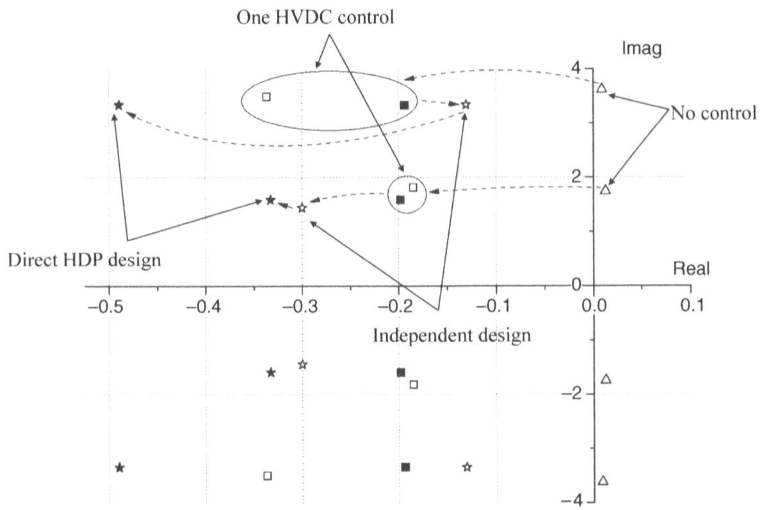

FIGURE 9.10 Closed loop pole locations for different control realizations to damp low-frquency oscillation.

9.6 SUMMARY

ADP is a class of promising design approaches to obtaining optimal controls for general nonlinear systems with uncertainties. The chapter focuses on examining the direct HDP learning controller from control theoretic perspectives. It illustrates the intuitive expectation that the direct HDP designs can result in large variations in the sensitivity maps corresponding to different initializations of the action and critic networks. Further, it demonstrates that among the many possible direct HDP realizations, there exists one that closely approximates an LQR design with desirable closed loop system performance in a classical control sense. The chapter introduces two cases to properly integrate the direct HDP design into the closed loop system and to guarantee system performance. In both cases, *a priori* knowledge about the system or a predesigned controller are important integral parts of the direct HDP design. As shown in the chapter that eventhough the so-designed direct HDP may require more information than the original online learning controller proposed in [11], guaranteed system performance is worth the additional effort. It is encouraging to know that in both cases, the direct HDP controls have outperformed traditional controllers.

ACKNOWLEDGMENT

The research was supported in part by NSF under Grant SECCS-0702057 and ECCS-1002391, and by NSFChina-50828701.

REFERENCES

1. J. Si, A.G. Barto, W.B. Powell, and D. Wunsch, editors. *Handbook of Learning and Approximate Dynamic Programming*. Wiley-IEEE Press, 2004.
2. D.P. Bertsekas and J.N. Tsitsiklis. *Neuro-Dynamic Programming*. Athena Scientific, Belmont, MA, 1996.
3. R.E. Bellman and S.E. Dreyfus. *Applied Dynamic Programming*. Princeton University Press, Princeton, NJ, 1962.
4. M.L. Puterman. *Markov Decision Processes—Discrete Stochastic Dynamic Programming*. John Wiley & Sons, Inc., New York, NY, 1994.
5. R.S. Sutton and A.G. Barto. *Reinforcement Learning: An Introduction*. The MIT Press, Boston, MA, 1998.
6. P.J. Werbos. A menu of design for reinforcement learning over time. In W.T. Miller III, R.S. Sutton, and P.J. Werbos editors. *Neural Networks for Control*, Chapter 3, MIT Press. Cambridge, MA, 1990, pp. 67–95.
7. P.J. Werbos. Approximate dynamic programming for real-time control and neural modeling. In D. White, and D. Sofge, editors. *Handbook of Intelligent Control*. Van Nostrand, New York, 1992, pp. 493–525.
8. D.V. Prokhorov, R.A. Santiago, and D.C. Wunsch. Adaptive critic designs: A case study for neurocontrol. *Neural Networks*, 8(9):1367–1372, 1995.
9. J. Dalton and S.N. Balakrishnan. A neighboring optimal adaptive critic for missile guidance. *Mathematical and Computer Modeling*, 23(1):175–188, 1996.
10. G.K. Venayagamoorthy, R.G. Harley, and D.C. Wunsch. Dual heuristic programming excitation neurocontrol for generators in a multimachine power system. *IEEE Transactions on Industry Applications*, 39(2):382–294, 2003.
11. J. Si and Y. Wang. Online learning control by association and reinforcement. *IEEE Transaction on Neural Networks*, 12(2):264–276, 2001.
12. A.G. Barto, R.S. Sutton, and C.W. Anderson. Neuron like adaptive elements that can solve difficult learning control problems. *IEEE Transactions On Systems, Man, and Cybernetics*, 13:834–847, 1983.
13. R. Enns and J. Si. Apache helicopter stabilization using neural dynamic programming. *AIAA Journal of Guidance, Control, and Dynamics*, 25(1):19–25, 2002.
14. R. Enns and J. Si. Helicopter trimming and tracking control using direct neural dynamic programming. *IEEE Transactions Neural Networks*, 14(4):929–939, 2003.
15. R. Enns, and J. Si. Helicopter flight-control reconfiguration for main rotor actuator failures, *AIAA Journal of Guidance, Control, and Dynamics*, 26(4):572–584, 2003.
16. C. Lu, J. Si, and X. Xie. Direct heuristic dynamic programming method for power system stability enhancement. *IEEE Transactions on Systems, Man, and Cybernetics, Part B*. 38(4):1008–1013, 2008.
17. L. Yang, J. Si, K.S. Tsakalis, and A.A. Rodreguez. Performance evaluation of direct heuristic dynamic programming using control-thoretic measures. *Jounral of Intelligent Robotic Systems*, 55:177–201, 2009.
18. K. Zhou, J.C. Doyle, and K. Glover. *Robust and Optimal Control*. Prentice-Hall, Upper Saddle River, NJ, 1995.

CHAPTER 10

Reinforcement Learning Control with Time-Dependent Agent Dynamics

KENTON KIRKPATRICK and JOHN VALASEK

Department of Aerospace Engineering, Texas A&M University, College Station, TX, USA

ABSTRACT

Reinforcement learning-based control systems have the advantage of being model-free. When used for control of dynamical systems, approximations of the time dynamics can be used to improve the policy development. This chapter proposes and develops a Q-Learning-based method for learning unknown information about the time scale properties of a dynamical system. Two candidate algorithm are proposed. The first determines optimal sample times for sampled data systems, and is demonstrated with a simulation of an inverted pendulum cart. The second algorithm learns a first-order approximation for a dynamical system, and uses the approximation to improve the control of a heterogenous multiagent system. This algorithm is demonstrated with a conceptual experiment involving such a system.

10.1 INTRODUCTION

In the field of machine learning, reinforcement learning (RL) has been a popular area of study for solving problems ranging from game playing to maze navigation, and for gain scheduling and the control of dynamical systems. Although many successful applications have been developed over the years, difficulties tend to arise when attempting to cast a dynamical system control problem as an RL problem. Theoretically, RL methods have been viewed as feasible for determining control policies because of the state-to-action mapping, and in the case of Q-Learning as an optimal control with provable convergence properties. It is the intrinsic capability of RL methods to learn this mapping without needing a model of the dynamics that makes them so appealing.

Reinforcement Learning and Approximate Dynamic Programming for Feedback Control, First Edition. Edited by Frank L. Lewis and Derong Liu.

However, in practice, they are usually difficult to implement and in many cases fails outright.

A closer look at problems with RL approaches reveals that failures often boil down to issues that have nothing to do with the fundamental approach of methods like Q-Learning. Rather, it is usually either a failure on the part of the engineer to properly represent the problem, for example, define the parameters, or to function approximation inaccuracies. In the case of many dynamical systems, the RL algorithm used is not sufficient for determining a proper control policy. For example, while the popular Watkins' Q-Learning algorithm is proven to converge to an optimal action-value function, it does not explicitly account for time. Any handling of time-dependent agent dynamics is either overlooked or handled outside of the learning process. This results in the optimal action-value function being incomplete. The action choices, and the set of possible actions themselves, assume that the dynamics only follow those precisely used by the system during the learning process. When considering sampled data systems, small adjustments to the sample time can have a large impact on the stability of the learned control policy.

One area of research that has received considerable study over recent years has been the application of Q-Learning-based algorithms to the control of cooperative multiagent systems. Learning control policies to allow multiple agents to work cooperatively toward achieving a goal is an attractive research topic with a lot of complexity. In some cases, an approach is taken that involves comparing the effect of using independent Q-Learning algorithms on each agent in a deterministic cooperative game to using Q-Learning-based algorithms that involve learning joint actions between agents [1]. Others have taken these ideas and extended them to conditions where the game is stochastic, and have tried treating the system as noncooperative by making the individual agents unaware of each other and not consider each other in action selection [2]. Considering systems where agents must coordinate but do not have knowledge of each other's actions has been an area of research deemed important for its general application to systems where agents cannot communicate with each other [3, 4]. Still, other research has explored improving the ability of Q-Learning to determine joint actions by considering mixed strategies and using Bayesian inference to estimate agent's strategies [5].

In most of these research scenarios, the agents are required to cooperate in games that do not have time-dependent dynamics. When approaching control of multiagent systems, one aspect of the problem that has not received much attention has been the control of systems that do exhibit real-time dynamics. One likely reason for this is that it is often difficult to learn control policies using RL approaches for even a single agent with time dynamics. A topic that needs to be addressed to make control of real time-based dynamical systems a possibility is the determination of an optimal sample time for a sampled data system. Since control of real, continuous systems using an RL-learned policy requires computer-based control, the sampled data system will require some sample time. This can become an issue when the policy is learned in simulation and then applied to an actual hardware experiment.

Another factor that needs to be addressed in the learning of dynamic multiagent system control policies is learning some approximation of the dynamics, and then using that approximation to improve the learning process. The main reason that RL has drawn the attention of the controls community has been its ability to learn a control policy without requiring a model of the dynamics. This is a benefit that can have great impact on the determination of control policies for systems that have unknown dynamics. However, it is often useful to determine some knowledge of the agent dynamics, and this is especially the case when dealing with heterogenous multiagent systems. When a multiagent system involves agents with different dynamics, the learning process can benefit by considering some approximation of the individual dynamics.

The rest of this chapter is organized as follows. In Section 10.2, the temporal-difference (TD) algorithm known as Q-Learning is explained in detail. Next is an explanation of the proposed sampled data Q-Learning algorithm in Section 10.3, along with an example applying this in a simulation with dynamics equations. This is proceeded by an explanation of a proposed method for incorporating learning of dynamics for the improvement of learning to control heterogeneous multiagent systems in Section 10.4. The chapter is then concluded with closing remarks in Section 10.5.

10.2 Q-LEARNING

There are multiple classes of algorithms that fall within the definition of RL. Currently, the RL algorithms that are most used in research are TD methods. TD methods are actually a conceptual combination from two other classes of algorithms known as dynamic programming (DP) and Monte Carlo [6]. Like Monte Carlo, TD methods use experience through interaction with the system to update the quality of the value function without the need of a model. Like DP, TD methods do not have to wait until the end to improve the value function, but rather update it along the way. This improves convergence time and also makes TD methods usable in online learning. In this section, the TD algorithm known as Q-Learning will be discussed, along with the limitations that lead to the development of extended Q-Learning-based algorithms for the use of learning in dynamical sampled data systems.

10.2.1 Q-Learning Algorithm

Of the various formulations of RL algorithms, Watkins' Q-Learning has been the most accepted and utilized algorithm for its proven convergence to the optimal action-value function [7]. Q-Learning is a TD method that learns the optimal action-value function in an off-policy manner [6]. This means that the policy used during a learning episode is not necessarily the same as the one that is updated at each time step. The Q-Learning algorithm is based upon an action-value update rule that uses a greedy policy to determine a predicted value for the state-action pair at the next (future) time step. The actual action selection may not be done using a greedy policy, and in fact it is typically better for optimality to include some degree of exploration in the policy. The rule used for updating the action-value function is as follows:

$$Q(s, a) \leftarrow Q(s, a) + \alpha\delta. \tag{10.1}$$

In (10.1), Q is the action-value function, s is the state at the current time step, a is the action selected for the current time step using the agent's policy (e.g., ε-greedy), α is the step-size parameter, and δ is the TD. The value of α is always in the interval (0,1), and can either be held constant or varied by some user-defined function. In some cases, α is designed to decrease within an episode according to how often the agent revisits the same state, essentially punishing the agent for repeating itself unnecessarily. The definition of δ is the main term that distinguishes Q-Learning from other TD algorithms, and it can be computed using (10.2).

$$\delta = r + \gamma \max_{a'} Q(s', a') - Q(s, a). \tag{10.2}$$

In (10.2), r is the reward received from the system, s' is the future state (due to taking action a), a' is the action that would be taken using a greedy policy when in state s', and γ is the weight for the future value. The selection of γ affects convergence time, and it is always within the range (0,1). The value of γ can either be kept constant throughout learning, or it can be chosen to vary episodically so that later learning episodes value the future prediction differently than early episodes. Combining (10.1) and (10.2) leads to the full Q-Learning update equation, shown in (10.3).

$$Q(s, a) \leftarrow Q(s, a) + \alpha[r + \gamma \max_{a'} Q(s', a') - Q(s, a)]. \tag{10.3}$$

This update rule is the backbone of the famous Watkins' Q-Learning algorithm. It is an off-policy TD algorithm with a user-determined policy for selecting actions at each time step, but uses a fully-greedy policy with the action-value function when updating the action-value function. Rather than utilizing past information to perform this update, this algorithm uses the predicted future state-action pair chosen greedily. The Watkins' Q-Learning algorithm is displayed in Algorithm 10.1.

Algorithm 10.1: Q-Learning [6]

- Initialize $Q(s,a)$ arbitrarily
- Repeat for each episode:
  - Initialize s
  - Repeat for each time step:
    * Choose a from s using policy derived from $Q(s,a)$ (e.g., ε-greedy)
    * Take action a, observe r, s'
    * $Q(s,a) \leftarrow Q(s,a) + \alpha\ [\ r + \gamma \max_{a'} Q(s',a')$ - $Q(s,a)\ ]$
    * $s \leftarrow s'$
  - Until s is terminal

10.2.2 ε-Greedy

When determining a policy for selecting actions during the learning process, there are any number of possibilities to choose. The most commonly used policies for the Q-Learning process are exploration, greedy, and ε-greedy methods. An exploration policy simply means that at any given state and any given time step, an action is selected completely at random from the possible action-space. This is demonstrated by (10.4), where A represents the full action-space.

$$a = \text{random} \in A. \tag{10.4}$$

Conversely, a fully greedy method requires that the action-value function be exploited for choosing an action at any given time step. This is demonstrated by (10.5).

$$a = \max_{a} Q(s, a). \tag{10.5}$$

Both of these extremes have limitations that can inhibit the learning process due to either slowing the convergence process or converging to paths that may not be optimal. It is often necessary to find a middle ground, and the ε-greedy method accomplishes just that [6].

An ε-greedy policy uses a user-defined probability, ε, that determines whether to choose actions randomly or according to the action-value function each time a new action must be taken. This speeds up the convergence time by reinforcing paths that have already been designated either good or bad while still allowing for new paths to be explored for optimality [8]. In the early episodes, the agent is required to explore due to lack of knowledge, regardless of the value assigned to ε. This method can be implemented during the action selection portion of the learning process according to Algorithm 10.2.

The chosen value of ε can have a drastic effect on the number of episodes required to achieve convergence, and ε can either be chosen as a constant value or a variable.

Algorithm 10.2: ε-Greedy

- Choose $\varepsilon \in [0\ 1]$
- Repeat for each action selection:
  - Generate random value $\beta \sim U(0, 1)$
  - If $\beta \geq 1 - \varepsilon$
    - $a \leftarrow$ random
  - If $\beta < 1 - \varepsilon$
    - $a \leftarrow \max_a Q(s, a)$

It is usually of benefit to choose a high probability of exploration in early episodes, and then anneal the value of ε in later episodes. Decreasing ε as more episodes are completed leads to encouraging exploration in early episodes and exploitation in later episodes. Maintaining a nonzero probability of exploration during late learning (e.g., $\varepsilon = 0.05$) typically helps the learning process by allowing the agent to discover a better path that may not have been discovered yet.

10.2.3 Function Approximation

When applying RL methods like Q-Learning, a key issue that causes problems with the final learned function is generalization [6]. The learning of an action-value function results in a policy that requires looking up a table of values. Tables are by nature discrete, so encountering every possible combination of state-action pairs is practically impossible when the system is continuous. The state- and action-spaces must be discretized to use a method like Watkins' Q-Learning, so the issue of generalization to state-action pairs not available in the final table must be addressed. It is for this reason that function approximation methods have been used in conjunction with RL.

Approximation of the action-value function can be accomplished through any number of methods. One possibility that has been explored is the use of least squares [9]. A drawback that can make using least squares function approximators unattractive is the need to design models or basis functions. Machine learning methods other than RL are most often used for the approximation of value functions. This can include artificial neural networks [10], genetic algorithms [11], support vector machines [12], and instance-based methods like k-Nearest Neighbor [13–15]. Of the various machine learning methods that have been implemented in practice for action-value function approximation, the simplest and most often used has been k-Nearest Neighbor.

The instance-based machine learning method known as k-Nearest Neighbor is derived from the basic assumption that each instance has the most in common with instances that are closest to it in Euclidean space [16]. In Euclidean space, n-dimensional information instances are given ordinates for determination of distance. The distances between these instances are determined according to the common distance formula for n-dimensional space, shown in (10.6).

$$d(x_i, x_j) = \sqrt{\sum_{c=1}^{n}(s_c(x_i) - s_c(x_j))^2} \tag{10.6}$$

In (10.6), the distance is denoted as d, with x_i and x_j being the ith and jth instances. The state dimension index is denoted by c, and the variables n and s are the dimension of the state-space and the state, respectively. Using (10.6), the Euclidean distance can be determined between any two points in n-dimensional space. The k-Nearest Neighbor algorithm then classifies an instance based on the average of the k-nearest points by Euclidean distance. This is an ideal way of handling function approximation of the action-value function due to the fact that no information is lost during the approximation. The Q-matrix holds a discrete table of values representing the

Algorithm 10.3: k-Nearest Neighbor

- Choose value of $k \in$ positive integers
- For the current state, s_t:
  - Find the k states in $Q(s,a)$ closest to s_t by (10.6)
  - Create new set from the k-nearest states, denoted $\bar{s}$
  - Repeat for each possible action, a:
    * $Q(s_t,a) = \frac{1}{k} \sum_{i=1}^{k} Q(\bar{s}_i, a)$

state-action pair values according to how the state- and action-spaces were discretized. This learned function is retained, and the k-Nearest Neighbor algorithm can be used to determine values for any states that fall between the discrete states listed in the Q-matrix. The k-Nearest Neighbor algorithm as it applies to action-value function approximation is shown in Algorithm 10.3.

10.3 SAMPLED DATA Q-LEARNING

When Algorithm 10.1 is used in dynamical systems, the problem of handling time-dependencies arises. Because it is much faster (and safer) to perform the learning in a computer simulation rather than online with a hardware system, researchers often turn to using dynamics models. The issue of ensuring an accurate model is obvious, but it is often overlooked that using the final control policy on a hardware model results in a sampled data system. The control policy is handled using a discrete computer, but the policy is learned assuming continuous dynamics. In some cases, the user realizes this and assumes a sampled data system in the simulated learning, but the Q-Learning algorithm will converge to a policy that assumes the same sample time will always be used. Here, an attempt to overcome this issue is addressed by wrapping the sample time into the learning process.

10.3.1 Sampled Data Q-Learning Algorithm

Incorporating the sample time, denoted T, into the learning process requires determining the value of individual sample times without adversely affecting the stability of the system during an episode. It would therefore be wise to not allow a sample time to change during a single episode. However, to determine optimal action-value functions for a range of sample times requires incorporating it into the state-space. It is therefore necessary to append the state-space with T while not allowing it to be affected by the action-space.

This gives rise to the question of how a particular sample time is to be selected when it cannot be affected by the action-space. It is necessary that a value be associated with each possible selection of T, but T must be held constant throughout an episode. It is therefore proposed that a state-value function for T be determined using

Monte Carlo-based learning [6]. The value function can be updated according to an every-visit Monte Carlo method, shown in (10.7).

$$V_1(T) \leftarrow V_1(T) + \alpha(R - V_1(T)). \tag{10.7}$$

This update rule will be the basis for sampled data Q-Learning. At the beginning of each episode, the sample time value-function, V_1, can be used to select T for the episode according to a user-defined policy. The sample time is appended to the system state vector, s, so that the normal Q-Learning update rule will determine separate control policies according to different values of T. When the reward for a given time step is determined, it is used to update both Q, and the total rewards for the episode are updated. At the end of the episode, V_1 is updated using the average return, R. This causes V_1 to be updated such that episodes which experience more positive rewards than negative rewards result in the value associated with that particular T increasing. Likewise, when an episode experiences more negative rewards than positive, the total value for T after the episode ends will have decreased. Using this new sampled data value function and update rule, the sampled data Q-Learning algorithm becomes as shown in Algorithm 10.4.

10.3.2 Example

To demonstrate the sampled data Q-Learning algorithm, an example of a dynamical system controlled by an RL method was constructed. In this example, the system

---

Algorithm 10.4: Sampled Data Q-Learning

- Initialize $Q(s,a)$ arbitrarily
- Initialize $V_1(T)$ arbitrarily
- Repeat for each episode:
  - Choose T using policy derived from $V_1(T)$ (e.g., ε-greedy)
  - Initialize $R = 0$
  - Initialize s, append T to s
  - Repeat for each sample time step, T:
    * Choose a from s using policy derived from $Q(s,a)$ (e.g., ε-Greedy)
    * Take action a, observe r, s'
    * $Q(s,a) \leftarrow Q(s,a) + \alpha\,[\, r + \gamma \max_{a'} Q(s',a') - Q(s,a)\,]$
    * $s \leftarrow s'$
  - Until s is terminal
  - R = average(r)
  - $V_1(T) \leftarrow V_1(T) + \alpha\,[\, R - V_1(T)\,]$

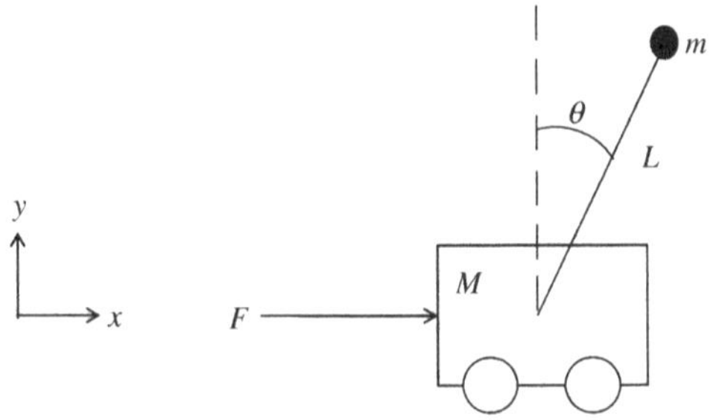

FIGURE 10.1 Inverted pendulum.

consists of an inverted pendulum on a cart. The pendulum is allowed to rotate freely about the point of connection, and the only control input available is the force applied to the cart along the x-axis. This system is shown in Figure 10.1.

10.3.2.1 Problem Representation The position-level states that describe this system's orientation are x and θ. To describe the dynamics of this system, the state-space is chosen as shown in (10.8). The full nonlinear dynamics can then be shown by (10.9).

$$\vec{x} = [x \quad \dot{x} \quad \theta \quad \dot{\theta}]^{\mathrm{T}} = [x_1 \quad x_2 \quad x_3 \quad x_4]^{\mathrm{T}}. \tag{10.8}$$

$$\begin{aligned} \dot{x_1} &= x_2 \\ \dot{x_2} &= \frac{F + mLx_4^2 \sin(x_3) - mg\sin(x_3)\cos(x_3)}{M + m\sin^2(x_3)} \\ \dot{x_3} &= x_4 \\ \dot{x_4} &= \frac{g\sin(x_3) - \dot{x_2}\cos(x_3)}{L} \end{aligned} \tag{10.9}$$

To determine the proper state-space for RL, one might assume that the full state vector should be used. However, in this case, it can be seen that including the position (x) and the cart velocity ($\dot{x}$) will not contribute much since it does not affect the pendulum orientation (θ) explicitly. In fact, it can be seen from the equations of motion in (10.9) that it is the acceleration of the cart that explicitly contributes to the pendulum orientation dynamics. Since the only states propagated and known for the cart at any given time interval are x and $\dot{x}$, an estimate of the acceleration is used for the RL state space according to (10.10).

$$\ddot{x}_{\mathrm{est}} = \frac{\dot{x}_t - \dot{x}_{t-1}}{T}. \tag{10.10}$$

To learn action-value functions for each sample time, T, the state-space must also include the sample time. The action-space must likewise be selected, but in this problem the only thing that makes sense is to use the force, F, as the action since it is the only control. The state- and action-spaces for the RL formulation of this problem are as shown in (10.11) and (10.12).

$$\vec{s} = [\ddot{x}_{\text{est}} \quad \theta \quad \dot{\theta} \quad T]^{\text{T}} \tag{10.11}$$

$$a = F. \tag{10.12}$$

The next step is to define the reward structure. The rewards depend upon both the goals of the system and the constraints placed on the system. For this example, the goal was determined to be the state where both $\theta = 0$ and $\dot{\theta} = 0$, with a tolerance of $\pm 1^\circ$ and $\pm 2^\circ$/s, respectively. If this goal was achieved, a positive reward of +20 was used to update the action-value function. The horizontal position and velocity of the cart were not considered for determination of the goal achievement. This structure provides the ranges for which positive rewards are achieved, but it is likewise necessary to define the constraints for which negative rewards are given. Again, the cart position and velocity do not contribute to the negative rewards by remaining unconstrained. The pendulum orientation was constrained so that it would not be allowed to move beyond $\pm 12^\circ$ without incurring negative rewards. The angular velocity was constrained so that it was not allowed to go beyond $\pm 20^\circ$/s without incurring negative rewards. If these constraints were breached, a negative reward of -20 was used to update the action-value function. All orientations that do not yield either a positive or negative reward receive a neutral reward of 0. The reward structure in place can be seen in (10.13)

$$r = \begin{cases} +20 & \text{if } |\theta| < 1^\circ \text{ and } |\dot{\theta}| < 2^\circ/\text{s} \\ -20 & \text{if } |\theta| > 12^\circ \text{ or } |\dot{\theta}| > 20^\circ/\text{s} \\ 0 & \text{otherwise} \end{cases}. \tag{10.13}$$

For a system where longer sample times are preferred, some function of the sample time would need to be included in the determination of the reward structure. For this simple system with no requirements placed on the sample time, it is expected that the optimal sample time will be the shortest available since it provides the highest control authority. With the RL representation determined, the simulation was performed. The parameters for the inverted pendulum cart were set as follows. The mass of the cart, M, was set to be 5 kg. The mass at the end of the pendulum, m, was set to be 1 kg, with the pendulum rod considered to be massless for simplicity. The length of the pendulum rod, L, was chosen to be 2 m. The total range of possible actions was chosen to be $F = -40$ N to 40 N in 1 N intervals. This system was then simulated continuously by using a fourth-order Runge–Kutta method with the full nonlinear dynamics from (10.9) and a small integration time step of $\Delta t = 0.01$ s. The system was treated as a sampled data system by sampling the states at a rate of T chosen by an ε-greedy policy from V_1 at the beginning of each episode. The value assigned to ε for this simulation was $\varepsilon = 0.7$. The total range of possible sample times from which to choose was $T = 0.01$–1.0 s in 0.01 s intervals. For this simulation, α was chosen as a constant 0.01 and γ was selected to be a constant 0.7. Each simulation was allowed to last until $t_{\text{f}} = 5$ s.

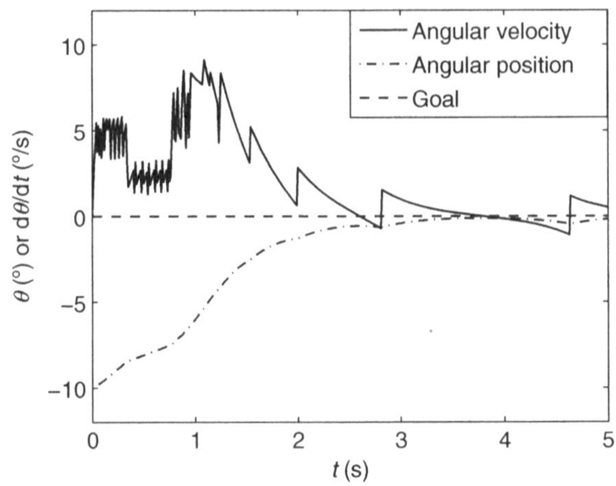

FIGURE 10.2 Inverted pendulum state time history.

When determining the length of an episode, there are many possibilities. Here, an episode is defined to be some significant portion of the learning process. For this particular problem, an episode was determined to last until either $t = t_{\mathrm{f}}$ or $|\theta| > 12°$. The simulation was then allowed to run for many episodes to converge to both the optimal value of T and the optimal action-value function, Q.

10.3.2.2 Results Using this representation of the inverted pendulum control problem, the simulation was allowed to learn for 200,000 episodes using MATLAB. This translated to roughly 12 h of computer time. After the learning was completed, the action-value function was tested, and an example of this is shown in the following figures. The sample time value function, V_1, converged to the highest value of the sample time being $T = 0.01$ s. This makes sense for this particular experiment as the wide range of total possible actions allow for fine tuning with short time intervals, and no constraints were placed on sample time length.

In Figure 10.2, the cart began stationary with no pendulum angular velocity and an initial orientation of $\theta = -10°$. As can be seen, by exploiting the learned action-value function the control policy was able to stabilize the pendulum to the goal within the 5 s timespan alloted. After reaching the goal region, the control policy was able to use occasional impulses to keep the angle within the goal region of $0° \pm 1°$ and the angular velocity within $0°/\mathrm{s} \pm 2°/\mathrm{s}$. These impulses can be seen in the force time history in Figure 10.3. As the phase portrait in Figure 10.4 demonstrates, the angular position and velocity successfully converge to the stable (0,0) state. This indicates a successful control policy learned using sampled data Q-Learning.

10.4 SYSTEM DYNAMICS APPROXIMATION

When using Q-Learning to determine a control policy for dynamical systems, the benefit of not needing to have a model of the dynamics can lead to neglecting

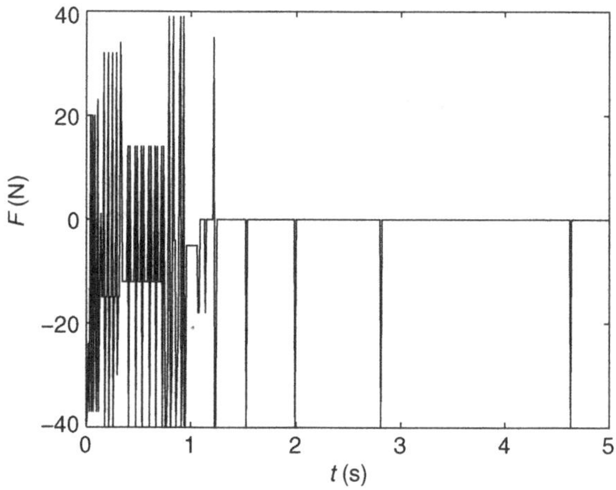

FIGURE 10.3 Inverted pendulum force time history.

the dynamics completely. Learning a control policy for these systems is necessary, but often it is desired to also determine some approximation of the dynamics. This is especially important when dealing with the control of heterogenous multiagent systems since coordinating the agents requires knowing how the individual agents respond differently to similar action inputs.

10.4.1 First-Order Dynamics Learning

One of the simplest ways to represent an approximation of agent dynamics is a first-order approximation. In a first-order system, the individual state transitions can be described by determining the time constant, τ, from the stable first-order ODE shown

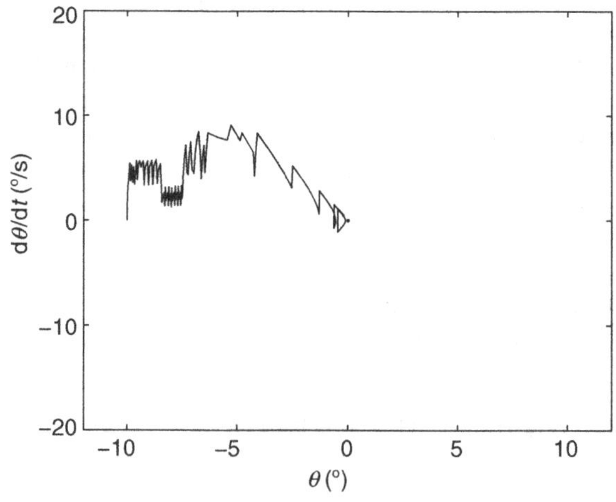

FIGURE 10.4 Inverted pendulum phase diagram.

in (10.14).

$$\tau \frac{\mathrm{d}x}{\mathrm{d}t} = -x \tag{10.14}$$

This ODE can be easily solved to get a solution for the state transition from x_1 to x_2 according to the increment in time, Δt. The derivation of this equation is shown in (10.15)–(10.17), with the final solution shown in (10.17).

$$\int_{x_1}^{x_2} \frac{\mathrm{d}x}{x} = -\int_{t_1}^{t_2} \frac{\mathrm{d}t}{\tau} \tag{10.15}$$

$$\ln(x_2) - \ln(x_1) = -\frac{t_2 - t_1}{\tau} = -\frac{\Delta t}{\tau} \tag{10.16}$$

$$x_2 = x_1 e^{-\Delta t/\tau}. \tag{10.17}$$

Equation (10.17) provides the solution to a first-order differential equation that can determine a state x_2 given the state x_1, Δt, and the time constant, τ. It can be seen from examining (10.17) that determining the opposite relationship can be accomplished by inverting the equation. The transition from state x_2 to x_1 can be thereby determined by (10.18).

$$x_1 = x_2 e^{\Delta t/\tau}. \tag{10.18}$$

Thus, the time constants are related by:

$$\tau_{2\to 1} = -\tau_{1\to 2}. \tag{10.19}$$

A first-order approximation is not an accurate general representation of real dynamical systems, which are typically of higher order. However, when considering transitions across individual time steps, first-order approximations can be used to accurately approximate the dynamics if using values of τ specifically designed for the current state-action pair. By considering the time step, Δt, an approximation for an individual state-transition from s to the approximate next state s^* can be performed according to (10.20).

$$s^* = s e^{-\Delta t/\tau}. \tag{10.20}$$

Equation (10.20) can be used to approximate a state transition for a single one-dimensional state. If more than one state dimension is to be approximated, multiple time constants would be needed. For instance, if this method were applied to the inverted pendulum example in Section 10.3.2, one might want to know the state transition dynamics of $\ddot{x}_{\text{est}}$, θ, and $\dot{\theta}$. This would simply require learning three time

Algorithm 10.5: First-Order Dynamics Learning

- Determine T and $Q(s,a)$ for system (e.g., sampled data Q-Learning)
- Initialize $V_2(s,a,\tau)$ arbitrarily
- Repeat for each episode:
  - Initialize s, append T to s
  - Repeat for each time step:
    * Choose a from s using greedy policy derived with $Q(s,a)$
    * Choose τ using policy derived from $V_2(s,a,\tau)$ (e.g., ε-greedy)
    * Predict next state, s^*, with s, a, and τ using first-order approximations
    * Take action a, observe actual next state, s'
    * Observe r shaped from observed s' and predicted s^*
    * $V_2(s,a,\tau) \leftarrow V_2(s,a,\tau) + \alpha\, [\, r - V_2(s,a,\tau)\,]$
    * $s \leftarrow s'$
  - Until s is terminal

constants, $\tau_{\ddot{x}_{\text{est}}}$, τ_{θ}, and $\tau_{\dot{\theta}}$, for each state-action pair, and approximating each independently at every time step. An example algorithm for how this may be achieved in a similar manner to sample-data Q-Learning is shown in Algorithm 10.5.

10.4.2 Multiagent System Thought Experiment

The learning of time constants for first-order dynamics approximations can be very beneficial for problems involving coordination of multiple heterogenous agents. Learning the time constants removes uncertainty about the time response behavior of individual agents. Figure 10.5 shows an example of a multiagent system that requires coordination to maximize goal achievement in minimal time. In this system there are four agents, although only three are pictured. The three pictured agents each have differing dynamics that cause them to transition from state-to-state in different amounts of time. While they each have different dynamics, each agent is omnidirectional and has the same time constant regardless of state position or direction of motion. The fourth, unseen agent, is the high-level agent that gives commands to the three low-level agents.

In this hypothetical 2D path planning scenario, there are two separate goals that each must be reached by any two of the three low-level agents. Minimizing time is considered to be of highest priority, so the high-level agent must take each low-level agent's dynamics into consideration. For instance, agent A_2 may be sufficiently close enough to goal G_2 to make the decision easy, but the relative positions of A_1 and A_2 to G_1 must be considered before deciding which agent to send. Although it may be the simplest approach to just send both agents and see who reaches the goal first, it is

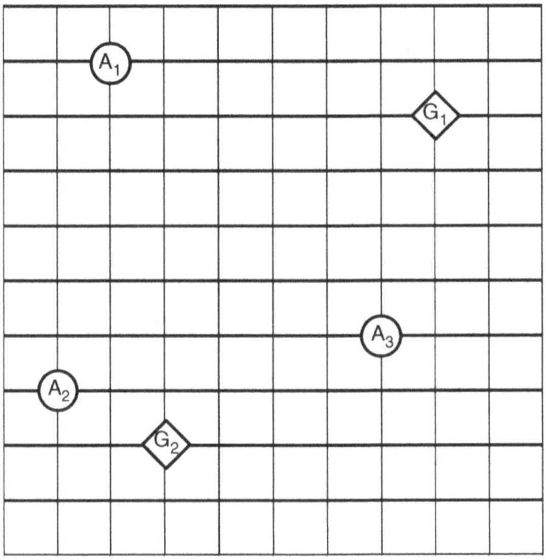

FIGURE 10.5 Multiagent system.

considered a waste of valuable resources to send two agents to achieve a goal that only takes one agent. For this scenario, let us assume that agent A_1 is significantly faster than A_3, and although A_3 is closer in distance it will actually take longer to achieve the goal. This may result in a set of high-level commands, as shown in Figure 10.6.

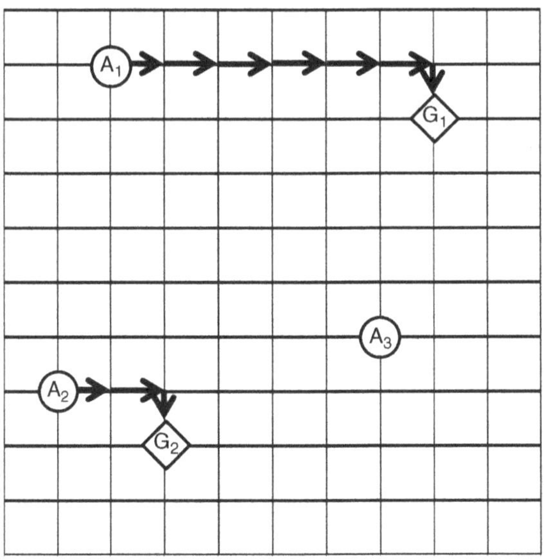

FIGURE 10.6 High-level agent commands.

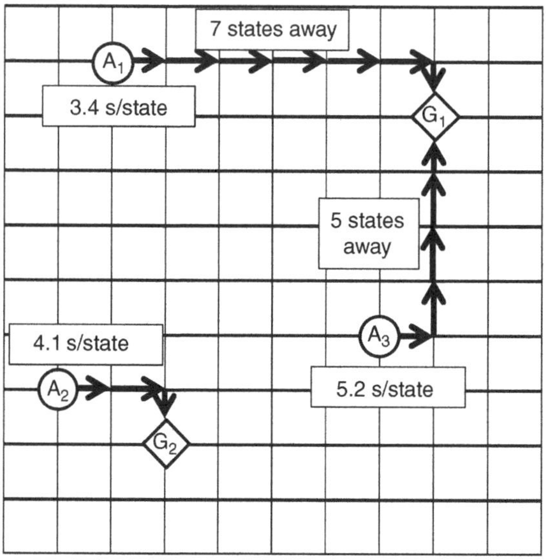

FIGURE 10.7 High-level agent decision-making.

In this scenario, the high-level agent gives commands to each of the low-level agents. For this to be possible, each low-level agent must have its own control policy based on the commands given. But if each agent has different dynamics and the same sample time, the discretization of the state-space may not be optimal for determining these commands accurately. For a continuous dynamical system that has been discretized for the purposes of Q-Learning, it makes the most sense to choose the individual sample times appropriately and then determine the state-to-state transition time constants for each agent. This would simply require that each agent learn its individual optimal sample time, T, using Algorithm 10.4, and then learn its time constant for performing state-to-state transitions using Algorithm 10.5. The high-level agent simply needs to sample each agent's positions at different intervals, according to the individual T associated with each. It can then use each agent's learned τ to make decisions regarding minimum time goal achievement. An example of how this might affect the decision making of the high-level agent from our thought experiment is shown in Figure 10.7.

10.5 CLOSING REMARKS

In this chapter, two ideas were presented that each account for some portion of the time domain for a dynamical system that has been cast as a RL problem. By using the proposed algorithm known as sampled data Q-Learning, the optimal value for sample time can be determined for a sampled data system. Depending on the possible inputs available in the action-space, the sample time that allows for the best control input intervals can be determined. The second method proposed an algorithm for first-order dynamics learning, in which a first-order approximation is found for a dynamical system, which has a previously determined sample time and optimal action-value function. Each state-action pair has an associated time constant that is used to predict the following state deterministically, resulting in a set of first-order approximations for each state-to-state transition. It was then shown in a thought experiment how this approach could be used for improving the learning process of a multiagent system controlled by a high-level agent.

Accounting for time-dependent dynamics in RL control is a complex issue that warrants further study. The approaches presented in this chapter are an initial look into accomplishing this goal for the purpose of making RL a more feasible method of determining control policies for dynamical systems. Learning an approximation of the system dynamics and using the approximation to improve the learning process may be the approach needed to make RL more feasible for dynamical system control.

REFERENCES

1. C. Claus and C. Boutilier. The dynamics of reinforcement learning in cooperative multiagent systems. In *Proceedings of National Conference on Artificial Intelligence*, AAAI-98, pp. 746–752, 1998.
2. J. Hu and M.P. Wellman. Multiagent reinforcement learning: Theoretical framework and an algorithm. In *Proceedings of the 15th International Conference on Machine Learning*, pp. 242–250, Madison, WI, 1998.
3. M. Lauer and M. Riedmiller. An algorithm for distributed reinforcement learning in cooperative multi-agent systems. In *Proceedings of the Seventeenth International Conference on Machine Learning*, ICML '00, Morgan Kaufmann Publishers Inc. San Francisco, CA, 2000, pp. 535–542.
4. S. Kapetanakis and D. Kudenko. Reinforcement learning of coordination in cooperative multi-agent systems. In *Proceedings of the Nineteenth National Conference on Artificial Intelligence*, AAAI-02, pp. 326–331, 2002.
5. G. Tesauro. Extending Q-learning to general adaptive multiagent systems. In S. Thrun, L. Saul and B. Schölkopf, editors. *Advances in Neural Information Processing Systems*, Vol. 16. Vancouver and Whistler, Canada, December 2003, pp. 8–13.
6. R. Sutton and A. Barto. *Reinforcement Learning: An Introductions*. The MIT Press, Cambridge, MA, 1998.
7. C. Watkins and P. Dayan. Q-learning. *Machine Learning*, 8(3):279–292, 1992.
8. S. Whiteson, M.E. Taylor, and P. Stone. Empirical studies in action selection with reinforcement learning. *Adaptive Behavior*, 15:33–50, 2007.
9. G. Taylor and R. Parr. Kernelized value function approximation for reinforcement learning. In *Proceedings of the 26th International Conference on Machine Learning*, Montreal, Canada, 2009.
10. M. Riedmiller. Neural fitted q iteration - first experiences with a data efficient neural reinforcement learning method. In *Machine Learning: ECML 2005*, Vol. 3720 of *Lecture Notes in Computer Science*. Springer Berlin/Heidelberg, Porto, Portugal, 2005, pp. 317–328.

11. P.L. Lanzi. Learning classifier systems from a reinforcement learning perspective. *Soft Computing - A Fusion of Foundations, Methodologies and Applications*, 6:162–170, 2002.
12. M. Martin. On-line support vector machines for function approximation. In *Technical Report, Universitat Politècnica de Catalunya, Software Department*, 2002.
13. K. Kirkpatrick and J. Valasek. Reinforcement learning for characterizing hysteresis behavior of shape memory alloys. *Journal of Aerospace Computing, Information, and Communication*, 6(3):227–238, 2009.
14. K. Kirkpatrick and J. Valasek. Active length control of shape memory alloy wires using reinforcement learning. *Journal of Intelligent Material Systems and Structures*, 22(14):1595–1604, 2011.
15. R.A. McCallum. Instance-based state identification for reinforcement learning. In G. Tesauro, D. S. Touretzky and T. K. Leen, editors. *Advances in Neural Information Processing Systems*, Vol. 7. MIT Press, Cambridge, MA, 1995.
16. T. M. Mitchell. *Machine Learning*. The McGraw-Hill Companies, Inc., Singapore, 1997.

CHAPTER 11

Online Optimal Control of Nonaffine Nonlinear Discrete-Time Systems without Using Value and Policy Iterations

HASSAN ZARGARZADEH,[1] QINMIN YANG,[2] and S. JAGANNATHAN[1]

[1]Embedded Systems and Networking Laboratory, Electrical & Computer Engineering Department, Missouri University of Science and Technology, Rolla, MO, USA
[2]State Key Laboratory of Industrial Control Technology, Department of Control Science and Engineering, Zhejiang University, Hangzhou, Zhejiang, China

ABSTRACT

Online optimal control of nonlinear discrete-time systems in a forward-in-time manner is a challenging problem due to a lack of closed-form solution to the Hamilton–Bellman–Jacobi (HJB) equation. Traditionally, value and policy iteration-based approximate optimal control schemes are developed in the literature by assuming that a significant number of iterations can be performed within a sampling interval which is not practical. By contrast, this book chapter introduces a novel online time-based optimal framework for nonlinear discrete-time systems by using both: (a) reinforcement learning and (b) online neural network approximation-based forward-in-time dynamic programming without using iterative methodology. The overall stability proof is provided and it is shown that the approximated control input converges to the optimal controller over time. Simulation results are provided to validate the proposed approach.

11.1 INTRODUCTION

Today, numerous methods [1] are available for designing controllers for nonlinear discrete-time systems with stability being the only consideration. However, optimal

Reinforcement Learning and Approximate Dynamic Programming for Feedback Control, First Edition. Edited by Frank L. Lewis and Derong Liu.

design based on a predefined cost function is preferred over stability for most applications. In other words, a controller scheme should not only achieve the stability of the closed-loop system, but also keep the cost or performance index as small as possible. To this end, recently a number of optimal control schemes [2–5] have been introduced for nonlinear systems.

Of the available methods, dynamic programming (DP) [6–8] technique has been extensively applied to generate optimal control law for nonlinear dynamic systems [6, 9] by utilizing the Bellman's Principle of Optimality. While this method allows optimization over time [10], the main drawback is the "curse of dimensionality" [11]. Therefore, to confront this issue, adaptive or approximate DP (or ADP) based iterative methods (e.g., [11–13]) have been introduced in the past couple of decades. Most of them use significant number of value and/or policy iterations to attain optimality and are implemented either by an offline manner [5] or require the dynamics of the nonlinear system to be known *a priori* [2]. Unfortunately, these requirements are often not practical, since the exact dynamics are not available in practice for real-world systems.

On the other hand, originated from animal behavior research and its interactions with the environment, reinforcement learning is proposed based on the common sense reasoning in that if an action is followed by a satisfactory outcome (reinforcement signal), then the tendency to repeat that action is strengthened, that is, reinforced [14–17]. Differing from the traditional supervised neural network (NN) learning, there is no desired behavior or training examples but an evaluation signal is employed with reinforcement learning schemes. Nevertheless, reinforcement learning is employed for optimal controller design, since the cost function can be viewed as a form of reinforcement signal [18].

Among the proposed reinforcement learning schemes, the temporal difference (TD) learning-based iterative schemes [14–17, 19] have found many applications in engineering since they does not require the knowledge of the system dynamics. However, to obtain a satisfactory reinforcement signal for an action, the approach must visit each system state by applying an action often enough [20]. This in turn requires the system under consideration to be time-invariant, or stationary in the case of stochastic system. Moreover, for implementation, discretization of the action and state space is invoked.

To overcome the offline methodology for real-time applications, several appealing online approximator (OLA)-based controller designs using value and/or policy iterations are introduced in [7, 10, 12, 13]. These methods are referred to as forward dynamic programming (FDP) or adaptive critic designs (ACD). The central theme of this approach is that the optimal control law and cost function are approximated respectively by online parametric structures, such as neural networks (NNs), which are trained over time by using the information that is fed back from the system response. Depending upon whether the NNs approximate the cost function or its derivative, or both, the ACDs are classified as: (1) heuristic dynamic programming (HDP), (2) dual heuristic dynamic programming (DHP), and (3) globalized dual heuristic dynamic programming (DHP). It is important to note that when the action is introduced as an additional input to the critic, then the ACD will be referred as action dependent (AD) version of the ACD.

Instead of finding the exact minimum, ACDs approximate the Bellman equation $J(x(k)) = \min_{u(k)} \{J(x(k+1)) + U(x(k), x(k+1))\}$ where $x(k)$ is the state and $u(k)$ the control at time step k. While the strategic utility function $J(x(k))$ represents the minimum cost or performance measure incurred in the range from k to final step N, $U(x(k), x(k+1))$ is the utility function denoting the cost incurred in the range from k to $k+1$ using control $u(k)$ and $J(k+1)$ is the minimum cost or performance measure incurred in the range from state $k+1$ to the final step N. The NNs are widely used for approximation in the ACD literature.

Numerous papers have appeared on ACDs using NN but stability is either not studied [2, 3, 21] or only addressed under a simplified binary reward or cost function [19, 21, 22] and for affine nonlinear systems. The common problem found with either the reinforcement learning [2] or ADP-based offline and online schemes [2–18] is the number of iterations needed for convergence. Under large number of iterations, convergence of the HDP algorithm is shown in [2] while closed-loop stability is not attempted. Inadequate number of iterations within a sampling interval for control applications would lead to instability as shown in [4, 23] and reproduced in this chapter through an example. Therefore, for real-time control applications, value and/or policy iterative schemes are not preferred due to both hardware challenges and closed-loop stability. While most of the ADP or reinforcement schemes deal with affine nonlinear systems [2–5], control of nonaffine systems is not adequately addressed since control schemes for affine nonlinear systems cannot be directly applicable to nonaffine systems.

The online approximator-based optimal methodology presented in this chapter uses both reinforcement learning and ADP with NNs as the fundamental learning block for the control of nonaffine nonlinear discrete systems with standard quadratic criterion as the cost function. In both the schemes, first the nonaffine nonlinear discrete-time system is transformed into an affine-like equivalent nonlinear system. The reinforcement learning based state feedback control scheme for the affine-like system consists of two online approximators (in this case two NNs): an action NN to derive the optimal (or near optimal) control signal to track not only the desired system output but also to minimize the long-term cost function; a critic NN to approximate the long-term cost function $J(x(k))$ and to tune the action NN weights. Subsequently, an online based ADP scheme is proposed wherein a NN is utilized to identify the input gain matrix and a second NN for approximating the cost function. The proposed methodology is generic enough that any online approximator such as radial basis functions (RBFs), splines, CMAC can be utilized. Value and policy iterations that are common in other techniques [2] are not needed here. Instead, the value function and the control input are updated once a sampling interval. Closed-loop stability is demonstrated and it is shown that the approximated control input approaches the optimal value over time.

Since the control signal is not used in the critic NN as an additional input, the proposed approach can be viewed as an online approximator-based HDP or neural HDP scheme. It is important to note that an uniformly ultimately boundedness (UUB) of the overall system is demonstrated in this paper even in the presence of approximation errors and bounded unknown disturbances unlike the existing adaptive critic works [4]. The NN weights are tuned online without any offline training phase ACD [5].

11.2 BACKGROUND

Of fundamental importance in NN closed-loop control applications is the universal function approximation property of NN having at least two layers. It is well recognized that on a compact set S, any continuous function $f(X) \in C^N(S)$ can be written as

$$f(X) = W^T \phi(V^T X) + \varepsilon(X), \tag{11.1}$$

with $\varepsilon(X)$ being a NN functional reconstruction error vector and $\phi(\bullet)$ is referred to as the basis function [1]. According to Stone–Weierstrass Theorem, the NN approximation or reconstruction error $\varepsilon(X)$ can be arbitrarily chosen small by selecting arbitrarily large number of hidden-layer neurons. In our design, the output layer weight matrix, W, is updated online while the input layer weight matrix, V, is initially selected at random and held fixed to form stochastic basis [24]. Here, novel weight tuning laws are proposed making the NN weights robust.

In this chapter, the optimal control of nonaffine nonlinear discrete-time systems will be considered. In the generic case, the nonaffine system is represented as $x(k+1) = f(x(k), u(k))$ where k is the sampling time instant, $x(k)$ the state, and $u(k)$ the system input. Due to the inclusion of the input within the unknown nonlinearity, controlling nonaffine systems in general is a challenging problem (e.g., [25]). To the best knowledge of the authors, the optimal control of such nonaffine systems is not done satisfactorily without using value and policy iterations and therefore attempted here. Next, the following definition is needed.

Definition 11.1 *[1]: An equilibrium point x_e is said to be uniformly ultimately bounded (UUB) if there exists a compact set $S \subset \Re^n$ so that for all initial states $x_0 \in S$ there exists a bound B and a time $T(B, x_o)$ such that $\|x(k) - x_e\| \leq B$ for all $k \geq k_0 + T$.*

The boundedness of the NN-based optimal controller instead of asymptotic result arises due to the inevitable upper bound on the functional approximation error $\varepsilon(X)$. Although in practice, this bound cannot be driven to zero, it can be arbitrarily made small by selecting certain design parameters and the number of hidden-layer neurons. It is important to note that a nonaffine nonlinear discrete-time system can be described in a variety of form as presented in this book chapter. Despite these various forms, first reinforcement learning based controller is developed for a class of single-input–single-output (SISO) nonaffine nonlinear discrete-time system and subsequently the time-based ADP scheme is introduced for multi-input and multi-output (MIMO) nonaffine system.

11.3 REINFORCEMENT LEARNING BASED CONTROL

11.3.1 Affine-Like Dynamics

Consider the following nonaffine single-input single-output (SISO) nonlinear discrete-time system with disturbance that is expressed in NARMAX form [26] as

$$y(k+\tau) = f(\bar{y}_k, \bar{u}_{k-1}, u(k), \bar{d}_{k+\tau-1}) = f(w_k, u(k), \bar{d}_{k+\tau-1}), \tag{11.2}$$

where $w_k = \left[\bar{y}_k^T, \bar{u}_{k-1}^T\right]^T$, $\bar{y}_k = [y(k), \ldots, y(k-n+1)]^T$, and $\bar{u}_{k-1} = [u(k-1), \ldots, y(k-n+1)]^T$ denotes a vector of past system inputs and outputs, respectively. The term $\bar{d}_{k+\tau-1} = [d(k+\tau-1), \ldots, d(k)]^T$ denotes the disturbance vector while τ represents the system delay, or the relative degree of the system [27]. In this chapter, the output $y(k)$ is considered measurable with initial values located in a compact set Ω_{y_0}. The objective is to design a control law to drive the system output $y(k)$ to track a desired trajectory $y_d(k)$, which is bounded. Next, several mild assumptions are needed to proceed.

Assumption 11.1 *The unknown nonlinear function $f(\bullet)$ in (11.2) is smooth.*

Assumption 11.2 *The disturbance $d(k)$ is bounded with a known bound $|d(k)| \le d_M$, and the partial derivative $|\partial f/\partial d(k)| \le D_M$ is also bounded, with D_M a positive constant.*

With Assumption 11.2, and by using Mean Value Theorem, Equation (11.2) can be rewritten as

$$\begin{aligned} y(k+\tau) &= f(w_k, u(k), \bar{d}_{k+\tau-1}) = f(w_k, u(k), 0) + \delta_f^T \bar{d}_{k+\tau-1} \\ &= f(w_k, u(k), 0) + \delta_{d_k}, \end{aligned} \tag{11.3}$$

where $\delta_f = \left[\left.\frac{\partial f}{\partial d(k+\tau-1)}\right|_{d(k+\tau-1)=d_\xi(k+\tau-1)}, \ldots, \left.\frac{\partial f}{\partial d(k)}\right|_{d(k)=d_\xi(k)} \right]^T$, $\delta_{d_k} = \delta_f^T \bar{d}_{k+\tau-1}$, and $d_\xi(k)$ is between 0 and $d(k)$, or $d_\xi(k) = 0 + \lambda(d(k) - 0)$, $\lambda \in [0, 1]$. Throughout this chapter, they are equivalent, and we will present by the former for simplicity.

Lemma 11.1 *δ_{d_k} is bounded by $\left|\delta_{d_k}\right| \le \tau D_M d_M$.*

Proof: Use (11.3) and Assumption 11.2.

The following virtual system is constructed first without disturbance as follows

$$y_n(k+\tau) = f(w_k, u(k), 0). \tag{11.4}$$

■

Assumption 11.3 *The term* $\partial f/\partial u(k) = g(k)$ *is bounded and satisfies* $0 < g_{\min} \leq g(k) \leq g_{\max}$*, where* $g_{\min}$ *and* $g_{\max}$ *are positive constants [1].*

Assumption 11.4 *Virtual system (11.4) is invertibly stable [27] which means bounded system output implies bounded system input.*

From Assumptions 11.3 and 11.4, we can draw the conclusion that for any output trajectory $y_n(k+\tau) = f(w_k, u(k), 0)$, there exists a unique and continuous (smooth) function $u(k) = f^{-1}(w_k, y_n(k+\tau), 0)$ [1]. Applying the Taylor series expansion of (3) with respect to $u(k)$ around $u(k-1)$ yields

$$\begin{aligned} y(k+\tau) &= f(w_k, u(k), 0) + \delta_{d_k} = f(w_k, u(k-1), 0) \\ &\quad + \frac{\partial f(w_k, u(k-1), 0)}{\partial u}\Delta u(k) + \frac{1}{2}\cdot\frac{\partial^2 f(w_k, u_\mu(k), 0)}{\partial u^2}\Delta u^2(k) + \delta_{d_k} \\ &= F(w_k, u(k)) + G(w_k)\Delta u(k) + \delta_{d_k}, \end{aligned} \tag{11.5}$$

where $F(w_k, u(k)) = f(w_k, u(k-1), 0) + \frac{1}{2}\cdot\partial^2 f(w_k, u_\mu(k), 0)/\partial u^2 \Delta u^2(k)$ and $G(w_k) = \partial f(w_k, u(k-1), 0)/\partial u$. Moreover, $u_\mu(k)$ is between $u(k)$ and $u(k+1)$ (or $u_\mu(k) = u(k+1) + \lambda(u(k+1) - u(k))$, $\lambda \in [0, 1]$) by using mean value theorem. In other words, there are no higher order terms in the Taylor series expansion missing, since they are incorporated into the second derivative. From (11.5), it is clear that it is similar to affine system as

$$y_n(k+\tau) = F(w_k, u(k)) + G(w_k)\Delta u(k). \tag{11.6}$$

Lemma 11.2 *Consider any desired system trajectory* $y_d(k) \in R$ *and corresponding nominal desired control input* $u_d(k) = f^{-1}(w_k, y_d(k+\tau), 0)$*, then there exists* $u_\xi(k)$ *between an input* $u_n(k)$ *and* $u_d(k)$ *such that*

$$\begin{aligned} F(w_k, u_n(k)) = F(w_k, u_d(k)) + &\frac{\partial F(w_k, u_\xi(k))}{\partial u}\cdot\frac{\partial f^{-1}(w_k, y_\xi(k+\tau), 0)}{\partial y}\cdot \\ &(y_n(k+\tau) - y_d(k+\tau)), \end{aligned} \tag{11.7}$$

where $u_\xi(k) = f^{-1}(w_k, y_\xi(k+\tau), 0)$.

Lemma 11.3 *Consider the output of the virtual system* $y_n(k+\tau) = f(w_k, u_n(k), 0)$ *for a given input* $u_n(k)$*, then there exists* $y_\varsigma(k+\tau)$ *between* $y_n(k+\tau)$ *and* $y_d(k+\tau)$

such that

$$\begin{aligned} u_n(k) &= f^{-1}(w_k, y_n(k+\tau), 0) = f^{-1}(w_k, y_d(k+\tau), 0) \\ &\quad + \frac{\partial f^{-1}(w_k, y_\varsigma(k+\tau), 0)}{\partial y}(y_n(k+\tau) - y_d(k+\tau)) \\ &= u_d(k) + \frac{\partial f^{-1}(w_k, y_\varsigma(k+\tau), 0)}{\partial y}(y_n(k+\tau) - y_d(k+\tau)). \end{aligned} \tag{11.8}$$

Proof: Lemmas 11.2 and 11.3 can be obtained by using the chain-rule and mean value theorem. Further, we have following lemma. ■

Lemma 11.4 *Consider system (11.6) with Lemmas 11.2 and 11.3, we have*

$$\frac{\partial F(w_k, u_\xi(k))}{\partial u} \cdot \frac{\partial f^{-1}(w_k, y_\xi(k+\tau), 0)}{\partial y} + G(w_k)\frac{\partial f^{-1}(w_k, y_\varsigma(k+\tau), 0)}{\partial y} = 1. \tag{11.9}$$

Proof:

i. If $y_n(k+\tau) = y_d(k+\tau)$, then $y_\xi(k+\tau) = y_\varsigma(k+\tau) = y_d(k+\tau)$. Therefore, (9) could be obtained by differentiating (11.66) with respect to $y_n(k+\tau)$,
ii. If $y_n(k+\tau) \neq y_d(k+\tau)$, then from (11.6), one has

$$\begin{aligned} y_n(k+\tau) &= F(w_k, u_n(k)) + G(w_k)(u_n(k) - u(k-1)) = F(w_k, u_d(k)) \\ &\quad + G(w_k)(u_d(k) - u(k-1)) \\ &\quad + \left(\frac{\partial F(w_k, u_\xi(k))}{\partial u} \frac{\partial f^{-1}(w_k, y_\xi(k+\tau), 0)}{\partial y} + \frac{\partial f^{-1}(w_k, y_\varsigma(k+\tau), 0)}{\partial y} \right) \\ &\quad \times (y_n(k+\tau) - y_d(k+\tau)). \end{aligned} \tag{11.10}$$

Substituting $y_d(k+\tau) = F(w_k, u_d(k)) + G(w_k)(u_d(k) - u(k-1))$ into Equation (11.10) yields $\partial F(w_k, u_\xi(k))/\partial u \cdot \partial f^{-1}(w_k, y_\xi(k+\tau), 0)/\partial y + G(w_k)\partial f^{-1}(w_k, y_\varsigma(k+\tau))/\partial y = 1$. ■

Lemma 11.5 *For any $y_\varsigma(k+\tau) \in S$ and corresponding control input $u_\varsigma(k) = f^{-1}(w_k, y_\varsigma(k+\tau), 0)$, the following statement holds*

$$\frac{\partial f(w_k, u_\varsigma(k), 0)}{\partial u} \cdot \frac{\partial f^{-1}(w_k, y_\varsigma(k+\tau))}{\partial y} = 1. \tag{11.11}$$

Proof: It can be straightforward to verify (11.11) by differentiating $y_\varsigma(k+\tau) = f(w_k, f^{-1}(w_k, y_\varsigma(k+\tau)), 0)$ with respect to $y_\varsigma(k+\tau)$.

Therefore, substituting (11.7) into (11.5) produces the system dynamics in terms of the tracking error as

$$\begin{aligned} e(k+\tau) &= y(k+\tau) - y_d(k+\tau) = F(w_k, u(k)) + G(w_k)\Delta u(k) \\ &\quad + \delta_{d_k} - y_d(k+\tau) = F(w_k, u_d(k)) \\ &\quad + \frac{\partial F(w_k, u_\xi(k))}{\partial u} \cdot \frac{\partial f^{-1}(w_k, y_\xi(k+\tau), 0)}{\partial y} \cdot (y(k+\tau) - y_d(k+\tau)) \\ &\quad + G(w_k)\Delta u(k) + \delta_{d_k} - y_d(k+\tau). \end{aligned} \tag{11.12}$$

Making use of Lemma 11.4, (11.12) can be written as

$$\begin{aligned} e(k+\tau) &= F(w_k, u_d(k)) + G(w_k)\Delta u(k) + \delta_{d_k} - y_d(k+\tau) \\ &\quad + \left(1 - G(w_k)\frac{\partial f^{-1}(w_k, y_\varsigma(k+\tau), 0)}{\partial y}\right) \cdot (y(k+\tau) - y_d(k+\tau)) \\ &= F(w_k, u_d(k)) + G(w_k)\Delta u(k) + \delta_{d_k} - y_d(k+\tau) \\ &\quad + \left(1 - G(w_k)\frac{\partial f^{-1}(w_k, y_\varsigma(k+\tau), 0)}{\partial y}\right) \cdot e(k+\tau). \end{aligned} \tag{11.13}$$

Combining (11.11) and (11.13), one has

$$e(k+\tau) = \frac{\partial f(w_k, u_\varsigma(k), 0)}{\partial u}\left(\frac{F(w_k, u_d(k)) + \delta_{d_k} - y_d(k+\tau)}{G(w_k)} + \Delta u(k)\right). \tag{11.14}$$

By defining $\partial f(w_k, u_\varsigma(k), 0)/\partial u = \kappa_k$, Equation (11.14) can be expressed as

$$\begin{aligned} e(k+\tau) &= \frac{\kappa_k}{G(w_k)}(F(w_k, u_d(k)) - y_d(k+\tau)) + \kappa_k \Delta u(k) + \frac{\kappa_k}{G(w_k)}\delta_{d_k} \\ &= F_a(w_k, y_d(k+\tau), \kappa_k) + \kappa_k \Delta u(k) + \delta_{\kappa_k} \\ &\quad F_a(w_k, y_d(k+\tau), \kappa_k) = \frac{\kappa_k}{G(w_k)}(F(w_k, u_d(k)) \\ &\quad - y_d(k+\tau)) \text{ and } \delta_{\kappa_k} = \frac{\kappa_k}{G(w_k)}\delta_{d_k}. \end{aligned} \tag{11.15}$$

Notice that $0 < g_{\min} \leq \kappa_k \leq g_{\max}$, $0 < g_{\min} \leq G(w_k) \leq g_{\max}$ due to Assumption 11.3. By referring to Lemma 11.1, one also observes that δ_{κ_k} is bounded above by $|\delta_{\kappa_k}| \leq g_{\max}\tau D_M d_M / g_{\min}$.

By rewriting the nonaffine SISO nonlinear discrete-time system into an equivalent affine-like representation (11.15) in terms of error dynamics and incorporating the higher order terms into the second derivative, the difficulty of designing controllers for nonaffine nonlinear discrete-time systems is avoided. ∎

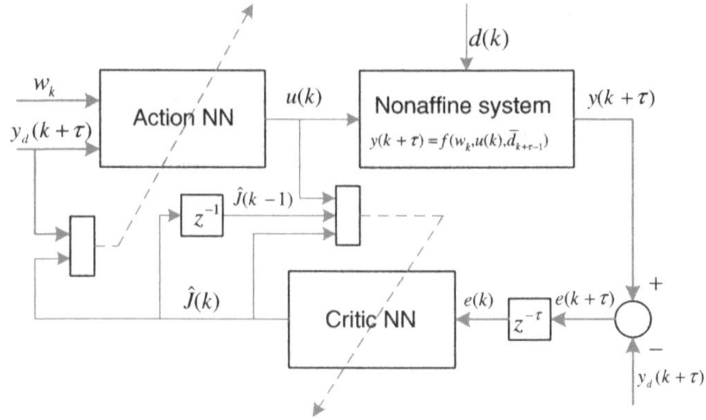

FIGURE 11.1 Online reinforcement learning neural controller structure.

11.3.2 Online Reinforcement Learning Controller Design

The purpose of this study is to design an online reinforcement learning NN controller for the equivalent error dynamics (11.15), such that (1) all the signals in the closed-loop system remain bounded, (2) the output $y(k)$ follows a desired trajectory $y_d(k) \in S$, and (3) the long-term cost function is minimized so that a near optimal control input can be generated [8]. Here, the "online" means the learning of the controller takes place "in real-time" by interacting with the plant, instead of in an offline or iterative manner. No value and policy iterations are needed. Instead, the value (or cost function) and policy (or control input) will be updated once per sampling interval.

The block diagram representation of the proposed controller is shown in Figure 11.1. Here the weights of two NN are initialized at zero and tuned online without any offline learning phase. In addition, no iterations are used.

11.3.3 The Action NN Design

Consider system (11.15), and to reduce the tracking error, a desired control law is given by

$$u_d(k) = u(k-1) - \frac{1}{\kappa_k} F_a(w_k, y_d(k+\tau), \kappa_k). \tag{11.16}$$

As a consequence, the tracking error will go to zero after τ steps if no disturbance is present. An action NN is employed to produce an approximation of control signal since the nonlinearity in Equation (11.16) is not known. From (11.16) and considering Assumptions 3 and 4, the desired control signal can be approximated by the action NN as

$$u_d(k) = w_a^T \phi_a(v_a^T s(k)) + \varepsilon_a(s(k)) = w_a^T \phi_a(k) + \varepsilon_a(k). \tag{11.17}$$

where $s(k) = \left[w_k^T, y_d(k+\tau)\right]^T$ is the action NN input vector. n_a is the number of neurons in the hidden layer, and $w_a \in R^{n_a \times 1}$, $v_a \in R^{2n \times n_a}$ denote the desired weight matrix of the output and hidden layer respectively with $\varepsilon_a(k) = \varepsilon_a(s(k))$ is the action NN approximation error. Since v_a is fixed, for convenience, the hidden layer activation function vector $\phi_a(v_a^T s(k)) \in R^{n_a}$ is written as $\phi_a(k)$. Meanwhile, the actual output can be expressed as

$$u(k) = \hat{w}_a^T(k)\phi_a(v_a^T s(k)) = \hat{w}_a^T(k)\phi_a(k). \tag{11.18}$$

where $\hat{w}_a(k) \in R^{n_a \times 1}$ is the actual weight matrix of the output layer at instant k.

Using the action NN output as the control signal, and substituting (11.17) and (11.18) into (11.15) yields

$$\begin{aligned} e(k+\tau) &= F_a(w_k, y_d(k+\tau), \kappa_k) + \kappa_k \Delta u(k) + \delta_{\kappa_k} = \kappa_k(u(k) - u_d(k)) + \delta_{\kappa_k} \\ &= \kappa_k(\tilde{w}_a^T(k)\phi_a(k) - \varepsilon_a(k)) + \delta_{\kappa_k} = \kappa_k \zeta_a(k) + d_a(k). \end{aligned} \tag{11.19}$$

where $\tilde{w}_a(k) = \hat{w}_a(k) - w_a$ is the weight estimation error, $\zeta_a(k) = \tilde{w}_a^T(k)\phi_a(k)$, and $d_a(k) = -\kappa_k \varepsilon_a(k) + \delta_{\kappa_k}$. Next the critic NN design is introduced.

11.3.4 The Critic NN Design

As stated earlier, a near optimal controller can stabilize the closed-loop system by minimizing the cost function. In this regard, a critic NN is employed to approximate the unknown long-term cost function $J(k)$.

First, the prediction error generated by the critic or the Bellman error [8] is defined as

$$e_c(k) = \gamma \hat{J}(k) - \hat{J}(k-1) + r(k), \tag{11.20}$$

where the subscript "c" stands for the "critic" and the approximated cost function is given by

$$\hat{J}(k) = \hat{w}_c^T(k)\phi_c(v_c^T e(k)) = \hat{w}_c^T(k)\phi_c(k), \tag{11.21}$$

with $\hat{J}(k) \in R$ being the critic NN output for approximating $J(k)$. $\hat{w}_c(k) \in R^{n_c \times 1}$ and $v_c \in R^{1 \times n_c}$ represent the actual weight matrices of the output and hidden layer respectively. The term n_c denotes the number of the neurons in the hidden layer. Similar to HDP, the tracking error $e(k)$ is selected as the critic NN input. The activation function vector of the hidden layer $\phi_c(v_c^T e(k)) \in R^{n_c}$ is denoted as $\phi_c(k)$ for convenience. Provided with enough number of hidden layer neurons, the optimal long-term cost function $J^*(k)$ can be approximated with arbitrarily small error $\varepsilon_c(k)$ as

$$J^*(k) = w_c^T \phi_c(v_c^T e(k)) + \varepsilon_c(e(k)) = w_c^T \phi_c(k) + \varepsilon_c(k), \tag{11.22}$$

Next, the critic NN weight estimation error can be defined as

$$\tilde{w}_c(k) = \hat{w}_c(k) - w_c, \tag{11.23}$$

where the approximation error is given by $\zeta_c(k) = \tilde{w}_c^T(k)\phi_c(k)$. Then the critic error can be rewritten as

$$\begin{aligned} e_c(k) = \gamma \hat{J}(k) - \hat{J}(k-1) + r(k) &= \gamma\zeta_c(k) + \gamma J^*(k) - \zeta_c(k-1) \\ &- J^*(k-1) + r(k) - \varepsilon_c(k) + \varepsilon_c(k-1). \end{aligned} \tag{11.24}$$

Next the weight tuning laws are introduced for both action and critic NNs.

11.3.5 Weight Updating Laws for the NNs

Following the discussion from the last section, the objective function to be minimized by the critic NN can be defined as a quadratic function of Bellman error as

$$E_c(k) = \frac{1}{2}e_c^T(k)\, e_c(k) = \frac{1}{2}e_c^2(k). \tag{11.25}$$

Using a standard gradient-based adaptation method, the weight updating scheme for the critic NN is given by

$$\hat{w}_c(k+\tau) = \hat{w}_c(k) + \Delta\hat{w}_c(k), \tag{11.26}$$

where $\Delta\hat{w}_c(k) = \alpha_c\left[-\partial E_c(k)/\partial \hat{w}_c(k)\right]$ with $\alpha_c \in R$ is the adaptation gain. Combining (11.20), (11.21), and (11.25) in $\Delta\hat{w}_c(k)$, the critic NN weight updating rule can be obtained by using the chain rule as

$$\begin{aligned} \Delta\hat{w}_c(k) &= -\alpha_c \frac{\partial E_c(k)}{\partial \hat{w}_c(k)} = -\alpha_c \frac{\partial E_c(k)}{\partial e_c(k)} \frac{\partial e_c(k)}{\partial \hat{J}(k)} \frac{\partial \hat{J}(k)}{\partial \hat{w}_c(k)} \\ &= -\alpha_c\gamma\phi_c(k)(\gamma\hat{J}(k) + r(k) - \hat{J}(k-1)). \end{aligned} \tag{11.27}$$

In the meantime, the objective for adapting the action NN is to track the desired output and to lower the cost function simultaneously. Therefore, the error for the action NN can be formed by combining the functional estimation error $\zeta_a(k)$, and the critic signal $\hat{J}(k)$. Let

$$e_a(k) = \sqrt{\kappa_k}\zeta_a(k) + \left(\sqrt{\kappa_k}\right)^{-1}\left(\hat{J}(k) - J_d(k)\right) = \sqrt{\kappa_k}\zeta_a(k) + \left(\sqrt{\kappa_k}\right)^{-1}\hat{J}(k), \tag{11.28}$$

where $\zeta_a(k)$ is defined in (11.22). The desired long-term cost function $J_d(k)$ is nominally defined and is considered to be zero ("0"), which means as low as possible [24].

Hence, the action NN weights $\hat{w}_a(k)$ are tuned to minimize the error

$$E_a(k) = \frac{1}{2} e_a^T(k) e_a(k). \tag{11.29}$$

Combining Equations (11.20), (11.22), (11.34), and (11.35) and using the chain rule yields

$$\begin{aligned}\Delta \hat{w}_a(k) &= -\alpha_a \frac{\partial E_a(k)}{\partial \hat{w}_a(k)} = -\alpha_a \frac{\partial E_a(k)}{\partial e_a(k)} \frac{\partial e_a(k)}{\partial \zeta_a(k)} \frac{\partial \zeta_a(k)}{\partial \hat{w}_c(k)} \\ &= -\alpha_a \phi_a(k)(\kappa_k \zeta_a(k) + \hat{J}(k))^T = -\alpha_a \phi_a(k)(e(k+\tau) - d_a(k) + \hat{J}(k))^T,\end{aligned} \tag{11.30}$$

where $\alpha_a \in R^+$ is the adaptation gain of the action NN. Since $d_a(k)$ is typically unavailable, and similar to the ideal case, the $d(k)$ and the mean value of $\varepsilon_a(k)$ over the compact subset of R are assumed to be zero [21]. Then the weight-updating scheme for the action NN will be given as

$$\hat{w}_a(k+\tau) = \hat{w}_a(k) - \alpha_a \phi_a(k)(e(k+\tau) + \hat{J}(k))^T. \tag{11.31}$$

11.3.6 Main Theoretic Results

Assumption 11.5 *Let the desired output layer weights for the action and critic NNs are unknown but bounded above such that $\|w_a\| \le w_{am}$, and $\|w_c\| \le w_{cm}$, where $w_{am} \in R^+$ and $w_{cm} \in R^+$ represent the bounds on the unknown target weights.*

Assumption 11.6 *The activation functions for the action and critic NNs are bounded by known positive values, such that $\|\phi_a(k)\| \le \phi_{am}$, $\|\phi_c(k)\| \le \phi_{cm}$ where ϕ_{am}, $\phi_{cm} \in R^+$ represent the upper bound. This bound is easily satisfied, since hyperbolic tangent sigmoid transfer function is normally chosen.*

Assumption 11.7 *The NN approximation errors and unmodeled dynamics $\varepsilon_a(k)$ and $\varepsilon_c(k)$ are assumed to be bounded above in the compact set$S \subset R$ byε_{am} and ε_{cm} [1].*

Lemma 11.6 *With the Assumptions 11.3 and 11.9, $d_a(k)$ in Equation (11.23) is bounded above in the compact set $S \subset R$ by*

$$|d_a(k)| \le d_{am} = g_{\max}\varepsilon_{am} + g_{\max}\tau D_M d_M / g_{\min}. \tag{11.32}$$

Combining Assumptions 11.1, 11.3, and 11.4 and others, the main result is introduced next.

Theorem 11.1 *Consider the nonlinear discrete-time system given by (11.2) whose dynamics can be expressed as (11.15). Let the Assumptions 11.1 through 11.7 hold with the disturbance bound d_M a known constant. Let the control input be provided by the action NN (11.19), with the critic NN (11.21). Further, let the weights of the action NN and the critic NN be tuned by (11.27) and (11.31), respectively. Then the tracking error $e(k)$, and the NN weight estimates of the action and critic NNs, $\tilde{w}_a(k)$, and $\tilde{w}_c(k)$ are UUB [22], provided the controller design parameters are selected as (a) $0 < \alpha_a \phi_a^2(k) < g_{\min}/g_{\max}^2$, (b) $0 < \alpha_c \phi_c^2(k) < 1/\gamma^2$, and (c) $\gamma > 2$ where α_a and α_c are NN adaptation gains, and γ is part of the* strategic *utility function.*

Proof: Define a Lyapunov candidate as

$$L(k) = \sum_{i=1}^{4} L_i = \frac{\gamma_1}{2} \sum_{j=0}^{\tau-1} e^2(k+j) + \frac{\gamma_2}{\alpha_a} \sum_{j=0}^{\tau-1} tr\left(\tilde{W}_a^T(k+j)\tilde{W}_a(k+j)\right)$$
$$+ \frac{\gamma_3}{\alpha_c} \sum_{j=0}^{\tau-1} tr\left(\tilde{W}_c^T(k+j)\tilde{W}_c(k+j)\right) + \gamma_4 \sum_{j=0}^{\tau-1} \zeta_c{}^2(k+j), \tag{11.33}$$

where $\gamma_i \in R^+$, $i = 1, 2, 3, 4$ are design parameters. Set $\gamma_2 = \gamma_2' \gamma_2''$, where $\gamma_2'' 1 - \alpha_a \phi_a^2(k) g_{\min}/g_{\min} - \alpha_a \phi_a^2(k) g_{\max}^2 \leq \frac{1}{2}$, therefore,

$$\begin{aligned}
\Delta L(k) \leq & -\left(\frac{\gamma_1}{2} - \frac{\gamma_3}{4} Q\right) e^2(k) - \left(\gamma_2 g_{\min} - \gamma_1 g_{\max}^2 - \frac{\gamma_3}{8} R\right) \zeta_a^2(k) \\
& - \left(\gamma_3 \gamma^2 - \gamma_2' - \gamma_4\right) \zeta_c^2(k) - \left(\gamma_4 - \frac{\gamma_3}{8}\right) \zeta_c^2(k-1) \\
& - \gamma_3 \left(1 - \alpha_c \gamma^2 \phi_c^2(k)\right) e_c^2(k) - \gamma_2 \left(g_{\min} - \alpha_a \phi_a^2(k) g_{\max}^2\right) \\
& \times \left(\zeta_a(k) + \frac{(1 - \alpha_a \phi_a^2(k) \kappa_k)}{g_{\min} - \alpha_a \phi_a^2(k) g_{\max}^2}\right)^2 + D_M^2,
\end{aligned} \tag{11.34}$$

where

$$D_M^2 = \left(\gamma_1 + \frac{\gamma_2'}{2}\right) d_{am}^2 + \left(\frac{\gamma_2'}{2} + \frac{\gamma_3}{8}(\gamma+1)^2\right) J_M^2 + \frac{\gamma_3}{8} R w_{am}^2 \phi_{am}^2 + \gamma_3 \varepsilon_{cm}^2, \tag{11.35}$$

From the standard Lyapunov analysis, $\Delta L \leq 0$ holds as long as $\|e(k)\| \geq D_M/(\sqrt{\gamma_1 - 4\gamma_3 Q})$, $\|\zeta_a(k)\| \leq D_M/\left(\sqrt{\gamma_2 g_{\min} - \gamma_1 g_{\max}^2} - 8\gamma_3 R\right)$, and $\|\zeta_c(k)\| \leq D_M/\left(\sqrt{\gamma_3 \gamma^2 - \gamma_2' - \gamma_4}\right)$. According to the standard Lyapunov extension theorem [22], the analysis earlier demonstrates that the tracking error $\|e(k)\|$ and the weights of the estimation errors are UUB. Further, the boundedness of $\|\zeta_a(k)\|$ and $\|\zeta_c(k)\|$ implies that the weight estimations $\|\hat{w}_a(k)\|$ and $\|\hat{w}_c(k)\|$ are also bounded. For detailed steps of proof, refer to [22]. ■

11.4 TIME-BASED ADAPTIVE DYNAMIC PROGRAMMING-BASED OPTIMAL CONTROL

This section addresses the forward-in-time optimal control of unknown nonaffine MIMO nonlinear discrete-time systems by transforming the nonaffine system into an affine-like equivalent system in the *input–output* form and by assuming that the higher order terms are bounded above. First, a NN based online identifier is proposed to learn the unknown input gain matrix (IGM) matrix of the affine-like equivalent system. Subsequently, the forward-in-time approach similar to [4] is introduced to the generic unknown affine-like equivalent system by using output feedback. Here value and policy iterations are not utilized whereas, the value function and the control input are updated once a sampling interval. Using an initial stabilizing control, an online approximator (OLA) is tuned to learn the cost function which is subsequently utilized along with the estimated IGM to generate the optimal control input. The nonaffine nature of the system, online identifier, and lack of system states complicate the stability analysis whereas the boundedness of all the closed-loop signals is demonstrated while demonstrating that the actual control input approaches its optimal value. It will be shown that given an initial stabilizing control, the proposed approach renders an optimal controller over time.

Subsequently, to mitigate the modeling errors due to higher order terms in the ultimate bounds, an auxiliary term is designed via fast dynamic inversion technique. The fast dynamic solver, along with the closed-loop system, forms a singularly perturbed system whose stability is shown to be guaranteed. The net result is the output feedback-based robust optimal controller design for nonaffine MIMO systems.

Consider a generic form of MIMO nonaffine system described by

$$x_{k+1} = f(x_k, u_k), \tag{11.36}$$

$$y_k = g(x_k, u_k), \tag{11.37}$$

where $u_k \in E_u \subset \mathbb{R}^m$, $x_k \in E_x \subset \mathbb{R}^n$, and $y_k \in E_y \subset \mathbb{R}^\ell$ represent the system input, states, and the outputs respectively, and $f(\bullet)$ and $g(\bullet)$ are assumed to be unknown continuous nonlinear functions with respect to x_k and u_k. The system representation (11.1) indicates a nonlinear system whose u_k output y_k is controlled u_k. In other words, $g(x_k, u_k)$ is a one-to-one mapping between y_k and x_k for a given u_k. In this case, there exists a one-to-one function g^{-1} such that $x_k = g^{-1}(y_k, u_k)$. Using Equations (11.36) and (11.37), we have

$$\begin{aligned} y_{k+1} &= g(x_{k+1}, u_{k+1}) = g(f(x_k, u_k), u_{k+1}) = h(x_k, u_k, u_{k+1}) \\ &= h(g^{-1}(y_k, u_k), u_k, u_{k+1}). \end{aligned} \tag{11.38}$$

Now, assume that the overall controller is designed such that the system input increment, denoted by Δu_k, is generated. This implies that the system input and the controller output are connected such that $u_{k+1} = u_k + \Delta u_k$. Using Equations (11.36) and (11.38) we can express the system dynamics as

$$y_{k+1} = h(y_k, u_k, \Delta u_k) \text{ or } X_{k+1} = H(X_k, \Delta u_k), \tag{11.39}$$

where $X_{k+1} = (u_k \;\; y_k)^T \subset \mathbb{R}^{m+\ell}$ and $H(X_k, \Delta u_k) = ((u_k + \Delta u_k)^T \;\; h(g^{-1}(y_k, u_k), u_k, u_k + \Delta u_k)^T)^T$. By applying the Taylor series expansion, (11.39) can be expanded as [28]

$$\begin{aligned} X_{k+1} = H(X_k, \Delta u_k) &= H(X_k) + \frac{\partial}{\partial \Delta u_k} H(X_k, \Delta u_k)|_{\Delta v_k=0} \Delta u_k \\ &+ \frac{1}{2} \frac{\partial}{\partial \Delta u_k} \left(\frac{\partial}{\partial \Delta u_k} H(X_k, \Delta u_k) \Delta u_k \right) |_{\Delta u_k=0} \Delta u_k + \cdots \\ &= \sum\nolimits_{i=0}^{q} F_i(X_k) \Delta u_k^i + O(X_k, \Delta u_k), \end{aligned} \tag{11.40}$$

where $O(X_k, \Delta u_k)$ represents the higher order terms of the Taylor series expansion. Moreover, Δu_k^i denotes a vector whose elements include all possible ith multiplication of the elements of Δu_k^i. For example

$$\Delta u_k^2 = \begin{pmatrix} \Delta u_k(1)\Delta u_k(1), \Delta u_k(1)\Delta u_k(2), \ldots, \Delta u_k(1)\Delta u_k(m), \Delta u_k(2)\Delta u_k(2), \Delta u_k(2)\Delta u_k(3), \ldots, \Delta u_k(2)\Delta u_k(m), \\ \ldots, \Delta u_k(m)\Delta u_k(m) \end{pmatrix}, \tag{11.41}$$

with $\Delta u_k^2 \in \mathbb{R}^{m \times (m+1)/2}$. Typically, for practical systems, the higher order terms, $O(X_k, \Delta u_k)$, can be considered small and negligible due to uniform convergence of Taylor series when $H(X_k, \Delta u_k)$ is differentiable [28]. Instead, the higher order modeling errors, $O(X_k, \Delta u_k)$, are assumed to be bounded above when $q = 1$ and later relaxed. In the sequel, we denote $O(X_k, \Delta u_k)$ by O_k for the sake of simplicity. Therefore, the unknown affine-like system representation of (11.1) takes the following input–output form as

$$X_{k+1} = F(X_k) + G(X_k)\Delta u_k + O_k, \tag{11.42}$$

where $F(X_k) = F_0(X_k)$ is the internal dynamics and $G(X_k) = F_1(X_k)$ being the input gain matrix (IGM). In Section 11.4.2, the optimal control scheme for the affine-like system (11.42) in the input–output form is introduced which requires $G(X_k)$ and the higher order terms O_k to be known which is a strong assumption while the information on internal dynamics, $F(X_k)$, is not required [4, 11]. Therefore, in the next subsection, a new online NN identifier is introduced first to approximate IGM and the higher order terms.

11.4.1 Online NN-Based Identifier

The objective of this section is to introduce a novel NN-based identifier for (11.42) when the system inputs and outputs are given. The NN-based identifier is novel in the sense that it can directly identify $F(X_k)$, IGM, and O_k by using a single NN without using states of the system. Instead of offline methods [1, 9, 5] the proposed identifier works online. Assume that a stabilizing input is applied to the system (11.42), the following expression can be used for approximation of the nonaffine system (11.42) at the instant k in a compact set Σ_1 as

$$X_k = \Psi(X_{k-1}, \Delta u_{k-1})^T W \Delta U_{k-1} + \bar{\varepsilon}_{k-1}, \tag{11.43}$$

where $\Psi(X_{k-1}, \Delta u_{k-1}) \in \mathbb{R}^{p\times(m+\ell)}$, $W \in \mathbb{R}^{p\times(m+\ell)}$, $\Delta U_k = \begin{pmatrix} 1 & \cdots & 1 & \Delta u_k^T \end{pmatrix}^T \in \mathbb{R}^{(m+\ell)}$, $\bar{\varepsilon}_k \in \mathbb{R}^{m+\ell}$, $\|\Psi(X_{k-1}, \Delta u_{k-1})\| \le \Psi_M$, $\|\Psi(X_{k-1}, \Delta u_{k-1})U_{k-1}\| \le \bar{\Psi}_M$ being the bounded NN activation function and $\bar{\varepsilon}_k$ is the estimation error satisfying $\|\bar{\varepsilon}_k\| < \bar{\varepsilon}_M$ [1]. Moreover, the following bound $U_m \le \|\Delta U_k\|$ holds due to the presence of constant values in the input vector.

It should be noted that the identification scheme has the advantage of identifying $F(X_k)$,$G(X_k)$, and O_k separately and online. Next, select $\Psi(X_k, \Delta u_k)^T = \left(\Psi_F(X_k)^T, \Psi_O(X_k, \Delta u_k)^T, \Psi_G(X_k)^T\right)$, and $W = \left(W_F^T, W_O^T, W_G^T\right)^T$ accordingly. Our goal next is to identify the NN weight matrix W denoted here as $\hat{W}_k$ by estimating the state vector X_k. Define the identification error as

$$\begin{aligned}\tilde{e}_k &= X_k - \hat{X}_k = \Psi(X_{k-1}, \Delta u_{k-1})^T (W - \hat{W}_{k-1}) \Delta U_{k-1} + \bar{\varepsilon}_{k-1} \\ &= \Psi(X_{k-1}, \Delta u_{k-1})^T \tilde{W}_{k-1} \Delta U_{k-1} + \bar{\varepsilon}_{k-1}.\end{aligned} \tag{11.44}$$

Define the update law for the actual NN weights $\hat{W}_k$ as

$$\hat{W}_k = \hat{W}_{k-1} + \alpha \tilde{E}_k \Psi(X_{k-1}, \Delta u_{k-1}) \Delta U_{k-1}^T / \left(\|\Delta U_{k-1}\|^2 + 1\right), \tag{11.45}$$

where

$$\tilde{E}_k = \mathrm{diag}(\tilde{e}_k) \in \mathbb{R}^{(m+\ell)\times(m+\ell)}, \tag{11.46}$$

with $\alpha > 0$ being the design parameter or NN learning rate. The error dynamics in weight update law can be written as

$$\tilde{W}_k = \tilde{W}_{k-1} - \alpha \tilde{E}_k \Psi(X_{k-1}, \Delta u_{k-1}) \Delta U_{k-1}^T / \left(\|\Delta U_{k-1}\|^2 + 1\right). \tag{11.47}$$

Now, by considering the structure of ΔU_k we have

$$X_{k+1} = W_F^T \Psi_F(X_k) + W_G^T \Psi_G(X_k) \Delta u_k + W_O^T \Psi_O(X_k, \Delta u_k) + \bar{\varepsilon}_k. \tag{11.48}$$

It should be noted that if the identification error $\|\tilde{e}_k\|$ and weight estimation error $\left\|\tilde{W}_k\right\|_F$ converges, the identification of $F(X_k)$,$G(X_k)$, and O_k is complete. Therefore, the following theorem can be stated to show the boundedness of the identification and the NN weight estimation errors. Here we will use Ψ_k instead of $\Psi(X_k, \Delta u_k)$ for the sake of brevity.

Theorem 11.2 *[25]: Assume that the proposed identifier in (11.43) with the update law in (11.47) is used to identify the system (11.45) and the design parameter is chosen as $0 < \alpha < 32U_{\mathrm{m}}^2\Psi_m^2/\left(4\bar{\Psi}_M^2 + \bar{\Psi}_M^2\Psi_M^2\right)$. Then the identification error $\tilde{e}_k$ and the NN weight estimation errors $\tilde{W}_k$ are uniformly ultimately bounded (UUB) [1] and converge to a basin of attraction with the bounds given by $\left\|\tilde{W}_k\right\| \leq W_B$ or $\|\tilde{e}_k\| \leq b_e$ where*

$$e_B \triangleq \bar{\varepsilon}_M \sqrt{1 + \frac{\Psi_M^2}{4} + \frac{\left(\Psi_M^2 + 2\alpha\bar{\Psi}_M^2\right)^2}{\alpha\Psi_M^2\left(8U_{\mathrm{m}}^2\Psi_m^2 - \alpha\Psi_M^2 U_M^2\left\{4 + \Psi_M^2\right\}\right)}}, \tag{11.49}$$

or

$$W_B \triangleq 2\bar{\varepsilon}_M\left(\Psi_M^2 + 2\alpha\bar{\Psi}_M^2\right)/\left\{8U_{\mathrm{m}}^2\Psi_m^2\Psi_M - \alpha\bar{\Psi}_M^2\left(4\Psi_M + \Psi_M^3\right)\right\}$$
$$+\bar{\varepsilon}_M \sqrt{1 + \frac{\Psi_M^2}{4} + \frac{\left(\Psi_M^2 + 2\alpha\bar{\Psi}_M^2\right)^2}{\alpha\Psi_M^2\left(8U_{\mathrm{m}}^2\Psi_m^2 - \alpha\bar{\Psi}_M^2\left(4 + \Psi_M^2\right)\right)}}. \tag{11.50}$$

11.4.2 Neural Network-Based Optimal Controller Design

The next step is to optimally stabilize the affine-like system by using the identifier and to guarantee that the closed-loop remains stable in the neighborhood around the origin while ensuring that the actual control input is bounded and close to the optimal control input. To this end, the affine representation (11.39) is considered with the cost function $J(k)$ such that

$$J(X_k) \equiv \sum_{i=0}^{\infty} r(k+i) = r(X_k, \Delta u_k) + J(X_{k+1}), \tag{11.51}$$

where $r(X_k, \Delta u_k) = Q(X_k) + \Delta u_k^T R \Delta u_k$ with $Q(X_k) \geq 0$, $Q(0) = 0$, and $R \in \mathbb{R}^{m\times m}$ as a positive definite matrix. In the sequel, we will denote $J(X_k)$ by J_k for the sake of simplicity. Next the following definition is needed in order to proceed.

Definition 11.2 *[29] The control input Δu_k is admissible with respect to the infinite horizon cost function (11.51) on a compact set Σ_2 provided the control action Δu_k is (a) continuous on Σ_2, (b) stabilizes (11.42) on Σ_2 with $\Delta u_k|_{X_k=0} = 0$, and (c) makes $J(X_0)$ finite (upper bounded) for all $X_0 \in \Sigma_2$.*

The objective is minimize J_k by starting with an admissible control law and modifying it with respect to the system dynamics so that the estimated cost function and control input converge to the optimal cost function J_k^* and control law Δu_k^*, respectively. By applying the stationarity condition [29] to (11.51) we have

$$\frac{\partial J_k}{\partial \Delta u_k} = \frac{\partial r(X_k, \Delta u_k)}{\partial \Delta u_k} + \frac{\partial J_{k+1}}{\partial \Delta u_k} = 2R\Delta u_k + \left(\frac{\partial X_{k+1}}{\partial \Delta u_k}\right)^T \frac{\partial J_{k+1}}{\partial X_{k+1}} = 0. \quad (11.52)$$

Using (11.52) and (11.42) we can write

$$\Delta u_k^* = -\frac{1}{2}R^{-1}G^T(X_k)\frac{\partial J_{k+1}^*}{\partial X_{k+1}} - \frac{1}{2}R^{-1}\left(\frac{\partial O_k}{\partial \Delta u_k}\right)^T \frac{\partial J_{k+1}^*}{\partial X_{k+1}}. \quad (11.53)$$

Since O_k is a function of Δu_k, solving (11.53) is impossible in general, therefore, we assume that O_k is upper bounded initially for the optimal control design. With this assumption, we will show in the next section that optimal controller $\Delta u_k^* = -0.5R^{-1}G^T(X_k)\partial J_{k+1}^*/\partial X_{k+1}$ for the affine system $X_{k+1} = F(X_k) + G(X_k)\Delta u_k$ can stabilize (11.42). However, the region of convergence will depend upon the bound of O_k which is not necessarily small. This will require the design of an auxiliary term, $\Delta\rho_k$, for robustness to be added to the optimal controller ($\Delta u_k = \Delta u_k^* + \Delta\rho_k$) in order to mitigate the effect of the higher order terms. The next sections are dedicated to the design of Δu_k^* and $\Delta\rho_k$ along with the assessment of the overall system stability.

11.4.3 Cost Function Approximation for Optimal Regulator Design

The objective of the optimal control law is to stabilize the system (11.42) while minimizing the cost-function (11.51). The cost-function (11.51) will be approximated by an OLA and written as

$$\hat{J}(k) = \hat{J}(X_k) = \hat{\Phi}_k^T \sigma(X_k) = \hat{\Phi}_k^T \sigma(k), \quad (11.54)$$

where $\hat{J}(k)$ represents an approximated value of the original cost function $J(k)$, $\hat{\Phi}_k$ is the vector of actual parameter vector for the target OLA parameter vector, Φ, and $\sigma(k) = \sigma_\ell(k)_1^{L:}$ is set of activation functions which are each chosen to be basis sets and thus are linearly independent. The basis function should satisfy $\|\sigma(0)\| = 0$ for $\|x\| = 0$ with $x \in \Re^n$. Selection of $\sigma(\bullet)$ in this way ensures $J(0) = 0$ [30]. For convenience, define the error in the cost function as

$$e_c(k) = r(k-1) + \hat{\Phi}_k^T \sigma(k) - \hat{\Phi}_k^T \sigma(k-1), \quad (11.55)$$

whose dynamics are given by

$$\Delta e_c(k+1) = r(k) + \hat{\Phi}_{k+1}^T \left(\sigma(k+1) - \sigma(k)\right). \quad (11.56)$$

Next, we define an auxiliary cost error vector as

$$E_c(k) = Y(k-1) + \hat{\Phi}_k^T \bar{X}(k-1) \in \Re^{1\times(1+j)}, \quad (11.57)$$

where $Y(k-1) = [r(k-1) \quad r(k-2) \ldots r(k-1-j)]$ and $\bar{X}(k-1) = [\Delta\sigma(k) \quad \Delta\sigma(k-1) \ldots \Delta\sigma(k-j)]$ with $\Delta\sigma(k) = \sigma(k) - \sigma(k-1)$, $0 < j < k-1 \in \boldsymbol{N}$ and $\boldsymbol{N}$ being the set of natural real numbers. It is useful to observe that (11.57) can be rewritten as $E_c(k) = [e_c(k|k) \quad e_c(k|k-1) \cdots e_c(k|k-j)]$ where the notation $e_c(k|k-1)$ means the cost error $e_c(k-1)$ re-evaluated at time k using the actual cost parameter matrix $\hat{\Phi}_k^T$. The dynamics of the auxiliary vector (11.57) are formed similar to (11.56) and revealed to be

$$E_c^T(k+1) = Y^T(k) + \bar{X}^T(k)\hat{\Phi}_{k+1}. \tag{11.58}$$

Examining the error dynamics in (11.58), it is observed that they closely resemble a nonlinear discrete-time system with $\hat{\Phi}_{k+1}$ being the control input, and $Y(k)$ and $X(k)$ being nonlinear vector fields. To proceed, the following technical results are needed.

Now define the cost function OLA parameter update to be

$$\hat{\Phi}_{k+1} = \bar{X}(k)\left(\bar{X}^T(k)\bar{X}(k)\right)^{-1}\left(\alpha_c E_c^T(k) - Y^T(k)\right), \tag{11.59}$$

where $0 < \alpha_c < 1$. Substituting (11.59) into (11.58) reveals

$$E_c^T(k+1) = \alpha_c E_c^T(k). \tag{11.60}$$

Remark 11.1 *It is interesting to observe that the parameter update law (11.59) resembles the least squares update rule commonly used in offline ADP [2, 29]; however, instead of summing over a mesh of training points, the update (11.59) represents a sum over the system's time history stored in $E_c(k)$. Thus, the update (11.59) uses data collected online instead of data formed offline. This update law relaxes the iterative approach.*

Remark 11.2 *As a result of (11.59), the matrix $\bar{X}^T(k)\bar{X}(k)$ is invertible provided $X(k) \neq 0$ which can be viewed as a persistency of excitation (PE) condition. Observing the definition of the cost function (11.51) and OLA approximation (11.54), it is evident that both become zero only when $X_k = 0$. To ensure the PE condition, an output measurement noise will be added to X_k.*

As a final step in the cost function OLA design, we define the parameter estimation error to be $\tilde{\Phi}_k = \Phi - \hat{\Phi}_k$, and rewrite (11.51) using the ideal OLA representation

$$J(X_k) = \Phi^T\sigma(X_k) + \varepsilon_c, \tag{11.61}$$

revealing $\Phi^T\sigma(X_k) + \varepsilon_c(k) = r(k) + \Phi^T\sigma(X_k) + \varepsilon_c(k+1)$ which can be rewritten as $r(k) = -\Phi^T\Delta\sigma(X_k) - \Delta\varepsilon_c(k)$ where $\Delta\varepsilon_c(k) = \varepsilon_c(k+1) - \varepsilon_c(k)$. Substituting $r(k)$ into (11.56) as well as utilizing (11.55) and $e_c(k+1) = \alpha_c e_c(k)$ from (11.60) yields

$$\tilde{\Phi}_{k+1}^T \Delta\sigma(X_k) = -\alpha_c(r(k-1) + \hat{\Phi}_k^T \Delta\sigma(k-1)) - \Delta\varepsilon_c(k). \tag{11.62}$$

In a similar manner as $r(k)$, we now form $r(k-1) = -\Phi^T \Delta\sigma(X_{k-1}) - \Delta\varepsilon_c(k-1)$ and substitute this expression into (11.62), revealing $\Delta\sigma^T(X_k)\tilde{\Phi}_{k+1} = \alpha_c \Delta\sigma^T(X_k)\tilde{\Phi}_k + \alpha_c \Delta\varepsilon_c(k-1) - \Delta\varepsilon_c(k)$. Then the OLA parameter estimation error dynamics are revealed to be

$$\begin{aligned}\tilde{\Phi}_{k+1}^T &= \alpha_c \Delta\sigma(X_k)\big(\Delta\sigma^T(X_k)\Delta\sigma(X_k)\big)^{-1} \Delta\sigma(X_{k-1})\tilde{\Phi}_k + \Delta\sigma(X_k)\\ &= \big(\Delta\sigma^T(X_k)\Delta\sigma(X_k)\big)^{-1}(\alpha_c \Delta\varepsilon_c(k-1) - \Delta\varepsilon_c(k)).\\ &= \big(\Delta\sigma^T(X_k)\Delta\sigma(X_k)\big)^{-1}(\alpha_c \Delta\varepsilon_c(k-1) - \Delta\varepsilon_c(k)).\end{aligned} \tag{11.63}$$

Next, the boundedness of the cost function error (11.55) and OLA estimation error (11.63) is demonstrated in the following theorem. To proceed, the following theorem is needed.

Theorem 11.3 *[4]:* (Boundedness of the Cost OLA Errors). *Let $u_0(x_k)$ be a fixed admissible control policy for the controllable system (11.42), and let the cost NN weight update law be given by (11.59). Then, given a positive constant δ_c satisfying $0 < \delta_c < 1/2$, there exists positive constant, α_c given by $\alpha_c = \sqrt{1/2 - \delta_c}$ such that the critic NN weight estimation error (11.48) is uniformly ultimately bounded (UUB) with ultimate bounds given by $\left\|\tilde{\Phi}_k\right\|_F \le b'_\Phi$ for a positive constant b'_Φ.*

Remark 11.3 *The results of* Theorem 3 *are drawn under the assumption of a fixed control policy or when the control input is not updated with time. These results will aid in the proof of* Theorem 4 *where only an initial admissible control is required. Moreover, the estimation of the cost function presented earlier is a generic approach that does not depend on the system structure. This approach was used in [4] for an affine system while here it is employed for the nonaffine system.*

11.4.4 Estimation of the Optimal Feedback Control Signal

To begin the development of the feedback control policy that minimizes (11.51), we define the OLA approximation of (11.53) to be

$$\Delta\hat{u}(k) = \Delta\hat{u}(X_k) = \hat{\Theta}_k^T \vartheta(X_k), \tag{11.64}$$

where $\hat{\Theta}_k$ is the estimated value of the ideal parameter matrix Θ and $\vartheta(\bullet)$ denotes the linearly independent basis function.

Next, the optimal control signal error is defined to be the difference between the feedback control applied to (11.42) and the optimal control signal, as

$$e_a(k) = \hat{\Theta}_k^T \vartheta(X_k) + R^{-1} G^T(k)(\partial\sigma(X_k)/\partial X_k)^T \hat{\Phi}_k/2, \tag{11.65}$$

The error dynamics is given by

$$e_a(k+1) = \hat{\Theta}_{k+1}^T \vartheta(X_k) + \frac{1}{2} R^{-1} G^T(k+1) \left(\frac{\partial\sigma(X_{k+2})}{\partial X_{k+2}} \right)^T \hat{\Phi}_{k+1}, \tag{11.66}$$

while the proposed control OLA parameter update is defined to be

$$\hat{\Theta}_{k+1} = \hat{\Theta}_k - \alpha_a \vartheta(k) e_a^T(k)/(\vartheta^T(k)\vartheta(k) + 1), \tag{11.67}$$

where $0 < \alpha_a < 1$ is a positive design parameter. Substituting the parameter update (11.65) into (11.66) yields

$$\begin{aligned} e_a(k+1) &= \hat{\Theta}_k^T \vartheta(X_{k+1}) - \alpha_a e_a(k) \vartheta^T(k) \vartheta(X_{k+1}) / \left(\vartheta^T(k)\vartheta(k) + 1 \right) \\ &\quad + \frac{R^{-1} G^T(k+1)}{2} \left(\frac{\partial\sigma(X_{k+2})}{\partial X_{k+2}} \right)^T \hat{\Phi}_c(k+1). \end{aligned} \tag{11.68}$$

Since the control policy $u(x(k))$ in (11.64) minimizes the cost function (11.51), from (11.53) we can write

$$\begin{aligned} 0 &= \varepsilon_A(k+1) + \frac{1}{2} R^{-1} G^T(k+1) \frac{\partial\varepsilon_c(k+2)^T}{\partial X_{k+2}} \\ &\quad + \frac{1}{2} R^{-1} G^T(k+1) \frac{\partial O_{k+2}^T}{\partial X_{k+2}} + \Theta^T \vartheta(k+1) \\ &\quad + \frac{1}{2} R^{-1} G^T(k+1) \frac{\partial\sigma(X_{k+2})^T}{\partial X_{k+2}} \Phi. \end{aligned} \tag{11.69}$$

Obviously $\partial O_k/\partial X_k$ is bounded due to smooth bounded function O_k. Define the control OLA parameter estimation error as $\tilde{\Theta}_k = \Theta - \hat{\Theta}_k$ and recall $\tilde{\Theta}_{k+1} = \Theta - \hat{\Theta}_{k+1}$. Subtracting (11.69) from (11.68) yields

$$\begin{aligned} e_a(k+1) &= -\alpha_a e_a(k) \vartheta^T(k) \vartheta(X_{k+1}) / \left(\vartheta^T(k)\vartheta(k) + 1 \right) \\ &\quad - \varepsilon_A(k+1) - \frac{1}{2} R^{-1} G^T(k+1) \frac{\partial\varepsilon_c(k+2)}{\partial X_{k+2}} - \frac{1}{2} R^{-1} G^T(k+1) \frac{\partial O_{k+2}^T}{\partial X_{k+2}} \\ &\quad - \frac{R^{-1} G^T(k+1)}{2} \frac{\partial\sigma(X_{k+2})^T}{\partial X_{k+2}} \tilde{\Phi}_{k+1} - \Theta_k^T \vartheta(X_{k+1}). \end{aligned} \tag{11.70}$$

As a final step, the parameter estimation error dynamics are given by

$$\tilde{\Theta}_{k+1} = \tilde{\Theta}_k + \alpha_a \vartheta(k) e_a^T(k) / (\vartheta^T(k)\vartheta(k) + 1). \tag{11.71}$$

Remark 11.4 *As mentioned before, in order to calculate $e_a(k)$ in (11.65) and implement the OLA parameter update (11.67), knowledge of IGM is required while the internal dynamics $F(k)$ is not needed. In the following theorem, it will be shown that by starting with an initial stabilizing control, the control OLA update (11.67) ensures all future control inputs are also admissible. The following corollary illustrates that for any admissible control policy, the system states as well as the cost and control input basis functions are bounded.*

Corollary 11.1 (Boundedness of OLA Basis Functions) *[4]. Let $\mu(k)$ be any admissible control for the controllable system (11.42). Then, there exists a positive constant $X_0 = X_{k=0}$ such that $X_0 \geq \|X_k\|$ for all $k > 0$. Moreover, there exists positive constants $\vartheta_M = \|\vartheta(x_0)\|$ and $\sigma_M = \|\sigma(X_o)\|$ such that $\vartheta_M \geq ||\vartheta(X_k)||$ and $\sigma_M \geq \|\sigma(X_k)\|$ for all $k > 0$.*

Now, we are ready to propose the main result that is the stability analysis of the optimal NN-based controller.

11.4.5 Convergence Proof

Theorem 11.4 (Overall Stability Proof) *Assume that the proposed online NN identifier in (11.43) with the weight update law in (11.47) is used to identify the affine-like system (11.45). Let the optimal control law proposed in (11.64) by using a NN is applied to the affine like representation of the nonaffine system (11.42) when the higher order term O_k is assumed to be bounded above with $\alpha_u/2 < \delta_J < 1/2$. Let the NN (11.54) is used to estimate the cost function. Then the state error X_k, the weight estimation error $\tilde{\Theta}_k$, the cost function weight estimation error $\tilde{\Phi}_k$, the identification error $\tilde{e}_k$, and NN weight estimation error $\tilde{W}_k$ for the identifier are UUB that is, $\left\|\tilde{W}_k\right\| \leq b_W, \|X_k\| \leq b_X$, $\|\Xi_{Ak}\| \leq b_\Xi$, $\|\tilde{e}_k\| \leq b_e$, and $\left\|\tilde{\Phi}_k\right\| \leq b_\Phi$, respectively.*

Proof: Consider the following Lyapunov function candidate

$$\begin{aligned} V_k = V_D(X_k) + V_u(\tilde{\Theta}_k) + V_J(\tilde{\Phi}_k) + V_I(\tilde{W}_k) = K_D \|X_k\| + tr\left\{\tilde{\Theta}_k^T \tilde{\Theta}_k\right\} \\ + tr\left\{\tilde{W}_k^T \tilde{W}_k\right\} + \left(\tilde{\Phi}_k^T \Delta\sigma(X_{k-1})\right)^2 + \alpha^2 \tilde{e}_k^T \tilde{e}_k, \end{aligned} \tag{11.72}$$

where $K_D \in \mathbb{R}^+$. Therefore, the first difference of the Lyapunov function can be denoted as follows

$$\Delta V_k = \Delta V_D(X_k) + \Delta V_u(\tilde{\Theta}_k) + \Delta V_J(\tilde{\Phi}_k) + \Delta V_I(\tilde{W}_k). \tag{11.73}$$

The first difference of the Lyapunov function candidate can be expressed as (see [25] for details)

$$\begin{aligned}\Delta V_k \le & -\alpha^2\tilde{e}_k^T\tilde{e}_k + \left\{\alpha_u\Pi_G\Psi_M^2 + \alpha\left(-2U_{\mathrm{m}}^2\Psi_m^2 + \alpha\bar{\Psi}_M^2\left\{1+\Psi_M^2/4\right\}\right)\right\}\left\|\tilde{W}_k\right\|^2 \\ & + \left(\alpha + 2\alpha^2\bar{\Psi}_M\right)\Psi_M\bar{\varepsilon}_M\left\|\tilde{W}_k\right\| - K_D(1-L)\left\|X_k\right\| \\ & - \frac{\alpha_u}{2(\vartheta_M^2+1)}(3-3\alpha_u)\left\|\Xi_{Ak}\right\|^2 + K_D\alpha_u G_M\left\|\Xi_{Ak}\right\| \\ & + \left(-2\delta_J\Delta_{\min}^2 + \alpha_u\Pi_A\right)\left\|\tilde{\Phi}_k\right\|^2 + \varepsilon_{SM},\end{aligned} \tag{11.74}$$

with

$$\begin{aligned}\varepsilon_{SM} = & \left(\alpha_u(5\alpha_u+4) + \alpha_u^2\left(\sigma_M' G_M\lambda_{\max}\left(R^{-1}\right)\right)^2\right)\left(\varepsilon_{uk}+\varepsilon_{uk}^*\right) \\ & + K_D\alpha_u G_M\left\|\varepsilon_{uM}\right\| + K_D O_M + \alpha^2\bar{\varepsilon}_M^2\left\{1+\Psi_M^2/4\right\} + 2\Delta\bar{\varepsilon}_{IM}^2.\end{aligned} \tag{11.75}$$

Here, it is assumed that $\|O_k\|_2 \le O_M$. It is important to note that O_k in a vast class of systems can be considered as the modeling error that is typically a uniformly bounded function of the states. Therefore, here, we assume that the systems dealt with in this approach have the property of $\|O_k\| \le O_M \in \mathbb{R}$ on the compact set $\Sigma = \Sigma_1 \cap \Sigma_2$ around the operating neighborhood. An upper bound on the higher order terms is the only requirement for the rest of the approach, since we will see that the effect of O_M on the convergence will be mitigated by an auxiliary robust term regardless of how large it is.

Thus, it is clear that ΔV_k is less than or equal to zero as long as the following is satisfied:

$$\left\|\tilde{W}_k\right\| > \frac{\bar{\varepsilon}_M\left(\Psi_M + 2\alpha\frac{\bar{\Psi}_M^2}{\Psi_M}\right) + \sqrt{\begin{array}{l}\alpha\left(-8U_{\mathrm{m}}^2\Psi_m^2 + \alpha\bar{\Psi}_M^2\left\{4+\Psi_M^2\right\}\right)\varepsilon_{SM} \\ +4\varepsilon_{SM}\alpha_u\Pi_G\Psi_M^2 - \left(\alpha+2\alpha^2\bar{\Psi}_M\right)^2\Psi_M^2\bar{\varepsilon}_M^2\end{array}}}{2\alpha_u\Pi_G\Psi_M^2 + 2\alpha\left(-2U_{\mathrm{m}}^2\Psi_m^2 + \alpha\bar{\Psi}_M^2\left\{1+\Psi_M^2/4\right\}\right)} \equiv b_W, \tag{11.76}$$

or

$$\|X_k\| > \varepsilon_{SM}/(1-L) \equiv b_X, \text{ or} \tag{11.77}$$

$$\left\|\Xi_{Ak}\right\| > \frac{K_D\alpha_u G_M + \sqrt{K_D^2\alpha_u^2 G_M^2 - \frac{4\varepsilon_{SM}\alpha_u}{2(\vartheta_M^2+1)}(3-3\alpha_u)}}{\frac{\alpha_u}{(\vartheta_M^2+1)}(3-3\alpha_u)} \equiv b_\Xi, \text{ or} \tag{11.78}$$

$$\|\tilde{e}_k\| > \sqrt{\varepsilon_{SM}}/\alpha \equiv b_e, \text{ or} \tag{11.79}$$

$$\left\|\tilde{\Phi}_k\right\| > \sqrt{\varepsilon_{SM}/\left(\alpha_u \Pi_A - 2\delta_c \Delta_{\min}^2\right)} \equiv b_\Phi. \tag{11.80}$$

Therefore, the closed-loop system is UUB with the bounds given by (11.76)–(11.80). One can observe the dependence of the convergence of the closed-loop system on the NNs learning rate parameters α_u,α, and α_c as well as the higher order terms. Although the convergence is guaranteed, finding proper learning rates for the best convergence rate remains a major challenge. ■

11.4.6 Robustness

Next to minimize the ultimate bounds due to modeling errors resulting from higher order terms O_k and to ensure robustness, an auxiliary term is designed. To this end, assume that the control law is enhanced by an auxiliary term that appears in the outer-loop as follows

$$X_{k+1} = \hat{F}(X_k) + \hat{G}(X_k)\Delta u_k + \hat{G}(X_k)\Delta\rho_k + \hat{O}_k + \bar{\varepsilon}_k + \tilde{e}_{k+1}, \tag{11.81}$$

where the robust control law $\Delta\rho_k$ is chosen such that

$$\hat{G}(X_k)\Delta\rho_k + \hat{O}_k\left(X_k, \Delta u_k + \Delta\rho_k\right) = 0, \tag{11.82}$$

provided a solution exists for (11.82). To this effect, define the estimation error as follows

$$\hat{G}(X_k)\Delta\rho_k + \hat{O}_k\left(X_k, \Delta u_k + \Delta\rho_k\right) = \zeta_k. \tag{11.83}$$

Our aim is to design the dynamics of the robust controller term $\Delta\rho_k$ such that $\|\zeta_{k+1}\|^2 \le \alpha_\zeta^2 \|\zeta_k\|^2$ for $\alpha_\zeta < 1$. In other words, ζ_k has to be exponentially stable. Now, consider the following update law for $\Delta\rho_k$ as

$$\Delta\rho_{k+1} = \Delta\rho_k + \delta\rho_k, \tag{11.84}$$

which makes (11.82) to have the following form

$$\begin{aligned} &\hat{G}(X_{k+1})\Delta\rho_{k+1} + \hat{O}\left(X_{k+1}, \Delta u_{k+1} + \Delta\rho_{k+1}\right) \\ &\quad = \hat{G}(X_{k+1})\left(\Delta\rho_k + \delta\rho_k\right) + \hat{O}\left(X_{k+1}, \Delta u_{k+1} + \Delta\rho_k + \delta\rho_k\right) = \zeta_{k+1}. \end{aligned} \tag{11.85}$$

By using Taylor series first order approximation we can write

$$\begin{aligned} &\hat{O}\left(X_{k+1}, \Delta u_{k+1} + \Delta\rho_k + \delta\rho_k\right) \\ &\quad \cong \hat{O}\left(X_{k+1}, \Delta u_{k+1} + \Delta\rho_k\right) \frac{\partial}{\partial \Delta u_k} \hat{O}\left(X_{k+1}, \Delta u_{k+1} + \Delta\rho_k\right)\delta\rho_k. \end{aligned} \tag{11.86}$$

Equation (11.86) is valid since the convergence of the outputs to a certain neighborhood of the operation point, which is the origin, is guaranteed by the proposed optimal controller. Equation (11.85) can be written as follow

$$\left\{\hat{G}(X_{k+1}) + \partial\hat{O}\left(X_{k+1}, \Delta u_{k+1} + \Delta\rho_k\right)/\partial\Delta u_k\right\}\delta\rho_k + \hat{G}(X_{k+1})\Delta\rho_k \\ + \hat{O}\left(X_{k+1}, \Delta u_{k+1} + \Delta\rho_k\right) = \zeta_{k+1}. \tag{11.87}$$

We desire to have the left hand side of the Equation (11.87) equal to $\alpha_\zeta\zeta_k$ with α_ζ being the rate of the fast dynamics which is known as ε in continuous-time singularly perturbed systems. Therefore, using pseudo-inversion to calculate the desired $\delta\rho_k$ as

$$\delta\rho_k = -\left(A_\zeta^T A_\zeta\right)^{-1} A_\zeta^T B_\zeta, \text{ with} \tag{11.88}$$

$$A_\zeta = \left\{\hat{G}(X_{k+1}) + \partial\hat{O}\left(X_{k+1}, \Delta u_{k+1} + \Delta\rho_k\right)/\partial\Delta u_k\right\}, \tag{11.89}$$

$$B_\zeta = \left\{\hat{G}(X_{k+1})\Delta\rho_k + \hat{O}\left(X_{k+1}, \Delta u_{k+1} + \Delta\rho_k\right) - \alpha_\zeta\zeta_k\right\}, \text{ and } 0 < \alpha_\zeta < 1.$$

Remark 11.5 *It is clear that by having an invertible* A_ζ, $\|\zeta_{k+1}\| = \alpha_\zeta\|\zeta_k\|$ *holds. However, in the case that* A_ζ *is not invertible,* $\|\zeta_{k+1}\| < \|\zeta_k\|$ *may not hold and therefore, the robust term should be chosen as* $\delta\rho_k = 0$ *which means a solution may not exist to reduce the convergence bound* ε_{SM}. *In the sequel, we assume* A_ζ *is invertible and therefore there exists* $L_\zeta < 1$ *such that*

$$\|\zeta_{k+1}\| - \|\zeta_k\| \leq -L_\zeta\|\zeta_k\|. \tag{11.90}$$

Remark 11.6 *When (11.82) holds, the identified higher order term* $\hat{O}_k$ *gets mitigated and therefore the systems behaves as an affine system. For guaranteeing the existence of solution for* $\Delta\rho_k$ *the* implicit function theorem *requires* $\hat{G}(X_{k+1}) + \partial\hat{O}\left(X_{k+1}, \Delta u_{k+1} + \Delta\rho_k\right)/\partial\Delta u_k$ *to be invertible. Therefore, the invertability for* A_ζ *is an inevitable condition. To the best knowledge of the authors, the only work available on affinization of nonaffine systems [31] requires the same assumption. Consequently, under the locally invertible assumption of* A_ζ *and after proving the boundedness of the closed-loop system by the optimal controller, the robust controller (11.88) improves the convergence by mitigating the modeling errors due to higher order terms of the Taylor series expansion.*

Therefore consider the following representation of a discrete singularly perturbed system

$$X_{k+1} = F(X_k) + G(X_k)\Delta u_k + G(X_k)\Delta\rho_k + O_k + \bar{\varepsilon}_k, \tag{11.91}$$

$$\Delta\rho_{k+1} = \Delta\rho_k + \delta\rho_k. \tag{11.92}$$

This system (91)–(92) is utilized for the robustness analysis which is introduced further.

Theorem 11.5 *(Stability of* Robust Optimal Controller*): Consider the nonlinear singularly perturbed discrete-time system (11.42) and (11.84) rewritten as (11.78) and (11.79). Let (11.82) be solvable and (11.64) be a control law stabilizing (11.42) with $O_k(X_k, \Delta u_k) \le O_M$. Assume that there exist $L_\zeta < 1$ so that (11.90) holds. Then the overall system (11.42) and (11.64) along with (11.84) is UUB and $\Delta\rho_k \to \Delta\bar{\rho}_k$ holds exponentially, where $\Delta\bar{\rho}_k$ is the solution of (11.82). In addition, the robust controller $\Delta u_k + \Delta\rho_k$ converges to the robust optimal controller $\Delta u_k^* + \Delta\bar{\rho}_k$ in (11.53) with an arbitrarily small error.*

Proof: Select the Lyapunov candidate $V_k^r = V_k + \|\zeta_k\|$ with $\zeta_k = \Delta\rho_k - \Delta\bar{\rho}_k$. Taking the first difference we have

$$\begin{aligned}
V_{k+1}^r - V_k^r &= \Delta V_k + \|\zeta_{k+1}\| - \|\zeta_k\| \\
&\le K_D \|F(X_k) + G(X_k)\Delta u_k + G(X_k)\Delta\rho_k + O_k\| - K_D\|X_k\| - L_\zeta \|\zeta_k\| \\
&\quad + \Delta V_u(\tilde{\Theta}_k) + \Delta V_J(\tilde{\Phi}_k) + \Delta V_I(\tilde{W}_k) \\
&= K_D \left\|F(X_k) + G(X_k)\Delta u_k + \zeta_k + \varepsilon_k^o\right\| - K_D\|X_k\| - L_\zeta\|\zeta_k\|^2 + \Delta V_u(\tilde{\Theta}_k) \\
&\quad + \Delta V_J(\tilde{\Phi}_k) + \Delta V_I(\tilde{W}_k)
\end{aligned},$$

where $O_k + G(X_k)\Delta\rho_k = \hat{O}_k + \hat{G}(X_k)\Delta\rho_k + \varepsilon_k^o$ that is ε_k^o is the identification error for $\hat{O}_k$. Therefore,

$$\begin{aligned}
V_{k+1}^r - V_k^r &\le K_D \|F(X_k) + G(X_k)\Delta u_k\| + K_D\|\zeta_k\| + K_D\left\|\varepsilon_k^o\right\| - K_D\|X_k\| \\
&\quad - L_\zeta\|\zeta_k\| + \Delta V_u(\tilde{\Theta}_k) + \Delta V_J(\tilde{\Phi}_k) + \Delta V_I(\tilde{W}_k) \\
&= K_D \|F(X_k) + G(X_k)\Delta u_k\| - K_D\|X_k\| + \Delta V_u(\tilde{\Theta}_k) + \Delta V_J(\tilde{\Phi}_k) + \Delta V_I(\tilde{W}_k) \\
&\quad + K_D\left\|\varepsilon_k^o\right\| + K_D\|\zeta_k\| - L_\zeta\|\zeta_k\| + K_D\|\zeta_k\| - L_\zeta\|\zeta_k\|.
\end{aligned} \tag{11.93}$$

Here for deriving (11.93), $\|\zeta_{k+1}\|^2 - \|\zeta_k\|^2 \le -L_\zeta\|\zeta_k\|^2$ is utilized which is guaranteed by exponential stability of (11.84) and (11.88). Since $\left\|\varepsilon_k^o\right\| \le \|\tilde{e}_k\|$, it is easy to observe the inequality (11.93) implies that the results by Theorem 11.5 can be enhanced if $K_D\|\zeta_k\| - L_\zeta\|\zeta_k\| < 0$. Therefore, K_D should be chosen such that $K_D < L_\zeta$. Therefore, it is easy to verify that with the first difference (11.93) the convergence bounds of the robust optimal controller $b_{\rho W}$, $b_{\rho X}$, $b_{\rho\Xi}$, $b_{\rho e}$, $b_{\rho\Phi}$, and $b_{\rho\zeta}$ are given as $b_{\rho W} = b_W|_{O_M=0}$, $b_{\rho X} = b_X|_{O_M=0}$, $b_{\rho\Xi} = b_\Xi|_{O_M=0}$, $b_{\rho e} = \left(K_D + \sqrt{K_D^2 - 4\alpha^2\varepsilon_{SM}}\right)/2\alpha^2\Big|_{O_M=0}$, $b_{\rho\Phi} = b_\Phi|_{O_M=0}$, and $b_{\rho\zeta} = \varepsilon_{SM}/|K_D - L_\zeta|\big|_{O_M=0}$.

So far, we showed that the closed-loop system remains uniformly ultimate bounded when the system is viewed as the singularly perturbed dynamics system (11.92) to mitigate the higher order terms. On the other hand, (11.84) is exponentially stable which means the effect of $\hat{O}_k$ can be neglected within a short time. This implies that the singularly perturbed controller can overcome the effect of the higher order terms.

As the last step, we need to study the convergence of the proposed control input (11.64), along with the robust term (11.84), to the optimal controller (11.53). In other words, we need to show that $\Delta u_k + \Delta \rho_k \to \Delta u_k^* + \Delta \bar{\rho}_k$ is valid with a negligible error. Now consider the input error as $\left\| \Delta u_k + \Delta \rho_k - \Delta u_k^* - \Delta \bar{\rho}_k \right\|$ which is shown to converge to $\left\| \Delta u_k - \Delta u_k^* \right\|$ when $k \to \infty$. To do this, we have proven that the robust term will mitigate the effect of the higher order terms leaving small error term that is the effect of $\hat{O}_k$ will be eliminated. Therefore,

$$\lim_{k\to\infty} \left\| \Delta u_k - \Delta u_k^* \right\| < \lim_{k\to\infty} \|\Xi_{Ak}\| + \varepsilon_A = b_\Xi|_{O_M=0} + \varepsilon_A \triangleq b_{\rho\Xi} + \varepsilon_A. \quad (11.94)$$

Thus according to (11.76), the actual control input approaches the optimal control input with small error which is a function of NN approximation errors. Thus, by selecting suitable number of NN activation functions, the identification error $\bar{\varepsilon}_k$, the cost function estimation error, the estimation error ε_c, and the optimal controller estimation error ε_A can be made arbitrarily small. This implies that $\varepsilon_{\rho SM} = \varepsilon_{SM}|_{O_M=0}$ in (11.75) can be made small by selecting the number of hidden-layer neurons. Therefore, from (11.80), $b_{\Xi 0} + \varepsilon_A$ can arbitrarily considered small. As a consequence, it can be concluded that the initial admissible controller converges to the optimal controller over time with a negligible error. ■

Figure 11.2 shows the standard HDP scheme while Figure 11.3 depicts the overall controller block diagram representation of the proposed scheme where value and policy iterations are not utilized. Here, the cost function is updated once every sampling interval. An initial admissible controller is given in terms of the initial value of $\hat{\Theta}_k$. After tuning, the initial admissible controller will become robust optimal controller.

11.5 SIMULATION RESULT

In this section, first the performance of a reinforcement-learning based controller is given. Subsequently, it will be shown that the standard HDP policy iteration based controller becomes unstable due to insufficient number of iteration by using a very simple SISO nonlinear system example. Finally, the time-based ADP-based controller will be applied to a complicated nonaffine MIMO dynamics of a HCCI engine.

11.5.1 Reinforcement-Learning-Based Control of a Nonlinear System

Consider the following second-order input–output representation with delay [22] described by

$$\begin{aligned} y(k+2) &= 0.2\cos(0.8(y(k)+y(k-1))) + 0.4\sin(0.8(y(k)+y(k-1)) \\ &\quad + 2u(k) + u(k-1)) + 0.1(9 + y(k) + y(k-1)) \\ &\quad + \frac{2(u(k)+u(k-1))}{1+\cos(y(k))} + d, \quad \text{for } k > 0. \end{aligned} \quad (11.95)$$

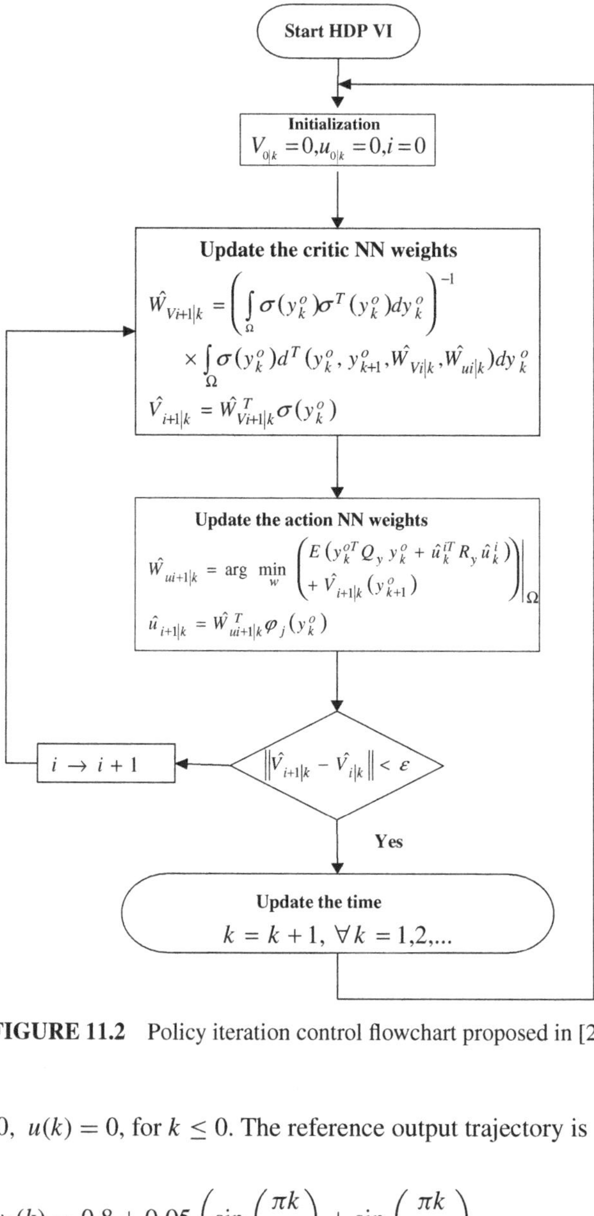

FIGURE 11.2 Policy iteration control flowchart proposed in [2].

with $y(k) = 0,\ u(k) = 0$, for $k \le 0$. The reference output trajectory is set to be

$$y_d(k) = 0.8 + 0.05 \left(\sin\left(\frac{\pi k}{50}\right) + \sin\left(\frac{\pi k}{100}\right) + \sin\left(\frac{\pi k}{150}\right) \right), \quad \text{for } k > 0, \quad \text{otherwise zero.} \tag{11.96}$$

The objective of our controller is to drive the system to track the reference trajectory. The sampling interval is set as 0.02 second and the uniformly distributed noise d

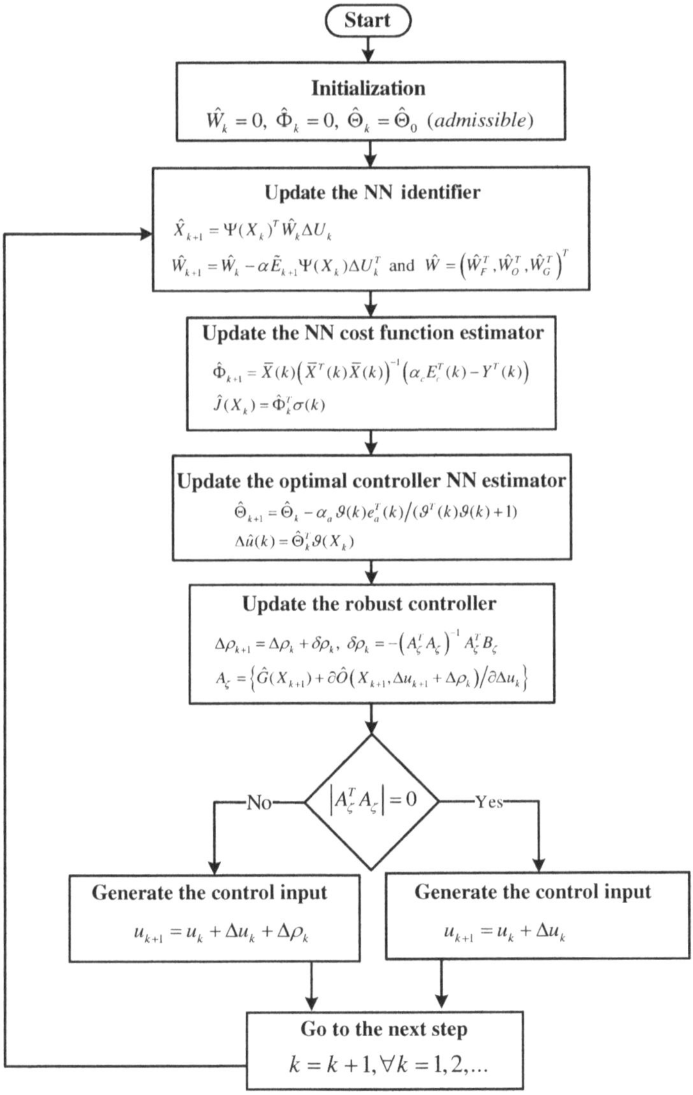

FIGURE 11.3 The proposed controller block diagram representation.

bounded by d_M is considered. Other parameters are listed as $R = 0.05$, $Q = 0.01$, $\gamma = 0.8$, $\alpha_c = 0.001$, $\alpha_a = 0.05$, $n_a = 10$, $n_c = 10$, and $d_M = 0.01$.

The tracking performance of the online reinforcement-learning controller is depicted in Figure 11.4. This figure demonstrates that a satisfactory performance with quick response is observed under the influence of noise. At the same time, the control signal and the norm of the action and critic NN weights are displayed in Figures 11.5 and 11.6, respectively, which illustrates the boundedness and convergence of the control input and the NN weights.

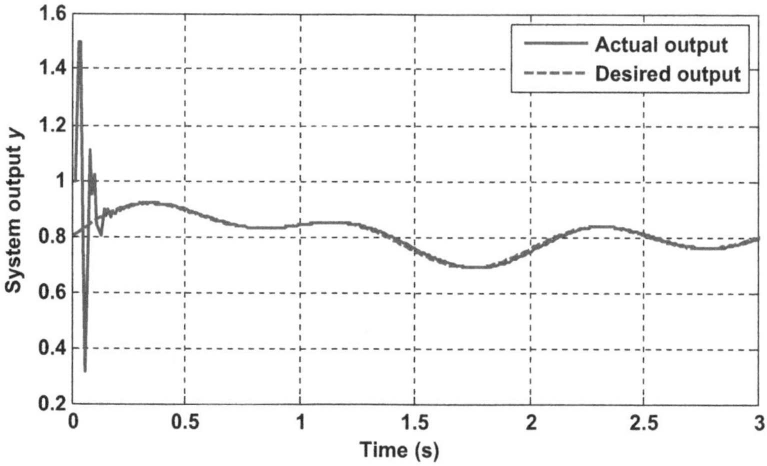

FIGURE 11.4 Tracking performance of the online learning controller.

11.5.2 The Drawback of HDP Policy Iteration Approach

Figure 11.2 demonstrates the HDP algorithm proposed for nonlinear affine discrete systems [2]. As it is obvious from the figure, the approach is based on estimating the value function during each iteration by fixing the time instant and iterating until $\left\|\hat{V}_{i+1|k} - \hat{V}_{i|k}\right\| < \varepsilon$. This makes the approach impractical when the number of iterations required is significant and if the computational issues are considered. Moreover, the convergence proof of the HDP algorithm is provided in the case of large number of iterations between sampling interval. To demonstrate this drawback, consider a simple system as $x_{k+1} = x_k + x_k^2 + (2 + \cos(x_k))u_k$ with x_k as the state and u_k as the input. Assume that $Q = 1$ and $R = 1$ are taken for the cost function. Figure 11.7

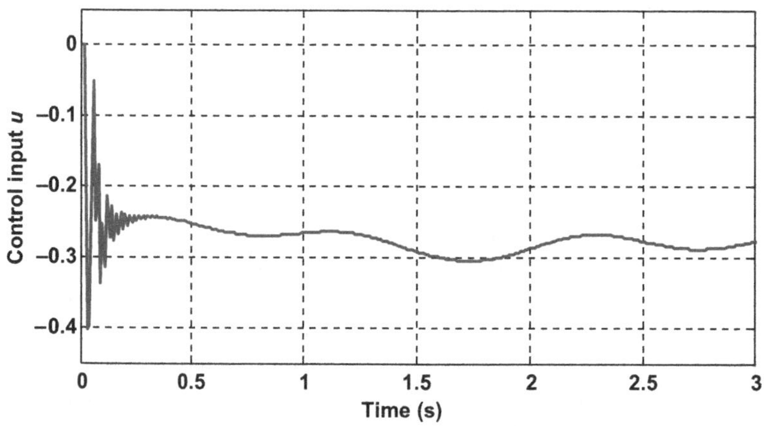

FIGURE 11.5 Input signal of the online learning controller.

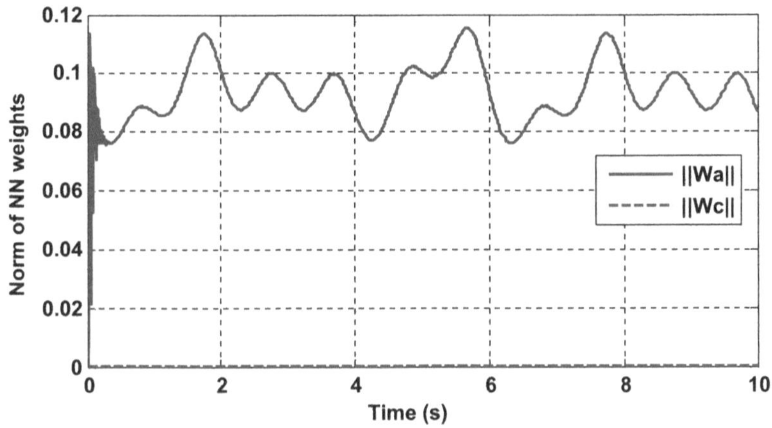

FIGURE 11.6 Hidden layer weights of the action and critic NN.

shows that the approach converges if we have 100 (enough) iteration per sampling interval while the overall system becomes unstable when 20 iterations are utilized. Selection of the number of iterations is not known. This problem is resolved by the proposed OLA-based optimal controller that is given next.

11.5.3 OLA-Based Optimal Control Applied to HCCI Engine

Low temperature combustion modes, such as homogeneous charge compression ignition (HCCI), represent a promising means to increase the efficiency and significantly reduce the emissions of internal combustion engines. Implementation and control are difficult, however, due to the dependence of the combustion event on chemical kinetics rather than an external trigger. In [32], the author outlined a nonlinear control-oriented

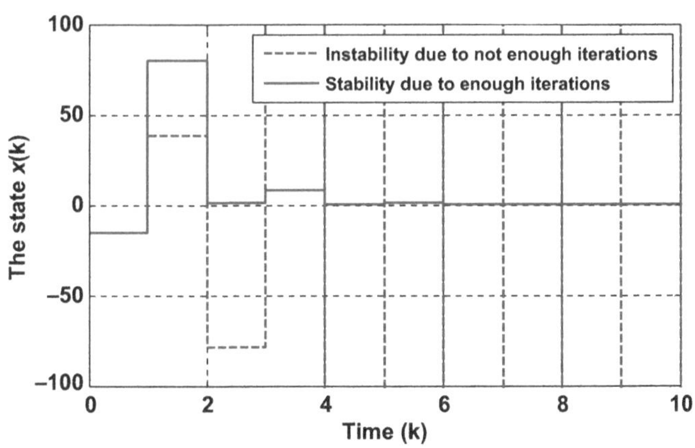

FIGURE 11.7 The stability of the value iteration method with respect to the number of the iterations per sampling interval.

FIGURE 11.8 Laboratory version of the HCCI engine.

model of a single cylinder HCCI engine, which is physically based on a five state thermodynamic cycle. This model is aimed at capturing the behavior of an engine.

Figure 11.8 illustrates the laboratory HCCI engine whose operating data are used for deriving the model. HCCI engine dynamics in discrete-time is well-developed in [32] and it is given as a MIMO nonaffine system. The system dynamics are derived from cycle-by-cycle information of the thermo-dynamical behavior of the engine. Therefore, the proposed controller from the previous sections can be applied to the HCCI engine representation. In this section, some simulation results are provided to show the significance of the proposed approach. It is required to have an admissible controller Δu_k^O. Then, the update law (11.45), (11.59), (11.67), and (11.84) will

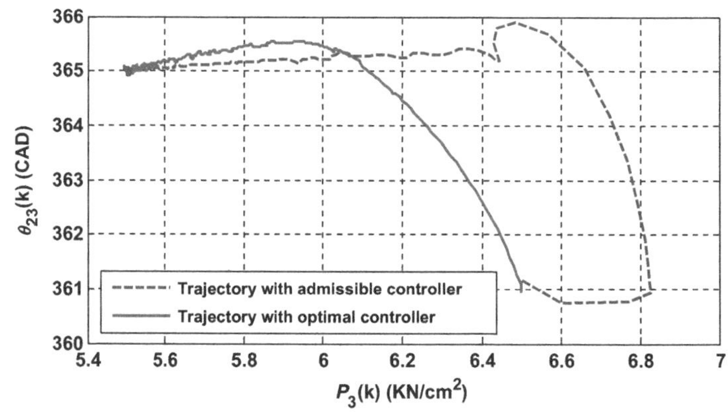

FIGURE 11.9 Closed-loop system convergence for $(\theta_{23,k}, P_3) = (365, 0.55)$ with admissible and optimal controllers.

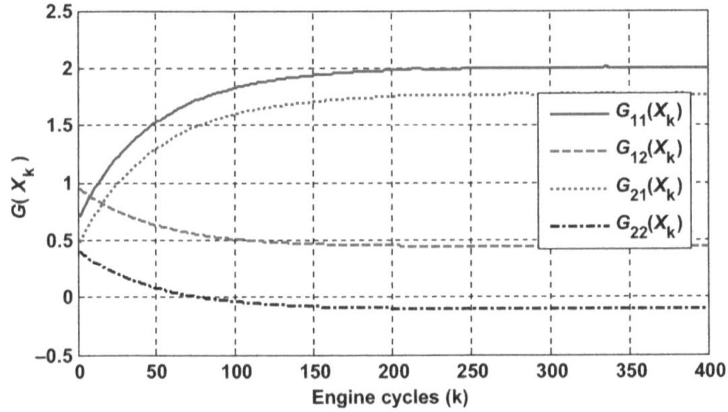

FIGURE 11.10 Convergence of the closed-loop system for $G(X_k)$ with admissible and optimal controllers.

converge to a optimal control signal Δu_k^*. The dispersion noise is injected to the system dynamics (as actuator and sensor noise) to provide persistence of excitation for the system identification and cost function estimation.

Now, we consider two different cases to observe the improvement made by the online near optimal controller. The first case is when (11.45), (11.59), (11.67), and (11.84) are not updated; and the second case is when they are updated and Δu_k^O converges to Δu_k^*. Figure 11.9 illustrates the convergence of the closed-loop system with the two earlier cases. Moreover, we investigate the index function (11.51) where $Q = 1$ and $R = 1$. Over all numerical examinations

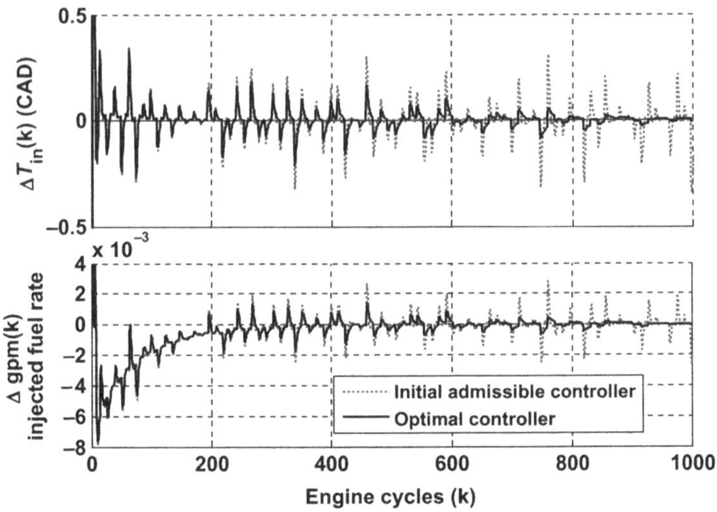

FIGURE 11.11 Comparison of the admissible and optimal controllers.

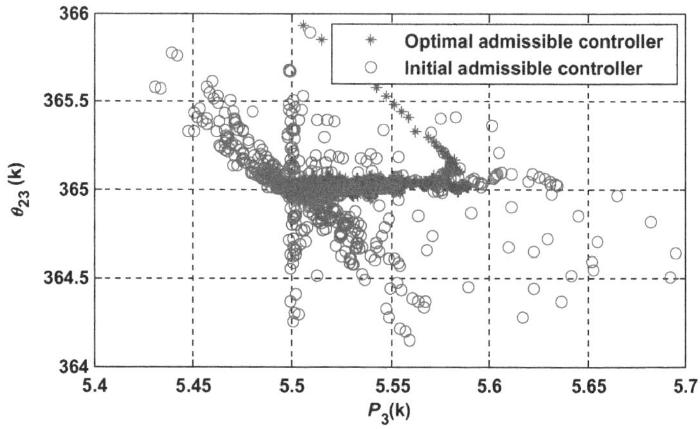

FIGURE 11.12 Comparison of admissible and optimal controllers in terms of cyclic dispersion.

$k = 0.008333$ s (the engine speed which is 188.49 rpm), $\alpha = 0.005$, $\alpha_u = 0.001$, $\sigma(X_k) = \{x_1^2, x_1x_2, x_2^2, x_1^4, x_1^3x_2, \ldots, x_2^6\}$, $\Psi(X_k) = \{1, \ldots, 1, \sin(X), \sin(2X), \ldots,$ $\tanh(X), \tanh(2X), \ldots\} \in \mathbb{R}^{16\times32}$, $\vartheta = \vartheta = \{\tanh(X), \tanh(2x), \ldots\} \in \mathbb{R}^{16\times32}$, $\alpha_\zeta = 0.1$, and the initial admissible controller is chosen as $-0.1 \times [(\theta_{23} - \theta_{23d}), (P_3 - P_{3d})]^T$. Figure 11.9 clearly shows that the trajectory selected by the online optimal controller is much shorter. Assume that both of the controllers have been operating at steady state mode. Under this scenario, the estimation of $G\,(X_k)$ is illustrated in Figure 11.10. Figure 11.11 makes a comparison between the control signals applied by the admissible and optimal controllers. Here the optimal controller starts with a same behavior as that of the admissible controller,

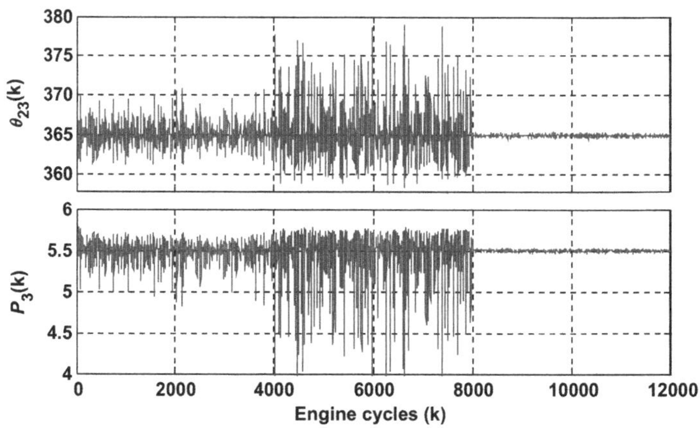

FIGURE 11.13 Comparison among open-loop, admissible, and the optimal controllers when the set point is ($P_3 = 0.55$, $\theta_{23} = 365$): the controller switches from open-loop to admissible at $k = 400$; then, to the optimal controller at $k = 800$.

whereas after about 200 engine cycles, the control effort begins to decrease in magnitude due to online learning.

The injected process and sensor noise will not allow the system to converge to its set point whereas the output variation can be viewed as the dispersion reduction ability. Figure 11.12 compares two controllers in terms of reduction in dispersion where $P_{3,k}$ is plotted versus $\theta_{23,k}$. Compared with the admissible controller, the optimal controller, after a transient behavior, shows significant smaller variation around the set point.

In another numerical experiment, we compare open-loop, admissible, and the optimal controllers in terms of their cyclic dispersion reduction abilities. To this end, we assume that initially the engine is considered to be running in the open-loop mode for 400 engine cycles, then it switches to an admissible controller for the same duration of run, and finally, to the optimal controller. Figure 11.13 shows the result of this experiment for setpoint ($P_3 = 0.55KN/\text{cm}^2, \theta_{23} = 365CAD$). It is obvious that the optimal controller significantly reduces the dispersion when compared with the open-loop and the admissible controllers. We conclude from Figure 11.13 that the proposed controller renders a significant reduction in output dispersion.

REFERENCES

1. S. Jagannathan. *Neural Network Control of Nonlinear Discrete-time Systems*, Taylor and Francis, Boca Raton, FL, 2006.
2. A. Al-Tamimi, F.L. Lewis, and M. Abu-Khalaf. Discrete-time nonlinear HJB solution using approximate dynamic programming: convergence proof, *IEEE Transactions on Systems, Man, and Cybernetics—Part B*, 38:943–949, 2008.
3. F.Y. Wang, N. Jin, D. Liu, and Q. Wei. Adaptive dynamic programming for finite-horizon optimal control of discrete-time nonlinear systems with ε-error bound, *IEEE Transactions on Neural Networks*, 22(1):24–36, 2011.
4. T. Dierks and S. Jagannathan. Online optimal control of nonlinear discrete-time systems using approximate dynamic programming, *Journal of Control Theory and Applications*, 9(3):361–369, 2011.
5. T. Dierks, B.T. Thumati, and S. Jagannathan. Optimal control of unknown affine nonlinear discrete-time systems using offline-trained neural networks with proof of convergence, *Neural Networks*, 22(5–6):851–860, 2008.
6. R. Bellman and S. Dreyfus. *Applied Dynamic Programming*, Princeton, NJ, Princeton University Press, 1962.
7. W.T. Miller, R.S. Sutton, and P.J. Werbos. *Neural Networks for Control*, Cambridge, MA, MIT Press, 1990.
8. D.P. Bertsekas and J.N. Tsitsiklis. *Neuro-Dynamic Programming*, MA, Athena Scientific, 1996.
9. F.L. Lewis, S. Jagannathan, and A. Yesildirek. *Neural Network Control of Robot Manipulators and Nonlinear Systems*, Taylor & Francis, PA, 1999.
10. J. Si, A.G. Barto, W.B. Powell, and D. Wunsch. *Handbook of Learning and Approximate Dynamic Programming*, Wiley-IEEE Press, 2004.

11. J.J. Murray, C.J. Cox, G.G. Lendaris, and R. Saeks. Adaptive dynamic programming, *IEEE Transactions on Systems, Man, and Cybernetics–Part C*, 32:140–153, 2002.
12. P.J. Werbos. "Approximate dynamic programming for real-time control and neural modeling," In White and Sofge, editors. *Handbook of Intelligent Control: Neural, Fuzzy, and Adaptive Approaches*, Van Nostrand Reinhold, NY, 1992, pp. 493–525.
13. P.J. Werbos. Stable adaptive control using new critics designs, In N. Karayiannis, M. Padgett, and L. Zadeh, editors. *Handbook Applied Computational Intelligence*, CRC, Boca Raton, FL, 2000.
14. A.G. Barto, R.S. Sutton, and C.W. Anderson. Neuron like adaptive elements that can solve difficult learning control problems, *IEEE Transactions on System, Man, Cybernetics*, 13:834–847, 1983.
15. R.S. Sutton. Learning to predict by the methods of temporal difference, *Machine Learning*, 3:9–44, 1988.
16. C.H. Watkins and P. Dayan. Q-Learning, *Machine Learning*, 8:257–277, 1992.
17. R.S. Sutton and A.G. Barto. *Reinforcement Learning: An Introduction*, the MIT Press, Cambridge, MA, 1998.
18. P.J. Werbos. A menu of designs for reinforcement learning over time. In W.T. Miller, R.S. Sutton, P. J. Werbos, editors. *Neural Networks for Control,* Cambridge: MIT Press, 1991, pp. 67–95.
19. P. He and S. Jagannathan. Reinforcement learning-based output feedback control of nonlinear systems with input constraints, *IEEE Transactions on System, Man, and Cybernetics*, 35:150–154, 2005.
20. G. Boone. Efficient reinforcement learning: model-based acrobat control. In *Proceedings of IEEE International Conference on Robotics and Automation*, Albuquerque, 1997, pp. 229–234.
21. J. Si and Y.T. Wang. On-line learning control by association and reinforcement, *IEEE Transactions on Neural Networks*, 12(2):264–276, 2001.
22. Q. Yang, J.B. Vance, and S. Jagannathan. "Control of nonaffine nonlinear discrete-time systems using reinforcement learning-based linearly parameterized neural networks," *IEEE Transaction on Systems, Man, and Cybernetics—Part B*, 38(4):994–1001, 2008.
23. H. Xu, and S. Jagannathan. Stochastic optimal control design for nonlinear networked control system via neurodynamics programming using input–output measurements, In *IEEE Conference on Decision and Control*. 2001.
24. B. Igelnik and Y.H. Pao. Stochastic choice of basis functions in adaptive function approximation and the functional-link net, *IEEE Transactions on Neural Networks*, 6(6):1320–1329, 1995.
25. H. Zargarzadeh, S. Jagannathan, and J. Drallmeier. Near optimal control of nonaffine nonlinear system with application to HCCI engines, In *Proceedings of the IEEE Symposium on Adaptive Dynamic Programming and Reinforcement Learning*, 2011.
26. O. Adetona, S. Sathananthan, and L.H. Keel. Robust adaptive control of nonaffine nonlinear plants with small input signal changes, *IEEE Transactions on Neural Networks*, 15(2):408–416, 2004.
27. M.S. Ahmed. Neural-net-based direct adaptive control for a class of nonlinear plants, *IEEE Transactions on Automatic Control*, 45(1):119–124, 2000.
28. P. Hartman. *Ordinary Differential Equations*, 2nd edition. SIAM, 2002.

29. Z. Chen, and S. Jagannathan. Generalized Hamilton–Jacobi–Bellman formulation based neural network control of affine nonlinear discrete-time systems, *IEEE Transactions on Neural Networks*, 19(3):90–106, 2008.

30. F.L. Lewis and V.L. Syrmos. 1995. *Optimal Control*, 2nd edition Hoboken, NJ: Wiley.

31. J. Stefanovski. Feedback affinization of nonlinear control systems, *Systems and Control Letters*, 46:207–217, 2002.

32. J. Bettis, J. Masse, J. Drallmeier, and S. Jagannathan. Thermodynamic Based modeling for nonlinear control of combustion phasing in HCCI, *Technical Meeting Champaign*, 21–23, 2010.

CHAPTER 12

An Actor–Critic–Identifier Architecture for Adaptive Approximate Optimal Control

S. BHASIN,[1] R. KAMALAPURKAR,[2] M. JOHNSON,[2]
K.G. VAMVOUDAKIS,[3] F.L. LEWIS,[3] and W.E. DIXON[2]

[1]Department of Electrical Engineering, Indian Institute of Technology, Delhi, India
[2]Department of Mechanical and Aerospace Engineering, University of Florida, Gainesville, FL, USA
[3]Automation and Robotics Research Institute, The University of Texas at Arlington, Fort Worth, TX, USA

ABSTRACT

An online adaptive reinforcement learning-based solution is developed for the infinite-horizon optimal control problem for continuous-time uncertain nonlinear systems. A novel actor-critic-identifier (ACI) is proposed to approximate the Hamilton-Jacobi-Bellman equation using three neural network (NN) structures—actor and critic NNs approximate the optimal control and the optimal value function, respectively, and a robust dynamic neural network identifier asymptotically approximates the uncertain system dynamics. An advantage of the using the ACI architecture is that learning by the actor, critic, and identifier is continuous and simultaneous, without requiring knowledge of system drift dynamics. Convergence of the algorithm is analyzed using Lyapunov-based adaptive control methods. A persistence of excitation condition is required to guarantee exponential convergence to a bounded region in the neighborhood of the optimal control and uniformly ultimately bounded (UUB) stability of the closed-loop system. Simulation results demonstrate the performance of the actor-critic-identifier method for approximate optimal control.

Reinforcement Learning and Approximate Dynamic Programming for Feedback Control, First Edition. Edited by Frank L. Lewis and Derong Liu.

12.1 INTRODUCTION

Reinforcement learning (RL) provides a way for learning agents to optimally interact with uncertain complex environments, and hence, can address problems from a variety of domains, including artificial intelligence, controls, economics, operations research and so on. Actor–critic (AC) architectures have been proposed as models of RL [1, 2]. Since AC methods are amenable to online implementation, they have become an important subject of research, particularly in the controls community [3–12]. Typically, the AC architecture consists of two neural networks (NNs)—an actor NN and a critic NN. The critic NN approximates the evaluation function, mapping states to an estimated measure of the value function, whereas the action NN approximates an optimal control law and generates actions or control signals. Following the works of Sutton [13], Barto [14], Watkins [15], and Werbos [16], current research focuses on the relationship between RL and dynamic programming (DP) [17] methods for solving optimal control problems. Owing to the curse of dimensionality associated with using DP [17], Werbos [3] introduced an alternate approximate dynamic programming (ADP) approach which gives an approximate solution to the DP problem, or the *Hamiltonian–Jacobi–Bellman* (HJB) equation for optimal control. A detailed review of ADP designs can be found in Ref. [4]. Various modifications to ADP algorithms have since been proposed [5, 18, 19].

Convergence of ADP algorithms for RL-based control is studied in Refs [5–8] and [20]. A policy iteration (PI) algorithm is proposed in Ref. [21] using Q-functions for the discrete-time LQR problem and convergence to the state feedback optimal solution is proven. In Ref. [22], model-free Q-learning is proposed for linear discrete-time systems with guaranteed convergence to the $\mathcal{H}_2$ and $\mathcal{H}_\infty$ state feedback control solution. Most of the previous work on ADP has focused on either finite state Markovian systems or discrete-time systems [23, 24]. The inherently iterative nature of the ADP algorithm has impeded the development of closed-loop controllers for continuous-time uncertain nonlinear systems. Extensions of ADP-based controllers to continuous-time systems entails challenges in proving stability, convergence, and ensuring the algorithm is online and model-free. Early solutions to the problem consisted of using a discrete-time formulation of time and state, and then applying an RL algorithm on the discretized system. Discretizing the state space for high dimensional systems requires a large memory space and a computationally prohibitive learning process. Convergence of PI for continuous-time LQR was first proved in Ref. [25]. Baird [26] proposed *Advantage Updating*, an extension of the Q-learning algorithm, which could be implemented in continuous-time and provided faster convergence. Doya [27] used an HJB framework to derive algorithms for value function approximation and policy improvement, based on a continuous-time version of the temporal difference (TD) error. Murray et al. [6] also used the HJB framework to develop a *stepwise stable* iterative ADP algorithm for continuous-time input-affine systems with an input quadratic performance measure. In Beard et al. [28], Galerkin's spectral method is used to approximate the solution to the generalized HJB (GHJB), using which a stabilizing feedback controller was computed offline. Similar to Ref. [28], Abu-Khalaf and Lewis [29] proposed a least squares (LS) successive approximation solution to the GHJB, where an NN is trained offline to learn the GHJB solution. Another continuous-time formulation of adaptive critic is proposed in Hanselman [10].

All of the aforementioned approaches for continuous-time nonlinear systems require complete knowledge of system dynamics. The fact that continuous-time ADP requires knowledge of the system dynamics has hampered the development of continuous-time extensions to ADP-based controllers for nonlinear systems. Recent result by Vrabie and Lewis [11] has made new inroads by addressing the problem for partially unknown nonlinear systems. A PI-based hybrid continuous-time/discrete-time sampled data controller is designed in Refs [11] and [30], where the feedback control operation of the actor occurs at faster time scale than the learning process of the critic. Vamvoudakis and Lewis [12] extended the idea by designing a model-based online algorithm called *synchronous PI*, which involved synchronous continuous-time adaptation of both actor and critic NNs. A contribution of this work is the use of a novel AC–identifier architecture, which obviates the need to know the system drift dynamics, and where the learning of the actor, critic, and identifier is continuous and simultaneous. Moreover, the AC–identifier method utilizes an identification-based online learning scheme, and hence is the first ever indirect adaptive control approach to RL.

In the developed method, the actor and critic NNs use gradient and LS-based update laws, respectively, to minimize the Bellman error, which is the difference between the exact and the approximate HJB equation. The identifier is a combination of a Hopfield-type [31] dynamic NN (DNN), in parallel configuration with the system [32], and a novel robust integral of sign of the error (RISE) component [33]. The identifier asymptotically estimates the state derivative, allowing the AC–identifier architecture to be implemented without the knowledge of system drift dynamics; however, knowledge of the input gain matrix is required to implement the control policy. Convergence of the AC–identifier-based algorithm and stability of the closed-loop system are analyzed using Lyapunov-based adaptive control methods, and a persistence of excitation (PE) condition is used to guarantee exponential convergence to a bounded region in the neighborhood of the optimal control and uniformly ultimately bounded (UUB) stability of the closed-loop system. The PE condition is equivalent to the exploration paradigm in RL [34] and ensures adequate sampling of the system's dynamics, required for convergence to the optimal policy.

12.2 ACTOR–CRITIC–IDENTIFIER ARCHITECTURE FOR HJB APPROXIMATION

Consider a continuous-time nonlinear system

$$\dot{x} = F(x, u),$$

where $x(t) \in \mathcal{X} \subseteq \mathbb{R}^n$, $u(t) \in \mathcal{U} \subseteq \mathbb{R}^m$ is the control input, $F : \mathcal{X} \times \mathcal{U} \to \mathbb{R}^n$ is Lipschitz continuous on $\mathcal{X} \times \mathcal{U}$ containing the origin, such that the solution $x(t)$ of the system is unique for any finite initial condition x_0 and control $u \in \mathcal{U}$. The optimal value function can be defined as

$$V^*(x(t)) = \min_{\substack{u(\tau)\in\Psi(\mathcal{X})\\ t\le\tau<\infty}} \int_t^{\infty} r(x(s), u(x(s)))\, ds, \tag{12.1}$$

where $\Psi(\mathcal{X})$ is a set of admissible policies, and $r(x, u) \in \mathbb{R}$ is the immediate or local cost, defined as

$$r(x, u) = Q(x) + u^{\mathrm{T}} R u, \tag{12.2}$$

where $Q(x) \in \mathbb{R}$ is continuously differentiable and positive definite, and $R \in \mathbb{R}^{m \times m}$ is a positive-definite symmetric matrix. For the local cost in (12.2), which is convex in the control, and control-affine dynamics of the form

$$\dot{x} = f(x) + g(x)u, \tag{12.3}$$

where $f(x) \in \mathbb{R}^n$ and $g(x) \in \mathbb{R}^{n \times m}$, the closed-form expression for optimal control is derived as [35]

$$u^*(x) = -\frac{1}{2} R^{-1} g^{\mathrm{T}}(x) \frac{\partial V^*(x)}{\partial x}^{\mathrm{T}}, \tag{12.4}$$

where it is assumed that the value function $V^*(x)$ is continuously differentiable and satisfies $V^*(0) = 0$.

The Hamiltonian of the system in (12.3) is given by

$$H(x, u, V_x^*) \triangleq V_x^* F_u + r_u,$$

where $V_x^* \triangleq \frac{\partial V^*}{\partial x} \in \mathbb{R}^{1 \times n}$ denotes the gradient of the optimal value function $V^*(x)$, $F_u(x, u) \triangleq f(x) + g(x)u \in \mathbb{R}^n$ denotes the system dynamics with control $u(x)$, and $r_u \triangleq r(x, u)$ denotes the local cost with control $u(x)$. The optimal value function $V^*(x)$ in (12.1) and the associated optimal policy $u^*(x)$ in (12.4) satisfy the HJB equation

$$H^*(x, u^*, V_x^*) = V_x^* F_{u^*} + r_{u^*} = 0. \tag{12.5}$$

Replacing $u^*(x)$, $V_x^*(x)$, and $F_{u^*}(x, u^*)$ in (12.5) by their approximations, $\hat{u}(x)$ (actor), $\hat{V}(x)$(critic), and $\hat{F}_{\hat{u}}(x, \hat{x}, \hat{u})$ (identifier), respectively, the approximate HJB equation is given by

$$\hat{H}^*(x, \hat{x}, \hat{u}, \hat{V}_x) = \hat{V}_x \hat{F}_{\hat{u}} + r_{\hat{u}}, \tag{12.6}$$

where $\hat{x}(t)$ denotes the state of the identifier. Using (12.5) and (12.6), the error between the actual and the approximate HJB equation is given by the Bellman residual error $\delta_{hjb}(x, \hat{x}, \hat{u}, \hat{V}_x)$, defined as

$$\delta_{hjb} \triangleq \hat{H}^*(x, \hat{x}, \hat{u}, \hat{V}_x) - H^*(x, u^*, V_x^*). \tag{12.7}$$

Since $H^*(x, u^*, V_x^*) \equiv 0$, the Bellman error can be written in a measurable form as

$$\delta_{hjb} = \hat{H}^*(x, \hat{x}, \hat{u}, \hat{V}_x) = \hat{V}_x \hat{F}_{\hat{u}} + r(x, \hat{u}). \tag{12.8}$$

The actor and critic learn based on the Bellman error $\delta_{hjb}(\cdot)$, whereas the identifier estimates the system dynamics online using the identification error $\tilde{x}(t) \triangleq x(t) - \hat{x}(t)$,

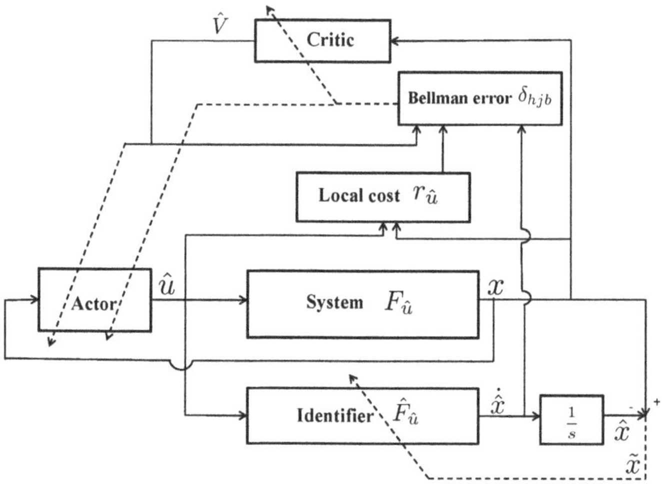

FIGURE 12.1 AC–identifier architecture to approximate the HJB.

and hence is decoupled from the AC design. The block diagram of the ACI architecture is shown in Figure 12.1.

The following assumptions are made about the control-affine system in (12.3).

Assumption 12.1 *The functions $f(x)$ and $g(x)$ are second-order differentiable.*

Assumption 12.2 *The input gain matrix $g(x)$ is known and bounded, that is, $0 < \|g(x)\| \leq \bar{g}$, where $\bar{g}$ is a known positive constant.*

Assuming the optimal control, the optimal value function and the system dynamics are continuous and defined on compact sets, and NNs can be used to approximate them [36, 37]. Some standard NN assumptions which will be used throughout the chapter are

Assumption 12.3 *Given a continuous function $\Upsilon : \mathbb{S} \to \mathbb{R}^n$, where $\mathbb{S}$ is a compact simply connected set, there exists ideal weights W, V such that the function can be represented by a NN as*

$$\Upsilon(x) = W^{\mathrm{T}}\sigma(V^{\mathrm{T}}x) + \varepsilon(x),$$

where $\sigma(\cdot)$ is the nonlinear activation function and $\varepsilon(x)$ the function reconstruction error.

Assumption 12.4 *The ideal NN weights are bounded by known positive constants, that is, $\|W\| \leq \bar{W}$, $\|V\| \leq \bar{V}$ [38].*

Assumption 12.5 *The NN activation function $\sigma(\cdot)$ and its derivative with respect to its arguments, $\sigma'(\cdot)$, are bounded.*

Assumption 12.6 *Using the NN universal approximation property [36, 37], the function reconstruction errors and its derivative with respect to its arguments are bounded [38] as* $\|\varepsilon(\cdot)\| \leq \bar{\varepsilon}$, $\left\|\varepsilon'(\cdot)\right\| \leq \bar{\varepsilon}'$.

12.3 ACTOR–CRITIC DESIGN

Using Assumption 12.3 and (12.4), the optimal value function and the optimal control can be represented by NNs as

$$V^*(x) = W^{\mathrm{T}}\phi(x) + \varepsilon_v(x),$$
$$u^*(x) = -\frac{1}{2}R^{-1}g^{\mathrm{T}}(x)(\phi'(x)^{\mathrm{T}}W + \varepsilon_v'(x)^{\mathrm{T}}), \tag{12.9}$$

where $W \in \mathbb{R}^N$ are unknown ideal NN weights, N is the number of neurons, $\phi(x) = [\phi_1(x)\ \phi_2(x)\ldots\phi_N(x)]^{\mathrm{T}} \in \mathbb{R}^N$ is a smooth NN activation function such that $\phi_i(0) = 0$ and $\phi_i'(0) = 0\ \ \forall i = 1\ldots N$, and $\varepsilon_v(\cdot) \in \mathbb{R}$ is the function reconstruction error.

Assumption 12.7 *The NN activation functions* $\{\phi_i(x)\ :\ i = 1\ldots N\}$ *are selected so that as* $N \to \infty$, $\phi(x)$ *provides a complete independent basis for* $V^*(x)$.

Using Assumption 12.7 and the Weierstrass higher order approximation theorem, both $V^*(x)$ and $\frac{\partial V^*(x)}{\partial x}$ can be uniformly approximated by NNs in (12.9), that is, as $N \to \infty$, the approximation errors $\varepsilon_v(x)$, $\varepsilon_v'(x) \to 0$ [29]. The critic $\hat{V}(x)$ and the actor $\hat{u}(x)$ approximate the optimal value function and the optimal control in (12.9), and are given by

$$\hat{V}(x) = \hat{W}_c^{\mathrm{T}}\phi(x); \qquad \hat{u}(x) = -\frac{1}{2}R^{-1}g^{\mathrm{T}}(x)\phi'^{\mathrm{T}}(x)\hat{W}_a, \tag{12.10}$$

where $\hat{W}_c(t) \in \mathbb{R}^N$ and $\hat{W}_a(t) \in \mathbb{R}^N$ are estimates of the ideal weights of the critic and actor NNs, respectively. The weight estimation errors for the critic and actor NNs are defined as $\tilde{W}_c(t) \triangleq W - \hat{W}_c(t) \in \mathbb{R}^N$ and $\tilde{W}_a(t) \triangleq W - \hat{W}_a(t) \in \mathbb{R}^N$, respectively.

Remark 12.1 *Since the optimal control is determined using the gradient of the optimal value function in (12.9), the critic NN in (12.10) may be used to determine the actor without using another NN for the actor. However, for ease in deriving weight update laws and subsequent stability analysis, separate NNs are used for the actor and the critic [12].*

The actor and critic NN weights are both updated based on the minimization of the Bellman error $\delta_{hjb}(\cdot)$ in (12.8), which can be rewritten by substituting $\hat{V}(x)$ from (12.10) as

$$\delta_{hjb} = \hat{W}_c^{\mathrm{T}}\omega + r(x, \hat{u}), \tag{12.11}$$

where $\omega(x, \hat{x}, \hat{u}) \triangleq \phi'(x)\hat{F}_{\hat{u}}(x, \hat{x}, \hat{u}) \in \mathbb{R}^N$ is the critic NN regressor vector. Let $E_c(\delta_{hjb}) = \int_0^t \delta_{hjb}^2(\tau)\, d\tau \in \mathbb{R}^+$ denote the integral-squared Bellman error for the critic. The normalized recursive LS update law for the critic is given by Ref. [39]

$$\dot{\hat{W}}_c = -\eta_c \Gamma \frac{\omega}{1 + \nu \omega^{\mathrm{T}} \Gamma \omega} \delta_{hjb}, \tag{12.12}$$

where $\nu, \eta_c \in \mathbb{R}$ are constant positive gains, and $\Gamma(t) \in \mathbb{R}^{N \times N}$ is a symmetric estimation gain matrix generated as

$$\dot{\Gamma} = -\eta_c \Gamma \frac{\omega \omega^{\mathrm{T}}}{1 + \nu \omega^{\mathrm{T}} \Gamma \omega} \Gamma; \quad \Gamma(t_r^+) = \Gamma(0) = \varphi_0 I, \tag{12.13}$$

where t_r^+ is the resetting time at which $\lambda_{\min}\{\Gamma(t)\} \leq \varphi_1, \varphi_0 > \varphi_1 > 0$. The covariance resetting ensures that $\Gamma(t)$ is positive-definite for all time and prevents its value from becoming arbitrarily small in some directions, thus avoiding slow adaptation in some directions (also called the covariance wind-up problem) [39]. From (12.13), it is clear that $\dot{\Gamma} \leq 0$, which means that the covariance matrix $\Gamma(t)$ can be bounded as

$$\varphi_1 I \leq \Gamma(t) \leq \varphi_0 I. \tag{12.14}$$

Unlike the critic weights, the actor weights appear nonlinearly in $\delta_{hjb}(\cdot)$, making it problematic to develop a LS update law. Hence, a gradient update law is developed for the actor, which minimizes the squared Bellman error $E_a(t) \triangleq \delta_{hjb}^2$, whose gradient is given by

$$\begin{aligned} \dot{\hat{W}}_a = \operatorname{proj}\Bigg[& -\frac{\eta_{a1}}{\sqrt{1+\omega^{\mathrm{T}}\omega}} \left(\hat{W}_c^{\mathrm{T}} \phi' \frac{\partial \hat{F}_{\hat{u}}}{\partial \hat{u}} \frac{\partial \hat{u}}{\partial \hat{W}_a} + \hat{W}_a^{\mathrm{T}} \phi' G \phi'^{\mathrm{T}} \right)^{\mathrm{T}} \delta_{hjb} \\ & - \eta_{a2}(\hat{W}_a - \hat{W}_c) \Bigg], \end{aligned} \tag{12.15}$$

where proj$\{\cdot\}$ is a projection operator used to bound the weight estimates [40], [41], $G(x) \triangleq g(x) R^{-1} g(x)^{\mathrm{T}} \in \mathbb{R}^{n \times n}$, $\eta_{a1}, \eta_{a2} \in \mathbb{R}$ are positive adaptation gains, $\frac{1}{\sqrt{1+\omega^{\mathrm{T}}\omega}}$ is the normalization term, and the last term in (12.15) is added for stability (based on the subsequent stability analysis).

12.4 IDENTIFIER DESIGN

The following assumption is made for the identifier design:

Assumption 12.8 *The control input is bounded, that is, $u(t) \in \mathcal{L}_\infty$. Using Assumptions 12.2, 12.5, and the projection algorithm in (12.15), this assumption holds for the control design $u(t) = \hat{u}(x)$ in (12.10).*

Using Assumption 12.3, the dynamic system in (12.3), with control $\hat{u}(x)$, can be represented using a multilayer NN as

$$\dot{x} = F_{\hat{u}}(x, \hat{u}) = W_f^{\mathrm{T}} \sigma(V_f^{\mathrm{T}} x) + \varepsilon_f(x) + g(x)\hat{u}, \tag{12.16}$$

where $W_f \in \mathbb{R}^{L_f+1\times n}$, $V_f \in \mathbb{R}^{n\times L_f}$ are the unknown ideal NN weights, $\sigma_f \triangleq \sigma(V_f^{\mathrm{T}}x) \in \mathbb{R}^{L_f+1}$ is the NN activation function, and $\varepsilon_f(x) \in \mathbb{R}^n$ is the function reconstruction error. The following multilayer dynamic neural network (MLDNN) identifier is used to approximate the system in (12.16)

$$\dot{\hat{x}} = \hat{F}_{\hat{u}}(x, \hat{x}, \hat{u}) = \hat{W}_f^{\mathrm{T}}\hat{\sigma}_f + g(x)\hat{u} + \mu, \tag{12.17}$$

where $\hat{x}(t) \in \mathbb{R}^n$ is the DNN state, $\hat{\sigma}_f \triangleq \sigma(\hat{V}_f^{\mathrm{T}}\hat{x}) \in \mathbb{R}^{L_f+1}$, $\hat{W}_f(t) \in \mathbb{R}^{L_f+1\times n}$, and $\hat{V}_f(t) \in \mathbb{R}^{n\times L_f}$ are weight estimates, and $\mu(t) \in \mathbb{R}^n$ denotes the RISE feedback term defined as [33, 42]

$$\mu \triangleq k\tilde{x}(t) - k\tilde{x}(0) + v, \tag{12.18}$$

where $\tilde{x}(t) \triangleq x(t) - \hat{x}(t) \in \mathbb{R}^n$ is the identification error, and $v(t) \in \mathbb{R}^n$ is the generalized solution (in Filippov's sense [43]) to

$$\dot{v} = (k\alpha + \gamma)\tilde{x} + \beta_1 sgn(\tilde{x}); \quad v(0) = 0,$$

where $k,\ \alpha,\ \gamma,\ \beta_1 \in \mathbb{R}$ are positive constant control gains, and $sgn\,(\cdot)$ denotes a vector signum function. The identification error dynamics can be written as

$$\dot{\tilde{x}} = \tilde{F}_u(x, \hat{x}, u) = W_f^{\mathrm{T}}\sigma_f - \hat{W}_f^{\mathrm{T}}\hat{\sigma}_f + \varepsilon_f(x) - \mu, \tag{12.19}$$

where $\tilde{F}_{\hat{u}}(x, \hat{x}, \hat{u}) \triangleq F_{\hat{u}}(x, \hat{u}) - \hat{F}_{\hat{u}}(x, \hat{x}, \hat{u}) \in \mathbb{R}^n$. A filtered identification error is defined as

$$r \triangleq \dot{\tilde{x}} + \alpha\tilde{x}. \tag{12.20}$$

Taking the time derivative of (12.20) and using (12.19) yields

$$\begin{aligned}\dot{r} = {} & W_f^{\mathrm{T}}\sigma'_f V_f^{\mathrm{T}}\dot{x} - \dot{\hat{W}}_f^{\mathrm{T}}\hat{\sigma}_f - \hat{W}_f^{\mathrm{T}}\hat{\sigma}'_f\dot{\hat{V}}_f^{\mathrm{T}}\hat{x} - \hat{W}_f^{\mathrm{T}}\hat{\sigma}'_f\hat{V}_f^{\mathrm{T}}\dot{\hat{x}} + \dot{\varepsilon}_f(x) - kr \\ & - \gamma\tilde{x} - \beta_1 sgn(\tilde{x}) + \alpha\dot{\tilde{x}}.\end{aligned} \tag{12.21}$$

Based on (12.21) and the subsequent stability analysis, the weight update laws for the DNN are designed as

$$\dot{\hat{W}}_f = \mathrm{proj}(\Gamma_{wf}\hat{\sigma}'_f\hat{V}_f^{\mathrm{T}}\dot{\hat{x}}\tilde{x}^{\mathrm{T}}), \quad \dot{\hat{V}}_f = \mathrm{proj}(\Gamma_{vf}\dot{\hat{x}}\tilde{x}^{\mathrm{T}}\hat{W}_f^{\mathrm{T}}\hat{\sigma}'_f), \tag{12.22}$$

where proj$(\cdot)$ is a smooth projection operator [40], [41], and $\Gamma_{wf} \in \mathbb{R}^{L_f+1\times L_f+1}$, $\Gamma_{vf} \in \mathbb{R}^{n\times n}$ are positive constant adaptation gain matrices. The expression in (12.21) can be rewritten as

$$\dot{r} = \tilde{N} + N_{B1} + \hat{N}_{B2} - kr - \gamma\tilde{x} - \beta_1 sgn(\tilde{x}), \tag{12.23}$$

where the auxiliary signals, $\tilde{N}(x, \tilde{x}, r, \hat{W}_f, \hat{V}_f, t)$, $N_{B1}(x, \hat{x}, \hat{W}_f, \hat{V}_f, t)$, and $\hat{N}_{B2}(\hat{x}, \dot{\hat{x}}, \hat{W}_f, \hat{V}_f, t) \in \mathbb{R}^n$ are defined as

$$\tilde{N} \triangleq \alpha\dot{\tilde{x}} - \dot{\hat{W}}_f^{\mathrm{T}}\hat{\sigma}_f - \hat{W}_f^{\mathrm{T}}\hat{\sigma}'_f\dot{\hat{V}}_f^{\mathrm{T}}\hat{x} + \frac{1}{2}W_f^{\mathrm{T}}\hat{\sigma}'_f\hat{V}_f^{\mathrm{T}}\dot{\tilde{x}} + \frac{1}{2}\hat{W}_f^{\mathrm{T}}\hat{\sigma}'_f V_f^{\mathrm{T}}\dot{\tilde{x}}, \tag{12.24}$$

$$N_{B1} \triangleq W_f^{\mathrm{T}}\sigma'_f V_f^{\mathrm{T}}\dot{x} - \frac{1}{2}W_f^{\mathrm{T}}\hat{\sigma}'_f\hat{V}_f^{\mathrm{T}}\dot{x} - \frac{1}{2}\hat{W}_f^{\mathrm{T}}\hat{\sigma}'_f V_f^{\mathrm{T}}\dot{x} + \dot{\varepsilon}_f(x), \tag{12.25}$$

$$\hat{N}_{B2} \triangleq \frac{1}{2}\tilde{W}_f^{\mathrm{T}}\hat{\sigma}'_f\hat{V}_f^{\mathrm{T}}\dot{\hat{x}} + \frac{1}{2}\hat{W}_f^{\mathrm{T}}\hat{\sigma}'_f\tilde{V}_f^{\mathrm{T}}\dot{\hat{x}}, \tag{12.26}$$

where $\tilde{W}_f \triangleq W_f - \hat{W}_f(t) \in \mathbb{R}^{L_f+1\times n}$ and $\tilde{V}_f \triangleq V_f - \hat{V}_f(t) \in \mathbb{R}^{n\times L_f}$. To facilitate the subsequent stability analysis, an auxiliary term $N_{B2}(\hat{x}, \dot{x}, \hat{W}_f, \hat{V}_f, t) \in \mathbb{R}^n$ is defined by replacing $\dot{\hat{x}}(t)$ in $\hat{N}_{B2}(\cdot)$ by $\dot{x}(t)$, and $\tilde{N}_{B2}(\hat{x}, \dot{\tilde{x}}, \hat{W}_f, \hat{V}_f, t) \triangleq \hat{N}_{B2}(\cdot) - N_{B2}(\cdot)$. The terms $N_{B1}(\cdot)$ and $N_{B2}(\cdot)$ are grouped as $N_B \triangleq N_{B1} + N_{B2}$. Using Assumptions 12.2, 12.4–12.6, and (12.20), (12.22), (12.25), and (12.26), the following bounds can be obtained

$$\left\|\tilde{N}\right\| \leq \rho_1(\|z\|)\,\|z\|, \tag{12.27}$$

$$\|N_{B1}\| \leq \zeta_1, \quad \|N_{B2}\| \leq \zeta_2, \quad \left\|\dot{N}_B\right\| \leq \zeta_3 + \zeta_4\rho_2(\|z\|)\,\|z\|, \tag{12.28}$$

$$\left\|\dot{\tilde{x}}^{\mathrm{T}}\tilde{N}_{B2}\right\| \leq \zeta_5\,\|\tilde{x}\|^2 + \zeta_6\,\|r\|^2, \tag{12.29}$$

where $z \triangleq \left[\tilde{x}^{\mathrm{T}}\; r^{\mathrm{T}}\right]^{\mathrm{T}} \in \mathbb{R}^{2n}$, $\rho_1(\cdot), \rho_2(\cdot) \in \mathbb{R}$ are positive, globally invertible, nondecreasing functions, and $\zeta_i \in \mathbb{R}$, $i = 1, ..., 6$ are computable positive constants. To facilitate the subsequent stability analysis, let $\mathcal{D} \subset \mathbb{R}^{2n+2}$ be a domain containing $y(t) = 0$, where $y(t) \in \mathbb{R}^{2n+2}$ is defined as

$$y \triangleq \left[\tilde{x}^{\mathrm{T}}\;\; r^{\mathrm{T}}\;\; \sqrt{P}\;\; \sqrt{Q}\right]^{\mathrm{T}}, \tag{12.30}$$

where the auxiliary function $P(z, t) \in \mathbb{R}$ is the generalized solution to the differential equation

$$\dot{P} = -L, \qquad P(0) = \beta_1 \sum_{i=1}^{n} |\tilde{x}_i(0)| - \tilde{x}^{\mathrm{T}}(0)\, N_B(0), \tag{12.31}$$

where the auxiliary function $L(z, t) \in \mathbb{R}$ is defined as

$$L \triangleq r^{\mathrm{T}}(N_{B1} - \beta_1 sgn(\tilde{x})) + \dot{\tilde{x}}^{\mathrm{T}} N_{B2} - \beta_2\rho_2(\|z\|)\,\|z\|\,\|\tilde{x}\|, \tag{12.32}$$

where $\beta_1, \beta_2 \in \mathbb{R}$ are chosen according to the following sufficient conditions to ensure $P(t) \geq 0$ [33]

$$\beta_1 > \max\left(\zeta_1 + \zeta_2,\; \zeta_1 + \frac{\zeta_3}{\alpha}\right), \qquad \beta_2 > \zeta_4. \tag{12.33}$$

The auxiliary function $Q(\tilde{W}_f, \tilde{V}_f) \in \mathbb{R}$ in (12.30) is defined as

$$Q \triangleq \frac{1}{4}\alpha\left[tr(\tilde{W}_f^{\mathrm{T}}\Gamma_{wf}^{-1}\tilde{W}_f) + tr(\tilde{V}_f^{\mathrm{T}}\Gamma_{vf}^{-1}\tilde{V}_f)\right],$$

where $tr(\cdot)$ denotes the trace of a matrix.

Theorem 12.1 *For the system in (12.3), the identifier developed in (12.17) along with the weight update laws in (12.22) ensures asymptotic identification of the state and its derivative, in the sense that*

$$\lim_{t\to\infty}\|\tilde{x}(t)\| = 0 \qquad \textit{and} \qquad \lim_{t\to\infty}\left\|\dot{\tilde{x}}(t)\right\| = 0,$$

provided the control gains k and γ are chosen sufficiently large based on the initial conditions of the states[1]*, and satisfy the following sufficient conditions*

$$\gamma > \frac{\zeta_5}{\alpha}, \qquad k > \zeta_6, \tag{12.34}$$

where ζ_5 and ζ_6 are introduced in (12.29), and β_1, β_2 introduced in (12.32), are chosen according to the sufficient conditions in (12.33).

Proof: Let $V_I(y) : \mathcal{D} \to \mathbb{R}$ be a Lipschitz continuous regular positive definite function defined as:

$$V_I \triangleq \frac{1}{2}r^{\mathrm{T}}r + \frac{1}{2}\gamma\tilde{x}^{\mathrm{T}}\tilde{x} + P + Q, \tag{12.35}$$

which satisfies the following inequalities:

$$U_1(y) \le V_I(y) \le U_2(y), \tag{12.36}$$

where $U_1(y), U_2(y) \in \mathbb{R}$ are continuous positive definite functions defined as

$$U_1 \triangleq \frac{1}{2}\min(1,\gamma)\|y\|^2 \qquad U_2 \triangleq \max(1,\gamma)\|y\|^2.$$

From (12.19), (12.22), (12.23), and (12.31), the differential equations of the closed-loop system are continuous except in the set $\{y|\tilde{x} = 0\}$. Using Filippov's differential inclusion [44, 45], the existence of solutions can be established for $\dot{y} = f(y)$, where $f(y) \in \mathbb{R}^{2n+2}$ denotes the right-hand side of the closed-loop error signals. Under Filippov's framework, a generalized Lyapunov stability theory can be used (see Refs [46–48] for further details) to establish strong stability of the closed-loop system. The generalized time derivative of (12.35) exists almost everywhere (a.e.), and $\dot{V}_I(y) \in^{a.e.}$

[1] See subsequent semiglobal stability analysis.

$\dot{\tilde{V}}_I(y)$ where

$$\dot{\tilde{V}}_I = \bigcap_{\xi \in \partial V_I(y)} \xi^{\mathrm{T}} K \left[\dot{r}^{\mathrm{T}} \ \dot{\tilde{x}}^{\mathrm{T}} \ \frac{1}{2} P^{-\frac{1}{2}} \dot{P} \ \frac{1}{2} Q^{-\frac{1}{2}} \dot{Q} \right]^{\mathrm{T}},$$

where ∂V_I is the generalized gradient of V_I [46], and $K[\cdot]$ is defined as [47, 48]

$$K[f](y) \triangleq \bigcap_{\delta > 0} \bigcap_{\mu M = 0} \overline{co} f(B(y, \delta) - M),$$

where $\bigcap_{\mu M=0}$ denotes the intersection of all sets M of Lebesgue measure zero, $\overline{co}$ denotes convex closure, and $B(y, \delta) = \{x \in \mathbb{R}^{2n+2} |\ \|y - x\| < \delta\}$. Since $V_I(y)$ is a Lipschitz continuous regular function,

$$\begin{aligned}
\dot{\tilde{V}}_I &= \nabla V_I^{\mathrm{T}} K \left[\dot{r}^{\mathrm{T}} \ \dot{\tilde{x}}^{\mathrm{T}} \ \frac{1}{2} P^{-\frac{1}{2}} \dot{P} \ \frac{1}{2} Q^{-\frac{1}{2}} \dot{Q} \right]^{\mathrm{T}} \\
&= \left[r^{\mathrm{T}} \ \gamma \tilde{x}^{\mathrm{T}} \ 2P^{\frac{1}{2}} \ 2Q^{\frac{1}{2}} \right] K \left[\dot{r}^{\mathrm{T}} \ \dot{\tilde{x}}^{\mathrm{T}} \ \frac{1}{2} P^{-\frac{1}{2}} \dot{P} \ \frac{1}{2} Q^{-\frac{1}{2}} \dot{Q} \right]^{\mathrm{T}}.
\end{aligned}$$

Using the calculus for $K[\cdot]$ from Ref. [48], and substituting the dynamics from (12.23) and (12.31), yields

$$\begin{aligned}
\dot{\tilde{V}}_I &\subset r^{\mathrm{T}}(\tilde{N} + N_{B1} + \hat{N}_{B2} - kr - \beta_1 K[sgn(\tilde{x})] - \gamma \tilde{x}) + \gamma \tilde{x}^{\mathrm{T}}(r - \alpha \tilde{x}) \\
&\quad - r^{\mathrm{T}}(N_{B1} - \beta_1 K[sgn(\tilde{x})]) - \dot{\tilde{x}}^{\mathrm{T}} N_{B2} + \beta_2 \rho_2(\|z\|) \|z\| \|\tilde{x}\| \\
&\quad - \frac{1}{2}\alpha \left[tr(\tilde{W}_f^{\mathrm{T}} \Gamma_{wf}^{-1} \dot{\hat{W}}_f) + tr(\tilde{V}_f^{\mathrm{T}} \Gamma_{vf}^{-1} \dot{\hat{V}}_f) \right] \\
&= -\alpha \gamma \tilde{x}^{\mathrm{T}} \tilde{x} - kr^{\mathrm{T}} r + r^{\mathrm{T}} \tilde{N} + \frac{1}{2} \alpha \tilde{x}^{\mathrm{T}} \tilde{W}_f^{\mathrm{T}} \hat{\sigma}'_f \hat{V}_f^{\mathrm{T}} \dot{\hat{x}} + \dot{\tilde{x}}^{\mathrm{T}} (\hat{N}_{B2} - N_{B2}) \\
&\quad + \frac{1}{2} \alpha \tilde{x}^{\mathrm{T}} \hat{W}_f^{\mathrm{T}} \hat{\sigma}'_f \tilde{V}_f^{\mathrm{T}} \dot{\hat{x}} + \beta_2 \rho_2(\|z\|) \|z\| \|\tilde{x}\| \qquad (12.37) \\
&\quad - \frac{1}{2} \alpha tr(\tilde{W}_f^{\mathrm{T}} \hat{\sigma}'_f \hat{V}_f^{\mathrm{T}} \dot{\hat{x}} \tilde{x}^{\mathrm{T}}) - \frac{1}{2} \alpha tr(\tilde{V}_f^{\mathrm{T}} \dot{\hat{x}} \tilde{x}^{\mathrm{T}} \hat{W}_f^{\mathrm{T}} \hat{\sigma}'_f),
\end{aligned}$$

where (12.22) and the fact that $(r^{\mathrm{T}} - r^{\mathrm{T}})_i SGN(\tilde{x}_i) = 0$ is used (the subscript i denotes the ith element), where $K[sgn(\tilde{x})] = SGN(\tilde{x})$ [48], such that $SGN(\tilde{x}_i) = 1$ if $\tilde{x}_i > 0$, $[-1, 1]$ if $\tilde{x}_i = 0$, and -1 if $\tilde{x}_i < 0$. Canceling common terms, substituting for $k \triangleq k_1 + k_2$ and $\gamma \triangleq \gamma_1 + \gamma_2$, using (12.27), (12.29), and completing the squares, the expression in (12.37) can be upper bounded as

$$\dot{\hat{V}}_I \leq -(\alpha\gamma_1 - \zeta_5)\left\|\tilde{x}\right\|^2 - (k_1 - \zeta_6)\left\|r\right\|^2 + \frac{\rho_1(\|z\|)^2}{4k_2}\left\|z\right\|^2 + \frac{\beta_2^2\rho_2(\|z\|)^2}{4\alpha\gamma_2}\left\|z\right\|^2. \tag{12.38}$$

Provided the sufficient conditions in (12.34) are satisfied, the expression in (12.38) can be rewritten as

$$\dot{\hat{V}}_I \leq -\lambda\left\|z\right\|^2 + \frac{\rho(\|z\|)^2}{4\eta}\left\|z\right\|^2 \leq -U(y) \qquad \forall y \in \mathcal{D}, \tag{12.39}$$

where $\lambda \triangleq \min\{\alpha\gamma_1 - \zeta_5,\ k_1 - \zeta_6\}$, $\rho(\|z\|)^2 \triangleq \rho_1(\|z\|)^2 + \rho_2(\|z\|)^2$, $\eta \triangleq \min\{k_2,\ \frac{\alpha\gamma_2}{\beta_2^2}\}$, and $U(y) = c\left\|z\right\|^2$, for some positive constant c, is a continuous, positive semidefinite function defined on the domain $\mathcal{D} \triangleq \left\{y(t) \in \mathbb{R}^{2n+2} \mid \|y\| \leq \rho^{-1}\left(2\sqrt{\lambda\eta}\right)\right\}$. The size of the domain $\mathcal{D}$ can be increased by increasing the gains k and γ. The result in (12.39) indicates that $\dot{V}_I(y) \leq -U(y)\ \forall\ \dot{V}_I(y) \in^{a.e.} \dot{\hat{V}}_I(y)$ $\forall y \in \mathcal{D}$. The inequalities in (12.36) and (12.39) can be used to show that $V_I(y) \in \mathcal{L}_\infty$ in $\mathcal{D}$; hence, $\tilde{x}(t), r(t) \in \mathcal{L}_\infty$ in $\mathcal{D}$. Using (12.20), standard linear analysis can be used to show that $\dot{\tilde{x}}(t) \in \mathcal{L}_\infty$ in $\mathcal{D}$, and since $\dot{x}(t) \in \mathcal{L}_\infty$, $\dot{\hat{x}}(t) \in \mathcal{L}_\infty$ in $\mathcal{D}$. Since $\hat{W}_f(t) \in \mathcal{L}_\infty$ from the use of projection in (12.22), $\hat{\sigma}_f(t) \in \mathcal{L}_\infty$ from Assumption 12.5, and $\hat{u}(t) \in \mathcal{L}_\infty$ from Assumption 12.8, $\mu(t) \in \mathcal{L}_\infty$ in $\mathcal{D}$ from (12.17). Using the above bounds and the fact that $\hat{\sigma}'_f(t)$, $\dot{\varepsilon}_f(t) \in \mathcal{L}_\infty$, it can be shown from (12.21) that $\dot{r}(t) \in \mathcal{L}_\infty$ in $\mathcal{D}$. Since $\tilde{x}(t)$,$r(t) \in \mathcal{L}_\infty$, the definition of $U(y)$ can be used to show that it is uniformly continuous in $\mathcal{D}$. Let $\mathcal{S} \subset \mathcal{D}$ denote a set defined as

$$\mathcal{S} \triangleq \left\{y(t) \subset \mathcal{D} \mid U_2(y(t)) < \frac{1}{2}\left(\rho^{-1}\left(2\sqrt{\lambda\eta}\right)\right)^2\right\}. \tag{12.40}$$

The region of attraction in (12.40) can be made arbitrarily large to include any initial conditions by increasing the control gain η (i.e., a semiglobal type of stability result), and hence

$$c\left\|z\right\|^2 \to 0 \qquad \text{as} \qquad t \to \infty \qquad \forall y(0) \in \mathcal{S},$$

and using the definition of $z(t)$

$$\left\|\tilde{x}(t)\right\|, \left\|\dot{\tilde{x}}(t)\right\|, \left\|r\right\| \to 0 \qquad \text{as} \qquad t \to \infty \qquad \forall y(0) \in \mathcal{S}.$$

■

Using the developed identifier in (12.17), the actor weight update law can now be simplified using (12.15) as

$$\dot{\hat{W}}_a = \text{proj}\left\{-\frac{\eta_{a1}}{\sqrt{1+\omega^{\mathrm{T}}\omega}}\phi' G \phi'^{\mathrm{T}}\left(\hat{W}_a - \hat{W}_c\right)\delta_{hjb} - \eta_{a2}(\hat{W}_a - \hat{W}_c)\right\}. \tag{12.41}$$

12.5 CONVERGENCE AND STABILITY ANALYSIS

The unmeasurable form of the Bellman error can be written using (12.5)–(12.8) and (12.11), as

$$\begin{aligned}\delta_{hjb} &= \hat{W}_c^{\mathrm{T}}\omega - W_c^{\mathrm{T}}\phi' F_{u^*} + \hat{u}^{\mathrm{T}} R\hat{u} - u^{*^{\mathrm{T}}} Ru^* - \varepsilon_v' F_{u^*}.\\ &= -\tilde{W}_c^{\mathrm{T}}\omega - W^{\mathrm{T}}\phi'\tilde{F}_{\hat{u}} + \frac{1}{4}\tilde{W}_a^{\mathrm{T}}\phi' G\phi'^{\mathrm{T}}\tilde{W}_a - \frac{1}{4}\varepsilon_v' G\varepsilon_v'^{\mathrm{T}} - \varepsilon_v' F_{u^*},\end{aligned} \tag{12.42}$$

where (12.9) and (12.10) are used. The dynamics of the critic weight estimation error $\tilde{W}_c(t)$ can now be developed by substituting (12.42) in (12.12), as

$$\begin{aligned}\dot{\tilde{W}}_c = -\eta_c\Gamma\psi\psi^{\mathrm{T}}\tilde{W}_c + \eta_c\Gamma\frac{\omega}{1+\nu\omega^{\mathrm{T}}\Gamma\omega}\Bigg[-W^{\mathrm{T}}\phi'\tilde{F}_{\hat{u}} + \frac{1}{4}\tilde{W}_a^{\mathrm{T}}\phi' G\phi'^{\mathrm{T}}\tilde{W}_a\\ - \frac{1}{4}\varepsilon_v' G\varepsilon_v'^{\mathrm{T}} - \varepsilon_v' F_{u^*}\Bigg],\end{aligned} \tag{12.43}$$

where $\psi(t) \triangleq \frac{\omega(t)}{\sqrt{1+\nu\omega(t)^{\mathrm{T}}\Gamma(t)\omega(t)}} \in \mathbb{R}^N$ is the normalized critic regressor vector, bounded as

$$\|\psi\| \le \frac{1}{\sqrt{\nu\varphi_1}}, \tag{12.44}$$

where φ_1 is introduced in (12.14). The error system in (12.43) can be represented by the following perturbed system

$$\dot{\tilde{W}}_c = \Omega_{\mathrm{nom}} + \Delta_{\mathrm{per}}, \tag{12.45}$$

where $\Omega_{\mathrm{nom}}(\tilde{W}_c, t) \triangleq -\eta_c\Gamma\psi\psi^{\mathrm{T}}\tilde{W}_c \in \mathbb{R}^N$, denotes the nominal system, and $\Delta_{\mathrm{per}}(t) \triangleq \eta_c\Gamma\frac{\omega}{1+\nu\omega^{\mathrm{T}}\Gamma\omega}\left[-W^{\mathrm{T}}\phi'\tilde{F}_{\hat{u}} + \frac{1}{4}\tilde{W}_a^{\mathrm{T}}\phi' G\phi'^{\mathrm{T}}\tilde{W}_a - \frac{1}{4}\varepsilon_v' G\varepsilon_v'^{\mathrm{T}} - \varepsilon_v' F_{u^*}\right] \in \mathbb{R}^N$ denotes the perturbation. Using Theorem 2.5.1 in Ref. [39], the nominal system

$$\dot{\tilde{W}}_c = -\eta_c\Gamma\psi\psi^{\mathrm{T}}\tilde{W}_c \tag{12.46}$$

is globally exponentially stable, if the bounded signal $\psi(t)$ is PE, that is,

$$\mu_2 I \ge \int_{t_0}^{t_0+\delta} \psi(\tau)\psi(\tau)^{\mathrm{T}} d\tau \ge \mu_1 I \qquad \forall t_0 \ge 0,$$

for some positive constants $\mu_1, \mu_2, \delta \in \mathbb{R}$. Since $\Omega_{\mathrm{nom}}(\tilde{W}_c, t)$ is continuously differentiable and the Jacobian $\frac{\partial\Omega_{\mathrm{nom}}}{\partial\tilde{W}_c} = -\eta_c\Gamma\psi\psi^{\mathrm{T}}$ is bounded for the exponentially stable system in (12.46), the converse Lyapunov Theorem 4.14 in Ref. [49] can be used to show that there exists a function $V_c : \mathbb{R}^N \times [0, \infty) \to \mathbb{R}$, which satisfies the following inequalities

$$
\begin{aligned}
c_1 \left\|\tilde{W}_c\right\|^2 \leq V_c(\tilde{W}_c, t) \leq c_2 \left\|\tilde{W}_c\right\|^2 \\
\frac{\partial V_c}{\partial t} + \frac{\partial V_c}{\partial \tilde{W}_c}\Omega_{\text{nom}}(\tilde{W}_c, t) \leq -c_3 \left\|\tilde{W}_c\right\|^2 \\
\left\|\frac{\partial V_c}{\partial \tilde{W}_c}\right\| \leq c_4 \left\|\tilde{W}_c\right\|,
\end{aligned}
\tag{12.47}
$$

for some positive constants $c_1, c_2, c_3, c_4 \in \mathbb{R}$. Using Assumptions 12.2, 12.4–12.6, and 12.8, the projection bounds in (12.15), the fact that $F_{u^*} \in \mathcal{L}_\infty$ (using (12.4), Assumptions 12.2–12.6, and (12.9)), and provided the conditions of Theorem 12.1 hold (required to prove that $\tilde{F}_{\hat{u}} \in \mathcal{L}_\infty$), the following bounds can be developed:

$$
\begin{aligned}
\left\|\tilde{W}_a\right\| \leq \kappa_1, \qquad \left\|\phi' G \phi'^{\mathrm{T}}\right\| \leq \kappa_2, \\
\left\|\frac{1}{4}\tilde{W}_a^{\mathrm{T}}\phi' G \phi'^{\mathrm{T}}\tilde{W}_a - \frac{1}{4}\varepsilon_v' G \varepsilon_v'^{\mathrm{T}} - W^{\mathrm{T}}\phi'\tilde{F}_{\hat{u}} - \varepsilon_v' F_{u^*}\right\| \leq \kappa_3, \\
\left\|\frac{1}{2}W^{\mathrm{T}}\phi' G \varepsilon_v'^{\mathrm{T}} + \frac{1}{2}\varepsilon_v' G \varepsilon_v'^{\mathrm{T}} + \frac{1}{2}W^{\mathrm{T}}\phi' G \phi'^{\mathrm{T}}\tilde{W}_a + \frac{1}{2}\varepsilon_v' G \phi'^{\mathrm{T}}\right\| \leq \kappa_4,
\end{aligned}
\tag{12.48}
$$

where $\kappa_1, \kappa_2, \kappa_3, \kappa_4 \in \mathbb{R}$ are computable positive constants.

Theorem 12.2 *If Assumptions 12.1–12.8 hold, the regressor* $\psi(t) \triangleq \frac{\omega}{\sqrt{1+\omega^{\mathrm{T}}\Gamma\omega}}$ *is PE, and provided (12.33), (12.34) and the following sufficient gain condition is satisfied*[2]

$$
\frac{c_3}{\eta_{a1}} > \kappa_1\kappa_2,
\tag{12.49}
$$

where $\eta_{a1}, c_3, \kappa_1, \kappa_2$ *are introduced in (12.15), (12.47), and (12.48), then the controller in (12.10), the AC weight update laws in (12.12)–(12.13) and (12.41), and the identifier in (12.17) and (12.22), guarantee that the state of the system* $x(t)$*, and the ARAC weight estimation errors* $\tilde{W}_a(t)$ *and* $\tilde{W}_c(t)$ *are UUB.*

Proof: To investigate the stability of (12.3) with control $\hat{u}(x)$, and the perturbed system in (12.45), consider $V_L : \mathcal{X} \times \mathbb{R}^N \times \mathbb{R}^N \times [0, \infty) \to \mathbb{R}$ as the continuously differentiable, positive-definite Lyapunov function candidate defined as

$$
V_L(x, \tilde{W}_c, \tilde{W}_a, t) \triangleq V^*(x) + V_c(\tilde{W}_c, t) + \frac{1}{2}\tilde{W}_a^{\mathrm{T}}\tilde{W}_a,
$$

[2]Because c_3 is a function of the critic adaptation gain η_c, η_{a1} is the actor adaptation gain, and κ_1, κ_2 are known constants, the sufficient gain condition in (12.49) can be easily satisfied.

where $V^*(x)$ (the optimal value function), is the Lyapunov function for (12.3), and $V_c(\tilde{W}_c, t)$ is the Lyapunov function for the exponentially stable system in (12.46). Because $V^*(x)$ is continuously differentiable and positive-definite from (12.1) and (12.2), there exist class $\mathcal{K}$ functions α_1 and α_2 defined on $[0,\ r]$, where $B_r \subset \mathcal{X}$(see Lemma 4.3 in Ref. [49]), such that

$$\alpha_1(\|x\|) \leq V^*(x) \leq \alpha_2(\|x\|) \qquad \forall x \in B_r. \tag{12.50}$$

Using (12.47) and (12.50), $V_L(x, \tilde{W}_c, \tilde{W}_a, t)$ can be bounded as

$$\begin{aligned}\alpha_1(\|x\|) + c_1 \left\|\tilde{W}_c\right\|^2 + \frac{1}{2}\left\|\tilde{W}_a\right\|^2 &\leq V_L(x, \tilde{W}_c, \tilde{W}_a, t) \\ &\leq \alpha_2(\|x\|) + c_2 \left\|\tilde{W}_c\right\|^2 + \frac{1}{2}\left\|\tilde{W}_a\right\|^2,\end{aligned}$$

which can be written as

$$\alpha_3(\|\tilde{z}\|) \leq V_L(x, \tilde{W}_c, \tilde{W}_a, t) \leq \alpha_4(\|\tilde{z}\|) \qquad \forall \tilde{z} \in B_s,$$

where $\tilde{z}(t) \triangleq [x(t)^{\mathrm{T}}\ \tilde{W}_c(t)^{\mathrm{T}}\ \tilde{W}_a(t)^{\mathrm{T}}]^{\mathrm{T}} \in \mathbb{R}^{n+2N}$, α_3 and α_4 are class $\mathcal{K}$ functions defined on $[0,\ s]$, where $B_s \subset \mathcal{X} \times \mathbb{R}^N \times \mathbb{R}^N$. Taking the time derivative of $V_L(\cdot)$ yields

$$\dot{V}_L = \frac{\partial V^*}{\partial x} f + \frac{\partial V^*}{\partial x} g\hat{u} + \frac{\partial V_c}{\partial t} + \frac{\partial V_c}{\partial \tilde{W}_c}\Omega_{\mathrm{nom}} + \frac{\partial V_c}{\partial \tilde{W}_c}\Delta_{\mathrm{per}} - \tilde{W}_a^{\mathrm{T}}\dot{\hat{W}}_a, \tag{12.51}$$

where the time derivative of $V^*(\cdot)$ is taken along the the trajectories of the system (12.3) with control $\hat{u}(\cdot)$ and the time derivative of $V_c(\cdot)$ is taken along the trajectories of the perturbed system (12.45). To facilitate the subsequent analysis, the HJB in (12.5) is rewritten as $\frac{\partial V^*}{\partial x} f = -\frac{\partial V^*}{\partial x} gu^* - Q(x) - u^{*^{\mathrm{T}}} Ru^*$. Substituting for $\frac{\partial V^*}{\partial x} f$ in (12.51), using the fact that $\frac{\partial V^*}{\partial x} g = -2u^{*^{\mathrm{T}}} R$ from (12.4), and using (12.15) and (12.47), (12.51) can be upper bounded as

$$\begin{aligned}\dot{V}_L \leq& -Q - u^{*^{\mathrm{T}}} Ru^* - c_3 \left\|\tilde{W}_c\right\|^2 + c_4 \left\|\tilde{W}_c\right\| \left\|\Delta_{\mathrm{per}}\right\| + 2u^{*^{\mathrm{T}}} R(u^* - \hat{u}) \\ &+ \eta_{a2}\tilde{W}_a^{\mathrm{T}}(\hat{W}_a - \hat{W}_c) + \frac{\eta_{a1}}{\sqrt{1+\omega^{\mathrm{T}}\omega}}\tilde{W}_a^{\mathrm{T}}\phi' G \phi'^{\mathrm{T}}(\hat{W}_a - \hat{W}_c)\delta_{hjb}.\end{aligned} \tag{12.52}$$

Substituting for u^*, $\hat{u}$, δ_{hjb}, and Δ_{per} using (12.4), (12.10), (12.42), and (12.45), respectively, and using (12.14) and (12.44) in (12.52), yields

$$\begin{aligned}
\dot{V}_L \leq & -Q - c_3 \left\|\tilde{W}_c\right\|^2 - \eta_{a2}\left\|\tilde{W}_a\right\|^2 + \frac{1}{2}W^{\mathrm{T}}\phi' G \varepsilon_v'^{\mathrm{T}} + \frac{1}{2}W^{\mathrm{T}}\phi' G \phi'^{\mathrm{T}}\tilde{W}_a \\
& + c_4 \frac{\eta_c \varphi_0}{2\sqrt{\nu\varphi_1}} \left\| -W^{\mathrm{T}}\phi' \tilde{F}_{\hat{u}} + \frac{1}{4}\tilde{W}_a^{\mathrm{T}}\phi' G \phi'^{\mathrm{T}}\tilde{W}_a - \frac{1}{4}\varepsilon_v' G \varepsilon_v'^{\mathrm{T}} - \varepsilon_v' F_{u^*}\right\| \left\|\tilde{W}_c\right\| \\
& + \frac{\eta_{a1}}{\sqrt{1+\omega^{\mathrm{T}}\omega}} \tilde{W}_a^{\mathrm{T}}\phi' G \phi'^{\mathrm{T}}(\tilde{W}_c - \tilde{W}_a)\left(-\tilde{W}_c^{\mathrm{T}}\omega - W^{\mathrm{T}}\phi'\tilde{F}_{\hat{u}} - \frac{1}{4}\varepsilon_v' G \varepsilon_v'^{\mathrm{T}}\right. \\
& \left. + \frac{1}{4}\tilde{W}_a^{\mathrm{T}}\phi' G \phi'^{\mathrm{T}}\tilde{W}_a - \varepsilon_v' F_{u^*}\right) + \frac{1}{2}\varepsilon_v' G \phi'^{\mathrm{T}}\tilde{W}_a + \eta_{a2}\left\|\tilde{W}_a\right\|\left\|\tilde{W}_c\right\| + \frac{1}{2}\varepsilon_v' G \varepsilon_v'^{\mathrm{T}}.
\end{aligned} \tag{12.53}$$

Using the bounds developed in (12.48), (12.53) can be further upper bounded as

$$\begin{aligned}
\dot{V}_L \leq & -Q - (c_3 - \eta_{a1}\kappa_1\kappa_2)\left\|\tilde{W}_c\right\|^2 - \eta_{a2}\left\|\tilde{W}_a\right\|^2 + \eta_{a1}\kappa_1^2\kappa_2\kappa_3 + \kappa_4 \\
& + \left(\frac{c_4\eta_c\varphi_0}{2\sqrt{\nu\varphi_1}}\kappa_3 + \eta_{a1}\kappa_1\kappa_2\kappa_3 + \eta_{a1}\kappa_1^2\kappa_2 + \eta_{a2}\kappa_1\right)\left\|\tilde{W}_c\right\|.
\end{aligned}$$

Provided $c_3 > \eta_{a1}\kappa_1\kappa_2$, and completing the square yields

$$\begin{aligned}
\dot{V}_L \leq & -Q - (1-\theta)(c_3 - \eta_{a1}\kappa_1\kappa_2)\left\|\tilde{W}_c\right\|^2 - \eta_{a2}\left\|\tilde{W}_a\right\|^2 + \eta_{a1}\kappa_1^2\kappa_2\kappa_3 + \kappa_4 \\
& + \frac{1}{4\theta(c_3 - \eta_{a1}\kappa_1\kappa_2)}\left[\frac{c_4\eta_c\varphi_0}{2\sqrt{\nu\varphi_1}}\kappa_3 + \eta_{a1}\kappa_1\kappa_2\kappa_3 + \eta_{a1}\kappa_1^2\kappa_2 + \eta_{a2}\kappa_1\right]^2
\end{aligned} \tag{12.54}$$

where $0 < \theta < 1$. Since $Q(x)$ is positive definite, Lemma 4.3 in Ref. [49] indicates that there exist class $\mathcal{K}$ functions α_5 and α_6 such that

$$\alpha_5(\|\tilde{z}\|) \leq Q + (1-\theta)(c_3 - \eta_{a1}\kappa_1\kappa_2)\left\|\tilde{W}_c\right\|^2 + \eta_{a2}\left\|\tilde{W}_a\right\|^2 \leq \alpha_6(\|\tilde{z}\|) \quad \forall v \in B_s,$$

which can be used to further upper bound the expression in (12.54) as

$$\begin{aligned}
\dot{V}_L \leq & -\alpha_5(\|\tilde{z}\|) + \frac{1}{4\theta(c_3 - \eta_{a1}\kappa_1\kappa_2)}\left[\frac{c_4\eta_c\varphi_0}{2\sqrt{\nu\varphi_1}}\kappa_3 + \eta_{a1}\kappa_1\kappa_2\kappa_3 + \eta_{a1}\kappa_1^2\kappa_2 + \eta_{a2}\kappa_1\right]^2 \\
& + \eta_{a1}\kappa_1^2\kappa_2\kappa_3 + \kappa_4,
\end{aligned}$$

which proves that $\dot{V}_L(\cdot)$ is negative whenever $\tilde{z}(t)$ lies outside the compact set $\Omega_{\tilde{z}} \triangleq \left\{\tilde{z} : \|\tilde{z}\| \leq \alpha_5^{-1}\left(\frac{1}{4\theta(c_3-\eta_{a1}\kappa_1\kappa_2)}\left[\frac{c_4\eta_c\varphi_0}{2\sqrt{\nu\varphi_1}}\kappa_3 + \eta_{a1}\kappa_1\kappa_2\kappa_3 + \eta_{a1}\kappa_1^2\kappa_2 + \eta_{a2}\kappa_1\right]^2 + \eta_{a1}\kappa_1^2\kappa_2\kappa_3 + \kappa_4\right)\right\}$, and hence, $\|\tilde{z}(t)\|$ is UUB (see Theorem 4.18 in Ref. [49]). The bounds in (12.48) depend on the actor NN approximation error ε_v', which can be reduced by increasing the number of neurons N, thereby reducing the size of the residual set $\Omega_{\tilde{z}}$. From Assumption 12.7, as the number of neurons of the actor and critic NNs $N \to \infty$, the reconstruction error $\varepsilon'_v \to 0$. ■

Remark 12.2 *Since the actor, critic, and identifier are continuously updated, the developed RL algorithm can be compared to fully optimistic PI in machine learning literature [50], where policy evaluation and policy improvement are done after every state transition, unlike traditional PI, where policy improvement is done after convergence of the policy evaluation step. Proving convergence of optimistic PI is complicated and is an active area of research in machine learning [50, 51]. By considering an adaptive control framework, this result investigates the convergence and stability behavior of fully optimistic PI in continuous-time.*

Remark 12.3 *The PE condition in Theorem 12.2 is equivalent to the exploration paradigm in RL which ensures sufficient sampling of the state space and convergence to the optimal policy [34].*

12.6 SIMULATION

The following nonlinear system is considered [12]

$$\dot{x} = \begin{bmatrix} -x_1 + x_2 \\ -0.5x_1 - 0.5x_2(1 - (\cos(2x_1) + 2)^2) \end{bmatrix} + \begin{bmatrix} 0 \\ \cos(2x_1) + 2 \end{bmatrix} u, \qquad (12.55)$$

where $x(t) \triangleq [x_1(t)\ x_2(t)]^{\mathrm{T}} \in \mathbb{R}^2$ and $u(t) \in \mathbb{R}$. The state and control penalties are chosen as

$$Q(x) = x^{\mathrm{T}} \begin{bmatrix} 1 & 0 \\ 0 & 1 \end{bmatrix} x; \qquad R = 1.$$

The optimal value function and optimal control for the system in (12.55) are known, and given by [12]

$$V^*(x) = \frac{1}{2}x_1^2 + x_2^2; \qquad u^*(x) = -(\cos(2x_1) + 2)x_2.$$

The activation function for the critic NN is selected with $N = 3$ neurons as

$$\phi(x) = [x_1^2 \quad x_1x_2 \quad x_2^2]^{\mathrm{T}},$$

while the activation function for the identifier DNN is selected as a symmetric sigmoid with $L_f = 5$ neurons in the hidden layer. The identifier gains are selected as

$$k = 800, \quad \alpha = 300, \quad \gamma = 5, \quad \beta_1 = 0.2, \quad \Gamma_{wf} = 0.1\mathbb{I}_{6\times6}, \quad \Gamma_{vf} = 0.1\mathbb{I}_{2\times2},$$

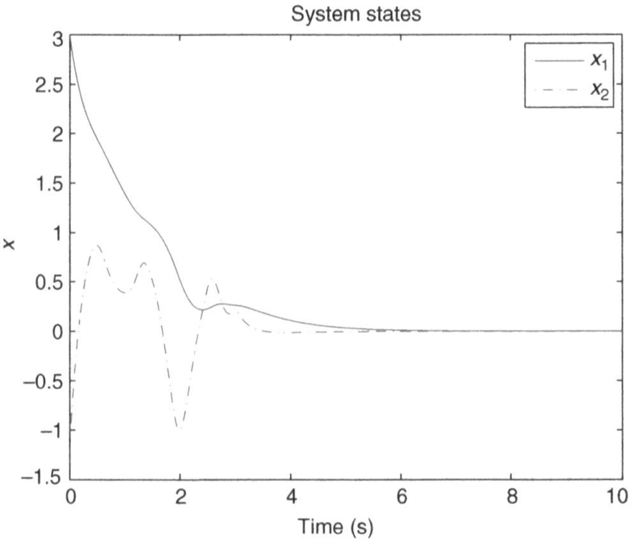

FIGURE 12.2 System states $x(t)$ with persistently excited input for the first 3 s.

and the gains for the AC learning laws are selected as

$$\eta_{a1} = 10, \quad \eta_{a2} = 50, \quad \eta_c = 20, \quad \nu = 0.005.$$

The covariance matrix is initialized to $\Gamma(0) = 5000$, all the NN weights are randomly initialized in $[-1,\ 1]$, and the states are initialized to $x(0) = [3,\ -1]$. An implementation issue in using the developed algorithm is to ensure PE of the critic regressor vector. Unlike linear systems, where PE of the regressor translates to sufficient richness of the external input, no verifiable method exists to ensure PE in nonlinear regulation problems. To ensure PE qualitatively, a small exploratory signal consisting of sinusoids of varying frequencies, $n(t) = \sin^2(t)\cos(t) + \sin^2(2t)\cos(0.1t) + \sin^2(-1.2t)\cos(0.5t) + \sin^5(t)$, is added to the control $u(t)$ for the first 3 s [12]. The evolution of states is shown in Figure 12.2. The identifier approximates the system dynamics, and the state derivative estimation error is shown in Figure 12.3. PE ensures that the weights converge to their optimal values of $W = [0.5\ 0\ 1]^{\mathrm{T}}$ in approximately 2 s, as seen from the evolution of actor–critic weights in Figures 12.4 and 12.5. The errors in approximating the optimal value function and optimal control at steady state ($t = 10$ s) are plotted against the states in Figures 12.6 and 12.7, respectively.

12.7 CONCLUSION

An AC–identifier architecture is proposed to learn the approximate solution to the HJB equation for infinite-horizon optimal control of uncertain nonlinear systems. The online method is the first ever indirect adaptive control approach to continuous-time RL.

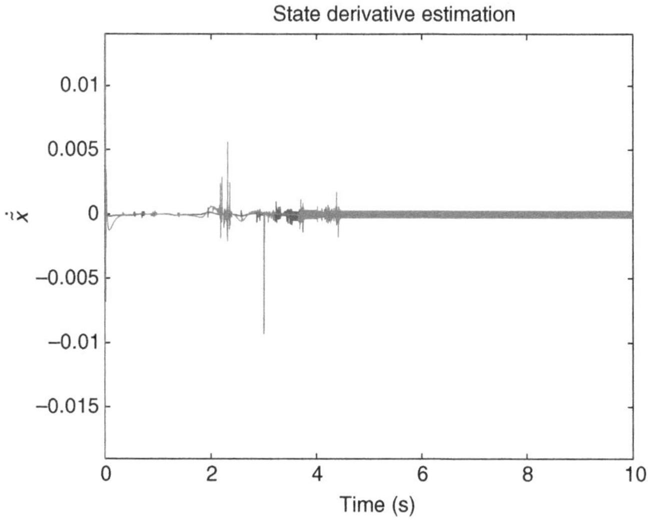

FIGURE 12.3 Error in estimating the state derivative $\dot{\tilde{x}}(t)$ by the identifier.

The learning by the actor, critic, and identifier is continuous and simultaneous, and the novel addition of the identifier to the traditional AC architecture eliminates the need to know the system drift dynamics. The actor and critic minimize the Bellman error using gradient and LS update laws, respectively, and provide online approximations to the optimal control and the optimal value function, respectively. The identifier estimates the system dynamics online and asymptotically converges to the system state and its

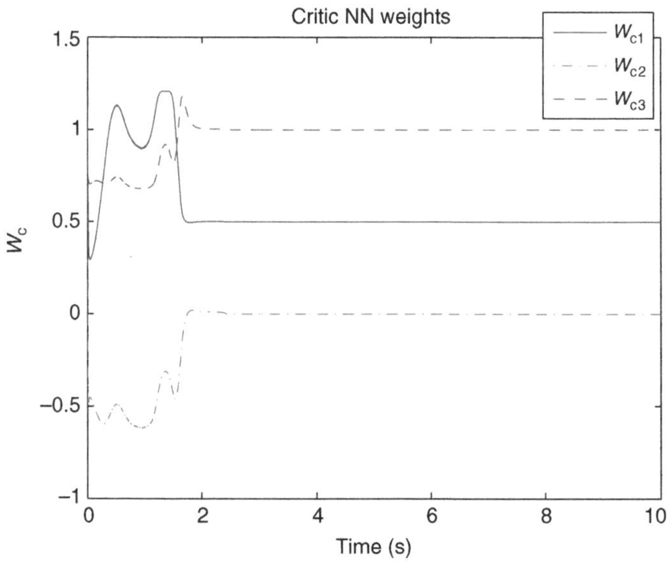

FIGURE 12.4 Convergence of critic weights $\hat{W}_c(t)$.

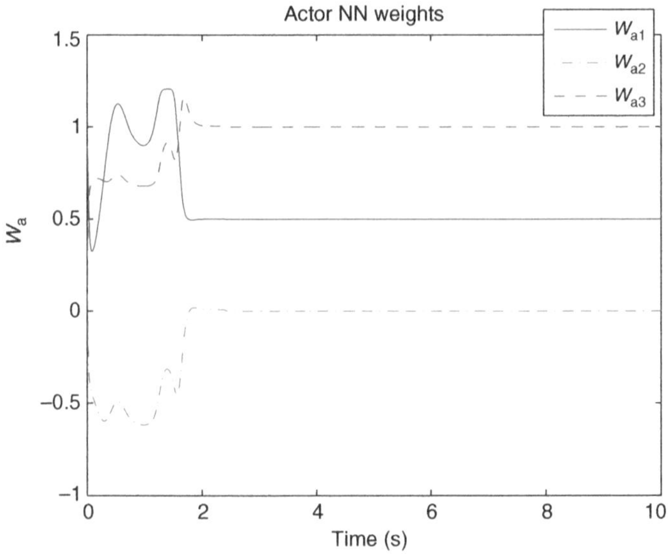

FIGURE 12.5 Convergence of actor weights $\hat{W}_a(t)$.

derivative. A PE condition is required to ensure exponential convergence to a bounded region in the neighborhood of the optimal control and UUB stability of the closed-loop system. Simulation results demonstrate the performance of the AC–identifier-based method. A limitation of the method is the requirement of the knowledge of the input gain matrix. Future efforts will investigate ways to overcome this limitation, for example, using methods similar to the model-free Q-learning methods [15, 21, 52].

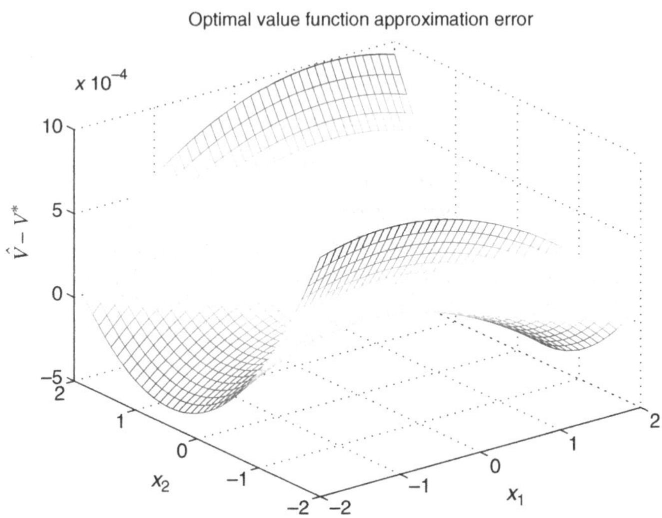

FIGURE 12.6 Error in approximating the optimal value function by the critic at steady state.

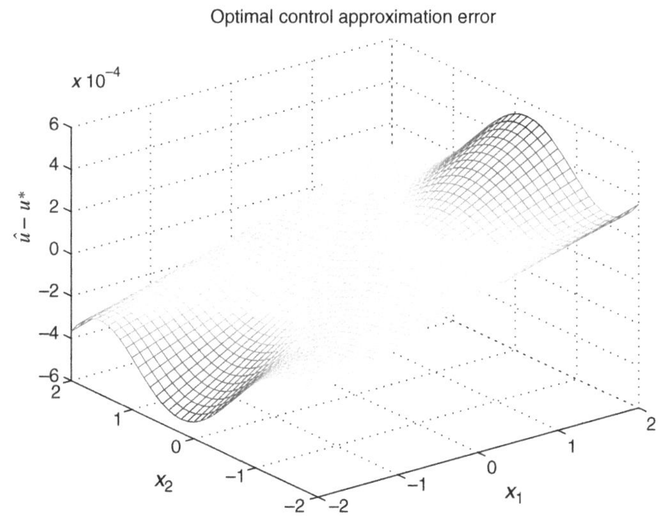

FIGURE 12.7 Error in approximating the optimal control by the actor at steady state.

REFERENCES

1. R.S. Sutton and A.G. Barto. *Introduction to Reinforcement Learning*. MIT Press, Cambridge, MA, USA, 1998.
2. B. Widrow, N.K. Gupta, and S. Maitra. Punish/reward: learning with a critic in adaptive threshold systems. *IEEE Transactions on Systems, Man, and Cybernetics*, 3(5):455–465, 1973.
3. P.J. Werbos. Approximate dynamic programming for real-time control and neural modeling. In D.A. White and D.A. Sofge, editors. *Handbook of Intelligent Control: Neural, Fuzzy, and Adaptive Approaches*. Van Nostrand Reinhold, New York, 1992.
4. D.V. Prokhorov and D.C. Wunsch, II. Adaptive critic designs. *IEEE Transactions Neural Networks*, 8:997–1007, 1997.
5. S. Ferrari and R.F. Stengel. An adaptive critic global controller. *American Control Conference*, Anchorage, AK, Vol. 4, 2002.
6. J.J. Murray, C.J. Cox, G.G. Lendaris, and R. Saeks. Adaptive dynamic programming. *IEEE Transactions on Systems, Man, and Cybernetics—Part C: Applications and Reviews*, 32(2):140–153, 2002.
7. X. Liu and S.N. Balakrishnan. Convergence analysis of adaptive critic based optimal control. *American Control Conference*, Chicago, IL, Vol. 3, 2000.
8. T. Dierks, B.T. Thumati, and S. Jagannathan. Optimal control of unknown affine nonlinear discrete-time systems using offline-trained neural networks with proof of convergence. *Neural Networks*, 22(5-6):851–860, 2009.
9. R. Padhi, S.N. Balakrishnan, and T. Randolph. Adaptive-critic based optimal neuro control synthesis for distributed parameter systems. *Automatica*, 37(8):1223–1234, 2001.
10. T. Hanselmann, L. Noakes, and A. Zaknich. Continuous-time adaptive critics. *IEEE Translations on Neural Networks*, 18(3):631–647, 2007.

11. D. Vrabie and F.L. Lewis. Neural network approach to continuous-time direct adaptive optimal control for partially unknown nonlinear systems. *Neural Networks*, 22(3):237–246, 2009.
12. K.G. Vamvoudakis and F.L. Lewis. Online actor-critic algorithm to solve the continuous-time infinite horizon optimal control problem. *Automatica*, 46:878–888, 2010.
13. R.S. Sutton. Learning to predict by the methods of temporal differences. *Machine Learning*, 3(1):9–44, 1988.
14. A.G. Barto, R.S. Sutton, and C.W. Anderson. Neuronlike adaptive elements that can solve difficult learning control problems. *IEEE Transactions on Systems Man. and Cybernetics*, 13(5):834–846, 1983.
15. C.J. C.H. Watkins and P. Dayan. Q-learning. *Machine Learning*, 8(3):279–292, 1992.
16. P.J. Werbos. Building and understanding adaptive systems: A statistical/numerical approach to factory automation and brain research. *IEEE Transactions on Systems, Man, and Cybernetics*, 17(1):7–20, 1987.
17. R. Bellman. *Dynamic Programming*. Dover Publications, Inc., 2003.
18. J. Si and Y.T. Wang. On-line learning control by association and reinforcement. *IEEE Transactions on Neural Networks*, 12(2):264–276, 2001.
19. J. Si, A. Barto, W. Powell, and D. Wunsch, editors. *Handbook of Learning and Approximate Dynamic Programming*. Wiley-IEEE Press, 2004.
20. T. Landelius. *Reinforcement Learning and Distributed Local Model Synthesis*. Ph.D. thesis, Linkoping University, Sweden, 1997.
21. S.J. Bradtke, B.E. Ydstie, and A.G. Barto. Adaptive linear quadratic control using policy iteration. *American Control Conference*, Baltimore, MD, pp. 3475–3479. 1994.
22. A. Al-Tamimi, F.L. Lewis, and M. Abu-Khalaf. Model-free q-learning designs for linear discrete-time zero-sum games with application to h-[infinity] control. *Automatica*, 43:473–481, 2007.
23. D.P. Bertsekas. *Dynamic Programming and Optimal Control*. Athena Scientific, 2007.
24. D.A. White and D.A. Sofge. *Handbook of Intelligent Control: Neural, Fuzzy, and Adaptive Approaches*. Van Nostrand Reinhold Company, 1992.
25. D. Kleinman. On an iterative technique for riccati equation computations. *IEEE Transactions on Automatic Control*, 13(1):114–115, 1968.
26. L.C. Baird. Advantage updating. Technical report, Wright Lab, Wright-Patterson Air Force Base, OH, 1993.
27. K. Doya. Reinforcement learning in continuous time and space. *Neural Computation*, 12(1):219–245, 2000.
28. R.W. Beard, G.N. Saridis, and J.T. Wen. Galerkin approximations of the generalized Hamilton–Jacobi–Bellman equation. *Automatica*, 33:2159–2178, 1997.
29. M. Abu-Khalaf and F.L. Lewis. Nearly optimal control laws for nonlinear systems with saturating actuators using a neural network HJB approach. *Automatica*, 41(5):779–791, 2005.
30. D. Vrabie, M. Abu-Khalaf, F.L. Lewis, and Y. Wang. Continuous-time ADP for linear systems with partially unknown dynamics. In *Proceedings of IEEE International Symposium on Approximate Dynamic Programming and Reinforcement Learning*, pp. 247–253, 2007.
31. J.J. Hopfield. Neurons with graded response have collective computational properties like those of two-state neurons. *Proceedings of the National Academy of Sciences, U.S.A.*, 81(10):3088, 1984.

32. A.S. Poznyak, E.N. Sanchez, and W. Yu. *Differential Neural Networks for Robust Nonlinear Control: Identification, State Estimation and Trajectory Tracking*. World Scientific Publishing Co Inc, 2001.
33. P.M. Patre, W. MacKunis, K. Kaiser, and W.E. Dixon. Asymptotic tracking for uncertain dynamic systems via a multilayer neural network feedforward and RISE feedback control structure. *IEEE Transactions on Automatic Control*, 53(9):2180–2185, 2008.
34. R.S. Sutton and A.G. Barto. *Reinforcement Learning: An Introduction*. MIT Press, 1998.
35. D. Kirk. *Optimal Control Theory: An Introduction*. Dover Publications, 2004.
36. K. Hornik, M. Stinchcombe, and H. White. Multilayer feedforward networks are universal approximators. *Neural Networks*, 2:359–366, 1985.
37. G. Cybenko. Approximation by superpositions of a sigmoidal function. *Mathematics Control Signals Systems*, 2:303–314, 1989.
38. F.L. Lewis, R. Selmic, and J. Campos. *Neuro-Fuzzy Control of Industrial Systems with Actuator Nonlinearities*. Society for Industrial and Applied Mathematics, Philadelphia, PA, USA, 2002.
39. S. Sastry and M. Bodson. *Adaptive Control: Stability, Convergence, and Robustness*. Prentice-Hall, Upper Saddle River, NJ, 1989.
40. W.E. Dixon, A. Behal, D.M. Dawson, and S. Nagarkatti. *Nonlinear Control of Engineering Systems: A Lyapunov-Based Approach*. Birkhäuser Boston, 2003.
41. M. Krstic, P.V. Kokotovic, and I. Kanellakopoulos. *Nonlinear and Adaptive Control Design*. John Wiley & Sons, 1995.
42. B. Xian, D.M. Dawson, M.S. de Queiroz, and J. Chen. A continuous asymptotic tracking control strategy for uncertain nonlinear systems. *IEEE Transactions on Automatic Control*, 49:1206–1211, 2004.
43. A.F. Fillipov. *Differential Equations with Discontinuous Righthand Sides*. Kluwer Academic Publishers, 1988, pp. 48–122.
44. A. Filippov. Differential equations with discontinuous right-hand side. *American Mathematical Society Translations*, 42(2):199–231, 1964.
45. G.V. Smirnov. *Introduction to the Theory of Differential Inclusions*. American Mathematical Society, 2002.
46. F.H. Clarke. *Optimization and Nonsmooth Analysis*. SIAM, 1990.
47. D. Shevitz and B. Paden. Lyapunov stability theory of nonsmooth systems. *IEEE Transactions on Automatic Control*, 39(9):1910–1914, 1994.
48. B. Paden and S. Sastry. A calculus for computing Filippov's differential inclusion with application to the variable structure control of robot manipulators. *IEEE Transactions on Circuits Systems*, 34(1):73–82, 1987.
49. H.K. Khalil. *Nonlinear Systems*, 3rd edition, Prentice Hall, 2002.
50. D.P. Bertsekas and J.N. Tsitsiklis. *Neuro-Dynamic Programming*. Athena Scientific, 1996.
51. L. Busoniu, R. Babuska, B. De Schutter, and D. Ernst. *Reinforcement Learning and Dynamic Programming Using Function Approximators*. CRC Press, 2010.
52. P. Mehta and S. Meyn. Q-learning and Pontryagin's minimum principle. *IEEE Conference on Decision and Control*, Shanghai, P.R. China, pp. 3598–3605, 2009.

CHAPTER 13

Robust Adaptive Dynamic Programming

YU JIANG and ZHONG-PING JIANG

Department of Electrical and Computer Engineering, Polytechnic Institute of New York University, Brooklyn, NY, USA

ABSTRACT

In this chapter, we propose a framework of robust adaptive dynamic programming (for short, robust-ADP), which is aimed at computing globally asymptotically stabilizing control laws with robustness to dynamic uncertainties, via off-line/on-line learning. It is shown that robust optimal control problems can be solved for higher-dimensional, partially linear composite systems by integration of ADP and modern nonlinear control design tools such as backstepping and ISS small-gain methods. Finally, the robust-ADP framework is applied to the load-frequency control for a power system and the controller design for a machine tool power drive system.

13.1 INTRODUCTION

Approximate/adaptive dynamic programming (ADP) together with reinforcement learning [1] is a methodology inspired by the learning behavior from biological systems, and it has become an effective computational approach for solving optimal control problems in recent years. The concept of ADP was pioneered by Werbos in Refs. [2–5]. Based on the specific ADP schemes: heuristic dynamic programming (HDP) [5] and action-dependent heuristic dynamic programming (ADHDP) [3] (or Q-learning [6]), various on-line ADP algorithms emerged. These algorithms have been studied both in theory and applications (see, for example, [7–13]).

Reinforcement Learning and Approximate Dynamic Programming for Feedback Control, First Edition. Edited by Frank L. Lewis and Derong Liu.

In the past literature, optimality is the main concern in developing ADP-based control algorithms and a common assumption is that the order of the system model and the state measurements are fully known. However, it is also important to notice that biological systems learn to achieve enhanced robustness (or greater chance of survival) through interacting with the unknown environment. Furthermore, due to the complexity of the real-world environment, biological systems may only be able to make decisions based on *partial-state* information. To capture and model these two features from biological learning, we propose a new concept of robust adaptive dynamic programming (robust-ADP), a natural extension of ADP to uncertain dynamic systems. It is worth noting that we focus on the presence of dynamic uncertainties with incomplete full-state information and unknown system order.

The first distinctive feature in robust-ADP is that the controller design issue is developed in the context of robust control with disturbance attenuation. Specifically, by means of the popular backstepping approach [14], we show that a robust controller, or adaptive critic, can be synthesized to yield an arbitrarily small gain with respect to the disturbance input. In addition, by studying the relationship between optimality and robustness, it is shown that in the absence of disturbance input, the robust controller also preserves optimality with respect to some iteratively constructed cost function. Then, we develop an off-line robust-ADP algorithm that computes a robust control law using only partial system knowledge. It should be mentioned that in Ref. [7] the theory of zero-sum games was employed in ADP design, but the gain from the disturbance input to the output cannot be made arbitrarily small, because the game algebraic Riccati equation (ARE) may not have solutions for arbitrarily small input–output gains.

Our study on the effects of dynamic uncertainties, or unmodeled dynamics, is motivated by engineering applications in situations where the exact mathematical model of a physical system is not easy to be obtained. In our robust-ADP framework, we decompose the uncertain environment into two parts: the *system model* (ideal environment) with known system order and fully accessible state, and the *dynamic uncertainties*, with unknown system order and dynamics, interacting with the ideal environment. To characterize dynamic uncertainties, we adopt the notion of Sontag's input-to-state stability (ISS) [15, 16], which has been proved to be an efficient tool for nonlinear system analysis and synthesis. Then, we develop an on-line strategy that can be used to compute suboptimal, globally asymptotically stabilizing control policies for the overall system in finite steps. We achieve the robust stability and suboptimality properties for the overall system, by means of Lyapunov and small-gain techniques [17, 18].

In the rest of this chapter, we investigate the relationship between optimality and robustness for a general class of partially linear, uncertain composite systems with a lower-triangular structure [19]. Then, we develop robust-ADP algorithms for both the L_2-gain disturbance attenuation robust control and the partial-state feedback control designs. Finally, we apply the proposed robust-ADP design methodology to two practical engineering problems: power system load-frequency control (LFC) and a machine tool power drive system. Concluding remarks are contained in the last section.

13.2 OPTIMALITY VERSUS ROBUSTNESS

In this section, we study the relationship between optimality and robustness. We will show that robust optimal controllers can be obtained from solving an optimal control problem with iteratively defined weighting matrices. First, we begin with linear systems with matched (nonlinear) disturbance input. Then, the result is extended to a general class of linear uncertain systems in the lower triangular form, where the matching condition is no longer assumed.

13.2.1 Systems with Matched Disturbance Input

Consider the system

$$\dot{x} = Ax + B(u + W\Delta), \tag{13.1}$$

$$y = Cx, \tag{13.2}$$

where $x \in \mathbb{R}^n$ is the system state, $u \in \mathbb{R}^m$ the control input, $\Delta \in \mathbb{R}^p$ the disturbance input; $A \in \mathbb{R}^{n\times n}$, $B \in \mathbb{R}^{n\times m}$, $C \in \mathbb{R}^{q\times n}$, $W \in \mathbb{R}^{m\times p}$ are constant matrices, with (A, B) controllable and (A, C) observable. In the remainder of this chapter, we assume that A is an unknown matrix.

Since the disturbance input is in the range of B, it is said to satisfy the *matching condition*. In such a case, the relationship between optimality and robustness for a given controller can be stated in Lemma 13.1.

Lemma 13.1 *Consider system (13.1)–(13.2). Assume the performance index*

$$J_0 = \int_0^{\infty} \left(x^{\mathrm{T}} Q x + u^{\mathrm{T}} R u\right) \mathrm{d}\tau, \tag{13.3}$$

is minimized for system (13.1) under the control law $u = -Kx$ with $\Delta = 0$, where $Q = Q^{\mathrm{T}} \geq \gamma^{-1} CC^{\mathrm{T}} + \epsilon_0 I_n$, $R = R^{\mathrm{T}} \geq \left(WW^{\mathrm{T}} + \epsilon_0 I_m\right)^{-1}$, $\gamma > 0$, $\epsilon_0 > 0$. Then, there exists a positive definite and radially unbounded function $V(x)$, such that

$$\dot{V} \leq -\gamma^{-1}\|y\|^2 - \epsilon_0\|x\|^2 + \|\Delta\|^2. \tag{13.4}$$

Proof: According to linear optimal control theory [20], we have

$$u = -Kx = -R^{-1}B^{\mathrm{T}}Px, \tag{13.5}$$

where $P = P^{\mathrm{T}} > 0$ is the solution of the ARE

$$PA + A^{\mathrm{T}}P + Q - PBR^{-1}B^{\mathrm{T}}P = 0. \tag{13.6}$$

Define $V = x^{\mathrm{T}} P x$. Then, along the solutions of the closed-loop system, we have

$$\begin{aligned}
\dot{V} &= x^{\mathrm{T}} P(A - BK)x + x^{\mathrm{T}}(A - BK)^{\mathrm{T}} Px + 2x^{\mathrm{T}} PBW\Delta \\
&\leq -x^{\mathrm{T}} Qx - x^{\mathrm{T}} PBWW^{\mathrm{T}} B^{\mathrm{T}} Px + 2x^{\mathrm{T}} PBW\Delta - \|\Delta\|^2 + \|\Delta\|^2 \\
&\leq -x^{\mathrm{T}} Qx - \|W^{\mathrm{T}} B^{\mathrm{T}} Px - \Delta\|^2 + \|\Delta\|^2 \\
&\leq -\gamma^{-1}\|y\|^2 - \epsilon_0\|x\|^2 + \|\Delta\|^2.
\end{aligned}$$

The proof is complete. ■

13.2.2 Adding One Integrator

Lemma 13.1 in the earlier section can be extended to systems when the disturbance input is not in the range of the control input, that is, when the matching condition does not hold. We will demonstrate this extension by means of the technique of adding an integrator, or the backstepping approach; see Ref. [14] and references therein.

To this end, let us consider the following cascade system:

$$\dot{x} = Ax + b(z_1 + w_0^{\mathrm{T}}\Delta), \tag{13.7}$$

$$\dot{z}_1 = u + w_1^{\mathrm{T}}\Delta, \tag{13.8}$$

$$y = Cx, \tag{13.9}$$

where $X_1 = \begin{bmatrix} x^{\mathrm{T}} & z_1 \end{bmatrix}^{\mathrm{T}} \in \mathbb{R}^{n+1}$ is the system state, $u \in \mathbb{R}$ the control input, $\Delta \in \mathbb{R}^p$ the disturbance. $A \in \mathbb{R}^{n\times n}$, $b \in \mathbb{R}^n$, and $C \in \mathbb{R}^{q\times n}$ are constant matrices with (A, b) controllable and (A, C) observable, and $w_0, w_1 \in \mathbb{R}^p$ are constant vectors.

Further, we use $P = P^{\mathrm{T}} > 0$ to denote the solution of the ARE

$$A^{\mathrm{T}} P + PA + Q - r_0^{-1} Pbb^{\mathrm{T}} P = 0, \tag{13.10}$$

where $Q = Q^{\mathrm{T}} \geq \gamma^{-1} C^{\mathrm{T}} C + \epsilon_0 I_n$, $r_0 = \left(\|w_0\|^2 + \epsilon_0\right)^{-1}$, $\gamma > 0$, and $\epsilon_0 > 0$.

In addition, we define

$$K = r_0^{-1} b^{\mathrm{T}} P. \tag{13.11}$$

Lemma 13.2 *Suppose a linear feedback control law of the form $u = -\bar{K}_1 X_1$ minimizes the following performance index for (13.7) and (13.8) with $\Delta \equiv 0$*

$$J_1 = \int_0^{\infty} \left[X_1^{\mathrm{T}} Q_1 X_1 + a_1^{-1}(u + s_1 X_1)^2\right] \mathrm{d}\tau, \tag{13.12}$$

where

$$X_1 = \begin{bmatrix} x^{\mathrm{T}} & z_1 \end{bmatrix}^{\mathrm{T}}, \; a_1 = \tfrac{1}{2}\left(\epsilon_1 + \|w_1^{\mathrm{T}} + Kbw_0^{\mathrm{T}}\|^2\right), \; \epsilon_1 > 0, \; T_1 = \begin{bmatrix} I_n & 0 \\ Kx & 1 \end{bmatrix},$$

$$Q_1 = T_1^{\mathrm{T}} \begin{bmatrix} Q + r_0 K^{\mathrm{T}} K & 0 \\ 0 & a_1 \end{bmatrix} T_1, \; A_0 = \begin{bmatrix} A & b \end{bmatrix}, \quad s_1 = KA_0 + \begin{bmatrix} b^{\mathrm{T}} P & 0 \end{bmatrix}.$$

Then, there exists a positive definite and radially unbounded function $V_1(X_1)$, such that under the same control law, the following holds

$$\dot{V}_1 \leq -\gamma^{-1}\|y\|^2 - \epsilon_0\|x\|^2 - \epsilon_1\|z_1 + Kx\|^2 + 2\|\Delta\|^2. \tag{13.13}$$

Proof: First, we introduce the following coordinate and input transformations

$$x_0 = x, \;\; x_1 = z_1 + Kx, \;\; v_1 = u + KA_0X_1 + b^{\mathrm{T}}Px.$$

In the new coordinates, we have

$$\dot{x}_0 = (A - bK)x_0 + bx_1 + bw_0^{\mathrm{T}}\Delta, \tag{13.14}$$

$$\dot{x}_1 = v_1 - b^{\mathrm{T}}Px_0 + \bar{w}_1^{\mathrm{T}}\Delta, \tag{13.15}$$

where $\bar{w}_1 = w_1 + w_0 b^{\mathrm{T}} K^{\mathrm{T}}$.

Notice that in the new coordinates the performance index becomes

$$J_1 = \int_0^{\infty} \left(\begin{bmatrix} x_0^{\mathrm{T}} & x_1 \end{bmatrix} \begin{bmatrix} Q + r_0 K^{\mathrm{T}} K & 0 \\ 0 & a_1 \end{bmatrix} \begin{bmatrix} x_0 \\ x_1 \end{bmatrix} + \frac{v_1^2}{a_1} \right) \mathrm{d}\tau. \tag{13.16}$$

By linear optimal control theory [20], the optimal controller is related to the ARE as follows:

$$\begin{aligned} 0 = P_1 &\begin{bmatrix} A - bK & b \\ -b^{\mathrm{T}}P & 0 \end{bmatrix} + \begin{bmatrix} A^{\mathrm{T}} - K^{\mathrm{T}}b^{\mathrm{T}} & -Pb \\ b^{\mathrm{T}} & 0 \end{bmatrix} P_1 \\ &+ \begin{bmatrix} Q + r_0K^{\mathrm{T}}K & 0 \\ 0 & a_1 \end{bmatrix} - a_1 P_1 \begin{bmatrix} 0 \\ 1 \end{bmatrix} \begin{bmatrix} 0 & 1 \end{bmatrix} P_1. \end{aligned} \tag{13.17}$$

It can be solved that

$$P_1 = \begin{bmatrix} P & 0 \\ 0 & 1 \end{bmatrix}, \quad v_1 = -a_1 \begin{bmatrix} 0 & 1 \end{bmatrix} P_1 \begin{bmatrix} x \\ x_1 \end{bmatrix} = -a_1 x_1.$$

To verify (13.13), define $V_1 = x^T P x + x_1^2$. Differentiating V_1 with respect to time gives:

$$\begin{aligned}\dot{V}_1 &\le -\gamma^{-1}\|y\|^2 - \epsilon_0\|x_0\|^2 + \|\Delta\|^2 + 2x_0^T Pbx_1 + 2x_1(u + \bar{w}_1^T\Delta + KA_0X_1) \\ &= -\gamma^{-1}\|y\|^2 - \sum_{i=0}^{1}\epsilon_i\|x_i\|^2 - \|\Delta\|^2 + 2x_1\bar{w}_1^T\Delta - \|\bar{w}_1\|^2 x_1^2 + 2\|\Delta\|^2 \\ &\le -\gamma^{-1}\|y\|^2 - \sum_{i=0}^{1}\epsilon_i\|x_i\|^2 + 2\|\Delta\|^2.\end{aligned}$$

The proof is complete. ■

Remark 13.1 *Notice that the cost function we introduced here is different from the ones used in game theory [7, 21], where the policy iterations are developed based on the game algebraic Riccati equation (GARE). The existence of a solution of the GARE cannot be guaranteed when the input–output gain is arbitrarily small. Therefore, a significant advantage of our method versus the game-theoretic approach of [7, 21] is that we are able to render the gain γ arbitrarily small.*

13.2.3 Systems in Lower-Triangular Form

Inspired by the adding one integrator technique, we are able to study a general class of linear uncertain systems in the lower-triangular form as follows:

$$\begin{aligned}\dot{x} &= Ax + b(z_1 + w_0^T\Delta), \\ \dot{z}_1 &= z_2 + w_1^T\Delta, \\ &\vdots \\ \dot{z}_m &= u + w_m^T\Delta, \\ y &= Cx,\end{aligned} \tag{13.18}$$

where $x \in \mathbb{R}^n$, $z_i \in \mathbb{R}$ with $i = 1, 2, \ldots, m$, are the state variables, and $w_i \in \mathbb{R}^p$ are known constant vectors with $i = 0, 1, \ldots, m$. $u \in \mathbb{R}$ is the control input, and $\Delta \in \mathbb{R}^p$ is the disturbance input. $(A, b) \in \mathbb{R}^{n\times n} \times \mathbb{R}^n$ is a pair of controllable matrices, and $(A, C) \in \mathbb{R}^{n\times n} \times \mathbb{R}^{q\times n}$ is a pair of observable matrices.

We introduce the following notations

$$A_i = \begin{bmatrix} A & b & 0 \\ 0 & 0 & I_i \end{bmatrix} \in \mathbb{R}^{(n+i)\times(n+1+i)}, \quad B_i = \begin{bmatrix} bw_0^T \\ w_1^T \\ w_2^T \\ \vdots \\ w_i^T \end{bmatrix} \in \mathbb{R}^{(n+i)\times p}, \quad i = 0, 1, \ldots, m,$$

$$X_i = \begin{bmatrix} x^{\mathrm{T}} & z_1 & z_2 & \cdots & z_i \end{bmatrix}^{\mathrm{T}}, \quad i = 1, 2, \ldots, m,$$

$$\bar{K}_i = \begin{cases} K, & i = 0, \\ \bar{K}_0 A_0 + a_1 \begin{bmatrix} \bar{K}_0 & 1 \end{bmatrix} + \begin{bmatrix} b^{\mathrm{T}} P & 0 \end{bmatrix}, & i = 1, \\ \bar{K}_{i-1} A_{i-1} + a_i \begin{bmatrix} \bar{K}_{i-1} & 1 \end{bmatrix} + \begin{bmatrix} \bar{K}_{i-2} & 1 & 0 \end{bmatrix}, & i = 2, 3, \ldots, m, \end{cases}$$

$$a_i = \begin{cases} r_0{}^{-1}, & i = 0, \\ \frac{1}{2}(\|w_i + B_{i-1}^{\mathrm{T}} \bar{K}_{i-1}^{\mathrm{T}}\|^2 + \epsilon_i), & i = 1, \ldots, m, \end{cases}$$

$$s_i = \begin{cases} 0, & i = 0, \\ \bar{K}_0 A_0 + [b^{\mathrm{T}} P \quad 0], & i = 1, \\ \bar{K}_{i-1} A_{i-1} + \begin{bmatrix} \bar{K}_{i-2} & 1 & 0 \end{bmatrix}, & i = 2, 3, \ldots, m, \end{cases}$$

$$Q_i = \begin{cases} Q, & i = 0, \\ T_i^{\mathrm{T}} \begin{bmatrix} Q + r_0 \bar{K}_0^{\mathrm{T}} \bar{K}_0 & 0 & 0 & \cdots & 0 \\ 0 & 2a_1 & 0 & \cdots & 0 \\ 0 & 0 & \ddots & \ddots & \vdots \\ \vdots & \vdots & \ddots & 2a_{i-1} & 0 \\ 0 & 0 & \cdots & 0 & a_i \end{bmatrix} T_i, & i = 1, 2, \ldots, m, \end{cases}$$

$$T_i = \begin{cases} I_n, & i = 0, \\ \begin{bmatrix} T_{i-1} & 0 \\ \bar{K}_{i-1} & 1 \end{bmatrix}, & i = 1, 2, \ldots, m, \end{cases}$$

where $\epsilon_i > 0$ is any small constant for all $i = 0, 1, \ldots, m$, and $P = P^{\mathrm{T}} > 0$ the solution of the ARE

$$PA + A^{\mathrm{T}} P + Q - r_0 K^{\mathrm{T}} K = 0, \tag{13.19}$$

with $Q = Q^{\mathrm{T}} \geq \gamma^{-1} C^{\mathrm{T}} C + \epsilon_0 I_n$, and $K = r_0^{-1} b^{\mathrm{T}} P$.

Theorem 13.1 *The control law $u = -\bar{K}_m X_m$ minimizes the following performance index for (13.18) with $\Delta \equiv 0$*

$$J_m = \int_0^{\infty} \left[X_m^{\mathrm{T}} Q_m X_m + a_m^{-1} (u + s_m X_m)^2 \right] \mathrm{d}\tau. \tag{13.20}$$

Also, there exists a positive definite and radially unbounded function $V_m(X_m)$, such that under the control law u, the following holds

$$\dot{V}_m \leq -\gamma^{-1} \|y\|^2 - \sum_{i=0}^{m} \epsilon_i \|x_i\|^2 + (m+1) \|\Delta\|^2, \tag{13.21}$$

where $x_0 = x$, and $x_i = z_i + \bar{K}_{i-1} X_{i-1}$, for $i = 1, 2, \ldots, m$.

Remark 13.2 *Theorem 13.1 can be proved by setting $V_m = x^{\mathrm{T}} P x + \sum_{i=1}^{m} x_i^2$ and iteratively following the adding-one-integrator technique as in the proof of Lemma 13.2.*

Remark 13.3 *Since we assume that A is not known, (13.19) cannot be directly used to compute the robust controller. In the next two sections, we will overcome this obstacle by introducing numerical algorithms in the spirit of ADP.*

Remark 13.4 *Notice that (13.21) implies*

$$\sqrt{\int_0^t \|y\|^2 \mathrm{d}t} \leq \sqrt{\gamma(m+1) \int_0^t \|\Delta\|^2 \mathrm{d}t} + \sqrt{V_m(0)}, \quad \forall\, t \geq 0. \tag{13.22}$$

Hence, the closed-loop system comprised of (13.18) and the control law $u = -\bar{K}_m X_m$ is finite-gain L_2 stable [22], with respect to Δ as the disturbance input and y as the output, and its L_2 gain is less than or equal to $\sqrt{\gamma(m+1)}$. Clearly, the L_2-gain can be made arbitrarily small because $\gamma > 0$ is arbitrary.

13.3 ROBUST-ADP DESIGN FOR DISTURBANCE ATTENUATION

In this section, we compute a robust optimal controller using the robust-ADP scheme. The proposed strategy consists of two subroutines: *horizontal learning* and *vertical learning*. In horizontal learning, we compute P that satisfies the ARE (13.19). In vertical learning, we compute $\bar{K}_i$, $i = 1, 2, \ldots, m$ iteratively. In both learning schemes, we will show that the knowledge of A is not required.

13.3.1 Horizontal Learning

In this phase, we focus on the computation of P without knowing A. First, recall the well-known Kleinman algorithm [23] that numerically solves AREs.

Lemma 13.3 *[23] Let $K_0 \in \mathbb{R}^{1\times n}$ be any constant matrix such that $A - bK_0$ is stable, and let $\{P_k\}$ be a sequence of real symmetric matrices, such that each P_k is the positive definite solution of the Lyapunov equation*

$$P_k(A - bK_k) + (A - bK_k)^{\mathrm{T}} P_k + Q + r_0 K_k^{\mathrm{T}} K_k = 0 \qquad (13.23)$$

where

$$K_{k+1} = r_0^{-1} b^{\mathrm{T}} P_k, \quad k = 0, 1, \ldots. \qquad (13.24)$$

Then, the following properties hold:

1. *$A - bK_k$ is Hurwitz,*
2. *$P \leq P_{k+1} \leq P_k$,*
3. *$\lim_{k\to\infty} K_k = K$, $\lim_{k\to\infty} P_k = P$.*

The Kleinman algorithm cannot be directly applied because the unknown A still appears in (13.23). To overcome this obstacle, we rewrite the x-subsystem in (13.18) with $\Delta = 0$ as

$$\dot{x} = Ax - bK_k x + b(z_1 + K_k x). \qquad (13.25)$$

Now, along the trajectories of (13.25), it follows that

$$\begin{aligned} & x(t)^{\mathrm{T}} P_k x(t) - x(t+\delta t)^{\mathrm{T}} P_k x(t+\delta t) \\ &= \int_t^{t+\delta t} x^{\mathrm{T}}(Q + r_0 K_k^{\mathrm{T}} K_k) x \mathrm{d}\tau - 2\int_t^{t+\delta t} x^{\mathrm{T}} P_k b(z_1 + K_k x)\mathrm{d}\tau. \qquad (13.26) \end{aligned}$$

Using Kronecker product representation [24], we have

$$x^{\mathrm{T}} P_k x = \left(x^{\mathrm{T}} \otimes x^{\mathrm{T}}\right) \mathrm{vec}(P_k),$$

$$\int_t^{t+\delta t} x^{\mathrm{T}}(Q + r_0 K_k^{\mathrm{T}} K_k) x \mathrm{d}\tau = \int_t^{t+\delta t} \left(x^{\mathrm{T}} \otimes x^{\mathrm{T}}\right) \mathrm{d}\tau \ \mathrm{vec}\left(Q + r_0 K_k^{\mathrm{T}} K_k\right),$$

$$\begin{aligned} \int_t^{t+\delta t} x^{\mathrm{T}} P_k b(z_1 + K_k x)\mathrm{d}\tau &= \int_t^{t+\delta t} (x^{\mathrm{T}} \otimes x^{\mathrm{T}})(K_k^{\mathrm{T}} b^{\mathrm{T}} \otimes I_n)\mathrm{d}\tau \ \mathrm{vec}(P_k) \\ &+ \int_t^{t+\delta t} z_1 x^{\mathrm{T}} (b^{\mathrm{T}} \otimes I_n)\mathrm{d}\tau \ \mathrm{vec}(P_k). \end{aligned}$$

Further, define

$$\delta_{xx} = \begin{bmatrix} x^{\mathrm{T}}(t_1) \otimes x^{\mathrm{T}}(t_1) - x^{\mathrm{T}}(t_1 + \delta t) \otimes x^{\mathrm{T}}(t_1 + \delta t) \\ x^{\mathrm{T}}(t_2) \otimes x^{\mathrm{T}}(t_2) - x^{\mathrm{T}}(t_2 + \delta t) \otimes x^{\mathrm{T}}(t_2 + \delta t) \\ \vdots \\ x^{\mathrm{T}}(t_l) \otimes x^{\mathrm{T}}(t_l) - x^{\mathrm{T}}(t_l + \delta t) \otimes x^{\mathrm{T}}(t_l + \delta t) \end{bmatrix},$$

$$I_{xx} = \begin{bmatrix} \int_{t_1}^{t_1+\delta t} \left(x^{\mathrm{T}} \otimes x^{\mathrm{T}}\right) \mathrm{d}\tau \\ \int_{t_2}^{t_2+\delta t} \left(x^{\mathrm{T}} \otimes x^{\mathrm{T}}\right) \mathrm{d}\tau \\ \vdots \\ \int_{t_l}^{t_l+\delta t} \left(x^{\mathrm{T}} \otimes x^{\mathrm{T}}\right) \mathrm{d}\tau \end{bmatrix}, \text{ and } I_{zx} = \begin{bmatrix} \int_{t_1}^{t_1+\delta t} z_1 x^{\mathrm{T}} \mathrm{d}\tau \\ \int_{t_2}^{t_2+\delta t} z_1 x^{\mathrm{T}} \mathrm{d}\tau \\ \vdots \\ \int_{t_l}^{t_l+\delta t} z_1 x^{\mathrm{T}} \mathrm{d}\tau \end{bmatrix},$$

where $[t_j, t_j + \delta t]$, $j = 1, 2, \ldots, l$ are some disjoint intervals.

Then, (13.26) implies the following matrix form of linear equations

$$\Theta_k \mathrm{vec}(P_k) = \Xi_k, \tag{13.27}$$

where

$$\Theta_k = \delta_{xx} + 2I_{zx}(b^{\mathrm{T}} \otimes I_n) + 2I_{xx}\left(K_k^{\mathrm{T}} b^{\mathrm{T}} \otimes I_n\right), \quad \Xi_k = I_{xx}\,\mathrm{vec}(Q + r_0 K_k^{\mathrm{T}} K_k).$$

It is clear to see if (13.27) has a unique solution, we are able to replace (13.23) with (13.27). In this way, the knowledge of A is no longer needed. Therefore, we introduce the following assumption:

Assumption 13.1 *Θ_k has full column rank, for all $k \in \mathbb{Z}_+$.*

Hence, starting from K_0, the optimal matrices P and K as shown in Lemma 13.3 can be obtained by iteratively using (13.24) and (13.27).

13.3.2 Vertical Learning

The horizontal learning computes the matrices P and K in (13.19). Now, to obtain the robust controller $u = -\bar{K}_m X_m$ as in Theorem 13.1, we need to iteratively solve $\bar{K}_i$ with $i = 1, 2, \ldots, m$. However, since the unknown matrix A is involved in each iteration step, (13.19) cannot be directly used. To overcome this obstacle, we develop the vertical learning scheme.

Define $\bar{K}_i = \begin{bmatrix} K_i^{(1)} & K_i^{(2)} \end{bmatrix}$, where $K_i^{(1)} \in \mathbb{R}^{1\times n}$, $K_i^{(2)} \in \mathbb{R}^{1\times i}$ with $i = 0, 1, \ldots, m$.

Then,

$$\bar{K}_i A_i = \begin{bmatrix} K_i^{(1)} A & K_i^{(1)} b & K_i^{(2)} \end{bmatrix}. \tag{13.28}$$

Hence, $L_i = K_i^{(1)} A$ with $i = 0, 1, \ldots, m$ is the unknown part that becomes an obstacle of directly using (13.19).

Inspired by the methodology of ADP, we compute this unknown vector via state information. Notice that

$$K_i^{(1)} [x(t+\delta t) - x(t)] = L_i \int_t^{t+\delta t} x \mathrm{d}\tau + K_i^{(1)} b \int_t^{t+\delta t} z_1 \mathrm{d}\tau. \tag{13.29}$$

Therefore, (13.29) implies the following matrix equation

$$L_i I_x^{\mathrm{T}} = K_i^{(1)} \left(\delta_x^{\mathrm{T}} - b I_z^{\mathrm{T}}\right), \tag{13.30}$$

where

$$I_x = \begin{bmatrix} \int_{t_1}^{t_1+\delta t} x^{\mathrm{T}} \mathrm{d}\tau \\ \int_{t_2}^{t_2+\delta t} x^{\mathrm{T}} \mathrm{d}\tau \\ \vdots \\ \int_{t_l}^{t_l+\delta t} x^{\mathrm{T}} \mathrm{d}\tau \end{bmatrix}, \quad I_z = \begin{bmatrix} \int_{t_1}^{t_1+\delta t} z_1 \mathrm{d}\tau \\ \int_{t_2}^{t_2+\delta t} z_1 \mathrm{d}\tau \\ \vdots \\ \int_{t_l}^{t_l+\delta t} z_1 \mathrm{d}\tau \end{bmatrix}, \text{ and } \delta_x = \begin{bmatrix} x^{\mathrm{T}}(t_1+\delta t) - x^{\mathrm{T}}(t_1) \\ x^{\mathrm{T}}(t_2+\delta t) - x^{\mathrm{T}}(t_2) \\ \vdots \\ x^{\mathrm{T}}(t_l+\delta t) - x^{\mathrm{T}}(t_l) \end{bmatrix}.$$

Similarly, to guarantee the uniqueness of L_i, we introduce another assumption

Assumption 13.2 *I_x has full column rank.*

Applying both horizontal and vertical learning schemes, the robust control law can be formulated without using the knowledge of the A.

13.3.3 Robust-ADP Algorithm for Disturbance Attenuation

We compute robust controllers for the system (13.18) by using robust-ADP based learning strategy. This is an off-line design scheme, so we first assume $\Delta \equiv 0$ when applying our algorithm. Then, we input exploration noise $u = e(t)$ to the system, and record corresponding variable information in I_x, I_z, I_{xx}, I_{zx}, δ_x, and δ_{xx}. The exploration noise can be random noise as used in Refs. [7, 8], or the sum of sinusoidal functions with different frequencies as in Refs. [9, 25]. The length l of these matrices can be set sufficiently large, so that Assumptions 13.1 and 13.2 can be satisfied.

The robust-ADP algorithm is generalized in the following pseudo-code:

Algorithm 13.1: Robust-ADP algorithm for disturbance attenuation

---

Input: I_x, I_z, I_{xx}, I_{zx}, δ_x, and δ_{xx},
Input: K_0, such that $A - bK_0$ is stable.

$k \leftarrow 0$
$P_{-1} \leftarrow -I_n$
while $\|P_k - P_{k-1}\| > \epsilon$ **do**
    $\Theta_k \leftarrow \delta_{xx} + 2I_{zx}\left(b^{\mathrm{T}} \otimes I_n\right) + 2I_{xx}\left(K_k^{\mathrm{T}} b^{\mathrm{T}} \otimes I_n\right)$
    $\Xi_k \leftarrow I_{xx}\mathrm{vec}\left(Q + r_0 K_k^{\mathrm{T}} K_k\right)$
    $\mathrm{vec}(P_k) \leftarrow (\Theta_k^{\mathrm{T}}\Theta_k)^{-1}\Theta_k^{\mathrm{T}}\Xi_k$
    $K_{k+1} \leftarrow r_0^{-1} b^{\mathrm{T}} P_k$
    $k \leftarrow k + 1$
end while
$\bar{K}_0 \leftarrow K_k$
for $i = 0$ to $i = m - 1$ **do**
    $L_i \leftarrow K_i^{(1)}(\delta x - bI_z) I_x^{\mathrm{T}} \left(I_x I_x^{\mathrm{T}}\right)^{-1}$
    $a_i \leftarrow \frac{1}{2}(\|w_i^{\mathrm{T}} + \bar{K}_{i-1}B_{i-1}\|^2 + \epsilon_i)$
    if $i = 0$ **then**
        $\bar{K}_{i+1} \leftarrow \left[L_0 K_i^{(1)} b\right] + a_1\left[\bar{K}_0 \quad 1\right] + \left[b^{\mathrm{T}} P \quad 0\right]$
    else
        $\bar{K}_{i+1} \leftarrow \left[L_i K_i^{(1)} b \quad K_i^{(2)}\right] + a_{i+1}\left[\bar{K}_i \quad 1\right] + \left[\bar{K}_{i-1} \quad 1 \quad 0\right]$
    end if
end for
Output: $\bar{K}_m$

13.4 ROBUST-ADP FOR PARTIAL-STATE FEEDBACK CONTROL

A common assumption in previous ADP-based control methods is that the plant to be controlled is of known system order, and its state is either directly measurable, or reconstructible from output measurements [9, 11]. However, the system order may be unknown because of the dynamic uncertainty. Therefore, as pointed out by Werbos [26], an important question in ADP is how to proceed learning and assure convergence using only limited information and partial-state measurements of a given system, the order of which is completely unknown. To perform stability analysis of the interconnected systems, we adopt the notion of input-to-state stability (ISS) [15, 16] in the robust-ADP framework. Then, we achieve the robust stability and suboptimality properties for the overall system, by means of Lyapunov and small-gain techniques [17, 18].

To be more specific, in this section we study systems in the following form

$$\begin{aligned}
\dot{\xi} &= f(\xi, y), \\
\dot{x} &= Ax + b\left(z_1 + w_0^{\mathrm{T}} \Delta(\xi, y)\right), \quad y = Cx, \\
\dot{z}_1 &= z_2 + w_1^{\mathrm{T}} \Delta(\xi, y), \\
&\vdots \\
\dot{z}_m &= u + w_m^{\mathrm{T}} \Delta(\xi, y),
\end{aligned} \tag{13.31}$$

where $x \in \mathbb{R}^n$, z_i, $i = 1, 2, \ldots, m$ are the state variables available for feedback design, and $\xi \in \mathbb{R}^{n_\xi}$ is the unmeasurable state vector. $A \in \mathbb{R}^{n\times n}$, $b \in \mathbb{R}^n$, $C \in \mathbb{R}^{q\times n}$, and $w_i \in \mathbb{R}^p$ with $i = 0, 1, \ldots, m$ are defined in the same way as in (13.18). Again, A is assumed to be unknown. The unknown functions $f : \mathbb{R}^{n_\xi} \times \mathbb{R}^q \to \mathbb{R}^{n_\xi}$ and $\Delta : \mathbb{R}^{n_\xi} \times \mathbb{R}^q \to \mathbb{R}^p$ are locally Lipschitz satisfying $f(0, 0) = 0$ and $\Delta(0, 0) = 0$.

13.4.1 The ISS Property

Consider the following control system having $x \in \mathbb{R}^n$ the state, $u \in \mathbb{R}^m$ the input, and $y \in \mathbb{R}^q$ the output:

$$\dot{x} = g(x, u), \quad y = h(x, u), \tag{13.32}$$

where g is a locally Lipschitz function and h is a continuous function.

Definition 13.1 (Ref. [15]) *System (13.32) is said to be input-to-state stable (ISS) with gain γ if, for any measurable locally essentially bounded input u and any initial condition $x(0)$, the solution $x(t)$ exists for every $t \geq 0$ and satisfies*

$$|x(t)| \leq \beta(\|x(0)\|, t) + \gamma\left(\sup_{0\leq\tau\leq t} \|u(\tau)\|\right), \tag{13.33}$$

where β, γ are of class $\mathcal{KL}$ and of class $\mathcal{K}$, respectively [22].

The following theorem gives necessary and sufficient conditions for the ISS property [16].

Theorem 13.2 *Consider the system*

$$\dot{x} = g(x, u), \tag{13.34}$$

where $x \in \mathbb{R}^n$ is the state, $u \in \mathbb{R}^m$ the input, and $g : \mathbb{R}^n \times \mathbb{R}^m \to \mathbb{R}^n$ a locally Lipschitz function.

Then, the system (13.34) is ISS if and only if there exists a continuously differentiable function $V : \mathbb{R}^n \to \mathbb{R}_+$, such that the following hold for all $(x, u) \in \mathbb{R}^n \times \mathbb{R}^m$:

$$\alpha_1(\|x\|) \leq V(x) \leq \alpha_2(\|x\|), \tag{13.35}$$

$$\frac{\partial V}{\partial x} f(x, u) \leq -\alpha_3(\|x\|) + \alpha_4(\|u\|), \tag{13.36}$$

where α_1, α_2, α_3 are of class $\mathcal{K}_\infty$, and α_4 is of class $\mathcal{K}$.

To achieve global asymptotic stability, let us make a few assumptions on system (13.31), which are often required in the literature of nonlinear control design [14, 27, 28].

Assumption 13.3 *The ξ-subsystem is ISS with respect to y as the input.*

Assumption 13.4 *There exist a continuously differentiable, positive definite and radially unbounded function $W : \mathbb{R}^{n_\xi} \to \mathbb{R}_+$, and two constants $c_1 > 0$, $c_2 \geq 0$, such that*

$$\dot{W} = \frac{\partial W(\xi)}{\partial \xi} f(\xi, y) \leq -c_1 \|\Delta(\xi, y)\|^2 + c_2\|y\|^2 \tag{13.37}$$

for all $\xi \in \mathbb{R}^{n_\xi}$ and $y \in \mathbb{R}^q$.

Theorem 13.3 *Under Assumptions 13.3 and 13.4, the control law $u = -\bar{K}_m X_m$ globally asymptotically stabilizes (13.31), if $\gamma \leq \dfrac{c_1}{c_2(m+1)}$.*

Proof: Define $U(\xi, x, z_1, z_2, \ldots, z_m) = V_m(x, z_1, z_2, \ldots, z_m) + \frac{m+1}{c_1} W(\xi)$. Along the trajectories of the closed-loop system comprised of (13.31) and $u = -\bar{K}_m X_m$, according to Theorem 13.1, we have

$$\begin{aligned}
\dot{U} &\leq -\gamma^{-1}\|y\|^2 - \sum_{i=0}^{m} \epsilon_i \|x_i\|^2 + (m+1)\|\Delta\|^2 - (m+1)\|\Delta\|^2 + \frac{c_2(m+1)}{c1}\|y\|^2 \\
&\leq -\gamma^{-1}\|y\|^2 - \sum_{i=0}^{m} \epsilon_i \|x_i\|^2 + \frac{c_2(m+1)}{c1}\|y\|^2 \\
&\leq -\sum_{i=0}^{m} \epsilon_i \|x_i\|^2.
\end{aligned}$$

Note that U is positive definite and radially unbounded, and $\sum_{i=0}^{m} \epsilon_i \|x_i\|^2 = 0$ if and only if $X_m = 0$. Thus, the global asymptotic stability of the trivial solution of the closed-loop system is proved. ■

13.4.2 Online Learning Strategy

Now we develop the online learning scheme of robust-ADP for partial-state feedback control. The on-line learning has to be proceeded in the presence of dynamic uncertainty. To this end, let us make the following assumption:

Assumption 13.5 *The variable $\omega(t) := w_0^{\mathrm{T}}\Delta(\xi(t), y(t))$, viewed as the output to the dynamic uncertainty, is assumed to be available, at least, on some disjoint intervals*

$$\bigcup_{j=0}^{\infty} \left[t_j, t_j + \delta t\right] \bigcap [0,\ t], \tag{13.38}$$

where $t \geq 0$ denotes the current time, $\delta t > 0$ is a small positive constant, $\{t_j\}$ is an increasing sequence satisfying $0 \leq t_j < t_j + \delta t < t_{j+1}$ for all $j = 1, 2, \ldots$

We rewrite the x-subsystem as

$$\dot{x} = Ax - bK_k x + b\left(K_k x + z_1 + \omega\right). \tag{13.39}$$

Now, along the trajectories of (13.39), it follows that

$$\begin{aligned} & x(t)^{\mathrm{T}} P_k x(t) - x(t+\delta t)^{\mathrm{T}} P_k x(t+\delta t) \\ &= \int_t^{t+\delta t} x^{\mathrm{T}}(Q + r_0 K_k^{\mathrm{T}} K_k) x_0 \mathrm{d}\tau - 2\int_t^{t+\delta t} x^{\mathrm{T}} P_k b(z_1 + \omega + K_k x)\mathrm{d}\tau. \end{aligned} \tag{13.40}$$

Then, (13.40) implies the following matrix form of linear equations

$$\Phi_k \mathrm{vec}(P_k) = \Xi_k, \tag{13.41}$$

where

$$\Phi_k = \delta_{xx} + 2I_{\omega x}(b^{\mathrm{T}} \otimes I_n) + 2I_{xx}\left(K_k^{\mathrm{T}} b^{\mathrm{T}} \otimes I_n\right),$$

$$I_{\omega x} = \begin{bmatrix} \int_{t_1}^{t_1+\delta t} (z_1 + \omega)\, x^{\mathrm{T}} \mathrm{d}\tau \\ \int_{t_2}^{t_2+\delta t} (z_1 + \omega)\, x^{\mathrm{T}} \mathrm{d}\tau \\ \vdots \\ \int_{t_l}^{t_l+\delta t} (z_1 + \omega)\, x^{\mathrm{T}} \mathrm{d}\tau \end{bmatrix},$$

and the other notations are defined in the same way as in the earlier section.

Similarly, to guarantee the uniqueness of P_k in each iteration step, we make the following assumption.

Assumption 13.6 *Φ_k has full column rank, for all $k \in \mathbb{Z}_+$.*

Next, we modify the vertical learning scheme for online purpose. Note that (13.29) becomes

$$\bar{K}_i^{(1)}\left[x(t+\delta t)-x(t)\right] = L_i \int_t^{t+\delta t} x \mathrm{d}\tau + \bar{K}_i^{(1)} b \int_t^{t+\delta t} (z_1+\omega)\,\mathrm{d}\tau, \qquad (13.42)$$

which implies the matrix equation

$$L_i I_x^{\mathrm{T}} = K_i^{(1)}\left(\delta_x^{\mathrm{T}} - b I_\omega^{\mathrm{T}}\right), \qquad (13.43)$$

where I_x and δ_x are defined as in the earlier section, and

$$I_\omega = \begin{bmatrix} \int_{t_1}^{t_1+\delta t} (z_1+\omega)\,\mathrm{d}\tau \\ \int_{t_2}^{t_2+\delta t} (z_1+\omega)\,\mathrm{d}\tau \\ \vdots \\ \int_{t_l}^{t_l+\delta t} (z_1+\omega)\,\mathrm{d}\tau \end{bmatrix}.$$

The online algorithm is similar to Algorithm 1.1. Hence, it is briefly sketched as follows:

1. employ exploration noise $e(t)$ as the input to the system,
2. compute $I_x, I_\omega, I_{\omega x}, I_{xx}, \delta_x, \delta_{xx}$ on enough time intervals such that Assumptions 13.5 and 13.6 are satisfied,
3. compute $\bar{K}_m$ by performing horizontal learning and vertical learning,
4. replace the input by $u = -\bar{K}_m X_m$.

Remark 13.5 *Clearly, it can be interpreted that in (13.27) and (13.41), two sets of quadratic basis functions Θ_k, Φ_k are employed to approximate the quadratic cost function $V_m(X_m)$. It is also worth noting that similar schemes can be developed to handle more general nonlinear systems, by extending Θ_k, Φ_k with other basis sets, such as sigmoid, hyperbolic tangent, and Gaussian radial basis functions.*

13.5 APPLICATIONS

13.5.1 Load-Frequency Control for a Power System

In electric power system design and operation, LFC is of importance. It is necessary to design a LFC system which deals with the control of loading of the generator depending on the frequency, to ensure the quality of the power supply. The model of the plant can be described using the following ordinary differential equations [29]:

$$\Delta \dot{f} = -\frac{1}{T_p}\Delta f + \frac{K_p}{T_p}\left(\Delta P_g - \Delta P_d\right), \tag{13.44}$$

$$\Delta \dot{P}_g = -\frac{1}{T_T}\Delta P_g + \frac{1}{T_T}\Delta X_g, \tag{13.45}$$

$$\Delta \dot{X}_g = -\frac{1}{R_G T_G}\Delta f - \frac{1}{T_G}\Delta X_g + \frac{1}{T_G}\Delta E + \frac{1}{T_G}u, \tag{13.46}$$

$$\Delta \dot{E} = K_E \Delta f, \tag{13.47}$$

where Δf is the incremental frequency deviation, ΔP_g the incremental change in generator output, and ΔX_g the incremental change in governor valve position. Parameters T_G, T_T, T_p, K_p, and R_G denote the governor time constant, the turbine time constant, the plant model time constant, the plant gain, and the speed regulation due to governor action, respectively. $u \in \mathbb{R}$ is the control input.

The main difficulty of the control problem comes from the fact that only the range of some parameters can be determined. In this example, except for $K_E > 0$ and $T_p > 0$ being assumed unknown, all other system parameters as well as the ratio $\frac{K_p}{T_p}$ are assumed to be known. The objective is to find a linear controller under which the closed-loop system is finite-gain L_2 stable, with respect to the disturbance input $\Delta P_d(t)$ and the output $\Delta f(t)$, with a gain less or equal to 60%.

Now, we are able to define the coordinates and input transformations:

$$x_1 = \Delta E, \quad x_2 = \Delta f, \quad z_1 = \Delta P_g, \quad z_2 = \frac{1}{T_T}\left(\Delta X_g - \Delta P_g\right),$$

$$\bar{u} = -\frac{1}{R_g T_G T_T}\Delta f - \frac{T_G + T_I}{T_G T_I^2}\Delta X_g + \frac{1}{T_G T_T}\Delta E + \frac{1}{T_T^2}\Delta P_g + \frac{1}{T_G T_T}u.$$

Thus, the original system is converted to

$$\dot{x} = \begin{bmatrix} 0 & K_E \\ 0 & -\frac{1}{T_p} \end{bmatrix} x + \begin{bmatrix} 0 \\ \frac{K_p}{T_p} \end{bmatrix} (z_1 - P_d), \tag{13.48}$$

$$\dot{z}_1 = z_2, \tag{13.49}$$

$$\dot{z}_2 = \bar{u}. \tag{13.50}$$

For simulation purpose, we set $T_T = 0.21$, $T_p = 15.0376$, $K_p = 120.3008$, $T_G = 0.0728$, and $R_G = 0.4579$. The initial conditions are randomly chosen as $\Delta f(0) = 1$, $\Delta P_g(0) = 0$, and $\Delta X_g(0) = 0$, $\Delta E(0) = 1$. The weighting matrices are

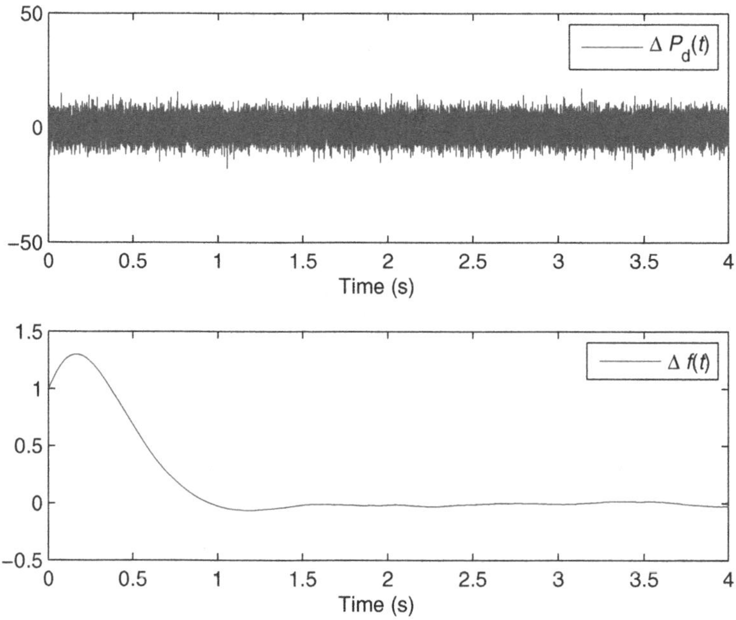

FIGURE 13.1 Robust performance.

set to be $Q = \mathrm{diag}\,(10, 0.01)$ and $R = 1$. We perform Algorithm 1.1 with $\Delta P_{\mathrm{d}}(t) \equiv 0$, and employ the following exploration noise

$$\bar{u}(t) = v(t) = 10\,(\sin 0.1t + \sin 0.3t + \sin 0.7t + \sin 1.1t)\,, \quad t \in [0, 1]\,.$$

Then, we record the matrices I_x, I_z, I_{xx}, I_{zx}, δ_x, and δ_{xx}, with $t_j = (j-1)\delta t$, $\delta t = 0.002s$, $j = 1, 2, \ldots, 500$. After updating the policy for 18 times in the horizontal learning, the iteration stops when the criterion $|P_k - P_{k-1}| \leq 10^{-4}$ is satisfied.

We now use the controller formulated from Algorithm 1.1 to simulate the closed-loop system with disturbance. The simulation is conducted using the data from the system at every $0.0001s$, where the disturbance P_{d} is simulated using white noise with zero mean and the variance is set to be 16.

The robust performance of the closed-loop system is shown in Figure 13.1. It is clear to see that the influence of the disturbance on Δf taken as the output is significantly attenuated.

13.5.2 Machine Tool Power Drive System

Consider a single axis of a machine tool in which a motor drives a lead screw through a reduction gear. The lead screw moves the workpiece under the cutter to produce the finished item. A computer sends desired positions of the table to a control system that attempts to place the table in the correct position as rapidly as possible. The equations of motion of the motor and load shaft can be written as [30]

$$\tau_{\rm m} - c_{\rm m}\omega_{\rm m} - T_{\rm l} = J_{\rm m}\ddot{\theta}_{\rm m}, \tag{13.51}$$

$$k\left(\frac{\theta_{\rm m}}{N} - \theta_t\right) - c_l\omega_l = J_l\ddot{\theta}_l \tag{13.52}$$

where $\theta_{\rm l}$ is the rotation angular of the lead-screw, $\theta_{\rm m}$ is the rotation angular of the motor, k is the torsional spring, $\tau_{\rm m}$ is the input motor torque. Details of the other parameters can be found in Ref. [30]. It is assumed the ISS gain for the ξ system is less than one.

Defining $\xi = \left[\theta_{\rm l}\ \dot{\theta}_{\rm l}\right]^{\rm T}$ and $x = \left[\theta_{\rm m}\ \dot{\theta}_{\rm m}\right]^{\rm T}$, (13.51) and (13.52) can be converted into

$$\dot{\xi} = \begin{bmatrix} 0 & 1 \\ \frac{-k}{J_{\rm l}} & \frac{-c_{\rm l}}{J_{\rm l}} \end{bmatrix}\xi + \begin{bmatrix} 0 \\ y \end{bmatrix}, \tag{13.53}$$

$$\dot{x} = \begin{bmatrix} 0 & 1 \\ \frac{-k}{J_{\rm m}N^2} & \frac{-c_{\rm m}}{J_{\rm m}} \end{bmatrix}x + \begin{bmatrix} 0 \\ \frac{1}{J_{\rm m}} \end{bmatrix}(\tau_{\rm m} + \Delta(\xi)), \tag{13.54}$$

where $y = \left[\frac{-k}{J_{\rm l}N}\ 0\right]x$, and $\Delta(\xi) = \left[\frac{k}{N}\ 0\right]\xi$.

Suppose the parameter $J_{\rm m}$ and the ratio $\frac{k}{N}$ are known, but all the other parameters are unknown. The state variables $\theta_{\rm m}$ and $\dot{\theta}_{\rm m}$ are available, $\theta_{\rm l}$ is measurable on the time interval of $[0,\ 1]$, $\dot{\theta}_{\rm l}$ is not available at all. We now apply the on-line robust-ADP algorithm to formulate a controller that globally asymptotically stabilizes the overall system. For simulation purpose, we set $J_{\rm m} = 0.1$ N·m·s^2/rad, $J_{\rm l} = 0.001$ N·m·s^2/rad, $c_{\rm l} = 0.01$ N·m·s/rad, $c_{\rm m} = 0.01$ N·m·s/rad, $k = 1$ N·m/rad, and $N = 2$. The initial conditions are randomly chosen to be $\theta_{\rm l}(0) = 1, \dot{\theta}_{\rm l}(0) = -2, \theta_{\rm m}(0) = -2, \dot{\theta}_{\rm m}(0) = 3$. The weighting matrices are set to be $Q = \text{diag}(1, 0.01)$, and $R = 1$.

During the first second, the following exploration noise is employed as the input motor torque

$$\tau_{\rm m}(t) = 0.01(\sin 0.1t + \sin 0.3t + \sin 0.7t + \sin 1.1t), \quad t \in [0,\ 1)$$

and corresponding matrices I_x, $I_{\omega x}$, I_ω, I_x, and I_{xx} are recorded, with length $l = 20$. At $t = 1s$, the control law is instantaneously updated eight times until convergence is attained. Then, the following controller is applied to the system from $t = 1s$ until the end of the simulation

$$\tau_{\rm m}(t) = -0.7809\theta_{\rm m} - 0.3978\dot{\theta}_{\rm m}, \quad t \in [1,\ 4]. \tag{13.55}$$

The global asymptotic stability of the trivial solution of the closed-loop system is illustrated in Figure 13.2. For comparison, we have also carried out simulations on

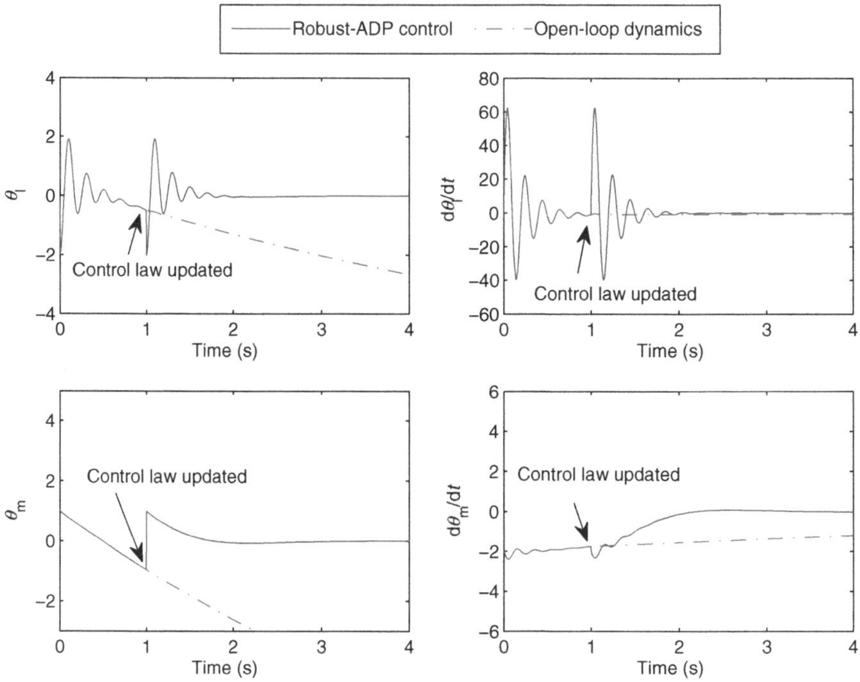

FIGURE 13.2 Trajectories of the state variables.

the open-loop system, which is not asymptotically stable due to the presence of a zero eigenvalue.

13.6 SUMMARY

In this chapter, we have proposed a framework of robust-ADP, which is aimed at computing globally asymptotically stabilizing control laws with robustness to dynamic uncertainties, via off-line/on-line learning. The robust-ADP framework has been applied to the LFC for a power system and a machine tool power drive system. In addition, it has been shown that robust optimal control problems can be solved for higher-dimensional, partially linear composite systems by integration of ADP and techniques in modern nonlinear control such as backstepping and ISS small-gain methods.

ACKNOWLEDGMENT

The authors would like to thank the National Science Foundation for financial support under Grants DMS-0906659 and ECCS-1101401.

REFERENCES

1. R. S. Sutton and A. G. Barto. *Reinforcement Learning: An Introduction*, MIT Press, 1998.
2. P.J. Werbos. *Beyond regression: new tools for prediction and analysis in the behavioural sciences*, Ph.D. Thesis, Harvard University, 1972.
3. P.J. Werbos. Neural networks for control and system identification, *Proceedings of IEEE Conference on Decision and Control*, 260–265, 1989.
4. P.J. Werbos. A menu of designs for reinforcement learning over time, *Neural Networks for Control*, ed. W. T. Miller, R. S. Sutton, P. J. Werbos, Cambridge: MIT Press, 1991, pp 67–95.
5. P. J. Werbos. Approximate dynamic programming for real-time control and neural modeling, *Handbook of Intelligent Control*, ed. D.A. White and D.A. Sofge, New York: Van Nostrand Reinhold, 1992.
6. C. Watkins. *Learning from delayed rewards*, Ph.D. thesis, King's College of Cambridge, UK, 1989.
7. A. Al-Tamimi, F.L. Lewis, and M. Abu-Khalaf. Model-free Q-learning designs for linear discrete-time zero-sum games with application to H-infinity control, *Automatica*, 43(3):473–481, 2007.
8. S.J. Bradtke, B.E. Ydstie, and A.G. Barto. Adaptive linear quadratic control using policy iteration, *Proceedings of American Control Conference*, 3:3475–3479, 1994.
9. Y. Jiang and Z. P. Jiang. Approximate dynamic programming for output feedback control, In *Proceedings of the 29th Chinese Control Conference*, 5815–5820, 2010.
10. Y. Jiang and Z. P. Jiang. Approximate dynamic programming for optimal stationary control with control-dependent noise, *IEEE Transactions on Neural Networks*, 22(12):2392–2398, 2011.
11. F. L. Lewis, K. G. Vamvoudakis. Reinforcement learning for partially observable dynamic processes: adaptive dynamic programming using measured output data, *IEEE Transactions Systems, Man, and Cybernetics, Part B*, SMCB-41(1):14–23, 2011.
12. D. Vrabie, O. Pastravanu, M. Abu-Khalaf, and F. L. Lewis. Adaptive optimal control for continuous-time linear systems based on policy iteration, *Automatica*, 45(2):477–484, 2009.
13. H. Zhang, Q. Wei, and D. Liu. An iterative adaptive dynamic programming method for solving a class of nonlinear zero-sum differential games, *Automatica*, 47(1):207–214, 2011.
14. M. Krstic, I. Kanellakopoulos, and P. V. Kokotovic. *Nonlinear and Adaptive Control Design*, John Wiley & Sons, 1995.
15. E. D. Sontag. Further facts about input to state stabilization, *IEEE Transactions on Automatic Control*, AC-35(4):473–476, 1990.
16. E. D. Sontag and Y. Wang. On characterizations of the input-to-state stability property, *Systems & Control Letters*, 24:351–359, 1995.
17. Z. P. Jiang, A. R. Teel, and L. Praly. Small-gain theorem for ISS systems and applications, *Mathematics of Control, Signals, and Systems*, 7(2):95–120, 1994.
18. Z. P. Jiang, I. Mareels and Y. Wang. A Lyapunov formulation of the nonlinear small gain theorem for interconnected ISS systems, *Automatica*, 32(8):1211–1215, 1996.
19. A. Saberi, P. V. Kokotovic, and H. J. Sussmann. Global stabilization of partially linear composite systems, *SIAM Journal on Control and Optimization*, 2(6):1491–1503, 1990.
20. F.L. Lewis and V.L. Syrmos. *Optimal Control*, John Wiley & Sons, 1995.

21. D. Vrabie, F. Lewis. Adaptive dynamic programming algorithm for finding online the equilibrium solution of the two-player zero-sum differential game, *The 2010 International Joint Conference on Neural Networks*, 1–8, 2010.
22. H. K. Khalil. *Nonlinear Systems*, 3rd edition, Prentice-Hall, 2002.
23. D. Kleinman. On an iterative technique for Riccati equation computations, *IEEE Transactions on Automatic Control*, AC-13(1):114–115, 1969.
24. R. A. Horn and C. R. Johnson. *Matrix Analysis*, Cambridge University Press, NY, 1985.
25. Y. Jiang and Z. P. Jiang. Robust approximate dynamic programming and global stabilization with nonlinear dynamic uncertainties, *Proceedings of the Joint 2011 IEEE Conference on Decision and Control and European Control Conference*, Orlando, FL, 115–120, 2011.
26. P.J. Werbos. Intelligence in the brain: a theory of how it works and how to build it, *Neural Networks*, 22(3):200–212, 2009.
27. A. Isidori. *Nonlinear Control Systems*. Vol. II, Springer-Verlag, 1999.
28. Z. P. Jiang and L. Praly. Design of robust adaptive controllers for nonlinear systems with dynamic uncertainties, *Automatica*, 34(7):825–840, 1998.
29. Y. Wang, R. Zhou, and C. Wen. Robust load-frequency controller design for power systems, *IEE Proceedings-Generation, Transmission and Distribution*, 104(1):11–16, 1993.
30. M. Driels. *Linear Control Systems Engineering*. The McGraw-Hill Companies, Inc., 1996.

PART II

LEARNING AND CONTROL IN MULTIAGENT GAMES

CHAPTER 14

Hybrid Learning in Stochastic Games and Its Application in Network Security

QUANYAN ZHU,[1] HAMIDOU TEMBINE,[2] and TAMER BAŞAR[1]

[1]Coordinated Science Laboratory, University of Illinois at Urbana-Champaign, Urbana, IL, USA
[2]Telecommunication Department, Supelec, Gif sur Yvette, France

ABSTRACT

In this chapter, we consider a class of two-player nonzero-sum stochastic games with incomplete information, which is inspired by recent applications of game theory in network security. We develop fully distributed reinforcement learning algorithms, which require for each player a minimal amount of information regarding the other player. At each time, each player can be in an active mode or in a sleep mode. If a player is in an active mode, the player updates the strategy and estimates of unknown quantities using a specific pure or hybrid learning pattern. The players' intelligence and rationality are captured by the weighted linear combination of different learning patterns. We use stochastic approximation techniques to show that, under appropriate conditions, the pure or hybrid learning schemes with random updates can be studied using their deterministic ordinary differential equation (ODE) counterparts. Convergence to state-independent equilibria is analyzed for special classes of games, namely, games with two actions, and potential games. Results are applied to network security games between an intruder and an administrator, where the noncooperative behaviors are characterized well by the features of distributed hybrid learning.

14.1 INTRODUCTION

In recent years, game-theoretic methods have been applied to study resource allocation problems in communication networks [1], security mechanisms for network security and privacy [2, 3], and economic pricing in power networks [4]. Most frameworks have

Reinforcement Learning and Approximate Dynamic Programming for Feedback Control, First Edition. Edited by Frank L. Lewis and Derong Liu.

assumed the rationality of the agents or the decision-makers as well as the complete information about their payoffs and strategies. However, in practice, due to the noise and the uncertainties in the environment, agents often have information limitations in their knowledge not only of other players' payoffs and strategies but also of their own. For this reason, we must consider the learning aspects of the decision-makers and address their estimation and assessment of the payoff and strategy based on the information accessible to them.

In this chapter, we consider a class of two-player nonzero-sum stochastic games with incomplete information. We develop combined fully distributed payoff and strategy reinforcement learning (CODIPAS-RL) algorithms, which require for each player a minimal amount of information regarding the other player. At each time, each player can be in an active mode or in a sleep mode. If a player is in an active mode, the player updates the strategy and estimates of unknown quantities using a specific pure or hybrid learning pattern. In contrast to the standard reinforcement learning algorithms which focus only on either strategy or payoff reinforcement for the equilibrium learning, the algorithm that couples the payoff reinforcement learning together with strategy-reinforcement learning allows an immediate prediction and updates the strategies by updated estimations based on recent experiences. The payoff reinforcement learning in our proposed algorithms bears a connection with the Q-learning algorithms in [5, 6], which have been commonly applied to learn the Q-functions of Markov decision processes (MDPs).

We specifically discuss five pure CODIPAS-RL algorithms and use stochastic approximation techniques to show that, under appropriate conditions, the pure or hybrid learning schemes with random updates can be studied using their deterministic ordinary differential equation (ODE) counterparts. Convergence to state-independent equilibria is analyzed under specific payoff functions such as those in games with two actions and Lyapunov games.

We apply the learning algorithms to a class of security games where an attacker and an intrusion detection system (IDS) strategically choose their actions to optimize their payoffs. Many forms of security games have been formulated to provide quantitative security and dependability analysis of networked systems [2, 3, 7]. However, technical difficulties in quantifying appropriate security metrics or payoff functions render it difficult to specify the utility functions for the attacker and the defender. In addition, the inevitable false positive and false negative errors in the detection often lead to incomplete information in a dynamic network environment. Our hybrid learning framework for the two-person game with incomplete information provides an appropriate theoretical basis for the online implementation of game-theoretic algorithms.

14.1.1 Related Work

Learning in games has been investigated in several papers in the recent literature. In Refs. [8, 9], a fictitious-play algorithm is used to find Nash equilibrium in a nonzero-sum game. Players observe opponents' actions and update their strategies in reaction to others' actions in a best-response fashion. Marden et al. [10] propose a modified version of the fictitious play called joint fictitious play with inertia for potential games, in which players alternate their updates at different time slots. In all these learning schemes, players have to monitor the actions of every other player and need to know their own payoff so as to find their optimal actions. In this chapter, we are interested in fully distributed learning procedures, where players do not need any information about the actions or payoffs of the other players, and, moreover, they do not need to have complete information of their own payoff structure.

Young proposes in [11] such a completely uncoupled learning rule called interactive trial and error learning. Players occasionally try out new actions and accept them if they lead to higher payoffs. If a player experiences a decrease in payoff due to strategy changes by some other players, the player initiates a random search for a new strategy and settles on one with a probability that increases monotonically with its realized payoff. When used by all players, the learning scheme yields pure-strategy Nash equilibrium behavior under an interdependency condition. However, in games without pure-strategy Nash equilibrium, it fails to yield Nash equilibrium strategies.

In Refs. [12, 13], strategy reinforcement learning in finite games has been studied. The ordinary differential equation (ODE) approximation of the learning algorithm is shown to be equivalent to an adjusted replicator dynamics [14]. In Ref. [15], a multiple time scale model-free algorithm is introduced, and it is shown to be asymptotically equivalent to the smooth fictitious play algorithm. In Refs. [7, 16], we introduce a class of combined distributed payoff and strategy reinforcement learning schemes (CODIPAS-RL) and propose a heterogeneous learning algorithm for two-person zero-sum stochastic games with incomplete information, where different players can adopt different learning schemes and learning rates. In Ref. [17], we propose a Q-learning algorithm for zero-sum stochastic games and apply it to dynamic configuration problems of intrusion detection systems.

14.1.2 Contribution

In this chapter, we consider a class of general-sum two-person games and introduce the new paradigm of *hybrid learning* under the frameworks of combined distributed payoff and strategy reinforcement learning (CODIPAS-RL), where in order to render the learning algorithm practical to implement in the context of network security, we introduce the following features of the game:

- (F1) In addition to exogenous environment uncertainties, we introduce inherent mode uncertainties in players. Each player can be in an *active* mode or in a *sleeping* mode. Players learn their strategies and average payoffs only when they are in an *active* mode.
- (F2) We allow the interaction between the players to occur at random times unknown to the players.

We use stochastic approximation techniques to show that the hybrid learning schemes with random updates can be studied using their deterministic ODE counterparts.

TABLE 14.1 Table of Notations

Symbol	Meaning
$a_{i,t} \in \mathcal{A}_i$	Action of player i (Pi) at time t
$\mathbf{x}_{i,t} \in \mathcal{X}_i$	Mixed strategy of Pi at t
$B_i \in \{0, 1\}$	Active or sleep mode of Pi
$s_i \in \mathcal{S}_i$	External state of Pi
$u_{i,t} \in \mathbb{R}$	Observed payoff by Pi at t
$\hat{\mathbf{u}}_{i,t} \in \mathbb{R}^{\vert\mathcal{A}_i\vert}$	Estimated payoff vector of Pi at t
$\mathbb{U}_i \in \mathbb{R}$	Mixed extension of the payoff U_i.
$\beta_i(\hat{\mathbf{u}}_{i,t}) \subseteq \mathcal{A}_i$	Best response
$\tilde{\beta}_{i,\epsilon}(\hat{\mathbf{u}}_{i,t}) \in \mathbb{R}^{\vert\mathcal{A}_i\vert}$	Boltzmann–Gibbs (B-G) strategy
$\tilde{\beta}_i^I(\mathbf{x}_{i,t}, \hat{\mathbf{u}}_{i,t}) \in \mathbb{R}^{\vert\mathcal{A}_i\vert}$	Imitative B-G strategy
$\tilde{\beta}_i^W(\mathbf{x}_{i,t}, \hat{\mathbf{u}}_{i,t}) \in \mathbb{R}^{\vert\mathcal{A}_i\vert}$	Weighted imitative B-G strategy
$\tilde{\beta}_i^F(\hat{\mathbf{u}}_{i,t}) \in \mathbb{R}^{\vert\mathcal{A}_i\vert}$	Weakened fictitious-play strategy
$\nu_{i,t} \in \mathbb{R}_+$	Payoff learning rates of Pi at t
$\lambda_{i,t} \in \mathbb{R}_+$	Strategy learning rates of Pi at t
$e_{a_i} \in \mathbb{R}^{\vert\mathcal{A}_i\vert}$	The unit vector with 1 at the position of a_i and 0 otherwise

The ODE obtained for hybrid learning is a linear combination of ODEs from pure learning schemes. We show the convergence properties of the learning algorithms for special classes of games, namely, games with two actions, and potential games, and demonstrate their applications in a network security environment.

14.1.3 Organization of the Chapter

The chapter is structured as follows. In Section 14.2, we formulate the two-player nonzero-sum stochastic game with incomplete information and introduce the solution concept of state-independent Nash equilibrium. In Section 14.3, we present a number of distinct learning schemes and discuss their properties. In Section 14.4, we present main results on learning for general-sum games. In Section 14.5, we apply the learning algorithms to a network security application. Section 14.6 concludes the chapter. In Table 14.1, we summarize the notation used in the chapter for reader's convenience.

14.2 TWO-PERSON GAME

In this section, we consider a finite two-person nonzero-sum game (NZSG) in which each player has stochastic payoffs and the interactions between the players are random. Let $\Xi := \langle \mathcal{N}, \{\mathcal{S}_i\}_{i\in\mathcal{N}}, \{\Omega_i\}_{i\in\mathcal{N}}, \{\mathcal{A}_i\}_{i\in\mathcal{N}}, \{U_i(s, B^2, .)\}_{s\in\mathcal{S}, b\in\mathcal{B}, i\in\mathcal{N}}\rangle$ be the stochastic NZSG, where $\mathcal{N} = \{1, 2\}$ is the set of players P1 and P2 who maximize their payoffs, and $\mathcal{A}_1, \mathcal{A}_2$ are the finite sets of actions available to players P1 and P2, respectively. The set $\mathcal{S}_i := [s_{i,1}, s_{i,2}, \ldots, s_{i,N_S^i}]$ comprises all possible N_S^i external states of Pi, which describes the environment where Pi resides. We assume that the state space $\mathcal{S} := \prod_{i \in \mathcal{N}} \mathcal{S}_i$ and the probability transition on the states are both unknown to the players. A state s_i is randomly and independently chosen at each time from the set $\mathcal{S}_i$. We assume that the action spaces are the same in each state.

In addition, players do not interact at all times. A player can be in one of the two modes: *active mode* or *sleep mode*, denoted by mode $B_i = 1$ and $B_i = 0$, respectively. Let $B_i, i \in \mathcal{N}$, be an i.i.d. random variable on $\Omega_i := \{0, 1\}$ whose probability mass function is given by

$$\rho_B^i = \begin{cases} p_i, & B^i = 1, \\ 1 - p_i, & B^i = 0 \end{cases}, i \in \mathcal{N}. \tag{14.1}$$

The player modes can be viewed as internal states that are governed by the inherent randomness of the player. The system mode $B^2 \in \Omega := \Omega_1 \times \Omega_2$ is a set of independent modes of the players and we denote by $\mathcal{B}^2 \subseteq \mathcal{N}$ as the set of active players corresponding to B^2.

The NZSG is characterized by utility functions $U_i : \mathcal{S} \times \Omega_i \times \mathcal{A}_1 \times \mathcal{A}_2 \to \mathbb{R}$. P$i$ collects a payoff $U_i(s, B^2, a_1, a_2)$ when P1 chooses $a_1 \in \mathcal{A}_1$ and P2 uses $a_2 \in \mathcal{A}_2$ at state $s \in \mathcal{S}$ and mode B^2.

The preceding game model can be viewed as a special class of stochastic games in which the state transitions are independent of the player actions as well as the current state.

We have slotted time, $t \in \{0, 1, \ldots\}$, when players pick their mixed strategies as functions of what has transpired in the past, to the extent the information available to them allows. Toward this end, we let $x_{i,t}(a_i)$ denote the probabilities of Pi choosing $a_i \in \mathcal{A}_i$ at time t, and let $\mathbf{x}_{i,t} = [x_{i,t}(a_i)]_{a_i \in \mathcal{A}_i}$ be the mixed strategies of Pi at time t, where more precisely

$$\mathbf{x}_{i,t} \in \mathcal{X}_i := \left\{ \mathbf{x}_i \in \mathbb{R}^{|\mathcal{A}_i|} : x_i(a_i) \in [0, 1], \sum_{a_i \in \mathcal{A}_i} x_i(a_i) = 1 \right\}.$$

In particular, we define $e_{a_i} \in \mathbb{R}^{|\mathcal{A}_i|}$, with $a_i \in \mathcal{A}_i$, as unit vectors of sizes $|\mathcal{A}_i|$, whose entry that corresponds to a_i is 1 while others are zeros. We assume that the mixed strategies of the players are independent of the current state s and the player mode B_i. For any given pair of mixed strategies, $(\mathbf{x}_1, \mathbf{x}_2) \in \mathcal{X}_1 \times \mathcal{X}_2$, and for a fixed $s_i \in S_i$, $B^2 \in \Omega$, we define the expected utility (as expected payoff to Pi) as

$$\mathbb{U}_i(s, B^2, \mathbf{x}_1, \mathbf{x}_2) := \mathbb{E}_{\mathbf{x}_1, \mathbf{x}_2} U_i(s, B^2, a_1, a_2),$$

where $\mathbb{E}_{\mathbf{x}_1, \mathbf{x}_2} U_i$ denotes expectation of U_i over the action sets of the players under the given mixed strategies. A further expectation of this quantity over the states s and B^2, denoted $\mathbb{E}_{s, B^2}$, yields the performance index of the *expected game*. We now define the equilibrium concept of interest for this game, that is, the equilibrium of the expected game.

Definition 14.1 (State-independent equilibrium) *A strategy profile* $(\mathbf{x}_1^*, \mathbf{x}_2^*) \in \mathcal{X}_1 \times \mathcal{X}_2$ *is a state-independent equilibrium of the game* Ξ *if it is equilibrium of the expected game, that is,* $\forall \mathbf{x}_1 \in \mathcal{X}_1, \mathbf{x}_2 \in \mathcal{X}_2$

$$\mathbb{E}_{s,B^2}\mathbb{U}_1(s, B^2, \mathbf{x}_1^*, \mathbf{x}_2^*) \geq \mathbb{E}_{s,B^2}\mathbb{U}_1(s, B^2, \mathbf{x}_1, \mathbf{x}_2^*),$$
$$\mathbb{E}_{s,B^2}\mathbb{U}_2(s, B^2, \mathbf{x}_1^*, \mathbf{x}_2^*) \geq \mathbb{E}_{s,B^2}\mathbb{U}_2(s, B^2, \mathbf{x}_1^*, \mathbf{x}_2).$$

Since the expected game is a two-player game with finite actions for each player, we can show the existence of the equilibrium using Nash's existence theorem [18] and state the following lemma.

Lemma 14.1 (Existence) *The stochastic NSZG* Ξ *with unknown states and changing modes admits a state-independent equilibrium.*

14.3 LEARNING IN NZSGs

In this section, we introduce different learning schemes for the stochastic NZSGs introduced in Section 14.2 and discuss the stochastic approximation of the learning schemes with ODEs.

14.3.1 Learning Procedures

In many practical applications, the players in the two-person NZSGs do not have the complete knowledge of their payoff functions and the state of their environment. Moreover, they do not know whether they interact with the other player or not. Hence, the equilibrium strategy has to be learned online by observing the payoffs at each time slot. A general learning procedure is outlined as follows. At each time slot $t \in \mathbb{Z}_+$, each player generates an internal mode B_i to determine whether to participate in the game or not. If both players are active, they interact and receive a payoff after the play. If only one of the players is active, then the active player receives a payoff as an outcome of the player's action at t only without the interaction with the other player. If players do not have the knowledge of their active mode probability p_i, then each player keeps count of its interaction with others by updating its vectors $\theta_{ij,t} \in \mathbb{R}^2$, $i, j \in \{1, 2\}$, as follows:

$$\theta_{ij,t} = \theta_{ij,t-1} + \mathbb{1}_{\{B_j=1\}},$$

where $\theta_{ij,t}$ is Pi's count of Pj's number of activities since $t \geq 0$ and the initial condition is given by $\theta_{ij} = 0, \forall i, j \in \{1, 2\}$. The active players choose an action $a_{i,t} \in \mathcal{A}_i$ at time t and observe or measure an output $u_{j,t} \in \mathbb{R}$ as an outcome of their actions. Players estimate their payoffs by updating the entry of the estimated payoff vector $\hat{\mathbf{u}}_{i,t+1} \in \mathbb{R}^{|\mathcal{A}_i|}$ that corresponds to the chosen action $a_{i,t}$. In a similar way, players update their strategy vectors $\mathbf{x}_{i,t+1}$ based on a specific learning scheme (to be introduced later). The update of the strategy vectors can exploit the payoff information $\hat{u}_{i,t}$ from the previous time step. In this case, we say the learning is coupled; otherwise, we say that it is uncoupled.

The general coupled learning updates on the strategy and utility vectors take the following form:

$$\begin{cases} \mathbf{x}_{i,t+1} = \mathbf{x}_{i,t} + \Pi_{i,t}(\lambda_{i,t}, a_{i,t}, u_{i,t}, \hat{\mathbf{u}}_{i,t}, \mathbf{x}_{i,t}), \\ \hat{\mathbf{u}}_{i,t+1} = \hat{\mathbf{u}}_{i,t} + \Sigma_{i,t}(\nu_{i,t}, a_{i,t}, u_{i,t}, \hat{\mathbf{u}}_{i,t}, \mathbf{x}_{i,t}), \end{cases} \tag{14.2}$$

where $\Pi_{i,t}$, $\Sigma_{i,t}$, $i \in \mathcal{N}$, are properly chosen functions for strategy and utility updates, respectively. The parameters $\lambda_{i,t}$, $\nu_{i,t}$ are learning rates indicating players' capabilities of information retrieval and update. The vectors $\mathbf{x}_{i,t} \in \mathcal{X}_i$ are mixed strategies of the players at time t. $\hat{\mathbf{u}}_{i,t}$, $i \in \mathcal{N}$, are estimated average payoffs updated at each iteration t, and $u_{i,t}$, $i \in \mathcal{N}$, are the observed payoffs received by players at time t. The learning rates $\lambda_{i,t}, \nu_{i,t} \in \mathbb{R}_+$ need to satisfy the conditions

(C1) $\sum_{t\geq 0} |\lambda_{i,t}|^2 < \infty$, $\sum_{t\geq 0} |\nu_{i,t}|^2 < \infty$.

(C2) $\sum_{t\geq 0} |\lambda_{i,t}| = +\infty$, $\sum_{t\geq 0} |\nu_{i,t}| = +\infty$.

The learning rate of Pi is relative to its frequency of activity. In general, the learning rates are functions of θ_{ii}, $i \in \mathcal{N}$ and can be written as $\lambda_{i,\theta_{ii}(t)}$, $\nu_{i,\theta_{ii}(t)}$. We need to adopt a time reference for the game using maximum learning rates among the active players, that is, $\lambda_t^* := \max_{i\in\mathcal{B}^2(t)} \lambda_{i,\theta_{ii}(t)}$, $\nu_t^* := \max_{i\in\mathcal{B}^2(t)} \nu_{i,\theta_{ii}(t)}$. It can be verified that the reference learning rates λ_t^*, ν_t^* satisfy (C1) and (C2) if $\lambda_{i,t}$, $\nu_{i,t}$ satisfy the conditions for every $i \in \mathcal{N}$. The learning rates λ_t^*, ν_t^*, as we will see later, affect the ODE approximation.

We call the learning in Equation (14.2) a combined distributed payoff and strategy reinforcement learning (CODIPAS-RL) [16]. The players can have different learning rates for their utility and strategy updates. The payoff learning rate is on a faster time scale than strategy learning rate if $\lambda_{i,t}/\nu_{i,t} \to 0$ as $t \to \infty$; it is on a slower time scale if $\nu_{i,t}/\lambda_{i,t} \to 0$ as $t \to \infty$. In the former case, the payoff learning can be seen as quasi-static compared to the strategy learning and *vice versa* for the latter.

14.3.2 Learning Schemes

We introduce different learning schemes in the form of (14.2) for the stochastic NZSG. Let $\mathcal{L} = \{\mathcal{L}_k, k = 1, 2, \ldots, 5\}$ be a set of five pure learning schemes. A player Pi chooses a learning schemes $\mathcal{P}_i$ from the set $\mathcal{L}$. We call the learning *homogenous* if both players use the same pure learning schemes and *heterogenous* if players use different learning schemes, that is, $\mathcal{P}_1 \neq \mathcal{P}_2$.

14.3.2.1 Bush–Mosteller-Based CODIPAS-RL $\mathcal{L}_1$ Let $\Gamma_i \in \mathbb{R}$ be a reference level of Pi and

$$\widetilde{\Gamma}_{i,t} := \frac{u_{i,t} - \Gamma_i}{\sup_{s,B^2,\mathbf{a}} |U_i(s, B^2, \mathbf{a}) - \Gamma_i|}. \tag{14.3}$$

The learning pattern $\mathcal{L}_1$ is given by

$$\begin{cases} x_{i,t+1}(a_i) &= x_{i,t}(a_i) + \lambda_{i,t} \mathbb{1}_{\{i \in \mathcal{B}^2(t)\}} \widetilde{\Gamma}_{i,t} \left(\mathbb{1}_{\{a_{i,t}=a_i\}} - x_{i,t}(a_i) \right), \\ \hat{u}_{i,t+1}(a_i) &= \hat{u}_{i,t}(a_i) + \nu_{i,t} \mathbb{1}_{\{a_{i,t}=a_i, i \in \mathcal{B}^2(t)\}} \left(u_{i,t} - \hat{u}_{i,t}(a_i) \right). \end{cases}$$

The updates on the strategy and the estimated payoff are decoupled. The strategy update does not exploit the knowledge of estimated payoff but only relies on the observed payoffs at each time slot. The strategy update of $\mathcal{L}_1$ is widely studied in machine learning and has been initially proposed by Bush and Mosteller in [19]. Combined with the payoff update, we obtain a COPIDAS-RL based on Bush–Mosteller learning. When $\Gamma_i = 0$, we obtain the learning schemes in [20, 21].

14.3.2.2 Boltzmann–Gibbs-Based CODIPAS-RL $\mathcal{L}_2$ Let $\tilde{\beta}_{i,\epsilon} : \mathbb{R}^{|\mathcal{A}_i|} \rightarrow \mathbb{R}^{|\mathcal{A}_i|}$ be the Boltzmann–Gibbs (B-G) strategy mapping given by

$$\tilde{\beta}_{i,\epsilon}(\hat{\mathbf{u}}_{i,t})(a_i) := \frac{e^{\frac{1}{\epsilon}\hat{u}_{i,t}(a_i)}}{\sum_{a_i' \in \mathcal{A}_i} e^{\frac{1}{\epsilon}\hat{u}_{i,t}(a_i')}}, \quad a_i \in \mathcal{A}_i. \tag{14.4}$$

It is also known as the soft-max function. When $\epsilon \rightarrow 0$, the B-G strategy yields a (pure) strategy that picks the maximum entry of the payoff vector $\hat{\mathbf{u}}_{i,t}$. The learning pattern $\mathcal{L}_2$ is given by

$$\begin{cases} x_{i,t+1}(a_i) &= x_{i,t}(a_i) + \lambda_{i,t} \mathbb{1}_{\{i \in \mathcal{B}^2(t)\}} \left(\tilde{\beta}_{i,\epsilon}(\hat{\mathbf{u}}_{i,t})(a_{i,t}) - x_{i,t}(a_i) \right), \\ \hat{u}_{i,t+1}(a_i) &= \hat{u}_{i,t}(a_i) + \nu_{i,t} \mathbb{1}_{\{a_{i,t}=a_i, i \in \mathcal{B}^2(t)\}} \left(u_{i,t} - \hat{u}_{i,t}(a_i) \right). \end{cases}$$

The strategy and the estimated payoff are updated in a coupled fashion. The numerical value of experiment is used in the estimation, and the estimated payoffs are used to built the strategy (here the estimations are crucial since a player does not know the numerical value of the payoff corresponding to the other actions that the player did not use). The strategy update is a B-G based reinforcement learning. Combined together one gets the B-G based CODIPAS-RL. The rest point $\mathcal{L}_2$ can be seen as the equilibrium for a modified game with the perturbed payoff $\mathbb{E}_{s,B^2}\mathbb{U}_i + \epsilon_i H_i$, where H_i is the extra entropy term as discussed in [8].

14.3.2.3 Imitative B-G CODIPAS-RL $\mathcal{L}_3$ Let $\beta^I_{i,\epsilon,t} : \mathcal{X}_i \times \mathbb{R}^{|\mathcal{A}_i|} \to \mathbb{R}^{|\mathcal{A}_i|}$ be the imitative B-G strategy mapping given by

$$\tilde{\beta}^I_{i,\epsilon,t}(\mathbf{x}_{i,t}, \hat{\mathbf{u}}_{i,t})(a_i) = \frac{x_{i,t}(a_i)e^{\frac{1}{\epsilon}\hat{u}_{i,t}(a_i)}}{\sum_{a'_i \in \mathcal{A}_i} x_{i,t}(a'_i)e^{\frac{1}{\epsilon}\hat{u}_{i,t}(a'_i)}}, \quad a_i \in \mathcal{A}_i. \tag{14.5}$$

The learning pattern $\mathcal{L}_3$ is given by

$$\begin{cases} x_{i,t+1}(a_i) &= x_{i,t}(a_i) + \lambda_{i,t} 1\!\!1_{\{i \in \mathcal{B}^2(t)\}} \left(\tilde{\beta}^I_{i,\epsilon,t}(\hat{\mathbf{u}}_{i,t})(a_i) - x_{i,t}(a_i)\right), \\ \hat{u}_{i,t+1}(a_i) &= \hat{u}_{i,t}(a_i) + \nu_{i,t} 1\!\!1_{\{a_{i,t}=a_i, i \in \mathcal{B}^2(t)\}} \left(u_{i,t} - \hat{u}_{i,t}(a_i)\right). \end{cases}$$

The imitative B-G learning weights the B-G strategy with the current strategy vector $\mathbf{x}_{i,t}$ and the strategy mapping $\tilde{\beta}^I_{i,\epsilon,t}$ is time dependent. It allows the learning strategies to be attained at the boundary of the simplex $\mathcal{X}_i$.

14.3.2.4 Weighted Imitative B-G CODIPAS-RL $\mathcal{L}_4$ Let $\tilde{\beta}^W_{i,t} : \mathcal{X}_i \times \mathbb{R} \times \mathbb{R}^{|\mathcal{A}_i|} \to \mathbb{R}^{|\mathcal{A}_i|}$ be the imitative weighted B-G strategy mapping given by

$$\tilde{\beta}^W_{i,t}(\mathbf{x}_{i,t}, \lambda_{i,t}, \hat{\mathbf{u}}_{i,t})(a_i) := \frac{x_{i,t}(a_i)(1+\lambda_{i,t})^{\hat{u}_{i,t}(a_i)}}{\sum_{a'_i \in \mathcal{A}_i} x_{i,t}(a'_i)(1+\lambda_{i,t})^{\hat{u}_{i,t}(a'_i)}}, \tag{14.6}$$

for every $a_i \in \mathcal{A}_i$. The learning pattern $\mathcal{L}_4$ is given by

$$\begin{cases} x_{i,t+1}(a_i) &= x_{i,t}(a_i) + 1\!\!1_{\{i \in \mathcal{B}^2(t)\}} \left(\tilde{\beta}^W_{i,t}(\mathbf{x}_{i,t}, \lambda_{i,t}, \hat{\mathbf{u}}_{i,t})(a_i) - x_{i,t}(a_i)\right), \\ \hat{u}_{i,t+1}(a_i) &= \hat{u}_{i,t}(a_i) + \nu_{i,t} 1\!\!1_{\{a_{i,t}=a_i, i \in \mathcal{B}^2(t)\}} \left(u_{i,t} - \hat{u}_{i,t}(a_i)\right). \end{cases}$$

Note that the exploitation function learning $\tilde{\beta}^W_{i,t}$ is time dependent in $\mathcal{L}_4$ and is independent of parameter ϵ. If the learning yields an interior point as the equilibrium, then it is the exact equilibrium of the expected game, while the equilibrium in $\mathcal{L}_2$ is an approximated one for the ϵ-perturbed game.

14.3.2.5 Weakened Fictitious-Play $\mathcal{L}_5$ Let $\tilde{\beta}^F_{i,t} : \mathbb{R}^{|\mathcal{A}_i|} \to 2^{\mathbb{R}^{|\mathcal{A}_i|}}$ be a point-to-set mapping (correspondence)

$$\tilde{\beta}^F_{i,t}(\hat{\mathbf{u}}_{i,t}) := (1-\epsilon)\delta_{\beta_i(\hat{\mathbf{u}}_{i,t})} + \frac{\epsilon}{|\mathcal{A}_i|}\mathbf{1}, \tag{14.7}$$

where $\mathbf{1} \in \mathbb{R}^{|\mathcal{A}_i|}$ is a vector with all its entries being 1; $\beta_i : \mathbb{R}^{|\mathcal{A}_i|} \to 2^{\mathcal{A}_i}$ is the best response correspondence:

$$\beta_i(\hat{\mathbf{u}}_{i,t}) \in \arg\max_{a'_i \in \mathcal{A}_i} \hat{u}_{i,t}(a'_i) \tag{14.8}$$

and $\delta_{\mathcal{Z}}$, $\mathcal{Z} \subseteq \mathcal{A}_i$, denotes a set of unit vectors $\{e_{a_i}, a_i \in \mathcal{Z}\}$.

The learning pattern $\mathcal{L}_5$ is given by

$$\begin{cases} x_{i,t+1}(a_i) &= x_{i,t}(a_i) + 1\!\!1_{\{i \in \mathcal{B}^2(t)\}} \left(\tilde{\beta}^F_{i,t}(\hat{\mathbf{u}}_{i,t}) - x_{i,t}(a_i)\right), \\ \hat{u}_{i,t+1}(a_i) &= \hat{u}_{i,t}(a_i) + \nu_{i,t} 1\!\!1_{\{a_{i,t}=a_i, i \in \mathcal{B}^2(t)\}} \left(u_{i,t} - \hat{u}_{i,t}(a_i)\right). \end{cases}$$

The weakened fictitious play $\mathcal{L}_5$ has been discussed in [10, 15]. Different from the classical fictitious play, a player does not observe the action played by the other player at the previous step and the payoff function is unknown. Each player estimates its payoff by updating $\hat{\mathbf{u}}_{i,t}$ using perceived payoffs. The strategy update equation is composed of two parts. A player chooses one of his optimal actions with probability $(1-\epsilon)$ by optimizing the up-to-date payoff estimate $\hat{u}_{i,t}$, and plays an arbitrary action with equal probability ϵ.

Remark 14.1 *We note that the average payoff-learning in the five pure learning schemes can be seen as the reinforcement learning of Q-functions in MDPs, which have been introduced in [5, 6]. Since we have considered a stochastic game with state transitions that are independent of the actions of the players, the Q-function in MDPs is reduced to the average-payoff function* $\hat{\mathbf{u}}_{i,t}$.

14.4 MAIN RESULTS

In this section, we introduce the new paradigm of hybrid learning, present the main results on learning in two-person general-sum games, and discuss their convergence properties for some special classes of games.

14.4.1 Stochastic Approximation of the Pure Learning Schemes

The pure learning schemes introduced in Section 14.3 share the same learning structure for average utility but differ in their strategy learning. Denote by $\Pi^{(l)}_{i,t}$ the strategy learning function for $l \in \mathcal{L}$ in the general form (14.2). Following the multiple time-scale stochastic approximation framework developed in [22–25], one can write the pure learning schemes into the form

$$\begin{cases} \mathbf{x}_{i,t+1} - \mathbf{x}_{i,t} &\in q_{i,t}\left(f_i^{(l)}(\mathbf{x}_{i,t}, \hat{\mathbf{u}}_{i,t}) + M^{(l)}_{i,t+1}\right) \\ \hat{\mathbf{u}}_{i,t+1} - \hat{\mathbf{u}}_{i,t} &\in \bar{q}_{i,t}\left(\mathbb{E}_{s,\mathbf{x}_{-i,t},\mathcal{B}^2} U_i - \hat{\mathbf{u}}_{i,t} + \bar{M}_{i,t+1}\right) \end{cases},$$

where $f_i^{(l)} = \mathbb{E}[\Pi^{(l)}_{i,t+1} | \mathcal{F}_t]$, $l \in \mathcal{L}$, is a learning pattern in the form of stochastic approximation. $q_{i,t}$ is a time-scaling factor which is a function of the learning rates $\lambda_{i,t}$ and the probability of Pi in active mode at time t, denoted by $\mathbb{P}(i \in \mathcal{B}^2(t))$; $\bar{q}_{i,t}$ is the time-scaling factor for $\hat{u}_{i,t}$. To use ODE approximation, we check first the conditions given in the appendix. The term $M^{(l)}_{i,t+1}$ is a bounded martingale difference because the strategies are in the product of simplices which are convex and compact, and the conditional expectation of $M_{i,t+1}$ given the sigma-algebra generated by the random variables $s_{t'}, \mathbf{x}_{t'}, u_{t'}, \hat{\mathbf{u}}_{t'},\ \ t' \leq t$, is zero. Similar properties hold for $\bar{M}_{t+1}$. The function f is a regular function, and hence Lipschitz over a compact set, which implies linear growth. Note that the case of constant learning rates can be analyzed under the same setting but the convergence result is weaker. Thus, the asymptotic pseudo-trajectories for the non vanishing time-scale ratio, that is, $\lambda_{i,t}/\nu_{i,t} \rightarrow \gamma_i$ for some $\gamma_i \in \mathbb{R}_{++}$ are

$$\begin{cases} \frac{\mathrm{d}}{\mathrm{d}t}\mathbf{x}_{i,t} & \in & g_{i,t}\left(f_i^{(l)}(\mathbf{x}_{i,t}, \hat{\mathbf{u}}_{i,t})\right) \\ \frac{\mathrm{d}}{\mathrm{d}t}\hat{\mathbf{u}}_{i,t} & = & \bar{g}_{i,t}\left(\mathbb{E}_{s,\mathbf{x}_{-i,t},\mathcal{B}^2}U_i - \hat{\mathbf{u}}_{i,t}\right) \end{cases},$$

where $g_{i,t}$ (resp. $\bar{g}_{i,t}$) are the asymptotic functions of $q_{i,t}, \lambda_t^*, p_i$ (resp. $\bar{q}_{i,t}, , \nu_t^*, p_i$).

If the learning rates have the vanishing ratio, that is, $\frac{\lambda_t}{\mu_t} \rightarrow 0$, the asymptotic pseudo-trajectories are

$$\begin{cases} \frac{\mathrm{d}}{\mathrm{d}t}\mathbf{x}_{i,t} & \in & g_{i,t}\left(f_i^{(l)}(\mathbf{x}_{i,t}, \mathbb{E}_{s,\mathbf{x}_{-i,t}}U_i)\right) \\ \hat{\mathbf{u}}_{i,t} & \longrightarrow & \mathbb{E}_{s,\mathbf{x}_{-i},\mathcal{B}^2}U_i. \end{cases}$$

14.4.2 Stochastic Approximation of the Hybrid Learning Scheme

Players can choose different patterns at different time slots. Consider the hybrid and switching learning

$$\begin{cases} \mathbf{x}_{i,t+1} - \mathbf{x}_{i,t} & \in & q_{i,t}\left(\sum_{l\in\mathcal{L}} \mathbb{1}_{\{l_{i,t}=l\}} f_i^{(l)}(\mathbf{x}_{i,t}, \hat{\mathbf{u}}_{i,t}) + M^{(l)}_{i,t+1}\right) \\ \hat{\mathbf{u}}_{i,t+1} - \hat{\mathbf{u}}_{i,t} & \in & \bar{q}_{i,t}\left(\mathbb{E}_{s,\mathbf{x}_{-i,t}}U_i - \hat{\mathbf{u}}_{i,t} + \bar{M}_{i,t+1}\right) \end{cases},$$

where $l_{i,t} \in \mathcal{L}$ is the learning pattern chosen by Pi at time t.

Theorem 14.1 *Assume that each player Pi, $i \in \mathcal{N}$, adopts one of the CODIPAS-RLs in $\mathcal{L}$ with probability $\omega_i = [\omega_{i,l'}]_{l'\in\mathcal{L}} \in \Delta(\mathcal{L})$ and the learning rates satisfy conditions (C1) and (C2). Then, the asymptotic pseudo-trajectories of the hybrid and switching learning can be written into the form*

$$\begin{cases} \frac{\mathrm{d}}{\mathrm{d}t}\mathbf{x}_{i,t} & \in & g_{i,t}\left(\sum_{l\in\mathcal{L}} \omega_{i,l} f_i^{(l)}(\mathbf{x}_{i,t}, \hat{\mathbf{u}}_{i,t})\right) \\ \frac{\mathrm{d}}{\mathrm{d}t}\hat{\mathbf{u}}_{i,t} & = & \bar{g}_{i,t}\left(\mathbb{E}_{s,\mathbf{x}_{-i,t}}U_i - \hat{\mathbf{u}}_{i,t}\right) \end{cases}$$

for the non vanishing time-scale learning ratio $\lambda_{i,t}/\nu_{i,t}$*; and,*

$$\begin{cases} \frac{d}{dt}\mathbf{x}_{i,t} & \in & g_{i,t}\left(\sum_{l\in\mathcal{L}}\omega_{i,l}f_i^{(l)}(\mathbf{x}_{i,t},\mathbb{E}_{s,\mathbf{x}_{-i,t},\mathcal{B}^2}U_i)\right) \\ \hat{\mathbf{u}}_{i,t} & \longrightarrow & \mathbb{E}_{s,\mathbf{x}_{-i},\mathcal{B}^2}U_i \end{cases}$$

for the vanishing learning ratio $\lambda_{i,t}/\nu_{i,t}$.

Proof: We first examine the strategy learning given by

$$\mathbf{x}_{i,t+1}-\mathbf{x}_{i,t}\in \mathbb{1}_{\{i\in\mathcal{B}^2(t)\}}\lambda_{i,t}\left(\sum_{l\in\mathcal{L}}\mathbb{1}_{\{l_{i,t}=l\}}f_i^{(l)}(\mathbf{x}_t)+M_{i,t+1}^{(l)}\right).$$

By taking λ_t as the reference learning rate, the drift (expected change in one step) can be computed via

$$\lim_{\lambda_{i,t}\longrightarrow 0}\mathbb{E}\left(\frac{\mathbf{x}_{i,t+1}-\mathbf{x}_{i,t}}{\lambda_{i,t}}\mid\mathcal{F}_t\right)=\mathbb{P}(i\in\mathcal{B}^2(t))\left(\sum_{l\in\mathcal{L}}\omega_{i,l}f_i^{(l)}(\mathbf{x}_t)\right),$$

where we used the fact that $\mathbb{E}\left(M_{i,t+1}^{(l)}\mid\mathcal{F}_t\right)=0$. The drift has the form

$$g_{i,t}\sum_{l\in\mathcal{L}}\omega_{i,l}f_i^{(l)}(\mathbf{x}_t).$$

We check that the assumptions A1-A4 given in the appendix are all met. The asymptotic pseudo-trajectory reduces to

$$\frac{d}{dt}\mathbf{x}_{i,t}=g_{i,t}\sum_{l\in\mathcal{L}}\omega_{i,l}f_i^{(l)}(\mathbf{x}_t).$$

For two time-scales CODIPAS-RL, we use the same lines as in [16, 22]. ■

In Table 14.2, we give the asymptotic pseudo-trajectory of the pure learning when the rate of payoff learning is faster than that of strategy learning.

TABLE 14.2 Asymptotic Pseudo-Trajectories of Pure Learning

Learning Patterns	Class of ODE
$\mathcal{L}_1$	Adjusted replicator dynamics
$\mathcal{L}_2$	Smooth best response dynamics
$\mathcal{L}_3$	Imitative B-G dynamics
$\mathcal{L}_4$	Time-scaled replicator dynamics
$\mathcal{L}_5$	Perturbed best response dynamics

Let $\overline{\mathbb{U}}_j(\mathbf{x}) := \mathbb{E}_{s,B^2}\mathbb{U}_j(s, B^2, \mathbf{x}),\ j \in \mathcal{N}$. In Table 14.2, the replicator dynamics are given by

$$\dot{x}_j(a_j) = q_j x_j(a_j)\left[\overline{\mathbb{U}}_j(e_{a_j}, \mathbf{x}_{-j}) - \sum_{a'_j \in \mathcal{A}_j} \overline{\mathbb{U}}_j(e_{a'_j}, \mathbf{x}_{-j}) x_j(a'_j)\right].$$

The smooth best response dynamics are given by

$$\dot{x}_j(a_j) = q_j \left(\frac{e^{\frac{1}{\epsilon}\overline{\mathbb{U}}_j(e_{a_j}, \mathbf{x}_{-j})}}{\sum_{a'_j} e^{\frac{1}{\epsilon}\overline{\mathbb{U}}_j(e_{a'_j}, \mathbf{x}_{-j})}} - x_j(a_j) \right).$$

The best response dynamics are given by

$$\dot{\mathbf{x}}_j \in q_j(\beta_j(\mathbf{x}_{-j}) - \mathbf{x}_j),$$

and the payoff dynamics are

$$\frac{\mathrm{d}}{\mathrm{d}t}\hat{u}_j(a_j) = \bar{q}_j x_j(a_j)(\overline{\mathbb{U}}_j(e_{a_j}, \mathbf{x}_{-j}) - \hat{u}_j(a_j)).$$

The imitative Boltzman–Gibbs dynamics are given by

$$\dot{x}_j(a_j) = q_j \left(\frac{x_j(a_j) e^{\frac{1}{\epsilon}\overline{\mathbb{U}}_j(e_{a_j}, \mathbf{x}_{-j})}}{\sum_{a'_j} x_j(a'_j) e^{\frac{1}{\epsilon}\overline{\mathbb{U}}_j(e_{a'_j}, \mathbf{x}_{-j})}} - x_j(a_j) \right).$$

14.4.3 Connection with Equilibria of the Expected Game

We study the convergence properties of the dynamics and their connection with the state-independent Nash equilibrium for three special classes of games.

14.4.3.1 Games with Two Actions For two-player games with two actions, that is, $\mathcal{A}_1 = \{a_1^1, a_1^2\}$, $\mathcal{A}_2 = \{a_2^1, a_2^2\}$, one can transform the system of ODEs of the strategy-learning into a planar system under the form

$$\dot{\alpha}_1 = Q_1(\alpha_1, \alpha_2),\ \dot{\alpha}_2 = Q_2(\alpha_1, \alpha_2), \tag{14.9}$$

where we let $\alpha_i = x_i(a_i^1)$. The dynamics for Pi can be expressed in terms of α_1, α_2 only as $x_1(a_1^2) = 1 - x_1(a_1^2)$, and $x_2(a_2^2) = 1 - x_2(a_2^2)$.

We use the Poincaré–Bendixson theorem and the Dulac criterion [26] to establish a convergence result for (14.9).

Theorem 14.2 ([26]) *For an autonomous planar vector field as in Equation (14.9), the Dulac's criterion states as follows: Let $\gamma(\cdot)$ be a scalar function defined on the unit square $[0, 1]^2$. If $\frac{\partial[\gamma(\alpha))\dot{\alpha}_1]}{\partial\alpha_1} + \frac{\partial[\gamma(\alpha)\dot{\alpha}_2]}{\partial\alpha_2}$ is not identically zero and does not change sign in $[0, 1]^2$, then there are no cycles lying entirely in $[0, 1]^2$.*

Corollary 14.1 *Consider a two-player two-action game. Assume that each of the players adopts the Boltzmann–Gibbs CODIPAS-RL with $\frac{\lambda_{i,t}}{\nu_{i,t}} = \frac{\lambda_t}{\nu_t} \longrightarrow 0$. Then, the asymptotic pseudo-trajectory reduces to a planar system in the form*

$$\dot{\alpha}_1 = \beta_{1,\epsilon}(u_1(e_{a_1}, \alpha_2)) - \alpha_1; \quad \dot{\alpha}_2 = \beta_{2,\epsilon}(u_2(\alpha_1, e_{a_2})) - \alpha_2.$$

Moreover, the system satisfies the Dulac's criterion.

Proof: We apply Theorem 14.2 with $\gamma(\cdot) \equiv 1$ and find the divergence to be equal to -2, which is strictly negative. Hence, the result follows. ∎

Note that for the replicator dynamics, the Dulac criterion reduces to

$$(1 - 2\alpha_1)(\overline{\mathbb{U}}_1(e_{a_1^1}, \alpha_2) - \overline{\mathbb{U}}_1(e_{a_1^2}, \alpha_2)) + (1 - 2\alpha_2)(\overline{\mathbb{U}}_2(\alpha_1, e_{a_2^1}) - \overline{\mathbb{U}}_2(\alpha_1, e_{a_2^2})),$$

which vanishes for $(\alpha_1, \alpha_2) = (1/2, 1/2)$. It is possible to have limit cycles in replicator dynamics and hence the Dulac criterion does not apply. However, the stability of the replicator dynamics can be directly studied in the two-action case by identifying the game as one of four types: coordination, anticoordination, prisoner's dilemma, and hawk-and-dove [27, 28].

The following corollary now follows from Corollary 14.1.

Corollary 14.2 *Consider a two-player two-action game.*

(CR1) Heterogeneous learning: If P1 is with Boltzmann–Gibbs CODIPAS-RL and P2's learning leads to replicator dynamics, then the convergence condition reduces to $(1 - 2\alpha_2)(u_2(\alpha_1, e_{a_2^1}) - u_2(\alpha_1, e_{a_2^2})) < 1$ for all (α_1, α_2).

(CR2) Hybrid learning: If the players use hybrid learning obtained by combining Boltzmann–Gibbs CODIPAS-RL with weight $\omega_{i,1}$ and the replicator dynamics with weight $1 - \omega_{i,1}$ then the Dulac criterion reduces to

$$\begin{array}{l} \omega_{1,2}[(1 - 2\alpha_1)(u_1(e_{a_1^1}, \alpha_2) - u_1(e_{a_1^2}, \alpha_2))] \\ + \omega_{2,2}[(1 - 2\alpha_2)(u_2(\alpha_1, e_{a_2^1}) - u_2(\alpha_1, e_{a_2^2}))] \end{array} < w_{1,1} + w_{2,2}$$

for all (α_1, α_2).

Remark 14.2 (Symmetric games with three actions) *If the expected game is a symmetric game with three actions per player, then, the symmetric game dynamics* *reduce to the two-dimensional dynamical system. This allows us to apply the Dulac criterion.*

14.4.3.2 Lyapunov Games We say that a game Ξ is a *Lyapunov game* under a given hybrid dynamics if the resulting dynamics has an associated Lyapunov function $V(\mathbf{x}) : \Delta \subseteq \mathbb{R}^{\sum_i |\mathcal{A}_i|} \rightarrow \mathbb{R}_+$. Note that a Lyapunov function $V(\mathbf{x})$ is positive definite on $\mathbb{R}^{\sum_i |\mathcal{A}_i|}$ for every $\mathbf{x} \neq \mathbf{x}^* \in \Delta$, and its time derivative is negative, $\frac{dV}{dt} < 0$, for all $\mathbf{x} \neq \mathbf{x}^*$, where x^* is a stationary point of the dynamics [29]. The Lyapunov function can also defined to be negative definite as in [30]; in this case, the time derivative will need to be positive.

Theorem 14.3 *Consider a Lyapunov game under the learning schemes $\mathcal{L}_1$ and $\mathcal{L}_4$. Then, the learning procedure has convergence to the set of equilibria of the expected robust game for all interior initial conditions.*

Proof: Lyapunov function V provides the stability of the set of rest points. Since the dynamics are positively correlated for adjusted replicator dynamics [27, 28], the state-independent equilibria are rest points of the dynamics obtained from $\mathcal{L}_1$ and $\mathcal{L}_4$. The stability of any convex combination of these dynamics follows. ■

Note that Theorem 14.3 states that starting from interior initial points, the hybrid dynamics lead to one of the equilibria, which we do not know which one in advance. The set can have either a finite number or a continuum of equilibria. This result holds also for n-player stochastic games with random updates.

14.4.3.3 Potential Games We say that the stochastic game Ξ is an *expected robust potential game* if the expected payoff derives from a potential function. Potential games are a special class of games where the payoff functions of the players are governed by a potential function $\Phi : \mathbb{R}^{\sum_{i \in \mathcal{N}} |\mathcal{A}_i|} \rightarrow \mathbb{R}$, i.e., $\mathbb{U}_i(e_{a_i}, \mathbf{x}_{-i}) = \frac{\partial \Phi(\mathbf{x})}{\partial x_i(a_i)}$, $i \in \mathcal{N}$, $a_i \in \mathcal{A}_i$. We use a Lyapunov approach to show the global convergence of hybrid learning for potential games.

Lemma 14.2 *Assume that the stochastic NZSG Φ has a potential function Φ. Then, there exists a Lyapunov function $V^R(\mathbf{x}_1, \mathbf{x}_2) : \mathbb{R}^{|\mathcal{A}_1|+|\mathcal{A}_2|} \rightarrow \mathbb{R}$ for learning schemes $\mathcal{L}_1$, $\mathcal{L}_4$-associated replicator dynamics and it is given by its potential $V^R = \Phi$. Hence, the replicator dynamics converge to a rest point. In addition, if starting from an interior point of the simplex, the dynamics converge to the Nash equilibrium of the game Ξ.*

Proof: Since the payoff matrix is bounded, we can study its strategically equivalent game [31, 32] by subtracting a certain offset from every matrix entries so that $U_i(\mathbf{a})$ is negative for every strategy pair $\mathbf{a}$, and hence $\mathbb{E}_{s,B^2}\mathbb{U}_i(e_{a_j}, x_{-i,t})$ is negative. Without loss of generality, we can assume the game payoff matrix or its strategically equivalent game payoff matrix is negative entry-wise. Therefore, $V^R = \Phi$ is negative. We take the time derivative of the Lyapunov function V^R as follows:

$$\frac{\mathrm{d}}{\mathrm{d}t}V^R(\mathbf{x}_{1,t},\mathbf{x}_{2,t}) = \sum_{i\in\mathcal{N}}\sum_{a_j\in\mathcal{A}_i}\left(\frac{\mathrm{d}x_{i,t}(a_j)}{\mathrm{d}t}\right)\left(\frac{\partial V^R}{\partial x_{i,t}(a_j)}\right),$$

which leads to the following set of inequalities

$$\begin{aligned}\frac{\mathrm{d}}{\mathrm{d}t}V^R(\mathbf{x}_{1,t},\mathbf{x}_{2,t}) &= \sum_{i\in\mathcal{N}} g_{i,t}\left\{\sum_{a_j\in\mathcal{A}_i} x_{i,t}(a_j)\left(\mathbb{E}_{s,B^2}\mathbb{U}_i(e_{a_j},\mathbf{x}_{-i,t})\right)^2\right.\\ &\quad\left.-\left(\sum_{a_j\in\mathcal{A}_i} x_{i,t}(a_j)\mathbb{E}_{s,B^2}\mathbb{U}_i(e_{a_j},\mathbf{x}_{-i,t})\right)^2\right\}\\ &\geq \sum_{i\in\mathcal{N}} g_{i,t}\left\{\sum_{a_j\in\mathcal{A}_i} x_{i,t}(a_j)\left(\mathbb{E}_{s,B^2}\mathbb{U}_i(e_{a_j},\mathbf{x}_{-i,t})\right)^2\right.\\ &\quad\left.-\sum_{a_j\in\mathcal{A}_i} x_{i,t}^2(a_j)\left(\mathbb{E}_{s,B^2}\mathbb{U}_i(e_{a_j},\mathbf{x}_{-i,t})\right)^2\right\}\geq 0.\end{aligned}$$

The last two inequalities result from Jensen's inequality and the positivity and the range of $\mathbf{x}_{i,t}$. We have $\frac{\mathrm{d}V^R}{\mathrm{d}t}\geq 0$ with equality only at the equilibrium. Hence, convergence to equilibria holds for all initial conditions in the interior of the simplex. ■

Lemma 14.3 *Let $V^B(\mathbf{x}_1,\mathbf{x}_2):\mathbb{R}^{|\mathcal{A}_1|+|\mathcal{A}_2|}\rightarrow\mathbb{R}$ be a Lyapunov function for learning pattern $\mathcal{L}_l$-associated replicator dynamics f^l, $l=2$, such that*

$$V^B(\mathbf{x}_1,\mathbf{x}_2)=\Phi(\mathbf{x}_1,\mathbf{x}_2)+\epsilon_1 H_1(\mathbf{x}_1)+\epsilon_2 H_2(\mathbf{x}_2),$$

where $H_i:\mathbb{R}^{|\mathcal{A}_i|}\rightarrow\mathbb{R}_+$ are strictly concave perturbation functions which can take different forms depending on the pure learning scheme l. The ODEs corresponding to the learning schemes converge to a set of perturbed equilibria of the game Ξ.

Proof: Using the same argument as in the proof of Lemma 14.2, we can assume Φ or its strategic equivalent form is positive without loss of generality, and hence V^B is non-negative. The Lyapunov function V^B has its critical points given by $\nabla_{\mathbf{x}_1}V^B=\nabla_{\mathbf{x}_2}V^B=0$, that is,

$$\nabla_{\mathbf{x}_i}\Phi+\epsilon_i\nabla_{\mathbf{x}_i}H_i=0,\, i=1,2. \tag{14.10}$$

The first-order condition (14.10) yields the perturbed equilibria of the B-G type of learning schemes. By taking the time derivative of V^B, we arrive at

$$\frac{\mathrm{d}V^B}{\mathrm{d}t} = \sum_{i\in\mathcal{N}} \sum_{a_j\in\mathcal{A}_i} \frac{\partial x_i(a_j)}{\partial t}\frac{\partial \Phi}{\partial x_i(a_j)} + \epsilon_i \frac{\partial x_i(a_j)}{\partial t}\frac{\partial H_i}{\partial x_i(a_j)}.$$

Denote the perturbed payoff function by $\widetilde{\mathbb{U}}_i(\mathbf{x}_1, \mathbf{x}_2) := \mathbb{E}_{s,B^2}\mathbb{U}_i(s, B^2, \mathbf{x}_1, \mathbf{x}_2) + \epsilon_i H_i(\mathbf{x}_i)$. The first-order condition for a maximum satisfies, for every $a_j \in \mathcal{A}_i$,

$$\mathbb{E}_{s,B^2}\mathbb{U}_i(s, B^2, e_{a_j}, \mathbf{x}_{-i}) + \epsilon_i \frac{\partial H_i(\bar{\beta}_i(\mathbf{x}_{-i}))}{\partial x_i(a_j)} = 0, \tag{14.11}$$

where $\bar{\beta}_i(\cdot)$ is a type of B-G strategy that corresponds to the learning type. Since the game is assumed to be a potential game, we have

$$\frac{\partial \Phi}{\partial x_i(a_j)} = -\epsilon_i \frac{\partial H_i(\bar{\beta}_i(\mathbf{x}_{-i}))}{\partial x_i(a_j)}. \tag{14.12}$$

Hence, we obtain from (14.11) and (14.12),

$$\frac{\mathrm{d}V^B}{\mathrm{d}t} = \sum_{i\in\mathcal{N}} \sum_{a_j\in\mathcal{A}_i} \epsilon_i g_{i,t} \left(\frac{\partial H_i(\bar{\beta}(\mathbf{x}_{-i}))(a_j)}{\partial x_i(a_j)} - \frac{\partial H_i(x_i(a_j))}{\partial x_i(a_j)} \right)$$
$$\cdot \left(\bar{\beta}_i\left(\mathbf{x}_{-i}\right)(a_j) - x_i(a_j)\right). \tag{14.13}$$

Due to the strict concavity of the perturbation functions H_i, we conclude that $\frac{\mathrm{d}V^B}{\mathrm{d}t} \leq 0$, with equality only at the equilibrium. Hence, the pure learning dynamics converge to the set of perturbed equilibria. ■

Theorem 14.4 *Assume that the stochastic NZSG Ξ has a potential function Φ. The hybrid learning with $\mathcal{L}_1$ and $\mathcal{L}_2$ converges locally to a perturbed state-independent Nash equilibrium $\mathbf{x}_1^*, \mathbf{x}_2^*$ of the potential game Ξ for sufficiently small ϵ_i.*

Proof: The Lyapunov functions for replicator and G-B dynamics share the same term Φ. For hybrid learning between these two dynamics, we can pick Φ as a Lyapunov function. For small ϵ_i close to zero, the Lyapunov function Φ for pure learning $\mathcal{L}_2$ yields a strictly positive time derivative for non equilibrium points due to continuity. Let $\mathbf{x}_\epsilon$ be a maximizer of $V^R(\mathbf{x}) = \Phi(\mathbf{x}) + \sum_i \epsilon_i H_i(\mathbf{x}_i)$. Then, there exists $\epsilon_h > 0$ such that $\tilde{V}(\mathbf{x}) = \Phi(\mathbf{x}) + \sum_i \epsilon_i' H_i(\mathbf{x}_i)$ is strictly positive in a neighborhood of the

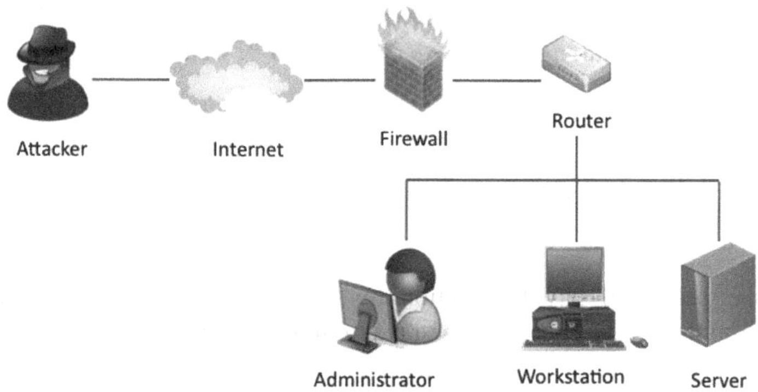

FIGURE 14.1 An illustration of the network security game scenario where an attacker attempts to breach the network security by compromising the servers and workstations, whereas the network administrator monitors the network activity to prevent possible intrusions.

considered nonequilibrium point with $\epsilon_i' = \min(\epsilon_h, \frac{\epsilon_i}{M_i})$, where M_i is the maximum of H_i over $\mathcal{X}_i$. Since the maximizer of $\tilde{V}$ is an ϵ'-equilibrium where $\epsilon' = \max_i \epsilon_i'$, there exists a subsequence of $\mathbf{x}_\epsilon$ converging to $\mathbf{x}^*$ which is an equilibrium and $\mathbf{x}^*$ maximizes Φ and this holds for any convergent subsequence. This means the time derivative of $\tilde{V}$ is strictly positive in all the neighborhood of $\mathbf{x}^*$ and vanishes only at $\mathbf{x}^*$. Thus, when $\epsilon' = \max \epsilon_i'$ goes to zero, one gets an equilibrium. Hence, in view of Lemma 14.3, we can conclude that for sufficiently small $\epsilon_i > 0$, we have local convergence of the hybrid learning. ■

Note that the equilibrium $\mathbf{x}_1^*, \mathbf{x}_2^*$ in Theorem 14.4 may not be unique, which depends on the rest point of the nonlinear hybrid dynamics.

14.5 SECURITY APPLICATION

In this section, we use the learning algorithm to study a two-person security game in a network between an intruder and an administrator. In Figure 14.1, we show a local network connected to the Internet where an attacker attempts to launch an internal denial-of-service attack to bring down a network server capture important data from a workstation. Let P1 and P2 denote the administrator and the intruder, respectively. An administrator P1 can use different levels of protection. The intruder P2 can launch an attack that can be of high or low intensity. Let the action sets for P1 and P2 be $\mathcal{A}_1 := \{H, L\}$ and $\mathcal{A}_2 := \{S, W\}$, respectively. The network administrator is assumed to be always on alert while the intruder attacks with a probability p. Hence, the set $\mathcal{B}^2(t)$ can be of two types, that is, (C1) {P1, P2} or (C2) {P1}. The former case (C1) corresponds to the scenario where the intruder and the administrator attack and defend, respectively, whereas the latter (C2) suggests that the administrator faces no threat.

We represent the payoff under these two scenarios by $\boldsymbol{M}_1$ and $\boldsymbol{M}_2$, respectively:

$$\boldsymbol{M}_1 := \begin{bmatrix} & S & W \\ H & 1,-1 & 1,0 \\ L & -2,1 & 2,0 \end{bmatrix}, \quad \boldsymbol{M}_2 := \begin{bmatrix} H & 1 \\ L & 2 \end{bmatrix}. \tag{14.14}$$

In (C1), a successful defense against attack yields a payoff of 2 for P1 whereas a failure results in a payoff of −2. A successful attack yields P2 a payoff of 1 whereas a failed attack yields a zero payoff. The employment of strong defense (H) or strong attack (S) costs an extra unit of effort as compared to the low defense (L) and the weak attack (W) for P1 and P2, respectively. In (C2), P1 stays secure without the threat from the intruder, and hence yields a payoff of 2. However, the high security level costs an extra unit of energy from the player.

The payoffs in $\boldsymbol{M}_1$ and $\boldsymbol{M}_2$ are subject to exogenous noise which varies in different environmental states s. The state-independent equilibrium of the game is found to be at $\mathbf{x}_1^* = [\frac{1}{2}, \frac{1}{2}]^{\mathrm{T}}, \mathbf{x}_2^* = [\frac{1}{3}, \frac{2}{3}]^{\mathrm{T}}$ and the optimal average payoffs are $\hat{\mathbf{u}}_1^* = [\frac{2}{3}, \frac{2}{3}]^{\mathrm{T}}, \hat{\mathbf{u}}_2^* = [0, 0]^{\mathrm{T}}$. In Figures 14.2 and 14.3, we show the payoffs and mixed strategies of both players when both players use the learning pattern $\mathcal{L}_1$. We can see that the replicator dynamics from $\mathcal{L}_1$ do not converge. However, the time average strategies $\lim_{T\to\infty} \frac{1}{T}\int_0^{\mathrm{T}} \mathbf{x}_{i,t} \mathrm{d}t$ converge to $\mathbf{x}_1^*, \mathbf{x}_2^*$, respectively, and, the time average payoffs $\lim_{T\to\infty} \frac{1}{T}\int_0^{\mathrm{T}} \hat{\mathbf{u}}_{i,t} \mathrm{d}t$ converge to $\hat{\mathbf{u}}_1^*, \hat{\mathbf{u}}_2^*$, respectively.

In Figures 14.4 and 14.5, we show the payoffs and mixed strategies of the players when they both adopt the learning pattern $\mathcal{L}_2$. We choose $\epsilon = 1/50$ and observe that the mixed strategies converge to $\bar{\mathbf{x}}_1 = [0.5277, 0.4723]^{\mathrm{T}}, \bar{\mathbf{x}}_1 = [0.3333, 0.6667]^{\mathrm{T}}$ and the payoffs converge to $\bar{\hat{\mathbf{u}}}_1 = [0.6667, 0.6667]^{\mathrm{T}}, \bar{\hat{\mathbf{u}}}_2 = [-0.027, 0]^{\mathrm{T}}$, which are in the close neighborhood of $\hat{\mathbf{u}}_1^*, \hat{\mathbf{u}}_2^*$.

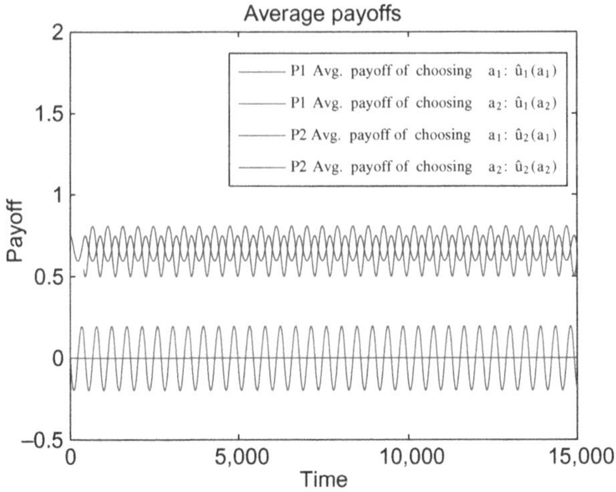

FIGURE 14.2 The payoffs to the players with both players using $\mathcal{L}_1$.

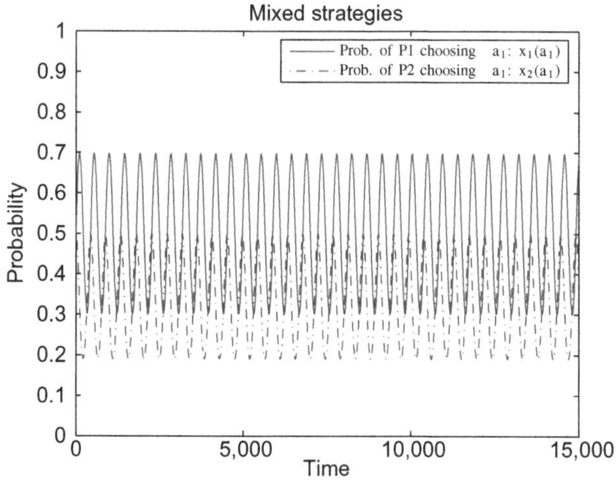

FIGURE 14.3 The mixed strategies of the players with both players using $\mathcal{L}_1$.

In Figures 14.6 and 14.7, we show the convergence of the heterogenous learning scheme where P1 uses $\mathcal{L}_1$ and P2 uses $\mathcal{L}_2$. With $\epsilon = 1/50$, we find the converging strategies at $\bar{\mathbf{x}}_1$, $\bar{\mathbf{x}}_1$ and the payoffs at $\bar{\hat{\mathbf{u}}}_1$, $\bar{\hat{\mathbf{u}}}_2$. We can see that the adoption of $\mathcal{L}_2$ by P2 in the heterogenous learning facilitates the convergence of the algorithm even though the learning exhibits high magnitude of oscillations at the beginning, which is mainly due to $\mathcal{L}_1$ learning pattern adopted by P1.

In Figures 14.8 and 14.9, we show the convergence of the hybrid learning scheme where P1 and P2 adopt $\mathcal{L}_1$ and $\mathcal{L}_2$ with equal weights. The strategies converge to

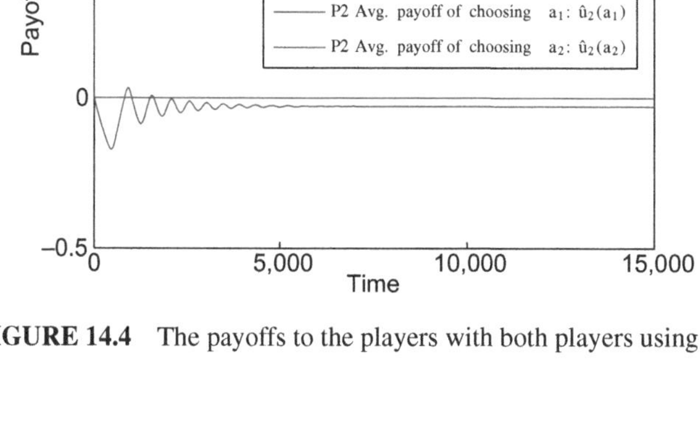

FIGURE 14.4 The payoffs to the players with both players using $\mathcal{L}_2$.

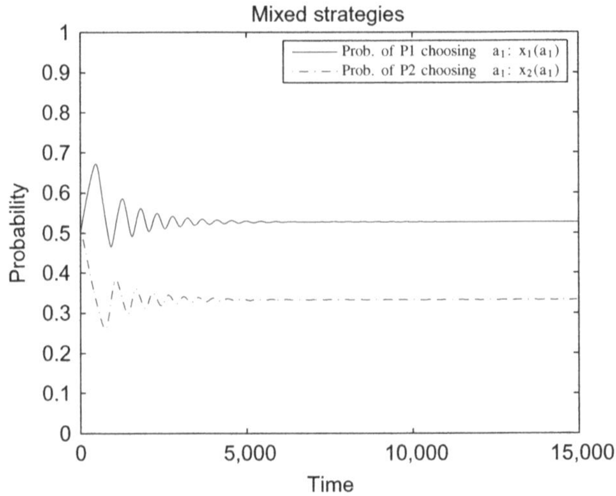

FIGURE 14.5 The mixed strategies of the players with both players using $\mathcal{L}_2$.

$[0.5145, 0.4855]^T$, $[0.3334, 0.6666]^T$ for P1 and P2, respectively, whereas the payoffs converge to $[0.6666, 0.6666]^T$, $[-0.01459, 0]^T$ for P1 and P2, respectively. We can see that the hybrid mixture of $\mathcal{L}_1$ and $\mathcal{L}_2$ learning patterns leads to convergence with smaller magnitude of oscillations in comparison to the ones shown in Figures 14.6 and 14.7.

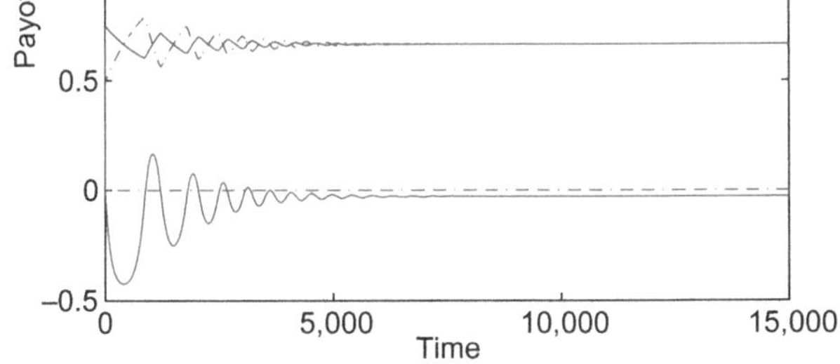

FIGURE 14.6 The payoffs to the heterogeneous players with P1 using $\mathcal{L}_1$ and P2 using $\mathcal{L}_2$.

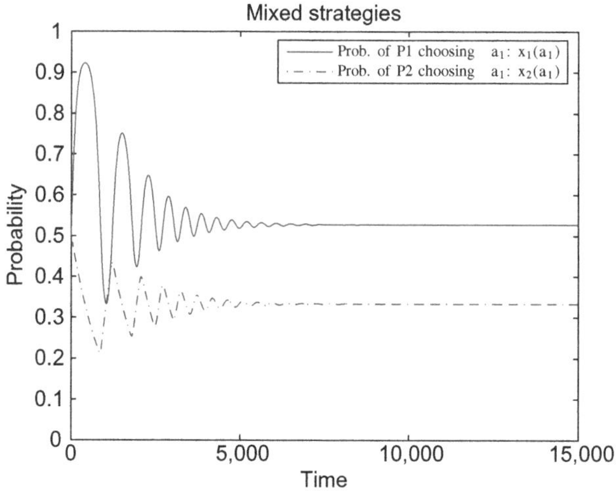

FIGURE 14.7 The mixed strategies of the heterogeneous players with P1 using $\mathcal{L}_1$ and P2 using $\mathcal{L}_2$.

14.6 CONCLUSIONS AND FUTURE WORKS

We have presented distributed hybrid strategic learning algorithms for a class of two-person nonzero-sum stochastic games along with their general convergence and non-convergence properties. The players are assumed to have information limitations in their knowledge not only of other players' payoff functions and strategies but also of their own. In addition, the interactions among the players occur at random times according to their modes. We have applied the framework to security games where the

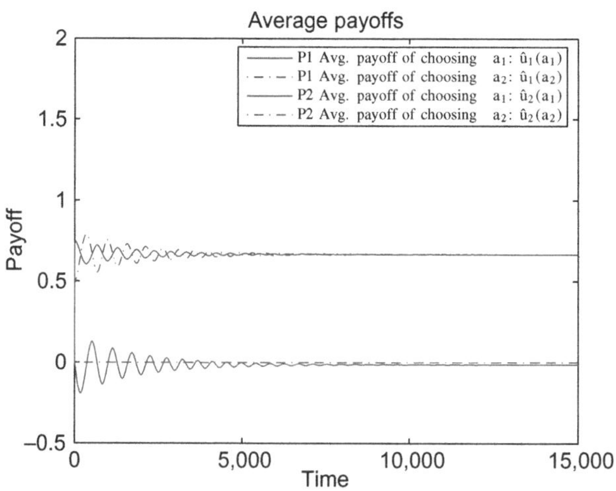

FIGURE 14.8 The payoffs to the players with both players using hybrid learning scheme with equal weights on $\mathcal{L}_1$ and $\mathcal{L}_2$.

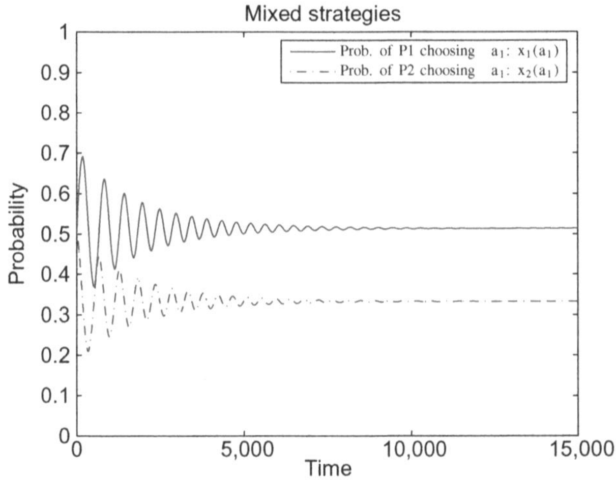

FIGURE 14.9 The mixed strategies of the players with both players using hybrid learning scheme with equal weights on $\mathcal{L}_1$ and $\mathcal{L}_2$.

noncooperative behaviors between an attacker and a defender are well characterized by the features of distributed hybrid learning. Interesting work that we leave for the future is to extend this learning framework to the case of an arbitrary (but still fixed) number of players, each of them adopting hybrid learning with a diffusion term leading to *stochastic differential equations*. It is also of our interest to capture the evolution of the players' rationality through a time-varying ordinary differential equation of the learning weights, which should be analyzed together with the hybrid learning dynamics.

APPENDIX: ASSUMPTIONS FOR STOCHASTIC APPROXIMATION

Consider the difference equation $\mathbf{x}_{t+1} = \mathbf{x}_t + \lambda_t(f(\mathbf{x}_t) + M_{t+1})$ in $\mathbb{R}^{\sum_i |\mathcal{A}_i|}$ and assume that

- **(A1)** f is Lipschitz.
- **(A2)** $\lambda_t \geq 0$, $\sum_{t\geq 0} \lambda_t = +\infty$, $\sum_{t\geq 0} \lambda_t^2 < \infty$.
- **(A3)** M_{t+1} is a martingale difference sequence with respect to the increasing family of sigma-fields $\mathcal{F}_t = \sigma(\mathbf{x}_{t'}, \hat{\mathbf{u}}_{t'}, M_{t'},\ t' \leq t)$, that is, $\mathbb{E}\left(M_{t+1} \mid \mathcal{F}_t\right) = 0$.
- **(A4)** M_t is square integrable and there is a constant $c > 0$ such that

$$\mathbb{E}\left(\| \mathbb{M}_{t+1} \|^2 \mid \mathcal{F}_t\right) \leq c(1+ \| \mathbf{x}_t \|^2)$$

  almost surely, for all $t \geq 0$.
- **(A5)** $\sup_t \| \mathbf{x}_t \| < \infty$ almost surely.

Then, the asymptotic pseudo-trajectory of the difference equation is given by the ordinary differential equation (ODE) [22, 25], $\dot{\mathbf{x}}_t = f(\mathbf{x}_t)$, with $\mathbf{x}_0$ fixed.

ACKNOWLEDGMENTS

This material is based on the work supported in part by the U.S. Air Force Office of Scientific Research (AFOSR) under grant number AFOSR MURI FA9550-09-1-0249.

REFERENCES

1. E. Altman, T. Boulogne, R. El-Azouzi, T. Jiménez, and L. Wynter. A survey on networking games in telecommunications. *Computers & Operations Research*, 33(2):286–311, 2006.
2. T. Alpcan and T. Başar. *Network Security: A Decision and Game Theoretic Approach*. Cambridge University Press, 2011.
3. M.H. Manshaei, Q. Zhu, T. Alpcan, T. Başar, and J.-P. Hubaux. Game theory meets network security and privacy. *ACM Computing Survey*, 45(3), September 2013 (to appear).
4. R.W. Ferrero, S.M. Shahidehpour, and V.C. Ramesh. Transaction analysis in deregulated power systems using game theory. *Power Systems, IEEE Transactions on*, 12(3):1340–1347, 1997.
5. C.J.C.H. Watkins. Learning from delayed rewards. Ph.D. thesis, 1989.
6. R.S. Sutton and A.G. Barto. *Reinforcement Learning: An Introduction*. MIT Press, 1998.
7. Q. Zhu, H. Tembine, and T. Başar. Distributed strategic learning with application network security. In *Proceedings of American Control Conference (ACC)*, pp. 4057–4062, 2011.
8. J.S. Shamma and G. Arslan. Dynamic fictitious play, dynamic gradient play, and distributed convergence to Nash equilibria. *IEEE Trans Automatic Control*, 50(3):312–327, 2005.
9. D. Fudenberg and D. Levine. *Learning in Games*. MIT Press, Cambridge, MA, 1998.
10. J.R. Marden, G. Arslan, and J.S. Shamma. Joint strategy fictitious play with inertia for potential games. In *Proceedings of 44th IEEE Conference Decision Control*, pp. 6692–6697, Dec. 2005.
11. H.P. Young. Learning by trial and error. *Games and Economic Behavior*, 65(2):626–643, 2009.
12. M.A.L. Thathachar, P.S. Sastry, and V.V. Phansalkar. Decentralized learning of Nash equilibria in multiperson stochastic games with incomplete information. *IEEE Transactions on System, Man, and Cybernetics* 24(5):769–777, 1994.
13. Y. Xing and R. Chandramouli. Stochastic learning solution for distributed discrete power control game in wireless data networks. *IEEE/ACM Transactions on Networking*, 16(4):932–944, 2008.
14. P.D. Taylor and L.B. Jonker. Evolutionarily stable strategies and game dynamics. *Mathematical Bioscience*, 40:145–156, 1978.
15. D. Leslie and E. Collins. Individual Q-learning in normal form games. *SIAM Journal on Controlled Optimization*, 44:495–514, 2005.

16. Q. Zhu, H. Tembine, and T. Başar. Heterogeneous learning in zero-sum stochastic games with incomplete information. In *49th IEEE Conference on Decision and Control, Atlanta, GA, USA*, pp. 219–224, 2010.
17. Q. Zhu and T. Başar. Dynamic policy-based IDS configuration. In *Proceedings of the 48th IEEE Conference on Decision and Control (CDC/CCC) 2009*, pp. 8600 –8605, Dec. 2009.
18. J. Nash. Equilibrium points in n-person games. *Proceedings of the National Academy of Sciences*, 36(1):48–49, 1950.
19. R. Bush and F. Mosteller. *Stochastic Models of Learning*. John Wiley and Sons, 1955.
20. T. Borgers and R. Sarin. Learning through reinforcement and replicator dynamics. *University College London Mimeo*, 1993.
21. W.B. Arthur. On designing economic agents that behave like human agents. *Journal of Evolutionary Economic*, 3:1–22, 1993.
22. V.S. Borkar. *Stochastic Approximation: a Dynamical Systems Viewpoint*. Cambridge University Press, 2008.
23. M. Benaïm and M. Faure. Stochastic approximations, cooperative dynamics and supermodular games. Available at http://members.unine.ch/michel.benaim/perso/papers1.html, 2010.
24. D.S. Leslie and E.J. Collins. Convergent multiple timescales reinforcement learning algorithms in normal form games. *The Annals of Applied Probability*, 13(4):1231–1251, 2003.
25. H.J. Kushner and D.S. Clark. *Stochastic Approximation Methods for Constrained and Unconstrained Systems*. Springer, New York, 1978.
26. J. Guckenheimer and P. Holmes. *Nonlinear Oscillations, Dynamical Systems, and Bifurcations of Vector Fields*. Springer-Verlag, New York, 1983.
27. J. Weibull. *Evolutionary Game Theory*. MIT Press, 1995.
28. W.H. Sandholm. *Population Games and Evolutionary Dynamics*. MIT Press, 2010.
29. H. K. Khalil. *Nonlinear Systems*, 3rd edition, Prentice Hall, 2001.
30. F.P. Kelly, A.K. Maulloo, and D.K.H. Tan. Rate control for communication networks: Shadow prices, proportional fairness and stability. *The Journal of the Operational Research Society*, 49(3):237–252, 1998.
31. T. Başar and G.J. Olsder. Dynamic noncooperative game theory. *SIAM Series in Classics in Applied Mathematics, Philadelphia*, January 1999.
32. O. Morgenstern and J. Von Neumann. *Theory of Games and Economic Behavior*. 3rd edition, Princeton University Press, 1980.

CHAPTER 15

Integral Reinforcement Learning for Online Computation of Nash Strategies of Nonzero-Sum Differential Games

DRAGUNA VRABIE[1] and F.L. LEWIS[2]

[1]United Technologies Research Center, East Hartford, CT, USA
[2]Automation and Robotics Research Institute, The University of Texas at Arlington, Fort Worth, TX, USA

ABSTRACT

This chapter presents an approximate dynamic programming (ADP) algorithm for determining online the Nash equilibrium solution for nonzero-sum linear quadratic differential games. Each of the two game players associates a value function with every given set of feedback control policies. Each player learns its value function and updates its policy online in real-time using data measured along the system trajectories. The procedure is an adaptive critic structure with each player having its own actor network and critic network. The players use a value iteration procedure based on integral reinforcement learning (IRL) for calculating online their value functions. IRL extends heuristic dynamic programming to continuous-time systems and allows the online solution of nonzero-sum games without knowing the full system dynamics. The proposed NZS game double actor–critic structure effectively solves a set of *coupled algebraic Riccati equations* (ARE) online in real time without knowing the system drift dynamics matrix.

The two participants in the continuous-time differential game compete in real time and their feedback Nash control strategies are determined based on data measured along the system trajectory. The algorithm is built on interplay between a *learning* phase, where each of the players learns online the value associated with the current set of play policies, and a policy *update* step, performed by each of the payers in order to decrease the value of the cost. The players learn concurrently based on interplay with

Reinforcement Learning and Approximate Dynamic Programming for Feedback Control, First Edition. Edited by Frank L. Lewis and Derong Liu.

each other and their prescribed performance indices. A computer simulation example shows that the online IRL nonzero-sum games ADP algorithm correctly learns the solution to coupled Riccati equations, and so finds the Nash solution in real time without knowing the system drift dynamics matrix.

15.1 INTRODUCTION

Nash differential games were originally introduced in [1]. Their study is highly relevant as they have a number of potential applications in control engineering and economics (see e.g., [2–4]). The underlying game theory formulation appears, for instance, in the study of coupled large scale systems [5] and networking and wireless communication systems [6]. Traditional game theory solutions are offline methods that rely on solving complex coupled matrix design equations. This chapter is concerned with presenting a continuous-time approximate dynamic programming (ADP) algorithm that calculates the Nash equilibrium solution for the linear two-player nonzero-sum (NZS) differential game online in real time by measuring data along the system trajectories. The algorithm solves coupled matrix Riccati design equations online without knowing the system plant matrix (i.e., the system drift dynamics).

Reinforcement learning (RL) algorithms provide an online approach to the solution of optimal control problems. Much work has developed actor–critic RL structures to solve optimal control problems for discrete-time systems in real time. Many of these online actor–critic learning controllers are in the approximate dynamic programming (ADP) class developed by Werbos [7, 8]. By now, actor–critic adaptive control structures have been developed for continuous-time systems [9]. Most previous research on RL for continuous-time systems assumed that the dynamical system is affected only by a single control strategy. In a game theory setup, the controlled system is affected by a number of control inputs, computed by different controllers that try to optimize their individual performance functions. In these situations, the control problem is formulated with the purpose of finding the set of control policies that are admissible, that is, control policies that guarantee the stability of the controlled dynamical system, and that minimize the individual performance functions in a Nash equilibrium sense. This kind of solution is characterized by the fact that any change in the policy of any given player (in the space of admissible policies) will result in a worse performance for that player, relative to the performance that it receives by means of the Nash equilibrium solution policy.

For the case when the system has linear dynamics and the cost indices are quadratic and have infinite horizon, it is known that finding the Nash equilibrium for the game problem is equivalent to calculating the solution of a set of coupled algebraic Riccati equations (ARE) (see e.g., [1–4]). The solution of the coupled ARE has been approached in [10–14] by means of iterative procedures. These algorithms construct sequences of cost functions, or their matrix kernels, that converge to the equilibrium solution. In the case of [10–13], convergence results for these procedures are still to be determined. All these algorithms require exact and complete knowledge of the system dynamics, and the solution is obtained by means of offline iterative computation procedures.

An ADP procedure that computes the solution to the Hamilton–Jacobi–Isaacs equation, associated with the two-player zero-sum nonlinear differential game, was introduced in [15]. This ADP algorithm involves calculation of two sequences of cost functions, the upper and lower performance indices, which converge to the saddle point solution of the game. The adaptive critic structure that is required for learning the saddle point solution is comprised of four action networks and two critic networks. The algorithm requires full knowledge of the system dynamics.

The result introduced in this chapter is the first reinforcement learning approach to find the saddle point solution of a two-player nonzero-sum differential game online in real time. We make use of the *integral reinforcement learning* (IRL) approach introduced for single-player optimal control in [9], so that complete knowledge of part of the system dynamics is not required. Specifically, the system drift dynamics, or A matrix, need not be known. To our knowledge, there exists no ADP algorithm that provides the Nash equilibrium solution of the two-player nonzero-sum game in an online fashion and without using complete information of the model of the dynamical system to be controlled.

The main traits of this new online procedure are the following:

- It involves the use of ADP techniques that determine the Nash equilibrium solution of the game in an online data-based procedure that does not require full knowledge of the system dynamics.
- It is an online version of a mathematical algorithm [14] that solves the underlying set of coupled algebraic Riccati equations of the game problem. The supporting algorithm [14] makes use of offline procedures and requires full knowledge of the system dynamics to determine the Nash equilibrium of the game.

In this ADP approach, both game players are actively learning and improving their policies. The algorithm is built on interplay between:

- a *learning* phase
- a *policy update* step.

During the learning phase, each of the players is learning the value function that it associates with the use of the current given pair of admissible policies. Both players are learning simultaneously. During the policy update step, both players are changing their feedback control policies in the sense of performance improvement. This means that each player changes its policy in such a way that it will minimize the player cost given the previous policy of their opponent.

For learning the value that each player associates with a given admissible pair of control policies, we will use the value function approximation (VFA) technique that uses the concept of minimization of the temporal difference error, described in [9]. VFA was used in ADP for discrete-time systems by Werbos [7, 8].

The objective of this chapter is to present an online algorithm that makes use of ADP techniques to provide the solution to the two-player differential nonzero-sum game. It is shown that the novel online procedure described here is equivalent to the offline procedure introduced in [14] that needs full knowledge of the system dynamics. The algorithm in [14] involves solving a sequence of Lyapunov equations in order to build a sequence of control policies that converges to the Nash equilibrium solution of the game. Herein, we will show how, by means of ADP techniques, the solution of these game optimal control problems can be obtained in an online fashion, using online measured data from the system, and without knowing the drift dynamics or system plant matrix A.

We begin our investigation by providing in Section 15.2 the formulation of the two-player nonzero-sum game problem. We describe the integral reinforcement learning (IRL) method [9] that can be used online to determine the value associated with a given pair of admissible control strategies. In Section 15.3 is given the algorithm that implements IRL for the single-player optimal control problem. This algorithm solves the algebraic Riccati equation online in real time using date measured along the system trajectories and without knowing the system plant matrix A. In Section 15.4, we describe the online method that provides the Nash equilibrium solution of the two-player nonzero-sum game. The adaptive actor–critic structure that implements the online solution of the game is discussed. It is important to note that in this case each of the two players makes use of a critic structure that uses reinforcement learning ideas to learn online the value that the player associates with a given admissible control strategy. Section 15.5 investigates the convergence properties of the online reinforcement learning algorithm. It is shown that the ADP procedure introduced in this chapter is theoretically equivalent to the offline iterative procedure introduced in [14], and thus has the same convergence properties. A formulation of the algorithm in the form of a quasi-Newton method is provided. Section 15.6 presents a simulation result.

15.2 TWO-PLAYER GAMES AND INTEGRAL REINFORCEMENT LEARNING

In this section, we define the two-player nonzero-sum games and present their solution [3]. Then, we review the integral reinforcement learning (IRL) method for applying approximate dynamic programming (ADP) to continuous-time systems [9]. IRL allows the solution of algebraic Riccati equations online using data measured along the system trajectories and without knowing the system plant matrix A.

15.2.1 Two-Player Nonzero-Sum Games and Nash Equilibrium

Consider the linear dynamical system described by the equation

$$\begin{aligned} \dot{x} &= Ax + B_1 u_1 + B_2 u_2 \\ x(t_0) &= x_0 \end{aligned}, \tag{15.1}$$

where $x \in R^n$, $u_i \in R^{m_i}$. This system has two inputs, termed players. Each player i, $i = 1, 2$, desires to determine the feedback control strategy $u_i = K_i x$ to minimize its own quadratic performance index or cost function

$$J_i = \frac{1}{2}\int_{t_0}^{\infty}(x^{\mathrm{T}}Q_i x + u_1^{\mathrm{T}}R_{i1}u_1 + u_2^{\mathrm{T}}R_{i2}u_2)\mathrm{d}\tau, \tag{15.2}$$

where $Q_i \geq 0,\ R_{ij} \geq 0(i \neq j),\ R_{ii}>0$. When the policies $u_i = K_i x$ are fixed, that is, when the feedback gains are given, the associated performance function is know as the value function.

Definition 15.1 *A feedback control policy pair (u_1, u_2) is admissible if the dynamics of the closed-loop system (15.1) are stable and the value functions (15.2) calculated for the given control pair have finite values.*

The set of admissible policies is denoted by $U_i \subset R^{m_i}\ i = 1, 2$.

A pair of policies (u_1^*, u_2^*) is said to be a Nash equilibrium if

$$\begin{aligned} J_1(u_1, u_2^*) &\geq J_1(u_1^*, u_2^*) \\ J_2(u_1^*, u_2) &\geq J_2(u_1^*, u_2^*) \end{aligned}, \tag{15.3}$$

for any admissible controls (u_1, u_2). This means that if either player changes its own policy while the other does not, that player will achieve worse performance in the sense of having a higher cost function or value. This is often stated as "any player that unilaterally changes its own policy achieves worse performance that if he remains in the Nash policy."

It is desired here to determine the pair of linear state-feedback control strategies (u_1^*, u_2^*) that corresponds to the Nash equilibrium of the two-player differential game.

Define the notation $S_i = B_i R_{ii}^{-1} B_i^{\mathrm{T}},\quad i = 1, 2,$ and $S_{ij} = B_j R_{jj}^{-1} R_{ij} R_{jj}^{-1} B_j^{\mathrm{T}},$ $i, j = 1, 2,\ j \neq i$. Then it is known that the closed-loop Nash control pair (u_1^*, u_2^*) has the state feedback form

$$u_i^* = -R_{ii}^{-1} B_i^{\mathrm{T}} P_i x = K_i x \quad i = 1, 2, \tag{15.4}$$

where the two matrices $P_i,\quad i = 1, 2$ are solutions to the coupled algebraic Riccati equations (AREs)

$$\begin{aligned} N_1(P_1, P_2) \triangleq A^{\mathrm{T}} P_1 + P_1 A + Q_1 + P_2 S_{12} P_2 \\ -P_1 S_1 P_1 - P_2 S_2 P_1 - P_1 S_2 P_2 = 0 \end{aligned}, \tag{15.5a}$$

and

$$\begin{aligned} N_2(P_1, P_2) \triangleq A^{\mathrm{T}} P_2 + P_2 A + Q_2 + P_1 S_{21} P_1 \\ -P_2 S_2 P_2 - P_1 S_1 P_2 - P_2 S_1 P_1 = 0 \end{aligned}. \tag{15.5b}$$

These are necessary conditions for finding the Nash equilibrium. For the derivations of these equations, see for instance [3].

15.2.2 Integral Reinforcement Learning for Two-Player Nonzero-Sum Games

The online iterative procedure for solving the two-player NZS game to be presented in Section 15.4 relies heavily on value function approximation (VFA) [7, 8]. Thus, the goal of this section is to briefly present the online algorithm that uses reinforcement learning ideas to find the value of the parameters of the infinite horizon cost associated with a quadratic cost function such as J_i, $i = 1, 2$. We refer to this online method as integral reinforcement learning (IRL). IRL was introduced in [9] and provides the means for extending to continuous-time systems the family of approximate dynamic programming controllers developed for discrete-time systems by Werbos [7, 8].

The procedure presented here is based on VFA where the value functions J_i, $i = 1, 2$ are given in terms of parameters. Then, the problem is not to find the value function, but the associated parameters. To bring the general theoretical concept into focus, let us formulate the following problem:

Given the dynamical system (15.1) and the admissible pair of control policies $(u_1, u_2) = (K_1 x, K_2 x)$ determine the parameters of the associated infinite horizon cost function J_i (15.2) of player i.

Before giving an online procedure for solving this problem, one must choose a parametric representation for the value function to be determined. In this particular case the cost functions are quadratic and the control policies have a linear state-feedback structure. Thus, for prescribed fixed feedback gains in $(u_1, u_2) = (K_1 x, K_2 x)$, a quadratic representation in the initial state can provide an exact representation for each of the two associated cost functions. That is, VFA can be taken in the quadratic form

$$J_i(x(t)) = x_t^{\mathrm{T}} P_i x_t = \frac{1}{2} \int_t^{\infty} x_\tau^{\mathrm{T}} \bar{Q}_i x_\tau \, \mathrm{d}\tau, \tag{15.6}$$

where $\bar{Q}_i = Q_i + K_1^{\mathrm{T}} R_{i1} K_1 + K_2^{\mathrm{T}} R_{i2} K_2$, P_i is a kernel matrix corresponding to J_i, and $x_t \equiv x(t)$ denotes the state of the system $\dot{x} = (A + B_1 K_1 + B_2 K_2)x$. The feedback control gains are the policies that are assumed fixed. The elements of matrix P_i are the parameters of the value function to be determined. To compute the value functions associated with fixed feedback policies one can integrate the closed-loop dynamics $\dot{x} = (A + B_1 K_1 + B_2 K_2)x$ and evaluate the infinite integrals in (15.2). An attractive alternative is to find a method to directly determine the parameters P_i by some other procedure.

Based on VFA of the form (15.6) one may determine the values of its parameters, namely, the elements of the kernel matrix P_i, as follows. The IRL algorithm for finding the value function parameters is based on the following IRL equation that is satisfied for every time t and interval $T_0 > 0$,

$$x_t^{\mathrm{T}} P_i x_t = \int_t^{t+T_0} x_\tau^{\mathrm{T}} x_\tau \mathrm{d}\tau + x_{t+T_0}^{\mathrm{T}} P_i x_{t+T_0}, \tag{15.7}$$

where x_τ denotes the state of the system described by $\dot{x} = (A + B_1K_1 + B_2K_2)x$ with initial condition x_t, and x_{t+T_0} is the value of the state at time $t + T_0$. The IRL (15.7) comes from the value function definition (15.2) by splitting the right-hand infinite integral into two parts, and using VFA Equation (15.6) for the remaining infinite-integral.

The IRL (15.7) holds for all times t and all integration intervals $T_0>0$. The integral

$$d_i(\bar{x}_t K_1, K_2) \equiv \int_t^{t+T_0} x_\tau^{\mathrm{T}} \bar{Q}_i x_\tau \, \mathrm{d}\tau \tag{15.8}$$

is known as the integral reinforcement for player i on time interval $[t, t + T_0]$. It depends on the prescribed feedback gains $(u_1, u_2) = (K_1x, K_2x)$. The IRL equation can be interpreted as a consistency relationship that holds for all t if P_i contains the correct parameters corresponding to VFA (15.6) for the prescribed feedback policies $(u_1, u_2) = (K_1x, K_2x)$.

IRL (15.7) can be considered as a *Bellman equation* for NZS games for continuous-time (CT) systems. It does not contain any of the system dynamics (A, B_1, B_2), yet it captures information that can be used as below to compute online the value function parameters P_i. It has the same form as the discrete-time (DT) Bellman equation used to developed the family of approximate dynamic programming algorithms by Werbos [7, 8].

An online algorithm that uses the CT NZS games Bellman equation (15.7) to compute the value function parameters is given next.

The solution of (15.7) is the matrix P_i that parameterizes the value function. The quadratic value function can be written as

$$J_i(x(t)) = x_t^{\mathrm{T}} P_i x_t = \bar{p}_i^{\mathrm{T}} \bar{x}_t, \tag{15.9}$$

where $\bar{x}_t$ denotes the Kronecker product [16] quadratic polynomial basis vector with the elements $x_i(t)x_j(t)_{i=1,n;j=i,n}$. Variable $\bar{p} = \nu(P)$, with $\nu(.)$, a vector-valued matrix function that acts on symmetric matrices and returns a column vector by stacking the elements of the diagonal and upper triangular part of the symmetric matrix into a vector. The off-diagonal elements are taken as $2P_{ij}$.

Based on these notations and structures, the Bellman (15.7) for player i is rewritten as the parameterized Bellman equation

$$\bar{p}_i^{\mathrm{T}}(\bar{x}_t - \bar{x}_{t+T_0}) = d_i(\bar{x}_t, K_1, K_2). \tag{15.10}$$

This is a scalar equation in a vector of unknown $\bar{p}_i$ that has the same form as equations appearing in system identification. The parameter vector $\bar{p}_i$ can be found by using least-squares. The equation can be solved online by measuring data along the system trajectories using methods such as recursive least-squares or batch least-squares. The quadratic bases vector $\bar{x}_t - \bar{x}_{t+T_0}$ acts as a regression vector. The right-hand side is a target integral reinforcement function that is measured based on the state trajectories over the time interval $[t, t + T_0]$. To solve (15.10) online, the data

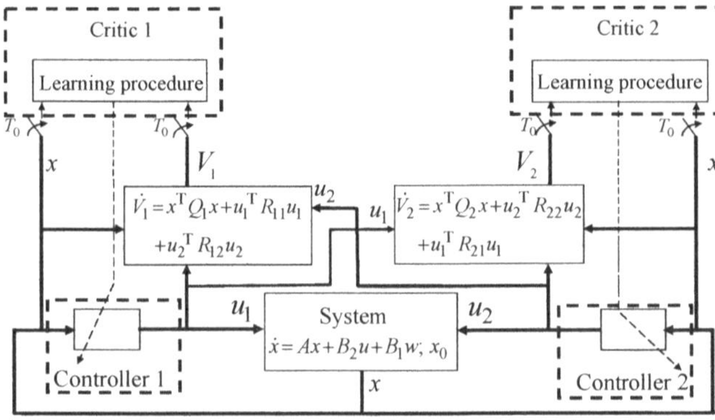

FIGURE 15.1 Double adaptive actor–critic structure for solving the NZS game with IRL.

measured by player i on each time interval $[t, t + T_0]$ is $(x(t), x(t + T_0), d_i(\bar{x}(t)))$, the initial state, the final state, and the integral reinforcement over the interval.

The integral reinforcement $d_i(\bar{x}(t))$ is computed using (15.8) knowing the trajectories over the time interval $[t, t + T_0]$. This is simplified by introducing a new state variable for each player defined as

$$\dot{V}_i = x_\tau^{\mathrm{T}} \bar{Q}_i x_\tau. \tag{15.11}$$

Then, with proper initialization on each time interval $[t, t + T_0]$, that is, $V_i(t) = 0$, the value of this state at the end of the interval $V_i(t + T_0)$ is equal to $d_i(\bar{x}(t))$ in Equation (15.9). This means that each player has a dynamic critic structure whose state $V_i(t)$ acts as a memory. This is shown in Figure 15.1. It is noted that $V_i(t)$ is the value function of player i evaluated over the time interval $[t, t + T_0]$.

15.3 CONTINUOUS-TIME VALUE ITERATION TO SOLVE THE RICCATI EQUATION

Based on the parameterized Bellman Equation (15.10), we will now develop an online algorithm for computing the NZS game value functions associated with a prescribed pair of feedback gains $(u_1, u_2) = (K_1 x, K_2 x)$. This is a data-based approach based on reinforcement learning ideas that does not require any knowledge of the system matrix A. The online algorithm for solving the NZS game in real time is presented in Section 15.4. It relies on repeated solutions of certain algebraic Riccati equations. Therefore, the goal of this section is to briefly present the online algorithm from [9], which uses reinforcement learning ideas to solve the single-player optimal control ARE

$$A^{\mathrm{T}} P + PA + Q - PBR^{-1} B^{\mathrm{T}} P = 0. \tag{15.12}$$

The next algorithm from [9] solves the ARE online by measuring data along the system trajectories. It is based on reinforcement learning and extends the idea of value iteration to CT systems. It is a continuous-time version of the heuristic dynamic programming (HDP) algorithm developed for discrete-time systems by Werbos [7, 8].

Algorithm 15.1: Continuous-Time Heuristic Dynamic Programming (CT-HDP)

Initialize. Let $P_0 = 0$ and let K_0 be a state-feedback control policy (not necessarily stabilizing).
Iterate between

Policy evaluation step

$$x_t^{\mathrm{T}} P_{i+1} x_t = \int_t^{t+T_0} x_\tau^{\mathrm{T}} (Q + K_i^{\mathrm{T}} R K_i) x_\tau \mathrm{d}\tau + x_{t+T_0}^{\mathrm{T}} P_i x_{t+T_0}. \tag{15.13}$$

Feedback gain update step

$$K_{i+1} = R^{-1} B^{\mathrm{T}} P_{i+1},$$

until convergence.

In this algorithm, at the ith iteration the old parameter matrix P_i appears on the right-hand side of (15.13), while the updated parameter matrix P_{i+1} appears on the left-hand side. Therefore, this is a value iteration algorithm. In contrast, a policy iteration algorithm would have the updated parameters on both sides of (15.13), as in Equation (15.10).

The online implementation of the algorithm is given next.

The solution of (15.13) consists of the value of the matrix P_{i+1} that parameterizes the cost function as in (15.9). The reinforcement over the time interval $[t, t + T_0]$ for a single-player is

$$d(\bar{x}_t, K_i) \equiv \int_t^{t+T_0} x^{\mathrm{T}}(\tau)(Q + K_i^{\mathrm{T}} R K_i) x(\tau) \mathrm{d}\tau. \tag{15.14}$$

Based on these notations and structures, Equation (15.13) is rewritten in parameterized form as

$$\bar{p}_{i+1}^{\mathrm{T}} \bar{x}_t = d(\bar{x}_t, K_i) + \bar{p}_i^{\mathrm{T}} \bar{x}_{t+T_0}. \tag{15.15}$$

This is a scalar equation for a vector of unknown parameters $\bar{p}_{i+1}$, and $\bar{x}_t$ acts as a regression vector. The right-hand side is computed based on measured state trajectories over the time interval $[t, t + T_0]$ and the state value at $t + T_0$, $\bar{x}_{t+T_0}$.

The parameter vector $\bar{p}_{i+1}$ is found by solving (15.15) the least-squares sense. The solution can be obtained online using data measured along the system trajectories by batch least-squares or by recursive least-squares algorithms, as is standard in system identification.

The online value iteration algorithm is an online data-based approach that uses reinforcement learning ideas to find the solution of the algebraic Riccati equation with sign definite quadratic term. This algorithm does not require any knowledge of the model of the controlled system's drift dynamics, that is, matrix A.

15.4 ONLINE ALGORITHM TO SOLVE NONZERO-SUM GAMES

In this section, we formulate an adaptive algorithm that finds the NZS game Nash solution online by measuring data along the system trajectories. The algorithm solves the coupled AREs (15.5) online without knowing the system drift dynamics A. This algorithm relies on repeated solutions of certain ARE equations of the form (15.12) using the CT-HDP Algorithm 15.1.

15.4.1 Finding Stabilizing Gains to Initialize the Online Algorithm

The online algorithm to solve the NZS game requires initial stabilizing policies $(u_1, u_2) = (K_1 x, K_2 x)$. This subsection shows how to compute such policies. The next standard necessary assumption is made.

Assumption 15.1 *The triples* $(A, B_i, \sqrt{Q_i})$, $i = 1, 2$ *are stabilizable and detectable.*

Under this assumption initial state feedback control strategies $u_i^{(0)} = K_i^{(0)} x \quad i = 1, 2$, *exist such that closed loop system matrix* $A - B_1 K_1^{(0)} - B_2 K_2^{(0)}$ *is Hurwitz. A procedure for finding such initial stabilizing gains is described next.*

The control pair $(u_1^{(0)}(x), u_2^{(0)}(x))$ is admissible, that is, $A - S_1 P_1^{(0)} - S_2 P_2^{(0)}$ is Hurwitz. Therefore, at this point we are in the possession of an initial admissible pair of feedback control strategies $(u_1^{(0)}, u_2^{(0)}) = (K_1^{(0)} x, K_2^{(0)} x)$, also represented by $(P_1^{(0)}, P_2^{(0)})$. It is worth noting that the step 1 above can also be executed with respect to player 2, followed by step 2 that is now relative to player 1. In this case also, a pair of admissible control policies will be obtained.

15.4.2 Online Partially Model-Free Algorithm for Solving the Nonzero-Sum Differential Game

In this subsection, we formulate an iterative algorithm that learns online the Nash equilibrium solution of the two-player zero-sum differential game. At every step of the iterative procedure, each player uses reinforcement learning to estimate the infinite horizon value function that it associates with the current admissible control pair. Following the value function estimation procedure, each of the two players makes

Algorithm 15.2: Finding Initial Stabilizing Policies

Step 1. Let player 2 use the zero control $u_2(x) = 0$ and determine the optimal control strategy of player 1 with respect to the cost index J_1. This is a classical linear quadratic regulation problem and the optimal control strategy has the form $u_1^{(0)}(x) = K_1^{(0)}x = -R_{11}^{-1}B_1^{\mathrm{T}}P_1^{(0)}x$, where $P_1^{(0)}$ is the solution of the ARE

$$A^{\mathrm{T}}P + PA + Q_1 - PB_1R_{11}^{-1}B_1^{\mathrm{T}}P = 0. \tag{15.16}$$

This ARE is of the standard form (15.12). It can be solved online without knowing the A matrix using CT-HDP Algorithm 15.1. The result is that the dynamics of the system (15.1) with the control pair $(u_1^{(0)}(x), 0)$ are stable, that is, $A - S_1P_1^{(0)}$ is Hurwitz.

Step 2. Let player 1 use the stabilizing control policy $u_1^{(0)}(x) = K_1^{(0)}x$ and determine the optimal control strategy of player 2 with respect to the cost index J_2. Again, this is a classical linear quadratic regulation problem and the optimal control strategy has the form $u_2^{(0)}(x) = K_2^{(0)}x = -R_{22}^{-1}B_2^{\mathrm{T}}P_2^{(0)}x$, where $P_2^{(0)}$ is the solution of the ARE

$$\begin{aligned}(A - S_1P_1^{(0)})^{\mathrm{T}}P + P(A - S_1P_1^{(0)}) + \\ +Q_2 + P_1^{(0)}S_{21}P_1^{(0)} - PB_2R_{22}^{-1}B_2^{\mathrm{T}}P = 0\end{aligned}. \tag{15.17}$$

This is of the standard ARE form (15.12), and it can be solved online without knowing the A matrix using CT-HDP Algorithm 15.1.

a decision to improve its control policy. The end result is the online Algorithm 15.3 that effectively solves the coupled matrix AREs (15.5) without knowing the system A matrix.

15.4.3 Adaptive Critic Structure for Solving the Two-Player Nash Differential Game

Algorithm 15.3 corresponds to a double adaptive actor–critic RL structure. It has value iteration steps followed by policy update steps for each player. The adaptive critic structure that represents the implementation of Algorithm 15.3 is given in Figure 15.1. There is a critic network and an actor network associated with each of the two players.

Each critic requires knowledge of its integral reinforcement (15.8) to solve the parameterized Bellman Equation (15.10). This is computed using the additional value state $V_i(t)$ defined by the state equation (15.11). Thus, each critic has a dynamic structure whose state is a memory that serves to keep track of the value function $V_i(t)$ on each time interval $[t, t + T_0]$. These extra critic states are shown in Figure 15.1.

Algorithm 15.3: IRL for Online Solution of Two-Player NZS Games

Initialize. Start with initial matrices $(P_1^{(0)}, P_2^{(0)})$ such that $A - S_1 P_1^{(0)} - S_2 P_2^{(0)}$ is Hurwitz. Use Algorithm 15.2 to find these matrices. Let $k = 0$.

For $k \geq 0, k \in N$. Use the integral reinforcement learning procedure described in connection with (15.10) to determine the value that each of the two players is associating with the control policies described by the matrix pair $(P_1^{(k)}, P_2^{(k)})$. Namely determine $P_i^{(k+1)}$, $i = 1, 2$ that satisfy

$$x_0^{\mathrm{T}} P_i^{(k+1)} x_0 = \frac{1}{2} \int_{t_0}^{\infty} x^{\mathrm{T}} \bar{Q}_1^{(k)} x \mathrm{d}\tau, \tag{15.18}$$

where $\bar{Q}_1^{(k)} = Q_i + P_i^{(k)} S_i P_i^{(k)} + P_j^{(k)} S_{ij} P_j^{(k)} \quad i = 1, 2, j \neq i.$

- Update the control policies of the two players such that the new control policy pair is characterized by $(P_1^{(k+1)}, P_2^{(k+1)})$.
- Stop the online algorithm when the following criterion is satisfied for a specified value of ε

$$\max(N_1(P_1^{(k+1)}, P_2^{(k+1)}), N_2(P_1^{(k+1)}, P_2^{(k+1)})) \leq \varepsilon. \tag{15.19}$$

An important aspect that is revealed by the adaptive critic structure is the fact that this ADP algorithm uses three time scales:

- The continuous time scale, represented by the full lines, which is connected with the continuous-time dynamics of the system and the continuous-time control policy computation performed by the two players.
- A discrete time scale defined by the integration interval T_0. This time scale is connected with the online value learning procedure and specifies at what times the data samples are collected along the system trajectories.
- A slower, discrete time scale that is a multiple of T_0. This time scale is connected with the policy evaluation step. It is the time required to solve the IRL Bellman equation for the values corresponding to the current policies of the two players. The policy update procedure is performed only after the value function learning procedure that uses integral reinforcement information has converged.

The values of the time periods T_0 can be variable and are controlled by the two learning critics. Each critic outputs a matrix $P_i^{(k)}$, $i = 1, 2$, in a synchronous fashion, after both online learning algorithms for the value functions have converged. Each controller uses the information from its corresponding critic to calculate a new control policy.

From the perspective of two-player games, the proposed online algorithm can be presented as follows.

Algorithm 15.4: Game Theory Interpretation of IRL NZS Game Solution

Initialization.

Let the initial policy of player 2 be zero $u_2^{(00)} = 0$.

Let the player 1 determine its optimal control policy $K_1^{(0)} = -R_{11}^{-1} B_1^T P_1^{(0)}$ in an online optimization procedure while player 2 is not playing the game.

Let the player 2 determine its optimal control policy $K_2^{(0)} = -R_{22}^{-1} B_2^T P_2^{(0)}$ in an online optimization procedure while player 1 is playing the game using $K_1^{(0)}$

Iterative procedure.

For $k \geq 0$, let both players determine online, using the integral reinforcement learning procedure, the values that they associate with the use of the policy pair $(K_1^{(k)}, K_2^{(k)})$, namely, the pair of matrices $(P_1^{(k+1)}, P_2^{(k+1)})$.

Let both players update their control policies using

$$K_i^{(k+1)} = -R_{ii}^{-1} B_i^T P_i^{(k+1)}. \tag{15.20}$$

Stop criterion

Let both players stop this iterative procedure when there is no change in the control policies at two successive steps (i.e., the Nash equilibrium has been obtained and both players can not further improve their cost function by changing their behavior policy).

15.5 ANALYSIS OF THE ONLINE LEARNING ALGORITHM FOR NZS GAMES

In this section, we provide an analysis for the online NZS Game Algorithm 15.3.

15.5.1 Mathematical Formulation of the Online Algorithm

Using the notation $A^{(k)} = A - S_1 P_1^{(k)} - S_2 P_2^{(k)}$, it can be shown that (15.18) can be written as

$$(A^{(k)})^T P_i^{(k+1)} + P_i^{(k+1)} A^{(k)} = -\bar{Q}_1^{(k)}. \tag{15.21}$$

Thus, the online procedure described in Algorithm 15.3 is equivalent with the following:

- start with initial matrices $(P_1^{(0)}, P_2^{(0)})$ such that $A - S_1 P_1^{(0)} - S_2 P_2^{(0)}$ is Hurwitz. Let $k = 0$;
- for $k \geq 0, k \in N$, solve (15.21);

- stop the online algorithm when the criterion (15.19) is satisfied for a user-specified value of ε.

This offline algorithm that uses iterations on Lyapunov equations has been proposed by [14], and its convergence has been discussed in [14] and further analyzed and [17]. Considering the mathematical equivalence between the algorithm introduced in [14] and the online procedure based on reinforcement learning that we proposed in Algorithm 15.3, we can conclude that the online, partially model free, algorithm that is presented herein has the same convergence properties.

It is interesting to see that, similarly to the Newton method proposed in [18] for solving the classical continuous-time algebraic Riccati equation, this algorithm relies on iterations on Lyapunov equations. However, the online procedure introduced here, and its underlying algorithm, is not a Newton method for finding the solution of the coupled AREs given in (15.5). This shall be clarified by means of the next two propositions. Compare to [5].

First, let us look at the formulation of the Newton method that determines the unique positive definite solution of the classical continuous-time algebraic Riccati equation

$$A^{\mathrm{T}}P + PA + Q - PBR^{-1}B^{\mathrm{T}}P = 0. \tag{15.22}$$

Denote by $Ric(P_i)$ the matrix valued function defined as

$$Ric(P_i) = A^{\mathrm{T}}P_i + P_iA + Q - P_iBR^{-1}B^{\mathrm{T}}P_i \tag{15.23}$$

and let Ric'_{P_i} denote the Frechet derivative of $Ric(P_i)$ taken with respect to P_i. The matrix function Ric'_{P_i} evaluated at a given matrix M is defined as

$$Ric'_{P_i}(M) = (A - BR^{-1}B^{\mathrm{T}}P_i)^{\mathrm{T}}M + M(A - BR^{-1}B^{\mathrm{T}}P_i) \tag{15.24}$$

Proposition 15.1 The unique positive solution of (15.23) can be determined by Newton's method given by

$$P_i = P_{i-1} - (Ric'_{P_{i-1}})^{-1}Ric(P_{i-1}) \tag{15.25}$$

provided that the initial matrix P_0 is such that $A - BR^{-1}B^{\mathrm{T}}P_0$ is Hurwitz; and considering that the regular conditions for existence and uniqueness of positive definite solution are satisfied.

Next, we will use the same mathematical tools to provide formulation to Algorithm 15.3. Consider the notations introduced in (15.5) for the two coupled algebraic Riccati equations, and let $N'_{1P_1^{(i)}}$ and $N'_{2P_2^{(i)}}$ denote the Frechet derivatives of $N_1(P_1^{(i)}, P_2^{(i)})$

and $N_2(P_1^{(i)}, P_2^{(i)})$ taken with respect to $P_1^{(i)}$ and, respectively, $P_2^{(i)}$, such that

$$\begin{aligned} N'_1 P_1^{(i)}(M) &= (A^{(i)})^{\mathrm{T}} M + M A^{(i)} \\ N'_2 P_2^{(i)}(M) &= (A^{(i)})^{\mathrm{T}} M + M A^{(i)} \end{aligned}, \tag{15.26}$$

with the notation $A^{(i)} = A - S_2 P_2^{(i)} - S_1 P_1^{(i)}$.

Proposition 15.2 Given that the regular conditions for existence and uniqueness of solution of the infinite horizon nonzero-sum differential game with quadratic performance are satisfied, the Nash equilibrium solution of (15.5) can be determined by a quasi-Newton method formulated as

$$\begin{aligned} P_1^{(k+1)} &= P_1^{(k)} - (N'_{1P_1^{(k)}})^{-1} N_1(P_1^{(k)}, P_2^{(k)}) \\ P_2^{(k+1)} &= P_2^{(k)} - (N'_{2P_2^{(k)}})^{-1} N_2(P_1^{(k)}, P_2^{(k)}) \end{aligned} \tag{15.27}$$

provided that the initial pair $(P_1{}^{(0)}, P_2{}^{(0)})$ is such that $A^{(0)} = A - S_2 P_2^{(0)} - S_1 P_1^{(0)}$ is Hurwitz.

Proof: One can show, after some manipulations, that the two equations

$$\begin{aligned} \left(A^{(k)}\right)^{\mathrm{T}} P_1^{(k+1)} + P_1^{(k+1)} A^{(k)} &= -\bar{Q}_1^{(k)} \\ \left(A^{(k)}\right)^{\mathrm{T}} P_2^{(k+1)} + P_2^{(k+1)} A^{(k)} &= -\bar{Q}_2^{(k)} \end{aligned}$$

can be written in the form of

$$\begin{aligned} &(P_1^{(k+1)} - P_1^{(k)})A^{(k)} + (A^{(k)})^{\mathrm{T}}(P_1^{(k+1)} - P_1^{(k)}) + \\ &+ N_1(P_1^{(k)}, P_2^{(k)}) = 0 \end{aligned}, \tag{15.28}$$

and

$$\begin{aligned} &(P_2^{(k+1)} - P_2^{(k)})A^{(k)} + (A^{(k)})^{\mathrm{T}}(P_2^{(k+1)} - P_2^{(k)}) + \\ &+ N_2(P_1^{(k)}, P_2^{(k)}) = 0 \end{aligned}. \tag{15.29}$$

Making use of the notation for the Frechet derivatives, Equation (15.28) and (15.29) can now be written in the quasi-Newton formulation (15.27). ■

15.6 SIMULATION RESULT FOR THE ONLINE GAME ALGORITHM

This section presents computer simulation results for the online IRL nonzero-sum games algorithm. It is verified that the algorithm learns the solution to the coupled AREs (15.5) without knowing the system A matrix, and hence finds the Nash equilibrium solution online in real time.

We consider the system used in Example 1 in [13]. The purpose of the design method is to allow the two players to determine by means of online measurements and reinforcement learning techniques the control strategies that satisfy the equilibrium characterized by (15.3). It is important to emphasize that the equilibrium result is obtained without making use of any knowledge of the drift dynamics system matrix A.

The matrices of the model of the plant are

$$A_{\text{nom}} = \begin{bmatrix} -0.0366 & 0.0271 & 0.0188 & -0.4555 \\ 0.0482 & -1.0100 & 0.0024 & -4.0208 \\ 0.1002 & 0.2855 & -0.7070 & 1.3229 \\ 0 & 0 & 1.0000 & 0 \end{bmatrix}$$

$$B_1 = \begin{bmatrix} 0.4422 & 3.0447 & -5.52 & 0 \end{bmatrix}^{\mathrm{T}}$$

$$B_2 = \begin{bmatrix} 0.1761 & -7.5922 & 4.99 & 0 \end{bmatrix}^{\mathrm{T}}.$$

The cost function parameters were chosen as $Q_1 = diag(3.5; 2; 4; 5)$, $Q_2 = diag(1.5; 6; 3; 1)$, $R_{11} = 1$, $R_{22} = 2$, $R_{12} = 0.25$, $R_{21} = 0.6$.

For the purpose of demonstrating the online learning algorithm, the closed loop system is excited with an initial condition, the initial state of the system being $x_0 = \begin{bmatrix} 0 & 0 & 0 & 1 \end{bmatrix}$. The simulation is conducted using data obtained from the system at every $T_0 = 0.2$ s. The value of the stop criterion ε was 10^{-6}.

Algorithm 15.3 is initialized using the two matrices $P_i^{(0)}$, $i = 1, 2$ calculated by the initialization procedure in Algorithm 15.1, and their associated admissible feedback control policies $K_i^{(0)}$, $i = 1, 2$. This is based on IRL and does not require knowledge of the system drift dynamics matrix A.

In order to solve online for the values of the $P_i^{(k)}$, $i = 1, 2$ in Algorithm 15.3, a least-squares problem based on (15.10) is set up before each iteration step in the online algorithm. The data measured by player i on each time interval $[t, t + T_0]$ is $(x(t), x(t + T_0), d_i(\bar{x}(t)))$, the initial state, the final state, and the integral reinforcement. The integral reinforcement is computed using the additional value state $V_i(t)$ defined by the critic state (15.11).

Since there are 10 independent elements in the 4×4 symmetric matrices $P_i^{(k)}$, $i = 1, 2$, the setup of the least-squares problem in (15.10) requires at least 10 measurements of the cost function associated with the given control policy and measurements of the system's states at the beginning and the end of each time interval, provided that there is enough excitation in the system. A least-squares problem is solved for the

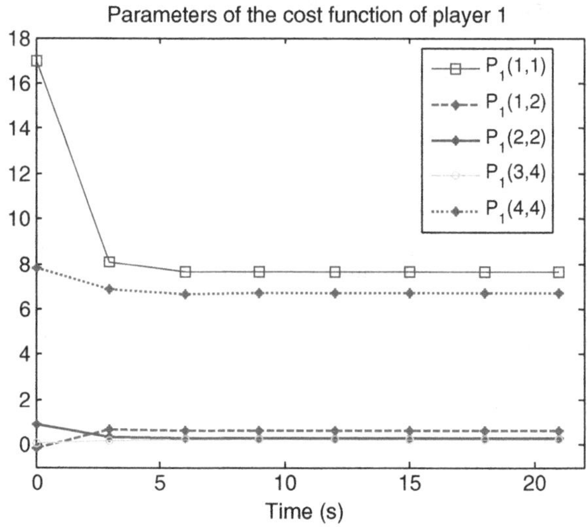

FIGURE 15.2 Convergence of the cost function of player 1 using the ADP method with integral reinforcement learning technique.

value function parameters after 15 data samples are acquired, and thus, the policy of the controller is updated every 3 s. This corresponds to one iteration step in Algorithm 15.3.

Figures 15.2 and 15.3 present the evolution of the parameters of the value of the game seen by player 1 and player 2, respectively.

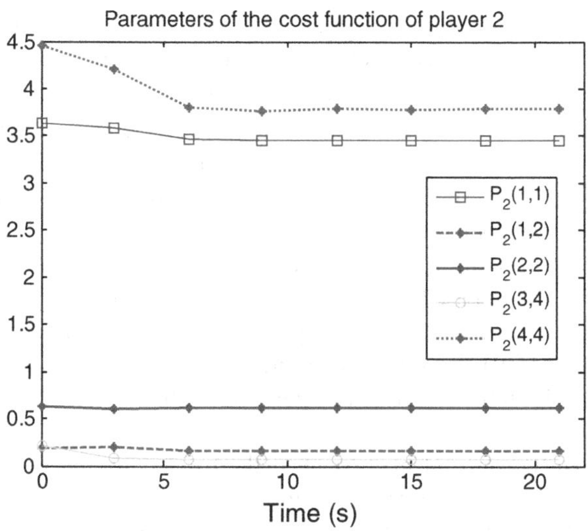

FIGURE 15.3 Convergence of the cost function of player 2 using the ADP method with integral reinforcement learning technique.

The kernel matrices that characterize the Nash equilibrium solution are obtained in simulation after seven iteration steps of 3 s each. These are

$$P_1^{(7)} = \begin{bmatrix} 7.6586 & 0.6438 & 0.6398 & -3.0831 \\ 0.6438 & 0.2878 & 0.2855 & -0.0945 \\ 0.6398 & 0.2855 & 0.5620 & 0.2270 \\ -3.0831 & -0.0945 & 0.2270 & 6.6987 \end{bmatrix}$$

and

$$P_2^{(7)} = \begin{bmatrix} 3.4579 & 0.1568 & 0.2047 & -1.8480 \\ 0.1568 & 0.6235 & 0.2889 & -0.0711 \\ 0.2047 & 0.2889 & 0.4014 & 0.0729 \\ -1.8480 & -0.0711 & 0.0729 & 3.7850 \end{bmatrix}.$$

The same matrices were obtained in [12] by a different method and can be verified to solve the coupled AREs (15.5).

The two Nash feedback control policies are

$$K_1^{(7)} = \begin{bmatrix} -1.8151 & 0.4150 & 1.9501 & 2.9041 \end{bmatrix}$$
$$K_2^{(7)} = \begin{bmatrix} -0.22 & 1.6323 & 0.0772 & -0.2891 \end{bmatrix}.$$

In summary, the ADP online gaming method described in Algorithm 15.3 uses measurements along the system trajectories and does not require any knowledge of the matrix A. Nonetheless, the resulting solution is close to the exact solution of the game problem (15.5) that can be obtained via numerical methods, which require an exact model of the system.

15.7 CONCLUSION

This chapter introduced an online data-based approach that makes use of reinforcement learning techniques to determine in an online fashion the Nash solution of the two-player nonzero-sum differential game with linear dynamics. The algorithm is suitable for real-time implementation and furthermore does not require exact knowledge of the system drift dynamics matrix A. The two players use a value iteration procedure based on integral reinforcement learning (IRL) for calculating online their value functions. IRL extends heuristic dynamic programming to continuous-time systems and allows the online solution of nonzero-sum games without knowing the full system dynamics.

In order to give a clear presentation, here we considered the infinite horizon, state-feedback, linear-quadratic case of the problem. Ideas related with the extension of this result to the more general case of a game with nonlinear dynamics will be pursued in detail in future work. Also, for the sake of clarity of the presentation, in this paper, we restricted the discussion to the case of two-player games. However, it is straightforward to formulate the ADP algorithm for the general case with N players.

REFERENCES

1. A. Starr and Y. Ho. Nonzero-sum differential games. *Journal of Optimization Theory and Applications*, 3(3):184–206, 1969.
2. H. Abou-Kandil, G. Freiling, and G. Jank. Necessary and sufficient conditions for constant solutions of coupled Riccati equations in Nash games. *Systems and Control Letters*, 21:295–306, 2003.
3. T. Basar and G.J. Olsder. *Dynamic Noncooperative Game Theory*, 2nd edition. (Classics in Applied Mathematics; 23), SIAM, 1999.
4. J.C. Engwerda. *LQ Dynamic Optimization and Differential Games*. Wiley, Chichester, 2005.
5. H. Mukaidani. Newton's method for solving cross-coupled sign-indefinite algebraic Riccati equations for weakly coupled large-scale systems. *Applied Mathematics and Computation*, 188:103–115, 2007.
6. V. Shah. Power control for wireless data services based on utility and pricing. M.S. thesis, Rutgers University, 1998.
7. P.J. Werbos. A menu of designs for reinforcement learning over time. In W.T. Miller, R.S. Sutton, and P.J. Werbos, editors. *Neural Networks for Control*. MIT Press, Cambridge, 1991, pp. 67–95.
8. P.J. Werbos. Approximate Dynamic Programming for real-time control and neural modeling. In D.A. White and D.A. Sofge, editors. *Handbook of Intelligent Control*. Van Nostrand Reinhold, New York, 1992.
9. D. Vrabie, O. Pastravanu, F. Lewis, and M. Abu-Khalaf. Adaptive optimal control for continuous-time linear systems based on policy iteration. *Automatica*, 45(2):477–484, 2009.
10. L. Cherfi, H. Abou-Kandil, and H. Bourles. Iterative method for general algebraic Riccati equation. In *Proceedings ACSE' 05*, 2005.
11. L. Cherfi, Y. Chitour, and H. Abou-Kandil. A new algorithm for solving coupled algebraic Riccati equations. In *Proceedings of CIMCA' 05*, pp.83–88, 2005.
12. G. Freiling, G. Jank, and H. Abou-Kandil. On global existence of solutions to coupled matrix Riccati equations in closed-loop Nash games. *IEEE Transaction on Automatic Control*, 41(2):264–269, 1996.
13. M. Jungers, E. De Pieri, and H. Abou-Kandil. Solving coupled Riccati equations for closed-loop Nash strategy, by lack of trust approach. *International Journal of Tomography and Statistics*, 7(F07):49–54, 2007.
14. T. Li and Z. Gajic. Lyapunov iterations for solving coupled algebraic Riccati equations of Nash differential games and algebraic Riccati equations of zero-sum games. In G. Olsder, editor. *New Trends in Dynamic Games*. Birkhauser, 1995, pp. 333–351.
15. Q. Wei and H. Zhang. A new approach to solve a class of continuous-time nonlinear quadratic zero-sum game using ADP. In *Proceedings IEEE International Conference on Networking, Sensing and Control (ICNSC'08)*, 6(8):507 –512, 2008.

16. J. W. Brewer. Kronecker products and matrix calculus in system theory. *IEEE Transaction on Circuit and System*, 25(9):772–781, 1978.

17. H. Mukaidani. Optimal numerical strategy for Nash games of weakly coupled large scale systems. *Dynamics of Continuous, Discrete and Impulsive Systems, Series B: Applications and Algorithms*, 13:249–268, 2006.

18. D. Kleinman. On an iterative technique for Riccati equation computations. *IEEE Transaction on Automatic Control*, 13:114–115, February, 1968.

CHAPTER 16

Online Learning Algorithms for Optimal Control and Dynamic Games

KYRIAKOS G. VAMVOUDAKIS[1] and FRANK L. LEWIS[2]

[1]Center for Control, Dynamical-Systems and Computation, University of California, Santa Barbara, CA, USA

[2]UTA Research Institute, University of Texas, Arlington, TX, USA

ABSTRACT

In this work, online adaptive learning algorithms are developed for optimal control and differential dynamic games using measurements along the trajectory. These algorithms are based on actor/critic schemes and involve simultaneous tuning of the actor/critic neural networks and provide online solutions to complex Hamilton–Jacobi equations, along with convergence and Lyapunov stability proofs.

16.1 INTRODUCTION

Optimal control deals with the problem of finding a control law for a given system such that a prescribed optimality criterion is achieved. Optimal control solutions can be derived using Pontryagin's minimum principle or by solving the Hamilton–Jacobi–Bellman equation. Major drawbacks of these solution procedures are that they are offline *a priori* methods that require full knowledge of the system dynamics. On the other hand, adaptive control allows stabilizing controllers to be learned online in real time for systems with unknown dynamics. Adaptive optimal controllers have been proposed by adding optimality criteria to an adaptive controller, or adding adaptive characteristics to an optimal controller.

In this chapter, online adaptive learning algorithms are developed for optimal control and differential dynamic games using measurements along the system trajectory. These algorithms are based on actor/critic schemes (reinforcement learning) and

Reinforcement Learning and Approximate Dynamic Programming for Feedback Control, First Edition. Edited by Frank L. Lewis and Derong Liu.

involve simultaneous tuning of the actor/critic neural networks and provide approximate local smooth solutions to complex Hamilton–Jacobi equations without explicitly solving them, along with convergence and Lyapunov stability proofs.

Reinforcement learning (RL) is a class of methods used in machine learning to methodically modify the actions of an agent based on observed responses from its environment [1–6]. The RL methods have been developed starting from learning mechanisms observed in mammals. Every decision-making organism interacts with its environment and uses those interactions to improve its own actions in order to maximize the positive effect of its limited available resources; this in turn leads to better survival chances. RL is a means of *learning optimal behaviors by observing the response from the environment to nonoptimal control policies*. In engineering terms, RL refers to the learning approach of an actor or agent that modifies its actions, or control policies, based on stimuli received in response to its interaction with its environment. This learning can be extended along two dimensions: (a) nature of interaction (competitive or collaborative) and (b) the number of decision makers (single or multi agent).

In view of the advantages offered by the RL methods, a recent objective of control systems researchers is to introduce and develop RL techniques which result in optimal feedback controllers for dynamical systems that can be described in terms of ordinary differential or difference equations.

Some of the methods involve a computational intelligence technique known as policy iteration (PI) [4, 6]. PI refers to a class of algorithms built as a two-step iteration: *policy evaluation* and *policy improvement*. Instead of trying a direct approach to solving the Hamilton–Jacobi (HJ) equations, the PI algorithm starts by evaluating the cost of a given initial admissible (in a sense to be defined herein) control policy. This is often accomplished by solving a nonlinear Lyapunov equation. The key to solving practically the nonlinear Lyapunov equations was in the use of neural networks (NN) [7], which can be trained to become approximate solutions of these equations. In fact, the PI algorithm for continuous time (CT) systems can be built on actor–critic structures that involve the use of neural networks: the critic NNs are trained to become approximations of the Lyapunov equations solution at the policy evaluation step, whereas the actor NNs are trained to approximate improving policies at the policy-improving step.

Game theory [8] has been very successful in modeling strategic behavior, where the outcome for each player depends on their actions and that of all the other players. Every player chooses a control to minimize independently from the others and their *own* performance objective. None has knowledge of the others' strategy. A lot of applications of optimization theory require the solution of coupled Hamilton–Jacobi equations [9, 10]. In games with N players, each player decides for the Nash equilibrium depending on Hamilton–Jacobi equations coupled through their quadratic terms [10, 11]. Each dynamic game consists of three parts: (a) players; (b) actions available for each player; and (c) costs for every player that depend on their actions.

The chapter starts by developing an online approximate local smooth solution, based on policy iteration (PI), for the infinite horizon optimal control problem for continuous-time nonlinear systems with known dynamics. We present an online adaptive algorithm that involves *simultaneous tuning* of both actor and critic neural networks (i.e., both neural networks are tuned at the same time). We term this algorithm "synchronous" policy iteration. This approach is an extremal version of the generalized Policy Iteration introduced in [6].

Then we developed an online learning algorithm to solve the continuous-time two-player zero-sum game with infinite horizon cost for nonlinear systems. The algorithm learns online in real time the solution to the game design Hamilton–Jacobi–Isaacs equation. This online gaming algorithm is called "synchronous" zero-sum game policy iteration.

One of the major outcomes of this chapter is the online learning algorithm to solve the continuous time multi player nonzero sum games with infinite horizon for linear and nonlinear systems. The adaptive algorithm learns online the solution of coupled Riccati and coupled Hamilton–Jacobi equations for linear and nonlinear systems, respectively. The optimal-adaptive algorithm is implemented as a separate actor–critic parametric network approximator structure for every player and involves simultaneous continuous-time adaptation of all the actor–critic networks.

16.2 OPTIMAL CONTROL AND THE CONTINUOUS TIME HAMILTON–JACOBI–BELLMAN EQUATION

The following section will provide the formulation of optimal control problem followed by the policy iteration algorithm that will give us the structure for the proposed online algorithm that follows.

16.2.1 Optimal Control and Hamilton–Jacobi–Bellman Equation

Consider the nonlinear time-invariant affine in the input dynamical system given by

$$\dot{x}(t) = f(x(t)) + g(x(t))\,u(x(t);\ x(0) = x_0 \tag{16.1}$$

with state $x(t) \in \mathbb{R}^n$, $f(x(t)) \in \mathbb{R}^n$, $g(x(t)) \in \mathbb{R}^{n\times m}$ and control input $u(t) \in \mathbb{R}^m$. We assume that, $f(0) = 0$, $f(x) + g(x)u$ is Lipschitz continuous on a set $\Omega \subseteq \mathbb{R}^n$ that contains the origin, and that the system is stabilizable on Ω, that is, there exists a continuous control function $u(t) \in U$ such that the system is asymptotically stable on Ω. The system dynamics $f(x)$, $g(x)$ are assumed known.

Define the infinite horizon integral cost

$$V(x_0) = \int_0^\infty r(x(\tau), u(\tau))\mathrm{d}\tau, \tag{16.2}$$

where $r(x, u) = Q(x) + u^{\mathrm{T}} Ru$ with $Q(x)$ positive definite, that is, $\forall x \neq 0$, $Q(x)>0$ and $x = 0 \Rightarrow Q(x) = 0$, and $R \in \mathbb{R}^{m\times m}$ a symmetric positive definite matrix.

Definition 16.1 *[7] (Admissible policy) A control policy $\mu(x)$ is defined as admissible with respect to (16.2) on Ω, denoted by $\mu \in \Psi(\Omega)$, if $\mu(x)$ is continuous on Ω, $\mu(0) = 0$, $u(x) = \mu(x)$ stabilizes (16.1) on Ω, and $V(x_0)$ is finite $\forall x_0 \in \Omega$.*

For any admissible control policy $\mu \in \Psi(\Omega)$, if the associated cost function

$$V^{\mu}(x_0) = \int_0^{\infty} r(x(\tau), \mu(x(\tau)))\mathrm{d}\tau \tag{16.3}$$

is C^1, then an infinitesimal version of (16.3) is the Bellman equation

$$0 = r(x, \mu(x)) + \left(V_x^{\mu}\right)^{\mathrm{T}}(f(x) + g(x)\mu(x)), \;\; V^{\mu}(0) = 0, \tag{16.4}$$

where V_x^{μ} denotes the partial derivative of the value function V^{μ} with respect to x. (Note that the value function does not depend explicitly on time)

We define the gradient here as a column vector, and use at times the alternative operator notation $\nabla \equiv \partial/\partial x$.

Equation (16.4) is a Lyapunov equation for nonlinear systems which, given a controller $\mu(x) \in \Psi(\Omega)$, can be solved for the value function $V^{\mu}(x)$ associated with it. Given that $\mu(x)$ is an admissible control policy, if $V^{\mu}(x)$ satisfies (16.4), with $r(x, \mu(x)) > 0$, then $V^{\mu}(x)$ is a Lyapunov function for the system (16.1) with control policy $\mu(x)$.

The optimal control problem can now be formulated: Given the continuous-time system (16.1), the set $\mu \in \Psi(\Omega)$ of admissible control policies and the infinite horizon cost functional (16.2), find an admissible control policy such that the cost index (16.2) associated with the system (16.1) is minimized.

Defining the Hamiltonian of the problem

$$H(x, u, V_x) = r(x(t), u(t)) + V_x^{\mathrm{T}}(f(x(t)) + g(x(t))\mu(t)) \tag{16.5}$$

the optimal cost function $V^*(x)$ defined by

$$V^*(x_0) = \min_{\mu \in \Psi(\Omega)} \left(\int_0^{\infty} r(x(\tau), \mu(x(\tau)))\mathrm{d}\tau \right)$$

with $x_0 = x$ is known as the *value function*, and satisfies the HJB equation

$$0 = \min_{\mu \in \Psi(\Omega)} [H(x, \mu, V_x^*)]. \tag{16.6}$$

Assuming that the minimum on the right-hand side of (16.6) exists and is unique then the optimal control function for the given problem is

$$\mu^*(x) = -\frac{1}{2}R^{-1}g^{\mathrm{T}}(x)V_x^*(x). \tag{16.7}$$

Inserting this optimal control policy in the nonlinear Lyapunov equation, we obtain the formulation of the HJB equation in terms of V_x^*

$$\begin{aligned} 0 &= Q(x) + V_x^{*\mathrm{T}}(x)f(x) - \frac{1}{4}V_x^{*\mathrm{T}}(x)g(x)R^{-1}g^{\mathrm{T}}(x)V_x^*(x) \\ V^*(0) &= 0. \end{aligned} \tag{16.8}$$

For the linear system case, considering a quadratic cost functional, the equivalent of this HJB equation is the well-known Riccati equation.

In order to find the optimal control solution for the problem one only needs to solve the HJB equation (16.8) for the value function and then substitute the solution in (16.7) to obtain the optimal control. However, due to the nonlinear nature of the HJB equation finding its solution is generally difficult or impossible.

16.2.2 Policy Iteration for Optimal Control

The approach of synchronous policy iteration used in this chapter is motivated by Policy iteration (PI) [6]. Therefore, in this section, we describe PI.

Policy iteration (PI) [6] is an iterative method of reinforcement learning for solving optimal control problems and consists of policy evaluation based on (16.4) and policy improvement based on (16.7). Specifically, the PI algorithm consists in solving iteratively the following two equations:

1. given $\mu^{(i)}(x)$, solve for the value $V^{\mu^{(i)}}(x(t))$ using the Bellman equation

$$\begin{aligned} 0 &= r\big(x, \mu^{(i)}(x)\big) + \big(\nabla V^{\mu^{(i)}}\big)^{\mathrm{T}}(f(x) + g(x)\mu^{(i)}(x)), \\ V^{\mu^{(i)}}(0) &= 0, \end{aligned} \tag{16.9}$$

2. update the control policy using

$$\mu^{(i+1)} = \underset{u\in\Psi(\Omega)}{\operatorname{argmin}}\left[H(x, u, \nabla V_X^{(i)})\right], \tag{16.10}$$

which explicitly is

$$\mu^{(i+1)}(x) = -\frac{1}{2}R^{-1}g^{\mathrm{T}}(x)\nabla V_x^{(i)}. \tag{16.11}$$

To ensure convergence of the PI algorithm an initial admissible policy $\mu^{(0)}(x(t)) \in \Psi(\Omega)$ is required. It is in fact required by the desired completion of the first step in the policy iteration: that is, finding a value associated with that initial policy (which needs to be admissible to have a finite value and for the nonlinear Lyapunov equation to have a solution). The algorithm then converges to the optimal control policy $\mu^* \in \Psi(\Omega)$ with corresponding cost $V^*(x)$. Proofs of convergence of the PI algorithm have been given in several references. See [4, 7, 12–18].

Policy iteration is a Newton method. In the linear time-invariant case, it reduces to the Kleinman algorithm [19] for solution of the Riccati equation, a familiar algorithm in control systems. Then, Equation (16.10) becomes a Lyapunov equation.

The policy iteration algorithm, as other reinforcement learning algorithms, can be implemented on an actor–critic structure which consists of two neural network structures to approximate the solutions of the two Equations (16.9) and (16.10) at each step of the iteration.

16.2.3 Online Synchronous Policy Iteration

The critic NN is based on value function approximation (VFA). In the following, it is desired to determine a rigorously justifiable form for the critic NN. We desire approximation in Sobolev norm, that is, approximation of the value $V(x)$ as well as its gradient. It is justified to assume there exist weights W_1 such that the value function $V(x)$ is approximated as

$$V(x) = W_1^{\mathrm{T}} \phi_1(x) + \varepsilon(x). \tag{16.12}$$

Then, $\phi_1(x) : \mathbb{R}^n \to \mathbb{R}^N$ is called the NN activation function vector, N the number of neurons in the hidden layer, and $\varepsilon(x)$ the NN approximation error.

Assumption 16.1 *The solution to (16.4) is smooth, that is,* $V(x) \in C^1(\Omega)$.

Assumption 16.2 *The solution to (16.4) is positive definite. This is guaranteed for stabilizable dynamics if the performance functional satisfies zero-state observability [20], which is guaranteed by the condition that* $Q(x)>0, x \in \Omega - 0; Q(0) = 0$ *be positive definite.*

The weights of the critic NN, W_1 which provide the best approximate solution for (16.5) are unknown. Therefore, the output of the critic neural network is

$$\hat{V}(x) = \hat{W}_1^{\mathrm{T}} \phi_1(x), \tag{16.13}$$

where $\hat{W}_1$ are the current estimated values of the ideal critic NN weights W_1. Recall that $\phi_1(x) : \mathbb{R}^n \to \mathbb{R}^N$ is the vector of activation functions, with N the number of neurons in the hidden layer. The approximate nonlinear Lyapunov equation is then

$$H(x, \hat{W}_1, u_2) = \hat{W}_1^{\mathrm{T}} \nabla\phi_1(f + gu_2) + Q(x) + u_2^{\mathrm{T}} R u_2 = e_1, \tag{16.14}$$

where the actor neural network is given as

$$u_2(x) = -\frac{1}{2}R^{-1}g^{\mathrm{T}}(x)\nabla\phi_1{}^{\mathrm{T}}\hat{W}_2. \tag{16.15}$$

Define the critic and the actor weight estimation errors as $\tilde{W}_1 = W_1 - \hat{W}_1$, $\tilde{W}_2 = W_1 - \hat{W}_2$.

It is desired to select $\hat{W}_1$ to minimize the squared residual error from (16.14)

$$E_1 = \frac{1}{2}e_1^{\mathrm{T}}e_1.$$

Definition 16.2 *[21] (UUB) A time signal $\zeta(t)$ is said to be uniformly ultimately bounded (UUB) if there exists a compact set $S \subset \mathbb{R}^n$ so that for all $\zeta(0) \in S$ there exists a bound B and a time $T(B, \zeta(0))$ such that $\|\zeta(t)\| \le B$ for all $t \ge t_0 + T$.*

Persistence of Excitation (PE) Assumption. Let the signal $\bar{\sigma}_{(\cdot)}$ be persistently exciting over the interval $[t, t+T]$, that is, there exist constants $\beta_1{>}0$, $\beta_2{>}0$, $T{>}0$ such that, for all t,

$$\beta_1 I \le S_0 \equiv \int_t^{t+T} \bar{\sigma}_{(\cdot)}(\tau)\bar{\sigma}_{(\cdot)}^{\mathrm{T}}(\tau)\mathrm{d}\tau \le \beta_2 I. \tag{16.16}$$

We now present the main theorem, which provides the tuning laws for the actor and critic neural networks that guarantee convergence of the synchronous online PI algorithm to the optimal controller, while guaranteeing closed-loop stability.

Theorem 16.1 *Let the dynamics be given by (16.1), the critic NN be given by (16.13) and the control input be given by actor NN (16.15). Let tuning for the critic NN be provided by the normalized gradient descent law*

$$\dot{\hat{W}}_1 = -a_1\frac{\sigma_2}{(\sigma_2^{\mathrm{T}}\sigma_2+1)^2}\left[\sigma_2^{\mathrm{T}}\hat{W}_1 + Q(x) + u_2^{\mathrm{T}}Ru_2\right], \tag{16.17}$$

where $\sigma_2 = \nabla\phi_1(f + gu_2)$, and assume that $\bar{\sigma}_2 = \sigma_2/(\sigma_2^{\mathrm{T}}\sigma_2 + 1)$ is persistently exciting. Let the actor NN be tuned as

$$_2 = -a_2\left\{(F_2\hat{W}_2 - F_1\bar{\sigma}_2^{\mathrm{T}}\hat{W}_1) - \frac{1}{4}\bar{D}_1(x)\hat{W}_2 m^{\mathrm{T}}(x)\hat{W}_1\right\}, \tag{16.18}$$

where $\bar{D}_1(x) \equiv \nabla\phi_1(x)g(x)R^{-1}g^{\mathrm{T}}(x)\nabla\phi_1^{\mathrm{T}}(x)$, $m \equiv \frac{\sigma_2}{(\sigma_2^{\mathrm{T}}\sigma_2+1)^2}$, and $F_1{>}0$ and $F_2{>}0$ are tuning parameters. Let Assumptions 16.1–16.2 hold, and the tuning parameters be selected appropriately as stated in [22]. Then there exists an N_0 such that, for the number of hidden layer units $N > N_0$ the closed-loop system state, the critic NN error $\tilde{W}_1$, and the actor NN error $\tilde{W}_2$ are UUB.

Proof: See [22]. ■

16.2.4 Simulation

Here we present a simulation of a nonlinear system to show that the game can be solved ONLINE by learning in real time and we converge to the approximate local smooth solution of HJB without solving it. Consider the following affine in control input nonlinear system, with a quadratic cost derived as in Refs. [18, 23],

$$\dot{x} = f(x) + g(x)u, \ x \in R^2,$$

where

$$f(x) = \begin{bmatrix} -x_1 + x_2 \\ -0.5x_1 - 0.5x_2(1 - (\cos(2x_1) + 2)^2) \end{bmatrix}$$

$$g(x) = \begin{bmatrix} 0 \\ \cos(2x_1) + 2 \end{bmatrix}.$$

We select $Q = \begin{bmatrix} 1 & 0 \\ 0 & 1 \end{bmatrix}$, $R = 1$.

Using the procedure in [23], the optimal value function is

$$V^*(x) = \frac{1}{2}x_1^2 + x_2^2$$

and the optimal control signal is

$$u^*(x) = -(\cos(2x_1) + 2)x_2.$$

We select the critic NN vector activation function as

$$\phi_1(x) = \left[x_1^2 \ x_1 x_2 \ x_2^2\right]^{\mathrm{T}},$$

Figure 16.1 shows the critic parameters, denoted by

$$\hat{W}_1 = [W_{c1} \ W_{c2} \ W_{c3}]^{\mathrm{T}}.$$

These converge after about 80 s to the correct values of

$$\hat{W}_1(t_f) = [0.5017 \ -0.0020 \ 1.0008]^{\mathrm{T}}.$$

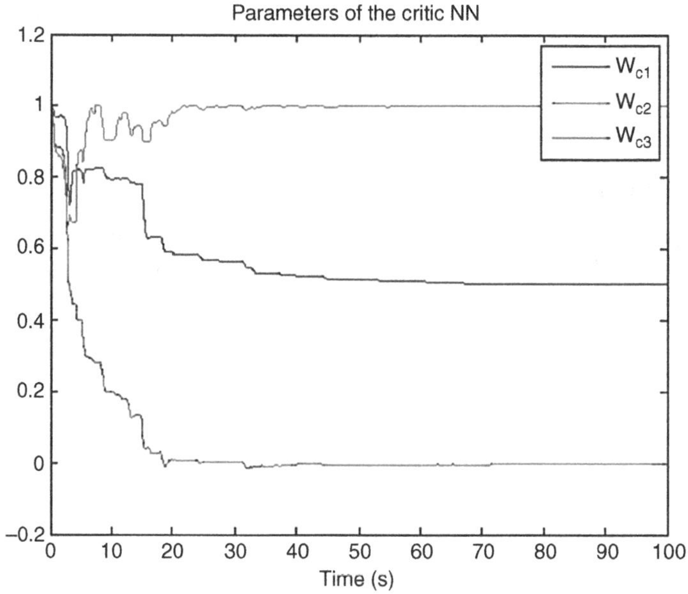

FIGURE 16.1 Convergence of the critic parameters, $\hat{W}_1 = [W_{c1}\ W_{c2}\ W_{c3}]^{\mathrm{T}}$.

The actor parameters after 80 s converge to the values of

$$\hat{W}_2(t_f) = [0.5017\ -0.0020\ 1.0008]^{\mathrm{T}}.$$

So that the actor NN (16.15)

$$\hat{u}_2(x) = -\frac{1}{2}R^{-1}\begin{bmatrix} 0 \\ \cos(2x_1)+2 \end{bmatrix}^{\mathrm{T}} \begin{bmatrix} 2x_1 & 0 \\ x_2 & x_1 \\ 0 & 2x_2 \end{bmatrix}^{\mathrm{T}} \begin{bmatrix} 0.5017 \\ -0.0020 \\ 1.0008 \end{bmatrix}$$

also converged to the optimal control.

The evolution of the system states is presented in Figure 16.2. We can see that after 80 s convergence of the NN weights in both critic and actor has occurred. This shows that the probing noise effectively guaranteed the PE condition. On convergence, the PE condition of the control signal is no longer needed, and the probing signal was turned off. After that, the states remain very close to zero, as required.

Figure 16.3 shows the 3D plot of the difference between the approximated value function, using the online algorithm, and the optimal one. This error is close to zero. Good approximation of the actual value function is being evolved.

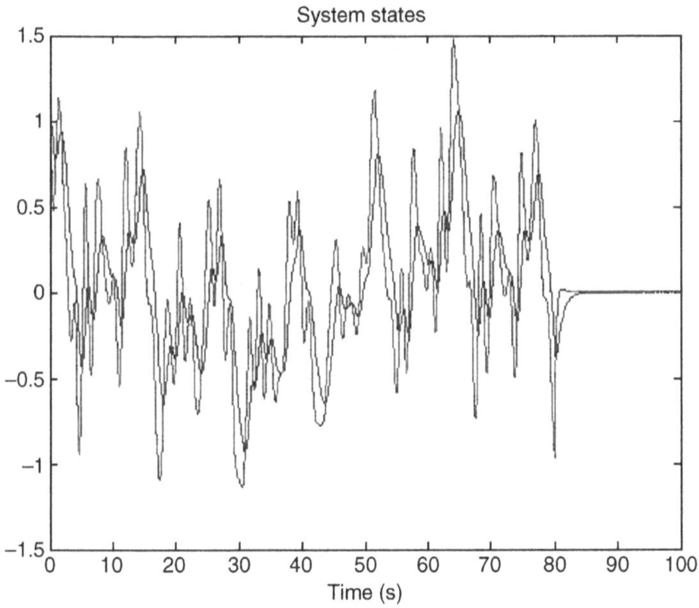

FIGURE 16.2 Evolution of the system states for the duration of the experiment.

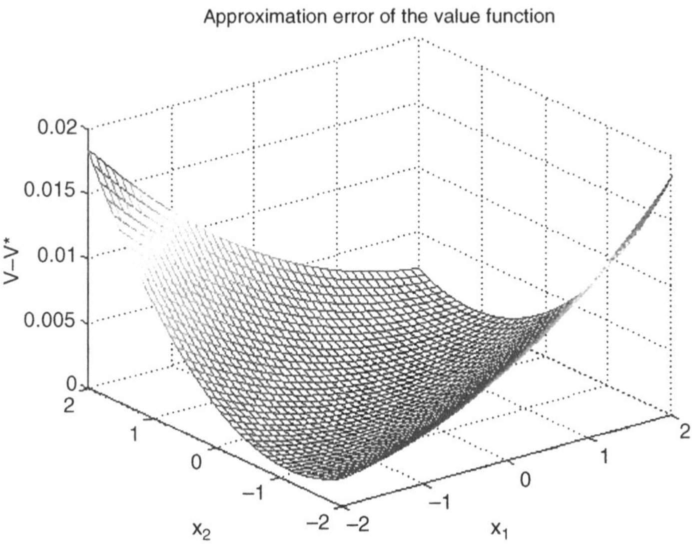

FIGURE 16.3 3D plot of the approximation error for the value function.

16.3 ONLINE SOLUTION OF NONLINEAR TWO-PLAYER ZERO-SUM GAMES AND HAMILTON–JACOBI–ISAACS EQUATION

The next section reviews the formulation of the two-player zero-sum differential game. A policy iteration algorithm is given to solve the Hamilton–Jacobi–Isaacs equation by successive solutions on nonlinear Lyapunov-like equations and will provide us with the structure for our online gaming algorithm that follows.

16.3.1 Zero-Sum Games and Hamilton–Jacobi–Isaacs Equation

Consider the nonlinear time-invariant affine in the input dynamical system given by

$$\dot{x} = f(x) + g(x)u(x) + k(x)d(x), \tag{16.19}$$

where state $x(t) \in \mathbb{R}^n$, $f(x(t)) \in \mathbb{R}^n$, $g(x(t)) \in \mathbb{R}^{nxm}$, control $u(x(t)) \in \mathbb{R}^m$, $k(x(t)) \in \mathbb{R}^{nxq}$ and disturbance $d(x(t)) \in \mathbb{R}^q$, Assume that $f(x)$ is locally Lipschitz, $f(0) = 0$ so that $x = 0$ is an equilibrium point of the system.

Define the performance index [24]

$$J(x(0), u, d) = \int_0^\infty \left(Q(x) + u^{\mathrm{T}} Ru - \gamma^2 \|d\|^2\right) \mathrm{d}t \equiv \int_0^\infty r(x, u, d)\, \mathrm{d}t, \tag{16.20}$$

for $Q(x) \geq 0$, $R = R^{\mathrm{T}} > 0$, $r(x, u, d) = Q(x) + u^{\mathrm{T}} Ru - \gamma^2 \|d\|^2$, and $\gamma \geq \gamma^* \geq 0$, where γ^* is the smallest γ for which the system is stabilized [20]. For feedback policies $u(x)$ and disturbance policies $d(x)$, define the value or cost of the policies as

$$V(x(t), u, d) = \int_t^\infty \left(Q(x) + u^{\mathrm{T}} Ru - \gamma^2 \|d\|^2\right) \mathrm{d}t. \tag{16.21}$$

When the value is finite, a differential equivalent to this is the Bellman equation

$$0 = r(x, u, d) + (\nabla V)^{\mathrm{T}}(f(x) + g(x)u(x) + k(x)d(x)), \; V(0) = 0, \tag{16.22}$$

where $\nabla V = \partial V / \partial x \in R^n$ is the (transposed) gradient and the Hamiltonian is

$$H(x, \nabla V, u, d) = r(x, u, d) + (\nabla V)^{\mathrm{T}}(f(x) + g(x)u(x) + k(x)d). \tag{16.23}$$

For feedback policies [25], a solution $V(x) \geq 0$ to (16.22) is the value (16.21) for given feedback policy $u(x)$ and disturbance policy $d(x)$.

Define the two-player zero-sum differential game [9, 25]

$$V^*(x(0)) = \min_u \max_d J(x(0), u, d) = \min_u \max_d \int_0^\infty \left(Q(x) + u^T R u - \gamma^2 \|d\|^2 \right) \mathrm{d}t \tag{16.24}$$

subject to the dynamical constraints (16.30). Thus, u is the minimizing player and d is the maximizing one. This two-player optimal control problem has a unique solution if a game theoretic saddle point exists, that is, if the Nash condition holds

$$\min_u \max_d J(x(0), u, d) = \max_d \min_u J(x(0), u, d). \tag{16.25}$$

To this game is associated the Hamilton–Jacobi–Isaacs (HJI) equation

$$\begin{aligned} 0 &= Q(x) + \nabla V^T(x) f(x) - \frac{1}{4} \nabla V^T(x) g(x) R^{-1} g^T(x) \nabla V(x) \\ &\quad + \frac{1}{4\gamma^2} \nabla V^T(x) k k^T \nabla V(x) \\ V(0) &= 0. \end{aligned} \tag{16.26}$$

Given a solution $V^*(x) \geq 0 : \mathbb{R}^n \to \mathbb{R}$ to the HJI (16.26), denote the associated control and disturbance as

$$u^* = -\frac{1}{2} R^{-1} g^T(x) \nabla V^*, \tag{16.27}$$

$$d^* = \frac{1}{2\gamma^2} k^T(x) \nabla V^*, \tag{16.28}$$

and write

$$\begin{aligned} 0 &= H(x, \nabla V, u^*, d^*) \\ &= Q(x) + \nabla V^T(x) f(x) - \tfrac{1}{4} \nabla V^T(x) g(x) R^{-1} g^T(x) \nabla V(x) + \tfrac{1}{4\gamma^2} \nabla V^T(x) k k^T \nabla V(x) \end{aligned} \tag{16.29}$$

16.3.2 Policy Iteration for Two-Player Zero-Sum Differential Games

The HJI equation (16.26) is usually intractable to solve directly. We can solve the HJI iteratively using one of several algorithms that are built on iterative solutions of the Lyapunov equation (16.22). Included are [26] which uses an inner loop with iterations on the control, and [20, 27, 28] which uses an inner loop with iterations on the disturbance. These are in effect extensions of Kleinman's algorithm [19] to nonlinear two-player games. The complementarity of these algorithms is shown in [16, 17]. Here, we shall use the latter algorithm (e.g., [20, 27, 28]).

The structure given in the algorithm below will be used as the basis for approximate online solution techniques in the next section.

Initialization: Start with a stabilizing feedback control policy u_0

1. for $j = 0, 1, \ldots$ given u_j;
2. for $i = 0, 1, \ldots$ set $d^0 = 0$, solve for $V_j^i(x(t))$, d^{i+1} using

$$0 = Q(x) + \nabla V_j^{iT}(x)(f + gu_j + kd^i) + u_j^{\mathrm{T}} R u_j - \gamma^2 ||d^i||^2, \quad (16.30)$$

$$d^{i+1} = \arg\max_d [H(x, \nabla V_j^i, u_j, d)] = \frac{1}{2\gamma^2} k^{\mathrm{T}}(x) \nabla V_j^i, \quad (16.31)$$

   on convergence, set $V_{j+1}(x) = V_j^i(x)$
3. update the control policy using

$$u_{j+1} = \arg\min_u [H(x, \nabla V_{j+1}), u, d] = -\frac{1}{2} R^{-1} g^{\mathrm{T}}(x) \nabla V_{j+1}, \quad (16.32)$$

   go to 1.

In practice, the iterations in i and j are continued until some convergence criterion is met, for example, $\left\| V_j^{i+1} - V_j^i \right\|$ or, respectively $\left\| V_{j+1} - V_j \right\|$ is small enough in some suitable norm.

16.3.3 Online Solution for Two-Player Zero-Sum Differential Games

The critic NN is based on value function approximation (VFA). Assume there exist NN weights W_1 such that the value function $V(x)$ is approximated as

$$V(x) = W_1^{\mathrm{T}} \phi_1(x) + \varepsilon(x), \quad (16.33)$$

with $\phi_1(x) : \mathbb{R}^n \to \mathbb{R}^N$ the NN activation function vector, N the number of neurons in the hidden layer, and $\varepsilon(x)$ the NN approximation error.

The ideal weights of the critic NN, W_1 which provide the best approximate solution for (16.29) are unknown. Therefore, the output of the critic neural network is

$$\hat{V}(x) = \hat{W}_1^{\mathrm{T}} \phi_1(x), \quad (16.34)$$

where $\hat{W}_1$ are the current estimated values of W_1. The approximate nonlinear Lyapunov-like equation is then

$$H(x, \hat{W}_1, \hat{u}, \hat{d}) = \hat{W}_1^{\mathrm{T}} \nabla \phi_1 (f + g\hat{u} + k\hat{d}) + Q(x) + \hat{u}^{\mathrm{T}} R \hat{u} - \gamma^2 \left\| \hat{d} \right\|^2 = e_1, \quad (16.35)$$

with the actor and the disturbance neural network defined as

$$\hat{u}(x) = -\frac{1}{2}R^{-1}g^{\mathrm{T}}(x)\nabla\phi_1^{\mathrm{T}}\hat{W}_2, \tag{16.36}$$

$$\hat{d}(x) = \frac{1}{2\gamma^2}k^{\mathrm{T}}(x)\nabla\phi_1{}^{\mathrm{T}}\hat{W}_3, \tag{16.37}$$

and e_1 a residual equation error.

Define the critic, actor, and disturbance weight estimation errors

$$\tilde{W}_1 = W_1 - \hat{W}_1,\ \tilde{W}_2 = W_1 - \hat{W}_2,\ \tilde{W}_3 = W_1 - \hat{W}_3,$$

where $\hat{W}_2,\ \hat{W}_3$ denotes the current estimated values of the ideal NN weights W_1.

It is desired to select $\hat{W}_1$ to minimize the squared residual error

$$E_1 = \frac{1}{2}e_1^{\mathrm{T}}e_1.$$

Select the tuning law for the critic weights as the normalized gradient descent algorithm.

Theorem 16.2 *Let the dynamics be given by (16.19), the critic NN be given by (16.34), the control input be given by actor NN (16.36) and the disturbance input be given by disturbance NN (16.47). Let tuning for the critic NN be provided by*

$$\dot{\hat{W}}_1 = -a_1\frac{\sigma_2}{(\sigma_2^{\mathrm{T}}\sigma_2+1)^2}[\sigma_2{}^{\mathrm{T}}\hat{W}_1 + Q(x) - \gamma^2\|\hat{d}\|^2 + \hat{u}^{\mathrm{T}}R\hat{u}], \tag{16.38}$$

where $\sigma_2 = \nabla\phi_1(f + g\hat{u} + k\hat{d})$. Let the actor NN be tuned as

$$\dot{\hat{W}}_2 = -a_2\left\{(F_2\hat{W}_2 - F_1\bar{\sigma}_2^{\mathrm{T}}\hat{W}_1) - \frac{1}{4}\bar{D}_1(x)\hat{W}_2m^{\mathrm{T}}(x)\hat{W}_1\right\} \tag{16.39}$$

and the disturbance NN be tuned as

$$\dot{\hat{W}}_3 = -a_3\left\{\left(F_4\hat{W}_3 - F_3\bar{\sigma}_2^{\mathrm{T}}\hat{W}_1\right) + \frac{1}{4\gamma^2}\bar{E}_1(x)\hat{W}_3m^{\mathrm{T}}\hat{W}_1\right\}, \tag{16.40}$$

where $\bar{D}_1(x) \equiv \nabla\phi_1(x)g(x)R^{-1}g^{\mathrm{T}}(x)\nabla\phi_1{}^{\mathrm{T}}(x)$, $\bar{E}_1(x) \equiv \nabla\phi_1(x)kk^{\mathrm{T}}\nabla\phi_1^{\mathrm{T}}(x)$, $m \equiv \frac{\sigma_2}{(\sigma_2^{\mathrm{T}}\sigma_2+1)^2}$, and $F_1>0$, $F_2>0$, $F_3>0$, $F_4>0$ are tuning parameters. Let $Q(x)>0$. Suppose that $\bar{\sigma}_2 = \sigma_2/(\sigma_2^{\mathrm{T}}\sigma_2 + 1)$ is persistently exciting. Let the tuning parameters F_1, F_2, F_3, F_4 in (16.39), and (16.40) be selected appropriately. Then there exists an N_0 such that, for the number of hidden layer units $N>N_0$ the closed-loop system state, the critic NN error $\tilde{W}_1$, the actor NN error $\tilde{W}_2$ and the disturbance NN error $\tilde{W}_3$ are UUB.

Proof: See [29]. ∎

16.3.4 Simulation

Here, we present a simulation of a nonlinear system to show that the game can be solved ONLINE by learning in real time and we converge to the approximate local smooth solution of HJI without solving it. Consider the following affine in control input nonlinear system, with a quadratic cost constructed as in [23]

$$\dot{x} = f(x) + g(x)u + k(x)d, x \in \mathbb{R}^2,$$

where

$$f(x) = \begin{bmatrix} -x_1 + x_2 \\ -x_1^3 - x_2^3 + 0.25x_2(\cos(2x_1) + 2)^2 - 0.25x_2\frac{1}{\gamma^2}(\sin(4x_1) + 2)^2 \end{bmatrix},$$

$$g(x) = \begin{bmatrix} 0 \\ \cos(2x_1) + 2 \end{bmatrix}, \quad k(x) = \begin{bmatrix} 0 \\ (\sin(4x_1) + 2) \end{bmatrix}.$$

We select $Q = \begin{bmatrix} 1 & 0 \\ 0 & 1 \end{bmatrix}$, $R = 1$, $\gamma = 8$.

Also $a_1 = a_2 = a_3 = 1$, $F_1 = I$, $F_2 = 10I$, $F_3 = I$, $F_4 = 10I$, where I is an identity matrix of appropriate dimensions.

The optimal value function is

$$V^*(x) = \frac{1}{4}x_1^4 + \frac{1}{2}x_2^2,$$

the optimal control signal is

$$u^*(x) = -\frac{1}{2}(\cos(2x_1) + 2)x_2$$

and

$$d^*(x) = \frac{1}{2\gamma^2}(\sin(4x_1) + 2)x_2.$$

We select the critic NN vector activation function as

$$\varphi_1(x) = \begin{bmatrix} x_1^2 \; x_2^2 \; x_1^4 \; x_2^4 \end{bmatrix}.$$

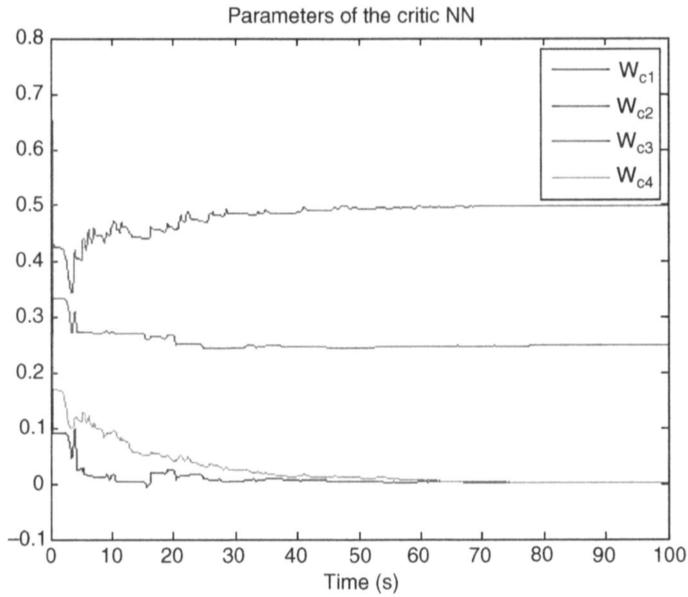

FIGURE 16.4 Convergence of the critic parameters, $\hat{W}_1 = [W_{c1}\ W_{c2}\ W_{c3}\ W_{c4}]^{\mathrm{T}}$.

Figure 16.4 shows the critic parameters, denoted by

$$\hat{W}_1 = \begin{bmatrix} W_{c1} & W_{c2} & W_{c3} & W_{c4} \end{bmatrix}^{\mathrm{T}}$$

using the synchronous zero-sum game algorithm. After convergence at about 50 s have

$$\hat{W}_1(t_f) = \begin{bmatrix} -0.0006 & 0.4981 & 0.2532 & 0.0000 \end{bmatrix}^{\mathrm{T}}.$$

The actor and disturbance parameters after 80 s converge to the values of

$$\hat{W}_3(t_f) = \hat{W}_2(t_f) = \hat{W}_1(t_f).$$

So that the actor NN

$$\hat{u}_2(x) = -\frac{1}{2}R^{-1}\begin{bmatrix} 0 \\ \cos(2x_1)+2 \end{bmatrix}^{\mathrm{T}} \begin{bmatrix} 2x_1 & 0 \\ 0 & 2x_2 \\ 4x_1^3 & 0 \\ 0 & 4x_2^3 \end{bmatrix}^{\mathrm{T}} \hat{W}_2(t_f)$$

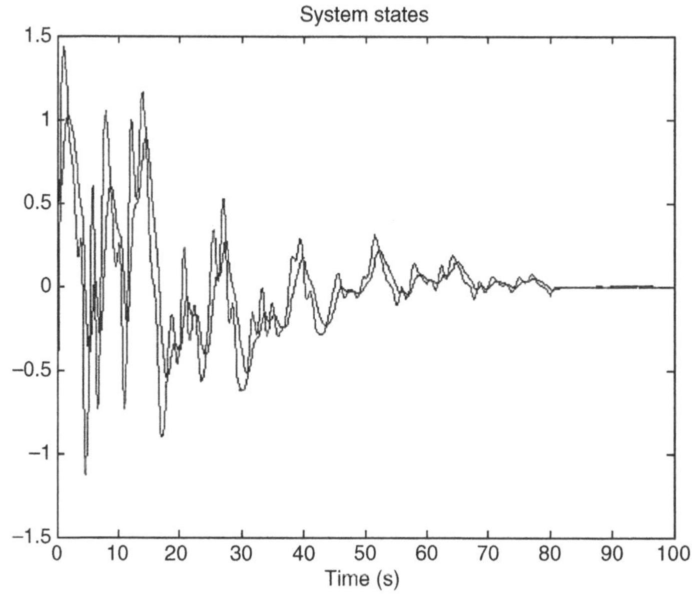

FIGURE 16.5 Evolution of the system states for critic parameters, $\hat{W}_1 = [W_{c1}\ W_{c2}\ W_{c3}\ W_{c4}]^{\mathrm{T}}$.

also converged to the optimal control, and the disturbance NN

$$\hat{d}(x) = \frac{1}{2\gamma^2}\begin{bmatrix} 0 \\ \sin(4x_1) + 2 \end{bmatrix}^{\mathrm{T}} \begin{bmatrix} 2x_1 & 0 \\ 0 & 2x_2 \\ 4x_1^3 & 0 \\ 0 & 4x_2^3 \end{bmatrix}^{\mathrm{T}} \hat{W}_3(t_f)$$

also converged to the optimal disturbance.

The evolution of the system states is presented in Figure 16.5.

16.4 ONLINE SOLUTION OF NONLINEAR NONZERO-SUM GAMES AND COUPLED HAMILTON–JACOBI EQUATIONS

The next section reviews the formulation of the nonzero sum differential games. A policy iteration algorithm is given to solve the coupled Hamilton–Jacobi equations by successive solutions on nonlinear Lyapunov-like equations and will provide us with the structure for our online algorithm that follows.

16.4.1 Nonzero Sum Games and Coupled Hamilton–Jacobi-Equations

Consider the N-player nonlinear time-invariant differential game on an infinite time horizon

$$\dot{x} = f(x) + \sum_{j=1}^{N} g_j(x)u_j, \tag{16.41}$$

where state $x(t) \in \mathbb{R}^n$, players or controls $u_j(t) \in \mathbb{R}^{m_j}$. Assume that $f(0) = 0$ and $f(x)$, $g_j(x)$ are locally Lipschitz.

The cost functionals associated with each player are

$$\begin{aligned} J_i(x(0), u_1, u_2, \ldots, u_N) &= \int_0^\infty \left(Q_i(x) + \sum_{j=1}^{N} u_j^{\mathrm{T}} R_{ij} u_j \right) \mathrm{d}t \\ &\equiv \int_0^\infty r_i(x(t), u_1, u_2, \ldots, u_N)\, \mathrm{d}t; \quad i \in N, \end{aligned} \tag{16.42}$$

where function $Q_i(x) \geq 0$ is generally nonlinear, and $R_{ii}>0$, $R_{ij} \geq 0$ are symmetric matrices.

We seek optimal controls among the set of feedback control policies with complete state information.

Given admissible feedback policies/strategies $u_i(t) = \mu_i(x)$, the value is

$$\begin{aligned} V_i(x(0), \mu_1, \mu_2, \ldots, \mu_N) &= \int_t^\infty \left(Q_i(x) + \sum_{j=1}^{N} \mu_j^{\mathrm{T}} R_{ij} \mu_j \right) \mathrm{d}\tau \\ &\equiv \int_t^\infty r_i(x(t), \mu_1, \mu_2, \ldots, \mu_N)\, \mathrm{d}\tau; \quad i \in N. \end{aligned} \tag{16.43}$$

Define the N-player game

$$V_i^*(x(t), \mu_1, \mu_2, \ldots, \mu_N) = \min_{\mu_i} \int_t^\infty \left(Q_i(x) + \sum_{j=1}^{N} \mu_j^{\mathrm{T}} R_{ij} \mu_j \right) \mathrm{d}\tau; \quad i \in N. \tag{16.44}$$

By assuming that all the players have the same hierarchical level, we focus on the so-called Nash equilibrium that is given by the following definition.

Definition 16.3 *[9] (Nash equilibrium strategies) An N-tuple of strategies* $\{\mu_1^*, \mu_2^*, \ldots, \mu_N^*\}$ *with* $\mu_i^* \in \Omega_i$, $i \in N$ *is said to constitute a Nash equilibrium* *solution for an N-player finite game in extensive form, if the following N inequalities are satisfied for all* $\mu_i^* \in \Omega_i$, $i \in N$:

$$J_i^* \triangleq J_i\left(\mu_1^*, \mu_2^*, \mu_i^*, \ldots, \mu_N^*\right) \leq J_i\left(\mu_1^*, \mu_2^*, \mu_i, \ldots, \mu_N^*\right), \; i \in N. \tag{16.45}$$

The *N-tuple* of quantities $\left\{J_1^*, J_2^*, \ldots, J_N^*\right\}$ is known as a Nash equilibrium outcome of the N-player game.

Differential equivalents to each value function are given by the following Bellman equations (different form from those given in [9])

$$0 = r(x, u_1, \ldots, u_N) + (\nabla V_i)^{\mathrm{T}} \left(f(x) + \sum_{j=1}^{N} g_j(x) u_j \right), \quad V_i(0) = 0, \quad i \in N, \tag{16.46}$$

where $\nabla V_i = \partial V_i / \partial x \in \mathbb{R}^{n_i}$ is the gradient vector (e.g., transposed gradient). Then, suitable nonnegative definite solutions to (16.46) are the values evaluated using the infinite integral (16.44) along the system trajectories. Define the Hamiltonian functions

$$H_i(x, \nabla V_i, u_1, \ldots, u_N) = r(x, u_1, \ldots, u_N) + (\nabla V_i)^{\mathrm{T}} \left(f(x) + \sum_{j=1}^{N} g_j(x) u_j \right), \; i \in N. \tag{16.47}$$

According to the stationarity conditions, associated feedback control policies are given by

$$\frac{\partial H_i}{\partial u_i} = 0 \Rightarrow \mu_i(x) = -\frac{1}{2} R_{ii}^{-1} g_i^{\mathrm{T}}(x) \nabla V_i, \quad i \in N. \tag{16.48}$$

Substituting (16.48) into (16.47) we obtain the N coupled Hamilton–Jacobi (HJ) equations

$$\begin{aligned} 0 &= (\nabla V_i)^{\mathrm{T}} \left(f(x) - \frac{1}{2} \sum_{j=1}^{N} g_i(x) R_{jj}^{-1} g_j^{\mathrm{T}}(x) \nabla V_j \right) \\ &\quad + Q_i(x) + \frac{1}{4} \sum_{j=1}^{N} \nabla V_j^{\mathrm{T}} g_j(x) R_{jj}^{-T} R_{ij} R_{jj}^{-1} g_j^{\mathrm{T}}(x) \nabla V_j \qquad (16.49) \\ V_i(0) &= 0. \end{aligned}$$

These coupled HJ equations are in "closed-loop" form. The equivalent "open-loop" form is

$$0 = \nabla V_i^{\mathrm{T}} f(x) + Q_i(x) - \frac{1}{2}\nabla V_i^{\mathrm{T}} \sum_{j=1}^{N} g_j(x) R_{jj}^{-1} g_j^{\mathrm{T}}(x) \nabla V_j$$

$$+ \frac{1}{4}\sum_{j=1}^{N} \nabla V_j^{\mathrm{T}} g_j(x) R_{jj}^{-T} R_{ij} R_{jj}^{-1} g_j^{\mathrm{T}}(x) \nabla V_j. \tag{16.50}$$

$$V_i(0) = 0.$$

In linear systems of the form $\dot{x} = Ax + \sum_{j=1}^{N} B_j u_j$, Equation (16.49) becomes the N coupled generalized algebraic Riccati equations

$$0 = P_i A_c + A_c^{\mathrm{T}} P_i + Q_i + \frac{1}{4}\sum_{j=1}^{N} P_j B_j R_{jj}^{-T} R_{ij} R_{jj}^{-1} B_j^{\mathrm{T}} P_j, \quad i \in N, \tag{16.51}$$

where $A_c = A - \frac{1}{2}\sum_{i=1}^{N} B_i R_{ii}^{-1} B_i^{\mathrm{T}} P_i$. It is shown in [9] that if there exist solutions to (16.74) and further satisfying the conditions that for each $i \in N$ the pair $\left(A - \frac{1}{2}\sum_{\substack{j \in N \\ j \neq i}} B_j R_{jj}^{-1} B_j^{\mathrm{T}} P_j \quad B_i\right)$ is stabilizable and the pair $\left(A - \frac{1}{2}\sum_{\substack{j \in N \\ j \neq i}} B_j R_{jj}^{-1} B_j^{\mathrm{T}} P_j,\ Q_i + \frac{1}{4}\sum_{\substack{j \in N \\ j \neq i}} P_j B_j R_{jj}^{-T} R_{ij} R_{jj}^{-1} B_j^{\mathrm{T}} P_j\right)$ is detectable, then the N-tuple of the stationary feedback policies $\mu_i^*(x) = -K_i x = -\frac{1}{2} R_{ii}^{-1} B_i^{\mathrm{T}} P_i x, \quad i \in N$ provides a Nash equilibrium solution for the linear quadratic N-player differential game among feedback policies with full state information. Furthermore, the resulting system dynamics, described by $\dot{x} = A_c x, \quad x(0) = x_0$ are asymptotically stable.

16.4.2 Policy Iteration for Nonzero Sum Differential Games

Equation (16.50) is difficult or impossible to solve. An iterative offline solution technique is given by the following policy iteration algorithm. It solves the coupled HJ equations by iterative solution of uncoupled nonlinear Lyapunov equations.

(a) Start with stabilizing initial policies $\mu_1^0(x), \ldots, \mu_N^0(x)$.

(b) Given the N-tuple of policies $\mu_1^k(x), \ldots, \mu_N^k(x)$, solve for the N-tuple of costs $V_1^k(x(t)), V_2^k(x(t)) \ldots V_N^k(x(t))$ using the Bellman equations

$$0 = r(x, \mu_1^k, \ldots, \mu_N^k) + (\nabla V_i^k)^{\mathrm{T}} \left(f(x) + \sum_{j=1}^{N} g_j(x)\mu_j^i\right), \quad V^k(0) = 0 \ \ i \in N. \tag{16.52}$$

(c) Update the N-tuple of control policies using:

$$\mu_i^{k+1} = \underset{u_i \in \Psi(\Omega)}{\operatorname{argmin}}[H_i(x, \nabla V_i, u_1, \ldots, u_N)] \quad i \in N, \tag{16.53}$$

which explicitly is

$$\mu_i^{k+1}(x) = -\frac{1}{2} R_{ii}^{-1} g_i^{\mathrm{T}}(x) \nabla V_i^k \quad i \in N. \tag{16.54}$$

A linear two-player version is given in [11] and can be considered as an extension of Kleiman's algorithm [19] to two-player games.

In the next section, we use PI Algorithm 1 to motivate the control structure for an online adaptive N-player game solution algorithm. Then it is proven that "optimal adaptive" control algorithm converges online to the solution of coupled HJs (16.50), while guaranteeing closed-loop stability.

16.4.3 Online Solution for Two-Player Nonzero Sum Differential Games

In this section, we show how to solve the two-player nonzero sum game online, the approach can easily be extended to more than two players.

Consider the nonlinear time-invariant affine in the input dynamical system given by

$$\dot{x} = f(x) + g(x)u(x) + k(x)d(x), \tag{16.55}$$

where state $x(t) \in \mathbb{R}^n$, first control input $u(x) \in \mathbb{R}^m$, and second control input $d(x) \in \mathbb{R}^q$. Assume that $f(0) = 0$ and $f(x)$, $g(x)$, $k(x)$ are locally Lipschitz, $\| f(x) \| < b_f \| x \|$.

This work uses nonlinear approximator structures for value function approximation (VFA) [30, 31, 32] to solve (16.50). Therefore, assume there exist constant neural network (NN) weights W_1 and W_2 such that the value functions $V_1(x)$ and $V_2(x)$ are approximated on a compact set Ω as

$$V_1(x) = W_1^{\mathrm{T}} \phi_1(x) + \varepsilon_1(x) \tag{16.56}$$

$$V_2(x) = W_2^{\mathrm{T}} \phi_2(x) + \varepsilon_2(x) \tag{16.57}$$

with $\phi_1(x) : \mathbb{R}^n \to \mathbb{R}^K$ and $\phi_2(x) : \mathbb{R}^n \to \mathbb{R}^K$ the NN activation function basis set vectors, K the number of neurons in the hidden layer, and $\varepsilon_1(x)$ and $\varepsilon_2(x)$ the NN approximation errors. From the approximation literature, the basis functions can be selected as sigmoids, tanh, polynomials, and so on.

Assuming current NN weight estimates $\hat{W}_1$ and $\hat{W}_2$, the outputs of the two critic NN are given by

$$\hat{V}_1(x) = \hat{W}_1^{\mathrm{T}}\phi_1(x) \tag{16.58}$$

$$\hat{V}_2(x) = \hat{W}_2^{\mathrm{T}}\phi_2(x). \tag{16.59}$$

The approximate Lyapunov-like equations are then

$$\begin{aligned} H_1(x, \hat{W}_1, u_3, d_4) &= Q_1(x) + u_3^{\mathrm{T}} R_{11} u_3 + d_4^{\mathrm{T}} R_{12} d_4 \\ &\quad + \hat{W}_1^{\mathrm{T}} \nabla\phi_1(f(x) + g(x)u_3 + k(x)d_4) \\ &= e_1, \end{aligned} \tag{16.60}$$

$$\begin{aligned} H_2(x, \hat{W}_2, u_3, d_4) &= Q_2(x) + u_3^{\mathrm{T}} R_{21} u_3 + d_4^{\mathrm{T}} R_{22} d_4 \\ &\quad + \hat{W}_2^{\mathrm{T}} \nabla\phi_2(f(x) + g(x)u_3 + k(x)d_4) \\ &= e_2. \end{aligned} \tag{16.61}$$

It is desired to select $\hat{W}_1$ and $\hat{W}_2$ to minimize the square residual error

$$E_1 = \frac{1}{2}e_1^{\mathrm{T}}e_1 + \frac{1}{2}e_2^{\mathrm{T}}e_2.$$

Then $\hat{W}_1(t) \to W_1$, $\hat{W}_2(t) \to W_2$ and we define the control policies in (16.60), (16.61) in the form of action neural networks that compute the control inputs in the structured form

$$u_3(x) = -\frac{1}{2}R_{11}^{-1}g^{\mathrm{T}}(x)\nabla\phi_1^{\mathrm{T}}\hat{W}_3, \tag{16.62}$$

$$d_4(x) = -\frac{1}{2}R_{22}^{-1}k^{\mathrm{T}}(x)\nabla\phi_2^{\mathrm{T}}\hat{W}_4, \tag{16.63}$$

where $\hat{W}_3$ and $\hat{W}_4$ denote the current estimated values of the ideal NN weights W_1 and W_2, respectively. Define the critic and the actor NN estimation errors, respectively, as

$$\tilde{W}_1 = W_1 - \hat{W}_1,\ \tilde{W}_2 = W_2 - \hat{W}_2,\ \tilde{W}_3 = W_1 - \hat{W}_3,\ \tilde{W}_4 = W_2 - \hat{W}_4. \tag{16.64}$$

The main Theorem of nonzero sum games is now given, which provide the tuning laws for the actor and critic neural networks that guarantee convergence of the two-player nonzero-sum game algorithm in real time to the Nash equilibrium solution, while also guaranteeing closed-loop stability.

Theorem 16.3 *Let the dynamics be given by (16.55), and consider the two-player nonzero sum game formulation described before. Let the critic NNs be given by (16.58)* *and (16.59), the control inputs be given by (16.62) and (16.63). Let tuning for the critic NNs be provided by*

$$\dot{\hat{W}}_1 = -a_1 \frac{\sigma_3}{(\sigma_3^T \sigma_3 + 1)^2} \left[\sigma_3^T \hat{W}_1 + Q_1(x) + u_3^T R_{11} u_3 + d_4^T R_{12} d_4\right], \tag{16.65}$$

$$\dot{\hat{W}}_2 = -a_2 \frac{\sigma_4}{(\sigma_4^T \sigma_4 + 1)^2} \left[\sigma_4^T \hat{W}_2 + Q_2(x) + u_3^T R_{21} u_3 + d_4^T R_{22} d_4\right], \tag{16.66}$$

where $\sigma_3 = \nabla\phi_1(f + gu_3 + kd_4)$ and $\sigma_4 = \nabla\phi_2(f + gu_3 + kd_4)$. Let the first actor NN (first player) be tuned as

$$\dot{\hat{W}}_3 = -a_3 \Big\{ \left(F_2 \hat{W}_3 - F_1 \bar{\sigma}_3^T \hat{W}\right)_1 - \frac{1}{4}\left(\nabla\phi_1 g(x) R_{11}^{-T} R_{21} R_{11}^{-1} g^T(x) \nabla\phi_1^T \hat{W}_3 m_2^T \hat{W}_2 + \bar{D}_1(x) \hat{W}_3 m_1^T \hat{W}_1\right)\Big\} \tag{16.67}$$

and the second actor (second player) NN be tuned as

$$\dot{\hat{W}}_4 = -a_4 \Big\{ \left(F_4 \hat{W}_4 - F_3 \bar{\sigma}_4^T \hat{W}_2\right) - \frac{1}{4}\left(\nabla\phi_2 k(x) R_{22}^{-T} R_{12} R_{22}^{-1} k^T(x) \nabla\phi_2^T \hat{W}_4 m_1^T \hat{W}_1 + \bar{D}_2(x) \hat{W}_4 m_2^T \hat{W}_2\right)\Big\}, \tag{16.68}$$

where
$\bar{D}_1(x) \equiv \nabla\phi_1(x) g(x) R_{11}^{-1} g^T(x) \nabla\phi_1^T(x)$, $\bar{D}_2(x) \equiv \nabla\phi_2(x) k R_{22}^{-1} k^T \nabla\phi_2^T(x)$, $m_1 \equiv \frac{\sigma_3}{(\sigma_3{}^T \sigma_3 + 1)^2}$, $m_2 \equiv \frac{\sigma_4}{(\sigma_4{}^T \sigma_4 + 1)^2}$, and $F_1 > 0$, $F_2 > 0$, $F_3 > 0$, $F_4 > 0$ are tuning parameters. Assume that $Q_1(x) > 0$ and $Q_2(x) > 0$. Suppose that $\bar{\sigma}_3 = \sigma_3/(\sigma_3^T \sigma_3 + 1)$ and $\bar{\sigma}_4 = \sigma_4/(\sigma_4^T \sigma_4 + 1)$ are persistently exciting. Let the tuning parameters be selected as detailed in [33]. Then there exists a K_0 such that, for the number of hidden layer units $K > K_0$ the closed-loop system state, the critic NN errors $\tilde{W}_1$ and $\tilde{W}_2$, the first actor NN error $\tilde{W}_3$ and the second actor NN error $\tilde{W}_4$ are UUB.

Proof: See [33]. ■

16.4.4 Simulation

Here, we present a simulation of a nonlinear system to show that the game can be solved online by learning in real time, using the method of this chapter. PE is needed to guarantee convergence to the Nash solution without solving the coupled

HJ equations. In this simulation, exponentially decreasing probing noise is added to the control inputs to ensure PE until convergence is obtained.

Consider the following affine in control input nonlinear system, with a quadratic cost constructed as in [23]

$$\dot{x} = f(x) + g(x)u + k(x)d, \; x \in \mathbb{R}^2,$$

where

$$f(x) = \begin{bmatrix} x_2 \\ -x_2 - \frac{1}{2}x_1 + \frac{1}{4}x_2(\cos(2x_1)+2)^2 + \frac{1}{4}x_2(\sin(4x_1^2)+2)^2 \end{bmatrix}$$

$$g(x) = \begin{bmatrix} 0 \\ \cos(2x_1)+2 \end{bmatrix}, \quad k(x) = \begin{bmatrix} 0 \\ \sin(4x_1^2)+2 \end{bmatrix}.$$

Select $Q_1 = 2Q_2$, $R_{11} = 2R_{22}$, and $R_{12} = 2R_{21}$, where Q_2,R_{22}, and R_{21} are identity matrices. Select $a_1 = a_2 = a_3 = a_4 = 1$ and $F_4 = F_3 = F_2 = F_1 = 100I$ for the constants on the tuning laws, where I is an identity matrix of appropriate dimensions.

The optimal value function for the first critic (player 1) is

$$V_1^*(x) = \frac{1}{2}x_1^2 + x_2^2$$

and for the second critic (player 2) is

$$V_2^*(x) = \frac{1}{4}x_1^2 + \frac{1}{2}x_2^2.$$

The optimal control signal for the first player is

$$u^*(x) = -2(\cos(2x_1)+2)x_2$$

and the optimal control signal for the second player is

$$d^*(x) = -(\sin(4x_1^2)+2)x_2.$$

We select the NN vector activation function for the critics as $\varphi_1(x) = \varphi_2(x) \equiv \begin{bmatrix} x_1^2 & x_1x_2 & x_2^2 \end{bmatrix}$ and uses the control algorithm presented in Theorem 16.3. Each player maintains two NN, a critic NN to estimate its current value and an action NN to estimate its current control policy.

Figure 16.6 shows the critic parameters for the first player, denoted by $\hat{W}_1 = \begin{bmatrix} W_{c1} & W_{c2} & W_{c3} \end{bmatrix}^{\mathrm{T}}$ using the proposed game algorithm. After convergence at about 150 s, we have $\hat{W}_1(t_f) = \begin{bmatrix} 0.5015 & 0.0007 & 1.0001 \end{bmatrix}^{\mathrm{T}}$. Figure 16.7 shows the critic parameters for the second player $\hat{W}_2 = \begin{bmatrix} W_{2c1} & W_{2c2} & W_{2c3} \end{bmatrix}^{\mathrm{T}}$. After

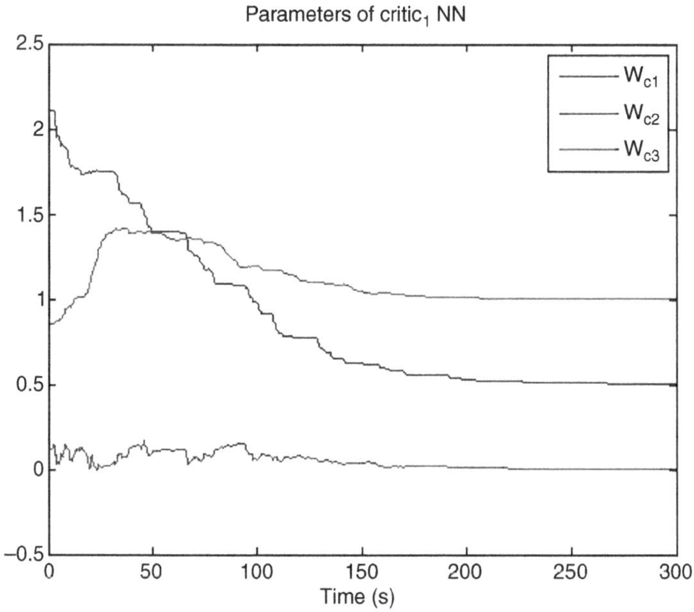

FIGURE 16.6 Convergence of the critic parameters for first player.

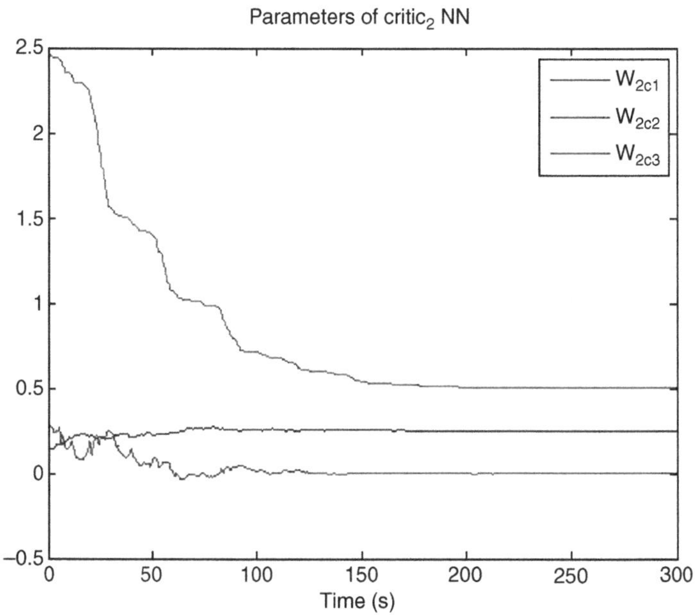

FIGURE 16.7 Convergence of the critic parameters for second player.

convergence at about 150 s, we have $\hat{W}_2(t_f) = \begin{bmatrix} 0.2514\ 0.0006\ 0.5001 \end{bmatrix}^{\mathrm{T}}$. The actor parameters for the first player after 150 s converge to the values of $\hat{W}_3(t_f) = \begin{bmatrix} 0.5015\ 0.0007\ 1.0001 \end{bmatrix}^{\mathrm{T}}$, and the actor parameters for the second player converge to the values of $\hat{W}_4(t_f) = \begin{bmatrix} 0.2514\ 0.0006\ 0.5001 \end{bmatrix}^{\mathrm{T}}$.

Therefore, the actor NN for the first player

$$\hat{u}(x) = -\frac{1}{2} R_{11}^{-1} \begin{bmatrix} 0 \\ \cos(2x_1) + 2 \end{bmatrix}^{\mathrm{T}} \begin{bmatrix} 2x_1 & 0 \\ x_2 & x_1 \\ 0 & 2x_2 \end{bmatrix}^{\mathrm{T}} \begin{bmatrix} 0.5015 \\ 0.0007 \\ 1.0001 \end{bmatrix}$$

also converged to the optimal control, and similarly for the second player

$$\hat{d}(x) = -\frac{1}{2} R_{22}^{-1} \begin{bmatrix} 0 \\ \sin(4x_1^2) + 2 \end{bmatrix}^{\mathrm{T}} \begin{bmatrix} 2x_1 & 0 \\ x_2 & x_1 \\ 0 & 2x_2 \end{bmatrix}^{\mathrm{T}} \begin{bmatrix} 0.2514 \\ 0.0006 \\ 0.5001 \end{bmatrix}.$$

The evolution of the system states is presented in Figure 16.8.

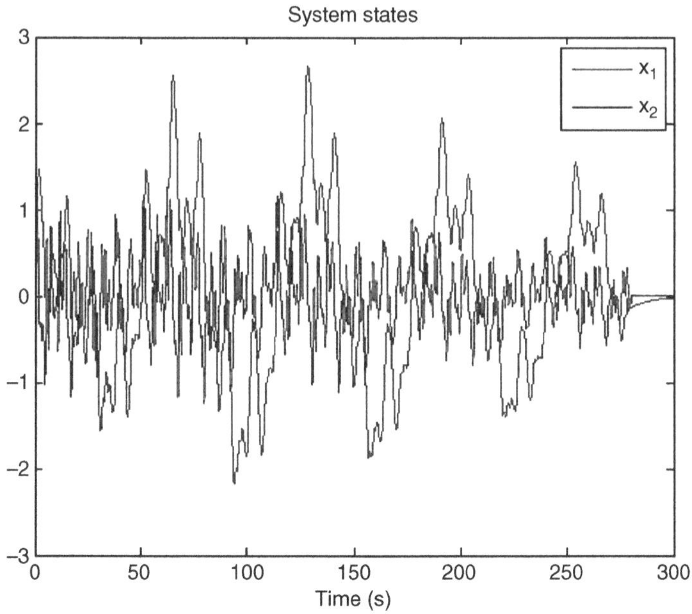

FIGURE 16.8 Evolution of the system states for critic parameters for first and second players.

ACKNOWLEDGMENTS

This work was supported by the National Science Foundation ECS-0801330, the Army Research Office W91NF-05-1-0314 and the Air Force Office of Scientific Research FA9550-09-1-0278.

REFERENCES

1. F.L. Lewis, G. Lendaris, and D. Liu. Special issue on approximate dynamic programming and reinforcement learning for feedback control. *IEEE Transactions Systems, Man, and Cybernetics, Part B*, 38(4): 2008.
2. K. Doya. Reinforcement learning in continuous time and space. *Neural Computation*, 12(1):219–245, 2000.
3. K. Doya, H. Kimura, and M. Kawato. Neural mechanisms of learning and control. *IEEE Control Systems Management*, 21(4):42–54, 2001.
4. R.A. Howard. *Dynamic Programming and Markov Processes*. MIT Press, Cambridge, MA, 1960.
5. F.L. Lewis, and K.G. Vamvoudakis. Reinforcement learning for partially observable dynamic processes: adaptive dynamic programming using measured output data. *IEEE Transactions Systems, Man, and Cybernetics, Part B*, 41(1):14–25, 2011.
6. R.S. Sutton and A.G. Barto. *Reinforcement Learning—An Introduction*. MIT Press, Cambridge, MA, 1998.
7. M. Abu-Khalaf and F.L. Lewis. Nearly optimal control laws for nonlinear systems with saturating actuators using a neural network HJB approach. *Automatica*, 41(5):779–791, 2005.
8. S. Tijs. *Introduction to Game Theory*. Hindustan Book Agency, India, 2003.
9. T. Başar and G.J. Olsder. *Dynamic Noncooperative Game Theory*. 2nd edition, SIAM, Philadelphia, PA, 1999.
10. G. Freiling, G. Jank, and H. Abou-Kandil. On global existence of solutions to coupled matrix Riccati equations in closed loop Nash games. *IEEE Transactions on Automatic Control*, 41(2):264–269, 2002.
11. Z. Gajic and T-Y. Li. Simulation results for two new algorithms for solving coupled algebraic Riccati equations. In *Third International Symposium On Differential Games, Sophia, Antipolis, France*, 1988.
12. L.C. III Baird. Reinforcement learning in continuous time: advantage updating. In *Proceedings of ICNN, Orlando, FL*, 1994.
13. R. Beard, G. Saridis, and J. Wen. Galerkin approximations of the generalized Hamilton–Jacobi–Bellman equation, *Automatica*, 33(12):2159–2177, 1997.
14. T. Hanselmann, L. Noakes, and A. Zaknich. Continuous-time adaptive critics. *IEEE Transactions on Neural Networks*, 18(3):631–647, 2007.
15. J.J. Murray, C.J. Cox, G.G. Lendaris, and R. Saeks. Adaptive dynamic programming. *IEEE Transactions on Systems, Man and Cybernetics*, 32(2):140–153. 2002.
16. D. Vrabie. *Online Adaptive Optimal Control for Continuous Time Systems*. Ph.D. thesis, Department of Electrical Engineering, University Texas at Arlington, Arlington, TX, USA, 2009.

17. D. Vrabie, O. Pastravanu, F.L. Lewis, and M. Abu-Khalaf. Adaptive optimal control for continuous-time linear systems based on policy iteration. *Automatica*, 45(2):477–484, 2009.

18. D. Vrabie, K. Vamvoudakis, and F. Lewis. Adaptive optimal controllers based on generalized policy iteration in a continuous-time framework. In *Proceedings of the IEEE Mediterranean Conference on Control and Automation*, 1402–1409, 2009.

19. D. Kleinman. On an iterative technique for Riccati equation computations. *IEEE Transactions on Automatic Control*, 13(1):114–115, 1968.

20. A.J. Van der Schaft. *L2*-gain analysis of nonlinear systems and nonlinear state feedback $H\infty$ control. *IEEE Transactions on Automatic Control*, 37(6):770–784, 1992.

21. F.L. Lewis, S. Jagannathan, and A. Yesildirek. *Neural Network Control of Robot Manipulators and Nonlinear Systems*. Taylor & Francis, 1999.

22. K.G. Vamvoudakis and F.L. Lewis. Online actor-critic algorithm to solve the continuous-time infinite horizon optimal control problem. *Automatica*, 46(5):878–888, 2010.

23. V. Nevistic and J.A. Primbs. Constrained nonlinear optimal control: a converse HJB approach, California Institute of Technology, Pasadena, CA 91125, Technical Report CIT-CDS 96-021, 1996.

24. F.L. Lewis, and V.L. Syrmos. *Optimal Control*. John Wiley, 1995.

25. T. Başar and P. Bernard. H_∞ *Optimal Control and Relate*d Minimax Design Problems. Birkhäuser, Boston, MA, 1995.

26. Y. Feng, B.D. Anderson, and M. Rotkowitz. A game theoretic algorithm to compute local stabilizing solutions to HJBI equations in nonlinear $H\infty$ control. *Automatica*, 45(4):881–888, 2009.

27. M. Abu-Khalaf and F.L. Lewis. Neurodynamic programming and zero-sum games for constrained control systems. *IEEE Transactions on Neural Networks*, 19(7):1243–1252, 2008.

28. M. Abu-Khalaf, F.L. Lewis, and Jie Huang. Policy iterations on the Hamilton–Jacobi–Isaacs equation for H_∞ state feedback control with input saturation. *IEEE Transactions on Automatic Control*, 51(12):1989–1995, 2006.

29. K.G. Vamvoudakis and F.L. Lewis. Online neural network solution of nonlinear two-player zero-sum games using synchronous policy iteration. *International Journal of Robust and Nonlinear Control*, 22(13):1460–1483, 2012.

30. D.P. Bertsekas and J.N. Tsitsiklis. *Neuro-Dynamic Programming*. Athena Scientific, MA, 1996.

31. P.J. Werbos. Beyond regression: new tools for prediction and analysis in the behavior sciences. Ph.D. thesis, 1974.

32. P.J. Werbos. Approximate dynamic programming for real-time control and neural modeling. In D.A. White and D.A. Sofge, editors. *Handbook of Intelligent Control*. Van Nostrand Reinhold, New York, 1992.

33. K.G. Vamvoudakis and F.L. Lewis. Multi-player non zero sum games: online adaptive learning solution of coupled Hamilton-Jacobi equations. *Automatica*, 47(8):1556–1569, 2011.

PART III

FOUNDATIONS IN MDP AND RL

CHAPTER 17

Lambda-Policy Iteration: A Review and a New Implementation

DIMITRI P. BERTSEKAS

Department of Electrical Engineering and Computer Science, Institute of Technology, Cambridge, MA, USA

ABSTRACT

In this chapter, we discuss λ-policy iteration, a method for exact and approximate dynamic programming. It is intermediate between the classical value iteration (VI) and the policy iteration (PI) methods, and it is closely related to optimistic (also known as modified) PI, whereby each policy evaluation is done approximately, using a finite number of VI. We review the theory of the method and associated questions of bias and exploration arising in simulation-based cost function approximation. We then discuss various implementations, which offer advantages over well-established PI methods that use LSPE(λ), LSTD(λ), or TD(λ) for policy evaluation with cost function approximation. One of these implementations is based on a new simulation scheme, called geometric sampling, which uses multiple short trajectories rather than a single infinitely long trajectory.

17.1 INTRODUCTION

Approximate dynamic programming (DP for short) has attracted substantial research interest, and has a wide range of applications, because of its potential to address large and complex problems that may not be treatable in other ways. The literature on the subject is very extensive and includes several textbooks, research monographs, and surveys that relate to the computational context of this chapter. For a nonexhaustive list, we mention the books by Bertsekas and Tsitsiklis [1], Sutton and Barto [2],

Reinforcement Learning and Approximate Dynamic Programming for Feedback Control, First Edition. Edited by Frank L. Lewis and Derong Liu.

Gosavi [3], Cao [4], Chang et al. [5], Meyn [6], Powell [7], Borkar [8], Haykin [9], Busoniu et al. [10], and the author's recent text [11]; the edited volumes and special issues by White and Sofge [12], Si et al. [13], Lewis et al. [14], and the 2007–2009 Proceedings of the IEEE Symposium on Approximate Dynamic Programming and Reinforcement Learning; and the recent surveys by Borkar [15], Lewis and Vrabie [16], Werbos [17], Szepesvari [18], and Bertsekas [19].

The purpose of this chapter is to critically review and extend a class of methods for exact and approximate DP, which are based on the λ-policy iteration (λ-PI) method, proposed by Bertsekas and Ioffe [20]. This method is intermediate between the classical value iteration (VI) and the policy iteration (PI) methods, and it is closely related to optimistic (also known as modified) PI, whereby each policy evaluation is done approximately, using a finite number of VI. It was originally used as the starting point for the development of approximate simulation-based DP methods of the temporal difference (TD) type, such as LSPE(λ) (see [20], and also [1], Sections 2.3.1 and 8.3). The emphasis in this chapter is on implementations of λ-PI, which provide alternatives to approximate PI methods that use other more established methods for policy evaluation.

We will focus on the α-discounted n-state Markovian decision problem (MDP), although the main ideas are more broadly applicable. The problem involves states $1, \ldots, n$, controls $u \in U(i)$ at state i, transition probabilities $p_{ij}(u)$, and cost $g(i, u, j)$ for transition from i to j under control u. A (stationary) policy μ is a function from states i to admissible controls $u \in U(i)$, and $J_\mu(i)$ is the cost starting from state i and using policy μ. It is well known (see e.g., Puterman [21] or Bertsekas [22]) that the vector $J_\mu \in \Re^n$, which has components $J_\mu(i)$,[1] is the unique fixed point of the mapping $T_\mu : \Re^n \mapsto \Re^n$, which maps $J \in \Re^n$ to the vector $T_\mu J \in \Re^n$ that has components

$$(T_\mu J)(i) = \sum_{j=1}^{n} p_{ij}\big(\mu(i)\big)\big(g(i, \mu(i), j) + \alpha J(j)\big), \qquad i = 1, \ldots, n. \tag{17.1}$$

Similarly, the optimal costs starting from $i = 1, \ldots, n$, are denoted $J^*(i)$, and the optimal cost vector $J^* \in \Re^n$, which has components $J^*(i)$, is the unique fixed point of the mapping $T : \Re^n \mapsto \Re^n$ defined by

$$(TJ)(i) = \min_{u \in U(i)} \sum_{j=1}^{n} p_{ij}(u)\big(g(i, u, j) + \alpha J(j)\big), \qquad i = 1, \ldots, n. \tag{17.2}$$

An important property is that T_μ and T are sup norm contractions. In particular, the iterations $J_{k+1} = T_\mu J_k$ and $J_{k+1} = TJ_k$ converge to J_μ and J^*, respectively, from any starting point J_0-this is the VI method.

[1] In our notation, $\Re^n$ is the n-dimensional Euclidean space, all vectors in $\Re^n$ are viewed as column vectors, and a prime denotes transposition. The identity matrix is denoted by I.

A major alternative to VI is PI. It produces a sequence of policies and associated cost functions through iterations that have two phases: *policy evaluation* (where the cost function of a policy is evaluated) and *policy improvement* (where a new policy is generated). In the exact form of the algorithm, the current policy μ is improved by finding $\bar{\mu}$ that satisfies $T_{\bar{\mu}} J_\mu = TJ_\mu$ (i.e., by minimizing in the right-hand side of Equation (17.2) with J_μ in place of J). The improved policy $\bar{\mu}$ is evaluated by solving the linear system of equations $J_{\bar{\mu}} = T_{\bar{\mu}} J_{\bar{\mu}}$, and $(J_{\bar{\mu}}, \bar{\mu})$ becomes the new cost vector–policy pair, which is used to start a new iteration. Thus, the exact form of PI can be succinctly defined as

$$T_{\mu_{k+1}} J_k = TJ_k, \qquad J_{k+1} = T_{\mu_{k+1}} J_{k+1}, \tag{17.3}$$

with the equation on the left describing the policy improvement and the equation on the right describing the evaluation of μ_{k+1}.

In a variant of the method, a policy μ_{k+1} is evaluated by a finite number of applications of $T_{\mu_{k+1}}$ to an approximate evaluation of the preceding policy. This is known as "optimistic" or "modified" PI, and its motivation is that in problems with a large number of states, the linear system $J_{k+1} = T_{\mu_{k+1}} J_{k+1}$ cannot be practically solved directly by matrix inversion, so it is best solved iteratively by VI. The method can be succinctly defined as

$$T_{\mu_{k+1}} J_k = TJ_k, \qquad J_{k+1} = T^{m_k}_{\mu_{k+1}} J_k. \tag{17.4}$$

If the number m_k of applications of $T_{\mu_{k+1}}$ is very large, the exact form of PI is essentially obtained, but practice has shown that it is most efficient to use a moderate value of m_k. In this case, the algorithm looks like a hybrid of VI and PI, involving a sequence of alternate applications of T and T_{μ_k}, with μ_k changing over time. Optimistic PI is generally believed to be more computationally efficient that either VI or PI. This is particularly so for problems where n is very large and implementation of exact PI is difficult due to the associated $n \times n$ matrix inversion, and also for problems with a large number of controls, where the overhead due to minimization over all controls $u \in U(i)$ in the mapping T [cf. Equation (17.2)] is substantial.

We note that the convergence properties of the optimistic PI method (17.4) are quite complicated and have been the subject of continuing research. The convergence $J_k \to J^*$ has been established by Rothblum [23] (see also the more recent work by Canbolat and Rothblum [24], which extends some of the results of [23]). On the other hand, when optimistic PI is implemented asynchronously (as it normally would be when simulation is used), it may oscillate as shown by the convergence counterexamples of Williams and Baird [25]. Recent work of Bertsekas and Yu [26–28] has developed convergent variants of synchronous and asynchronous optimistic PI and Q-learning, based on a new way to perform policy evaluation: by solving approximately an optimal stopping problem rather than a system of linear equations.

The λ-PI method is a form of optimistic PI, given by

$$T_{\mu_{k+1}} J_k = T J_k, \qquad J_{k+1} = T^{(\lambda)}_{\mu_{k+1}} J_k, \tag{17.5}$$

where for any μ and $\lambda \in [0, 1)$, $T^{(\lambda)}_{\mu}$ is the linear mapping given by

$$T^{(\lambda)}_{\mu} = (1 - \lambda) \sum_{\ell=0}^{\infty} \lambda^{\ell} T^{\ell+1}_{\mu}. \tag{17.6}$$

Note that the mapping $T^{(\lambda)}_{\mu}$ is central in much recent research on approximate DP, simulation-based PI, and TD methods, as will be discussed in the sequel.

To compare the optimistic PI method (17.4) and the λ-PI method (17.5), note that both mappings $T^{m_k}_{\mu_{k+1}}$ and $T^{(\lambda)}_{\mu_{k+1}}$ appearing in Equations (17.4) and (17.5), involve multiple applications of the VI mapping $T_{\mu_{k+1}}$: a fixed number m_k in the former case (with $m_k = 1$ corresponding to VI and $m_k \to \infty$ corresponding to PI), and a geometrically weighted number in the latter case (with $\lambda = 0$ corresponding to VI and $\lambda \to 1$ corresponding to PI). Thus, optimistic PI and λ-PI are similar: they just control the accuracy of the approximation $J_{k+1} \approx J_{\mu_{k+1}}$ by applying VI in different ways. In a classical DP/nonsimulation-based setting, λ-PI is far more complicated relative to optimistic PI, since exact computations using the mapping $T^{(\lambda)}_{\mu}$ are unwieldy. However, this advantage of optimistic PI is dissipated in a simulation context, where computations involving $T^{(\lambda)}_{\mu}$ can be performed conveniently, as extensive analytical and experimental work with TD methods has demonstrated.

Recent research on DP has focused on the use of simulation, in order to deal with model-free situations where the transition probabilities and/or the cost per stage are not known explicitly, and also to deal with the associated high-dimensional linear algebra operations. For problems with very large number of states, the evaluation of various fixed points of mappings, such as T_{μ} or $T^{(\lambda)}_{\mu}$, is typically done by approximation with a vector Φr from the subspace $S = \{\Phi r \mid r \in \Re^s\}$ that is spanned by the columns of an $n \times s$ matrix Φ. In this chapter, we will focus on the *projected equation approach*, whereby given a generic mapping $L : \Re^n \mapsto \Re^n$ (such as T_{μ}) we approximate its fixed point by solving the equation

$$\Phi r = \Pi L(\Phi r),$$

where Π denotes projection onto the subspace S. The projection is with respect to a Euclidean norm $\|\cdot\|_{\xi}$, weighted by a suitable vector ξ of positive weights. An alternative possibility is to solve instead the equation

$$\Phi r = \Pi L^{(\nu)}(\Phi r), \tag{17.7}$$

where similar to Equation (17.6),

$$L^{(\nu)} = (1 - \nu) \sum_{\ell=0}^{\infty} \nu^{\ell} L^{\ell+1},$$

and $\nu \in [0, 1)$ is a parameter [not necessarily the same as the λ parameter in Equations (17.5 and 17.6)]. In our context, we will encounter several different types of mappings L, and in all cases L is a contraction with respect to the projection norm $\| \cdot \|_\xi$, with fixed point $\hat{J}$, whereas $\Pi L^{(\nu)}$ are contractions with respect to $\| \cdot \|_\xi$ for all $\nu \in [0, 1)$. It is well known that the fixed point of $\Pi L^{(\nu)}$, denoted $\Phi r(\nu)$, converges to $\Pi \hat{J}$ as $\nu \to 1$. The norm of the difference $\Phi r(\nu) - \Pi \hat{J}$ is known as the *bias*. Its size/norm depends on ν and is generally smaller as ν gets closer to 1 (see [1, 29, 30] for error bound analyses).

A common example of fixed point approximation in PI is when $L = T_\mu$ for a policy μ, in which case the fixed point of ΠL or $\Pi L^{(\nu)}$ is an approximation to the fixed point of T_μ, that is, the cost vector J_μ. If the Markov chain corresponding to μ is irreducible and ξ is the corresponding steady-state distribution vector, the mapping $\Pi T_\mu^{(\lambda)}$ is a contraction with respect to $\| \cdot \|_\xi$ for all $\lambda \in [0, 1)$, and is unique fixed point, denoted $\Phi r_\mu(\lambda)$, converges to ΠJ_μ as $\lambda \to 1$. Generally, the projected equation $\Phi r = \Pi T_\mu^{(\lambda)}(\Phi r)$ is solved by a simulation process that generates a sequence of states according to a sampling scheme to be discussed later, and then by matrix inversion (this is the least squares temporal differences [LSTD(λ)] method, proposed by Bradtke and Barto [31]), or by iteration, using the TD(λ) method, proposed by Sutton [56] and analyzed by Tsitsiklis and VanRoy [29] among others, or the least squares policy evaluation [LSPE(λ)] method, proposed by Bertsekas and Ioffe [20].[2] These methods are extensively discussed in the literature, and exhibit complex and sometimes pathological behavior, particularly when embedded within PI (see [34–36] for some notable failures, and [37] for a recent assessment). Moreover, matrix inversion and iterative methods, such as TD(λ), LSTD(λ), and LSPE(λ), can be used for solving not only the projected equation $\Phi r = \Pi T_\mu^{(\lambda)}(\Phi r)$, but also the more general equation $\Phi r = \Pi L^{(\nu)}(\Phi r)$ of Equation (17.7), as long as L is a linear mapping that is convenient for the use of simulation (and in the case of TD(λ) and LSPE(λ), $\Pi L^{(\nu)}$ is a contraction; see [38] or [39]).

[2]The paper [20] as well as the book [1] used the name "λ-policy iteration" for both the lookup table and the compact representation versions of the method described here, and tested a compact representation version on the game of tetris, a challenging SSP problem. The name "LSPE" was first used in the subsequent paper by Nedić and Bertsekas [32] to describe a specific iterative implementation of the λ-PI method with cost function approximation for discounted MDP (essentially the discounted version of the implementation used in [20] and [1] for the aforementioned tetris case study). Reference [32] proved convergence of the LSPE(λ) method, as described in Section 17.3.1, for the case of a diminishing stepsize. Convergence for a stepsize equal to 1 was proved shortly afterwards by Bertsekas et al. [33]. The use of two different names for essentially the same method has been a source of some confusion. While in practical implementations these two names refer to algorithms that are closely related, we reserve the name "λ-policy iteration" for the more abstract form Equation (17.5–17.6), and we will view LSPE(λ) as an implementation of λ-PI (see Section 17.4.1).

In this chapter, we will review some of the basic issues in approximate PI using the projected equation approach, thereby setting the stage for assessing the relative strengths and weaknesses of the λ-PI methodology. We will then focus on three alternative implementations of λ-PI, which involve simulation and cost function approximation. The first is basically the LSPE(λ) method as implemented in [20]. The second is an interesting recent proposal by Thiery and Scherrer [40], who gave extensive and quite successful computational results, as well as error bounds [41]. The third implementation is new and may have some advantages over the first two. We will argue that it deals better with the combined issues of bias and exploration. This implementation embodies a new idea for λ-methods: a simulation scheme, called *geometric sampling*, that uses multiple short trajectories with random geometrically distributed length, and exploration-enhanced restart, rather than a single infinitely long trajectory.

The three implementations are described in Section 17.4, following a discussion of the generic properties of exact λ-PI in Section 17.2, and the LSTD(λ) and LSPE(λ) methods in Section 17.3. In our description, these implementations are model-based and use cost function approximation, but there are versions that are model-free and use Q-factor approximation; these can be straightforwardly constructed by the reader.

17.2 LAMBDA-POLICY ITERATION WITHOUT COST FUNCTION APPROXIMATION

We first recall a central result from [20]. It provides a helpful characterization of the λ-PI method (17.5), which will later become the basis for cost function approximations.

Proposition 17.1 Given $\lambda \in [0, 1)$, J_k, and μ_{k+1}, consider the mapping W_k defined by

$$W_k J = (1 - \lambda) T_{\mu_{k+1}} J_k + \lambda T_{\mu_{k+1}} J. \tag{17.8}$$

(a) W_k is a sup norm contraction of modulus $\lambda\alpha$.

(b) The vector $J_{k+1} = T^{(\lambda)}_{\mu_{k+1}} J_k$ generated next by the λ-PI method (17.5) is the unique fixed point of W_k.

Proof: (a) For any two vectors J and $\bar{J}$, using the definition (17.8) of W_k, we have

$$\|W_k J - W_k \bar{J}\| = \left\|\lambda(T_{\mu_{k+1}} J - T_{\mu_{k+1}} \bar{J})\right\| = \lambda \|T_{\mu_{k+1}} J - T_{\mu_{k+1}} \bar{J}\| \leq \lambda\alpha \|J - \bar{J}\|,$$

where $\|\cdot\|$ denotes the sup norm, so W_k is a sup norm contraction with modulus $\lambda\alpha$.

(b) We have

$$J_{k+1} = T^{(\lambda)}_{\mu_{k+1}} J_k = (1-\lambda)\sum_{\ell=0}^{\infty} \lambda^\ell T^{\ell+1}_{\mu_{k+1}} J_k,$$

so the fixed point property to be shown, $J_{k+1} = W_k J_{k+1}$, is written as

$$(1-\lambda)\sum_{\ell=0}^{\infty} \lambda^\ell T^{\ell+1}_{\mu_{k+1}} J_k = (1-\lambda) T_{\mu_{k+1}} J_k + \lambda T_{\mu_{k+1}} (1-\lambda)\sum_{\ell=0}^{\infty} \lambda^\ell T^{\ell+1}_{\mu_{k+1}} J_k,$$

and evidently holds. ■

From part (b) of the preceding proposition, we see that $J_{k+1} = W_k J_{k+1}$, or equivalently

$$J_{k+1}(i) = \sum_{j=1}^{n} p_{ij}\big(\mu_{k+1}(i)\big)\Big(g\big(i, \mu_{k+1}(i), j\big) + (1-\lambda)\alpha J_k(j) + \lambda\alpha J_{k+1}(j)\Big), \qquad i = 1, \ldots, n. \tag{17.9}$$

The solution of this fixed point equation can be obtained by viewing it as Bellman's equation for two different but equivalent MDP.

(a) *As Bellman's equation for an infinite-horizon $\lambda\alpha$-discounted MDP*, where μ_{k+1} is the only policy, and the cost per stage is

$$g\big(i, \mu_{k+1}(i), j\big) + (1-\lambda)\alpha J_k(j).$$

(b) *As Bellman's equation for an infinite-horizon stopping problem*, where μ_{k+1} is the only policy. In particular, J_{k+1} is the cost vector of policy μ_{k+1} in a stopping problem that is derived from the given α-discounted problem by introducing transitions from each state j to an artificial termination state as follows: at state i we first make a transition to j with probability $p_{ij}\big(\mu_{k+1}(i)\big)$ and transition cost $g\big(i, \mu_{k+1}(i), j\big)$; then we either stay at j and wait for the next transition (this occurs with probability λ), or else we move from j to the termination state with an additional termination cost $\alpha J_k(j)$ (this occurs with probability $1-\lambda$). All transition costs as well as the termination cost are discounted by an additional factor α with each transition.

The convergence and rate of convergence of the λ-PI method (17.5) was given in [20] and also in [1], Proposition 17.8. We will simply quote the results for completeness.

Proposition 17.2 Assume that $\lambda \in [0, 1)$, and let $\{J_k, \mu_k\}$ be the sequence generated by the λ-PI method (17.5). Then J_k converges to J^*. Furthermore, for all k greater than some index $\bar{k}$, μ_k is optimal.

Proposition 17.3 Let the assumptions of Proposition 17.2 hold and let $\bar{k}$ be the index such that for all $k \geq \bar{k}$, μ_k is optimal. The sequence $\{J_k\}$ generated by the λ-PI method (17.5) satisfies for all $k > \bar{k}$

$$\|J_{k+1} - J^*\| \leq \frac{\alpha(1-\lambda)}{1-\lambda\alpha} \|J_k - J^*\|, \tag{17.10}$$

where $\|\cdot\|$ denotes the sup norm.

Note that the convergence rate estimate (17.3) holds only for $k \geq \bar{k}$, essentially after an optimal policy has been identified, as per Proposition 17.2. Nonetheless, this rate estimate is qualitatively correct and supports the empirical observation that the iterates (J_k, μ_k) generated by λ-PI converge faster as λ increases. Indeed in the limit, as $\lambda \to 1$, λ-PI becomes exact PI, and converges to the optimum in a finite number of iterations. On the other hand, the computation of $J_{k+1} = T^{(\lambda)}_{\mu_{k+1}} J_k$ [cf. Equation (17.5)] becomes more time-consuming as λ increases, particularly when simulation is used, because the simulation-based calculation of $T^{(\lambda)}_{\mu_{k+1}} J_k$ involves more simulation noise as λ gets larger.

We finally note that Propositions 17.2 and 17.3 apply to synchronous implementations of λ-PI. When implemented asynchronously, λ-PI has similar convergence difficulties to optimistic PI. To see this, note that asynchronous implementations of these two methods essentially coincide when $m_k = 1$ in Equation (17.5) and $\lambda = 0$ in Equation (17.4), and the counterexamples of Williams and Baird [25] apply. Thus, the development of convergent versions of asynchronous λ-PI is an open research question.

17.3 APPROXIMATE POLICY EVALUATION USING PROJECTED EQUATIONS

In PI methods with cost function approximation, we evaluate μ by approximating J_μ with a vector Φr_μ from the subspace $S = \{\Phi r \mid r \in \Re^s\}$, spanned by the columns of an $n \times s$ matrix Φ, which may be viewed as basis functions. We generate an "improved" policy $\bar{\mu}$ using the formula $T_{\bar{\mu}}(\Phi r_\mu) = T(\Phi r_\mu)$, that is,

$$\bar{\mu}(i) \in \arg\min_{u \in U(i)} \sum_{j=1}^{n} p_{ij}(u)\big(g(i, u, j) + \alpha\phi(j)' r_\mu\big), \qquad i = 1, \ldots, n,$$

where $\phi(j)'$ is the row of Φ that corresponds to state j [the method terminates with μ if $T_\mu(\Phi r_\mu) = T(\Phi r_\mu)$]. We then repeat with μ replaced by $\bar{\mu}$. For the purposes of this chapter, we assume that Φ has rank s, and that the Markov chain corresponding to μ is irreducible.

As noted earlier, in the projected equation approach to approximate PI, we approximate J_μ with a vector of the form $\Phi r_\mu(\lambda)$ that solves the fixed point problem

$$\Phi r = \Pi T_\mu^{(\lambda)}(\Phi r). \tag{17.11}$$

Here Π denotes projection onto the subspace S with respect to a weighted Euclidean norm $\|\cdot\|_\xi$, where $\xi = (\xi_1, \ldots, \xi_n)$ is a probability distribution with positive components (i.e., $\|J\|_\xi^2 = \sum_{i=1}^n \xi_i x_i^2$, where $\xi_i > 0$ for all i). In nonoptimistic PI methods, the projected Equation (17.11) is solved exactly, whereas in optimistic PI methods it is solved approximately. We note that this approach has a long history in the context of Galerkin methods for the approximate solution of high-dimensional or infinite-dimensional linear equations (partial differential, integral, inverse problems, and so on see e.g., [42, 43]). In fact some of the policy evaluation theory referred to in this chapter applies to general projected equations arising in contexts beyond DP (see [38, 39, 44–46]). However, Monte Carlo simulation is not part of the Galerkin methodology, as currently practiced in the numerical analysis field. For this reason, much of the extensive available knowledge about Galerkin methods does not apply to the approximate DP context, which is primarily simulation-oriented.

We now discuss some of the issues relating to projected equations. While we focus on Equation (17.11), much of our discussion also applies to the more general projected equations.

17.3.1 Exploration–Contraction Trade-off

An important choice in the projected equation approach is the distribution ξ that defines the projection norm $\|\cdot\|_\xi$. This distribution is sometimes chosen to be the steady-state probability vector ξ_μ of the Markov chain corresponding to μ, in which case the mapping $\Pi T_\mu^{(\lambda)}$ can be shown to be a contraction with respect to $\|\cdot\|_{\xi_\mu}$ with modulus

$$\alpha_\lambda = \frac{\alpha(1-\lambda)}{1-\lambda\alpha} \tag{17.12}$$

(see [1], Lemma 6.6, or [22], Proposition 6.3.3).

On the other hand, the choice of ξ is related to *exploration*, that is, the need to collect an adequately rich set of samples from a broad and representative set of states. This is a critical issue in simulation-based PI, and results in a well-known trade-off: to evaluate a policy μ, we may need to generate cost samples using μ, but this may affect the simulation results by underrepresenting states that are unlikely to occur under μ (more weight is placed on states that are visited more frequently under μ). As a result, the cost-to-go estimates of the underrepresented states may be highly inaccurate, causing potentially serious errors in the calculation of the improved control policy.

A well-known approach for exploration is to choose ξ to be a mixture of the form

$$\xi = (1-\beta)\xi_\mu + \beta\tilde{\xi}, \tag{17.13}$$

where $\beta \in (0, 1)$ and $\tilde{\xi}$ is another distribution (often referred to as the *off-policy* distribution), which is added to enhance exploration (see the discussion of Section 17.1). Unfortunately, with such a choice the contraction property of $\Pi T_\mu^{(\lambda)}$ comes into doubt: it depends on the size of the parameters λ and β [it can be shown that $\Pi T_\mu^{(\lambda)}$ is a contraction for any $\beta \in [0, 1)$ provided λ is close enough to 1, and it is a contraction for any $\lambda \in [0, 1)$ provided β is close enough to 0]. This is important because for convergence of iterative methods such as TD(λ) and some forms of LSPE(λ), it is critical that $\Pi T_\mu^{(\lambda)}$ be a contraction. Thus, there is a trade-off between exploration enhancement using the mixture distribution (17.13) and ability to use a broader range of methods for solution of the projected equation.

17.3.2 Bias

Although the Bellman equation $J = T_\mu^{(\lambda)} J$ has the same fixed point J_μ for all $\lambda \in [0, 1)$, the fixed point $\Phi r_\mu(\lambda)$ of the projected version (17.11) depends on λ. The difference of $\Phi r_\mu(\lambda)$ and the closest point of S to J_μ, $\Phi r_\mu(\lambda) - \Pi J_\mu$, is generally nonzero. Its norm, the bias, tends to decrease to 0 as $\lambda \uparrow 1$ and tends to increase as $\lambda \downarrow 0$. It is known that the bias can be very large and may seriously degrade the practical value of the approximate policy evaluation for small values of λ; see [34] for some examples.

The following is a well-known error bound for the case $\xi = \xi_\mu$:

$$\|J_\mu - \Phi r_\mu(\lambda)\|_{\xi_\mu} \leq \frac{1}{\sqrt{1-\alpha_\lambda^2}} \|J_\mu - \Pi J_\mu\|_{\xi_\mu}, \tag{17.14}$$

where α_λ is given by Equation 17.12, and $\|\cdot\|_{\xi_\mu}$ is the weighted Euclidean norm corresponding to $\xi = \xi_\mu$, the steady-state probability vector of the Markov chain corresponding to μ. Thus, the error bound becomes worse as λ decreases (and α_λ increases), suggesting a larger size of bias. While the bound is rather conservative, the paper by Yu and Bertsekas [30] (see also Scherrer [47]) derives sharper error bounds, which also apply to cases where $\xi \neq \xi_\mu$ and $\Pi T_\mu^{(\lambda)}$ is not a contraction. These error bounds and the bound (17.14) are consistent in suggesting that the bias increases as λ decreases, and they are also largely consistent with the results of computational experimentation.

17.3.3 Bias-Variance Trade-off

In simulation-based methods for solving the projected Equation (17.11), one must deal with the effects of simulation error. Generally as λ increases, the methods become more vulnerable to simulation noise, and hence require more sampling for good performance. Indeed, the noise in a simulation sample of an ℓ-stages cost vector $T_\mu^\ell J$ tends to be larger as ℓ increases, and from the formula

$$T_\mu^{(\lambda)} = (1-\lambda)\sum_{\ell=0}^{\infty} \lambda^\ell T_\mu^{\ell+1},$$

it can be seen that simulation samples of $T_\mu^{(\lambda)}(\Phi r_k)$ tend to contain more noise as λ increases. This is consistent with practical experience, and gives rise to the so called bias-variance trade-off: a large value of λ to reduce bias results in slower and less reliable computation because of higher simulation noise (and consequently, a larger number of samples to achieve the same accuracy of various simulation-based estimates). Generally, there is no rule of thumb for selecting λ, which is usually chosen with some trial and error.

In summary, the preceding discussion suggests that if simulation noise is not an issue (i.e., one can afford many simulation samples) one should choose large values of λ, since then the bias is reduced and one may afford greater exploration without losing the contraction property of $\Pi T_\mu^{(\lambda)}$. In the contrary case, however, the degradation of the estimate of J_μ due to simulation noise may offset whatever bias/contraction benefits a large value of λ may bring.

17.3.4 TD Methods

Most of the simulation-based methods for solving the projected equation use explicitly or implicitly the notion of temporal difference (TD), which originated in reinforcement learning with the works of Samuel [48, 49] on a checkers-playing program. The first TD method is TD(λ), which can be viewed as an iterative stochastic approximation-type algorithm. The LSTD(λ) method is based on batch simulation: it first generates a batch of state and cost samples, it approximates the projected equation $\Phi r = \Pi T_\mu^{(\lambda)}(\Phi r)$ using these samples, and then solves the equation directly by matrix inversion. Another TD method is LSPE(λ), which while being more iterative, shares much of the simulation philosophy of LSTD(λ).

To describe more specifically the LSTD(λ) and LSPE(λ) methods, we first note that the orthogonality condition that characterizes the projection in the projected equation $\Phi r = \Pi T_\mu^{(\lambda)}(\Phi r)$ is

$$\Phi'\Xi\big(\Phi r - T_\mu^{(\lambda)}(\Phi r)\big) = 0, \tag{17.15}$$

where Ξ is the diagonal matrix with the vector ξ along the diagonal (see e.g., [22]). Thus, the projected equation (17.11) is equivalent to the lower-dimensional Equation (17.15), which can in turn be written in matrix form as

$$C^{(\lambda)} r = d^{(\lambda)}, \tag{17.16}$$

with

$$C^{(\lambda)} = \Phi'\Xi\big(I - P_\mu^{(\lambda)}\big)\Phi, \qquad d^{(\lambda)} = \Phi'\Xi g_\mu^{(\lambda)}, \tag{17.17}$$

and

$$P_\mu^{(\lambda)} = (1-\lambda)\sum_{\ell=0}^{\infty} \lambda^\ell \alpha^{\ell+1} P_\mu^{\ell+1}, \qquad g_\mu^{(\lambda)} = \sum_{\ell=0}^{\infty} \lambda^\ell \alpha^\ell P_\mu^\ell g_\mu, \tag{17.18}$$

where P_μ and g_μ are the transition probability matrix and expected single-stage cost vector corresponding to μ. The LSTD(λ) and LSPE(λ) methods use simulation-based approximations of $C^{(\lambda)}$ and $d^{(\lambda)}$. This is done by simulating a state sequence $(i_0, \ldots, i_t)$ and corresponding transition cost sequence, using the current policy μ (perhaps with exploration enhancement, as discussed earlier). Then after each simulated state i_ℓ, $\ell = 0, \ldots, t$, is generated, estimates $C_\ell^{(\lambda)}$ and $d_\ell^{(\lambda)}$ are obtained using the simulation samples up to time ℓ, using formulas that we will not give here, as they are not important for our purposes. Such formulas, in various alternative forms, can be found in several sources, including the textbooks cited earlier. The papers [32, 38, 45, 46] discuss the conditions for the convergence $\lim_{\ell\to\infty} C_\ell^{(\lambda)} = C^{(\lambda)}$, $\lim_{\ell\to\infty} d_\ell^{(\lambda)} = d^{(\lambda)}$ to hold with probability 1.

The LSTD(λ) method is based on simple matrix inversion: after the last state i_t of the simulation trajectory is generated, it computes the solution

$$\hat{r} = \big(C_t^{(\lambda)}\big)^{-1} d_t^{(\lambda)} \tag{17.19}$$

of the corresponding simulation-based approximation to Equation (17.16),

$$C_t^{(\lambda)} r = d_t^{(\lambda)}, \tag{17.20}$$

and approximates the cost vector J_μ by $\Phi\hat{r}$. An important point is that $\hat{r}$ can be obtained regardless of whether $\Pi T_\mu^{(\lambda)}$ is a contraction. It is only required that $C_t^{(\lambda)}$ is invertible, a much less restrictive condition.

One version of the LSPE(λ) method consists of iterative solution of the system Equation (17.20). It approximates the cost vector J_μ by Φr_{t+1}, where r_{t+1} is obtained at the last step of the iteration

$$r_{\ell+1} = r_\ell - \gamma G_\ell\big(C_\ell^{(\lambda)} r_\ell - d_\ell^{(\lambda)}\big), \qquad \ell = 0, \ldots, t, \tag{17.21}$$

where r_0 is some initial vector, likely the vector obtained from the preceding policy evaluation, γ is a positive stepsize, G_ℓ is the matrix

$$G_\ell = \left(\frac{1}{\ell}\sum_{m=0}^{\ell-1} \phi(i_m)\phi(i_m)'\right)^{-1}, \qquad \ell = 0, \ldots, t, \tag{17.22}$$

and as earlier, $\phi(i)'$ denotes the ith row of the matrix Φ. In the original proposal of [20], the stepsize is $\gamma = 1$; convergence of Φr_t to the fixed point of $\Pi T_\mu^{(\lambda)}$ for this stepsize was shown in [33]. The matrix G_ℓ is a simulation-based approximation of $(\Phi' \Xi \Phi)^{-1}$ (alternative choices of G_ℓ have been discussed recently in [19, 39]). There is also an equivalent implementation of this iteration, which is based on solution of a least squares problem (see Section 17.4.1).

The choice (17.22) for G_ℓ and the use of $\gamma = 1$ are based on a view of the method as an approximation to the projected value iteration method

$$\Phi r_{\ell+1} = \Pi T_\mu^{(\lambda)}(\Phi r_\ell),$$

which after some calculation can be written as

$$\Phi r_{\ell+1} = \Phi\big(r_\ell - (\Phi' \Xi \Phi)^{-1}\big(C^{(\lambda)} r_\ell - d^{(\lambda)}\big)\big),$$

or equivalently, since Φ has full rank, as

$$r_{\ell+1} = r_\ell - (\Phi' \Xi \Phi)^{-1}\big(C^{(\lambda)} r_\ell - d^{(\lambda)}\big);$$

cf. Equations (17.21–17.22) with $\gamma = 1$.

Note that the matrix inversion in Equation (17.22) is not so onerous, because it can be formed incrementally, with a rank-one correction as each sample becomes available. On the other hand, contrary to LSTD(λ) [and similar to TD(λ)], the LSPE(λ) method (17.21–17.22) requires that $\Pi T_\mu^{(\lambda)}$ be a contraction for convergence. Indeed, if the simulation is performed using the steady-state distribution ξ_μ, it can be shown that $\Pi T_\mu^{(\lambda)}$ is a contraction, but if the simulation is performed using a mixture/off-policy distribution (17.13) for the purpose of exploration-enhancement, the contraction property may be lost and repeated iterations of the form (17.2) may diverge.

We finally note that in iteration (17.21) the underlying assumption is that we update r as simulation samples are collected and used to form ever improving approximations to C and d. An alternative is to use batch simulation, such as in LSTD: first simulate to obtain $C_t^{(\lambda)}$, $d_t^{(\lambda)}$, and G_t, and then solve the system $C_t^{(\lambda)} r = d_t^{(\lambda)}$ iteratively rather than through the direct matrix inversion (17.19), using any number of iterations of the type (17.21). In fact, we may use *only one* iteration, in which case the method takes the form

$$r_1 = r_0 - \gamma G_t\big(C_t^{(\lambda)} r_0 - d_t^{(\lambda)}\big). \tag{17.23}$$

A single (or very few) iterations may be sufficient if λ is close to 1, since then the contraction modulus of $\Pi T_\mu^{(\lambda)}$ is close to 0 (see e.g., [1], Lemma 6.6, or [22], Proposition 6.3.3), so a single iteration with $\Pi T_\mu^{(\lambda)}$ is very effective, yielding a vector that is close to its fixed point. We will return to this variant of the method later.

17.3.5 Comparison of LSTD(λ) and LSPE(λ)

There has been speculation about the relative merits of LSTD(λ) and LSPE(λ). In general, it is difficult to reach definitive conclusions, as there are several complex factors to consider, such as the length of the simulation sequence $(i_0, \ldots, i_t)$, and the potential near-singularity of $C^{(\lambda)}$, which affects the error in the matrix inversion in the LSTD(λ) formula (17.19). As an illustration, consider a few different situations:

(a) Assume, as an idealization, that an infinite number of samples is collected. Then both methods yield in the limit the same result, the fixed point of the projected equation $J = \Pi T_\mu^{(\lambda)} J$. However, in contrast to LSTD(λ), in order to guarantee convergence, LSPE(λ) requires that $\Pi T_\mu^{(\lambda)}$ is a contraction, which interferes with the freedom to do exploration, as discussed earlier.

(b) Assume that $C^{(\lambda)}$ is invertible, but is nearly singular. Then the matrix inversion in the LSTD(λ) formula (17.19) may require a very large number of samples to yield a reasonably accurate solution of $C^{(\lambda)} r = d^{(\lambda)}$.[3] To correct the sensitivity of LSTD(λ) to simulation noise, it may be necessary to turn it into an iterative method through some form of regularization, which then brings it close to a form of LSPE(λ) (see [11, 19, 39, 44, 50] for such regularization methods and their connection to LSPE). Of course, the situation becomes even more complex if $C^{(\lambda)}$ is singular, perhaps due to inadvertent rank deficiency of Φ (see [51, 52] for a discussion of this possibility).

(c) When LSTD(λ) and LSPE(λ) are embedded within a PI framework, the number of samples collected using any one policy is often relatively small. Then the behavior of the two methods becomes very complicated, and it is hard to reach any kind of reliable conclusion [37]. Computational studies indicate that LSPE(λ) being an iterative method, is less sensitive to the matrix inversion errors that afflict LSTD(λ) in the presence of high simulation noise.

[3]It is well known from fundamental error analyses of linear equation solvers that small errors in a nearly singular matrix $C^{(\lambda)}$ will cause large errors in the solution of $C^{(\lambda)} r = d^{(\lambda)}$. Near-singularity of $C^{(\lambda)}$ may be due either to the columns of Φ being nearly linearly dependent or to the matrix $\Xi(I - \alpha P^{(\lambda)})$ being nearly singular [cf. Equation (17.17)]. Near-linear dependence of the columns of Φ will not affect the error in the solution of the high-dimensional projected equation, which can be written as $\Phi C^{(\lambda)} r = \Phi d^{(\lambda)}$. The reason is that this error depends only on the subspace S and not its representation in terms of the matrix Φ. In particular, if we replace Φ with a matrix ΦB, where B is an $s \times s$ invertible scaling matrix, the subspace S will be unaffected and the error in the solution of the projected equation will also be unaffected. On the other hand, near singularity of the matrix $I - \alpha P^{(\lambda)}$ may affect significantly the error. Note that $I - \alpha P^{(\lambda)}$ is nearly singular in the case where α is very close to 1, or in the corresponding undiscounted case where $\alpha = 1$ and P is substochastic with some eigenvalues very close to 1. Large variations in the size of the diagonal components of Ξ may also affect significantly the error, although this dependence is complicated by the fact that Ξ appears not only in the formula $C^{(\lambda)} = \Phi' \Xi (I - \alpha P^{(\lambda)}) \Phi$ but also in the formula $d^{(\lambda)} = \Phi' \Xi g^{(\lambda)}$.

The preceding discussion is also relevant to the implementations of λ-PI to be discussed in the next section, since these implementations bear strong relations to both LSTD(λ) and LSPE(λ).

17.4 LAMBDA-POLICY ITERATION WITH COST FUNCTION APPROXIMATION

We saw in Section 17.2 that the policy evaluation portion of λ-PI,

$$J_{k+1} = T^{(\lambda)}_{\mu_{k+1}} J_k, \tag{17.24}$$

[cf. Equation (17.5)] can be implemented in two ways:

(1) by computing $T^{(\lambda)}_{\mu_{k+1}} J_k$;

(2) by finding the fixed point of the mapping W_k [cf. Equation (17.8)] through solution of the equation

$$J = W_k J, \tag{17.25}$$

which can be viewed as Bellman's equation associated with the current policy for the two equivalent DP problems discussed in Section 17.2 [cf. Equation (17.9)]: a $\lambda\alpha$-discounted problem and a stopping problem.

Let us now consider approximation of λ-PI on the subspace $S = \{\Phi r \mid r \in \Re^s\}$. A natural possibility is to introduce projection in the preceding approaches. In particular, we may approximate the λ-PI iterate J_{k+1} of Equation (17.24) by Φr_{k+1} in three ways:

(a) Using a single projected value iteration for the original α-discounted problem,

$$\Phi r_{k+1} = \Pi T^{(\lambda)}_{\mu_{k+1}} (\Phi r_k). \tag{17.26}$$

This is the original proposal of [20]. It is the variant of the LSPE(λ) method (17.21–17.22), which involves just the last iteration.

(b) By solving a projected version of Equation (17.25), viewing it as Bellman's equation for the $\lambda\alpha$-discounted problem of Section 17.2, and setting r_{k+1} equal to its solution. This is the proposal of [40], and implements by simulation the solution of this projected equation, essentially by applying LSTD(0) to Bellman's equation for the $\lambda\alpha$-discounted problem formulated in Section 17.2.

(c) By solving a projected version of Equation (17.25), viewing it as Bellman's equation for the stopping problem formulated in Section 17.2, and setting r_{k+1} equal to its solution.

In the following three subsections, we will describe three alternative implementations of λ-PI corresponding to the possibilities (a)–(c) above. Of course when linear cost function approximation of the form Φr_k is used to represent J_k, the λ-PI method need not converge, and the cost vectors J_{μ_k} of the generated policies typically oscillate within some suboptimality threshold from J^*. We do not address this issue, but we note that related error bounds, which also apply to other forms of optimistic PI are given by Bertsekas and Yu [26], Thiery and Scherrer [41], and Scherrer [53].

17.4.1 The LSPE(λ) Implementation

A variant of the LSPE(λ) method (17.21–17.22) is to form batches of simulation samples and perform iteration (17.21) at the end of each batch. In an extreme case, we treat the entire simulation trajectory $(i_0, \ldots, i_t)$ as a single simulation batch, and we perform a *single* iteration (17.21), for $\ell = t$, yielding the method

$$r_{k+1} = r_k - G_t\big(C_t^{(\lambda)} r_k - d_t^{(\lambda)}\big), \tag{17.27}$$

where Φr_k is the approximate evaluation of the cost vector of the preceding policy μ_k [cf. Equation (17.23)]. As $t \to \infty$ and the simulation becomes exact in the limit, that is,

$$\lim_{t\to\infty} C_t^{(\lambda)} = C^{(\lambda)}, \qquad \lim_{t\to\infty} d_t^{(\lambda)} = d^{(\lambda)},$$

and if G_t is given by the formula (17.22), it can be verified that

$$\Phi r_{k+1} \to \Pi T_{\mu_{k+1}}^{(\lambda)}(\Phi r_k). \tag{17.28}$$

Thus, the method (17.27) with G_t given by Equation (17.22) can be viewed as a simulation-based implementation of Equation (17.26), the projected version of λ-PI, which becomes exact in the limit as $t \to \infty$. In practice of course t is finite, and one may consider variants of the method, whereby multiple iterations of the form (17.27) are performed, with each iteration using additional simulation samples.

We note a mathematically equivalent description of this method, which is given in terms of a least-squares optimization (see [22], Section 6.3.3 for a more detailed textbook account): we set

$$r_{k+1} = \arg\min_{r\in\Re^s} \sum_{\ell=0}^{t} \left(\phi(i_\ell)' r - \phi(i_\ell)' r_k - \sum_{m=\ell}^{t} (\lambda\alpha)^{m-\ell} q(i_m, i_{m+1}) \right)^2, \tag{17.29}$$

where $q(i_m, i_{m+1})$ is the temporal difference

$$q(i_m, i_{m+1}) = g\big(i_m, \mu_{k+1}(i_m), i_{m+1}\big) + \alpha\phi(i_{m+1})'r_k - \phi(i_m)'r_k, \qquad m = 0, \ldots, t. \tag{17.30}$$

In fact this is how the method was originally described in [20] and [1].

A positive aspect of this method is that it approximates directly $\Pi T^{(\lambda)}_{\mu_{k+1}}(\Phi r_k)$, so it is not subject to bias in the evaluation of the fixed point of W_k; cf. Equation (17.28). However, in the form given here, the method does not address the issue of exploration. Despite this fact, this implementation [in the form (17.29)] has been successful in several challenging computational studies, including the one involving the game of tetris in the original paper [20] and some follow-up works, and a recent one by Foderaro et al. [54] involving the game of pac-man, a benchmark problem of pursuit-evasion.

17.4.2 λ-PI(0)—An Implementation Based on a Discounted MDP

This implementation, suggested and tested by Thiery and Scherrer [40, 41], is based on the fixed point property of J_{k+1} [cf. Proposition 17.1(b)]. It produces an approximation Φr_{k+1} to J_{k+1} within the subspace S, by solving the projected equation

$$\Phi r = \Pi W_k(\Phi r),$$

with W_k given by

$$W_k J = (1-\lambda)T_{\mu_{k+1}}(\Phi r_k) + \lambda T_{\mu_{k+1}} J, \qquad J \in \Re^n. \tag{17.31}$$

We may find the solution r_{k+1} of this equation using an LSTD(0)-like simulation approach. In particular, r_{k+1} satisfies the orthogonality condition

$$Cr = d(k),$$

where

$$C = \Phi'\Xi(I - \lambda\alpha P_{\mu_{k+1}})\Phi, \qquad d(k) = \Phi'\Xi\big(g_{\mu_{k+1}} + (1-\lambda)\alpha P_{\mu_{k+1}}\Phi r_k\big),$$

so that

$$r_{k+1} = C^{-1}d(k). \tag{17.32}$$

We refer to this method as λ-PI(0) to distinguish it notationally from the method of the next subsection (the name LSλPI was introduced for this method in [40]).

In a simulation-based implementation, the matrix C and the vector $d(k)$ are approximated by estimates C_t and $d_t(k)$. Thus, this method does not require that $\Pi T^{(\lambda)}_{\mu_{k+1}}$ is a contraction, and like LSTD, it can deal well with the issue of exploration. The simulation samples need not depend on the policy μ_{k+1} being evaluated, so they can be generated only once within a PI process. On the other hand, the objective of the implementation is to approximate the next iterate of λ-PI, that is, $T^{(\lambda)}_{\mu_{k+1}}(\Phi r_k)$, and it is not clear that it is doing this well. To see this, suppose that the iteration (17.32), or equivalently $\Phi r_{k+1} = \Pi W_k(\Phi r_k)$, is repeated an infinite number of times so it converges to a limit $\bar{r}$, which must satisfy $\Phi\bar{r} = \Pi W_k(\Phi\bar{r})$. Then using Equation (17.31), we have

$$\Phi\bar{r} = (1-\lambda)\Pi T_{\mu_{k+1}}(\Phi\bar{r}) + \lambda\Pi T_{\mu_{k+1}}(\Phi\bar{r}),$$

which shows that $\Phi\bar{r} = \Pi T_{\mu_{k+1}}(\Phi\bar{r})$. Thus λ-PI(0) aims at $\bar{r}$, which is the limit of TD(0) independent of the value of λ. Indeed as $\lambda \to 1$, ΠW_k tends to $\Pi T_{\mu_{k+1}}$ [cf. Equation (17.31)], so its fixed point Φr_{k+1} tends to the fixed point of $\Pi T_{\mu_{k+1}}$, that is, the limit of TD(0). It follows that while this implementation deals well with the issue of exploration, it may be subject to significant bias-related error.

17.4.3 λ-PI(1)—An Implementation Based on a Stopping Problem

The third implementation is based on the property mentioned in Section 17.2: the fixed point equation $J = W_k J$ [or equivalently, Equation (17.9)] is Bellman's equation for the policy μ_{k+1} in the context of a stopping problem. Here, there is an artificial termination state 0, and for all states j, there is probability $1-\lambda$ that a transition to j will be followed by an immediate transition to state 0, with cost $\alpha J_k(j)$, cf. Equation (17.9). Note that if λ is not too close to 1, the trajectories of this problem tend to be short, and in fact if $\lambda = 0$ all trajectories consist of a single transition.

To compute an approximation Φr_{k+1} to the fixed point of W_k using the stopping problem, we may use any policy evaluation algorithm with cost function approximation over the subspace $S = \{\Phi r \mid r \in \Re^s\}$. An interesting choice is to use the LSPE(1) method, which consists of a least squares fit of Φr to the simulated costs of the trajectories of the stopping problem whose Bellman equation mapping is W_k. The use of LSPE(1) not only involves minimum bias relative to all LSPE(ν) methods with $\nu \in [0, 1]$ but also leads to a simple least squares implementation.

To this end, we introduce a simulation procedure, called *geometric sampling*, which departs from the single infinitely long simulation trajectory format of the implementation of Section 17.4.1, and has the following characteristics:

(a) It uses multiple relatively short simulation trajectories.
(b) The initial state of each trajectory is chosen essentially as desired, thereby allowing flexibility to generate a richer mixture of state visits.
(c) The length of each trajectory is random and is determined by a λ-dependent geometric distribution [a probability $(1-\lambda)\lambda^{\ell}$ that the number of transitions is $\ell + 1$].

In particular, given the current representation Φr_k of J_k and the current policy μ_{k+1}, we update the parameter vector from r_k to r_{k+1} after generating t simulated trajectories. The states of a trajectory are generated according to the transition probabilities $p_{ij}\big(\mu_{k+1}(i)\big)$, the transition cost is discounted by an additional factor α with each transition, and following each transition to a state j, the trajectory is terminated with probability $1-\lambda$ and with an extra cost $\alpha\phi(i)'r_k$. Once a trajectory is terminated, an initial state for the next trajectory is chosen according to a fixed probability distribution $\zeta_0 = \big(\zeta_0(1), \ldots, \zeta_0(n)\big)$, and the process is repeated. Note that the sequence of restart states need not depend on the policy being evaluated, so that it can be simulated only once within a PI process. Of course, the simulated trajectories have to be recalculated for each new policy. The details are as follows.

Let the mth trajectory, $m = 1, \ldots, t$, have the form $(i_{0,m}, i_{1,m}, \ldots, i_{N_m,m})$, where $i_{0,m}$ is the initial state, and $i_{N_m,m}$ is the state at which the trajectory is completed (the last state prior to termination). For each state $i_{\ell,m}$, $\ell = 0, \ldots, N_m - 1$, of the mth trajectory, the simulated cost is

$$c_{\ell,m}(r_k) = \alpha^{N_m-\ell}\phi(i_{N_m,m})'r_k + \sum_{q=\ell}^{N_m-1} \alpha^{q-\ell} g(i_{q,m}, u_{q,m}, i_{q+1,m}), \qquad (17.33)$$

where

$$u_{q,m} = \mu_{k+1}(i_{q,m}), \qquad m = 1, \ldots, t, \; q = 0, \ldots, N_m - 1.$$

Once the costs $c_{\ell,m}(r_k)$ are computed for all states $i_{\ell,m}$ of the mth trajectory and all trajectories $m = 1, \ldots, t$, the vector r_{k+1} is obtained by a least squares fit of these costs:

$$r_{k+1} = \arg\min_{r\in\Re^s} \sum_{m=1}^{t} \sum_{\ell=0}^{N_m-1} \big(\phi(i_{\ell,m})'r - c_{\ell,m}(r_k)\big)^2, \qquad (17.34)$$

cf. Equations (17.29–17.30). Equivalently, we can write the solution of the least squares problem explicitly as

$$r_{k+1} = \left(\sum_{m=1}^{t} \sum_{\ell=0}^{N_m-1} \phi(i_{\ell,m})\phi(i_{\ell,m})' \right)^{-1} \sum_{m=1}^{t} \sum_{\ell=0}^{N_m-1} \phi(i_{\ell,m}) c_{\ell,m}(r_k). \qquad (17.35)$$

We refer to the resulting implementation as λ-PI(1).

Note the extreme special case when $\lambda = 0$. Then all the simulated trajectories consist of a single transition, and there is a restart at every transition. This means that the simulation samples are from states that are generated independently according to the restart distribution ζ_0.

Convergence of the Simulation Process

We will now show that in the limit, as $t \to \infty$, the vector r_{k+1} of Equation (17.35) satisfies

$$\Phi r_{k+1} = \hat{\Pi} T^{(\lambda)}_{\mu_{k+1}}(\Phi r_k), \tag{17.36}$$

where $\hat{\Pi}$ denotes projection with respect to the weighted sup norm $\|\cdot\|_\zeta$ with weight vector $\zeta = \big(\zeta(1), \ldots, \zeta(n)\big)$, where

$$\zeta(i) = \frac{\hat{\zeta}(i)}{\sum_{j=1}^{n} \hat{\zeta}(j)}, \qquad i = 1, \ldots, n, \tag{17.37}$$

and

$$\hat{\zeta}(i) = \sum_{\ell=0}^{\infty} \zeta_\ell(i),$$

with $\zeta_\ell(i)$ being the probability of the state being i after ℓ transitions of a randomly chosen simulation trajectory. This is the underlying norm in TD methods such as LSTD, LSPE, and TD, as applied to SSP problems (see [1], Section 17.6.3.4). Note that $\zeta(i)$ is the long-term occupancy probability of state i during the simulation process. We assume that the restart distribution ζ_0 is chosen so that $\zeta(i) > 0$ for all $i = 1, \ldots, n$, implying that $\|\cdot\|_\zeta$ is a legitimate norm [this is guaranteed if we require that $\zeta_0(i) > 0$ for all i].

Indeed, let us view $T^{\ell+1}_{\mu_{k+1}} J$ as the vector of total discounted costs over a horizon of $(\ell + 1)$ stages with the terminal cost function being J, and write

$$T^{\ell+1}_{\mu_{k+1}} J = \alpha^{\ell+1} P^{\ell+1}_{\mu_{k+1}} J + \sum_{q=0}^{\ell} \alpha^q P^q_{\mu_{k+1}} g_{\mu_{k+1}},$$

where $P_{\mu_{k+1}}$ and $g_{\mu_{k+1}}$ are the transition probability matrix and cost vector, respectively, under μ_{k+1}. As a result the vector $T^{(\lambda)}_{\mu_{k+1}} J = (1-\lambda)\sum_{\ell=0}^{\infty} \lambda^\ell T^{\ell+1}_{\mu_{k+1}} J$ can be expressed as

$$\begin{aligned}\big(T^{(\lambda)}_{\mu_{k+1}} J\big)(i) = \sum_{\ell=0}^{\infty}(1-\lambda)\lambda^\ell E\Bigg\{ & \alpha^{\ell+1} J(i_{\ell+1}) \\ & + \sum_{q=0}^{\ell} \alpha^q g\big(i_q, \mu_{k+1}(i_q), i_{q+1}\big) \,\Big|\, i_0 = i \Bigg\}, \qquad i = 1, \ldots, n.\end{aligned}$$

Thus, $\big(T_{\mu_{k+1}}^{(\lambda)} J\big)(i)$ may be viewed as the expected value of the $(\ell+1)$-stages cost of policy μ_{k+1} starting at state i, with *the number of stages being random and geometrically distributed with parameter* λ [probability of $\kappa+1$ transitions is $(1-\lambda)\lambda^\kappa$, $\kappa = 0, 1, \ldots$]. It follows that the cost samples $c_{\ell,m}(r_k)$ of Equation (17.33), produced by the simulation process described earlier, can be used to estimate $\big(T_{\mu_{k+1}}^{(\lambda)}(\Phi r_k)\big)(i)$ for all i by Monte Carlo averaging. The estimation formula is

$$D_t(i) = \frac{1}{\sum_{m=1}^{t}\sum_{\ell=0}^{N_m-1}\delta(i_{\ell,m}=i)} \cdot \sum_{m=1}^{t}\sum_{\ell=0}^{N_m-1}\delta(i_{\ell,m}=i)c_{\ell,m}(r_k), \qquad (17.38)$$

where $\delta(i_{\ell,m}=i)=1$ if $i_{\ell,m}=i$ and $\delta(i_{\ell,m}=i)=0$ otherwise, and we have

$$\big(T_{\mu_{k+1}}^{(\lambda)}(\Phi r_k)\big)(i) = \lim_{t\to\infty} D_t(i), \qquad i = 1, \ldots, n,$$

(see also the discussion on the consistency of Monte Carlo simulation for policy evaluation in [1], Section 17.5.2).

Let us now compare the λ-PI iteration (17.36) with the simulation-based implementation (17.37). Using the definition of projection, Equation (17.36) can be written as

$$r_{k+1} = \arg\min_{r\in\Re^s}\sum_{i=1}^{n}\zeta(i)\Big(\phi(i)'r - \big(T_{\mu_{k+1}}^{(\lambda)}(\Phi r_k)\big)(i)\Big)^2,$$

or equivalently

$$r_{k+1} = \left(\sum_{i=1}^{n}\zeta(i)\phi(i)\phi(i)'\right)^{-1}\sum_{i=1}^{n}\zeta(i)\phi(i)\big(T_{\mu_{k+1}}^{(\lambda)}(\Phi r_k)\big)(i). \qquad (17.39)$$

Let $\tilde{\zeta}(i)$ be the empirical relative frequency of state i during the simulation, given by

$$\tilde{\zeta}(i) = \frac{1}{N_1+\cdots+N_t}\sum_{m=1}^{t}\sum_{\ell=0}^{N_m-1}\delta(i_{\ell,m}=i). \qquad (17.40)$$

Then the simulation-based estimate (Equation (17.35)) can be written as

$$\begin{aligned}
r_{k+1} &= \left(\sum_{m=1}^{t}\sum_{\ell=0}^{N_m-1} \phi(i_{\ell,m})\phi(i_{\ell,m})'\right)^{-1} \sum_{m=1}^{t}\sum_{\ell=0}^{N_m-1} \phi(i_{\ell,m})c_{\ell,m}(r_k) \\
&= \left(\sum_{i=1}^{n}\sum_{m=1}^{t}\sum_{\ell=0}^{N_m-1} \delta(i_{\ell,m}=i)\phi(i)\phi(i)'\right)^{-1} \sum_{i=1}^{n}\sum_{m=1}^{t}\sum_{\ell=0}^{N_m-1} \delta(i_{\ell,m}=i)\phi(i)c_{\ell,m}(r_k) \\
&= \left(\sum_{i=1}^{n} \tilde{\xi}(i)\phi(i)\phi(i)'\right)^{-1} \sum_{i=1}^{n} \frac{1}{N_1+\cdots+N_t}\cdot\phi(i)\cdot\sum_{m=1}^{t}\sum_{\ell=0}^{N_m-1}\delta(i_{\ell,m}=i)c_{\ell,m}(r_k) \\
&= \left(\sum_{i=1}^{n} \tilde{\xi}(i)\phi(i)\phi(i)'\right)^{-1} \sum_{i=1}^{n} \frac{\sum_{m=1}^{t}\sum_{\ell=0}^{N_m-1}\delta(i_{\ell,m}=i)}{N_1+\cdots+N_t}\cdot\phi(i)\cdot \\
&\qquad \cdot\frac{1}{\sum_{m=1}^{t}\sum_{\ell=0}^{N_m-1}\delta(i_{\ell,m}=i)}\cdot\sum_{m=1}^{t}\sum_{\ell=0}^{N_m-1}\delta(i_{\ell,m}=i)c_{\ell,m}(r_k)
\end{aligned}$$

and finally, using Equations (17.38) and (17.40),

$$r_{k+1} = \left(\sum_{i=1}^{n} \tilde{\xi}(i)\phi(i)\phi(i)'\right)^{-1} \sum_{i=1}^{n} \tilde{\xi}(i)\phi(i)D_t(i). \tag{17.41}$$

We can now compare the λ-PI iteration (17.39) and the simulation-based implementation (17.41). Since $\big(T^{(\lambda)}_{\mu_{k+1}}(\Phi r_k)\big)(i) = \lim_{t\to\infty} D_t(i)$ and $\zeta(i) = \lim_{t\to\infty}\tilde{\xi}(i)$, we see that these two iterations asymptotically coincide.

The expression (17.41) provides some insight on how λ-PI(1) approximates the λ-PI iteration (17.39) [or equivalently $\Phi r_{k+1} = \hat{\Pi} T^{(\lambda)}_{\mu_{k+1}}(\Phi r_k)$; cf. Equation (17.36)]. Generally, the simulation process of λ-PI(1) (many short trajectories) involves more noise than the simulation process of the other implementations (a single long trajectory), because the length of each simulation trajectory is random (exponentially distributed). This can be seen from iteration (17.41), which involves considerable simulation noise due to the presence of $\tilde{\xi}(i)$ and $D_t(i)$. However, we will argue that from a practical point of view much of this noise does not play a significant role. To see this, first note that the deviation of $\tilde{\xi}(i)$ from $\zeta(i)$, is not important since $\tilde{\xi}(i)$ simply redefines the projection norm. Next note that $D_t(i)$ can be written as

$$D_t(i) = \sum_{\ell=0}^{\infty} \tilde{f}_\ell(i)\tilde{E}_\ell(i), \tag{17.42}$$

where $\tilde{f}_\ell(i)$ and $\tilde{E}_\ell(i)$ are the following empirical averages over the entire simulation process:

(a) $\tilde{f}_\ell(i)$ is the empirical relative frequency of cost samples that start at state i, and correspond to trajectories consisting of $\ell + 1$ transitions. As $t \to \infty$, it converges to $(1 - \lambda)\lambda^\ell$ based on the way the simulation is structured.

(b) $\tilde{E}_\ell(i)$ is the Monte Carlo estimate of the cost of trajectories that start at state i, consist of $\ell + 1$ transitions, and have terminal cost vector Φr_k. As $t \to \infty$, it converges to $T^{\ell+1}_{\mu_{k+1}}(\Phi r_k)(i)$.

While both $\tilde{f}_\ell(i)$ and $\tilde{E}_\ell(i)$ contribute to the variance of $D_t(i)$, only $\tilde{E}_\ell(i)$ has practical significance. To see this note that based on Equation (17.42), $D_t(i)$ can also be viewed as an estimate of

$$\tilde{T}_{\mu_{k+1}}(\Phi r_k)(i) = \sum_{\ell=0}^{\infty} \tilde{f}_\ell(i) T^{\ell+1}_{\mu_{k+1}}(\Phi r_k)(i). \tag{17.43}$$

Thus, iteration (17.41) may also be viewed as a simulation-based implementation of the optimistic PI method

$$\Phi r_{k+1} = \tilde{\Pi} \tilde{T}_{\mu_{k+1}}(\Phi r_k),$$

where $\tilde{\Pi}$ is projection with respect to the weighted sup norm defined by $\tilde{\zeta}$. From a practical point of view, this iteration and the λ-PI iteration $\Phi r_{k+1} = \hat{\Pi} T^{(\lambda)}_{\mu_{k+1}}(\Phi r_k)$ perform similarly: there is only a difference in the projection norm ($\tilde{\Pi}$ rather than $\hat{\Pi}$), and a difference in the weights of the terms $T^{\ell+1}_{\mu_{k+1}}$ [$\tilde{f}_\ell(i)$ rather than $(1 - \lambda)\lambda^\ell$]; compare $\tilde{T}_{\mu_{k+1}}(\Phi r_k)(i)$ as given by Equation (17.43) with

$$T^{(\lambda)}_{\mu_{k+1}}(\Phi r_k)(i) = \sum_{\ell=0}^{\infty}(1 - \lambda)\lambda^\ell T^{\ell+1}_{\mu_{k+1}}(\Phi r_k)(i),$$

the definition of $T^{(\lambda)}_{\mu_{k+1}}$. Neither difference should affect significantly the quality of the obtained approximation Φr_{k+1}.

In conclusion, with the λ-PI(1) implementation (17.33–17.35), as $t \to \infty$, we obtain in the limit the λ-PI iteration, with comparable performance degradation due to simulation noise as for the LSPE(λ) implementation of Section 17.4.1. A key characteristic of the implementation is that it deals with the issue of exploration flexibly and effectively. Since a trajectory of the stopping problem is completed at each transition with the potentially large probability $1 - \lambda$, a restart with a new initial state i_0 is frequent and the length of each of the simulated trajectories is relatively small. The restart mechanism can be used as a "natural" form of exploration, by choosing appropriately the restart distribution ζ_0 so that $\zeta(i)$ reflects a "substantial" weight for all states i. Thus, λ-PI(1) is like LSPE(λ) (Section 17.4.1) but with built-in exploration enhancement. Compared to λ-PI(0) (Section 17.4.2) it involves reduced bias since it aims to find the limit point of TD(λ), not TD(0). In particular, as $\lambda \to 1$, it produces an evaluation Φr_{k+1} that tends to the fixed point of TD(1), i.e., the projection $\hat{\Pi} J_{\mu_{k+1}}$.

17.4.4 Comparison with Alternative Approximate PI Methods

The preceding λ-PI implementations are in direct competition with approximate PI methods that use LSTD(λ) for policy evaluation. A popular method, often referred to as LSPI (Lagoudakis and Parr [55]), can be simply described as approximate PI combined with LSTD(0) for policy evaluation. The LSPI and λ-PI(0) methods have been compared in [40] in terms of four characteristics.

(a) *Bias*: Both methods are subject to qualitatively similar bias [they aim to find the limit point of TD(0)].

(b) *Sample Efficiency*: Both methods can reuse the same set of sample state trajectories over all policies. (In the model-free case where Q-factors are approximated, again the set of sample state-control trajectories is reusable.)

(c) *Exploration*: Both methods provide the same options for exploration, since the validity of these methods does not depend on whether the simulation trajectories are obtained using the current policy [in fact these trajectories are reusable as per (b) above].

(d) *Optimistic Operation*: Since λ-PI(0) has an iterative character (r_{k+1} depends on r_k), it is less susceptible to simulation noise and has an advantage over LSPI in the case where the number of samples per policy is low. Indeed, this assertion is made by Thiery and Scherrer [40] based on experimentation, who also found that the effect of the choice of λ is more pronounced in this case.

Note that (b) and (c) are the advantages of LSPI and λ-PI(0) over the LSPE(λ) implementation of Section 17.4.1 (which in turn involves less bias because of the use of $\lambda > 0$, and also has an optimistic character).

Let us now compare λ-PI(1) with LSPI and λ-PI(0) in terms of the characteristics (a)-(d) above. It has better bias characteristics as noted earlier. It has worse sample efficiency as it cannot reuse simulation trajectories (it can only reuse the restart state sequence). It deals with exploration about as well, thanks to the restart mechanism of the SSP formulation. Finally, similar to λ-PI(0), λ-PI(1) has an optimistic character, and has a similar advantage over LSPI in this regard, cf. (d).

17.4.5 Exploration-Enhanced LSTD(λ) with Geometric Sampling

The geometric sampling idea underlying the λ-PI(1) implementation of Equations (17.33–17.35) may also be modified to obtain an exploration-enhanced version of LSTD(λ). In particular, we use the same simulation procedure, and in analogy to Equation (17.33) we define

$$c_{\ell,m}(r) = \alpha^{N_m-\ell}\phi(i_{N_m,m})'r + \sum_{q=\ell}^{N_m-1} \alpha^{q-\ell} g(i_{q,m}, u_{q,m}, i_{q+1,m}).$$

We then obtain an approximation $\Phi\hat{r}$ to the solution of the projected equation

$$\Phi r = \hat{\Pi} T^{(\lambda)}_{\mu_{k+1}}(\Phi r),$$

[cf. Equation (17.36)] by finding $\hat{r}$ such that

$$\hat{r} = \arg\min_{r \in \Re^s} \sum_{m=1}^{t} \sum_{\ell=0}^{N_m-1} \big(\phi(i_{\ell,m})' r - c_{\ell,m}(\hat{r})\big)^2. \tag{17.44}$$

By writing the optimality condition

$$\sum_{m=1}^{t} \sum_{\ell=0}^{N_m-1} \phi(i_{\ell,m})\big(\phi(i_{\ell,m})'\hat{r} - c_{\ell,m}(\hat{r})\big) = 0$$

for the least squares minimization in Equation (17.44) and solving for $\hat{r}$, we obtain the following implementation of LSTD(λ):

$$\hat{r} = \hat{C}^{-1}\hat{d}, \tag{17.45}$$

where

$$\hat{C} = \sum_{m=1}^{t} \sum_{\ell=0}^{N_m-1} \phi(i_{\ell,m})\big(\phi(i_{\ell,m}) - \alpha^{N_m-\ell}\phi(i_{N_m,m})\big)', \tag{17.46}$$

and

$$\hat{d} = \sum_{m=1}^{t} \sum_{\ell=0}^{N_m-1} \phi(i_{\ell,m}) \sum_{q=\ell}^{N_m-1} \alpha^{q-\ell} g(i_{q,m}, u_{q,m}, i_{q+1,m}). \tag{17.47}$$

For a large number of trajectories t, the exploration-enhanced LSTD(λ) method (17.44) [or equivalently Equations (17.45–17.47) and λ-PI(1) [cf. Equation (17.35) yield similar results, particularly when $\lambda \approx 1$. However, λ-PI(1) has an iterative character (r_{k+1} depends on r_k), so it is reasonable to expect that it is less susceptible to simulation noise in an optimistic PI setting where the number of samples per policy is low.

As an example, when $\lambda = 0$, all the simulation trajectories consist of a single transition, so $N_m = 1$ for all $m = 1, \ldots, t$. Then, using Equations (17.45) and (17.47), the equation $\hat{C}r = \hat{d}$ becomes

$$\sum_{m=1}^{t} \phi(i_{0,m})\big(\phi(i_{0,m}) - \alpha\phi(i_{1,m})\big)' r = \sum_{m=1}^{t} \phi(i_{0,m}) g(i_{0,m}, u_{0,m}, i_{1,m}).$$

It yields the same vector $\hat{r} = \hat{C}^{-1}\hat{d}$ as the LSTD(0) method that simulates t independent transitions according to the restart distribution ζ_0, rather than simulating a single long trajectory. In fact this is the policy evaluation process in the LSPI method mentioned in Section 17.4.4. The geometric sampling procedure described here allows exploration-enhancement for any λ.

17.5 CONCLUSIONS

We discussed a few implementations of λ-PI with linear cost function approximation, which have different strengths and weaknesses with respect to dealing with the critical issues of bias and exploration. Out of the three implementations, the one of Section 17.4.3, λ-PI(1), is new and seems capable of dealing well with both issues, although it has worse sample complexity than the λ-PI(0) implementation of Section 17.4.2.

On the other hand, our discussion has been somewhat speculative, and our assessments, while relying on past computational experience, still require supportive experimentation. Moreover, the λ-PI implementations should be compared to other approximate PI methods based on projected equations, such as the exploration-enhanced LSTD(λ) method for policy evaluation, discussed in Section 17.3, and the LSPI method discussed in Section 17.4.4. A computational comparison of λ-PI(0) with this latter method is given in [40], and a similar comparison with λ-PI(1) would be desirable.

Fundamentally, λ-PI(1) is based on geometric sampling, a new simulation idea for λ-methods that uses multiple short trajectories with exploration-enhanced restart, rather than a single infinitely long trajectory. This idea can also be applied to LSTD(λ), thereby obtaining a new exploration-enhanced version of this method, which has been described in Section 17.4.5. The method of geometric sampling has been generalized recently with a method called *free-form sampling*, where more general than geometric distributions are used to generate sequences of states; see Bertsekas [11] and Yu and Bertsekas [59].

ACKNOWLEDGMENTS

This research was supported by NSF Grant ECCS-08015-19 and by the Air Force Grant FA9550-10-1-0412. The author thanks Bruno Scherrer for helpful comments, Huizhen Yu for related collaboration, and Mengdi Wang for assistance with computational experimentation.

REFERENCES

1. D.P. Bertsekas and J.N. Tsitsiklis. *Neuro-Dynamic Programming*. Athena Scientific, Belmont, MA, 1996.
2. R.S. Sutton and A.G. Barto. *Reinforcement Learning*. MIT Press, Cambridge, MA, 1998.

3. A. Gosavi. *Simulation-Based Optimization Parametric Optimization Techniques and Reinforcement Learning*. Springer-Verlag, New York, 2003.
4. X.R. Cao. *Stochastic Learning and Optimization: A Sensitivity-Based Approach*. Springer, New York, 2007.
5. H.S. Chang, M.C. Fu, J. Hu, and S.I. Marcus. *Simulation-Based Algorithms for Markov Decision Processes*. Springer, New York, 2007.
6. S. Meyn. *Control Techniques for Complex Networks*. Cambridge University Press, New York, 2007.
7. W.B. Powell. *Approximate Dynamic Programming: Solving the Curses of Dimensionality*. Wiley, New York, 2007.
8. V.S. Borkar. *Stochastic Approximation: A Dynamical Systems Viewpoint*. Cambridge University Press, 2008.
9. S. Haykin. *Neural Networks and Learning Machines*, 3rd edition. Prentice-Hall, Englewood-Cliffs, New York, 2008.
10. L. Busoniu, R. Babuska, B. De Schutter, and D. Ernst. *Reinforcement Learning and Dynamic Programming Using Function Approximators*. CRC Press, New York, 2010.
11. D.P. Bertsekas, *Dynamic Programming and Optimal Central, Vol. II: Approximate Dynamic Programming*, Athena Scientific, 2012.
12. D. White and D. Sofge. *Handbook of Intelligent Control*. Van Nostrand Reinhold, New York, 1992.
13. J. Si, A. Barto, W. Powell, and D. Wunsch, editors. *Learning and Approximate Dynamic Programming*. IEEE Press, New York, 2004.
14. F.L. Lewis, G.G. Lendaris, and D. Liu. Special issue on adaptive dynamic programming and reinforcement learning in feedback control. *IEEE Transactions on Systems, Man, and Cybernetics*, 38, 2008.
15. V.S. Borkar. Reinforcement learning: a bridge between numerical methods and monte carlo. *World Scientific Review*, 9(Chap. 4), 2009.
16. F.L. Lewis and D. Vrabie. Reinforcement learning and adaptive dynamic programming for feedback control. *IEEE Circuits and Systems Magazine*, 3rd Q. Issue, 2009.
17. P.J. Werbos. Intelligence in the brain: a theory of how it works and how to build it. *Neural Networks*, 22:200–212, 2009.
18. C. Szepesvari. *Reinforcement Learning Algorithms for MDPs*. Morgan and Claypool Publishers, 2010.
19. D.P. Bertsekas. Approximate policy iteration: a survey and some new methods. *Journal of Control Theory and Applications*, 9:310–335, 2011.
20. D.P. Bertsekas and S. Ioffe. Temporal differences-based policy iteration and applications in neuro-dynamic programming. Lab. for Information and Decision Systems Report LIDS-P-2349, MIT, 1996.
21. M.L. Puterman. *Markov Decision Processes: Discrete Stochastic Dynamic Programming*. Wiley, New York, 1994.
22. D.P. Bertsekas. *Dynamic Programming and Optimal Control*, 3rd edition, Vol. II. Athena Scientific, Belmont, MA, 2007.
23. U.G. Rothblum. Iterated successive approximation for sequential decision processes. In J.W.B. van Overhagen and H.C. Tijms, editor. *Stochastic Control and Optimization*. Vrije University, Amsterdam, 1979.

24. P.G. Canbolat and U.G. Rothblum. (Approximate) Iterated Successive Approximations Algorithm for Sequential Decision Processes. Technical Report, The Technion - Israel Institute of Technology, May 2011.

25. R.J. Williams and L.C. Baird. Analysis of some incremental variants of policy iteration: first steps toward understanding actor-critic learning systems. Report NU-CCS-93-11, College of Computer Science, Northeastern University, Boston, MA, 1993.

26. D.P. Bertsekas and H. Yu. Q-Learning and enhanced policy iteration in discounted dynamic programming, Laboratory for Information and Decision Systems Report LIDS-P-2831, MIT, 2010; to appear in Mathematics of Operations Research.

27. D.P. Bertsekas and H. Yu. Asynchronous distributed policy iteration in dynamic programming. In *Proceedings of Allerton Conference on Information Sciences and Systems*, 2010, pp. 1368–1375.

28. H. Yu and D.P. Bertsekas. Q-Learning and policy iteration algorithms for stochastic shortest path problems. Laboratory for Information and Decision Systems Report LIDS-P-2871, MIT, 2011; to appear in Annals of Operations Research.

29. J.N. Tsitsiklis and B. Van Roy. An analysis of temporal-difference learning with function approximation. *IEEE Transactions on Automatic Control*, 42:674–690, 1997.

30. H. Yu, and D.P. Bertsekas. Error bounds for approximations from projected linear equations. *Mathematics of Operations Research*, 35:306–329, 2010.

31. S.J. Bradtke and A.G. Barto. Linear least-squares algorithms for temporal difference learning. *Machine Learning*, 22:33–57, 1996.

32. A. Nedić and D.P. Bertsekas. Least squares policy evaluation algorithms with linear function approximation. *Discrete Event Dynamic Systems: Theory and Applications*, 13:79–110, 2003.

33. D.P. Bertsekas, V.S. Borkar, and A. Nedić. Improved temporal difference methods with linear function approximation. In J. Si, A. Barto, W. Powell, and D. Wunsch, Editors. *Learning and Approximate Dynamic Programming*. IEEE Press, New York, 2004.

34. D.P. Bertsekas. A Counterexample to temporal differences learning. *Neural Computation*, 7:270–279, 1995.

35. I. Szita and A. Lorinz. Learning tetris using the noisy cross-entropy method. *Neural Computation*, 18:2936–2941, 2006.

36. C. Thiery and B. Scherrer. Improvements on learning tetris with cross-entropy. *International Computer Games Association Journal*, 32:23–33, 2009.

37. D.P. Bertsekas. Pathologies of temporal difference methods in approximate dynamic programming. In *Proceedings 2010 IEEE Conference on Decision and Control*, 2010, pp. 3034–3039.

38. D.P. Bertsekas and H. Yu. Projected equation methods for approximate solution of large linear systems, *Journal of Computational and Applied Mathematics*, 227:27–50, 2009.

39. D.P. Bertsekas. Temporal difference methods for general projected equations. *IEEE Trans. on Aut. Control*, 56:2128–2139, 2011.

40. C. Thiery and B. Scherrer. Least-squares policy iteration: bias-variance trade-off in control problems. In *Proceedings of 2010 ICML*. Haifa, Israel, 2010.

41. C. Thiery and B. Scherrer. Performance bound for approximate optimistic policy iteration. Technical Report, INRIA, 2010.

42. M.A. Krasnoselskii, et al. *Approximate Solution of Operator Equations*. Translated by D. Louvish. Wolters-Noordhoff Pub., Groningen, 1972.

43. C.A.J. Fletcher. *Computational Galerkin Methods*. Springer-Verlag, New York, 1984.

44. D.P. Bertsekas. Projected equations, variational inequalities, and temporal difference methods. Laboratory for Information and Decision Systems Report LIDS-P-2808, MIT, 2009.

45. H. Yu. Least squares temporal difference methods: an analysis under general conditions. Technical report C-2010-39, Department Computer Science, University of Helsinki, 2010.

46. H. Yu. Convergence of least squares temporal difference methods under general conditions. *Proceedings of the 27th ICML*, Haifa, Israel, 2010.

47. B. Scherrer. Should one compute the temporal difference fix point or minimize the Bellman residual? the unified oblique projection view. In *ICML'10: Proceedings of the 27th Annual International Conference on Machine Learning*, 2010.

48. A.L. Samuel. Some studies in machine learning using the game of checkers. *IBM Journal of Research and Development*, 210–229, 1959.

49. A.L. Samuel. Some studies in machine learning using the game of checkers. II – recent progress. *IBM Journal of Research and Development*, 601–617, 1967.

50. M. Wang, N. Polydorides and D.P. Bertsekas. Approximate simulation-based solution of large-scale least squares problems. Laboratory for Information and Decision Systems Report LIDS-P-2819, MIT, 2009.

51. M. Wang and D.P. Bertsekas. Stabilization of simulation-based iterative methods for singular and nearly singular linear systems. Laboratory for Information and Decision Systems Report LIDS-P-2878, MIT, 2011.

52. M. Wang and D.P. Bertsekas. On the convergence of iterative simulation-based methods for singular linear systems. Laboratory for Information and Decision Systems Report LIDS-P-2879, MIT, 2011.

53. B. Scherrer. Performance bounds for lambda policy iteration and application to the game of tetris. Report RR-6348, INRIA, 2011.

54. G. Foderaro, V. Raju, and S. Ferrari. A model-based approximate λ-policy iteration approach to online evasive path planning and the video game Ms. Pac-Man. *Journal of Control Theory and Applications*, 9:391–399, 2011.

55. M.G. Lagoudakis and R. Parr. Least-squares policy iteration. *Journal of Machine Learning Research*, 4:1107–1149, 2003.

56. R.S. Sutton. Learning to predict by the methods of temporal differences. *Machine Learning*, 3:9–44, 1988.

57. H. Yu and D.P. Bertsekas. Convergence results for some temporal difference methods based on least squares. *IEEE Transation on Automatic Control*, 54:1515–1543, 2009.

58. A.G. Barto, R.S. Sutton, and C.W. Anderson. Neuronlike elements that can solve difficult learning control problems. *IEEE Trans. on Systems, Man, and Cybernetics*, 13:835–846, 1983.

59. H. Yu and D.P. Bertsekas. Weighted Bellman Equations and their Applications in Dynamic Programming. Laboratory for Information and Decision Systems Report LIDS-P-2876, MIT, 2012.

CHAPTER 18

Optimal Learning and Approximate Dynamic Programming

WARREN B. POWELL[1] and ILYA O. RYZHOV[2]

[1]Department of Operations Research and Financial Engineering, Princeton University, NJ, USA
[2]Department of Decision, Operations, and Information Technologies Robert H. Smith School of Business, University of Maryland, College Patk, MD, USA

18.1 INTRODUCTION

Approximate dynamic programming (ADP) has emerged as a powerful tool for tackling a diverse collection of stochastic optimization problems. Reflecting the wide diversity of problems, ADP (including research under names such as reinforcement learning, adaptive dynamic programming, and neuro-dynamic programming) has become an umbrella for a wide range of algorithmic strategies. Most of these involve learning functions of some form using Monte Carlo sampling.

A recurring theme in these algorithms involves the need to not just learn policies, but to learn them quickly and effectively. Learning arises in both offline settings (training an algorithm within the computer) and online settings (where we have to learn as we go). Learning also arises in different ways within algorithms, including learning the parameters of a policy, learning a value function, and learning how to expand the branches of a tree.

The focus of this chapter is to discuss different strategies for learning policies in an efficient way. These "strategies" are themselves referred to as policies, as they represent rules for guiding the collection of information. Passive learning policies will collect information and update beliefs about functions, without making any explicit attempt at collecting information in a way that would accelerate the learning policy. Active learning refers to policies where we are willing to make suboptimal actions explicitly because the information gained will add value later in the process. Most of the literature has focused on simple heuristics, but in some cases these heuristics have provable suboptimality bounds. Our presentation will focus primarily on the

Reinforcement Learning and Approximate Dynamic Programming for Feedback Control, First Edition. Edited by Frank L. Lewis and Derong Liu.

knowledge gradient policy, which maximizes the rate of learning, and offers both theoretical and practical features in a Bayesian setting which focuses on minimizing expected opportunity cost.

To describe the different ways in which active learning arises, we need to first understand different problem settings and, most importantly, the major classes of policies. Section 18.2 provides a compact modeling framework and highlights some important problem classes. Section 18.3 identifies the major classes of policies, which we then use to identify opportunities for optimal learning.

18.2 MODELING

There are five core components of any stochastic, dynamic system. These include

- *The State Variable* S_t: The state variable captures the information available at time t.
- *Decisions, Actions, or Controls*: We use the notation a_t to represent discrete action spaces, although many problems exhibit vector-valued decisions or controls.
- *Exogenous Information*: We let W_t represent any random information that becomes known for the first time at discrete time t.
- *The Transition Function*: Also known as the system model, plant model, or just model, we represent it using the function $S^M(\cdot)$ where we will write

$$S_{t+1} = S^M(S_t, a_t, W_{t+1})$$

  to represent the evolution of the state variable over time due to the effect of an action a_t and new information W_{t+1}.
- *Objective Function*: We let $C(S_t, a_t)$ be the contribution (if maximizing) or cost (if minimizing) that depends on the state S_t and action a_t. In some applications, the one-period contribution may be random at time t, in which case we would write it as $C(S_t, a_t, W_{t+1})$. In this case, we assume that $C(S_t, a_t)$ is the expected contribution.

For our purposes, we need to further identify three perspectives of a state variable:

- *The Physical State*: Sometimes called the resource state, there are many problems where we are managing a resource such as robots, a fleet of vehicles, a chemical process, or financial investments. The resource state, R_t, captures the state of physical entities that are being managed.
- *The Information State*: Widely used in control theory, we let I_t capture information used to make a decision, which would include R_t but may also include information such as prices and the weather.
- *The Knowledge (or Belief) State*: The knowledge state captures our belief about unobservable parameters. This is a particularly important dimension of the prob-

lems we consider in this chapter, and represents one of the more subtle and difficult dimensions of dynamic programming.

For the moment, we do not address the problem of how we make decisions, other than to assume there is some rule, or *policy* for making decisions. The research literature will typically represent a policy as π. We prefer the notation that if we are determining an action a, our policy would be written as a function $A^\pi(S_t)$ where π parameterizes both the class of function (examples are given below), and any parameters that determine the behavior of the function. If we use decision x, our decision function would be written $X^\pi(S_t)$.

We need an objective function that defines the best policy. We assume that we are maximizing expected total (possibly discounted) contributions, recognizing that there are other objective functions. The problem of finding the best policy can be written as

$$\max_\pi \mathbb{E}^\pi \sum_{t=0}^{T} \gamma^t C(S_t, A^\pi(S_t)). \tag{18.1}$$

Equation (18.1) is the formal definition of a dynamic program, which is a sequential decision process. We are generally not able to find a policy that solves (18.1), so we have to resort to finding an approximate policy. This is the essence of ADP.

18.3 THE FOUR CLASSES OF POLICIES

The literature on stochastic optimization can sometimes be viewed as a jungle of modeling and algorithmic strategies. Terms like "policy search," "reinforcement learning," "stochastic programming," "simulation-optimization," "model predictive control," "Monte Carlo tree search," and "actor–critic methods" (to name a few) are often used without a clear definition.

Without question, perhaps one of the most confusing and misunderstood terms in dynamic programming is the word *policy*. Stated simply, a policy is a mapping from state to action. These mappings are typically written as functions, and are sometimes confusingly described as rules.

The research community has created a complex array of functions for making decisions. These can generally be boiled down to four fundamental classes, with many opportunities for creating hybrid policies. We list these because the structure of each policy introduces specific learning challenges.

18.3.1 Myopic Cost Function Approximation

There is the rare problem that lends itself to a simple myopic policy of the form

$$A^M(S_t) = \arg\max_a C(S_t, a).$$

One example involves balancing financial portfolios when there are no transaction costs. We can set up a model where we maximize one-period returns and a penalty for the aggregate volatility of the account. Without transaction costs, we can move funds from one asset to another with no cost. As a result, a policy that optimizes the one-period utility is optimal.

Myopic policies are often used as heuristics in specific applications. We might have a resource i (such as a truck) that we are looking to assign to task j (such as a load). Let c_{ij} be the cost of assigning truck i to load j. A myopic policy would be to solve

$$A^M(S_t) = \arg\min_a \sum_i \sum_j c_{ij} a_{ij},$$

subject to

$$\sum_i a_{ij} \leq 1,$$
$$\sum_j a_{ij} \leq 1,$$
$$a_{ij} \geq 0.$$

We further assume that unassigned drivers and loads are held to the next time period. This policy is actually used in some commercial dispatch systems in the truckload trucking industry, but it is hardly optimal. For example, it ignores how long a load has been held. A proper dynamic program would capture how long drivers and loads have been sitting, and would make a decision that balances minimizing costs and service delays over an entire horizon. Since this is hard to solve, a simple heuristic is to introduce a bonus for moving loads that have been waiting. Letting τ_i be the length of time that a load has been waiting, the policy would be given by

$$A^M(S_t|\theta) = \arg\min_a \sum_i \sum_j (c_{ij} - \theta\tau_i) a_{ij}. \tag{18.2}$$

Here, we have created a family of myopic policies parameterized by θ.

18.3.2 Lookahead Policies

Lookahead policies come in many forms, but they all involve optimizing explicitly over multiple time periods, but then just retaining the decision that we implement now. The most common type of lookahead uses a point forecast of any random information, which transforms the model into a deterministic optimization problem of the form

$$A^L(S_t) = \arg\max_t \left(c_t a_t + \sum_{t'=t+1}^{t+\bar{t}} c_{t'} a_{t'} \right).$$

Here, we define $\arg\max_t$ as the operator that maximizes over the vector $a_t, \ldots, a_{t+\bar{t}}$, but only returns a_t. This strategy is often referred to as a rolling horizon procedure, because we optimize over a horizon of length $\bar{t}$, implement the decision for time t, and then roll forward, updating any parameters due to new information.

There are several communities that work to incorporate uncertainty into the lookahead process, such as decision trees (for problems with small action spaces) or stochastic programs (for vector-valued decisions; see [1]). Monte Carlo tree search (MCTS) has become a popular field of research in the reinforcement learning community, contributing to breakthroughs on difficult problems such as computer Go [2–4].

18.3.3 Policy Function Approximation

We use the term "policy function approximation" to refer to direct mappings from states to actions, without the need to solve an embedded optimization problem. Examples of policy function approximations include

- If a patient has a set of symptoms s, choose drug d (for every state, we have to have an associated drug).
- If the inventory in a store is less than q, order enough to bring it up to Q.
- If the reservoir in a lake is of level h, set the rate of outflow from the dam to $x = \theta_0 + \theta_1 h$.
- There is a set of discrete actions a with contribution $C(s, a)$ when we are in state s. We are going to choose action a with probability

$$p(a, s|\theta) = \frac{e^{\theta C(s,a)}}{\sum_{a'} e^{\theta C(s,a)}},$$

  where θ is a tunable parameter. As θ is increased, we increase the likelihood that we will choose action a.

In the drug example, we have a parameter (the choice of drug d) for each state. Let θ_s be the choice of drug if we are in state s. In the inventory example, we might write $\theta = (q, Q)$. In the lake example, the decision is determined by a function parameterized by $\theta = (\theta_0, \theta_1)$. Note that while the second and third policies have only two parameters, the drug policy has a parameter for every state. The fourth policy (which is widely used in reinforcement learning) is parameterized by a scalar θ.

18.3.4 Policies Based on Value Function Approximations

It is well known that we can mathematically characterize an optimal policy using Bellman's optimality equation which we write using

$$V_t(S_t) = \max_a \left(C(S_t, a) + \gamma \sum_{s'} p(s'|S_t, a) V_{t+1}(s') \right) \tag{18.3}$$

$$= \max_a \left(C(S_t, a) + \gamma \mathbb{E}\{V_{t+1}(S^M(S_t, a, W_{t+1}))|S_t\} \right). \tag{18.4}$$

We use the time-dependent version because it makes the modeling clearer (in addition, there are many problems where policies are time dependent). Equation (18.3) is the form most widely used in texts on Markov decision processes (such as [5]), while Equation (18.4) is known as the expectation form, where the expectation is over the random variable W_{t+1} given that we are in state S_t.

There is a rich literature that has evolved around the idea of using Bellman's equation to derive optimal (or at least good) policies. We face several challenges when going down this path. First, for most applications, we cannot compute the expectation. We circumvent this by using the concept of the post-decision state S_t^a, which is the state after making a decision but before any new information has arrived. Second, we cannot compute the value function exactly, we replace it with a statistical approximation. While we can tap a wide array of approximation strategies (see, for example, [6]), the most popular strategy is a simple linear model of the form

$$\bar{V}_t(S_t^a|\theta_t) = \sum_{f\in\mathcal{F}} \theta_{tf}\phi_f(S_t^a). \tag{18.5}$$

where $(\phi_f(S_t^a))_{f\in\mathcal{F}}$ is a vector of user-specified *basis functions* and $(\theta_f)_{f\in\mathcal{F}}$ a vector of regression parameters. This allows us to write our policy as

$$A_t^{\pi,n}(S_t|\theta_t) = \arg\max_a \left(C(S_t^n, a) + \gamma \sum_{f\in\mathcal{F}} \theta_{tf}^n \phi_f(S^a(S_t^n, a)) \right). \tag{18.6}$$

Despite the elegance and simplicity of this general strategy, it requires decisions about a range of algorithmic choices. Provably convergent algorithms are rare and generally require fairly strong assumptions (e.g., the policy may be fixed rather than optimized; see [7], for a recent example). There are different strategies for generating sample estimates of the value of being in a state, estimating the value of a policy versus updating a policy, applying on-policy versus off-policy learning, and state sampling. One issue that arises is that the information we collect to update the value function approximation depends on what state we visit, and we have some control over the state.

For in-depth discussions of the use of value function approximations, we refer the reader to Chapter 6 in Bertsekas [8], Bertsekas [9], Bertsekas and Tsitsiklis [10], Sutton and Barto [11], Szepesvari [12] and Chapters 8–10 of Powell [13], and the many references cited there.

18.3.5 Learning Policies

We have outlined above four fundamental classes of policies. Needless to say, it is possible to create a host of additional policies by mixing and matching these strategies. One popular strategy is to use a limited tree search with a value function approximation at the end of the tree. Cost function approximations are widely used to improve myopic policies, but just as often will be merged with a rolling horizon procedure (again, a form of tree search).

Three of these fundamental policies involve some sort of functional approximation: approximating a cost function, approximating a function that specifies the policy, and approximating a value function which in turn is used to define a policy. There are three ways of approximating functions: lookup tables, parametric models, and nonparametric models. With a lookup table, specifying a discrete state returns an action (policy function approximation) or the value of being in a state (value function approximation). Parametric models, used for all three types of functional approximations, represent any analytical function with a predetermined set of parameters to be determined. Nonparametric models have also been suggested for both policies (primarily for engineering applications with continuous actions) and value functions.

Special challenges arise when we have to learn value function approximations. There is a major difference between learning the parameters θ that govern the behavior of a policy, and learning value functions. As we have shown in Section 18.4, the first can be modeled as a dynamic program with a pure belief state, whereas the second requires that we handle the presence of a physical state.

Section 18.4 provides an introduction to the basic problem of tuning the parameters of a policy and introduces several popular heuristic policies. Then, Section 18.5 presents optimal policies for learning, including a characterization of the optimal policy for learning as a dynamic program with a pure belief state. Finally, Section 18.6 closes with a discussion of optimal learning in the presence of a physical state, which is the challenge we face in ADP.

18.4 BASIC LEARNING POLICIES FOR POLICY SEARCH

All four classes of policies above can be represented as a function $A^\pi(S_t|\theta)$ if we wish to choose a discrete action a_t. The index π specifies the class of function, while θ captures all the tunable parameters that determine the behavior of the function within the class of policy (we should write the tunable parameter vector as θ^π to express its dependence on the class of policy, but we suppress this bit of additional notation for simplicity). We emphasize that $A^\pi(S_t|\theta)$ may be *any* of the four fundamental classes of policies, including policies based on value function approximations.

Once we have chosen the class of policy π, the problem of finding the best vector θ requires solving

$$\min_\theta F^\pi(\theta) = \mathbb{E}\sum_{t=0}^{T} C(S_t, A^\pi(S_t|\theta)). \tag{18.7}$$

Equation (18.7) is a classical problem in stochastic optimization that was first addressed in the seminal paper by Robbins and Monro [14]. Recognizing that the expectation cannot be computed for any but the most specialized problems, we use our ability to simulate a policy, or to observe a policy being used in a real situation. Given a sample path ω, let $F^\pi(\theta, \omega)$ be a simulation of the value of policy π. If we further assume that we can compute the gradient $\nabla_\theta F^\pi(\theta, \omega)$ (along with some other assumptions), the classic stochastic approximation procedure would allow us to find θ using

$$\theta^n = \theta^{n-1} - \alpha_{n-1} \nabla_\theta F^\pi(\theta^{n-1}, \omega^n). \tag{18.8}$$

Equation (18.8) is a simple, elegant optimization algorithm which offers provable asymptotic convergence (for reviews, see [15, 16]). At the same time, there are no promises on rate of convergence: while this algorithm can work quite well in practice, it can be slow. Finally, there are many problems where we cannot find the gradient. There are hundreds of papers addressing the many variations of this fundamental problem. For reviews, see Spall [17], Ruszczynski and Shapiro [18], and Shapiro et al. [19].

Our focus is not just learning the best policy, but learning it quickly, with the ultimate (if unachievable) goal of learning it at an optimal rate. Rate of convergence is particularly useful when computing a sample realization of $F^\pi(\theta^{n-1}, \omega^n)$ is slow. It is not hard to find industrial models where calculating $F^\pi(\theta^{n-1}, \omega^n)$ can take anywhere from 30 min to a day or more. There are settings where $F^\pi(\theta^{n-1}, \omega^n)$ is not a computer simulation but rather is a physical system. We might be interested in a sales policy where we would fix θ and then observe sales for a week or more.

The literature on optimal learning typically focuses on solving a problem of the form

$$\min_x \mathbb{E} F(x, W), \tag{18.9}$$

where we have to find the best value $x \in \mathcal{X}$ within a finite measurement budget. In our setting, it is more natural to use θ to parameterize the policy (in some communities, x is used as the decision variable within the dynamic program). It should be apparent that Equations (18.7) and (18.9) are equivalent. We are going to let θ_x be a particular value of θ corresponding to $x \in \mathcal{X}$, so that there is a one-to-one relationship between x and θ, allowing us to bridge two notational systems.

In practice x (or θ_x) may be (a) a small, discrete set, (b) a large, discrete set (that is, too large to enumerate), (c) a continuous scalar, or (d) a vector of any form. We do not have the space to describe specialized algorithms for all these cases, but we are going to cover algorithms that can handle a wide range of applications that arise in practice, including problems where x is continuous.

18.4.1 The Belief Model

We proceed by first forming a belief model about $F(x) = F^\pi(\theta_x)$. If $x \in \{1, 2, \ldots, M\}$, it is standard practice to use a Bayesian model where μ_x is the unknown true value of $F(x)$ where $x = \theta_x$. We treat μ_x as a random variable which is normally distributed with mean μ_x^0 and precision $\beta_x^0 = 1/(\sigma_x^0)^2$, where the precision is the inverse of the variance. We refer to (μ_x^0, β_x^0) as our prior (in addition to the assumption of normality). We have to specify these parameters based on some domain knowledge (or preliminary simulations). They reflect our belief about the likely values of μ_x.

We are going to start by assuming that we have a lookup table representation of $F(\theta)$, which is to say that there is a true value $\mu_x = \mathbb{E}F(\theta_x)$ giving us how well the policy performs if $\theta = \theta_x$. For our lookup table representation, we assume that we are searching for the best value of θ among a finite set of vectors $(\theta_x)_{x \in \mathcal{X}}$, $\mathcal{X} = \{1, 2, \ldots, M\}$. We let (μ_x^n, β_x^n) be our belief about $F(\theta_x)$ after n observations. For a Bayesian model with a normal prior and normally distributed observations W_x^n of $F(x)$, beliefs are updated using

$$\mu_x^{n+1} = \frac{\beta_x^n \mu_x^n + \beta^W W_x^{n+1}}{\beta_x^n + \beta^W}, \tag{18.10}$$

$$\beta_x^{n+1} = \beta_x^n + \beta^W. \tag{18.11}$$

Here, $\beta^W = 1/\sigma_W^2$ is the precision (inverse variance) of a measurement W_x^n. We assume that β^W is independent of x, but this is easily generalized.

18.4.2 Objective Functions for Offline and Online Learning

Assume that we have a budget of N measurements to find the best value of x. Let μ_x^N be our belief about $F(x) = F(\theta_x)$ after N measurements which were chosen according to policy π, and let

$$x^\pi = \arg\max_x \mu_x^N$$

be our choice of the optimal solution while using policy π. This problem is widely known as the ranking and selection problem but we prefer the term offline learning, where we do not care about rewards earned while we are trying to learn the best choice.

Our goal is to find a policy π that solves the following maximization problem

$$\max_\pi \mathbb{E}_\mu \mathbb{E}_W \mu_{x^\pi}. \tag{18.12}$$

The objective function in (18.12) reflects the fact that there are two sets of random variables we have to deal with. First, the true values μ_x are normally distributed with mean μ^0 and precision β^0. Second, the observations $W^1, \ldots, W^N$ are observed from distributions determined by μ and our policy π, and will affect the outcome of x^π.

If we are learning as we progress (known as online learning), our objective function becomes

$$\max_\pi \mathbb{E}_\mu \mathbb{E}_W \sum_{n=1}^{N} \gamma^n \mu_{x^n}, \tag{18.13}$$

where $x^n = \arg\max_{x'} \mu_{x'}^{n-1}$ and $0 < \gamma \leq 1$ is a discount factor. Online learning is widely referred to as the multiarmed bandit problem in the research literature (see [20], for a thorough review). It has drawn significant attention in the applied probability and computer science communities. If we are tuning a policy, most of the time we are doing this in an offline fashion within a computer, but we may have to design policies in real time, incurring costs, and rewards as we are learning the policy.

18.4.3 Some Heuristic Policies

Some of the best known heuristic policies include:

- *Pure Exploitation*: This policy uses

$$x^n = \arg\max_{x \in \mathcal{X}} \mu_x^n.$$

  Pure exploitation makes the most sense for online learning problems, where there is value in getting the best solution at each iteration. However, this is widely used in response-surface methods, where at each iteration we update a parametric approximation of $F(x)$.
- *Pure Exploration*: This policy chooses x at random. This approach makes sense in offline settings, where we are using observations to fit a parametric approximation of $F(x)$, which is then optimized to find the best value of x.
- *Epsilon-Greedy Exploration*: This is a hybrid, where we choose x at random (explore) with probability ϵ, and choose what appears to be the best (exploit) with probability $1 - \epsilon$. Typically, ϵ declines with the number of iterations. Epsilon-greedy focuses more energy on evaluating the option that appears to be the best, but still provides for a level of exploration in case we are missing what is truly the best option. See Sutton and Barto [11] for an introduction to epsilon-greedy and Singh et al. [21] for an analysis of convergence properties.
- *Boltzmann Exploration*: Here, we choose x with probability

$$p_x^n = \frac{\exp(\rho \mu_x^n)}{\sum_{x' \in \mathcal{X}} \exp(\rho \mu_{x'}^n)}.$$

  This policy overcomes a major limitation of epsilon-greedy, which does not treat the second best option any differently than the worst. If $\rho = 0$, Boltzmann-exploration is equivalent to pure exploration, while as ρ increases, it approaches pure exploitation. Boltzmann exploration is very popular in the reinforcement learning community as a parameterized random policy, where it is sometimes referred to as a soft-max policy.
- *Interval Estimation*: For each x, we compute

$$\nu_x^{IE,n} = \mu_x^n + z_\alpha \sigma_x^n,$$

where z_α is a tunable parameter (often set to around 2 or 3), and where $(\sigma_x^n)^2 = 1/\beta_x^n$ is the variance of our beliefs about μ_x. IE was introduced by Kaelbling [22].

- *Upper Confidence Bounding*: A UCB policy, designed specifically for normally distributed measurements, is calculated using

$$\nu_x^{\text{UCB},n} = \mu_x^n + 4\sigma_W \sqrt{\frac{\log n}{N_x^n}}. \tag{18.14}$$

  Here, N_x^n is the number of times we have observed x through time n.

Pure exploration and exploitation are heuristics with obvious weaknesses, but exploration is often used in offline learning where you observe the function at random locations, fit a function and then choose the best. By contrast, pure exploitation is widely used in practice for online learning problems, where choosing what appears to be the best is an understandable if imperfect policy. Epsilon-greedy and Boltzmann exploration are both hybrid exploration–exploitation policies.

Interval estimation is the first policy that uses both the current estimate of our belief μ_x^n about $F(x)$, and the uncertainty in this belief. Here, n refers to the number of times we have observed x. It is important to tune the parameter z_α, but properly tuned, IE has been found to work well in practice.

Upper confidence bounding, known generally as UCB policies, is similar to interval estimation in that there is a term added to the estimate μ_x^n that captures how good μ_x might be. UCB policies have attracted considerable interest because it is possible to prove that there is a logarithmic bound in N on suboptimality (or *regret*) in online learning. Regret bounds on UCB policies were first introduced by Lai and Robbins [23]; the UCB policy presented here is given in Auer et al. [24]. Researchers have started to develop similar policies for offline learning problems with similar regret bounds, but as of this writing, experimental evaluations of these policies are limited (see e.g., [25]).

We note that classical stochastic gradient methods such as the one in Equation (18.8) do not use any form of learning. After observing a stochastic gradient, we update our estimate of the best value of θ, but we do not otherwise update any belief models. In contrast, we assume that when using any of the policies listed in this section, we would use Equations (18.10)–(18.11) to update our beliefs about the choice we just observed. Pure exploitation makes no effort to collect information purely for the value of learning, while all the remaining policies use some form of explicit exploration, where we are willing to observe a value x that we do not believe is best, because it is possible that it might be the best. In our own experiments, interval estimation seems to work extremely well if z_α is properly tuned (and this tuning can have a real impact on performance). By contrast, we have had less success with UCB policies, but these policies are supported by rigorous bounds. However, these bounds also grow with N, so empirical performance may vary.

18.5 OPTIMAL LEARNING POLICIES FOR POLICY SEARCH

Ideally, we would have a way to find the policy π that produces the best possible result, whether it is offline or online. It turns out we can characterize the optimal policy by formulating the problem as a dynamic program, first presented in DeGroot [26]. Assume that everything is still normally distributed. Then, our state variable is our state of belief, which is given by $S^n = (\mu_x^n, \beta_x^n)_x$. Bellman's equation then characterizes the optimal policy using

$$V^n(S^n) = \max_x \left(C(S^n, x) + \gamma \mathbb{E}\{V^{n+1}(S^{n+1}(s, x))|S^n\}\right), \tag{18.15}$$

where S^{n+1} is given by Equations (18.10) and (18.11). Unfortunately, (18.15) is computationally intractable. It is important to emphasize that the problem setting of learning the parameters of a policy can be formulated as a dynamic program with just a belief state. Later, we consider the much more difficult problem of combining a belief state with a physical state.

For this basic model, the optimal policy can be characterized using Gittins indices, given by

$$\nu_x^{\text{Gitt},n}\left(\mu_x^n, \sigma_x^n, \sigma_W, \gamma\right) = \mu_x^n + \sigma_W G\left(\frac{\sigma_x^n}{\sigma_W}, \gamma\right). \tag{18.16}$$

This theory, described in-depth by Gittins et al. [20], is very elegant, but the functions $G\left(\frac{\sigma_x^n}{\sigma_W}, \gamma\right)$, known as Gittins indices, are hard to compute. Chick and Gans [27] has developed a simple analytical approximation to overcome this problem, but Gittins indices are only optimal within a relatively simple class of problems that does not arise often in practice. As an alternative, we propose the use of the knowledge gradient, a simple concept that can be extended to handle more general settings.

18.5.1 The Knowledge Gradient for Offline Learning

Let

$$V^n(S^n) = \max_{x'} \mu_{x'}^n$$

be the current value of our problem given the state of knowledge S^n after n iterations. Then let $S^{n+1}(x)$ be the belief state if we choose to measure $x^n = x$ and observe W_x^{n+1} (which is still random since we have not actually performed the observation), which means that $V^{n+1}(S^{n+1}(x))$ is a random variable. The knowledge gradient is a policy where we choose to measure the alternate x that gives us the greatest value of information, and is given by

$$\nu_x^{\text{KG},n} = \mathbb{E}\left[V^{n+1}(S^{n+1}(x)) - V^n(S^n)|S^n\right]. \tag{18.17}$$

Here, S^n is our current state of knowledge (that is, the beliefs about all the alternatives), and $S^{n+1}(x)$ is the state of knowledge if we choose to observe x, but before we have actually made the observation. The concept is simple; the trick is computing it.

Till now, we have focused on problems with independent beliefs, which is to say that if we choose to measure x and observe W_x^n, we do not learn anything about alternatives $x' \neq x$. For this problem, computing the knowledge gradient is quite simple. We first have to find the conditional *change* in the variance of an estimate given a measurement, given by

$$\tilde{\sigma}_x^{2,n} = \text{Var}^n \left[\mu_x^{n+1} - \mu_x^n\right].$$

It is possible to show that this is given by

$$\begin{aligned}\tilde{\sigma}_x^{2,n} &= \sigma_x^{2,n} - \sigma_x^{2,n+1} &(18.18)\\ &= \left(\beta_x^n\right)^{-1} - \left(\beta_x^n + \beta^W\right)^{-1}. &(18.19)\end{aligned}$$

We next compute ζ_x^n which is given by

$$\zeta_x^n = -\left|\frac{\mu_x^n - \max_{x' \neq x} \mu_{x'}^n}{\tilde{\sigma}_x^n}\right|. \tag{18.20}$$

Now let

$$f(\zeta) = \zeta\Phi(\zeta) + \phi(\zeta), \tag{18.21}$$

where $\Phi(\zeta)$ and $\phi(\zeta)$ are, respectively, the cumulative standard normal distribution and the standard normal density.

We can now express the knowledge gradient as

$$\nu_x^{\text{KG},n} = \tilde{\sigma}_x^n f\left(\zeta_x^n\right). \tag{18.22}$$

The knowledge gradient policy chooses to measure the value x, which maximizes $\nu_x^{\text{KG},n}$ over all x.

Compared to solving the dynamic program in Equation (18.15), the knowledge gradient is simply a one-step lookahead policy. However, in contrast with problems with a physical state, this policy seems to work extremely well for pure learning problems. For offline learning problems, it offers several nice theoretical properties:

(1) The knowledge gradient is always positive, $\nu_x^{\text{KG},n} \geq 0$ for all x. Thus, if the knowledge gradient of an alternative is zero, that means we won't measure it.

(2) The knowledge gradient policy is optimal (by construction) if we are going to make exactly one measurement.

(3) If there are only two choices, the knowledge gradient policy is optimal for any measurement budget N.

(4) If N is our measurement budget, the knowledge gradient policy is guaranteed to find the best alternative as N is allowed to be big enough. That is, if x^N is the solution we obtain after N measurements, and

$$x^* = \arg\max \mu_x$$

is the true best alternative, then $x^N \to x^*$ as $N \to \infty$. This property is known as asymptotic optimality or consistency.

(5) There are many heuristic policies that are asymptotically optimal (for example, pure exploration, mixed exploration–exploitation, epsilon-greedy exploration and Boltzmann exploration). But none of these heuristic policies are myopically optimal. The knowledge gradient policy is the only pure policy (a more precise term would be to say it is the only stationary policy) that is both myopically and asymptotically optimal.

(6) The knowledge gradient has no tunable parameters.

Myopic optimality and asymptotic optimality are the two properties that suggest that the knowledge gradient will have good convergence over intermediate budgets.

The one issue that requires some caution is known as the S-curve effect, that arises when the marginal value of information is nonconcave in the number of measurements. This is easy to detect, because all we have to do is to compute the knowledge gradient assuming that we measure alternate x a total of n_x times. The value of information can be nonconcave when the precision of a measurement β^W is low, which means that a single measurement does not contain enough information to accurately indicate whether an alternative is good. Frazier and Powell [28] identifies this problem and proposes a solution, called the KG(*) policy, where we find n_x that maximizes the *average* value of information over n_x consecutive measurements. We then choose the alternative x with the highest average value, but we may only measure it once.

The knowledge gradient for independent beliefs is appealing because it works well, is easy to compute and there are no tunable parameters such as the parameter z_α for interval estimation. There is, of course, the need to specify a prior, although it is always possible to create a prior using an initial sample (a strategy known as "empirical Bayes").

18.5.2 The Knowledge Gradient for Correlated Beliefs

The real power of the knowledge gradient arises in its ability to handle correlations between beliefs. This means that if we observe the value of x, we learn something about x'. Correlated beliefs might arise when $F(x)$ is continuous in x, so that the values of x and x' are similar when x and x' are closer together.

Let Σ^n be the matrix where element (x, x') gives $Cov^n(\mu_x, \mu_{x'})$. Let

$$B^n = (\Sigma^n)^{-1},$$

where B^n is the matrix version of our precision. It is straightforward to show that

$$\mu^{n+1} = \left(B^{n+1}\right)^{-1}\left(B^n\mu^n + \beta^W W^{n+1} e_{x^n}\right), \tag{18.23}$$
$$B^{n+1} = \left(B^n + \beta^W e_{x^n}(e_{x^n})^{\mathrm{T}}\right), \tag{18.24}$$

where e_x is a vector of 0's with a 1 in position x. We next compute the matrix form of the conditional change in the variance if we measure alternate x using

$$\tilde{\Sigma}^n(x) = \Sigma^n - \Sigma^{n+1} \tag{18.25}$$
$$= \frac{\Sigma^n e_x (e_x)^{\mathrm{T}} \Sigma^n}{\Sigma^n_{xx} + \lambda^W}, \tag{18.26}$$

where $\lambda^W = 1/\beta^W$ is the variance of a measurement. This matrix is analogous to the variance reduction in (18.19). Let

$$\tilde{\sigma}^n(x) = \frac{\Sigma^n e_x}{\sqrt{\Sigma^n_{xx} + \lambda^W}}, \tag{18.27}$$

be the column vector reflecting the change in the variances across all alternatives if we choose to measure x. The knowledge gradient in the presence of correlated beliefs is given by

$$X^{\mathrm{KG}}(s) = \arg\max_x \mathbb{E}\left[\max_i \mu_i^{n+1} \mid S^n = s, x^n = x\right] \tag{18.28}$$
$$= \arg\max_x \mathbb{E}\left[\max_i \left(\mu_i^n + \tilde{\sigma}_i(x^n) Z^{n+1}\right) \mid S^n, x^n = x\right],$$

where Z^{n+1} is a scalar, standard normal variable with mean 0, variance 1.

There is a way to compute the expectation exactly. We start by defining

$$h(\mu^n, \tilde{\sigma}(x)) = \mathbb{E}\left[\max_i \left(\mu_i^n + \tilde{\sigma}_i(x^n) Z^{n+1}\right) \mid S^n, x^n = x\right]. \tag{18.29}$$

Substituting (18.29) into (18.28) gives us

$$X^{\mathrm{KG}}(s) = \arg\max_x h(\mu^n, \tilde{\sigma}(x)). \tag{18.30}$$

Let $h(a, b) = \mathbb{E} \max_i (a_i + b_i Z)$, where $a = \mu_i^n$, $b = \tilde{\sigma}_i(\Sigma^n, x^n)$, and Z is our standard normal deviate. The lines $a_i + b_i Z$ can be used to determine the values of Z for which alternative i would become the best, but we have to recognize that some alternatives are simply dominated by all the other alternatives. We first assume that the lines are sorted in order of increasing b_i. It is then necessary to identify these dominated alternatives and eliminate them. This procedure is described by

Frazier et al. [29] and Powell and Ryzhov [30]. Once this step is complete, we can compute the knowledge gradient by using

$$h(a, b) = \sum_{i=1}^{M} (b_{i+1} - b_i) f(-|c_i|).$$

Correlated beliefs arise in a variety of settings. For the purpose of tuning parameters of a function, continuity is the most common motivation because an observation of x changes our belief about nearby points. For a problem where the measurement vector x (or θ) is multidimensional, we can randomly generate a large number of potential points and search over these. We can handle problems with several thousand alternatives, even if we have a measurement budget that is much smaller.

18.5.3 The Knowledge Gradient for Online Learning

Up to now, we have focused on developing the knowledge gradient for offline learning problems. It turns out that if we can compute the knowledge gradient for offline learning problems, there is a simple way to adapt this to give us the knowledge gradient for online learning problems. We are not going to duplicate the derivation (given in [31]). If we have a measurement budget of N measurements and we have already completed n, the knowledge gradient for an undiscounted problem is given by

$$\nu_x^{\text{online},n} = \mu_x^n + (N - n)\nu^{\text{KG},n}(x),$$

where $\nu^{\text{KG},n}(x)$ is the offline knowledge gradient for alternate x. If we have a discount factor γ, the knowledge gradient is given by

$$X^{\text{KG},n} = \arg\max_x \mu_x^n + \gamma \frac{1 - \gamma^{N-n}}{1 - \gamma} \nu_x^{\text{KG},n}.$$

Taking $N \to \infty$, we obtain the knowledge gradient rule for infinite-horizon problems,

$$X^{\text{KG},n} = \arg\max_x \mu_x^n + \frac{\gamma}{1 - \gamma} \nu_x^{\text{KG},n}.$$

18.5.4 The Knowledge Gradient for a Parametric Belief Model

There are many applications where it is simply not practical to use a lookup representation of a belief model. Often, it is useful to use a parametric model. Assume that we would like to approximate $F(x)$ using

$$F(x) \approx \sum_{f \in \mathcal{F}} \eta_f \phi_f(x).$$

Here, instead of one parameter $\mu_x = F(x)$ for each alternate x, we have a vector of parameters $(\eta_f)_{f \in \mathcal{F}}$, where the number of parameters is dramatically smaller than the number of alternatives x (which may in fact be continuous).

It is possible to compute the knowledge gradient when the vector η is the parameters of a linear model. The idea is the same as the knowledge gradient for correlated beliefs using a lookup table. The difference is that rather than computing and storing the matrix Σ^n, which may be computationally prohibitive (imagine a matrix with tens of thousands of rows and columns), we can compute $\tilde{\sigma}^n(x)$ in a much more compact way.

First, create a matrix X^n where each row is made up of $(\phi_1(x^n), \ldots, \phi_f(x^n), \ldots, \phi_F(x^n))$, and where there is a row for each measurement $x^1, \ldots, x^n$. Let Y^n be the column vector of the corresponding observations of the function $F(x)$. We can compute the parameter vector θ using the normal equations, given by

$$\theta^n = [(X^n)^{\mathrm{T}} X^n]^{-1} X^n Y^n.$$

Let $\Sigma^{\eta,n}$ be the covariance matrix for our beliefs about the true values of the vector θ after n observations, which is given by

$$\Sigma^{\eta,n} = [(X^n)^{\mathrm{T}} X^n]^{-1} \sigma_\epsilon^2,$$

where σ_ϵ^2 is the variance of a measurement. It is easy to show that

$$\Sigma^n = X^n \Sigma^{\eta,n} (X^n)^{\mathrm{T}}.$$

However, we do not need to compute the entire matrix Σ^n—we only need a single row corresponding to a particular alternative that we are measuring. So, while this row will still have many elements, at least we do not have to do an entire matrix.

18.5.5 Discussion

We now have a fairly powerful set of tools for performing policy search. We can treat each possible θ in (18.7) as an alternative in an optimal learning problem, and apply the knowledge gradient or other learning policies. We can obtain a realization $W(\omega)$ of the policy value $F^\pi(\theta)$ by simulating the exogenous information driving the dynamic program over a time horizon T, while using the particular vector θ to lock in all decisions via (18.6).

We emphasize that we have not made any attempt to survey the substantial literature that already exists for this problem class, but the knowledge gradient offers some uniquely attractive features: a fast rate of convergence (since it maximizes the value of each measurement) and the ability to handle correlated beliefs, even among large numbers of alternatives. This said, there is a cost to this power. While the knowledge gradient for independent beliefs is quite easy to compute, the knowledge gradient for correlated beliefs requires executing an algorithm. This introduces a level of overhead that will make sense for some applications, but not for others.

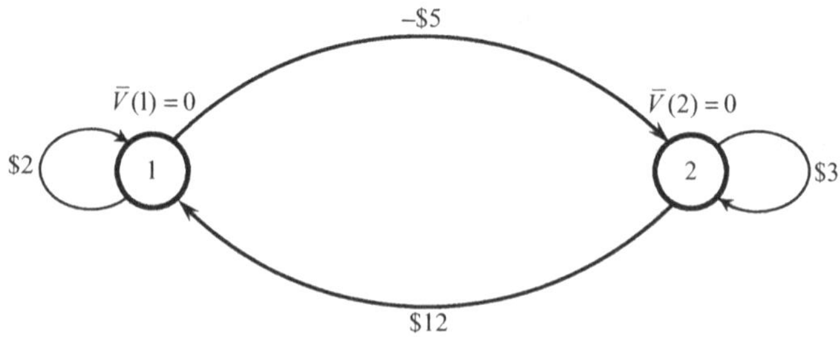

FIGURE 18.1 A two-state dynamic program, where the estimated value of being in each state is currently zero.

18.6 LEARNING WITH A PHYSICAL STATE

The challenge of optimal learning has been well-known in the ADP community for decades, where it is widely referred to as the "exploration vs. exploitation" problem. The problem is easily illustrated using the two-stage dynamic program in Figure 18.1, where we start by approximating the value of each state as zero. If we start in state 1, it appears that it is best to stay in state 1 because transitioning to state 2 would produce an immediate loss of $5, with an approximate future value of zero when we land in state 2. However, only by visiting state 2 do we realize that the approximation is not very good, and we would earn $12 when we make the transition from state 2 to state 1.

Although we previously formulated our learning problem as a dynamic program in Section 18.5, our success with a myopic policy depended on the property that the state variable consisted purely of a belief state. Now, we are considering problems where S_t represents a physical state. In this setting, an action a_t may bring information that allows us to learn, but it also has the effect of moving us from S_t to S_{t+1}, where we now face a different set of actions, and where the information we learned by taking action a_t from state S_t may no longer be useful (we may never return to S_t). The presence of a physical state is so common that references to "dynamic programming" typically imply the existence of a physical state.

We have to be specific about what we are learning in our new dynamic programming setting. These might include

- *Costs or Rewards*: This is what we were doing in our previous discussion of learning about policies. After each action, we may observe a cost, which allows us to update our belief about the costs.
- *Transition Functions*: Given a state s and action a, we may transition to a state s' with probability $p(s'|s, a)$. However, we may not be certain about this transition matrix, and as we observe the actual transition, we may update our belief about the transition matrix.
- *Value Functions*: Learning optimal policies may involve learning the value of being in a state. With each transition, we may update our belief about the value of being in a state.

We focus our discussion here on the last category, which is perhaps the one most widely associated with the exploration/exploitation problem.

18.6.1 Heuristic Policies

A number of heuristic policies have been suggested to assist with the exploration–exploitation problem in ADP. For example, all of the heuristic policies presented in Section 18.4.3 for problems without a physical state can and have been applied to problems with a physical state.

Several policies have been designed specifically for learning problems with a physical state. One is known as the R-max policy (developed by [32]), which starts with an upper bound on the value of visiting any state which has been visited fewer than a specified number of times. Needless to say, this policy is limited to problems with relatively small numbers of discrete states and actions.

The E^3 policy ("Explicit Explore or Exploit") is a variation of epsilon-greedy [33]. Upon reaching state S^n, our decision is made as follows. If S^n is a state that we have never visited before, we make a random decision. If we have visited S^n before, but fewer than m times, we make the decision that we have tried the fewest number of times among all our previous visits to the state. Finally, if we have visited S^n more than m times, we follow an exploitation policy. Once again, the policy reduces to pure exploitation once we have sufficiently explored the state space.

A more sophisticated policy is the local bandit approximation, an adaptation of the Gittins index policy proposed by Duff and Barto [34] and developed more fully in Ryzhov et al. [35]. If we are in a state s and considering an action a that takes us to a state s', we evaluate this action by estimating the reward earned between starting in s' and then finally terminating back in s, divided by the time required (this is an equivalent formulation of the Gittins index).

18.6.2 The Knowledge Gradient with a Physical State

The knowledge gradient has been adapted to problems with a physical state in Ryzhov and Powell [36] and Ryzhov and Powell [37], where the focus is on learning a parametric model of the value function. The idea is as follows. Imagine that we are in state s and considering an action a that might take us to state s' (a random variable given s and a). Once in s' (remember that we have not yet decided if we want to take action a), we identify the best action a' which takes us to another downstream state s''. In the process, we obtain a sample estimate $\hat{v}'$ of the value of being in state s'. If we had obtained this observation, we could use it to update a linear model of the value function, as in (18.5).

We do not have the space to fully derive the knowledge gradient policy with a physical state, but it is useful to take a look at the final result, which is given by

$$X^{\mathrm{KG},n}(S^n) = \max_x \Big(C(S^n, x) + \gamma \bar{V}^n(S^{x,n}) + \sum_{S^{n+1}} p(S^{n+1}|S^n, x) \nu^{\mathrm{KG},n}(S^{x,n}, S^{n+1}) \Big). \tag{18.31}$$

TABLE 18.1 Performance Values in a Benchmark Energy Storage Problem (from [37]).

	Offline Objective		Online Objective	
	Mean	Avg. SE	Mean	Avg. SE
Offline KG (basis functions)	1136.20	3.54	−342.23	19.96
Online KG (basis functions)	871.13	3.15	44.58	27.71
Offline KG (lookup)	210.43	0.33	−277.38	15.90
Online KG (lookup)	79.36	0.23	160.28	5.43
Epsilon-greedy (param.)	−475.54	2.30	−329.03	25.31

The policy consists of three terms. The first two include the one-period contribution and the value of the downstream state, just as we would have with any exploitation policy for dynamic programming. The third term $\nu^{\mathrm{KG},n}(S^{x,n}, S^{n+1})$ captures the value of information arising from being in post-decision state $S^{x,n}$ and then making a stochastic transition to S^{n+1} (in this setting, both $S^{x,n}$ and S^{n+1} represent physical states). This is multiplied by the probability of the transition from $S^{x,n}$ to S^{n+1}, which can be approximated using Monte Carlo simulation if necessary.

Equation (18.31) seems like a natural extension of a basic exploitation policy for ADP to include a value of information term. It is most suitable for online learning problems, which combine the reward being earned (contribution plus value of the next physical state) with the value of the information gained from the subsequent transition. If we compare the knowledge gradient for offline and online settings, we see that the knowledge gradient policy for offline learning would be given by

$$X^{Off,n}(S^n) = \arg\max_x \sum_{S^{n+1}} p\big(S^{n+1}|S^n, x\big)\nu^{\mathrm{KG},n}\big(S^{x,n}, S^{n+1}\big). \tag{18.32}$$

Table 18.1 presents comparisons between a series of algorithms in both an offline setting (where we use training iterations to learn a value function approximation, and then evaluate the policy using a separate set of testing iterations) and an online setting (where we accumulate rewards as we go). We compare both online and offline KG, using both a lookup table representation as well as basis functions, against an epsilon-greedy policy. We note the online version of KG works best when evaluated using an online objective function (as we would expect), while the offline version of KG works best when we use an offline objective function. Both outperform by a significant margin the epsilon-greedy policy when using a parametric value function approximation.

REFERENCES

1. J.R. Birge and F. Louveaux. *Introduction to Stochastic Programming*. Springer Verlag, New York, 1997.
2. D. Silver. Reinforcement learning and simulation-based search in computer go, PhD thesis, University of Alberta, 2009.

3. S. Gelly and Y. Wang. Exploration exploitation in go: UCT for Monte-Carlo go, In *Twentieth Annual Conference on Neural Information Processing Systems (NIPS 2006)*, Citeseer, 2006.
4. R. Coulom. Efficient selectivity and backup operators in monte-carlo tree search, In *Proceedings of the 5th International Conference on Computers and Games*, pp. 72–83, 2006.
5. M.L. Puterman. *Markov Decision Processes*, 2nd edition. John Wiley and Sons, Hoboken, NJ, 2005.
6. T. Hastie, R. Tibshirani, and J. Friedman. *The Elements of Statistical Learning: Data Mining, Inference and Prediction*. Springer, New York, 2009.
7. R. Sutton, H. Maei, D. Precup, S. Bhatnagar, D. Silver, C. Szepesvári, and E. Wiewiora. Fast gradient-descent methods for temporal-difference learning with linear function approximation, In *Proceedings of the 26th International Conference on Machine Learning*, pp. 993–1000, 2009.
8. D.P. Bertsekas. Approximate dynamic programming, in *Dynamic Programming and Optimal Control*, 3rd edition, Vol. II. Athena Scientific, Belmont, MA, chapter 6, 2011a.
9. D.P. Bertsekas. Approximate policy iteration: a survey and some new methods, *Journal of Control Theory and Applications*, 9(3):310–335, 2011b.
10. D.P. Bertsekas and J.N. Tsitsiklis. *Neuro-Dynamic Programming*. Athena Scientific, Belmont, MA, 1996.
11. R.S. Sutton and A.G. Barto. *Reinforcement Learning*, Vol. 35. MIT Press, Cambridge, MA, 1998.
12. C. Szepesvari. Algorithms for reinforcement learning. *Synthesis Lectures on Artificial Intelligence and Machine Learning*, 4(1):1–103, 2010.
13. W.B. Powell. *Approximate Dynamic Programming: Solving the Curses of Dimensionality*. 2nd edition. John Wiley & Sons, Hoboken, NJ, 2011.
14. H. Robbins and S. Monro. A stochastic approximation method. *The Annals of Mathematical Statistics*, 22(3):400–407, 1951.
15. H.S. Chang, M.C. Fu, J. Hu, and S.I. Marcus. *Simulation-based Algorithms for Markov Decision Processes*, Springer, 2007.
16. M.C. Fu. What you should know about simulation and derivatives, *Naval Research Logistics*, 55(8):723–736, 2008.
17. J.C. Spall. *Introduction to Stochastic Search and Optimization: Estimation, Simulation and Control*. John Wiley & Sons, Hoboken, NJ, 2003.
18. A. Ruszczynski and A. Shapiro. *Handbooks in Operations Research and Management Science: Stochastic Programming*, Vol. 10. Elsevier, Amsterdam, 2003.
19. A. Shapiro, D. Dentcheva, and A. Ruszczynski. *Lectures on Stochastic Programming: Modeling and Theory*. SIAM, Philadelphia, 2009.
20. J. Gittins, K. Glazebrook, and R.R. Weber. *Multi-Armed Bandit Allocation Indices*. John Wiley & Sons, New York, 2011.
21. S.P. Singh, T. Jaakkola, C. Szepesvari, and M. Littman. Convergence results for single-step on-policy reinforcement-learning algorithms. *Machine Learning*, 38(3):287–308, 2000.
22. L.P. Kaelbling. *Learning in Embedded Systems*. MIT Press, Cambridge, MA, 1993.
23. T.L. Lai and H. Robbins. Asymptotically efficient adaptive allocation rules. *Advances in Applied Mathematics*, 6:4–22, 1985.

24. P. Auer, N. Cesa-bianchi, and P. Fischer. Finite time analysis of the multiarmed bandit problem. *Machine Learning*, 47(2–3):235–256, 2002.

25. S. Bubeck, R. Munos, and G. Stoltz. Pure exploration in multi-armed bandits problems. In *Proceedings of the 20th International Conference on Algorithmic Learning Theory*, 2009.

26. M.H. DeGroot. *Optimal Statistical Decisions*. John Wiley and Sons, 1970.

27. S.E. Chick and N. Gans. Economic analysis of simulation selection problems. *Management Science*, 55(3):421–437, 2009.

28. P.I. Frazier and W.B. Powell. Paradoxes in learning and the marginal value of information, *Decision Analysis*, 7(4):378–403, 2010.

29. P.I. Frazier, W.B. Powell, and S. Dayanik, The knowledge-gradient policy for correlated normal beliefs, *INFORMS Journal on Computing*, 21(4):599–613, 2009.

30. W.B. Powell and I.O. Ryzhov. *Optimal Learning*. John Wiley & Sons Inc., 2012.

31. I.O. Ryzhov, W.B. Powell, and P.I. Frazier. The knowledge gradient algorithm for a general class of online learning problems. *Operations Research* in press, 2011.

32. R.I. Brafman and M. Tennenholtz. R-max – A general polynomial time algorithm for near-optimal reinforcement learning. *Journal of Machine Learning Research*, 3, 213–231, 2003.

33. M. Kearns and S. Singh. Near-optimal reinforcement learning in polynomial time. *Machine Learning* 49(2):209–232, 2002.

34. M. Duff and A. Barto. Local bandit approximation for optimal learning problems, In M. Mozer, M. Jordan, and T. Pesche, editors. *Advances in Neural Information Processing Systems*, Vol. 9. MIT Press, Cambridge, MA, 1996, pp. 1019–1025.

35. I.O. Ryzhov, M.R. Valdez-Vivas, and W.B. Powell. Optimal learning of transition probabilities in the two-agent newsvendor problem, In B. Johansson, S. Jain, J. Montoya-Torres, J. Hugan, and E. Yücesan, editors. *Proceedings of the 2010 Winter Simulation Conference*, pp. 1088–1098, 2010.

36. I.O. Ryzhov and W.B. Powell. Approximate dynamic programming with correlated Bayesian beliefs, In *Proceedings of the 48th Allerton Conference on Communication, Control, and Computing*, pp. 1360–1367, 2010.

37. I.O. Ryzhov and W.B. Powell. Bayesian active learning with basis functions, In *Proceedings of the 2011 IEEE Symposium on Adaptive Dynamic Programming and Reinforcement Learning*, pp. 143–150, 2011.

CHAPTER 19

An Introduction to Event-Based Optimization: Theory and Applications

XI-REN CAO,[1] YANJIA ZHAO,[2] QING-SHAN JIA,[2] and QIANCHUAN ZHAO[2]

[1]Shanghai Jiaotong University, Shanghai, China

[2]Department of Automation, Tsinghua University, Beijing, China

ABSTRACT

In this chapter, we review the basic idea of event-based optimization (EBO), which is specifically suitable for policy optimization of discrete event dynamic system (DEDS). With decisions based on certain events instead of on the system state, the number of potentials to be estimated is usually much smaller than the size of the state space. Performance difference and derivative formulas are developed for event-based policies. Under certain assumptions, policy iteration algorithms for EBO can be developed and they may converge to a globally optimal event-based policy. However, gradient-based optimization algorithms can always be applied and they usually converge to a local optimum in the parameterized policy space. We illustrate the EBO method by a material handling (MH) problem. We hope this chapter may bring insights for the study of event-based control, decision-making, and optimization in more general situations.

19.1 INTRODUCTION

Many systems nowadays follow not only physical laws but also man-made rules, such as smart grid, transportation, manufacturing, and financial systems. These systems are known as discrete event dynamic systems (DEDSs) [1]. Simulation is usually the only faithful way to accurately describe the dynamics of such systems, which is time-

Reinforcement Learning and Approximate Dynamic Programming for Feedback Control, First Edition. Edited by Frank L. Lewis and Derong Liu.

consuming and only provides noisy performance estimation. Policy optimization of DEDS has been a long-standing challenging problem due to the discrete and large state space, the simulation-based performance evaluation, and the pervasive randomness in the system. Since these systems are driven by events, and the number of events may increase only linearly when the system scale increases, event-based decision-making has specific advantage for DEDS than traditional state-based decision-making. Event-based optimization (EBO) provides a general framework for the optimization of event-based policies.

The rest of this chapter is organized as follows. In Section 19.2, the related literature is briefly reviewed. In Section 19.3, the mathematical formulation of EBO is presented. In Section 19.4, the performance difference and derivative formulas for EBO are presented, which then naturally lead to event-based policy iteration. The optimality condition is shown, which ensures the iteration converges to global optimal policy. In Section 19.5, the EBO is applied to a material handling (MH) problem to show that even when the optimality condition is not satisfied, event-based policy iteration may provide a good policy. A brief conclusion and discussion is presented in Section 19.6.

19.2 LITERATURE REVIEW

Markov decision process (MDP) [2, 3] has provided a general framework for many control, decision-making, and optimization problems, including the policy optimization of DEDS. Policy iteration and value iteration are well-known procedures to solve MDPs, both of which, however, suffer from the large state space and large action space when addressing practical systems. There exist substantial literatures on how to solve large-scale MDPs. One example is the various methods included in this book. Depending on whether the solution is accurate or approximate, the existing methods can be classified into two types. The first type is to utilize the problem information to provide simple structures of the optimal policies. State aggregation [4, 5], time aggregation [6, 7], and action elimination [8, 9] belong to this type. These methods require certain structural property of the system and are thus restrictive. The second type is to solve MDPs approximately. Neuro-dynamic programming [10–12], reinforcement learning [13], approximate dynamic programming [14–16], the rollout method [17], the approximate receding horizon approach [18], selective approximation [19], and some other simulation-based algorithms for MDPs [20] belong to this type. Approximations of the value functions or policies are usually used in these methods. Heuristics and experiences are usually important to find good approximation functions in practice.

Event-based control and decision-making have gained a lot of attention recently due to the broad applications in networked systems. There have been studies on event-based sampling [21–25], event-triggered data transmission [26], and event-based sensing [27–31] in the networked control systems. Most of these studies assume closed-form expressions for system dynamics, and consider stability issues and performance optimization in the LQG framework.

A general framework, called EBO, has been developed to provide a unified view for solutions to performance optimization of DEDSs and many other nonstandard problems, including POMDPs [32–34], hierarchical control [35], options [36], singular perturbation analysis [37, 38], and customer-average performance optimization in queueing systems[39]. Different events can be defined to capture the special features in these different problems. In EBO, an event is modeled as a set of state transitions. This captures the main features of many real world events and covers the special cases cited above. The basic idea of EBO was developed by Cao in [40, 41] and later on summarized in [42]. Existing studies mainly consider the average reward over infinite number of stages [42]. Potential-based policy iteration has been developed to use simulation-based evaluation models, which is suitable for applications in DEDSs. We will review these results in the following sections.

19.3 PROBLEM FORMULATION

We use average-reward over infinite-stage EBO as an example to introduce notations. Notations for discounted reward over infinite stages and total reward over finite stages can be defined similarly. Let $\mathcal{S} = \{1, 2, \ldots, S\}$ be the system state space. A single event is defined as a one-step state transition $\langle i, j\rangle$, $i, j \in \mathcal{S}$. The space of all the single events is denoted as $\mathcal{E} = \{\emptyset, \langle i, j\rangle | i, j \in \mathcal{S}\}$, with $\emptyset$ being a null event. A set of single events is called an event. Define $\mathcal{E}_o = \{e_o(k_1), k_1 = 1, \ldots, k_o\}, \mathcal{E}_c = \{e_c(k_2), k_2 = 1, \ldots, k_c\}$, and $\mathcal{E}_t = \{e_t(k_3), k_3 = 1, \ldots, k_t\}$ as the set of observable, controllable, and natural transition events, where $e_o(k_1), e_c(k_2), e_t(k_3) \subset \mathcal{E}$ are sets of state transitions. It is usually assumed that $\mathcal{E}_o$, $\mathcal{E}_c$, and $\mathcal{E}_t$ each provide a mutually exclusive decomposition of $\mathcal{E}$. An observable event has two features. First, we can tell whether the event occurs at any time instant from the system behavior; second, the event contains some information about the system, which can be used to determine the control actions. A controllable event is an event that we can control the probability of its occurrence by taking actions based on the information obtained from an observable event that has just happened. A natural transition event is an event whose corresponding transitions are governed by nature. Thus, the probability of the occurrence of a natural transition event cannot be controlled.

The event-based decision-making process is as follows. Suppose the system is at state $X = i$. The decision maker observes $E_o = e_o(k_1)$ with probability (w.p.) $\mu(e_o(k_1)|i)$, picks an action $\alpha \in \mathcal{A}(e_o(k_1))$, and receives an immediate reward $f(i, \alpha)$. This action will make a controllable event $e_c(k_2)$ appear w.p. $p^\alpha[e_c(k_2)|e_o(k_1)]$. Then the natural transition event $e_t(k_3)$ appears w.p. $p[e_t(k_3)|e_o(k_1), e_c(k_2)]$. The system state will then transit to $j \in O_i[e_o(k_1) \cap e_c(k_2) \cap e_t(k_3)]$, where $O_i[e] \equiv \{j | j \in \mathcal{S}, \langle i, j\rangle \in e\}$ represents the output state of i under event e, and $\langle i, j\rangle$ represents a one-step state transition from i to j. Note that by appropriate definition of $\mathcal{E}_t$, it is always possible to make $|O_i[e_o(k_1) \cap e_c(k_2) \cap e_t(k_3)]| = 1$ for all i, k_1, k_2, k_3. One way to do this is to let every element in $\mathcal{E}_t$ be a single event. So without loss of generality, we can write $j = O_i[e_o(k_1) \cap e_c(k_2) \cap e_t(k_3)]$. Similarly, we can define the input state of j under event e as $I_j[e] \equiv \{i | i \in \mathcal{S}, \langle i, j \rangle \in e\}$, and define $I[e] \equiv \cup_{j \in \mathcal{S}} I_j[e]$. Then, the one-step transition probability function is

$$p(j|i, \alpha) = \sum_{k_1, k_2, k_3} \mu(e_o(k_1)|i) p^{\alpha}[e_c(k_2)|e_o(k_1)] p[e_t(k_3)|e_o(k_1), e_c(k_2)] \times \mathcal{I}[j = O_i[e_o(k_1) \cap e_c(k_2) \cap e_t(k_3)]], \quad (19.1)$$

where $\mathcal{I}[A] = 1$ (or 0) if event A is true (or false). Define the following notations. $\Pi(X_0)$ is the probability distribution of the initial system state, which summarizes all the *a priori* information; E_0 is the initial observable event; $\mathcal{H}_n = \{\Pi(X_0), E_0, \alpha_0, \ldots, E_n\}$ is the entire history of observable events and all the actions that have been taken; $d_n : \mathcal{H}_n \to \Pi(\mathcal{A})$ is a history-dependent randomized event-based decision rule; $L = \{d_0, \ldots\}$ is called a history-dependent randomized policy; and $\mathcal{L}_{HR}$ is the set of all such policies. We want to maximize the average reward over an infinite horizon, that is,

$$\max_{L \in \mathcal{L}_{HR}} \eta^L(\Pi(X_0), E_0) = \max_{L \in \mathcal{L}_{HR}} \lim_{T \to \infty} \frac{1}{T+1} E\left[\sum_{n=0}^{T} f_n(X_n, d_n(\mathcal{H}_n)) | \Pi(X_0), E_0\right], \quad (19.2)$$

where the expectation is taken over all the sample paths. Note that the event-action sequence is not Markovian, that is, $\Pr\{E_n | \Pi(X_0), E_0, \ldots, E_{n-1}, \alpha_{n-1}\} \neq \Pr\{E_n | E_{n-1}, \alpha_{n-1}\}$. Discussions on how to solve EBOs for this general case can be found in [43] and [44]. Examples also exist in which the event-action sequences do obey the Markovian property and thus the optimal policy can be found within deterministic Markovian policies. To focus on the main idea of the EBO approach, we assume there exist deterministic Markovian optimal policies in the following discussions, and present methods to find such a policy. Under this assumption, a policy can be represented by its decision rule d. Thus, when there is no confusion, we may also refer d as the event-based policy in the following.

19.4 POLICY ITERATION FOR EBO

An important step toward policy iteration is to compare the performances of two policies. Thus, we introduce the performance difference formula and the performance derivative formula in Section 19.4.1, and then naturally lead to the policy iteration for EBO in Section 19.4.2.

19.4.1 Performance Difference and Derivative Formulas

As aforementioned, we only study ergodic policies. An event-based policy $d[e_o(k_1)]$, $k = 1, \ldots, k_o$, is said to be ergodic, if the Markov chain with transition probabilities $P^d(j|i)$, $i, j \in \mathcal{S}$ is ergodic, where

$$P^d(j|i) = \sum_{k_1,k_2,k_3} \mu(e_o(k_1)|i) p^{d[e_o(k_1)]}[e_c(k_2)|e_o(k_1)] p[e_t(k_3)|e_o(k_1), e_c(k_2)] \\ \times \mathcal{I}\left[j = O_i[e_o(k_1) \cap e_c(k_2) \cap e_t(k_3)]\right].$$

Let $\mathcal{D}_e$ be the set of all the event-based policies. We assume all the policies in $\mathcal{D}_e$ are ergodic.

Let $\pi_s^d = \left(\pi_s^d(1), \ldots, \pi_s^d(S)\right)$ be the steady-state distribution under policy d and let $\pi_e^d = \left(\pi_e^d(e_o(1), \ldots, \pi_e^d(e_o(k_o)))\right)$ be the steady distribution of observable events under policy d. Define the potential of state X under policy d as

$$g^d(X) = \lim_{T \to \infty} E\left\{ \sum_{n=0}^{T} \sum_{k_1} \left[f(X_n, d[e_o(k_1)]) - \eta^d \right] \Big| X_0 = X \right\}. \tag{19.3}$$

It is known that the average-reward difference formula for ergodic chain is

$$\eta^h - \eta^d = \pi_s^h \left[\left(f_s^h + P^h g^d \right) - \left(f_s^d + P^d g^d \right) \right], \tag{19.4}$$

where $f_s^d = \left(f_s^d(1), \ldots, f_s^d(S)\right)^\tau$, τ represents the transpose, and $f_s^d(i) = \sum_{k_1} \mu(e_o(k_1)|i) f(i, d[e_o(k_1)])$. Note that the vectors on the right-hand-side (RHS) of the above equation all have dimensionality of the state space. We would like to rewrite the equation to emphasize its relationship with the events rather than the states. Thus, we first rewrite Equation (19.4) as

$$\eta^h - \eta^d = \pi_s^h \left(P^h - P^d \right) g^d + \pi_s^h \left(f_s^h - f_s^d \right). \tag{19.5}$$

For the first term on the RHS of Equation (19.5), we have

$$\begin{aligned}
&\pi_s^h \left(P^h - P^d \right) g^d \\
&= \sum_{i=1}^{S} \pi_s^h(i) \sum_{j=1}^{S} \left[P^h(j|i) - P^d(j|i) \right] g^d(j), \\
&= \sum_{i=1}^{S} \pi_s^h(i) \sum_{k_1=1}^{k_o} \sum_{k_2=1}^{k_c} \sum_{k_3=1}^{k_t} \mu(e_o(k_1)|i) \\
&\quad \left\{ p^{h[e_o(k_1)]}[e_c(k_2)|e_o(k_1)] - p^{d[e_o(k_1)]}[e_c(k_2)|e_o(k_1)] \right\} \\
&\quad \cdot p[e_t(k_3)|e_c(k_2), e_o(k_1)] g^d(O_i[e_o(k_1) \cap e_c(k_2) \cap e_t(k_3)]),
\end{aligned}$$

$$= \sum_{k_1=1}^{k_o} \sum_{k_2=1}^{k_c} \sum_{k_3=1}^{k_t} \sum_{i=1}^{S} \pi_s^h(i)\mu(e_o(k_1)|i)$$

$$\cdot \left\{ p^{h[e_o(k_1)]}[e_c(k_2)|e_o(k_1)] - p^{d[e_o(k_1)]}[e_c(k_2)|e_o(k_1)] \right\}$$

$$p[e_t(k_3)|e_c(k_2), e_o(k_1)] g^d(O_i[e_o(k_1) \cap e_c(k_2) \cap e_t(k_3)]),$$

$$= \sum_{k_1=1}^{k_o} \sum_{k_2=1}^{k_c} \sum_{k_3=1}^{k_t} \sum_{i=1}^{S} \pi_e^h(e_o(k_1))\pi^h(i|e_o(k_1))$$

$$\cdot \left\{ p^{h[e_o(k_1)]}[e_c(k_2)|e_o(k_1)] - p^{d[e_o(k_1)]}[e_c(k_2)|e_o(k_1)] \right\}$$

$$\cdot p[e_t(k_3)|e_c(k_2), e_o(k_1)] g^d(O_i[e_o(k_1) \cap e_c(k_2) \cap e_t(k_3)]),$$

$$= \sum_{k_1=1}^{k_o} \pi_e^h(e_o(k_1)) \sum_{k_2=1}^{k_c} \left\{ p^{h[e_o(k_1)]}[e_c(k_2)|e_o(k_1)] - p^{d[e_o(k_1)]}[e_c(k_2)|e_o(k_1)] \right\}$$

$$\sum_{i \in I[e_o(k_1)]} \sum_{k_3=1}^{k_t} \pi^h(i|e_o(k_1)) p[e_t(k_3)|e_c(k_2), e_o(k_1)] g^d(j), \tag{19.6}$$

with $j = O_i[e_o(k_1) \cap e_c(k_2) \cap e_t(k_3)]$, and

$$\pi^h(i|e_o(k_1)) = \frac{\pi_s^h(i)\mu(e_o(k_1)|i)}{\sum_j \pi_s^h(j)\mu(e_o(k_1)|j)}. \tag{19.7}$$

Define

$$g^{d,h}(k_1, k_2) \equiv \sum_{i \in I[e_o(k_1)]} \sum_{k_3=1}^{k_t} \left\{ \pi^h(i|e_o(k_1)) p[e_t(k_3)|e_o(k_1), e_c(k_2)] g^d(j) \right\}. \tag{19.8}$$

Then we can rewrite Equation (19.6) as

$$\pi_s^h \left(P^h - P^d \right) g^d$$

$$= \sum_{k_1=1}^{k_o} \pi_e^h(e_o(k_1)) \sum_{k_2=1}^{k_c} \left\{ p^{d[e_o(k_1)]}[e_c(k_2)|e_o(k_1)] - p^{d[e_o(k_1)]}[e_c(k_2)|e_o(k_1)] \right\} g^{d,h}(k_1, k_2). \tag{19.9}$$

For the second term on the RHS of Equation (19.5), we have

$$\begin{aligned}
\pi_s^h f_s^h &= \sum_{i=1}^{S} \pi_s^h(i) f_s^h(i), \\
&= \sum_{i=1}^{S} \left\{ \sum_{k_1=1}^{k_o} \left[\pi^h(i|e_o(k_1)) \pi_e^h(e_o(k_1)) \right] f_s^h(i) \right\}, \\
&= \sum_{k_1=1}^{k_o} \left\{ \pi_e^h(e_o(k_1)) f_e^h(k_1) \right\},
\end{aligned} \tag{19.10}$$

where $f_e^h(k_1) \equiv \sum_{i=1}^{S} \left[\pi^h(i|e_o(k_1)) f_s^h(i) \right]$. Similarly, we have

$$\pi_s^h f_s^d = \sum_{k_1=1}^{k_o} \left[\pi_e^h(e_o(k_1)) f_e^{d,h}(k_1) \right], \tag{19.11}$$

where $f_e^{d,h}(k_1) \equiv \sum_{i=1}^{S} \left[\pi^h(i|e_o(k_1)) f_s^d(i) \right]$. Therefore, combining Equations (19.6)–(19.11), we have

$$\begin{aligned}
&\eta^h - \eta^d \\
&= \sum_{k_1=1}^{k_o} \pi_e^h(e_o(k_1)) \sum_{k_2=1}^{k_c} \left\{ p^{h[e_o(k_1)]}[e_c(k_2)|e_o(k_1)] - p^{d[e_o(k_1)]}[e_c(k_2)|e_o(k_1)] \right\} g^{d,h}(k_1, k_2) \\
&+ \sum_{k_1=1}^{k_o} \left\{ \pi_e^h(e_o(k_1)) \left[f_e^h(k_1) - f_e^{d,h}(k_1) \right] \right\}, \\
&= \sum_{k_1=1}^{k_o} \pi_e^h(e_o(k_1)) \left\{ \left[f_e^h(k_1) + \sum_{k_2=1}^{k_c} \left\{ p^{h[e_o(k_1)]}[e_c(k_2)|e_o(k_1)] g^{d,h}(k_1, k_2) \right\} \right] \right. \\
&\left. - \left[f_e^{d,h}(k_1) + \sum_{k_2=1}^{k_c} \left\{ p^{d[e_o(k_1)]}[e_c(k_2)|e_o(k_1)] g^{d,h}(k_1, k_2) \right\} \right] \right\}.
\end{aligned} \tag{19.12}$$

Equation (19.12) is the performance difference formula. Note that $\pi^h(i|e_o(k_1))$ appears in $g^{d,h}(k_1, k_2)$, which cannot be estimated in the sample path of policy d. The average-reward gradient along any direction, however, can be expressed in aggregated potentials that depend only on the original systems, as shown below.

Assume the conditional transition probabilities of the controllable events and the reward function depend on a continuous parameter $\theta \in \Theta \subset \mathbb{R}$ and are denoted as

$p_\theta[e_c(k_2)|e_o(k_1)]$, $k_1 = 1, \ldots, k_o$, $k_2 = 1, \ldots, k_c$, and $f_\theta(i)$, $i \in \mathcal{S}$, respectively. Taking $p_{\theta_2}[e_c(k_2)|e_o(k_1)]$ as $p^h[e_c(k_2)|e_o(k_1)]$ and $p_{\theta_1}[e_c(k_2)|e_o(k_1)]$ as $p^d[e_c(k_2)|e_o(k_1)]$ in Equation (19.12), where $\theta_1, \theta_2 \in \Theta$, we have

$$\eta(\theta_2) - \eta(\theta_1) = \sum_{k_1=1}^{k_o} \pi^{\theta_2}(e_o(k_1))$$
$$\times \left\{ \left[f_e^{\theta_2}(k_1) - f_e^{\theta_2,\theta_1}(k_1) \right] + \sum_{k_2=1}^{k_c} \left\{ p^{\theta_2}[e_c(k_2)|e_o(k_1)] - p^{\theta_1}[e_c(k_2)|e_o(k_1)] \right\} g^{\theta_1,\theta_2}(k_1,k_2) \right\}, \tag{19.13}$$

where

$$g^{\theta_1,\theta_2}(k_1,k_2) = \sum_{i \in I[e_o(k_1)]} \sum_{k_3=1}^{k_t} \left\{ \pi^{\theta_2}(i|e_o(k_1)) p[e_t(k_3)|e_o(k_1), e_c(k_2)] g^{\theta_1}(j) \right\}, \tag{19.14}$$

with

$$f_e^{\theta_2}(k_1) \equiv \sum_{i=1}^{S} \pi^{\theta_2}(i|e_o(k_1)) f_s^{\theta_2}(i),$$
$$f_e^{\theta_2,\theta_1}(k_1) \equiv \sum_{i=1}^{S} \pi^{\theta_2}(i|e_o(k_1)) f_s^{\theta_1}(i).$$

Dividing both sides with $\theta_2 - \theta_1$ and letting $\theta_2 \to \theta_1$ and assuming that the derivative exists, we obtain

$$\frac{\mathrm{d}\eta(\theta)}{\mathrm{d}\theta}\Big|_{\theta=\theta_1}$$
$$= \sum_{k_1=1}^{k_o} \pi_e^{\theta_1}(e_o(k_1)) \left\{ \sum_{i=1}^{S} \pi^{\theta_1}(i|e_o(k_1)) \frac{\mathrm{d}f^\theta(i)}{\mathrm{d}\theta}\Big|_{\theta=\theta_1} + \sum_{k_2=1}^{k_c} \left\{ \frac{d}{\mathrm{d}\theta} p^\theta[e_c(k_2)|e_o(k_1)] \right\} \Big|_{\theta=\theta_1} g^{\theta_1}(k_1,k_2) \right\},$$
$$= \sum_{i=1}^{S} \pi_s^{\theta_1}(i) \frac{\mathrm{d}f^\theta(i)}{\mathrm{d}\theta}\Big|_{\theta=\theta_1} + \sum_{k_1=1}^{k_o} \pi_e^{\theta_1}(e_o(k_1)) \left\{ \sum_{k_2=1}^{k_c} \left\{ \frac{d}{\mathrm{d}\theta} p^\theta[e_c(k_2)|e_o(k_1)] \right\} \Big|_{\theta=\theta_1} g^{\theta_1}(k_1,k_2) \right\}, \tag{19.15}$$

where

$$g^{\theta_1}(k_1,k_2) = \sum_{i \in I[e_o(k_1)]} \sum_{k_3=1}^{k_t} \left\{ \pi^{\theta_1}(i|e_o(k_1)) p[e_t(k_3)|e_o(k_1), e_c(k_2)] g^{\theta_1}(j) \right\}, \tag{19.16}$$

with $j = O_i[e_o(k_1) \cap e_c(k_2) \cap e_t(k_3)]$. Equation (19.15) is known as the performance derivative formula. Note that the aggregated potential $g^{\theta_1}(k_1, k_2)$ depend only on the original system with parameter θ_1, and thus can be estimated in the sample path of the original system with θ_1.

19.4.2 Policy Iteration for EBO

Note that in the performance diference formula in Equation (19.12) the aggregated potential $g^{d,h}(k_1, k_2)$ depends on both policies h and d. Such a quantity cannot be used in sample-path-based policy iteration, where all quantities should be estimated based on the sample path of the original system with policy d. However, under some special situations, $g^{d,h}(k_1, k_2)$ depends only on the original system and therefore can be estimated from the sample path of the original system.

First, for some systems, the following equation holds for any $h, d \in \mathcal{D}_e$,

$$\pi^h(i|e_o(k_1)) = \pi^d(i|e_o(k_1)), \text{ for all } i \in I[e_o(k_1)], k_1 = 1, 2, \ldots, k_o. \quad (19.17)$$

For example, in financial problem, it is a common assumption that the market behavior, thus the price distribution, does not depend on another one's control policy. Another example can be found in admission control [40]. Then Equation (19.8) becomes

$$g^d(k_1, k_2) = \sum_{i \in I[e_o(k_1)]} \sum_{k_3=1}^{k_t} \left\{ \pi^d(i|e_o(k_1)) p[e_t(k_3)|e_o(k_1), e_c(k_2)] g^d(j) \right\}, \quad (19.18)$$

which depends only on the original policy d.

Second, for some other systems, for any state $i \in I[e_o(k_1)]$, if $e_c(k_2)$ occurs (no matter what action is taken) and if the same natural transition event $e_t(k_3)$ is followed, then the next state will be the same state j. This essentially resets the system randomly by nature, after the action is taken. Then Equation (19.8) becomes

$$g^d(k_1, k_2) = \sum_{k_3=1}^{k_t} \left\{ p[e_t(k_3)|e_o(k_1), e_c(k_2)] g^d(j) \right\}, \quad (19.19)$$

with $j = O_i[e_o(k_1) \cap e_c(k_2) \cap e_t(k_3)]$ not explicitly depending on i.

Now we present policy iteration under the assumption that $g^d(k_1, k_2)$ can be estimated from the sample path of the original system with policy d as shown in Algorithm 19.1. We know that policy iteration terminates within finite iterations. In general, there are $k_o \times k_c$ aggregated potentials $g^d(k_1, k_2)$ compared with S potentials in the standard MDPs. In many systems, $k_o \times k_c \ll S$. Then, the policy iteration for EBO provides a computationally feasible way to solve the problem while policy iteration for standard MDP may not be feasible.

Note that when $g^{d,h}(k_1, k_2)$ depends on both d and h, we can always use gradient-based policy iteration, which converges to local optimum in the parameterized policy space. The algorithm can be developed similarly.

Algorithm 19.1: Policy Iteration for EBO

Step 1. Pick an initial policy $d^0 \in \mathcal{D}_e, k = 0$.
Step 2. Estimate $g^{d^k}(k_1, k_2)$ for all $k_1 = 1, \ldots, k_o$ and $k_2 = 1, \ldots, k_c$.
Step 3. For each observable event $e_o(k_1)$, let

$$d^{k+1}[e_o(k_1)] = \arg \max_{\alpha \in \mathcal{A}(k_1)} \left\{ \sum_{k_2=1}^{k_c} p^{\alpha}[e_c(k_2)|e_o(k_1)] g^{d^k}(k_1, k_2) + f(k_1, \alpha) \right\}.$$

Step 4. If $d^{k+1}[e_o(k_1)] = d^k[e_o(k_1)]$ for all $k_1 = 1, \ldots, k_o$, stops. Otherwise, $k = k + 1$ and go to step 2.

19.5 EXAMPLE: MATERIAL HANDLING PROBLEM

In this section, we consider an optimization problem in an MH system of a general assembly (GA) line motivated by a practical car manufacturing system.

19.5.1 Problem Formulation

As Figure 19.1 illustrates, the GA line consists of a conveyor and series of assembling stations [45]. Semiproducts (say car bodies) are loaded onto the conveyor, going through all the stations where different parts (such as engine, seats, tires, etc.) are assembled, and then final products are produced. The conveyor is called a paced-moving conveyor since it moves step-by-step in every period of time (called cycle time) and transfers semiproducts downstream synchronously when all stations complete their assembly operations. To have smooth production, the parts are supplied by a driver of the MH system from a central docking area to the lineside buffers with a 2-dolly train. Each dolly can hold parts for only one lineside buffer, so that the driver can supply at most two buffers in each trip.

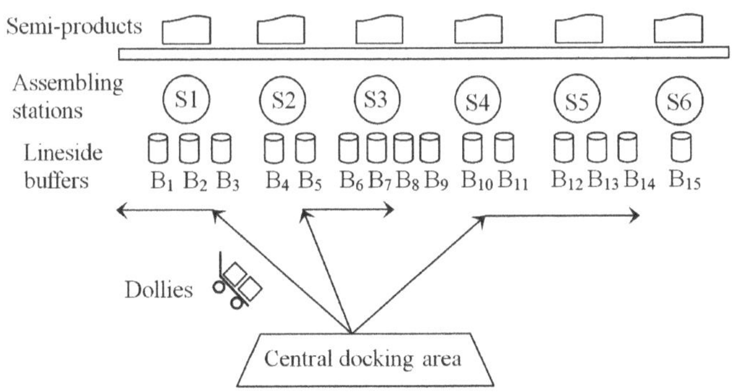

FIGURE 19.1 Sketched general assembly line with material handling.

This practical problem can be formulated as an EBO problem [46]. To simplify the discussion, we first assume the following assumptions.

1. There is one driver serving I lineside buffers with a 2-dolly train. The assembling operation at each station finishes within unit time, that is, the cycle time of all stations equals to one.
2. The number of parts consumed at each buffer at each unit time has the Bernoulli distribution, and the average consumption rates are constants, which may be different for different buffers.
3. To simplify the expression, the inventory level of lineside buffers will be increased within the same time unit as the replenishment action is taken; however, the driver may take several time units to finish the entire trip and come back to the central docking area.

Note that all these assumptions have been approved by our industry partners according to the practical requirements in factories.

We choose decisions at each time unit, that is, $t = 0, 1, 2, \cdots$, and the system state at decision epoch t is defined as $\mathbf{s_t} = (\mathbf{n_t}, \mathbf{y_t})^\tau$, where $\mathbf{n_t}$ is the stack vector of inventory levels with components $n_{i,t}$ for lineside buffer i, $i = 1, \cdots, I$; y_t is the status of the driver, that is, the time units left for the driver to come back to the central docking area.

The action of the system at decision epoch t is defined as $\mathbf{a_t} = (\mathbf{a_t}, \mathbf{a_t'})^\tau$, where a_t and a_t' denote the indexes of the buffer to supply with the first and the second dollies, respectively, $a_t, a_t' \in \{0, 1, \cdots, I\}$. If the driver is in traveling, that is, $y_t > 0$, there is no choice except for $a_t = a_t' = 0$. If the driver is idle at the central docking area, that is, $y_t = 0$, the feasible actions are $a_t, a_t' = 0, 1, \cdots, I$. In other words, the feasible action set at time t with system state $\mathbf{s_t}$ is denoted as

$$\mathcal{A}(\mathbf{s_t}) = \begin{cases} \left\{ \left(a_t, a_t'\right)^\tau \middle| a_t, a_t' \in \{0, 1, \cdots, I\} \right\}, & \text{if } y_t = 0, \\ \left\{ \left(a_t, a_t'\right)^\tau \middle| a_t = a_t' = 0 \right\}, & \text{if } y_t > 0. \end{cases} \tag{19.20}$$

The system dynamic is depicted as follows. For all $i = 1, \cdots, I$, and $t = 0, 1, \cdots$,

$$n_{i,t+1} = \max\{\min\{n_{i,t} + Q_i(\mathcal{I}[a_t = i] + \mathcal{I}[a_t' = i]), C_i\} - \omega_t \xi_{i,t}, 0\}, \tag{19.21}$$

$$y_{t+1} = \max\{y_t + \lceil T(0, a_t) + T(a_t, a_t') + T(a_t', 0) \rceil - 1, 0\}, \tag{19.22}$$

where Q_i denotes the package size for lineside buffer i, that is, the number of parts supplied to lineside buffer i with one dolly in each trip; C_i is the inventory capacity of lineside buffer i; $T(i, j)$ denotes the travel time from point i to j, where $i, j = 1, 2, \cdots, I$ denote the lineside buffers; $i, j = 0$ denotes the central docking area; $\lceil \cdot \rceil$ is a ceiling function which rounds a number upward to the nearest integer;

$\lceil T(0, a_t) + T(a_t, a_t') + T(a_t', 0) \rceil$ denotes the round trip time with actions a_t, and a_t'; ω_t denotes whether or not a starvation for parts happens at epoch t:

$$\omega_t = \prod_{i=1}^{I} [\mathcal{I}(n_{i,t} \geq 1)]; \tag{19.23}$$

$\xi_{i,t}$ is a random variable, indicating the consumption amount of parts at lineside buffer i between time t and $t + 1$. Under assumption 2, we have the following probabilities:

$$\Pr\left\{\xi_{i,t} = 1\right\} = U_i, \Pr\left\{\xi_{i,t} = 0\right\} = 1 - U_i, \text{ with } 0 \leq U_i \leq 1. \tag{19.24}$$

Equations (19.21) and (19.22) clearly demonstrate the coupling of the delivery process in the MH system and the assembly process in the GA line.

From the above description of the system dynamics, we can get the states transition probability at time t under any given dispatching policy $\mathcal{L}$. Nevertheless, the promising event-based solution framework provides an easy way to do optimization without knowing the exact transition probability matrix.

The one-step cost function contains three aspects—starvation penalty cost, driver transportation cost, and buffers inventory cost—which is calculated as:

$$f(\mathbf{s_t}) = w_1 \cdot \sum_{i=1}^{I} \mathcal{I}(n_{i,t} = 0) + w_2 \cdot y_t + w_3 \cdot \sum_{i=1}^{I} n_{i,t}, \tag{19.25}$$

where weighing coefficients indicate the priorities of different costs, which are predetermined based on practical requirements as $w_1 \gg w_2 \gg w_3$. We intend to minimize the average dispatching cost of the MH system over a long horizon. That is, the problem is formulated as to minimize the long-run average cost

$$\min_{\mathcal{L}} \eta^{\mathcal{L}} = \min_{\mathcal{L}} \limsup_{T \to \infty} \frac{1}{T} E\left(\sum_{t=0}^{T-1} f(\mathbf{s_t}) \middle| \mathcal{L}\right). \tag{19.26}$$

From this formulation, we could see the challenge of the large-size problem. The size of the state space in every epoch is

$$|\mathcal{S}| = \prod_{i=1}^{I} (C_i + 1) \cdot (Y + 1), \tag{19.27}$$

where Y denotes the possible upper bound of travel time for one trip; $|\cdot|$ indicates the cardinality of the set argument. The size of the action space is

$$|\mathcal{A}| = I(I + 1) + 1. \tag{19.28}$$

With typical data from a practical system [45], the size of the state space is larger than 10^{26}, and the size of the action space is larger than 10^2. Thus, the size of the policy space, $|\mathcal{A}|^{|\mathcal{S}|}$, is extremely huge, which makes it difficult to optimize the dispatching policies through traditional approaches.

19.5.2 Event-Based Optimization for the Material Handling Problem

We first define the Estimated Remaining Lifetime (ERL) $l_{i,t}$ for lineside buffer i at decision epoch t as the expected time length that buffer i can maintain without replenishment since time t:

$$l_{i,t} = \max\left\{ \left\lfloor \frac{n_{i,t}}{U_i} - T(0,i) \right\rfloor, 0 \right\}, \forall i = 1, \cdots, I, t = 0, \cdots, T-1, \quad (19.29)$$

where $\lfloor \cdot \rfloor$ is a floor function which rounds a number downward to the nearest integer. We sort the series of ERLs in ascending order as $l_{(1),t}, l_{(2),t}, ..., l_{(I),t}$, and focus on the first two numbers, that is, $l_{(1),t}$ and $l_{(2),t}$, when making decisions.

In controlling the system, we first observe whether the driver is available at the central docking area at decision epoch t. If not, we do nothing. Otherwise, that is, $y_t = 0$, we suppose the first and the second minimum ERLs are $l_{(1),t}$ and $l_{(2),t}$, respectively. We say that an *observable event* happens at time t and denote it as $e(l_{(1)}, l_{(2)})$. After the observation of event $e(l_{(1)}, l_{(2)})$ at time t, there are three types of *controllable events*: $e(l_{(1)}, l_{(2)}, +2)$ if action $(a_t = (1), a'_t = (2))$ (i.e., a_{+2}) is taken; $e(l_{(1)}, l_{(2)}, +1)$ if action $(a_t = (1), a'_t = 0)$ (i.e., a_{+1}) is taken; $e(l_{(1)}, l_{(2)}, +0)$ if action $(a_t = a'_t = 0)$ (i.e., a_{+0}) is taken. Finally, after the controllable event, a *natural transition event* happens and determines the consumption process of buffers according to system dynamics (19.21) and (19.22).

An observable event $e(l_{(1)}, l_{(2)})$ for this MH problem is a set of state transitions:

$$e(l_{(1)}, l_{(2)}) = \{\langle \mathbf{s}, \mathbf{x}\rangle | \forall \mathbf{s}, \mathbf{x} \in \mathcal{S}, \text{ and the first and second minimum ERL of } \mathbf{s} \text{ equal to } l_{(1)}, l_{(2)}\}. \quad (19.30)$$

The set of observable events is

$$\mathcal{E}_o = \left\{ e(l_{(1)}, l_{(2)}) \,\middle|\, \forall\, l_{(1)} = 0, 1, \cdots, L,\ l_{(2)} = l_{(1)},\ l_{(1)} + 1, \cdots, L \right\}. \quad (19.31)$$

The size of the event space becomes $|\mathcal{E}_o| = L^2$. Comparing with state space (19.27), the event space is much reduced.

An event-based policy for this MH problem is a mapping from the event space (19.31) to action space $\mathcal{A}' = \{a_{+0}, a_{+1}, a_{+2}\}$. Unfortunately, in this problem, Equation (19.17) does not hold, and there is not any reset event. So policy iteration will need $g^{d,h}(k_1, k_2)$ which cannot be estimated in the sample path of policy d. So, we focus on gradient-based policy iteration. We consider a parameterized event-based policy $\mathcal{L}_a$ in this section, which is defined as follows. When observable event $e(l_{(1)}, l_{(2)})$ happens: $\mathcal{L}_1(l_{(1)}, l_{(2)})$ is the probability of taking action a_{+1}; $\mathcal{L}_2(l_{(1)}, l_{(2)})$ is the probability of taking action a_{+2}; $1 - \mathcal{L}_1(l_{(1)}, l_{(2)}) - \mathcal{L}_2(l_{(1)}, l_{(2)})$ is the probability of taking action a_{+0}. Obviously, $\mathcal{L}_a$ belongs to a convex set Γ_a:

$$\begin{aligned}\Gamma_a = \{\mathcal{L}_a |\ & 0 \le \mathcal{L}_1(l_{(1)}, l_{(2)}) \le 1,\ 0 \le \mathcal{L}_2(l_{(1)}, l_{(2)}) \le 1 \\ & 0 \le \mathcal{L}_1(l_{(1)}, l_{(2)}) + \mathcal{L}_2(l_{(1)}, l_{(2)}) \le 1 \\ & \forall\, l_{(1)} = 0, 1, \cdots, L,\ l_{(2)} = l_{(1)}, l_{(1)} + 1, \cdots, L\}. \end{aligned} \tag{19.32}$$

On the basis of Equation (19.15) in Section 19.4, we have the performance sensitive formula for the MH problem as follows: $\forall \mathcal{L}_1(l_{(1)}, l_{(2)}),\ \mathcal{L}_2(l_{(1)}, l_{(2)})$,

$$\begin{aligned}&\frac{\partial \eta}{\partial \mathcal{L}_1(l_{(1)}, l_{(2)})} \\ &= \pi(e(l_{(1)}, l_{(2)})) \cdot [g^{\mathcal{L}_a}(e(l_{(1)}, l_{(2)}), e(l_{(1)}, l_{(2)}, +1)) - g^{\mathcal{L}_a}(e(l_{(1)}, l_{(2)}), e(l_{(1)}, l_{(2)}, +0))],\end{aligned} \tag{19.33}$$

$$\begin{aligned}&\frac{\partial \eta}{\partial \mathcal{L}_2(l_{(1)}, l_{(2)})} \\ &= \pi(e(l_{(1)}, l_{(2)})) \cdot [g^{\mathcal{L}_a}(e(l_{(1)}, l_{(2)}), e(l_{(1)}, l_{(2)}, +2)) - g^{\mathcal{L}_a}(e(l_{(1)}, l_{(2)}), e(l_{(1)}, l_{(2)}, +0))],\end{aligned} \tag{19.34}$$

where

$$g^{\mathcal{L}_a}(e(l_{(1)}, l_{(2)}), e(l_{(1)}, l_{(2)}, +n)) = \sum_{\mathbf{s} \in \mathcal{I}(e(l_{(1)}, l_{(2)}))} \pi(\mathbf{s} | e(l_{(1)}, l_{(2)})) \cdot g(\mathbf{s}_{+n}), n = 0, 1, 2, \tag{19.35}$$

where $\mathbf{s}_{+n}$ denote the states after actions a_{+n} is taken under state $\mathbf{s}$, $n = 0$, 1, and 2.

The potentials for events in (19.35) can be estimated based on a sample path of the system under the original policy. Then, a gradient-based approach can be applied to optimize the MH policies based on the performance sensitivity formulas (19.33) and (19.34) as follows:

- *Step 1 (Initialization)*: Uniformly sample in set Γ_a to generate the initial policy $\mathcal{L}_a^0$; $j = 0$.
- *Step 2 (Policy Evaluation)*: Let $\mathcal{L}_a^j \in \Gamma_a$ denote the policy used in iteration j. Estimate the potentials of events in Equation (19.35) based on a sample path under policy $\mathcal{L}_a^j$.
- *Step 3 (Policy Improvement)*: Obtain the gradient $\nabla \eta^{\mathcal{L}_a^j}$ through performance derivative formulas (19.33) and (19.34). Update the policies according to the gradient projection methods in [47] with the steepest descent direction and diminishing step size:

$$\mathcal{L}_a^{j+1} = [\mathcal{L}_a^j - \beta^j \cdot \nabla \eta^{\mathcal{L}_a^j}]^+, \tag{19.36}$$

where $[\cdot]^+$ denotes projection on the set Γ_a as in [47]; $\alpha^j \in (0, 1]$ is a step size, $\beta^j = 1/j$.

- *Step 4 (Stopping Criterion)*: If

$$\frac{||\nabla \eta^{\mathcal{L}_a^j}||}{||\nabla \eta^{\mathcal{L}_a^0}||} \leq \epsilon, \tag{19.37}$$

stop; otherwise, let $j = j + 1$ and go to step 2.

19.5.3 Numerical Results

First, we demonstrate the effectiveness of the EBO method by comparing the sizes of state space $\mathcal{S}$, state-based action space $\mathcal{A}$, observable event space $\mathcal{E}_o$, and event-based action space $\mathcal{A}'$ of this MH problem. We use 16 sets of practical data from our industry partners, and the comparison results are listed in Table 19.1.

Comparing with the sizes of the state space $|\mathcal{S}|$ and state-based action space $|\mathcal{A}|$, the size of the observable event space $|\mathcal{E}_o|$ is significantly smaller, and the size of the event-based action space $|\mathcal{A}'|$ is considerably shrunk. Thus, the size of the event-based policy space, $|\mathcal{A}'|^{|\mathcal{E}|}$, has been greatly reduced. It is important to note that the size of event space is proportional to the system size despite that the size of the state space grows exponentially. This outstanding feature provides a possibility for the EBO theory to overcome or to alleviate the computational difficulty.

Second, we compare the optimization result from EBO with practical MH policy currently used in industry. The practical MH policy applied in automotive industry is a threshold policy, called reorder-point (RP) policy. For lineside buffer i, r_i denotes a

TABLE 19.1 The Effectiveness of the EBO Theory

Data Set	$\|\mathcal{S}\|$	$\|\mathcal{A}\|$	$\|\mathcal{E}_o\|$	$\|\mathcal{A}'\|$
#1	2.84E+31	4.84E+02	2.89E+05	3
#2	6.54E+64	1.23E+03	5.34E+06	3
#3	1.51E+37	5.29E+02	7.31E+05	3
#4	1.55E+32	4.84E+02	3.37E+05	3
#5	1.02E+45	7.29E+02	1.30E+06	3
#6	4.37E+30	3.61E+02	4.68E+05	3
#7	1.36E+39	6.25E+02	6.94E+05	3
#8	5.16E+20	1.96E+02	1.23E+05	3
#9	1.01E+31	4.00E+02	3.89E+05	3
#10	1.05E+20	1.96E+02	1.05E+05	3
#11	1.29E+25	2.89E+02	1.98E+05	3
#12	3.58E+27	3.24E+02	2.57E+05	3
#13	2.91E+26	3.24E+02	2.23E+05	3
#14	3.66E+25	2.89E+02	1.98E+05	3
#15	2.77E+29	3.61E+02	3.50E+05	3
#16	4.36E+19	1.96E+02	8.81E+04	3

TABLE 19.2 Comparison with Practical Policy in Industry

Data Set	RP Policy	EBO Policy	Relative Saving %
#1	172.56	157.24	8.87
#2	687.36	627.42	8.72
#3	121.57	113.47	6.66
#4	330.24	288.26	12.71
#5	200.20	192.61	3.79
#6	110.16	104.32	5.30
#7	153.43	142.63	7.04
#8	135.02	112.13	16.95
#9	136.44	125.52	8.00
#10	69.61	64.44	7.43
#11	119.39	108.87	8.82
#12	241.74	187.49	22.44
#13	107.01	96.80	9.54
#14	179.34	146.23	18.46
#15	157.86	140.93	10.73
#16	80.76	69.38	14.09
Mean			10.58

predefined threshold (RP). When the inventory of buffer i at decision epoch t is less than this threshold, that is, $n_{t,i} < r_i$, it generates a request to the driver. The driver serves the requests with the First Come First Serve (FCFS) principle. Here, we use 16 sets of practical data from a world-famous automotive manufactory, and compare the EBO performance with the RP policy that is currently used in factories. The numerical results are shown in Table 19.2.

From Table 19.2, it can be seen that the system saves about 10.58% costs on average with the dispatching policy that is found by our EBO approach, comparing with the currently used threshold policy in automotive industry. This result demonstrates the effectiveness and efficiency of the EBO theory introduced above.

Third, we test the robustness of the EBO theory by randomly generating the parameters of the MH system within a given range around the practical data. Specifically, we generate the testing-data x' within a small range, that is, $[0.8x, 1.2x]$, where x indicates the practical parameter in industry. The parameters consists of $T(i, j)$, C_i, Q_i, and U_i. Then, we compare the performance of the EBO approach with the currently used RP policy in industry, and show the results in Table 19.3.

In another group, we generate the testing-data x' within a large range, that is, $[0.5x, 1.5x]$, and show the comparison results of the EBO theory and current RP policy in Table 19.4.

From Tables 19.3 and 19.4, we can see that the performance of policies obtained from the EBO approach outperforms that of the currently used RP policy in industry. This result demonstrates that our EBO approach is not sensitive with respect to the system parameters, and can be extendedly applied in various systems robustly and effectively.

TABLE 19.3 Performance with Random Parameters in Small Range

Test Index	RP	EBO	Saving %
1	188.12	174.01	7.50
2	208.87	193.76	7.23
3	179.32	165.42	7.75
4	140.68	131.85	6.28
5	207.11	187.03	9.70
6	207.53	195.46	5.81
7	153.92	143.87	6.53
8	214.31	195.03	9.00
9	162.19	147.64	8.97
10	188.40	172.75	8.31
Mean			7.71
Var			1.65

TABLE 19.4 Performance with Random Parameters in Large Range

Test Index	RP	EBO	Saving %
1	1785.06	1668.34	6.54
2	3041.68	2831.34	6.92
3	221.56	204.79	7.57
4	157.57	143.33	9.03
5	201.58	192.72	4.39
6	119.17	110.96	6.89
7	115.30	107.31	6.93
8	147.48	137.60	6.70
9	154.26	137.25	11.02
10	144.95	131.04	9.60
Mean			7.56
Var			3.49

19.6 CONCLUSIONS

In this chapter, we review the basic idea of EBO, which is specifically suitable for policy optimization of DEDS. By making decisions based on certain events instead of on the entire system state, the number of event potentials to estimate usually is much smaller than the size of the state space. Performance difference and derivative formulas are developed. Under certain assumptions, policy iteration for EBO is developed to converge to globally optimal event-based policy. The gradient-based policy iteration can always be applied to converge to a local optimum in the parameterized policy space. The EBO method is demonstrated on a MH problem. More details of the EBO method can be found in [42]. As for future research, it is interesting to quantify the performance loss of event-based policies when the event sequence is not Markovian. Some work toward this goal can be found in [43] and [44]. Also note that an observable event contains information both on which state the system was and on which state the system is going to. It is an interesting research topic to study how to explore this information on the future to find better policies. Due to the broad applications in networked systems, the EBO method has gain more and more attention recently. We hope this chapter could bring insights for the study of event-based control, decision-making, and optimization in more general situations.

ACKNOWLEDGMENTS

This work was supported in part by the National Science Foundation of China under grants (Nos. 60704008, 60736027, 90924001, 61222302, and 61174072), the Specialized Research Fund for the Doctoral Program of Higher Education (No. 20070003110), the National 111 International Collaboration Project (No. B06002), and the Tsinghua National Laboratory for Information Science and Technology (TNList) Cross-discipline Foundation.

REFERENCES

1. C.G. Cassandras and S. Lafortune. *Introduction to Discrete Event Systems*, 2nd edition. Springer, New York, NY, 2008.
2. D.P. Bertsekas. *Dynamic Programming and Optimal Control*, 3rd edition, Vols. 1 and 2. Athena Scientific, Belmont, MA, 2007.
3. M.L. Puterman. *Markov Decision Processes: Discrete Stochastic Dynamic Programming*. John Wiley and Sons, Inc., New York, NY, 1994.
4. Q.-S. Jia. On state aggregation on approximate complex value functions in large-scale markov decision processes. *IEEE Transactions on Automatic Control*, 56(2):333–344, 2011.
5. Z. Ren and B.H. Krogh. State aggregation in markov decision processes. In *Proceedings of the 41st IEEE Conference on Decision and Control*, pp. 3819–3824, Las Vegas, NV, December 10–13, 2002.
6. X.R. Cao, Z.Y. Ren, S. Bhatnagar, M. Fu, and S. Marcus. A time aggregation approach to markov decision processes. *Automatica*, 38(6):929–943, 2002.
7. T. Sun, Q.C. Zhao, and P.B. Luh. Incremental value iteration for time aggregated markov decision processes. *IEEE Transactions on Automatic Control*, 52(11):2177–2182, 2007.
8. Q.-S. Jia. A structural property of optimal policies for multi-component maintenance problems. *IEEE Transactions on Automation Science and Engineering*, 7(3):677–680, 2010.
9. L. Xia, Q. Zhao, and Q.-S. Jia. A structure property of optimal policies for maintenance problems with safety-critical components. *IEEE Transactions on Automation Science and Engineering*, 5(3):519–531, 2008.
10. D.P. Bertsekas and J.N. Tsitsiklis. *Neuro-Dynamic Programming*. Athena Scientific, Belmont, MA, 1996.
11. P.J. Werbös. Approximate dynamic programming for real-time control and neural modeling. In D.A. White and D.A. Sofge, editors. *Handbook of Intelligent Control*. Van Nostrand Reinhold, New York, NY, 1992.

12. P.J. Werbös. Neurocontrol and supervised learning: An overview and valuation. In D.A. White and D.A. Sofge, editors. *Handbook of Intelligent Control*. Van Nostrand Reinhold, New York, NY, 1992.

13. R.S. Sutton and A.G. Barto. *Reinforcement Learning: An Introduction*. MIT Press, Cambridge, MA, 1998.

14. C. Cervellera and M. Muselli. Efficient sampling in approximate dynamic programming algorithms. *Computational Optimization and Applications*, 38(3):417–443, 2007.

15. C. Cervellera, A. Wen, and V.C.P. Chen. Neural network and regression spline value function approximations for stochastic dynamic programming. *Computers and Operations Research*, 34:70–90, 2007.

16. W.B. Powell. *Approximate Dynamic Programming: Solving the Curse of Dimensionality*. Wiley-Interscience, 2007.

17. D.P. Bertsekas and D.A. Castañon. Rollout algorithms for stochastic scheduling problems. *Journal of Heuristics*, 5:89–108, 1999.

18. H.S. Chang and S.I. Marcus. Approximate receding horizon approach for markov decision processes: Average reward case. *Journal of Mathematical Analysis and Applications*, 286(2):636–651, 2003.

19. Q.-S. Jia and Q.-C. Zhao. Strategy optimization for controlled markov process with descriptive complexity constraint. *Science in China Series F: Information Sciences*, 52(11):1993–2005, 2009.

20. H.S. Chang, M.C. Fu, J. Hu, and S.I. Marcus. *Simulation-based Algorithms for Markov Decision Processes*. Springer, New York, NY, 2007.

21. K. Arzen. A simple event-based pid controller. In *Proceedings of 1999 IFAC World Congress*, pp. 423–428, Beijing, China, July 5–9, 1999. Pergamon, Oxford, UK.

22. K. Aström. Event based control. In A. Astolfi and L. Marconi, editors. *Analysis and Design of Nonlinear Control Systems*. pp. 127–147, Springer-Verlag, Berlin, 2008.

23. K. Aström and B. Bernhardsson. Comparison of riemann and lebesgue sampling for first order stochastic systems. In *Proceedings of the 2002 IEEE Conference on Decision and Control*, pp. 2011–2016, Las Vegas, NV, December 10–13, 2002.

24. W.P. M.H. Heemels, J. Sandee, and P.P.J. Van Den Bosch. Analysis of event-driven controllers for linear systems. *International Journal of Control*, 81(4):571–590, 2007.

25. Y.-K. Xu and X.-R. Cao. Lebesgue-sampling-based optimal control problems with time aggregation. *IEEE Transactions on Automatic Control*, 56(5):1097–1109, 2011.

26. X. Wang and M.D. Lemmon. Event-triggering in distributed networked control systems. *IEEE Transactions on Automatic Control*, 56(3):586–601, 2011.

27. R. Cogill, S. Lall, and J.P. Hespanha. A constant factor approximation algorithm for event-based sampling. In *Proceedings of the 2007 American Control Conference*, pages 305–311, New York, NY, July 9–13, 2007.

28. O.C. Imer and T. Basar. Optimal estimation with limited measurements. In *Proceedings of the 44th IEEE Conference on Decision and Control*, pp. 1029–1034, Seville, Spain, December 12–15, 2005.

29. A. Molin and S. Hirche. On lqg joint optimal scheduling and control under communication constraints. In *Proceedings of the 48th IEEE Conference on Decision and Control and the 28th Chinese Control Conference*, pp. 5832–5838, Shanghai, China, December 15–18, 2009.

30. W. Wu and A. Arapostathis. Optimal control of stochastic systems with costly observations - the general markovian model and the lqg problem. In *Proceedings of the 2005 American Control Conference*, pp. 294–299, Portland, OR, June 8–10, 2005.
31. Y. Xu and J. Hespanha. Optimal communication logics in networked control systems. In *Proceedings of the 43rd IEEE Conference on Decision and Control*, pp. 3527–3532, The Bahamas, December 14–17, 2004.
32. L.P. Kaelbling, M.L. Littman, and A.R. Cassandra. Planning and acting in partially observable stochastic domains. *Artificial Intelligence*, 101:99–134, 1998.
33. R.D. Smallwood and E.J. Sondik. The optimal control of partially observable markov processes over a finite horizon. *Operations Research*, 21(5):1071–1088, 1973.
34. E.J. Sondik. The optimal control of partially observable markov processes over the infinite horizon: Discouted costs. *Operations Research*, 26(2):282–304, 1978.
35. J.A. Filar and A. Haurie. A two-factor stochastic production model with two time scales. *Automatica*, 37:1505–1513, 2001.
36. A. Barto and S. Mahadevan. Recent advances in hirerarchical reinforcement learning. *Discrete Event Dynamic Systems: Theory and Applications*, 13(1–2):41–77, 2003. Special Issue on Reinforcement Learning.
37. T.R. Bielecki and J.A. Filar. Singularly perturbed markov control problems: limiting average cost. *Annals of Operations Research*, 28:153–168, 1991.
38. J.A. Filar, V. Gaitsgory, and A.B. Haurie. Control of singularly perturbed hybrid stochastic systems. *IEEE Transactions on Automatic Control*, 46(2):179–190, 2001.
39. L. Xia, X. Chen, and X.R. Cao. Policy iteration of customer-average performance in queueing systems. *Automatica*, 45:1639–1648, 2009.
40. X.R. Cao. Basic ideas for event-based optimization of markov systems. *Discrete Event Dynamic Systems: Theory and Applications*, 15(2):169–197, 2005.
41. X.-R. Cao and J. Zhang. Event-based optimization of markov systems. *IEEE Transactions on Automatic Control*, 53(4):1076–1082, 2008.
42. X.R. Cao. *Stochastic Learning and Optimization: A Sensitivity-Based Approach*. Springer, New York, NY, 2007.
43. Q.-S. Jia. On solving event-based optimization with average reward over infinite stages. *IEEE Transactions on Automatic Control*, 56(12):2912–2917, 2011.
44. Q.-S. Jia. On solving optimal policies for finite-stage event-based optimization. *IEEE Transactions on Automatic Control*, 56(9):2195–2200, 2011.
45. Y. Zhao, C.B. Yan, Q. Zhao, N. Huang, J. Li, and X. Guan. Efficient simulation method for general assembly systems with material handling based on aggregated event-scheduling. *IEEE Transactions on Automation Science and Engineering*, 7(4):762–775, 2010.
46. Y. Zhao, Q. Zhao, and X. Guan. Stochastic optimal control for a class of manufacturing systems based on event-based optimization. In J. Huang, K.-Z. Liu, and Y. Ohta, editors. *Theory and Applications of Complex Systems and Robust Control*. Tsinghua University Press, Beijing, China, 2010, pp. 63–83.
47. D.P. Bertsekas. *Nonlinear Programming*. Athena Scientific, 1999.

CHAPTER 20

Bounds for Markov Decision Processes

VIJAY V. DESAI,[1] VIVEK F. FARIAS,[2] and CIAMAC C. MOALLEMI[3]

[1]Industrial Engineering and Operations Research, Columbia University, New York, NY, USA
[2]Sloan School of Management, Massachusetts Institute of Technology, Cambridge, MA, USA
[3]Graduate School of Business, Columbia University, New York, NY, USA

ABSTRACT

We consider the problem of producing lower bounds on the optimal cost-to-go function of a Markov decision problem. We present two approaches to this problem: one based on the methodology of approximate linear programming (ALP) and another based on the so-called martingale duality approach. We show that these two approaches are intimately connected. Exploring this connection leads us to the problem of finding "optimal" martingale penalties within the martingale duality approach which we dub the pathwise optimization (PO) problem. We show interesting cases where the PO problem admits a tractable solution and establish that these solutions produce tighter approximations than the ALP approach.

20.1 INTRODUCTION

Markov decision processes (MDPs) provide a general framework for modeling sequential decision-making under uncertainty. A large number of practical problems from diverse areas can be viewed as MDPs and can, in principle, be solved via dynamic programming. However, for many problems of interest, the state space of the corresponding dynamic program is intractably large. This phenomenon, referred to as the *curse of dimensionality*, renders exact approaches to solving Markov decision problems impractical.

Reinforcement Learning and Approximate Dynamic Programming for Feedback Control, First Edition. Edited by Frank L. Lewis and Derong Liu.

Solving an MDP may be viewed as equivalent to the problem of computing an *optimal cost-to-go function*. As such, approximation algorithms for solving MDPs whose state space is intractably large frequently treat the task of computing an approximation to this optimal cost-to-go function as the key algorithmic task; given such an approximation, the greedy policy with respect to the approximation is a canonical candidate for an approximate policy. The collective research area devoted to the development of such algorithms is frequently referred to as *approximate dynamic programming*; see Van Roy [1] or Bertsekas ([2], Chapter 6) for brief surveys of this area of research. Now consider a Markov decision problem wherein we wish to minimize expected costs, discounted over an infinite time horizon and consider the problem of producing upper and lower bounds on the costs incurred under an optimal policy starting at a specific state (the "cost-to-go" of that state). By simulating an arbitrary feasible policy starting at that state, we obtain an upper bound on the cost-to-go of the state. Given a complementary *lower* bound on the cost-to-go of this state, one may hope to construct a "confidence interval" of sorts on the cost-to-go of the state in question.[1] The task of finding a lower bound on the cost-to-go of a state is not quite as straightforward. Moreover, we are interested in *good* bounds. The literature offers us two seemingly disparate alternatives to serve this end:

- *Lower Bounds via Approximate Linear Programming (ALP)*: This approach was introduced by Schweitzer and Seidmann [3] and later developed and analyzed by Farias and Van Roy [4, 5]. Given a set of "basis functions," the ALP produces an approximation to the optimal cost-to-go function spanned by these basis functions that is provably a pointwise lower bound to the optimal cost-to-go function. The quality of the cost-to-go function approximation produced by the ALP can be shown to compete, in an appropriate sense, with the best possible approximation afforded by the basis function architecture. The ALP approach is attractive for two reasons: first, from a practical standpoint, the availability of reliable linear programming solvers allows the solution of large ADP problems. Second, the structure of the linear program allows strong theoretical guarantees to be established.
- *Lower Bounds via Martingale Duality*: A second approach to computing lower bounds, which constitutes an active area of research, relies on "information relaxations." As a trivial example, consider giving the optimizer *a priori* knowledge of all randomness that will be realized over time; clearly this might be used to compute a "clairvoyant" lower bound on the optimal cost-to-go. These approaches introduce, in the spirit of Lagrangian duality, a penalty for relaxing the restrictions on information available to the controller. The penalty function is itself a stochastic process and, frequently, is a martingale adapted to the natural

[1]Equivalently, in problems where reward is maximized, the quantity of interest is the value of rewards achieved under an optimal policy, starting from a specific state. Lower bounds are available from the simulation of suboptimal policies, and we might seek complimentary upper bounds. We will choose between the objectives of cost minimization and reward maximization in this chapter, according to what is most natural to the immediate setting.

filtration of the MDP; hence the nomenclature martingale duality. An important application of these approaches can be found in the context of pricing high dimensional American options following the work of Rogers [6] and Haugh and Kogan [7]. Generalizations of this approach to control problems other than optimal stopping, have also been studied (see, e.g., [8, 9]).

The two approaches are, at least superficially, fairly distinct from each other. Computing a good cost-to-go function approximation via the ALP relies on finding a good set of basis functions. The martingale duality approach on the other hand requires that we identify a suitable martingale to serve as the penalty process. The purpose of this chapter is to present a simple unified view of the two approaches through the lens of, what we call, the *pathwise optimization* (PO) method. This method was introduced in the context of high-dimensional optimal stopping problems by Desai et al. [10] and later extended to a larger class of problems (optimizing convex cost functionals subject to linear system dynamics) in Desai et al. [11].

We will shortly present a brief literature review. Following that, the remainder of the chapter is organized as follows: In Section 20.2, we formulate our problem and state the Bellman equation. Sections 20.3 and 20.4 introduce the ALP and martingale duality approaches, respectively, for the problem. The PO approach is described in Section 20.5 and its applications to optimal stopping and linear convex systems are described in Section 20.6.

20.1.1 Related Literature

The landscape of ADP algorithms is rich and varied; we only highlight some of the literature related to ALP. Bertsekas and Tsitsiklis [12] and Powell [13] are more detailed references on the topic. The ALP approach was introduced by Schweitzer and Seidmann [3] and further developed by Farias and Van Roy [4, 5] who established approximation guarantees for this approach. This method has seen a number of applications, which includes scheduling in queueing networks [14–16], revenue management [17–19], portfolio management [20], inventory problems [21, 22], and algorithms for solving stochastic games [23], among others.

Martingale duality methods for the pricing of American and Bermudan options, which rely on Doob's decomposition to generate the penalty process, were introduced by Rogers [6] and Haugh and Kogan [7]. Andersen and Broadie [24] showed how to compute martingale penalties using stopping rules and are able to obtain tight bounds. An alternative "multiplicative" approach to duality was introduced by Jamshidian [25] and its connections with the above "additive" duality approaches was explored in Chen and Glasserman [26]. Beyond stopping problems, these methods are applicable for general control problems as discussed in Rogers [8] and Brown et al. [9]. Further, Brown et al. [9] considered a broader class of information relaxations than the typical case of a perfect information relaxation. Applications of these methods were considered in portfolio optimization [27] and valuation of natural gas storage [28, 29], among others.

20.2 PROBLEM FORMULATION

Consider a discounted, infinite horizon problem with state space $\mathcal{X}$ and action set $\mathcal{A}$. At time t, given state x_t and action a_t, the per stage cost is given by $g(x_t, a_t)$. The state evolves according to

$$x_{t+1} = h(x_t, a_t, w_t),$$

where $\{w_t\}$ are independent and identically distributed random variables taking values in the set $\mathcal{W}$. Let $\mathcal{F} \triangleq \{\mathcal{F}_t\}$ be the natural filtration generated by the process $\{w_t\}$, that is, for each time t, $\mathcal{F}_t \triangleq \sigma(w_0, w_1, \ldots, w_t)$. So as to avoid discussion of technicalities which are not central to our main ideas, for ease of exposition, we assume finite state and control spaces.

A stationary policy $\mu\colon \mathcal{X} \to \mathcal{A}$ maps the state space $\mathcal{X}$ to the set of actions $\mathcal{A}$. In other words, given a state x_t, the action taken at that state under policy μ is $a_t = \mu(x_t)$. The cost-to-go function J_μ associated with a stationary policy μ is given by

$$J_\mu(x) = \mathsf{E}\left[\sum_{t=0}^{\infty} \alpha^t g(x_t, \mu(x_t)) \,\middle|\, x_0 = x \right],$$

where α is the discount factor.

We define the Bellman operator associated with policy μ according to

$$(T_\mu J)(x) \triangleq g(x, \mu(x)) + \alpha \mathsf{E}[J(h(x, \mu(x), w))].$$

Given this defintion, J_μ is given as the unique solution to the Bellman's equation $T_\mu J = J$. We further define the optimal cost-to-go function J^* according to $J^*(x) = \min_\mu J_\mu(x),\ \forall\ x \in \mathcal{X}$. J^* may be computed as the unique solution to *Bellman's equation*. In particular, define the Bellman operator $T : \mathbb{R}^{|\mathcal{X}|} \to \mathbb{R}^{|\mathcal{X}|}$ according to $TJ = \min_\mu T_\mu J$. Bellman's equation is simply the fixed point equation $TJ = J$.

Given the optimal cost-to-go function, the optimal policy is obtained by acting greedily with respect to the optimal cost-to-go function, that is,

$$\mu^*(x) \in \operatorname*{argmin}_a\ g(x, a) + \alpha \mathsf{E}[J^*(h(x, a, w))]. \tag{20.1}$$

The Problem: Computing J^* is in general intractable for state spaces $\mathcal{X}$ that are intractably large. As such our goal in this chapter will be to compute *lower bounds* to the optimal cost-to-go function of a specific state x, $J^*(x)$. We will particularly be interested in issues of tractability and the tightness of the resulting bounds.

20.3 THE LINEAR PROGRAMMING APPROACH

This section describes an approximate dynamic programming approach (dubbed approximate linear programming) to solving the above problem. The approach relies on solving a linear program motivated largely by a certain "exact" linear program for the exact solution of Bellman's equation. We begin by describing the exact linear program.

20.3.1 The Exact Linear Program

Given any vector $\nu \in \mathbb{R}^{|\mathcal{X}|}$ with positive components, the exact linear program, credited to Manne [30], is given by:

$$\begin{aligned} \underset{J}{\text{maximize}} \quad & \nu^\top J \\ \text{subject to} \quad & J \leq TJ. \end{aligned} \tag{20.2}$$

Although the Bellman operator T is nonlinear, this program can be easily transformed into a linear program. Consider a state $x \in \mathcal{X}$, the constraint $J(x) \leq (TJ)(x)$ is equivalent to $|\mathcal{A}|$ linear constraints given by

$$J(x) \leq g(x, a) + \alpha \mathsf{E}[J(h(x, a, w))], \quad \forall\, a \in \mathcal{A}.$$

Using this transformation, the exact linear program has as many variables as the state space size $|\mathcal{X}|$ and as many constraints as $|\mathcal{X} \times \mathcal{A}|$.

We recall the following basic properties of the Bellman operator T. The interested reader is referred to Bertsekas [31] for details of the proof.

Proposition 20.1 Let $J, J' \in \mathbb{R}^{|\mathcal{X}|}$.

1. (Monotonicity) If $J \geq J'$, then $TJ \geq TJ'$.
2. (Max-norm contraction) $\|TJ - TJ'\|_\infty \leq \alpha \|J - J'\|_\infty$.

The following theorem establishes that the program (20.2) yields, as its unique optimal solution, the optimal cost to go J^*. We provide a proof of this fact for completeness.

Theorem 20.1

1. *For all $J \in \mathbb{R}^{|\mathcal{X}|}$ such that $J \leq TJ$, we have $J \leq J^*$.*
2. *J^* is the unique optimal solution to the exact linear program* (20.2).

Proof: Now by the monotonicity of T, for any J satisfying $J \leq TJ$, we must also have $J \leq TJ \leq \cdots \leq T^k J$, for any integer $k \geq 1$. Since T is a contraction mapping, however, we have that, as $k \to \infty$, $T^k J \to J^*$, the unique fixed point of the operator T. It follows that any feasible solution to (20.2), J, satisfies $J \leq J^*$. This is the first part of the theorem. Further, since J^* is itself a feasible solution, and since the components of ν are strictly positive, we have the second part of the theorem. ■

Of course, the exact linear program has $|\mathcal{X}|$ variables and $|\mathcal{X} \times \mathcal{A}|$ constraints and, as such, we must still contend with the curse of dimensionality. This motivates an effort to reduce the dimensionality of the problem by permitting approximations to the cost-to-go function.

20.3.2 Cost-to-Go Function Approximation

Cost-to-go function approximations address the curse of dimensionality through the use of parameterized function approximations. In particular, it is common to focus on linear parameterizations. Consider a collection of *basis functions* $\{\phi_1, \ldots, \phi_K\}$ where each $\phi_i : \mathcal{X} \to \mathbb{R}$ is a real-valued function on the state space. ADP algorithms seek to find linear combinations of the basis functions that provide good approximations to the optimal cost-to-go function. In particular, we seek a vector of weights $r \in \mathbb{R}^K$ so that

$$\Phi r(x) \triangleq \sum_{\ell=1}^{K} \phi_\ell(x) r_\ell \approx J^*(x).$$

Here, we define $\Phi \triangleq [\phi_1 \; \phi_2 \; \ldots \; \phi_K]$ to be a matrix with columns consisting of the basis functions. Given such an approximation to the cost-to-go function, a natural policy to consider is simply the policy that acts greedily with respect to the cost-to-go function approximation. Such a policy is given by:

$$\mu^r(x) \in \underset{a \in \mathcal{A}}{\operatorname{argmin}} \; g(x, a) + \mathsf{E}[\Phi r(h(x, a, w))]. \tag{20.3}$$

Notice that such a policy is eminently implementable. In contrast with the optimal policy which would generally require a lookup table for the optimal cost-to-go function (and consequently, storage space on the order of the size of the state space), the policy μ^r simply requires that we store K numbers corresponding to the weights r and have access to an oracle that for a given state x computes the basis functions at that state. The approximations to the cost-to-go function can then be computed online, as and when needed.

20.3.3 The Approximate Linear Program

In light of the approximation described earlier, a natural idea would be to restrict attention to solutions of the exact linear program that lie in the lower dimensional space spanned by the basis functions (i.e., span(Φ)). The approximate linear program (ALP) does exactly this:

$$\begin{aligned} \underset{r}{\text{maximize}} \quad & \nu^\top \Phi r \\ \text{subject to} \quad & \Phi r \leq T \Phi r. \end{aligned} \tag{20.4}$$

Notice that the above program continues to have a large number of constraints but a substantially smaller number of variables, K.

For any feasible solution r to this program, we must have, by Theorem 20.1, that the approximation implied by r provides a lower bound to the optimal cost-to-go. That is, $\Phi r \leq J^*$. This observation also allows us to rewrite ALP as

$$\begin{aligned} \underset{r}{\text{minimize}} \quad & \|J^* - \Phi r\|_{1,\nu} \\ \text{subject to} \quad & \Phi r \leq T\Phi r, \end{aligned} \tag{20.5}$$

where the weighted 1-norm in the objective is defined by

$$\|J^* - \Phi r\|_{1,\nu} \triangleq \sum_{x \in \mathcal{X}} \nu(x)|J^*(x) - \Phi r(x)|.$$

This representation of the ALP makes it clear that ν can be used to emphasize regions of the state space where we would like a good approximation and consequently, the components of ν are referred to as the state-relevance weights.

Now, for a fixed state $x \in \mathcal{X}$, the best lower bound to $J^*(x)$ we might compute using this approach simply calls for us to choose the state-relevance weights such that $\nu(x)$ is large. Moreover, if J^* is in the linear span of Φ, then it is clear from (20.5) that the approximation error would be zero. Apart from obtaining lower bounds, the cost-to-go function approximation obtained by solving the ALP can be used to generate policies, simply by acting greedily with respect to the approximation as shown in (20.3).

20.4 THE MARTINGALE DUALITY APPROACH

Every feasible solution to the ALP constitues a lower bound to the optimal cost-to-go function; the quality of this bound is determined largely by our choice of basis functions. A different approach to obtaining lower bounds is via an information relaxation. The idea is to allow policies to have knowledge of all future randomness and "penalize" this relaxation in the spirit of Lagrangian duality. The penalties are themselves stochastic processes, and typically martingales. We describe this approach next.

Let $\mathcal{P}$ be the space of real-valued functions defined on $\mathcal{X}$. Intuitively, we can think of this as the space of cost-to-go functions. Let us begin with defining the martingale difference operator Δ that maps a function $J \in \mathcal{P}$ to a real-valued function ΔJ on $\mathcal{X} \times \mathcal{X} \times \mathcal{A}$ according to

$$(\Delta J)(x_{t+1}, x_t, a_t) \triangleq J(x_{t+1}) - \mathsf{E}[J(x_{t+1})|x_t, a_t].$$

We are interested in computing lower bounds by considering a perfect information relaxation. Let $\mathcal{A}^\infty$ be the set of infinite sequences of elements of $\mathcal{A}$. For an arbitrary sequence of actions $\mathbf{a} \in \mathcal{A}^\infty$, define the process $M_t^{\mathbf{a}}(J)$ by

$$M_0^{\mathbf{a}}(J) \triangleq 0, \qquad M_t^{\mathbf{a}}(J) \triangleq \sum_{s=1}^{t} \alpha^s \Delta J(x_s, x_{s-1}, a_{s-1}), \quad \forall\, t \geq 1.$$

Clearly $M_t^{\mathbf{a}}(J)$ is adapted to the filtration $\mathcal{F}$. Further, if actions are chosen according to $\mathbf{a}$, then $M_t^{\mathbf{a}}(J)$ is a martingale. Using the fact that the state space $\mathcal{X}$ and action space $\mathcal{A}$ are finite, there exists a constant C_J such that

$$|\Delta J(x_s, x_{s-1}, a_{s-1})| < C_J, \quad \forall\, (x_s, x_{s-1}, a_{s-1}) \in \mathcal{X} \times \mathcal{X} \times \mathcal{A}.$$

It then follows from the orthogonality of martingale increments that

$$\mathsf{E}\left[M_t^{\mathbf{a}}(J)^2\right] = \sum_{s=1}^{t} \alpha^{2s} \mathsf{E}\left[|\Delta J(x_s, x_{s-1}, a_{s-1})|^2\right] < \frac{C_J^2 \alpha^2}{1-\alpha^2}.$$

Thus, $M_t^{\mathbf{a}}(J)$ is a $\mathcal{L}^2$-martingale. By the martingale convergence theorem, the limit

$$M_\infty^{\mathbf{a}}(J) \triangleq \sum_{s=1}^{\infty} \alpha^s \Delta J(x_s, x_{s-1}, a_{s-1}), \tag{20.6}$$

is well defined.

We now define the *martingale duality* operator $F : \mathcal{P} \to \mathcal{P}$ according to:

$$(FJ)(x) \triangleq \mathsf{E}\left[\inf_{\mathbf{a} \in \mathcal{A}^\infty} \sum_{t=0}^{\infty} \alpha^t g(x_t, a_t) - M_\infty^{\mathbf{a}}(J) \,\middle|\, x_0 = x\right], \tag{20.7}$$

where the expectation is with respect the infinite sequence of disturbances $(w_0, w_1, \ldots)$. The deterministic minimization problem embedded inside the expectation will be referred to as the *inner problem.*

Given any $J \in \mathcal{P}$, $FJ(x)$ can be used to obtain lower bounds on the optimal cost-to-go function $J^*(x)$. Moreover, there exists $J \in \mathcal{P}$ for which the lower bounds are tight, and one such choice of J is the optimal cost-to-go function J^*. The following theorem justifies these claims.

Theorem 20.2

(i) (Weak duality) For any $J \in \mathcal{P}$ and all $x \in \mathcal{X}$, $FJ(x) \leq J^*(x)$.

(ii) (Strong duality) For all $x \in \mathcal{X}$, $J^*(x) = FJ^*(x)$.

Proof: (i) For each state $x \in \mathcal{X}$,

$$\begin{aligned}
J^*(x) &= \min_{\mu} \mathsf{E}\left[\sum_{t=0}^{\infty} \alpha^t g(x_t, \mu(x_t)) \,\middle|\, x_0 = x\right] \\
&\overset{\text{(a)}}{=} \min_{\mu} \mathsf{E}\left[\sum_{t=0}^{\infty} \alpha^t g(x_t, \mu(x_t)) - M_{\infty}^{\mu}(J) \,\middle|\, x_0 = x\right] \\
&\overset{\text{(b)}}{\geq} \mathsf{E}\left[\inf_{\mathbf{a} \in \mathcal{A}^{\infty}} \sum_{t=0}^{\infty} \alpha^t g(x_t, a_t) - M_{\infty}^{\mathbf{a}}(J) \,\middle|\, x_0 = x\right] \\
&= FJ(x).
\end{aligned}$$

Here (a) follows from the fact that $M_{\infty}^{\mu}(J)$ is zero mean, and (b) follows from that fact that the objective value can only be decreased given knowledge of the entire sample path of disturbances.

(ii) From (i), we have that $FJ^*(x) \leq J^*(x)$. We will establish the result by showing $FJ^*(x) \geq J^*(x)$. Using the definition of $FJ^*(x)$, we have

$$\begin{aligned}
FJ^*(x) &= \mathsf{E}\left[\inf_{\mathbf{a} \in \mathcal{A}^{\infty}} \sum_{t=0}^{\infty} \alpha^t \big(g(x_t, a_t) - \alpha \Delta J^*(x_{t+1}, x_t, a_t)\big) \,\middle|\, x_0 = x\right] \\
&= \mathsf{E}\left[\inf_{\mathbf{a} \in \mathcal{A}^{\infty}} \sum_{t=0}^{\infty} \alpha^t \left(g(x_t, a_t) + \alpha \mathsf{E}\big[J^*(x_{t+1}) | x_t, a_t\big] - \alpha J^*(x_{t+1})\right) \,\middle|\, x_0 = x\right] \\
&= \mathsf{E}\left[\inf_{\mathbf{a} \in \mathcal{A}^{\infty}} \sum_{t=0}^{\infty} \alpha^t \left(g(x_t, a_t) + \alpha \mathsf{E}\big[J^*(x_{t+1}) | x_t, a_t\big] - J^*(x_t)\right)\right. \\
&\qquad \left. + J^*(x_0) \,\middle|\, x_0 = x\right] \\
&\geq J^*(x).
\end{aligned}$$

The last inequality follows from the fact that J^* satisfies the Bellman equation, thus $J^*(x) \leq g(x, a) + \alpha \mathsf{E}[J^*(x_{t+1}) | x_t = x, a]$ for all $a \in \mathcal{A}$ and $x \in \mathcal{X}$. ■

We can succinctly state the above result as:

$$J^*(x) = \sup_{J \in \mathcal{P}} FJ(x), \tag{20.8}$$

which we refer to as the *dual problem*. Although this dual problem is typically thought of as an optimization over an appropriate space of martingales, our exposition suggests that as an alternative, we may think of the dual problem as optimizing over the space of cost-to-go functions. This view will be crucial in unifying the ALP and martingale duality approaches. The optimization over the space $\mathcal{P}$ will be referred to as the *outer problem* to distinguish it from the inner problem, which is a deterministic minimization problem embedded inside the F operator.

The dual problem is challenging for various reasons. In particular, optimizing over $\mathcal{P}$ is nontrivial when the state space is high-dimensional. It has nevertheless inspired heuristic methods for computing lower bounds. Given a cost-to-go function approximation J, we can use Monte Carlo simulation to estimate $FJ(x)$ and this serves as a lower bound on $J^*(x)$. The approximation J itself could be the product of an ADP method. Alternatively, it could be obtained by simplifying the original problem with the goal of being able to compute a surrogate to the cost-to-go function. These approaches have been successfully applied in a wide variety of settings. In the context of American option pricing, for example, Andersen and Broadie [24] used regression based approaches to obtain a cost-to-go function approximation, which can then be used to construct martingale penalties which yield remarkably tight bounds. Beyond American option pricing problem, such approaches have been used in portfolio optimization [27] and the valuation of natural gas storage [28], among other applications.

20.5 THE PATHWISE OPTIMIZATION METHOD

Observe that the dual problem entails optimization over a very high-dimensional space (namely, $\mathcal{P} \triangleq \mathbb{R}^{|\mathcal{X}|}$). This is reminiscent of the challenge with the exact linear program. Analogous to our derivation of the ALP then, we are led to restrict the optimization problem to a lower dimensional subspace. In particular, given a set of basis functions, Φ, define $\hat{\mathcal{P}} \triangleq \{\Phi r \; : \; r \in \mathbb{R}^K\} \subset \mathcal{P}$. We consider finding a good approximation to the cost-to-go function of the form FJ, with $J \in \hat{\mathcal{P}}$ restricted to the subspace spanned by the basis. To accomplish this, given a state-relevance vector $\nu \in \mathbb{R}^{|\mathcal{X}|}$ with positive components, we define the *pathwise optimization* (PO) problem by

$$\sup_r \; \nu^\top F\Phi r \triangleq \sup_r \sum_{x \in \mathcal{X}} \nu(x) F\Phi r(x). \tag{20.9}$$

Several remarks are in order. Observe that from Theorem 20.2, for any r, $F\Phi r(x) \leq J^*(x)$ for all states x. Therefore, the PO program (20.9) is equivalent to

$$\inf_r \; \|J^* - F\Phi r\|_{1,\nu}.$$

Thus, the PO program will seek to find $\Phi r \in \hat{\mathcal{P}}$, so that the resulting lower bound $F\Phi r(x)$ will be close to the true optimal cost-to-go $J^*(x)$, measured on average across states x according to the state-relevance weight ν.

Similar to the ALP, if J^* is in the span of Φ, it is clear that the optimal solution to the above problem will yield the optimal cost-to-go function J^*. In addition, as the following theorem establishes, the PO problem is a *convex* optimization problem[2] over a low-dimensional space.

Theorem 20.3 *The function $r \mapsto \nu^\top F\Phi r$ is concave in $r \in \mathbb{R}^K$.*

Proof: Observe that, as a function of r, $\nu^\top F\Phi r$ is a nonnegative linear combination of a set of pointwise infima of affine functions of r, and hence must be concave in r as each of these operations preserves concavity. ■

The PO problem puts the martingale duality and ALP approaches on a common footing: both approaches can now be seen to require a set of basis function Φ whose span ideally contains a good approximation to the optimal cost-to-go function. Given such a set of basis functions, both approaches require the solution of a convex optimization problem over a low-dimensional space of weight vectors $r \in \mathbb{R}^K$: (20.4) for the ALP, and (20.9) for the pathwise approach. Given an optimal solution r, both methods can produce a lower bound on the optimal cost-to-go $J^*(x)$ at an arbitrary state x: $\Phi r(x)$ for the ALP, and $F\Phi r(x)$ for the pathwise approach. The natural question one might then ask is: how do these approaches relate to each other in terms of the lower bounds they produce? We answer this question next:

Theorem 20.4 *Let r be any feasible solution to the ALP, that is, r satisfies $\Phi r \leq T\Phi r$. Then, for all $x \in \mathcal{X}$,*

$$\Phi r(x) \leq F\Phi r(x) \leq J^*(x).$$

Proof: Using the weak duality result in Theorem 20.2,

$$\begin{aligned} J^*(x) \geq F\Phi r(x) &= \mathsf{E}\left[\inf_{\mathbf{a} \in \mathcal{A}^\infty} \sum_{t=0}^{\infty} \alpha^t \left(g(x_t, a_t) - \alpha \Delta \Phi r(x_{t+1}, x_t, a_t) \right) \,\middle|\, x_0 = x \right] \\ &= \mathsf{E}\left[\inf_{\mathbf{a} \in \mathcal{A}^\infty} \sum_{t=0}^{\infty} \alpha^t \left(g(x_t, a_t) + \alpha \mathsf{E}\left[\Phi r(x_{t+1}) | x_t, a_t \right] - \Phi r(x_t) \right) \right. \\ &\qquad \left. + \, \Phi r(x_0) \,\middle|\, x_0 = x \right] \\ &\geq \Phi r(x), \end{aligned}$$

where the final inequality follows since r is feasible for the ALP. ■

Theorem 20.5 establishes a strong relationship between the lower bounds arising from the ALP and PO methods. For *any* feasible candidate weight vector r, the

[2]Here, we refer to an optimization problem as convex if it involves the minimization of a convex function over a convex feasible set, or, equivalently, the maximization of a concave function over a convex feasible set.

corresponding ALP lower bound $\Phi r(x)$ is dominated by the PO lower bound $F\Phi r(x)$, at *every* state x. Since the PO program (20.9) further considers a large set of feasible r, it immediately follows that the optimal solution of the PO program will provide an lower bound that is, in an appropriately weighted sense, tighter than that of the ALP method. The fact that the PO method provably dominates the ALP method is the content of the following theorem.

Theorem 20.5 *Suppose that* r_{PO} *is an optimal solution to the PO program* (20.9), *while* r_{ALP} *is an optimal solution to the ALP* (20.4). *Then,*

$$\|J^* - F\Phi r_{\mathsf{PO}}\|_{1,\nu} \leq \|J^* - \Phi r_{\mathsf{ALP}}\|_{1,\nu}.$$

Proof: Note that, using Theorems 20.2 and 20.5,

$$\begin{aligned}\|J^* - F\Phi r_{\mathsf{PO}}\|_{1,\nu} &= \nu^\top J^* - \nu^\top F\Phi r_{\mathsf{PO}} \leq \nu^\top J^* - \nu^\top F\Phi r_{\mathsf{ALP}} \\ &\leq \nu^\top J^* - \nu^\top \Phi r_{\mathsf{ALP}} = \|J^* - \Phi r_{\mathsf{ALP}}\|_{1,\nu}.\end{aligned}$$ ■

20.6 APPLICATIONS

The results of the prior section establish that the PO method is a convex optimization problem over a low-dimensional space that delivers provably stronger bounds than the ALP approach. However, challenges remain in implementing the PO method. The PO objective in (20.9) is the expectation of a complicated random variable, namely, the objective value of the inner optimization problem. We can use a sample average approximation to estimate the outer expectation. However, for each sample path, the inner optimization problem will correspond to a potentially high dimensional *deterministic* dynamic program. This program may be no easier to solve than the original *stochastic* dynamic program. In particular, for example, solution of the deterministic problem via exact dynamic programming would be subject to the same curse-of-dimensionality as the stochastic problem. Hence, we expect that solving the PO problem in a tractable fashion is likely to call for additional problem structure. In this section, we present two broad classes of problems whose structure admits a tractable PO problem.

Our discussion thus far has focused on the infinite horizon, discounted case. We chose to do so for two reasons: simplicity on the one hand, and the fact that in such a setting, results such as Theorem 20.5 demonstrate that the approximations produced by the PO method inherit approximation guarantees established for the ALP in the discounted, infinite horizon setting. In what follows we will consider two concrete classes of problems that are more naturally studied in a *finite* horizon setting. As it turns out, the PO problem has a natural analog in such a setting and the following examples will serve to illustrate this analog in addition to specifying broad classes of problems where the PO approach is tractable.

20.6.1 Optimal Stopping

Optimal stopping problems are a fundamental class of stochastic control problems. The problem of valuing American options is among the more significant examples of such a control problem. It is most natural to consider dealing with the finite horizon case here. As such, time becomes a relevant state variable and the PO method as stated earlier needs to be adapted. Further, our problem formulation and development of the ALP and PO method, so far, has been couched in a discounted infinite horizon setting where one seeks to minimize cost. However, these methods are equally applicable to the finite horizon case where we seek to maximize reward. Motivated by the application of option pricing, we will consider this latter setting in the context of optimal stopping.

In particular, consider a discounted problem over the finite horizon $\mathcal{T} \triangleq \{0, 1, \ldots, T\}$. The state evolves as a Markov process, so that

$$x_{t+1} = h(x_t, w_t),$$

where w_t is an i.i.d. disturbance. The action at each time step is either to stop or to continue and thus $\mathcal{A} \triangleq \{\mathsf{STOP}, \mathsf{CONTINUE}\}$. On choosing to stop at time t in state x_t, the discounted reward is $\alpha^t g(x_t)$, where α is the discount factor. An exercise policy $\mu \triangleq \{\mu_t,\ t \in \mathcal{T}\}$, is a sequence of functions where each $\mu_t \colon \mathcal{X} \to \{\mathsf{STOP}, \mathsf{CONTINUE}\}$ specifies the stopping decision at time t, as a function of state x_t. We require that stopping occur at some time in $\mathcal{T}$, and our goal is to obtain an exercise policy that maximizes the expected discounted reward.

In principle, J^* may be computed via the following dynamic programming backward recursion

$$J_t^*(x) \triangleq \begin{cases} \max\ \left\{g(x),\ \alpha \mathsf{E}\left[J_{t+1}^*(x_{t+1}) \mid x_t = x\right]\right\} & \text{if } t < T, \\ g(x) & \text{if } t = T, \end{cases} \tag{20.10}$$

for all $x \in \mathcal{X}$ and $t \in \mathcal{T}$, The corresponding optimal stopping policy μ^* that acts "greedily" with respect to J^* is given by

$$\mu_t^*(x) \triangleq \begin{cases} \mathsf{CONTINUE} & \text{if } t < T \text{ and } g(x) < \alpha \mathsf{E}[J_{t+1}^*(x_{t+1}) \mid x_t = x], \\ \mathsf{STOP} & \text{otherwise.} \end{cases} \tag{20.11}$$

20.6.1.1 *The Martingale Duality Approach* Let $\mathcal{S}$ be the space of real-valued functions defined on the state space $\mathcal{X}$, that is, functions of the form $V \colon \mathcal{X} \to \mathbb{R}$. Define $\mathcal{P}$ to be the set of functions $J \colon \mathcal{X} \times \mathcal{T} \to \mathbb{R}$ of state and time, and, for notational convenience, denote $J_t \triangleq J(\cdot, t)$. One can think of $\mathcal{P}$ as the space of value functions. We begin by defining the *martingale difference operator* Δ. The operator Δ maps a function $V \in \mathcal{S}$ to the function $\Delta V \colon \mathcal{X} \times \mathcal{X} \to \mathbb{R}$ according to

$$(\Delta V)(x_{t+1}, x_t) \triangleq V(x_{t+1}) - \mathsf{E}[V(x_{t+1})|x_t].$$

Given an arbitrary function $J \in \mathcal{P}$, and a time $\tau \in \mathcal{T}$, define the process

$$M_t^{(\tau)}(J) \triangleq \sum_{s=1}^{t\wedge\tau} \alpha^s (\Delta J_s)(x_s, x_{s-1}), \quad \forall\, t \in \mathcal{T}. \tag{20.12}$$

Then, $M^{(\tau)}$ is a martingale adapted to the filtration $\mathcal{F}$. Next, we define the *martingale duality operator* $F: \mathcal{P} \to \mathcal{S}$ according to:

$$(FJ)(x) \triangleq \mathsf{E}\left[\max_{t\in\mathcal{T}} \alpha^t g(x_t) - M_T^{(t)}(J) \,\middle|\, x_0 = x \right]. \tag{20.13}$$

Observe that the martingale penalty (20.12) is a natural analog of the penalty (20.6) introduced earlier. In the stopping problem, the sequence of actions simply corresponds to a choice of time $t \in \mathcal{T}$ at which to stop. Beyond that time, the optimal value function will take the value zero. Hence, when constructing a martingale penalty according to an optimal value function surrogate, it is not necessary to consider times after the stopping time. With these observations, it is clear that the penalty (20.6) simplifies to the penalty (20.12) for a stopping problem, and hence the operator (20.13) is a natural generalization of the operator (20.7).

For any given $J \in \mathcal{P}$ and a state $x_0 \in \mathcal{X}$, an analog to Theorem 20.2 establishes that $FJ(x_0)$ provides an upper bound on the optimal value $J_0^*(x_0)$. With the intention of optimizing the bound $FJ(x_0)$ over a parameterized subspace $\hat{\mathcal{P}} \subset \mathcal{P}$, we introduce the collection of K basis functions

$$\Phi \triangleq \{\phi_1, \phi_2, \ldots, \phi_K\} \subset \mathcal{P}.$$

Each vector $r \in \mathbb{R}^K$ determines a value function approximation of the form

$$(\Phi r)_t(x) \triangleq \sum_{\ell=1}^{K} \phi_\ell(x, t) r_\ell, \quad \forall\, x \in \mathcal{X},\ t \in \mathcal{T}.$$

Thus, the PO problem of finding the tightest upper bound of the form $F\Phi r(x_0)$ can be defined as

$$\inf_r \; F\Phi r(x_0). \tag{20.14}$$

The problem (20.14) is an unconstrained convex optimization problem over a low-dimensional space. However, the challenge is that the objective involves expectation over an inner optimization problem. Further, the inner optimization problem, in its use of the Δ operator, implicitly relies on the ability to take one-step conditional expectation of the basis functions. We approximate these expectations by sample averages.

In particular, consider sampling a set of S *outer* sample paths denoted by $x^{(i)} \triangleq \{x_s^{(i)},\ s \in \mathcal{T}\}$ for $i = 1, 2, \dots, S$, each sampled independently, conditional on the initial state x_0. Along each of these sample paths, we approximate the Δ operator by generating one-step *inner* samples. In particular, for each time $p \in \{1, \dots, T\}$, we generate I independent *inner* samples $\{x_p^{(i,j)},\ j = 1, \dots, I\}$, conditional on $x_{p-1} = x_{p-1}^{(i)}$, resulting in the approximation

$$\hat{\Delta}(\Phi r)_p\big(x_p^{(i)}, x_{p-1}^{(i)}\big) \triangleq (\Phi r)_p\big(x_p^{(i)}\big) - \frac{1}{I}\sum_{j=1}^{I}(\Phi r)_p\big(x_p^{(i,j)}\big). \tag{20.15}$$

Having thus replaced the expectations by their empirical counterparts, we obtain the following *nested* Monte Carlo approximation to the objective:

$$\hat{F}^{S,I}\Phi r(x_0) \triangleq \frac{1}{S}\sum_{i=1}^{S}\max_{0\le s\le d}\left(\alpha^s g\big(x_s^{(i)}\big) - \sum_{p=1}^{s}\alpha^p\hat{\Delta}(\Phi r)_p\big(x_p^{(i)}, x_{p-1}^{(i)}\big)\right). \tag{20.16}$$

Consequently, the sampled variant of PO is given by

$$\inf_r\ \hat{F}^{S,I}\Phi r(x_0),$$

which is equivalent to the following linear program

$$\begin{aligned}
\underset{r,u}{\text{minimize}}\quad & \frac{1}{S}\sum_{i=1}^{S}u_i\\
\text{subject to}\quad & u_i + \sum_{p=1}^{s}\alpha^p\hat{\Delta}(\Phi r)_p\big(x_p^{(i)}\big) \ge \alpha^s g\big(x_s^{(i)}\big), \quad \forall\, 1 \le i \le S, \quad 0 \le s \le d,\\
& r \in \mathbb{R}^K,\ u \in \mathbb{R}^S.
\end{aligned} \tag{20.17}$$

Desai et al. [10] establish the convergence of this sampled LP, as the number of samples (S, I) tend to infinity.

The linear program (20.17) has $K + S$ variables and $S(d + 1)$ constraints. Since the u_i variables appear only "locally," the Hessian corresponding to the logarithmic barrier function can be inverted in $O(K^2S)$ floating point operations (see, e.g., [32]). Therefore, one may argue that the complexity of solving this LP via an interior point method essentially scales linearly with the number of outer sample paths S.

The PO method is a specific instance of a method that uses value function approximations to compute the martingale penalty. Further, the method can be shown to enjoy strong approximation guarantees. The quality of the upper bound produced by the PO method depends on three parameters: the error due to the *best possible* approximation afforded by the chosen basis function architecture, the square root of the effective time horizon, and a certain measure of the "predictability" of the underlying Markov process. The latter parameter provides valuable insight on aspects of the underlying Markov process that make a particular pricing problem easy or hard. This result, described in Desai et al. [10], also makes precise the intuition that the PO method produces good price approximations if the chosen basis function architecture contains a good approximation to the value function.

20.6.2 Linear Convex Control

In this section, we consider yet another class of MDPs, which we refer to as *linear convex control problems*. These problems essentially call for the minimization of some convex cost function of the state trajectory subject to linear dynamics and, potentially, convex constraints on the control sequence. A number of interesting problems ranging from inventory control to portfolio optimization to network revenue management can be addressed using this framework.

Consider an MDP over the finite time horizon $\mathcal{T} \triangleq \{0, 1, \ldots, T\}$. For the purpose of this section, we assume that the state space $\mathcal{X} \triangleq \mathbb{R}^m$, the action space $\mathcal{A} \triangleq \mathbb{R}^n$, and the disturbance space $\mathcal{W} \triangleq \mathbb{R}^m$. The cost of taking some action a in state x at time t is given by a function $g_t \colon \mathcal{X} \times \mathcal{A} \to \mathbb{R}$ that is assumed jointly convex in its arguments. Further, the dynamics governing the evolution of x_t are assumed to be linear:

$$x_{t+1} = h(x_t, a_t, w_t) = A_t x_t + B_t a_t + w_t,$$

where $A_t \in \mathbb{R}^{m\times m}$ and $B_t \in \mathbb{R}^{m\times n}$ are deterministic matrices that govern the system dynamics, and $w_t \in \mathbb{R}^m$ is an i.i.d. disturbance. We allow for constraints on controls of the form $a_t \in \mathcal{K}_t$, where $\mathcal{K}_t \subset \mathbb{R}^n$ is a convex set. While we do not discuss this here, both the cost function and the nature of the constraints can be substantially relaxed: we may consider cost functions that are general convex functionals of the state and control trajectories and, under some technical conditions, can permit general convex constraints on the sequence of control actions employed; see Desai et al. [11] for further details.

Let the sequence of policies, actions, states, and disturbances be denoted by $\boldsymbol{\mu}_T \triangleq (\mu_0, \mu_1, \ldots, \mu_T)$, $\mathbf{a}_T \triangleq (a_0, a_1, \ldots, a_T)$, $\mathbf{x}_T \triangleq (x_0, x_1, \ldots, x_T)$, and $\mathbf{w}_T \triangleq (w_0, \ldots, w_{T-1})$, respectively. Define the set of feasible nonanticipative policies by

$$\mathcal{A}^{\mathbb{F}} \triangleq \{\boldsymbol{\mu}_T \ : \ \mu_t \in \mathcal{K}_t, \ \forall\, t \in \mathcal{T}, \text{ and } \boldsymbol{\mu}_T \text{ is adapted to filtration } \mathcal{F}\}.$$

We are interested in the following undiscounted, finite horizon optimization problem

$$\inf_{\boldsymbol{\mu}_T \in \mathcal{A}^{\mathbb{F}}} \mathsf{E}\left[\sum_{t=0}^{T} g_t(x_t, \mu_t)\right]. \tag{20.18}$$

Under mild technical conditions (for details, see [11]) the optimal cost-to-go function J^* satisfies the Bellman equation

$$J_t^*(x) = \begin{cases} \inf\limits_{a_t \in \mathcal{K}_t} g_t(x, a_t) + \mathsf{E}\left[J_{t+1}^*(x_{t+1}) \middle| x_t = x,\ a_t \right] & \text{if } t < T, \\ \inf\limits_{a_T \in \mathcal{K}_T} g_T(x, a_T) & \text{if } t = T. \end{cases} \tag{20.19}$$

20.6.2.1 The Martingale Duality Approach Let $\mathcal{S}$ be the space of real-valued functions defined on state space $\mathbb{R}^m$ and $\mathcal{P}$ be the space of real-valued functions on $\mathbb{R}^m \times \mathcal{T}$, such that $J_t \triangleq J(\cdot, t)$ belongs to $\mathcal{S}$. Define the *martingale difference operator* Δ that maps a function $V \in \mathcal{S}$ to the function $\Delta V \colon \mathbb{R}^m \times \mathbb{R}^m \times \mathbb{R}^n \to \mathbb{R}$ according to

$$(\Delta V)(x_{t+1}, x_t, a_t) \triangleq V(x_{t+1}) - \mathsf{E}[V(x_{t+1}) | x_t, a_t].$$

We are interested in computing lower bounds by considering a perfect information relaxation. Define $\mathbf{K} \triangleq \mathcal{K}_0 \times \mathcal{K}_1 \times \cdots \times \mathcal{K}_T$ to be the set of all feasible control sequences $\mathbf{a}_T$. Given a feasible sequence of actions $\mathbf{a}_T \in \mathbf{K}$ and a function $J \in \mathcal{P}$, define the martingale $M_t^{\mathbf{a}_T}(J)$ by

$$M_0^{\mathbf{a}_T}(J) \triangleq 0, \qquad M_t^{\mathbf{a}_T}(J) \triangleq \sum_{s=1}^{t} \Delta J_s(x_s, x_{s-1}, a_{s-1}), \quad \forall\, 1 \le t \le T.$$

Then, we can define the *martingale duality operator* $F \colon \mathcal{P} \to \mathcal{S}$ according to:

$$(FJ)(x) \triangleq \mathsf{E}\left[\inf_{\mathbf{a}_T \in \mathbf{K}} \sum_{t=0}^{T} g_t(x_t, a_t) - M_T^{\mathbf{a}_T}(J) \,\middle|\, x_0 = x \right]. \tag{20.20}$$

In order for the deterministic inner optimization problem in (20.20) to be tractable, we need to impose special structure on the function J. To this end, given a sequence of matrices $\mathbf{\Gamma} \triangleq (\Gamma_1, \ldots, \Gamma_T)$, define the function $J^{\mathbf{\Gamma}} \in \mathcal{P}$ by

$$J_0^{\mathbf{\Gamma}}(x) \triangleq 0, \qquad J_t^{\mathbf{\Gamma}}(x) \triangleq x^\top \Gamma_t x, \quad \forall\, 1 \le t \le T.$$

Denote by $\mathcal{C} \subset \mathcal{P}$ the set of all functions of the form $J^{\mathbf{\Gamma}}$. The following theorem establishes that, for this class of quadratic functions, the inner optimization problem in (20.20) is a convex optimization problem, and therefore is tractable:

Theorem 20.6 *For all $J \in \mathcal{C}$, the inner optimization problem of* (20.20) *is a convex optimization problem.*

Proof: Suppose that $J = J^{\mathbf{\Gamma}} \in \mathcal{C}$. For each time t, apply the martingale difference operator Δ to $J_t^{\mathbf{\Gamma}}$ to obtain

$$\begin{aligned}\Delta J_t(x_t, x_{t-1}, a_{t-1}) = 2w_{t-1}^{\top}\Gamma_t(A_{t-1}x_{t-1} + B_{t-1}a_{t-1}) + w_{t-1}^{\top}\Gamma_t w_{t-1} \\ - \mathsf{E}\left[w_{t-1}^{\top}\Gamma_t w_{t-1}\right].\end{aligned} \tag{20.21}$$

Observe that the quantity $w_{t-1}^{\top}\Gamma_t w_{t-1} - \mathsf{E}\left[w_{t-1}^{\top}\Gamma_t w_{t-1}\right]$ is zero mean and independent of the control $\mathbf{a}_T$. We may consequently eliminate those terms from the inner optimization problem. In particular, given a fixed sequence of disturbances $\mathbf{w}_T$, the inner optimization problem becomes:

$$\begin{aligned}\underset{\mathbf{a}_T, \mathbf{x}_T}{\text{minimize}} \quad & g_0(x_0, u_0) + \sum_{t=1}^{T}\left\{g_t(x_t, a_t) - 2w_{t-1}^{\top}\Gamma_t(A_{t-1}x_{t-1} + B_{t-1}a_{t-1})\right\} \\ \text{subject to} \quad & x_{t+1} = A_t x_t + B_t a_t + w_t, \quad \forall\, 0 \le t \le T-1, \\ & a_t \in \mathcal{K}_t, \quad \forall\, 0 \le t \le T.\end{aligned} \tag{20.22}$$

This is clearly a convex optimization problem. ■

Theorem 20.6 suggested that for a quadratic[3] cost-to-go function surrogate $J^{\mathbf{\Gamma}} \in \mathcal{C}$, the lower bound $FJ^{\mathbf{\Gamma}}(x)$ on the optimal cost-to-go $J_0^*(x)$ can be efficiently computed. Finding the *tightest* such lower bound suggests the optimization problem

$$\sup_{\mathbf{\Gamma}}\ FJ^{\mathbf{\Gamma}}(x). \tag{20.23}$$

We now establish that this is also a convex optimization problem:

Theorem 20.7 *$FJ^{\mathbf{\Gamma}}(x)$ is concave in $\mathbf{\Gamma}$.*

Proof: Using the definition of the F operator given by (20.20) and the expression for $\Delta J(x_t, x_{t-1}, a_{t-1})$ given by (20.21), we obtain

$$\begin{aligned}FJ^{\mathbf{\Gamma}}(x) = \mathsf{E}\Bigg[\inf_{a_T \in \mathbf{K}}\ & g_0(x_0, a_0) \\ & + \sum_{t=1}^{T}\left\{g_t(x_t, a_t) - 2w_{t-1}^{\top}\Gamma_t(A_{t-1}x_{t-1} + B_{t-1}a_{t-1})\right\} \Bigg|\, x_0 = x\Bigg].\end{aligned}$$

[3] In fact, a broader class of cost-to-go functions including constant and linear terms could also be considered. However, such constant and linear terms are eliminated in the evaluation of the martingale difference operator in (20.21). Hence, they do not enter into the lower bound and can be ignored.

Observe that $FJ^{\mathbf{\Gamma}}(x)$ is given by nonnegative linear combinations of infima of affine functions of $\mathbf{\Gamma}$. Since each of these operations preserves concavity, we obtain the desired result. ■

The PO problem given by (20.23) can be viewed as a stochastic optimization problem. This suggests two methods of solution:

- Iterative methods based on stochastic gradient descent can be used to solve (20.23). Starting from an initial guess for $\mathbf{\Gamma}$, the gradient of $FJ^{\mathbf{\Gamma}}(x)$ can be estimated along a single sample path $\mathbf{w}_T$ of random disturbances. The stochastic gradient estimate is then used to update the choice of $\mathbf{\Gamma}$, and the procedure is repeated until convergence. These steps together give rise to a simple online method that can be used to handle large problems with a low memory requirement.
- Alternatively, a sample average approximation can be used. Here, the objective function $FJ^{\mathbf{\Gamma}}(x)$ is approximated with a sample average over sequences of random disturbances $\mathbf{w}_T$. For a given realization of this sequence, the objective cost-to-go of the inner optimization problem, (20.22) can be expressed (using the appropriate conjugate functions) as a convex function of Γ. In several special cases this representation allows us to rewrite the overall optimization problem in a form suitable for direct optimization.

The details of both these approaches, along with application to a high-dimensional financial application, namely, an optimal execution problem, can be found in Desai et al. [11].

Observe that the classic convex *linear quadratic control* (LQC) problem is an example of a linear convex problem. It is well known that the optimal cost-to-go function for the convex LQC problem takes a positive semi definite quadratic form and can be computed recursively (and efficiently) by solving the so-called Ricatti equation (see, e.g., [33]). This tractability breaks down under seemingly innocuous constraints such as requiring nonnegative control actions. Loosely speaking, the PO method bootstraps our ability to solve convex LQC problems to the task of producing good approximations to linear convex problems. It does so by seeking martingale penalty functions derived from quadratic approximations to the cost-to-go function. In particular, if convex quadratic forms are likely to provide a reasonable approximation to the cost-to-go function of the linear convex problem at hand, then one can expect the PO method to produce good lower bounds.

20.7 CONCLUSION

This chapter set out with the task of producing lower bounds on the optimal cost-to-go for high-dimensional Markov decision problems. We considered two seemingly disparate approaches to this task: the approximate linear programming (ALP) methodology and an approach based on finding martingale "penalties" in a certain dual problem. In concluding, we observe that these two methodologies are intimately connected:

1. We have observed that given an approximation architecture for the ALP approach, one is naturally led to consider a corresponding family of martingale penalties derived from the same architecture. This consideration suggests an optimization problem that produces a martingale penalty yielding the tightest lower bound possible within the corresponding family of martingale penalties. We referred to this problem as the pathwise optimization (PO) problem.
2. We established that solving the PO problem yields approximations to the cost-to-go that are no worse than those produced by the ALP approach. This provided an elegant unification of the two approaches.
3. Finally, we demonstrated the algorithmic value of the PO method in the context of two broad classes of MDPs.

Moving forward, we believe that much remains to be done in developing the pathwise optimization approach described in this chapter. In particular, developing the approach successfully for a given class of problems requires that we first identify a suitable approximation architecture for that class of problems. This architecture should admit tractable PO problems and simultaneously be rich enough that it captures essential features of the true cost-to-go function. A number of problems from areas such as financial engineering, revenue management, and inventory management are ripe for precisely this sort of study.

On an orthogonal note, while we have not studied this issue here, much remains to be done in using the solution of the PO problem to generate good heuristic *policies*. Desai et al. [10] discuss this in the context of optimal stopping, and demonstrate in numerical examples that PO-derived policies can be superior to policies derived from more conventional ADP methods. In general, some careful thought is needed here since optimal solutions to the PO problem are not unique. For example, in linear convex setting, the optimal solutions are only identified up to affine translations.

REFERENCES

1. B. Van Roy. Neuro-dynamic programming: overview and recent trends. In A. Shwartz and E. Feinberg, editors, *Handbook of Markov Decision Processes*. Kluwer, Boston, 2002.
2. D.P. Bertsekas. *Dynamic Programming and Optimal Control*, 3rd edition, Volume 2. Athena Scientific, Belmont, MA, 3rd edition, 2007.
3. P. Schweitzer and A. Seidmann. Generalized polynomial approximations in Markovian decision processes. *Journal of Mathematical Analysis and Applications*, 110:568–582, 1985.
4. D.P. de Farias and B. Van Roy. The linear programming approach to approximate dynamic programming. *Operations Research*, 51(6):850–865, 2003.

5. D.P. de Farias and B. Van Roy. On constraint sampling in the linear programming approach to approximate dynamic programming. *Mathematics of Operations Research*, 293(3):462–478, 2004.
6. L.C.G. Rogers. Monte Carlo valuation of American options. *Mathematical Finance*, 12(3):271–286, 2002.
7. M.B. Haugh and L. Kogan. Pricing American options: a duality approach. *Operations Research*, 52(2):258–270, 2004.
8. L.C.G. Rogers. Pathwise stochastic optimal control. *SIAM Journal on Control and Optimization*, 46(3):1116–1132, 2008.
9. D.B. Brown, J.E. Smith, and P. Sun. Information relaxations and duality in stochastic dynamic programs. *Operations Research*, 58(4):785–801, 2010.
10. V.V. Desai, V.F. Farias, and C.C. Moallemi. Pathwise optimization for optimal stopping problems. Management Science, 2012.
11. V.V. Desai, V.F. Farias, and C.C. Moallemi. Pathwise optimization for linear convex systems. *Working Paper*, 2011.
12. D.P. Bertsekas and J.N. Tsitsiklis. *Neuro-Dynamic Programming*. Athena Scientific, Belmont, MA, 1996.
13. W.B. Powell. *Approximate Dynamic Programming: Solving the Curses of Dimensionality*. John Wiley and Sons, 2007.
14. J.R. Morrison and P.R. Kumar. New linear program performance bounds for queueing networks. *Journal of Optimization Theory and Applications*, 100(3):575–597, 1999.
15. M.H. Veatch. Approximate dynamic programming for networks: fluid models and constraint reduction. Working Paper, 2005.
16. C.C. Moallemi, S. Kumar, and B. Van Roy. Approximate and data-driven dynamic programming for queueing networks. Working Paper, 2008.
17. D. Adelman. Dynamic bid prices in revenue management. *Operations Research*, 55(4):647–661, 2007.
18. V.F. Farias and B. Van Roy. An approximate dynamic programming approach to network revenue management. Working Paper, 2007.
19. D. Zhang and D. Adelman. An approximate dynamic programming approach to network revenue management with customer choice. Working Paper, 2008.
20. J. Han. Dynamic portfolio management—an approximate linear programming approach. Ph.D. thesis, Stanford University, 2005.
21. D. Adelman. A price-directed approach to stochastic inventory/routing. *Operations Research*, 52(4):499–514, 2004.
22. D. Adelman and D. Klabjan. Computing near optimal policies in generalized joint replenishment. Working Paper, January 2009.
23. V.F. Farias, D. Saure, and G.Y. Weintraub. An approximate dynamic programming approach to solving dynamic oligopoly models. Working Paper, 2011.
24. L. Andersen and M. Broadie. Primal-dual simulation algorithm for pricing multidimensional American options. *Management Science*, 50(9):1222–1234, 2004.
25. F. Jamshidian. Minimax optimality of Bermudan and American claims and their Monte-Carlo upper bound approximation. Technical Report, NIB Capitial, The Hague, 2003.
26. N. Chen and P. Glasserman. Additive and multiplicative duals for American option pricing. *Finance and Stochastics*, 11(2):153–179, 2007.

27. D.B. Brown and J.E. Smith. Dynamic portfolio optimization with transaction costs: heuristics and dual bounds. *Management Science*, 2011.

28. G. Lai, F. Margot, and N. Secomandi. An approximate dynamic programming approach to benchmark practice-based heuristics for natural gas storage valuation. *Operations Research*, 58(3):564–582, 2010a.

29. G. Lai, M.X. Wang, S. Kekre, A. Scheller-Wolf, and N. Secomandi. Valuation of storage at a liquefied natural gas terminal. *Operations Research*, 2011.

30. A.S. Manne. Linear programming and sequential decisions. *Management Science*, 60(3):259–267, 1960.

31. D.P. Bertsekas. *Dynamic Programming and Optimal Control*, 3rd edition, Volume 2. Athena Scientific, Belmont, MA, 2006.

32. S. Boyd and L. Vandenberghe. *Convex Optimization*. Cambridge University Press, Cambridge, UK, 2004.

33. D.P. Bertsekas. *Dynamic Programming and Optimal Control*. Athena Scientific, Belmont, MA, 1995.

CHAPTER 21

Approximate Dynamic Programming and Backpropagation on Timescales

JOHN SEIFFERTT and DONALD WUNSCH

Department of Electrical and Computer Engineering, Missouri University of Science and Technology, Rolla, MO, USA

ABSTRACT

Timescales calculus allows signals to have both continuous-time and discrete-time properties. It has become an emerging key topic due to many multidisciplinary applications. This calculus is extended to approximate dynamic programming (ADP) and backpropagation, the most fundamental tools for material covered in this book. The core backwards induction algorithm of dynamic programming is extended from its traditional discrete case to all isolated timescales. Hamilton–Jacobi–Bellman (HJB) equations, the solution of which is the fundamental problem in the field of dynamic programming, are motivated and proven on timescales. Additionally, the chain rule for ordered derivatives and the backpropagation learning algorithm are established.

This chapter contains results first presented in Refs [1–3].

21.1 INTRODUCTION: TIMESCALES FUNDAMENTALS

The timescales calculus is concerned with functions defined on domains, which can be a mixture of the continuous and discrete. This section presents enough of the theory of such *dynamic equations* to support the results in this chapter. In addition to the established mathematics, new contributions are presented: the definition of σ_n-complete differentiability and a chain rule for n-variables. The best reference work on this material is currently [4].

Dynamic equations are defined on mathematical structures called *timescales*. Formally, a timescale (T, α) is an ordered pair such that $\mathbb{T} \subset \mathbb{R}$ is nonempty,

Reinforcement Learning and Approximate Dynamic Programming for Feedback Control, First Edition. Edited by Frank L. Lewis and Derong Liu.

every Cauchy sequence in $\mathbb{T}$ converges to either a point within $\mathbb{T}$ or to a finite infimum or supremum of $\mathbb{T}$, and α is a function from $\mathbb{T}$ into $\mathbb{T}$. This function α, along with the nature of the timescale under consideration, will dictate the properties of the chosen dynamic derivative and the resulting dynamic equations.

Sets which qualify as timescales include the integers $\mathbb{Z}$, the natural numbers $\mathbb{N}$, the scaled integers $h\mathbb{Z}$ (for $h>0$), the quantum calculus timescale $q^{\mathbb{N}_0}$ for $q>1$, limiting sets such as $\{0\} \cup \left\{\frac{1}{n} : n = 1, 2, \ldots\right\}$, finite unions of intervals such as $[0, 1] \cup [2, 3] \cup [5, 6]$, and unions of intervals and discrete points such as $\{1, 2, 3, 4, 5\} \cup [5, 8]$. More exotic sets such as the Cantor set are also technically timescales. Furthermore, well-known sets such as the rationals $\mathbb{Q}$, the irrationals $\mathbb{R}/\mathbb{Q}$, and open intervals such as $(0, 1)$ are *not* timescales.

Four characteristic functions are defined on timescales: two of which scroll through the timescale and two of which detail the degree of continuity of the timescale. The *forwards* and *backwards jump operators* $\sigma(t)$ and $\rho(t)$ give us the "next" and "previous" elements, and are defined as follows:

$$\begin{aligned} \sigma(t) &= \inf\{x \in \mathbb{T} : x>t\}, \\ \rho(t) &= \sup\{x \in \mathbb{T} : x<t\}. \end{aligned} \tag{21.1}$$

For $\mathbb{T} = \mathbb{R}$, $\sigma(t) = \rho(t) = t$ and for $\mathbb{T} = \mathbb{Z}$, $\sigma(t) = t + 1$, and $\rho(t) = t - 1$. More complicated timescales admit more involved jump operators; for example, the q-timescale studied in the quantum calculus: $\mathbb{T} = q^{\mathbb{N}_0}$(where $q>1$). The forward and backward jump operators are given by $\sigma(t) = q^{t+1}$ and $\rho(t) = q^{t-1}$, respectively. Also, it is standard to denote $f(\sigma(t))$ as $f^{\sigma}(t)$ and $f(\rho(t))$ as $f^{\rho}(t)$.

These jump operators are a critical determinant of the character of a dynamic equation. For example, to work with the commonly used delta derivative, consider $\mathbb{T}$ to be any nonempty closed subset of the real line and set α equal to the forward jump operator σ.

The remaining two characteristic functions are referred to as the *forward and backwards* step functions μ and ν. These functions trace the level of continuity of a timescale and are defined as follows:

$$\begin{aligned} \mu(t) &= \sigma(t) - t, \\ \nu(t) &= t - \rho(t). \end{aligned} \tag{21.2}$$

For continuous domains, the step functions are $\mu(t) = \nu(t) = 0$ and for the integers they are $\mu(t) = \nu(t) = 1$. For the q-timescale, $\mu(t) = q^t(q - 1)$. The forward step function $\mu(t)$ is often called the *graininess function*.

21.1.1 Single-Variable Calculus

The usual derivative of the timescales calculus is defined as follows. Let $f : \mathbb{T} \to \mathbb{R}$ be a function. Then the delta derivative $f^{\Delta}(t)$ of f at a point $t \in \mathbb{T}^{\kappa}$, where $\mathbb{T}^{\kappa}$ coincides with $\mathbb{T}$ except at a left-scattered maximum, if one should exist, is defined to be the number such that given $\varepsilon>0$ there is a neighborhood U of t such that

$$\left|[f(\sigma(t)) - f(s)] - f^{\Delta}(t)[\sigma(t) - s]\right| \leq \varepsilon\,|\sigma(t) - s|. \qquad (21.3)$$

for all $s \in U$, where neighborhood is defined such that $U = (t - \delta, t + \delta)$ for some $\delta>0$. Note that this follows the classical definition of the derivative, with the traditional $x + h$ increment replaced by the forward jump operator $\sigma(t)$. When $\mathbb{T} = \mathbb{R}$, the delta derivative reduces to the traditional derivative $f'(t)$. The terms $\sigma(t) - s$ and $\rho(t) - s$ take the place of the $x - h$ construction in the classical calculus.

For all of these dynamic derivatives, many of the standard rules for derivatives obtain. However, one crucial result from classical analysis does not hold for dynamic derivatives: the chain rule. Instead of a single formula to handle differentiation of compositions of functions, dynamic equations rely on a suite of different rules, each dependent on the character of the functions. The most general of these for functions of a single variable, and the one which will be used in the formulation of the Hamilton–Jacobi–Bellman (HJB) equation and the backpropagation update equations, is due to Potzsche [5] and is stated as follows: Let $f: \mathbb{R} \to \mathbb{R}$ be continuously differentiable and let $g: \mathbb{T} \to \mathbb{R}$ be delta differentiable. Then $f(g(t))$ is delta differentiable, with $(f \circ g)^{\Delta}(t)$ given by the following integral equation:

$$(f \circ g)^{\Delta}(t) = g^{\Delta}(t) \int_0^1 f'(g(t) + h\mu(t)\, g^{\Delta}(t))\mathrm{d}h. \qquad (21.4)$$

When $\mathbb{T} = \mathbb{R}$, then $\mu(t) = 0$ and the equation reduces to our expected chain rule. Thus, the forward step function (or graininess) of the dynamic domain under consideration has a great effect on the emergent calculus. It turns out that many results of standard analysis, including the HJB equation and backpropagation equations, are dependent on the use of the traditional chain rule. It is for this reason that our extension of this chain rule is required.

21.1.2 Calculus of Multiple Variables

This section uses a definition of partial derivatives on timescales given in [6]. Let $\mathbb{T}_1, \mathbb{T}_2, \ldots, \mathbb{T}_n$ be timescales, set $\mathbb{T} = \mathbb{T}_1 \times \mathbb{T}_2 \times \cdots \times \mathbb{T}_n$, and let $f: \mathbb{T} \to \mathbb{R}$ be a function. Define the operators on $\mathbb{T}$ as $\sigma(t) = (\sigma(t_1), \sigma(t_2), \ldots, \sigma(t_n))$ and $\rho(t) = (\rho(t_1), \rho(t_2), \ldots, \rho(t_n))$. Also define $\mathbb{T}^{\kappa} = \mathbb{T}_1^{\kappa} \times \mathbb{T}_2^{\kappa} \times \cdots \times \mathbb{T}_n^{\kappa}$, $f^{\sigma_i(t)} = f(t_1, t_2, \ldots, t_{i-1}, \sigma_i(t_i), t_i, \ldots, t_n)$, $f^{\rho_i(t)} = f(t_1, t_2, \ldots, t_{i-1}, \rho_i(t_i), t_i, \ldots, t_n)$, and $f_i^s = f(t_1, t_2, \ldots, t_{i-1}, s, t_i, \ldots, t_n)$.

The partial delta derivative of f at t with respect to t_i is the number f^{Δ_i}, provided it exists, such that given any $\varepsilon>0$ there exists a neighborhood U of t_i for $\delta>0$ such that

$$\left|[f^{\sigma_i}(t) - f_i^s(t)] - f^{\Delta_i}(t)[\sigma_i(t) - s]\right| \leq \varepsilon\,|\sigma_i(t) - s|, \qquad (21.5)$$

for all $s \in U$, where neighborhood is defined such that $U = (t_i - \delta, t_i + \delta) \cap \mathbb{T}_i$.

Necessary for the proof of the HJB equation on timescales will be the chain rule for partial derivatives given by Bohner and Guseinov. Let $t \in \mathbb{T}$, $x : \mathbb{T} \to \mathbb{R}$, $y : \mathbb{T} \to \mathbb{R}$, $x(\mathbb{T}) = \mathbb{T}_x$, $y(\mathbb{T}) = \mathbb{T}_y$, and $F(x(t), y(t))$. Assume $x(\sigma(t)) = \sigma_x(x(t))$ and $y(\sigma(t)) = \sigma_y(t)$. If $F = f(x(t), y(t))$ is σ_x-completely differentiable and x and y are differentiable, then

$$F^{\Delta}(t) = f^{\Delta_x}(x(t), y(t))\, x^{\Delta}(t) + f^{\Delta_y}(\sigma_x(x(t)), y(t))\, y^{\Delta}(t). \tag{21.6}$$

For more details on partial derivatives on timescales, including a complete treatment of σ_x-complete differentiability, the reader is directed to [7] and [8].

21.1.3 Extension of the Chain Rule

Extending this chain rule to the case of n variables is required. As a preamble, the important definition of σ_1- delta differentiability is discussed.

One fundamental part of the definition, which will be needed is as follows. In order for a function $f : \mathbb{T} \to \mathbb{R}$ to be σ_1-delta differentiable the following condition must hold:

$$\begin{aligned} f\left(\sigma_1\left(t_1^0\right), \sigma_2\left(t_2^0\right)\right) - f(t_1, t_2) &= A_1\left[\sigma_1\left(t_1^0\right) - t_1\right] + B\left[\sigma_2\left(t_2^0\right) - t_2\right] \\ &\quad + \gamma_1\left[\sigma_1\left(t_1^0\right) - t_1\right] + \gamma_2\left[\sigma_2\left(t_2^0\right) - t_2\right], \end{aligned} \tag{21.7}$$

where B is the number $f^{\Delta_{t_2}}\left(\sigma_1\left(t_1^0\right), t_2^0\right)$ and the $\gamma_i \to 0$ as $t \to t_0$.

Now, for the statement of the Bohner–Guseinov chain rule. Let $t \in \mathbb{T}$, $x : \mathbb{T} \to \mathbb{R}$, $y : \mathbb{T} \to \mathbb{R}$, $x(\mathbb{T}) = \mathbb{T}_x$, $y(\mathbb{T}) = \mathbb{T}_y$, and $F(x(t), y(t))$. Assume $x(\sigma(t)) = \sigma_x(x(t))$ and $y(\sigma(t)) = \sigma_y(y(t))$. If $F = f(x(t), y(t))$ is σ_1-completely differentiable and x and y are differentiable, then

$$F^{\Delta}(t) = f^{\Delta_x}(x(t), y(t))\, x^{\Delta}(t) + f^{\Delta_y}(\sigma_x(x(t)), y(t))\, y^{\Delta}(t). \tag{21.8}$$

The condition $x(\sigma(t)) = \sigma_x(x(t))$ is referred to as *forward jump commutativity* and is called upon often in the chain rules for partial derivatives on timescales.

For an extension of this theorem to the case of n variables, it is first needed to define the notion of σ_1-delta differentiability for n variables instead of just two. The condition (21.7) becomes

$$\begin{aligned} f\left(\sigma_1\left(t_1^0\right), \ldots, \sigma_n\left(t_n^0\right)\right) - f(t_1, \ldots, t_n) &= A_1\left[\sigma_1\left(t_1^0\right) - t_1\right] \\ &\quad + \sum_{i=2}^{n} B\left[\sigma_i\left(t_i^0\right) - t_i\right] + \sum_{i=1}^{n} \gamma_i\left[\sigma_1\left(t_1^0\right) - t_1\right]. \end{aligned} \tag{21.9}$$

Armed with this definition, the following theorem can be proven.

Theorem 21.1 (Chain Rule for Functions of n-variables): *Assume that* $x_i : \mathbb{T} \to \mathbb{T}_i$, $x_i\left(t^0\right) = t_i^0$, $t^0 \in \mathbb{T}$, *and* $x_i\left(\sigma\left(t^0\right)\right) = \sigma_i(x_i\left(t^0\right))$. *Let* $\left(t^0\right) = f\left(x_1\left(t^0\right), \ldots, x_n(t^0)\right)$. *Then*

$$F^{\Delta}(t_0) = f_{t_1}^{\Delta}\left(t_1^0, \ldots, t_n^0\right) x_1^{\Delta}\left(t^0\right) + \sum_{i=2}^{n} f_{t_i}^{\Delta}\left(\sigma_1\left(t_1^0\right), t_2^0, \ldots, t_n^0\right) x_i^{\Delta}(t^0). \tag{21.10}$$

Proof: We proceed by constructing the definition of $F^{\Delta}(t^0)$ and using our hypotheses:

$$F\left(\sigma\left(t^0\right)\right) - F(t) = f\left(x_1(\sigma\left(t^0\right), \ldots, x_n(\sigma\left(t^0\right)\right) - f\left(x_1(t), \ldots, x_n(t)\right). \tag{21.11}$$

Now applying the forward jump commutativity condition, the right-hand side becomes

$$f\left(\sigma_1(x_1\left(t^0\right), \ldots, \sigma_n(x_n\left(t^0\right)\right) - f\left(x_1(t), \ldots, x_n(t)\right), \tag{21.12}$$

which, upon substitution, reduces to

$$f\left(\sigma_1(x_1\left(t^0\right), \ldots, \sigma_n(x_n\left(t^0\right)\right) - f(t_1, \ldots, t_n). \tag{21.13}$$

Using the definition of σ_1-differentiability with $A_1 = f_{t_i}^{\Delta}\left(t_1^0, \ldots, t_n^0\right)$ and $B_i = f_{t_i}^{\Delta}\left(\sigma_1\left(t_1^0\right), t_2^0, \ldots, t_n^0\right)$, gives

$$f_{t_i}^{\Delta}\left(t_1^0, \ldots, t_n^0\right)\left(\sigma_1\left(t_1^0\right) - t\right) + \sum_{i=2}^{n} f_{t_i}^{\Delta}\left(\sigma_1\left(t_1^0\right), t_2^0, \ldots, t_n^0\right)\left(\sigma_i\left(t_i^0\right) - t\right) + \sum_{i=1}^{n} \gamma_i\left[\sigma_i\left(t_i^0\right) - t\right]. \tag{21.14}$$

Using the fact that $t_i^0 = x_i(t^0)$, we arrive at

$$f_{t_i}^{\Delta}\left(t_1^0, \ldots, t_n^0\right)\left(\sigma_1\left(x_1(t^0) - t\right) + \sum_{i=2}^{n} f_{t_i}^{\Delta}\left(\sigma_1\left(t_1^0\right), t_2^0, \ldots, t_n^0\right)(\sigma_i(x_i(t_0) - t) + \sum_{i=1}^{n} \gamma_i\left[\sigma_i(x_i(t^0) - t\right]. \tag{21.15}$$

Again with the forward jump commutativity, we see that

$$f_{t_i}^{\Delta}\left(t_1^0,\ldots,t_n^0\right)\left(x_1\left(\sigma_1(t^0\right)-t\right)+\sum_{i=2}^{n} f_{t_i}^{\Delta}\left(\sigma_1\left(t_1^0\right),t_2^0,\ldots,t_n^0\right)(x_i(\sigma_i(t_0)-t)$$
$$+\sum_{i=1}^{n}\gamma_i\left[x_i(\sigma_i(t^0)-t\right]. \tag{21.16}$$

Finally, dividing through by $\sigma\left(t^0\right)-t$ and taking the limit as $t_0 \to t$ yields

$$F^{\Delta}(t_0)=f_{t_1}^{\Delta}\left(t_1^0,\ldots,t_n^0\right)x_1^{\Delta}\left(t^0\right)+\sum_{i=2}^{n} f_{t_i}^{\Delta}\left(\sigma_1\left(t_1^0\right),t_2^0,\ldots,t_n^0\right)x_i^{\Delta}(t^0), \tag{21.17}$$

which is our desired result. ■

Note that, as a technical matter, it is important for the x_i's all to have the same domain $\mathbb{T}$ so that in the final step the limit can be taken on a single t. Without this restriction, the limit would have to carry across multiple t_i's and this would present a less tractable situation.

21.1.4 Induction on Timescales

A form of backwards induction exists on timescales. Let $t_0 \in \mathbb{T}$ and $S(t)$ be a statement for each $t \in [-\infty, t_0)$ such that the following four conditions hold:

1. $S(t_0)$ is true.
2. $S(t)$ being true at a left-scattered t forces $S(\rho(t))$ to be true.
3. $S(t)$ being true at a left-dense t forces $S(t')$ to be true for all t' in a left-neighborhood of t.
4. $S(t')$ being true for all $t' \in (t, t_0]$ when t is right-dense forces $S(t)$ to be true.

Then it can be concluded that $S(t)$ is true for all $t \in [t_0, \infty)$. There is also a forward version involving right-scattered and left-dense intervals and the forward jump operator $\sigma(t)$, but it is this backwards form which we use in Section 21.2.

21.2 DYNAMIC PROGRAMMING

Dynamic programming involves various methods for generating optimal solutions for multistage decision processes. The standard algorithm for dynamic programming involves a computationally intensive backwards induction update rule. In practice, many engineers working on industrial-scale applications have turned to approximation methods based on the optimal ones. Much research has been dedicated to the task of finding these suboptimal policies. The field of approximate dynamic programming (ADP) considers these approaches [9, 10].

ADP, united with the field of reinforcement learning, concerns itself with solving the HJB equation of dynamic programming. In discrete time, backwards induction is often used. For continuous time domains, the HJB equation takes on the form of a second order partial differential equation. This chapter extends the HJB equation to the timescales calculus, therefore considering both discrete and continuous domains as well as mixed domains, which are neither discrete nor continuous, and formulate it using the very general alpha derivatives.

21.2.1 Dynamic Programming Overview

The requirements of the dynamic programming framework considered herein are as follows: a timescale $\mathbb{T}$ in which our decision points lie, controls $c(x(t), t)$, a stochastic disturbance $w(t)$, states $x(t)$ which evolve according to a rule $f(x(t), c(x(t), t), w(t), t)$, and a cost/reward $r(x(t), c, w(t), t)$ where the cost at a terminal decision point T is piecewise defined as $r_T(x(T))$. A policy π is a set of state-control pairs for each point in $\mathbb{T}$ such that each control is valid for both the state and time. Denote by π^t the tail of the policy π beginning with time step t. Also introduced is a cost-to-go function J given by

$$J_\pi(x(t_0), t_0) = E\left\{\int_{t_0}^{T} r(x(\tau), c(x(\tau), \tau), w(\tau), \tau)\Delta\tau\right\}, \tag{21.18}$$

which measures the expected cost of a policy π. All values are assumed finite and well-defined.

Consider the following state-space dynamical system defined on a timescale:

$$x^\alpha(t) = f(x(t), c(x(t), t), w(t), t), \tag{21.19}$$

where $t \in T$ and $x^\alpha(t)$ indicate an interval taken using the alpha derivative. The task is to calculate a policy π which minimizes J_π. Call such a π an *optimal policy* and denote the optimal cost-to-go as $J^*(x(t_0), t_0) = \min_\pi J_\pi(x(t_0), t_0)$, where the minimum is considered over all policies.

Use Bellman's principle of optimality in the solution to the optimization problem. This principle can be framed in the following way. Let π^* be an optimal policy. Then, the optimal policy for the tail problem starting at time n, which is to minimize

$$E\left\{\int_{n}^{T} r(x(\tau), c(x(\tau), \tau), w(\tau), \tau)\Delta\tau\right\}, \tag{21.20}$$

is equal to the portion of π^*, which overlaps π^n. To justify this principle, note that if it were not true then the tail of π^* could be replaced by a more optimal π^n, thus contradicting the claim of optimality of π^*.

The dynamic programming algorithm, a form of backwards induction, involves stochastic optimization of control selection starting from the terminal time point T. Beginning with setting $J(x(T), T) = r_T(x(T))$, the algorithm proceeds via the following update rule:

$$J(x(t), t) = \min_c E\Big\{r(x(t), c(x(t), t), w(t), t) \\ + J(f(x(t), c((x(t), t), w(t), t), \sigma(t))\Big\}, \tag{21.21}$$

for t $\in \mathbb{T}$. This rule says that the cost-to-go of the current state $x(t)$ under a control $c(x(t), t)$ equals the expected value of the immediate cost $r(x(t), c(x(t), t), w(t), t)$ plus the future costs $J(f(x(t), c((x(t), t), w(t), t), \sigma(t))$. The symbol $\sigma(t)$ represents the "next" point in our timescale $\mathbb{T}$ and the use of this forward-jump operator is one of the hallmarks of the timescales calculus's ability to combine discrete and continuous analysis.

It is the goal in this work to move beyond this dynamic programming algorithm and to establish the HJB equation in full on general domains using the alpha derivative equation.

21.2.2 Dynamic Programming Algorithm on Timescales

Bellman's principle of optimality aids in the solution to the above optimization problem. This principle can be stated in the following way. Let π^* be an optimal policy. Then the optimal policy for the tail problem starting at time n, which is to minimize

$$E\left\{\int_n^T r(x(\tau), c(x(\tau), \tau), w(\tau), \tau)\Delta\tau\right\}, \tag{21.22}$$

is equal to the portion of π^* which overlaps π^n. To justify this principle, note that if it were not true then the tail of π^* could be replaced by a more optimal π^n, thus contradicting the claim of optimality of π^*.

The dynamic programming algorithm, a form of backwards induction, involves stochastic optimization of control selection starting from the terminal time point T. Beginning with setting $J(x(T), T) = r_T(x(T))$, the algorithm proceeds via the following update rule:

$$J(x(t), t) = \min_c E\Big\{r\big(x(t), c(x(t), t), w(t), t\big) \\ + J\big(f\big(x(t), c((x(t), t), w(t), t\big), \sigma(t)\big)\Big\}, \tag{21.23}$$

for t $\in \mathbb{T}$. This rule says that the cost-to-go of the current state $x(t)$ under a control $c(x(t), t)$ equals the expected value of the immediate cost $r(x(t), c(x(t), t), w(t), t)$ plus the future costs $J(f(x(t), c((x(t), t), w(t), t), \sigma(t))$. Recall that $\sigma(t)$ is the "next" point in our timescale $\mathbb{T}$.

It is standard to discuss this algorithm's optimality in terms which assume convergence. This proof, following Bertsekas declares controls optimal if they minimize the update rule.

The classical version of this update rule is true for the discrete timescale $t = \{1, 2, \ldots, T\}$. This chapter extends this result to any isolated timescale $\mathbb{T}$.

Theorem 21.2 (Dynamic Programming Algorithm, Delta Derivatives): *If $c^*(x(t), t)$ minimizes the update expression given above for each state and for all $t \in \mathbb{T}$, then the policy $c^*(x(t), t)$ is optimal.*

Proof: Set $J^*(x(T), T) = r_T(x(T))$ and proceed via timescales induction to show that application of the dynamic programming algorithm's recursive update equations yields the optimal policy at each stage, that is, that $J^*(x(t), t) = J(x(t), t))$ for all $\mathrm{t} \in \mathbb{T}$.

Letting $t = T$ yields, by definition,

$$J^*(x(T), T) = r_T(x(T)) = J(x(T), T)). \tag{21.24}$$

Now assume $J^*(x(t), t) = J(x(t), t)$ for some time point $t \in \mathbb{T}$ and all states $x(t)$. To apply the backwards induction algorithm, recall that in a timescale $\rho(t)$ is the point that comes just "before" the point t. Therefore, the quantity $\rho(t)$ plays a central role in our discussion. To wit, the following equation gives the immediate one-step application of our update rule:

$$\begin{aligned} J^*(x(\rho(t)), \rho(t)) = \min_{c,\pi} E\Bigg\{ & r(x(t), c(x(t), t), w(t), \rho(t)) \\ & + \int_t^T r(x(\tau), c(x(\tau), \tau), w(\tau), \tau)\Delta\tau \Bigg\}. \end{aligned} \tag{21.25}$$

The integral represents the value of the cost-to-go function J at the "next" time step after $\rho(t)$, which is t. Also note that the minimization is taken term-by-term over all controls and policies, respectively.

Now use the principle of optimality to distribute the *min* through the expectation, as the tail problem is indeed an optimal policy for the problem contained within the tail. This yields the following:

$$\begin{aligned} J^*(x(\rho(t)), \rho(t)) = \min_{c} E\Bigg\{ & r\big(x(t), c(x(t), t), w(t), \rho(t)\big) \\ & + \min_{\pi} E\bigg\{ \int_t^T r\big(x(\tau), c(x(\tau), \tau), w(\tau), \tau\big)\Delta\tau \bigg\}\Bigg\}. \end{aligned} \tag{21.26}$$

Using the definition of $J^*(x(t), t)$, which subsumes the term minimized over the policy, we reduce to

$$J^*(x(\rho(t)), \rho(t)) = \min_{c} E\left\{ r(x(t), c(x(t), t), w(t), \rho(t)) + J^*(x(t), t)\right\}. \tag{21.27}$$

By the induction hypothesis, we know $J^*(x(t), t)$ is equivalent to the approximation $J(x(t), t)$ due to the dynamic programming algorithm. Thus, we write $J^*(x(\rho(t)), \rho(t))$ as

$$\min_c E\{r(x(t), c(x(t), t), w(t), \rho(t)) + J(x(t), t)\}, \tag{21.28}$$

which, by definition, is simply

$$J^*(x(\rho(t)), \rho(t)) = J(x(\rho(t)), \rho(t)). \tag{21.29}$$

We have now satisfied conditions 1 and 2 of the principle of backwards induction on timescales given in the preliminaries section. Since we assume $\mathbb{T}$ to be isolated, conditions 3 and 4 do not apply and we conclude that, by backwards induction on timescales, we have proven our claim. ■

Thus, the dynamic programming algorithm is expanded to timescales.

21.2.3 HJB Equation on Timescales

Consider the dynamical system given by

$$x^{\Delta}(t) = f(x(t), c(t)), \tag{21.30}$$

where x represents states and c the control. Let $t \in \mathbb{T}$, $x : \mathbb{T} \to \mathbb{R}$, and $x(\mathbb{T}) = \mathbb{T}_x$. The cost-to-go function $J : \mathbb{T}_x \times \mathbb{T} \to \mathbb{R}$ is given by

$$J(x(t_0), t_0) = \int_{t_0}^{T} r(x(t), u(t))\Delta t, \tag{21.31}$$

where t_0 is the initial decision point and $r(x(t), u(t))$ the cost. Assume J is delta-differentiable and x is σ_x-completely delta differentiable. Furthermore, require x to satisfy

$$x(\sigma(t)) = \sigma_x(x(t)). \tag{21.32}$$

Then, the HJB equation on timescales is given by

$$0 = \min_c \left(r(x(t), c(x(t), t)) + J^{\Delta_t}(x(t), t) + J^{\Delta_x}(x(t), \sigma(t)) f(x(t), t)\right). \tag{21.33}$$

This is an equation that any optimal policy of the minimization problem must satisfy. Since precious few industrial-scale applications admit an analytic solution of this equation, ADP is used to develop approximation techniques for this purpose. The proof of this equation is the next theorem.

Theorem 21.3 (HJB Equation): *Let $V(x(t), t)$ be a solution to Equation (21.31) such that*

$$0 = \min_{c} \left(r(x(t), c(x(t), t)) + V^{\Delta_t}(x(t), t) + V^{\Delta_x}(x(t), \sigma(t)) f(x(t), t) \right). \tag{21.34}$$

Assume the boundary condition $V(x(T), T) = r_T(x(T))$ and $\hat{x}(t_0) = x(t_0)$ and suppose $c^*(x(t), t)$ attains the minimum called for in Equation (21.34) for all states and all time. Let $x^*(t)$ be the state trajectory, subject to the condition $x^*(t_0) = x(t_0)$, that corresponds to applying the controls $c^*(x(t), t)$ at each decision point t.

Then the function $V(x(t), t)$ is the optimal cost-to-go function $J^*(x(t), t)$ and the control $c^*(x(t), t)$ is optimal.

Proof: Let $\hat{c}(x(t), t)$ be a control policy with corresponding state trajectory $\hat{x}(t)$. We will show that the policy $c^*(x(t), t)$ achieves a cost no greater than this arbitrary $\hat{c}(x(t), t)$, thus forcing $c^*(x(t), t)$ to be our optimal control. We begin by invoking Equation (21.34) to give us

$$0 \le r(\hat{x}(t), \hat{c}(x(t), t)) + V^{\Delta_t}(\hat{x}(t), t) + V^{\Delta_x}(\hat{x}(t), \sigma(t)) f(\hat{x}(t), t). \tag{21.35}$$

Noting that, via (21.28), we have $x^{\Delta}(t) = f(x(t), c(t))$, we see that

$$0 \le r(\hat{x}(t), \hat{c}(x(t), t)) + V^{\Delta_t}(\hat{x}(t), t) + V^{\Delta_x}(\hat{x}(t), \sigma(t)) x^{\Delta}(t). \tag{21.36}$$

and, by reversing the chain rule implicit in this formulation, we arrive at

$$0 \le r(\hat{x}(t), \hat{c}(x(t), t)) + V^{\Delta}(t). \tag{21.37}$$

Integrating over our time horizon yields

$$0 \le \int_{t_0}^{T} r(\hat{x}(t), \hat{x}(x(t), t))\, \Delta t + \int_{t_0}^{T} V^{\Delta}(t) \Delta t. \tag{21.38}$$

Using the fundamental theorem we arrive at

$$0 \le \int_{t_0}^{T} r(\hat{x}(t), \hat{c}(x(t), t))\, \Delta t + V(\hat{x}(T), T) - V(\hat{x}(t_0), t_0). \tag{21.39}$$

Substituting in our boundary conditions $V(x(T), T) = r_T(x(T))$ and $\hat{x}(t_0) = x(t_0)$ gives us

$$0 \le \int_{t_0}^{T} r(\hat{x}(t), \hat{c}(x(t), t))\, \Delta t + r_T(x(T)) - V(x(t_0), t_0), \tag{21.40}$$

which is equal to

$$V(x(t_0), t_0) \le \int_{t_0}^{T} r(\hat{x}(t), \hat{c}(x(t), t))\, \Delta t + r_T(x(T)). \tag{21.41}$$

From our hypothesis, we assume the controls $c^*(x(t), t)$ and their corresponding state trajectory $x^*(t)$ minimize the value function $V(x(t), t)$. Using this information and the initial condition $x^*(t_0) = x(t_0)$, we can replace the inequality with equality in the case of these quantities:

$$V(x(t_0), t_0) = \int_{t_0}^{T} r\left(x^*(t), c^*(x(t), t)\right) \Delta t + r_T\left(x^*(T)\right). \tag{21.42}$$

Combining with the previous equation, we have

$$\begin{aligned} &\int_{t_0}^{T} r\left(x^*(t), c^*(x(t), t)\right) \Delta t + r_T\left(x^*(T)\right) \\ &\le \int_{t_0}^{T} r(\hat{x}(t), \hat{c}(x(t), t))\, \Delta t + r_T(x(T)). \end{aligned} \tag{21.43}$$

This tells us that the cost of the policy $c^*(x(t), t)$ is less than or equal to the cost of any admissible policy $\hat{c}(x(t), t)$. We conclude the policy $c^*(x(t), t)$ is optimal and that, since $\hat{c}(x(t), t)$ is arbitrary, we have $V(x(t), t) = J^*(x(t), t)$. Therefore, any optimal policy must satisfy the HJB equation. ■

21.3 BACKPROPAGATION

Backpropagation is the most widely used neural network learning technique. It is based on the mathematical notion of an ordered derivative. This section presents a formulation of ordered derivatives and the backpropagation training algorithm for the timescales calculus. To do so, we develop a new multivariate chain rule of this calculus, define ordered derivatives on timescales, prove a key theorem about them, and derive the backpropagation weight update equations for a feedforward multilayer neural network architecture. By drawing together the timescales calculus and the area of neural network learning, the first connection of timescales and neural network learning is presented.

A model of the backpropagation algorithm based on this timescales calculus is presented. The variables are allowed to take on either discrete or continuous values and a rule to update the weights in this general form is derived. The original version of backpropagation was discrete in nature and researchers have since developed it in the continuous case.

Ordered derivatives, invented by Paul Werbos to track the sensitivities of variables within systems often studied by social scientists, are also treated in this general timescales case. Key to the development of the backpropagation algorithm is the notion of a chain rule for ordered derivatives. This chain rule in the timescales calculus is proven and what an ordered derivative means within this new mathematics is defined.

21.3.1 Ordered Derivatives

In his PhD dissertation [11, 12], Paul Werbos discussed the idea of an *ordered derivative* distinct from the notion of a traditional direct partial derivative in common use in the analysis of physical systems. The use of these ordered derivatives is motivated by the particular character possessed by social systems as compared to those systems, which operate under strictly physical dynamics. In these systems, it makes sense to model using a series of variables which are each the consequence of a variable appearing earlier in the system. Hence, these are called *ordered variables*.

Werbos shows the traditional chain rule for total derivatives fails to provide an adequate framework for the calculations needed for the investigation not only of such an integrated socio-economic system, but also for a wide array of other mathematically related connectionist systems such as the problem of political forecasting described in detail in his dissertation work. However, he goes on to prove that a chain rule for ordered derivatives is in fact capable of driving the necessary calculation engine for studying these sorts of systems. While he does not term this method *backpropagation*, the algorithm he lays out is the one which today powers most of the neural network learning approaches.

To complete the theoretical foundation of backpropagation, Werbos proved a special chain rule for these ordered derivatives that worked for the applications in which the traditional chain rule of continuous analysis broke down. This section contains the definition of the ordered (delta) derivative on timescales and the proof that the chain rule obtained by Werbos holds true in this new environment.

Following Werbos, define the following variables: input $\boldsymbol{X}(\boldsymbol{t})$ and output $\bar{\boldsymbol{Y}}$ which approximates the target output $\boldsymbol{Y}(\boldsymbol{t})$. The dimension of the outputs is denoted by n. The relationship between inputs and outputs is determined via a series of adaptive weights W. It is the goal of backpropagation, and hence the ordered derivative calculations, to tune W in such a way as to reduce an error measure between $\bar{Y}$ and Y. The implicit timescale in Werbos's work is an isolated subset of the integers $\mathbb{Z}$. This section considers timescales of higher generality and will assume $\boldsymbol{t} \in \mathbb{T}$, where $\mathbb{T}$ is not restricted to a subset of $\mathbb{Z}$. Let N denote the number of inputs or the number of time steps the system is allowed to compute.

Define the following measure of system performance/error:

$$J = \int_1^{\sigma(N)} J(\tau)\,\Delta\tau = \int_1^{\sigma(N)} \sum_{i=1}^{n} \frac{1}{2}\left(\bar{Y}_l(\tau) - Y_i(\tau)\right)^2 \Delta\tau. \tag{21.44}$$

This form represents the standard least-squares error measure. Note that the integral takes on the role of the generalized summation operation and is appropriate for any timescale $\mathbb{T}$ free of restriction to an isolated case. The proper summation that remains represents the dimension of the output vector, which is a fixed scalar value and not related to our timescale $\mathbb{T}$ and is thus not subsumed by an integral.

The form of this measure J when defined on different timescales can be investigated.

Consider a function $f : \mathbb{T} \to \mathbb{R}$ where $\mathbb{T} = \{t_0, t_1, \ldots\}$. Then $\int_{t_0}^{t} f(\tau)\,\Delta\tau = \sum_{\tau \in [t_0,t)} f(\tau)\,\mu(\tau)$, where t_0 is the starting point of the timescale $\mathbb{T}$ and μ is the graininess function of $\mathbb{T}$ as defined in Section 21.4. So, if $\mathbb{T} = h\mathbb{Z}$, then the performance measure J becomes

$$J = \int_{t_0}^{N} J(\tau)\,\Delta\tau\sigma = \sum_{t\in[t_0,t)} J(\tau)\,h = h \sum_{t\in[t_0,t)} \sum_{i=1}^{n} \frac{1}{2}\left(\bar{Y}_l(\tau) - Y_i(\tau)\right)^2. \quad (21.45)$$

Similarly, for the case of the quantum calculus timescale $q^{\mathbb{N}_0}$, $q>1$, the following error obtains:

$$J = \int_{t_0}^{q^n} J(\tau)\,\Delta\tau = \sum_{i=0}^{n-1} J\left(q^i\right) q^i (q-1) = (q-1) \sum_{i=0}^{n-1} \left(\frac{q^i}{2}\right) \left(\bar{Y}_l\left(q^i\right) - Y_i(q^i)\right)^2. \quad (21.46)$$

This error term can be calculated for any timescale. Regardless of the choice of timescale, which informs the construction of the error measure, the network equations remain to be analyzed. It is here that the chain rule for ordered derivatives is needed, as this is the tool used to turn the network equations into a computation unit for the proper updating of the adaptive weights via backpropagation. As a preamble to the chain rule theorem, it is first necessary to discuss the notion of ordered derivative.

To discuss the structure of ordered derivatives, consider the following system:

$$\begin{aligned} x_4 &= 2x_1 + 5x_2 + x_3, \\ x_3 &= 4x_2 + x_1, \\ x_2 &= 7x_1. \end{aligned}$$

Distinguish between *direct* and *indirect* effects. The direct effect of a change in x_1 on x_4, denoted $\partial x_4/\partial x_1$, is equal to 2. Note that in this case the ordering of the variables is not taken into account. Unlike a physical system, where the evaluations of all the variables x_i would be assumed to take place simultaneously so that the calculation of $\partial x_4/\partial x_1$ would require the substitution of lower valued x_i's into the equation for x_4, in this *ordered system* the direct effect is calculated wholly from the equation for x_4 without recourse to the equations representing lower, or previous, levels of activity. The total derivative effect in such an ordered system is called the indirect effect and is denoted by $\partial^+ x_4/\partial x_1$. In this case, this indirect effect is equal to 66.

Furthermore, this notation corresponds to the continuous case. For the general timescale case, denote the ordered derivative of the performance measure J with respect to an ordered variable x_i as

$$J^{\Delta^+_{x_i}}.$$

The following text develops the necessary definitions to make the above notation precise.

Let $x_1(t), x_2(t), \ldots, x_n(t)$ be an ordered sequence of variables with $t \in \mathbb{T}$. These variables represent stages of a larger calculation (e.g., layers, in a sense of a multilayer perceptron network). What sets them apart as *ordered* is that they follow a recursion given by

$$x_i = f(x_1, x_2, \ldots, x_{i-1}). \tag{21.47}$$

In this way, can the mathematics speak meaningfully of causation as a basis for the relationship among the x_i's. The interest is in determining the way in which the performance term J changes with respect to one of the x_i's, that is, it is desired to calculate $J^{\Delta_{x_i}}$. Following Werbos, set up the error, which itself is actually the final variable in the extended ordered set, as a sequence of recursive functions, which is required to be σ_1-completely (delta) differentiable, such that

$$J_n(x_1, \ldots, x_n) = x_n \tag{21.48}$$

and

$$J_{i-1}(x_1, \ldots, x_{i-1}) = J_i(x_1, \ldots, x_{i-1}, f_i(x_i, \ldots, x_{i-1})). \tag{21.49}$$

These recursive functions describe the final node in the network, x_n, as a function of earlier variables in the ordering successively calculated. These functions J_i will provide the machinery needed to discuss the chain rule for ordered derivatives. In application, J_n, also equal to the final variable in the ordering x_n, is the network performance/value measure, which is to be optimized via the updating of adaptive weights.

Now, the performance can be defined by reference to these recursive J_i functions. Define the ordered (delta) derivative of x_n with respect to a previous variable x_i in the ordering by

$$x_n^{\Delta^+_{x_i}} = J_i^{\Delta_{x_i}}, \tag{21.50}$$

where the functional relationship among the x_i's is repressed (i.e., all variables $x_1, x_2, \ldots, x_{i-1}$ are held constant) when calculating $J_i^{\Delta_{x_i}}$ so as to properly represent the ordered nature of these variables (that is, the derivative $J_i^{\Delta_{x_i}}$ gives us only the *direct* effect of a change in x_i on J_i.)

With these definitions, the following chain rule may be proven.

Theorem 21.4 (The Chain Rule for Ordered Derivatives):

$$J_j^{\Delta_{x_i}} = \sum_{k=j+1}^{n} x_n(\sigma_1(x_1), x_2, \ldots, x_{n-1})^{\Delta^+_{x_k}} x_k^{\Delta_{x_i}}. \tag{21.51}$$

Proof: We proceed via reverse induction on j. We will start with $j = n - 1$ and end with $j = 1$. Note that with $j<i$, the nature of the definitions of the f_i 's and x_i's force the terms $f_k^{\Delta x_i}$ to equal 0 and, therefore, these terms do not contribute to the summation. So, it will suffice to consider j in the range $n - 1$ to i.

Let $j = n - 1$. Then our hypothesis becomes

$$J_n^{\Delta x_i} = x_n(\sigma_1(x_1), x_2, \ldots, x_{n-1})^{\overset{+}{\Delta} x_k} x_n^{\Delta x_i} = x_n^{\Delta x_i}. \tag{21.52}$$

Since by definition $J_n = x_n$, we have an identity and the claim is proven.

Now assume our hypothesis holds for $j + 1<n$. We will proceed to show it is true for $j>i$. Noting the recursive definition of the J_i's make them all equal, we have that $J_j^{\Delta x_i} = J_{j+1}^{\Delta x_i}$. This expression now requires use of our chain rule for n variables proven in Section 21.4. Write J_{j+1} explicitly as a function of functions of x_i:

$$J_{j+1}\left(f_1(x), \ldots, f_{j-1}\left(x_1, \ldots, x_{j-1}\right), f_{j+1}\left(x_1, \ldots, x_j\right)\right). \tag{21.53}$$

Due to the nature of the ordered system, $x_k^{\Delta x_i} = 0$ for $i \neq 0$ and $i<j$ so that the expansion given by our n variable chain rule

$$J_{j+1}^{\Delta x_i} = J_{j+1}^{\Delta x_1} x_1^{\Delta x_i} + \sum_{k=2}^{j+1} J_{j+1}(\sigma_1(x_1), x_2, \ldots, x_n)^{\Delta x_k} x_k^{\Delta x_i} \tag{21.54}$$

reduces to

$$J_{j+1}^{\Delta x_i} = J_{j+1}^{\Delta x_{j+1}} x_{j+1}^{\Delta x_{j+1}} + J_{j+1}^{\Delta x_i} x_i^{\Delta x_i} \tag{21.55}$$

so that

$$J_j^{\Delta x_i} = J_{j+1}^{\Delta x_{j+1}} x_{j+1}^{\Delta x_i} + J_{j+1}^{\Delta x_i}. \tag{21.56}$$

Our induction hypothesis gives us

$$J_{j+1}^{\Delta x_i} = \sum_{k=j+2}^{n} x_n(\sigma_1(x_1), x_2, \ldots, x_{n-1})^{\overset{+}{\Delta} x_k} x_k^{\Delta x_i}. \tag{21.57}$$

Combining these results yields

$$J_{j+1}^{\Delta x_i} = J_{j+1}^{\Delta x_{j+1}} x_{j+1}^{\Delta x_i} + \sum_{k=j+2}^{n} x_n(\sigma_1(x_1), x_2, \ldots, x_{n-1})^{\overset{+}{\Delta} x_k} x_k^{\Delta x_i}, \tag{21.58}$$

which is equal to

$$x_n^{\Delta^+_{x_{j+1}}} x_{j+1}^{\Delta_{x_i}} + \sum_{k=j+2}^{n} x_n(\sigma_1(x_1), x_2, \ldots, x_{n-1})^{\Delta^+_{x_k}} x_k^{\Delta_{x_i}}, \tag{21.59}$$

and which reduces to

$$\sum_{k=j+2}^{n} x_n(\sigma_1(x_1), x_2, \ldots, x_{n-1})^{\Delta^+_{x_k}} x_k^{\Delta_{x_i}}. \tag{21.60}$$

Thus, our claim, the chain rule for ordered derivatives, is proven. ■

The usual form of this theorem is given by the following corollary:

Theorem 21.5 (Chain Rule for Ordered Derivatives Corollary):

$$\begin{aligned} x_n(\sigma_1(x_1), x_2, \ldots, x_{n-1})^{\Delta^+_{x_i}} &= x_n(\sigma_1(x_1), x_2, \ldots, x_{n-1})^{\Delta_{x_i}} \\ &\quad + \sum_{k=i+1}^{n-1} x_n(\sigma_1(x_1), x_2, \ldots, x_{n-1})^{\Delta^+_{x_i}} x_k^{\Delta_{x_i}}. \end{aligned} \tag{21.61}$$

This corollary follows immediately from the chain rule theorem. Note that in practice J is used as the final variable in the ordered system, x_n, so that this corollary will take the form

$$J^{\Delta^+_{x_i}} = J^{\Delta_{x_i}} + \sum_{k=i+1}^{n-1} J^{\Delta^+_{x_k}} x_k^{\Delta_{x_i}}. \tag{21.62}$$

Note that the timescales version of the chain rule for ordered derivatives contains the term $\sigma_1(x_1)$. This is due to the requirement of σ_1-complete (delta) differentiability in our n-variable chain rule. Further versions of both chain rules can be developed which require σ_i differentiability, where $i \neq 1$. With this chain rule the weight update rules for the backpropagation algorithm for neural network learning can be derived.

21.3.2 The Backpropagation Algorithm on Timescales

Before the development of backpropagation as the method of choice for calculating derivatives to update the weights of neural networks, a direct differentiation method was used. This method demands much in terms of computational resources when the number of weights is larger than the number of neurons. Since multilayer feedforward systems typically satisfy this criteria, the more efficient backpropagation is preferred.

To derive the backpropagation algorithm on timescales, work from the assumption that the node activity of our network takes on the following form:

$$x_i = f\left(\sum_{j<i} w_{ij} w_j\right), \tag{21.63}$$

where the w_{ij} are the adaptive weights. Note that these weights are part of the ordered system model for the neural network.

From this, the change in an ordered variable with respect to changes in an earlier node using the timescales chain rule can be calculated:

$$x_j^{\Delta_{x_i}} = w_{ij} \int_0^1 f'\left(\sum_{j>i} w_{ij} x_j + h\mu(t) w_{ij}\right) \mathrm{d}h. \tag{21.64}$$

Now, the rate of change of the error measure J with respect to an activity node x_i is given by

$$J^{\Delta^+_{x_i}} = -J^{\Delta_{x_i}} + \sum_{k=i+1}^{n-1} J^{\Delta^+_{x_k}} x_j^{\Delta_{x_i}}, \tag{21.65}$$

which, via the chain rule may be written as

$$J^{\Delta^+_{x_i}} = -J^{\Delta_{x_i}} + \sum_{k=i+1}^{n-1} J^{\Delta^+_{x_k}} w_{ij} \int_0^1 f'\left(\sum_{j>i} w_{ij} x_j + h\mu(t) w_{ij}\right) \mathrm{d}h \tag{21.66}$$

and reduced to

$$J^{\Delta^+_{x_i}} = -J^{\Delta_{x_i}} + \sum_{k=i+1}^{n-1} w_{ij} \delta_i, \tag{21.67}$$

where

$$\delta_i = J^{\Delta^+_{x_k}} \int_0^1 f'\left(\sum_{j>i} w_{ij} x_j + h\mu(t) w_{ij}\right) \mathrm{d}h \tag{21.68}$$

is the *local error* term which measures the change in the performance measure with respect to a change at the level of a local neuron.

The gradient descent *delta rule* on timescales is given by

$$w_{ij}^{\Delta}(t) = -\beta \delta_i x_j(t), \tag{21.69}$$

which is the change in the weights w_{ij} as a change in the error function J modified by a learning rate β.

The final step in the derivation of the backpropagation algorithm is the integration of the delta learning rule with the equations obtained previously:

$$w_{ij}^{\Delta}(t) = -\beta J^{\Delta_{w_{ij}}^{+}}. \tag{21.70}$$

This form, which holds for nodes such that $x : \mathbb{T} \to \mathbb{R}$, can account for networks in which the input is either continuous or discrete using a single theoretical framework and update equation.

21.4 CONCLUSIONS

The timescales calculus is an increasingly relevant and developed area of mathematics with wide-ranging opportunities for application. In addition to showing its applicability to backpropagation, this chapter shows that the dynamic programming algorithm, derived from Bellman's principle of optimality, obtains on timescales. Also derived is the HJB equation on timescales.

Furthermore, ordered derivatives and the backpropagation update rule have been established using the emerging mathematical field of timescales calculus. This calculus unifies the discrete and continuous domains, so our results provide a complete theoretical framework for discussing learning in connectionist systems, which can admit input signals of any type.

While the study of timescales can provide a concise theoretical unification of control theory in the discrete and continuous case, it can also provide so much more than that. There are important application areas where dealing simultaneously with discrete and continuous variables is critical [13], and the timescales calculus provides a natural and powerful framework for such exploration.

ACKNOWLEDGMENTS

Partial support for this research from the National Science Foundation, the Missouri University of Science & Technology Intelligent Systems Center, and the Mary K. Finley Missouri Endowment are gratefully acknowledged.

REFERENCES

1. J. Seiffertt, S. Sanyal, and D. Wunsch. Hamilton–Jacobi–Bellman equations and approximate dynamic programming on time scales. *IEEE Transactions on Systems, Man, and Cybernetics, Part B*, 38(4):918–923, 2008.
2. J. Seiffertt and D. Wunsch. Backpropagation and ordered derivatives in the time scales calculus. *IEEE Transactions on Neural Networks*, 21(8):1262–1269, 2010.

3. J. Seiffertt and D. Wunsch. *Unified Computational Intelligence for Complex Systems*. Springer-Verlag, 2010.
4. M. Bohner and A. Peterson. *Dynamic Equations on Time Scales: An Introduction with Applications*. Birkhauser, Boston, MA, 2001.
5. C. Potzsche. Chain rule and invariance principle on measure chains. *Journal of Computational and Applied Mathematics* 141:249–254, 2002.
6. B. Jackson. Partial dynamic equations on time scales. *Journal of Computational and Applied Mathematics*, 186:391–415, 2006.
7. M. Bohner and G. Guseinov. Multiple integration on time scales. *Dynamic Systems and Applications*, 14(3–4):579–606, 2005.
8. M. Bohner and G. Guseinov. Partial differentiation on time scales. *Dynamic Systems and Applications*, 13:351–379, 2004.
9. W. Powell. *Approximate Dynamic Programming: Solving the Curses of Dimensionality*. Wiley Series in Probability and Statistics, Hoboken, NJ, 2007.
10. J. Si, A. Barto, W. Powell, and D. Wunsch. *Handbook of Learning and Approximate Dynamic Programming*. IEEE Press Series on Computational Intelligence, 2004.
11. P. Werbos. Beyond regression. Ph.D. dissertation, Harvard University, 1974.
12. P. Werbos. *The Roots of Backpropagation: From Ordered Derivatives to Neural Networks and Political Forecasting*. Wiley, New York, NY, 1994.
13. P. Werbos. Backwards differentiation in AD and neural nets: past links and new opportunities. In M. Bucker, G. Corliss, P. Hovland, U. Naumann, and N. Boyana, editors. *Automatic Differentiation: Applications, Theory, and Implementations*. Springer, New York, NY, 2005.

CHAPTER 22

A Survey of Optimistic Planning in Markov Decision Processes

LUCIAN BUŞONIU[1], RÉMI MUNOS[2], and ROBERT BABUŠKA[3]

[1]Université de Lorraine, CRAN, UMR 7039 and CNRS, CRAN, UMR 7039, Vandœuvre-lès-Nancy, France
[2]SequeL Team, INRIA Lille – Nord Europe, Villeneuve d'Ascq, France
[3]Delft Center for Systems and Control, Delft University of Technology, Delft, The Netherlands

ABSTRACT

We review a class of online planning algorithms for deterministic and stochastic optimal control problems, modeled as Markov decision processes. At each discrete time step, these algorithms maximize the predicted value of planning policies from the current state, and apply the first action of the best policy found. An overall receding-horizon algorithm results, which can also be seen as a type of model-predictive control. The space of planning policies is explored optimistically, focusing on areas with largest upper bounds on the value—or upper confidence bounds, in the stochastic case. The resulting *optimistic planning* framework integrates several types of optimism previously used in planning, optimization, and reinforcement learning, thereby obtaining several intuitive algorithms with good performance guarantees. We describe in detail three recent such algorithms, outline the theoretical guarantees on their performance, and illustrate their behavior in a numerical example.

22.1 INTRODUCTION

This chapter considers online algorithms for problems in which a nonlinear, possibly stochastic dynamic system must be optimally controlled in discrete time. Optimality is

Work performed in part while L. Buşoniu was with SequeL Team, INRIA Lille. He is also associated with the Automation Department, Technical University of Cluj-Napoca, Romania.

Reinforcement Learning and Approximate Dynamic Programming for Feedback Control, First Edition. Edited by Frank L. Lewis and Derong Liu.

measured by a cumulative reward signal which must be maximized: the return. Such problems arise in many fields, including artificial intelligence, automatic control, operations research, economics, medicine, and so on. They are often modeled as Markov decision processes (MDPs).

In an MDP, the system is described by a state signal x varying in the state space X, and can be influenced by actions u in the action space U. For the simplicity of notation, X will be considered countable here, but uncountable (e.g., continuous) spaces can easily be handled. Most of the algorithms that will be described in this chapter require that U consists of a small number of discrete actions. The probability that next state x' is reached after action u is taken in state x is $f(x, u, x')$, where $f : X \times U \times X \rightarrow [0, 1]$ is a transition probability function describing the dynamics of the system. After the transition to x', a reward $r' = \rho(x, u, x')$ is received, where $\rho : X \times U \times X \rightarrow \mathbb{R}$ is the reward function. A control policy $\pi : X \rightarrow U$ indicates how the controller should choose actions to interact with the system. Denoting by k the discrete time index, the expected infinite-horizon discounted return of state x under a policy π is

$$V^{\pi}(x) = \mathrm{E}_{x_{k+1} \sim f(x_k, \pi(x_k), \cdot)} \left\{ \sum_{k=0}^{\infty} \gamma^k r_{k+1} \right\}, \tag{22.1}$$

where $x_0 = x$, $r_{k+1} = \rho(x_k, \pi(x_k), x_{k+1})$, $\gamma \in (0, 1)$ is the discount factor, and the notation $x_{k+1} \sim f(x_k, \pi(x_k), \cdot)$ means that x_{k+1} is drawn from the distribution $f(x_k, \pi(x_k), \cdot)$ over next states.[1] Other types of return can also be used, such as finite-horizon or averaged over time. We call $V^{\pi} : X \rightarrow \mathbb{R}$ a value function. The goal is to control the system using an optimal policy π^*, so that the value function is maximized for every $x \in X$. This maximal, optimal value function is denoted by V^* and is unique, so it does not depend on the particular optimal policy. It is also useful to consider an action-dependent optimal value function, the optimal Q-function: $Q^*(x, u) = \mathrm{E}_{x' \sim f(x,u,\cdot)} \left\{ \rho(x, u, x') + \gamma V^*(x') \right\}$. Note that optimal control problems are often stated so that a cost is minimized, rather than a return being maximized—but the two formulations are equivalent.

We consider a class of online model-based algorithms that, at each step of interaction, look at the current system state x_k and employ the model to predict the system's response to various *planning policies*—which may be, for example, simple open-loop sequences of actions or more complex, closed-loop selection rules. Note that a planning policy is a different object from the interaction policy π. Exploiting the predictions, an action u_k that is as good as possible is applied, which results in a new state x_{k+1}. The entire cycle then repeats, in a receding-horizon fashion. In systems and control, such algorithms are known as model-predictive control [1], while in computer science they are called planning [2]. In this chapter, we mainly draw inspiration from the computer science literature, and to emphasize that we mean

[1] Note the "prime" notation, as in, for example, x', is used to generically indicate variables at the next step, whereas if the actual time step is important, it is included as a subscript, for example, x_{k+1}.

the online, feedback control over trajectories consisting of many steps, we call the algorithms online planning.

The quality of action u_k returned at step k by the planning algorithm is measured by the simple regret, defined as follows:

$$\mathcal{R}(x_k) = \max_{u \in U} Q^*(x_k, u) - Q^*(x_k, u_k), \tag{22.2}$$

that is, the loss incurred by choosing u_k and then acting optimally, with respect to acting optimally immediately, from step k. Note the regret is always nonnegative, and an optimal action choice attains a regret of 0. If an algorithm achieves a regret of ε for every state, the overall suboptimality in terms of discounted return is at most $\frac{\varepsilon}{1-\gamma}$ [3]. This property motivates the use of the simple regret as a performance measure.

The online planning approach is very different from the standard methods for solving MDPs considered in dynamic programming and reinforcement learning. The latter methods usually seek a global solution, whereas online planning finds actions on demand, locally for each state where they are needed. Online planning is therefore much less dependent on the state space size. It is generally suboptimal, but useful bounds can be placed on its suboptimality, as will be outlined later in the chapter. While global methods do achieve optimality in some restricted settings, in realistic problems they must also use approximation, thereby sacrificing optimality, as seen elsewhere in this book and specifically reviewed in [4].

We focus on a class of online planning algorithms based on the principle of optimism in the face of uncertainty: given the choice between planning policies having uncertain values, more promising policies are explored first. This optimistic principle has applications in many fields, among which those closest to our focus are optimization, where a representative optimistic method is branch-and-bound; classical planning, with algorithms such as A* [2]; and the exploration–exploitation dilemma in reinforcement learning [5]. In classical planning, optimism relies on *deterministic* upper bounds on the values of incomplete action sequences, whereas in exploration–exploitation, upper *confidence* bounds on stochastic values are used to guide action choices [6]. Upper confidence bounds have also been recently applied to planning [7, 8], but typically without making the connection with the deterministic optimism of classical planning. In this chapter, we integrate both types of optimism into a single framework, in the context of MDPs. To this end, planning is cast as the problem of optimizing returns over planning policies from the current state. This interpretation leads to several intuitive planning algorithms with strong performance guarantees, which have been developed over the last few years [3, 9, 10] and which we review in detail.

Next, Section 22.2 describes the foundations of optimism in online optimization. Section 22.3 forms the core of the chapter, and exploits the ideas of Section 22.2 to introduce the family of optimistic planning algorithms. Methods for three types of system dynamics are reviewed: deterministic, general stochastic, and stochastic but where each transition can only end up in a finite number of states. Related algorithms from the literature are discussed in relation to optimistic planning in Section 22.4. The behavior of the three methods is illustrated in a numerical example in Section 22.5.

22.2 OPTIMISTIC ONLINE OPTIMIZATION

The idea of optimism can be understood by looking at the problem of maximizing an unknown function $v : \mathcal{H} \to \mathbb{R}$ online, while observing possibly noisy values of this function at chosen points. At every round $t = 1, 2, \ldots$, the algorithm chooses a point $h_t \in \mathcal{H}$, and is provided with a random sample $\tilde{v}_t = v(h_t) + w_t$ where w is zero-mean noise, drawn independently for each t. The goal is not just to find an optimal point $h^* \in \arg\max_{h \in \mathcal{H}} v(h)$, but due to the online setting, to find it while sampling points of as high a quality as possible. We will not provide formal guarantees for optimistic optimization in terms of this requirement (although they are available), because we are only interested in these methods as stepping stones for optimistic planning. In that context, a different requirement will be sufficient, which only asks the algorithm to return the best point possible after expending some number of samples, without looking directly at the quality of these samples.

We first consider two special cases: noisy evaluations but small discrete set $\mathcal{H}$; and infinite set $\mathcal{H}$ but deterministic evaluations. We then proceed to the general case.

22.2.1 Bandit Problems

In the case of bandit problems, the observed values are noisy, but $\mathcal{H}$ is small and discrete: $\mathcal{H} = \{h^1, \ldots, h^M\}$. The name comes from the fact that this case models the optimization of the payoff achieved by pulling the arms of an M-armed slot machine (bandit), where each h^j is an arm. Online optimization is a way to solve the exploration–exploitation dilemma: given the information available so far, should the algorithm exploit the arm that seems the best (has the best average observed value), or should another arm be explored to achieve better confidence in its value? This formalism is used, for example, in reinforcement learning algorithms.

There are many ways to solve bandit problems, and here we will only consider the *upper confidence bound* (UCB) algorithm. At every round t, this algorithm defines a b-value for every h^j:

$$b(h^j) = \frac{\sum_{i=1}^{n(h^j)} \tilde{v}_i^j}{n(h^j)} + \sqrt{\frac{2 \log t}{n(h^j)}}, \tag{22.3}$$

where $n(h^j)$ is the number of instances h^j has been chosen up to round t, and $\tilde{v}_i^j$ is the random value (sample) observed at the ith such instance (the b-values of unsampled points, for which $n(h^j) = 0$, are by convention positive infinite). Each b-value is an UCB on $v(h^j)$, that is, with high probability it is an upper bound. Then the algorithm simply makes an *optimistic* choice $\arg\max_{h^j} b(h^j)$ for the next sample. It must be emphasized that the count $n(h^j)$ and thus the b-value $b(h^j)$ depend, of course, on the round t. Throughout the chapter, we leave such dependencies on the current round implicit to avoid cluttering the notation.

Such an algorithm is optimistic because it always explores further the point with the largest UCB. The optimistic principle has strong theoretical support in bandit problems [6].

22.2.2 Lipschitz Functions and Deterministic Samples

When the set $\mathcal{H}$ is infinite (or just very large), it is impossible to sample every point. Consider the deterministic case, where the algorithm is given the exact value $v(h)$ for each point h it chooses. Let $\ell(h, \bar{h})$ be a metric over $\mathcal{H}$, and assume v is Lipschitz:

$$\left|v(h) - v(\bar{h})\right| \leq \ell(h, \bar{h}) \quad \forall h, \bar{h} \in \mathcal{H},$$

where for convenience we do not make the Lipschitz constant explicit. Then for any set of points $H \subseteq \mathcal{H}$ and any point $h \in H$ we have the following upper bound on the supremum of v over H (i.e., this upper bound is larger than $v(\bar{h})$ for any $\bar{h} \in H$):

$$v(h) + \operatorname{diam}(H),$$

where $\operatorname{diam}(H) = \sup_{h,\bar{h} \in H} \ell(h, \bar{h})$ denotes the diameter of the set. If multiple samples are available in H, say their number is $n(H)$, then the bound is refined to:

$$b_{\text{init}}(H) = \min_{i=1,\dots,n(H)} \left[v(h_i) + \operatorname{diam}(H)\right]. \tag{22.4}$$

Using this property, a tree-based algorithm for optimizing v can be derived: Algorithm 22.1, called *optimistic optimization for deterministic functions* (OOD). This method builds a tree $\mathcal{T}$ of policy sets, where each node contains some set H and its children correspond to a partition of H. No difference is made between the notation of a node and its corresponding set, and the children of some H are denoted by $\mathcal{C}(H)$. OOD starts building the tree from a root node containing the entire space $\mathcal{H}$, and at each step, chooses an *optimistic* leaf by navigating the tree along the maximum-b-value path. The optimistic leaf is expanded, that is, the corresponding set is split (partitioned) into several child subsets, and an arbitrary point is sampled in each subset. The node splitting procedure must ensure that the diameter decreases with the depth. For example, if each node corresponds to a hyperbox in a Euclidean space, then the optimistic leaf could be split halfway along the longest edge.

B-values b_{init} at the leaves are computed with (22.4) and propagated upwards in the tree, where each parent node gets the largest bound from its children. The minimization in (22.4) is useful because samples inherited by the leaf from its parents may provide

Algorithm 22.1: Optimistic optimization of deterministic functions

initialize tree: $\mathcal{T} \leftarrow \{\mathcal{H}\}$
for $t = 1, 2, \ldots$ **do**
  $H \leftarrow$ root $\mathcal{H}$; **while** H is not leaf **do** $H \leftarrow \arg\max_{H' \in C(H)} b(H')$ **end while**
  optimistic leaf found: $H^{\dagger} \leftarrow H$
  expand $H^{\dagger}$ into several child subsets H', adding them to $\mathcal{T}$
  sample a point in every new child H'
  update b-values upwards in the tree, starting from the leaves:

$$b(H) \leftarrow \begin{cases} b_{\text{init}}(H) \text{ from (22.4)} & \text{if } H \text{ is leaf} \\ \max_{H' \in C(H)} b(H') & \text{otherwise} \end{cases}$$

end for

a better bound than its own, newly obtained sample. Due to the propagation rule, navigating the tree along the optimistic path is equivalent to directly choosing the leaf with the largest b-value. The optimistic path is nevertheless made explicit to emphasize the connection with later algorithms.

Note the similarity of this algorithm with branch-and-bound optimization. One difference is that branch-and-bound also uses lower bounds to eliminate some nodes from consideration. In Algorithm 22.1, all the nodes remain candidates for expansion, but the algorithm always focuses on the most promising node.

22.2.3 Lipschitz Functions and Random Samples

A general algorithm, which works for the noisy optimization of Lipschitz functions defined over arbitrary metric spaces, will combine the two types of optimism highlighted in Algorithm 22.1: upper confidence bounds due to the stochastic samples, with diameter-related bounds due to the nonzero size of the sets considered. In particular, at round t, a high-probability upper bound on the supremum of v over some set H is

$$b_{\text{init}}(H) = \frac{\sum_{i=1}^{n(H)} \tilde{v}_i}{n(H)} + \sqrt{\frac{2\log(t)}{n(H)}} + \text{diam}(H), \tag{22.5}$$

where $n(H)$ is as before the number of samples falling inside set H, and $\tilde{v}_i$ are the random values of these samples.

The resulting algorithm is called *hierarchical optimistic optimization* (HOO) [11] and is shown in Algorithm 22.2. Like for OOD, the splitting procedure should decrease the diameter. The only difference from OOD is that a parent either inherits the b-value from one of its children, or it uses its own b-value b_{init} if it is better (smaller). The latter situation could not arise in OOD, because the diameters decrease with the depth and sample values are deterministic. Figure 22.1 illustrates how HOO develops the tree and computes the b-values.

Algorithm 22.2: Hierarchical optimistic optimization

initialize tree: $\mathcal{T} \leftarrow \{\mathcal{H}\}$
for $t = 1, 2, \ldots$ **do**
  $H \leftarrow$ root $\mathcal{H}$; **while** H is not leaf **do** $H \leftarrow \arg\max_{H' \in \mathcal{C}(H)} b(H')$ **end while**
  optimistic leaf found: $H^\dagger \leftarrow H$
  expand $H^\dagger$ into several child subsets H', adding them to $\mathcal{T}$
  sample a point in every new child H'
  update b-values upwards in the tree, starting from the leaves:

$$b(H) = \begin{cases} b_{\text{init}}(H) \text{ from (22.5)} & \text{if } H \text{ is leaf} \\ \min\left\{b_{\text{init}}(H), \max_{H' \in \mathcal{C}(H)} b(H')\right\} & \text{otherwise} \end{cases}$$

end for

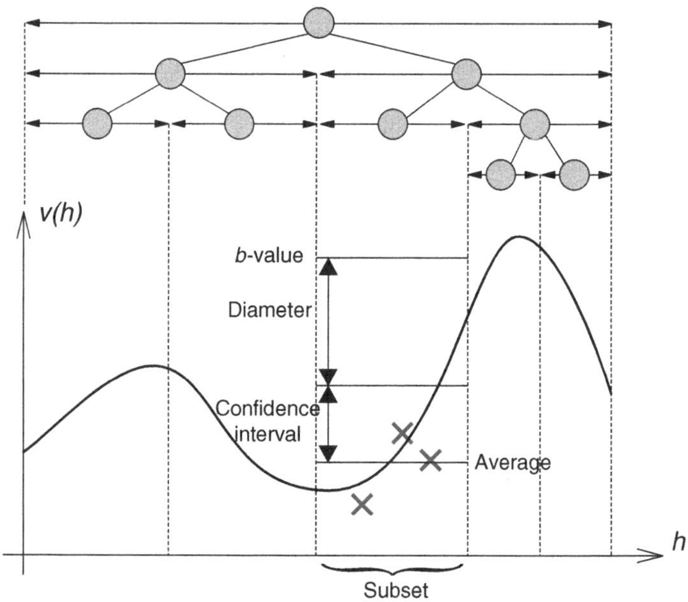

FIGURE 22.1 Illustration of HOO. Each tree node corresponds to a subset along the horizontal axis. For each leaf subset H, the b-value is computed as the sum of the average of the observed samples $(\sum_{i=1}^{n(H)} \tilde{v}_i)/n(H)$, the confidence interval $\sqrt{2\log(t)/n(H)}$, and the subset diameter $\text{diam}(H)$.

22.3 OPTIMISTIC PLANNING ALGORITHMS

In this section, the ideas of optimistic optimization are applied to planning. As already outlined in the introduction, the planning algorithms we consider work online, in a receding-horizon fashion, and use a model of the MDP in the form of the functions f and ρ. At each step k, the model is employed to predict possible behaviors of the system (specifically, state trajectories and ensuing rewards) starting from the current state x_k and responding to various planning policies. Using these predictions, the algorithm returns an action u_k that is as close to optimal as possible, where near-optimality is measured by the simple regret (22.2). This action is applied, the system transits to x_{k+1}, and the cycle repeats.

To place this problem in the context of optimistic optimization, let us isolate one such step of interaction with the system, and by convention, relabel the current time from k to 0, so that the system state is x_0. Planning should then maximize the expected return starting from x_0:

$$v(h) = \mathrm{E}_{x_{d+1} \sim f(x_d, h(d, x_d), \cdot)} \left\{ \sum_{d=0}^{\infty} \gamma^d r_{d+1} \right\}, \tag{22.6}$$

where h is a rule describing how actions are chosen for each state occurring along future trajectories. This equation immediately clarifies the link with the optimization problem discussed in Section 22.2: the function v to optimize is the expected return, and the space $\mathcal{H}$ over which we optimize consists of all possible h for x_0.

The time step along planned trajectories is denoted d instead of k, for two reasons: to emphasize the difference between planned and real interaction with the system, and because d will be related to the depth in certain trees considered later. The rule h is different from the policy π used while *interacting* with the system, which is the overall result of applying the planning algorithm at each step. That is why we call h a *planning* policy. Since in the sequel we are mostly interested in planning policies, they will often simply be called "policies" for the sake of brevity. When planning policies are discussed together with interaction policies, we will use the long, explicit names of both, to make things clear. Note it is not strictly necessary to have a dependence on time d in the planning policy, since for the discounted infinite-horizon MDPs considered in this chapter, a policy that only depends on the state is sufficient to obtain optimal behavior. Nevertheless, this is the general form of the policy sought by planning algorithms.

Usually, an amount n of computational resources will be specified, and the planning algorithm must return a u_0 with the smallest simple regret given this constraint. The computational units can be number of evaluations of the model functions, number of basic arithmetic operations, and so on. In practice, n may often not be specified in advance but may depend for example, on real-time constraints. This motivates an anytime behavior, which returns the best action possible after any amount n of units spent, without using the actual value of n. The exploration–exploitation tradeoff still appears in the weaker sense of judiciously prioritizing the sampling of planning policies to minimize the final, simple regret.

Three planning algorithms will be considered in detail in this section: optimistic planning for deterministic systems (OPD) [3], open loop optimistic planning (OLOP) [9], and optimistic planning for sparsely stochastic systems (OPSS) [10]; the latter two algorithms work for stochastic systems.

All these algorithms require discrete actions, so that $U = \{u^1, ..., u^M\}$. OPSS also requires that each random transition ends up in one of a finite number of next states. Thus it works for finite-state MDPs, as well as infinite-state MDPs that satisfy the condition, which we call "sparsely stochastic." Furthermore, for all algorithms it is assumed the rewards are bounded in the interval [0, 1], i.e., $\rho(x, u, x') \in [0, 1]$ for any x, u, x'. This assumption is not overly restrictive since, at least for problems that do not have terminal states,[2] any bounded reward function can be normalized to [0, 1] by translation and scaling, without changing the optimal policies.

22.3.1 Optimistic Planning for Deterministic Systems

In the deterministic case, the planning policy is sufficiently specified by an infinite sequence of actions $h = [u_0, u_1, \ldots, u_d, \ldots]$ with $u_d \in U$, since this sequence completely determines the system trajectory. Thus $\mathcal{H}$ is the infinitely dimensional space U^∞. The value (return) of a policy h is

$$v(h) = \sum_{d=0}^{\infty} \gamma^d r_{d+1} = \sum_{d=0}^{\infty} \gamma^d \rho(x_d, u_d, x_{d+1}). \tag{22.7}$$

Define the metric (distance between policies) over $\mathcal{H}$:

$$\ell(h, \bar{h}) = \frac{\gamma^{\delta(h,\bar{h})}}{1-\gamma},$$

where $\delta(h, \bar{h}) = \min \{d \mid u_d \neq \bar{u}_d\}$, that is, the smallest index where the two policies are different. Intuitively, this distance is the largest difference between the returns provided by the two policies in any possible MDP. Because the rewards are in [0, 1], the returns can differ by at most $\sum_{\bar{d}=\delta(h,\bar{h})}^{\infty} \gamma^{\bar{d}} \cdot 1 = \frac{\gamma^{\delta(h,\bar{h})}}{1-\gamma}$.

Optimistic planning for deterministic systems (OPD) is built starting from OOD (Algorithm 22.1). The sets considered are defined by incomplete sequences of length d, so that each policy in the set matches the sequence up to index $d-1$ and is free to choose any action afterwards: $H_d = \{[u_0, u_1, \ldots, u_{d-1}, \star, \star, \ldots]\}$ with '$\star$' taken to mean "any action." Note $H_0 = \mathcal{H}$ since it imposes no constraints on the policy. A set H_d is split by making the choices for u_d definite, so that we obtain M different children sets H_{d+1}, one for every value of $u_d \in \{u^1, \ldots, u^M\}$. Note that, under the chosen metric, the diameter of each such set is $\frac{\gamma^d}{1-\gamma}$ since any two policies in H_d can differ at the earliest at index d; thus also $\mathrm{diam}(\mathcal{H}) = \frac{1}{1-\gamma}$.

We have not specified how to sample a value in a leaf set H_d. In fact, since this value is an infinite sum of rewards, it is not possible to obtain an exact sample. This is fortunately not a problem, since an upper bound on the values of policies in some

[2]A terminal state is one from which the system can no longer escape, and from which all actions always obtain zero rewards. Such states can be used to represent, for example, "goal achieved" and "failure" situations.

leaf set H_d can still be obtained from the sequence of rewards simulated so far:

$$b_{\text{init}}(H_d) = \nu(H_d) + \text{diam}(H_d) = \nu(H_d) + \frac{\gamma^d}{1-\gamma},$$

$$\text{where } \nu(H_d) = \sum_{\bar{d}=0}^{d-1} \gamma^{\bar{d}} \rho(x_{\bar{d}}, u_{\bar{d}}, x_{\bar{d}+1}).$$

Note that $\nu(H_d)$ is a *lower* bound on the values of all policies in H_d. These bounds hold because the rewards that each policy in H_d obtains up to depth d are fixed, producing the partial return $\nu(H_d)$, while the rewards below d are all between 0 and 1. So, expanding a node and computing the rewards obtained for reaching each of its children replaces the usual sampling process.

After performing n expansions, OPD chooses the set that maximizes $\nu(H_d)$ amongst all the sets considered so far, and returns the action u_0 for this set. Notice that in contrast to the optimistic criterion used to build the tree, at the end OPD makes a safe choice, based on lower bounds. With this, the instantiation of OPD as a variant of Algorithm 22.1 is complete. Rather than making this instantiation explicit, we choose to present the algorithm in an alternative, equivalent way that contributes additional intuition and helps in understanding OPSS later.

A one-to-one mapping exists between the tree of sets H_d and a corresponding tree of *states*, Figure 22.2. A node s_d at depth d in this latter tree is labeled by an actual state $x(s_d)$, with the root s_0 labeled by the initial state x_0. An arc (s_d, s_{d+1}) corresponds to a transition from $x(s_d)$ to $x(s_{d+1})$ and is labeled by the associated action u_d and reward $r_{d+1} = \rho(x(s_d), u_d, x(s_{d+1}))$. A path through the tree gives the state trajectory obtained by applying a corresponding sequence $u_0, u_1, \ldots, u_{d-1}$; thus each node s_d

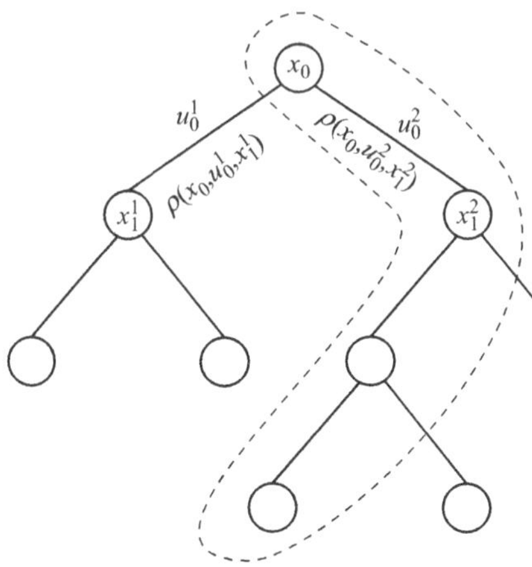

FIGURE 22.2 Illustration of an OPD tree. Nodes are labeled by states, and arcs correspond to transitions. Subscripts are depths, superscripts index the M possible actions/transitions from each node ($M = 2$ in this example). The dashed outline encloses a possible optimistic path.

Algorithm 22.3: Optimistic planning for deterministic systems

initialize tree: $\mathcal{T} \leftarrow \{s_0\}$
for $t = 0, \ldots, n$ **do**
  $s \leftarrow$ root s_0; **while** s is not leaf **do** $s \leftarrow \arg\max_{s' \in C(s)} b(s')$ **end while**
  optimistic leaf found: $s^\dagger \leftarrow s$
  expand $s^\dagger$: simulate all M transitions from $x(s^\dagger)$,
    add to $\mathcal{T}$ nodes s' containing the resulting states x'
  update b-values upwards in the tree, starting from the leaves:

$$b(s) = \begin{cases} b_{\text{init}}(s) \text{ from (22.8)} & \text{if } s \text{ is leaf} \\ \max_{s' \in C(s)} b(s') & \text{otherwise} \end{cases}$$

end for
output u_0, the first action along path with largest ν

corresponds to a set H_d. It is important to note that a certain state x may appear several times in the tree; the variant of OPD presented here does not merge information from these duplicates.

Expanding a node s is done by simulating the transitions from $x(s)$ for all M actions, and adding nodes for all the resulting next states x' as children $C(s)$ to the tree. The b-value of a leaf node is

$$b_{\text{init}}(s) = \nu(s) + \frac{\gamma^{d}(s)}{1-\gamma}, \tag{22.8}$$

where $\nu(s)$ is the return along the path from the root s_0 to s. Algorithm 22.3 summarizes OPD using this second type of tree.

22.3.2 Open-Loop Optimistic Planning

In the stochastic case, there are multiple state outcomes for a given action at some step, so a simple sequence of actions is no longer sufficient to represent a full planning policy. Such a policy would need to depend on the state, having for example, the form $h(d, s_d)$ in (22.6), which specifies an action choice for each possible state outcome s_d at every step d. Representing such a policy is unfortunately impossible in general, since there can be an infinite number of outcomes. Instead, the *open-loop optimistic planning* (OLOP) algorithm chooses to still optimize only over plain sequences of actions, without explicitly considering the underlying state outcomes—hence its "open-loop" nature.[3]

OLOP will be described using the tree-of-sets formalism, since a tree of states does not exist in this case. The space $\mathcal{H}$ and metric are the same as in the deterministic case, and splitting has the same meaning of initializing all action choices for the next

[3]Instead of the theoretical OLOP algorithm of [9], we present here a variant more directly related to the HOO template, and also more amenable to practical implementation.

step. However, the value of a policy h is now the expected return:

$$v(h) = \mathrm{E}_{x_{d+1} \sim f(x_d, u_d, \cdot)} \left\{ \sum_{d=0}^{\infty} \gamma^d \rho(x_d, u_d, x_{d+1}) \right\}, \tag{22.9}$$

so we are dealing with a noisy optimization problem: summing up the rewards along some simulated trajectory only gives a *sample* return. Moreover, to obtain independent return samples, whenever traveling down the optimistic action path in the tree a *new* trajectory must be simulated, rather than relying on the already existing reward samples as in OPD. Formally, each set (tree node) H_d is associated with a cumulative reward $r(H_d)$ and a count $n(H_d)$ of how many reward samples are available for it, with both quantities initialized to 0. Simulating a trajectory $(x_{\bar{d}}, u_{\bar{d}}, x_{\bar{d}+1}, r_{\bar{d}+1})$ for $\bar{d} = 0, \ldots, d-1$ provides new reward samples for all the sets $H_{\bar{d}+1} = \{[u_0, u_1, \ldots, u_{\bar{d}}, \star, \star, \ldots]\}$:

$$\begin{aligned} r(H_{\bar{d}+1}) &\leftarrow r(H_{\bar{d}+1}) + r_{\bar{d}+1}, \\ n(H_{\bar{d}+1}) &\leftarrow n(H_{\bar{d}+1}) + 1. \end{aligned}$$

OLOP can be seen as a variant of HOO (Algorithm 22.2) that uses the b-values:

$$b_{\mathrm{init}}(H_d) = \sum_{\bar{d}=0}^{d-1} \gamma^{\bar{d}} \left[\frac{r(H_{\bar{d}+1})}{n(H_{\bar{d}+1})} + \sqrt{\frac{2\log(t)}{n(H_{\bar{d}+1})}} \right] + \frac{\gamma^d}{1-\gamma}, \tag{22.10}$$

for some set H_d, where t is the number of *trajectories* simulated so far. These b-values are different from what the "vanilla" application of HOO would give, see (22.5): they cleverly exploit the additive structure of the value (return) to compute separate UCBs at each step, whereas vanilla HOO would compute a single UCB for the entire trajectory. The diameter component $\gamma^d/(1-\gamma)$ remains the same as in the deterministic case.

Algorithm 22.4 summarizes OLOP, in a variant where the computational unit consists of simulating a single transition (rather than M as in OPD), and the final action choice is based on the most-often sampled first action (rather than the largest ν). There is a clear parallel between OLOP as an application of HOO, and OPD as an application of OOD. For instance, OLOP requires a minimum with the b-value b_{init} of the current node when propagating b-values up the tree.

22.3.3 Optimistic Planning for Sparsely Stochastic Systems

When each random state transition can only end up in a small number N of next states, it is in fact possible to represent full, closed-loop planning policies $h(d, s_d)$, which also depend on the state node and hence on the underlying state. More specifically, a

Algorithm 22.4: Open-loop optimistic planning

initialize tree: $\mathcal{T} \leftarrow \{\mathcal{H}\}$
for $t = 1, 2, ...,$ stopping when n transitions have been exceeded **do**
  $H \leftarrow$ root $\mathcal{H}$; **while** H is not leaf **do** $H \leftarrow \arg\max_{H' \in \mathcal{C}(H)} b(H')$ **end while**
  optimistic leaf found: $H^\dagger \leftarrow H$
  expand $H^\dagger$ into M child subsets H', one for each action, adding them to $\mathcal{T}$
  simulate a trajectory from x_0, using all the actions in $H^\dagger$, reaching some x_d
  from x_d, simulate a transition for each action
  update b-values upwards in the tree, starting from the leaves:

$$b(H) = \begin{cases} b_{\text{init}}(H) \text{ from (22.10)} & \text{if } H \text{ is leaf} \\ \min\left\{b_{\text{init}}(H), \max_{H' \in \mathcal{C}(H)} b(H')\right\} & \text{otherwise} \end{cases}$$

end for
output u_0, a first action that was sampled most often

policy can be represented as a vector:

$$h = [u_0, u_1^1, \ldots, u_1^N, u_2^{1,1}, \ldots, u_2^{N,N}, \ldots],$$

because at step 1 it must provide an action for each of the possible N outcomes; at step 2, for each of the N^2 outcomes, and so on. Exploiting this idea, a deterministic algorithm can be built on the foundations of OOD, even though the system is stochastic. We call systems with the property above sparsely stochastic, so the algorithm is *optimistic planning for sparsely stochastic systems* (OPSS). This class includes all finite-state MDPs (where good performance can be expected when N is small), as well as important classes of continuous-state MDPs such as those that arise by combining deterministic dynamics with discrete random variables, or by discretizing in time some continuous-time stochastic systems.

Rather than use the h notation which is cumbersome in this case, OPSS is described using a tree of states, such as OPD earlier. This tree $\mathcal{T}$ is built starting from a root s_0 labeled by x_0 and iteratively expanding nodes. Each expansion consists of generating and adding all the N one-step successor states of the node-to-expand, for all M actions, see Figure 22.3, left. The b-values of leaf nodes are computed with:

$$b_{\text{init}}(s) = \nu(s) + \frac{\gamma^{d}(s)}{1 - \gamma}, \tag{22.11}$$

where $\nu(s)$ is the discounted sum of rewards along the path from s_0 to s. The b-values of nodes higher in the tree maximize *expected* b-values under the children's probabilities: $b(s) = \max_{u \in U} \sum_{s' \in \mathcal{C}(s,u)} f(x(s), u, x(s'))b(s')$. Here, $\mathcal{C}(s, u)$ denotes the set of children reachable from s after performing action u.

Rather than an optimistic path as in OPD and OLOP, OPSS builds an *optimistic subtree* $\mathcal{T}^\dagger$ by starting from the root and selecting at each node s *all* the children

Algorithm 22.5: Optimistic planning for sparsely stochastic systems

initialize tree: $\mathcal{T} \leftarrow \{s_0\}$
for $t = 1, \ldots, n$ **do**
    build optimistic subtree $\mathcal{T}^\dagger$
    choose node to expand: $s^\dagger \leftarrow \arg\max_{s \in \mathcal{L}^\dagger} \mathrm{P}(s) \frac{\gamma^{d(s)}}{1-\gamma}$
    expand $s^\dagger$: simulate all NM transitions from $x(s^\dagger)$,
        add to $\mathcal{T}$ nodes s' containing the resulting states x'
    update b-values upwards in the tree, starting from the leaves:

$$b(s) = \begin{cases} b_{\text{init}}(s) \text{ from (22.11)} & \text{if } s \text{ is leaf} \\ \max_{u \in U} \sum_{s' \in \mathcal{C}(s,u)} f(x(s), u, x(s'))b(s') & \text{otherwise} \end{cases}$$

end for
output u_0 maximizing ν

associated to an optimistic action:

$$u^\dagger(s) = \arg\max_{u \in U} \sum_{s' \in \mathcal{C}(s,u)} f(x(s), u, x(s'))b(s').$$

Among the leaves $\mathcal{L}^\dagger$ of this subtree, the node to expand is selected using

$$s^\dagger = \arg\max_{s \in \mathcal{L}^\dagger} \mathrm{P}(s) \frac{\gamma^{d(s)}}{1-\gamma}, \tag{22.12}$$

where $\mathrm{P}(s)$ is the probability to reach s, that is the product of all probabilities along the path to s: $\mathrm{P}(s) = \sum_{d=0}^{d(s)-1} f(x(s_d), u_d, x(s_{d+1}))$. After n expansions, the algorithm selects at the root an action:

$$u_0 = \arg\max_{u \in U} \sum_{s' \in \mathcal{C}(s_0,u)} f(x_0, u, x(s'))\nu(s'),$$

where ν-values are propagated up the tree just like b-values, starting from their definition at the leaves given above.

Algorithm 22.5 summarizes OPSS. The computational unit consists in this case of generating all the NM children of a particular state.

To better understand the choices made in OPSS, let us informally return to the tree-of-policy-sets interpretation. To keep the parallel clear, we denote the tree of sets by $\boldsymbol{\mathcal{T}}$ in contrast to the tree of states $\mathcal{T}$, and always specify to which tree each entity belongs. Selecting the children of one particular action at each state node in $\mathcal{T}$, recursively until reaching some set of leaves $\bar{\mathcal{L}} \in \mathcal{T}$, corresponds to selecting some *single* leaf subset $H \in \mathcal{T}$, with the b-value:

$$b(H) = \sum_{s \in \bar{\mathcal{L}}} \mathrm{P}(s) \left[v(s) + \frac{\gamma^{d(s)}}{1-\gamma} \right]$$

$$= \sum_{s \in \bar{\mathcal{L}}} \mathrm{P}(s)\, v(s) + \sum_{s \in \bar{\mathcal{L}}} \mathrm{P}(s) \frac{\gamma^{d(s)}}{1-\gamma} = v(H) + \mathrm{diam}(H),$$

the standard formula for the b-value. The lower bound $v(H)$ is the sum of truncated returns up to the leaves in $\mathcal{T}$, weighted by their probabilities. A similar weighted formula holds for the diameter. Among all these leaf policy sets in $\mathcal{T}$, the optimistic action choices in $\mathcal{T}$ lead to the optimistic set $H^\dagger \in \mathcal{T}$, which has the largest b-value, and expanding the optimistic node $s^\dagger \in \mathcal{T}$ splits $H^\dagger \in \mathcal{T}$ into M children *along the longest edge*, which corresponds to $s^\dagger$.

A subtle but essential property of OPSS is that, contrary to all algorithms so far, this expansion *also splits many additional sets*: those corresponding to all policies that reach $s^\dagger$ with nonzero probability. Figure 22.3, right illustrates this effect, and more generally how expanding leaves in $\mathcal{T}$ corresponds to splitting sets in $\mathcal{T}$. Due to this multiple-splitting property, there is no one-to-one mapping between $\mathcal{T}$ and $\mathcal{T}$.

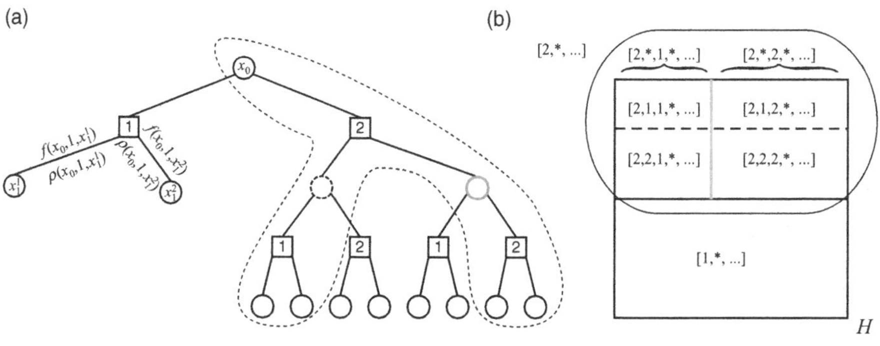

FIGURE 22.3 (a) Illustration of an OPSS tree after three expansions (root, gray node, and dashed node), for $N = M = 2$. The circles are state nodes, and the actions are explicitly represented as square, choice nodes. Transition arcs to next states are labeled by probabilities and rewards. The dashed outline encloses a possible optimistic subtree. (b) Corresponding partition of the policy space $\mathcal{H}$. Expanding the root splits $\mathcal{H}$ in two sets containing policies of the form $[1, \star, \ldots]$ and $[2, \star, \ldots]$, respectively (recall a policy has the structure $h = [u_0, u_1^1, u_1^2, \ldots]$). Next, expanding the gray node—which corresponds to the second random outcome of action 2—splits $\{[2, \star, \ldots]\}$ into $\{[2, \star, 1, \star, \ldots]\}$ and $\{[2, \star, 2, \star, \ldots]\}$; note the action for the first random outcome (dashed node) remains undefined. Expanding the dashed node then makes the action definite, thereby splitting *both* sets $\{[2, \star, 1, \star, \ldots]\}$ and $\{[2, \star, 2, \star, \ldots]\}$, since the dashed node is reached with nonzero probability by both classes of policies.

22.3.4 Theoretical Guarantees

The core theoretical question for optimistic planning algorithms concerns the quality of the action they produce after expending the budget n, in terms of the simple regret (22.2). Available guarantees rely on a measure of the complexity of the problem that can best be understood in terms of a branching factor of some related trees.

For OPD, this tree consists of near-optimal policy sets H: $\mathcal{T}^* = \{H_d \mid v^* - w(H_d) \leq \gamma^d/(1-\gamma)\}$, where v^* is the optimal return achievable from x_0, and where $w(H_d)$ is the infinite-horizon return obtained by first applying the actions defined in H_d and then acting optimally. So, $\mathcal{T}^*$ includes at every level d the policy sets for which it is impossible to tell, from the rewards achieved up that level, whether they contain an optimal policy. The larger the number of such sets, the more difficult the problem is for planning. Denoting the asymptotic (as $d \to \infty$) branching factor of $\mathcal{T}^*$ by κ, the simple regret of OPD is

$$\mathcal{R}(x_0) = \begin{cases} O(n^{-\frac{\log 1/\gamma}{\log \kappa}}) & \text{if } \kappa > 1 \\ O(\gamma^{nc}) & \text{if } \kappa = 1, \end{cases}$$

where c is a specific constant [3]. In the ideal case $\kappa = 1$, which means there is a single near-optimal set at every level, and a single optimal planning policy. A regret that decreases exponentially with n is obtained in this case. As κ grows to K, the regret decreases more slowly with n.

OLOP defines the branching factor κ with a similar formula, but allows for technical reasons the sets to be suboptimal by up to $2\gamma^d/(1-\gamma)$ rather than $\gamma^d/(1-\gamma)$. Then, roughly speaking (the actual technical statement is more involved) [9]:

$$\mathcal{R}(x_0) \approx \begin{cases} O(n^{-\frac{\log 1/\gamma}{\log \kappa}}) & \text{if } \gamma\sqrt{\kappa} > 1 \\ O(n^{-1/2}) & \text{if } \gamma\sqrt{\kappa} \leq 1 \end{cases}.$$

Remarkably, when $\gamma\sqrt{\kappa} > 1$ the regret has the same form as for OPD.

Similar guarantees hold for OPSS [12], but due to the additional intricacies caused by the multiple-splitting property explained earlier, introducing these guarantees would require extensive preliminaries that fall beyond the scope of this chapter.

22.4 RELATED PLANNING ALGORITHMS

A recent algorithm in the optimistic planning family is UCB applied to trees (UCT) [8], which has made a great impact in the planning field due to its competitive performance in games such as Go [7]. UCT basically applies UCB at every node in a planning tree similar to that of Figure 22.3. It travels an optimistic path along the tree choosing actions with maximal UCBs (22.3) on the returns obtained from the current node, and sampling states independently. An issue with UCT is that its b-values do not consider the size of the policy sets, which means they are not true upper confidence bounds; equivalently, it can be said that UCT assumes infinite smoothness of the value function it optimizes. A first attempt to address this problem, by introducing a so-called smoothness coefficient in place of the set diameter, was made in [13]. This "bandit algorithm for smooth trees" can be considered a precursor of OLOP. In [14] a continuous-action variant of UCT is introduced, which tackles continuous actions by using HOO instead of UCB in the tree nodes.

On the other hand, optimistic planning has a strong relationship to classical planning algorithms such as A* and AO* [2, 15], which also expand planning trees by optimistically prioritizing nodes. By using upper bounds on the values of incomplete action sequences, A* and AO* do consider diameters of policy sets and obey the principles of optimistic optimization (more specifically OOD). OPD can be seen as an extension of A* from goal-based problems to more general, infinite-horizon discounted MDPs, and OPSS as a similar extension of AO*, with the specific rule (22.12) for choosing the particular node to expand. Exploiting UCBs is a strong innovation brought by UCT and OLOP to classical planning.

The key observation here is that, in the context of MDPs, all these algorithms belong to a general framework for optimistic optimization over planning policies. This insight guides essential choices in the algorithms—for instance, the need to always include diameters in the b-values, and the node expansion criterion (22.12) in OPSS, intuitively seen as splitting a set along the longest edge. If these choices are made correctly, strong theoretical guarantees on the performance are achieved in terms of the simple regret, as outlined in Section 22.3.4.

An early, nonoptimistic sample-based planning algorithm for MDPs was given by [16]. This algorithm builds a uniform, symmetric planning tree by sampling a fixed number of states for each action, at each depth up to some horizon. It is called sparse sampling due to the finite number of outcomes sampled, so its sparseness is of a different nature than in OPSS. Sparse sampling was extended by [17] to consider an adaptive horizon. "Forward-search sparse sampling" [18] is an optimistic extension, which does not expand the tree uniformly, but along paths obtained by choosing maximum b-value actions, and then among the successors choosing the one with the maximum *gap* between the b-value and a lower-bound on the state value. Disturbance trees are very similar to the planning trees used here, and planning algorithms based on them were given for example by [19].

22.5 NUMERICAL EXAMPLE

To understand the behavior of OPD, OLOP, and OPSS in practice, they will be applied to the problem of swinging up an underactuated inverted pendulum—a rather simple problem commonly used in the literature on solving MDPs.

The inverted pendulum consists of a mass attached to an actuated link (a rod) that rotates in a vertical plane. The available power is taken insufficient to push the pendulum up in a single rotation from every initial state. Instead, from certain states (e.g., pointing down), the pendulum needs to be swung back and forth to gather energy, prior to being pushed up and stabilized. Finding such a solution is challenging for planning algorithms, because the swing-up must be planned over a longer horizon, and solutions that seem optimal over a short horizon will not work, instead just pushing the pendulum in one direction.

A continuous-time model of the pendulum dynamics is

$$\ddot{\alpha} = 1/J \cdot [mgl \sin(\alpha) - b\dot{\alpha} - K^2\dot{\alpha}/R + Ku/R],$$

where we have inertia $J = 1.91 \times 10^{-4}\,\text{kgm}^2$, mass $m = 0.055\,\text{kg}$, gravitational acceleration $g = 9.81\,\text{m/s}^2$, length $l = 0.042\,\text{m}$, damping $b = 3 \times 10^{-6}\,\text{Nms/rad}$, torque constant $K = 0.0536\,\text{Nm/A}$, and resistance $R = 9.5\,\Omega$. The angle α varies in the interval $[-\pi, \pi]$ rad, with $\alpha = 0$ pointing up, and "wraps around" so that for example a rotation of $3\pi/2$ corresponds to $\alpha = -\pi/2$. The state is $x = [\alpha, \dot{\alpha}]^\top$. The velocity $\dot{\alpha}$ is restricted to $[-15\pi, 15\pi]$ rad/s, using saturation. The control action $u \in [-3, 3]$ V may be affected by noise as described later in this chapter. It is discretized into the set $U = \{-3, 0, 3\}$, so that $M = 3$. The sampling time is $T_s = 0.05$ s.

The goal is to stabilize the pendulum in the unstable equilibrium $x = 0$ (pointing up), and is expressed by the *unnormalized* quadratic rewards:

$$r = \rho_{\text{unnorm}}(x, u, x') = -5\alpha^2 - 0.1\dot{\alpha}^2 - u^2.$$

Using the known bounds on the state and action variables, the rewards are normalized (scaled and translated) to the interval [0, 1]. The discount factor is $\gamma = 0.95$.

Two versions of the problem are considered. The first variant has deterministic (noiseless) actions, and is used to exemplify the behavior of OPD. In the second variant, an unreliable actuator is modeled that only applies the intended action u with probability 0.6, and applies an action with smaller magnitude, $0.7u$, with probability 0.4 (when the intended action is 0 it always remains 0). This corresponds to a sparsely stochastic MDP with $N = 2$, a more difficult problem than the deterministic variant. The results we will present for the stochastic inverted pendulum are similar to those in [10].

To obtain a global performance measure, all algorithms are applied in an offline fashion, to find actions for the states on the grid $X_0 = \left\{-\pi, \frac{-150\pi}{180}, \frac{-120\pi}{180}, \ldots, \pi\right\} \times \{-15\pi, -14\pi, \ldots, 15\pi\}$. Since an exact optimal Q-function for the inverted pendulum problem is not known, to approximate the simple regret a near-optimal Q-function is computed instead using interpolation-based approximate value iteration [20, 21] with an accurate approximator.

Baseline planning solutions are obtained using uniform planning, which always expands a node having the smallest depth instead of an optimistic node, thus developing uniform, symmetric trees.

Results with OPD. Figure 22.4 shows the results obtained by OPD for the deterministic inverted pendulum, in comparison to uniform planning. The budget n varies in the set $\{50, 100, 200, 300, \ldots, 900\}$. As expected, OPD works better (obtains smaller regret) than uniform planning, since it expands the planning trees in a

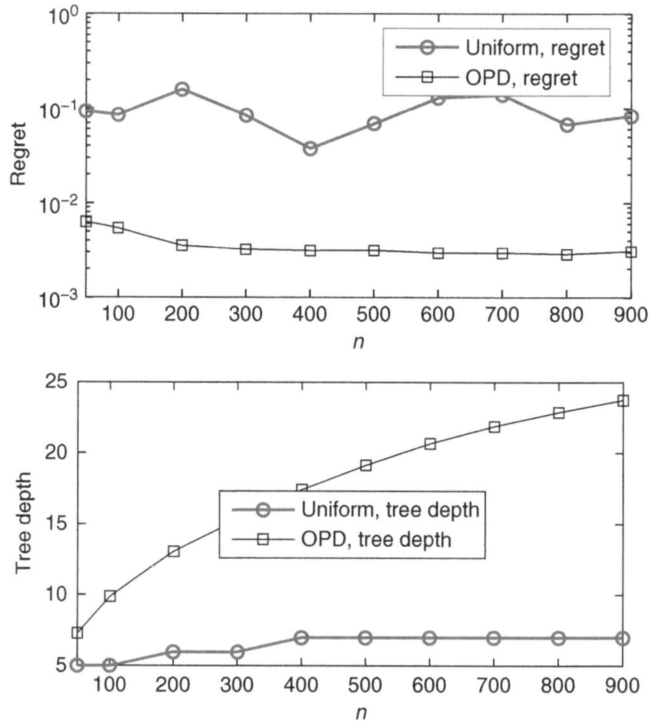

FIGURE 22.4 Results of OPD, compared to uniform planning: average regret over X_0 (left), average tree depth over X_0 (right).

smart way, resulting in much deeper trees. The regret of OPD decreases with n but quickly plateaus after $n = 300$, which indicates this budget is already sufficient.[4]

Results with OLOP and OPSS. For OLOP and OPSS, the sparsely stochastic variant of the inverted pendulum is used. For OPSS, the computational budget n varies in the set $\{50, 100, 200, 300, \ldots, 900\}$ like for OPD above. Each node expansion requires in principle $NM = 6n$ transitions—in practice, since action 0 has a deterministic outcome, the number is in fact $5n$. Nevertheless, because the computational unit of OLOP consists of simulating a single transition, $6n$ transitions are allowed (and the algorithm may use slightly more to finish the last trajectory).

[4]Note that the regret does not decrease monotonically with n. This is because regret can vary nonmonotonically with the planning horizon. For a concrete example, consider a deterministic MDP with state space $\{1, 2, \ldots, 6\}$, two actions $-1, 1$, and additive dynamics $x' = \max(1, \min(6, x + u))$. The rewards obtained upon reaching one of the six states are, respectively, 4, 0, 0, 1, -10, 100, and the discount factor is 0.5. Then, the optimal action for $x = 3$ is 1. If a fully expanded planning tree of depth (horizon) 1 is used to compute ν-values and choose an action, the action returned will be 1. If the depth is 2, the action changes to -1 due to the -10 reward. For depths larger than 2, due to the 100 reward, the action is again optimal, i.e., 1. The simple regret for $x = 3$ is thus clearly nonmonotonic with respect to the depth.

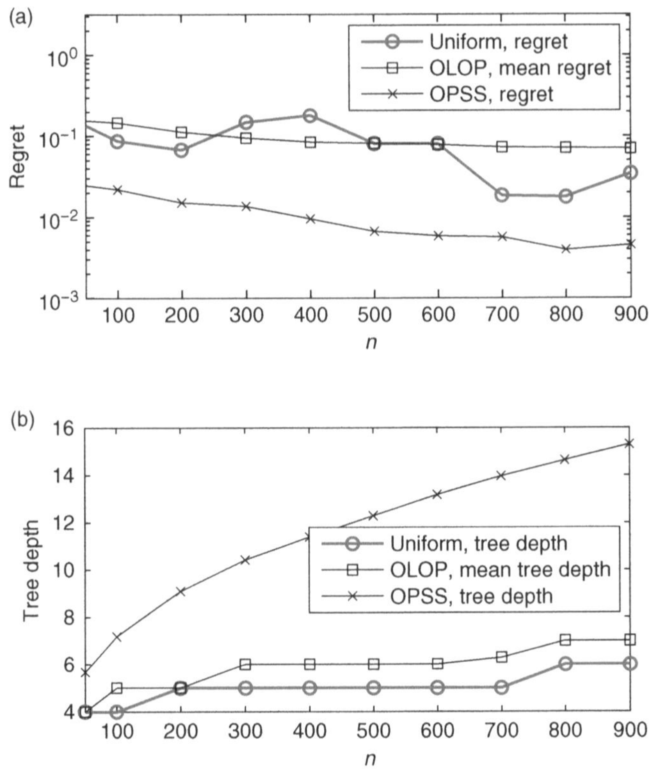

FIGURE 22.5 Comparison between OPSS and uniform planning: average regret over X_0 (a), average tree depth over X_0 (b). As the results of OLOP depend on particular realizations of stochastic trajectories, this algorithm is run 10 times and mean results are reported (the 95% confidence regions are too narrow to be visible at this scale).

Figure 22.5 shows the results obtained, compared to the uniform planning baseline. The relationship between OPSS and uniform planning mirrors that between OPD and uniform planning in the deterministic case, although here the performance of OPSS continues to improve as n increases, reflecting the more difficult nature of the stochastic problem. Less expected is that OLOP works poorly, similar to uniform planning. This likely happens because the computational budgets considered do not allow OLOP to decrease the upper confidence bounds on the returns to informative levels; OLOP may work better with a larger budget.

To illustrate the online control performance of optimistic planning, OPSS is applied in a receding-horizon fashion starting from the stable equilibrium of the pendulum (pointing down), and $n = 600$. Figure 22.6 shows the resulting trajectory, compared to that given by uniform planning. OPSS swings up the pendulum in one go, whereas uniform planning manages it only after a few attempts.

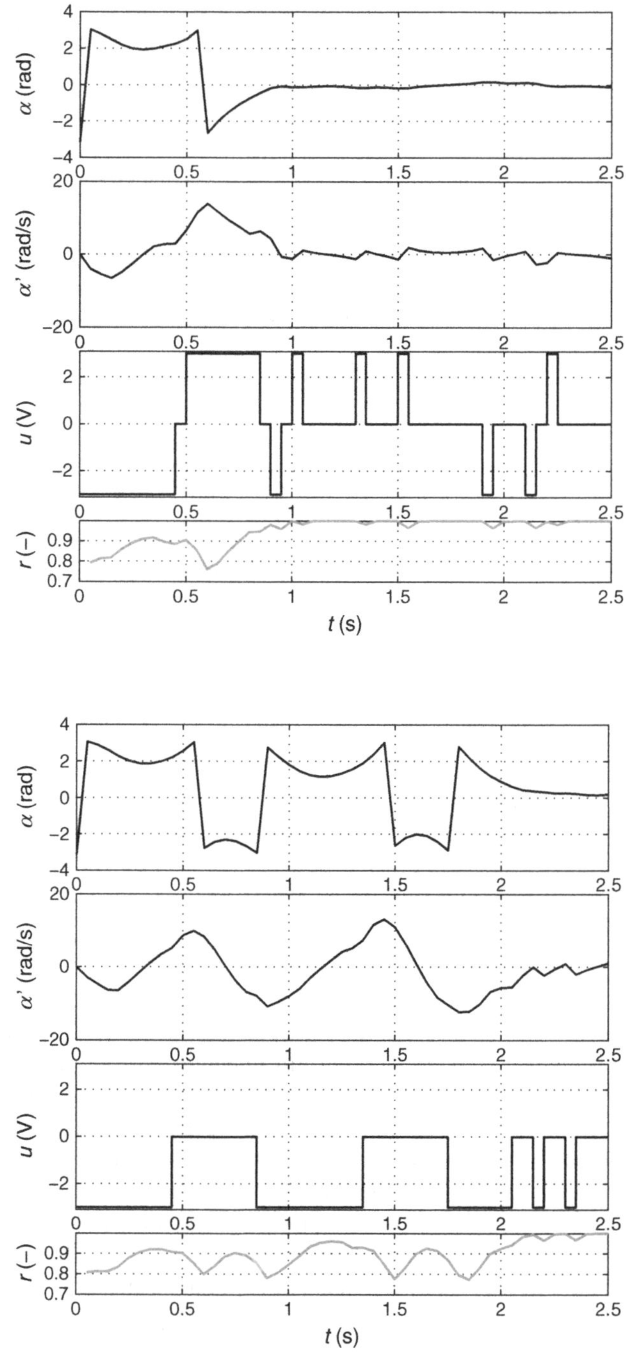

FIGURE 22.6 Online control results of OPSS and uniform planning.

REFERENCES

1. J.M. Maciejowski. *Predictive Control with Constraints*. Prentice Hall, 2002.
2. S.M. La Valle. *Planning Algorithms*. Cambridge University Press, 2006.
3. J.-F. Hren and R. Munos. Optimistic planning of deterministic systems. In *Proceedings 8th European Workshop on Reinforcement Learning (EWRL-08)*, pp. 151–164, Villeneuve d'Ascq, France, June 30 to July 3, 2008.
4. L. Buşoniu, R. Babuška, B. De Schutter, and D. Ernst. *Reinforcement Learning and Dynamic Programming Using Function Approximators*. Automation and Control Engineering. Taylor & Francis, CRC Press, 2010.
5. R.S. Sutton and A.G. Barto. *Reinforcement Learning: An Introduction*. MIT Press, 1998.
6. P. Auer, N. Cesa-Bianchi, and P. Fischer. Finite-time analysis of the multiarmed bandit problem. *Machine Learning*, 47(2–3):235–256, 2002.
7. S. Gelly, Y. Wang, R. Munos, and O. Teytaud. Modification of UCT with patterns in Monte-Carlo Go. Technical report, INRIA, 2006.
8. L. Kocsis and C. Szepesvári. Bandit based Monte-Carlo planning. In *Proceedings 17th European Conference on Machine Learning (ECML-06)*, pp. 282–293, Berlin, Germany, September 18–22, 2006.
9. S. Bubeck and R. Munos. Open loop optimistic planning. In *Proceedings 23rd Annual Conference on Learning Theory (COLT-10)*, pp. 477–489, Haifa, Israel, June 27–29, 2010.
10. L. Buşoniu, R. Munos, B. De Schutter, and R. Babuška. Optimistic planning for sparsely stochastic systems. In *Proceedings 2011 IEEE International Symposium on Adaptive Dynamic Programming and Reinforcement Learning (ADPRL-11)*, pp. 48–55, Paris, France, April 11–15, 2011.
11. S. Bubeck, R. Munos, G. Stoltz, and C. Szepesvári. Online optimization in X-armed bandits. In D. Koller, D. Schuurmans, Y. Bengio, and L. Bottou, editors, *Advances in Neural Information Processing Systems 21*, MIT Press, 2009, pp. 201–208.
12. L. Buşoniu and R. Munos. Optimistic planning for Markov decision processes. In *15th International Conference on Artificial Intelligence and Statistics (AISTATS-12)*, 2012.
13. P.-A. Coquelin and R. Munos. Bandit algorithms for tree search. In *Proceedings 23rd Conference on Uncertainty in Artificial Intelligence (UAI-07)*, pp. 67–74, Vancouver, Canada, July 19–22, 2007.
14. C. Mansley, A. Weinstein, and M.L. Littman. Sample-based planning for continuous action Markov decision processes. In *Proceedings 21st International Conference on Automated Planning and Scheduling*, pp. 335–338, Freiburg, Germany, June 11–16, 2011.
15. N.J. Nilsson. *Principles of Artificial Intelligence*. Tioga Publishing, 1980.
16. M.J. Kearns, Y. Mansour, and A.Y. Ng. A sparse sampling algorithm for near-optimal planning in large Markov decision processes. *Machine Learning*, 49(2-3):193–208, 2002.
17. L. Péret and F. Garcia. On-line search for solving Markov decision processes via heuristic sampling. In *Proceedings 16th European Conference on Artificial Intelligence, ECAI'2004*, pp. 530–534, Valencia, Spain, August 22–27, 2004.
18. T.J. Walsh, S. Goschin, and M.L. Littman. Integrating sample-based planning and model-based reinforcement learning. In *Proceedings 24th AAAI Conference on Artificial Intelligence (AAAI-10)*, Atlanta, US, July 11–15, 2010.

19. B. Defourny, D. Ernst, and L. Wehenkel. Planning under uncertainty, ensembles of disturbance trees and kernelized discrete action spaces. In *Proceedings 2009 IEEE International Symposium on Adaptive Dynamic Programming and Reinforcement Learning (ADPRL-09)*, pp. 145–152, Nashville, US, March 30 to April 2, 2009.
20. L. Buşoniu, D. Ernst, B. De Schutter, and R. Babuška. Approximate dynamic programming with a fuzzy parameterization. *Automatica*, 46(5):804–814, 2010.
21. J.N. Tsitsiklis and B. Van Roy. Feature-based methods for large scale dynamic programming. *Machine Learning*, 22(1–3):59–94, 1996.

CHAPTER 23

Adaptive Feature Pursuit: Online Adaptation of Features in Reinforcement Learning

SHALABH BHATNAGAR,[1] VIVEK S. BORKAR,[2] and L.A. PRASHANTH[1]

[1]Department of Computer Science and Automation, Indian Institute of Science, Bangalore, India
[2]Department of Electrical Engineering, Indian Institute of Technology, Powai, Mumbai, India

ABSTRACT

We present a novel feature adaptation scheme based on temporal difference (TD) learning for the problem of prediction. The scheme suitably combines aspects of *exploitation* and *exploration* by (a) finding the worst basis vector in the feature matrix at each stage and replacing it with the current best estimate of the normalized value function and (b) replacing the second worst basis vector with another vector chosen randomly that would result in a new subspace of basis vectors getting picked. We apply our algorithm to a problem of prediction in traffic signal control and observe good performance over two different network settings.

Keywords: online feature adaptation, temporal difference learning, traffic signal control.

23.1 INTRODUCTION

Markov decision process (MDP) [1–3] is a general framework for solving stochastic control problems. Classical solution approaches for MDP such as policy and value iteration solve the associated control problem precisely by identifying an optimal action to pick in each state. These approaches typically suffer from two major problems:

Reinforcement Learning and Approximate Dynamic Programming for Feedback Control, First Edition. Edited by Frank L. Lewis and Derong Liu.

(a) they require precise knowledge of the transition probabilities, that is, the system model and (b) the amount of computation required to obtain a solution using these approaches grows exponentially with the size of the state and/or action space. Reinforcement learning (RL) [4–6] provides efficient solutions for both problems above. Many RL algorithms are incremental update stochastic approximation algorithms that work directly with real or simulated data. These algorithms make use of the averaging property of stochastic approximation as a result of which they work efficiently even when the system transition model is not known. For problems where the cardinality of the state/action space is so large that precise solutions cannot be obtained easily, one resorts to certain approximation methods. Often the value function is approximated as a function of certain parameter(s). The functional form of the approximator is also called an architecture. Architectures based on linear function approximators have been well studied in the literature because algorithms such as temporal difference (TD) learning [7] have been shown convergent when linear architectures are used [8, 9]. Here, the value of a given state is approximated using the scalar product of the parameter vector with the feature associated with the state. The feature usually quantifies important state attributes. On the other hand, TD with nonlinear function approximators has been seen to diverge in some cases. An important problem that we address in this chapter is to design an efficient scheme to adaptively select the "best features" when linear function approximation is used.

The problem of feature adaptation has been studied recently for problems of prediction as well as control. In [10], features are assumed parameterized and the problem considered is of finding the optimum feature parameter when TD learning is used. Two algorithms, one based on a gradient approach and the other based on a cross entropy method are then presented for tuning the feature parameter. Certain generalizations of the parameterized basis adaptation approach of [10] have been presented in [11]. The basis adaptation scheme there involves low-order calculations. In [12], a procedure based on state aggregation and neighborhood component analysis is used for constructing basis functions assuming a linear approximation architecture. In [13], Krylov subspace basis functions, involving powers of the transition probability matrix of the associated Markov chain are used. Noisy samples of the basis functions are obtained as they cannot be computed exactly. In [14], a Laurent series expansion of the value function is used for basis construction. Another interesting work is [15] where through a recursive procedure, an "ideal" basis function that is a representative of the value function is obtained. Whereas references [15] deal with basis adaptation for the problem of prediction, in [16], the problem of control with adaptive bases is considered. Multiscale actor-critic algorithms are developed where on the "slowest" timescale, the feature parameters are updated. Algorithms based on TD error, mean square Bellman error and mean square projected Bellman error are presented there. All of the above works consider features to be parameterized and the goal in these references is to find optimal parameters and thus the optimal features within the specific parameterized feature classes. In [17], feature adaptation for a problem of prediction is considered. Unlike in the above references, the basis functions there are not assumed parameterized. The Bellman error is included as an additional basis in each iteration, thereby increasing the dimension of the subspace at each iteration. In [18], an *a priori* approximation of the value function based on a simplified model as one of the features has been considered. In [19], an approach that makes use of both state aggregation and linear function approximation is presented. While the feature matrix is kept fixed there, state aggregation is done adaptively on a slower timescale using estimates of the approximate value function. In this case, we obtain locally optimal state clusters asymptotically.

In this chapter, we propose a new algorithm for feature adaptation. We consider the problem of prediction (i.e., estimating the value function corresponding to a given policy) under the infinite horizon discounted cost criterion and use the TD learning scheme with linear function approximation for this purpose. We let all feature components to take values only between 0 and 1. Upon convergence of the TD scheme for a fixed set of features, we consider the converged weights in the weight vector (i.e., the parameter whose scalar product with the state-feature approximates the value of that state). We then replace two of the columns of the feature matrix that correspond to the two smallest components of the weight vector as follows. The column of the feature matrix corresponding to the smallest component of the weight vector is replaced by the most recent estimate of the value function (suitably normalized so that all its entries are between 0 and 1) obtained after running the TD scheme for a given large number of iterations using the feature matrix of the previous step, while the column corresponding to the second smallest component of the weight vector is replaced by independent and uniformly generated random numbers between 0 and 1. Thus, while the worst performer (among the columns of the feature matrix) is replaced by the normalized value function estimate, the second-worst performer is replaced with a random search direction to aid in exploration of better features and thereby a better subspace than the previous. The remaining columns of the feature matrix are left unchanged. The TD algorithm is then run again with the new feature matrix and the process repeated.

As an application setting, we consider the problem of estimating the value of a policy in a problem of traffic signal control [20]. The general control problem in this domain is to obtain a policy for signal switching across various sign configurations so that traffic flow is maximized and levels of congestion and hence delays at the traffic junctions are minimized. In [20], Q-learning with function approximation for a given feature matrix is applied for this problem. By keeping the optimal policy (obtained from the Q-learning algorithm in [20]) corresponding to the original feature matrix fixed, we apply our algorithm for feature adaptation and observe consistent improvement in performance over "episodes" during each of which the feature matrix is held fixed and TD is applied. We study the performance of our scheme on two different network settings and observe performance improvements in both.

The rest of the chapter is organized as follows: In Section 23.2, we give the framework. The feature adaptation scheme is presented in Section 23.3. In Section 23.4, we present a partial analysis of the convergence behavior of our scheme. In Section 23.5, the results of numerical experiments in the traffic signal control setting are shown. Finally, Section 23.6 presents the concluding remarks.

23.2 THE FRAMEWORK

An MDP is a stochastic process $\{X_n\}$ taking values in a set S (called the state space) that is governed by a control sequence $\{Z_n\}$. Let $A(i)$ be the set of feasible controls or actions in state i and $A \triangleq \cup_{i \in S} A(i)$ be the set of all actions or the action space. We assume that both S and A are finite sets. The process $\{X_n\}$ satisfies the controlled Markov property

$$P(X_{n+1} = j \mid X_m, Z_m, m \leq n) = p(X_n, Z_n, j) \text{ a.s.},$$

where $p : S \times A \times S \to [0, 1]$ is a given function for which $\sum_{j \in S} p(i, a, j) = 1, \forall a \in A(i), i \in S$.

A policy is a decision rule for selecting actions. We call the sequence $\bar{\pi} \triangleq \{\mu_0, \mu_1, \ldots\}$ of maps $\mu_n : S \to A$ an admissible policy when $\mu_n(i) \in A(i) \, \forall i \in S$. In other words, each map μ_n in an admissible policy assigns feasible controls to any state. When $\mu_n \equiv \mu, \forall n \geq 0$, where μ is independent of n, we call $\bar{\pi}$ or by abuse of notation, μ itself a stationary deterministic policy (SDP).

In what follows, we consider the MDP $\{X_n\}$ to be governed by a given (fixed) SDP μ. In such a case, $\{X_n\}$ is in fact a Markov chain taking values in S. Let $c(i, a)$ denote the single-stage cost when the Markov chain is in state i and action a is picked. Given $\gamma \in (0, 1)$, let

$$V^{\mu}(i) = E\left[\sum_{m=0}^{\infty} \gamma^m c(X_m, \mu(X_m)) \mid X_0 = i\right] \tag{23.1}$$

denote the value function under SDP μ when the initial state of $\{X_n\}$ is i. The value function can be obtained by solving the system of equations

$$V^{\mu}(i) = c(i, \mu(i)) + \gamma \sum_{j \in S} p(i, \mu(i), j) V^{\mu}(j), \;\; i \in S, \tag{23.2}$$

or in vector notation,

$$V^{\mu} = c^{\mu} + \gamma P^{\mu} V^{\mu}, \tag{23.3}$$

where $P^{\mu} = [[p(i, \mu(i), j)]]_{i, j \in S}$ is the transition probability matrix of $\{X_n\}$, $c^{\mu} = (c(i, \mu(i)), i \in S)^{\mathrm{T}}$ is the vector of single-stage costs, and $V^{\mu} = (V^{\mu}(i), i \in S)^{\mathrm{T}}$ is the vector of "values" over individual states respectively, under policy μ. The system of Equation (23.3) yields the solution

$$V^{\mu} = (I - \gamma P^{\mu})^{-1} c^{\mu}, \tag{23.4}$$

where I denotes the $(|S| \times |S|)$–identity matrix. When the size of the state space ($|S|$) is large, obtaining the inverse of the matrix $(I - \gamma P^{\mu})$ is computationally hard. An alternative is to solve (23.3) using a value iteration procedure. However, since (23.3) corresponds to a linear system of $|S|$ equations, we expect such a procedure to be slow as well for large $|S|$.

Another problem in using a purely numerical approach is that in most real life situations, the transition probabilities $p(i, \mu(i), j)$, $i, j \in S$ are in fact not known. A way out is to use a combination of value function approximation (via a so-called approximation architecture) and stochastic approximation. The former helps to make the computational complexity manageable, whereas the latter helps to learn the (approximated) value function with only real or simulated measurements and without any knowledge of the transition probabilities.

We approximate $V^{\mu}(j) \approx \phi(j)^{\mathrm{T}}\theta$, where $\phi(j) = (\phi_1(j), \ldots, \phi_d(j))^{\mathrm{T}}$ is a d-dimensional feature associated with state i. Also, $\theta = (\theta(1), \ldots, \theta(d))^{\mathrm{T}}$ is an associated parameter. Let Φ denote the $|S| \times d$ feature matrix with $\phi(j)^{\mathrm{T}}$, $j \in S$, as its rows. Thus, $\Phi = [[\phi_k(s)]]_{k=1,\ldots,d, s \in S}$. Let $\phi_k \stackrel{\triangle}{=} (\phi_k(s), s \in S)^{\mathrm{T}}$ denote the kth column of Φ, $k \in \{1, \ldots, d\}$.

We make the following assumptions.

Assumption 23.1 *The Markov chain $\{X_n\}$ under SDP μ is irreducible.*

Assumption 23.2 *The d columns of the matrix Φ, that is, $\phi_1, \ldots, \phi_d$ are linearly independent. Further, $d \leq |S|$.*

From Assumption 23.1, $\{X_n\}$ is also positive recurrent (since $|S| < \infty$). Let $d^{\mu}(i)$, $i \in S$ be the stationary distribution of $\{X_n\}$ under policy μ. Further, let D^{μ} be a diagonal matrix with entries $d^{\mu}(i)$, $i \in S$ along the diagonal. The regular TD(0) algorithm is as follows.

23.2.1 The TD(0) Learning Algorithm

$$\theta_{n+1} = \theta_n + a(n)\delta_n\phi(X_n), \tag{23.5}$$

where δ_n is the TD term that is defined according to

$$\delta_n = (c(X_n, \mu(X_n)) + \gamma\phi(X_{n+1})^{\mathrm{T}}\theta_n - \phi(X_n)^{\mathrm{T}}\theta_n), \ n \geq 0.$$

Further, $a(n)$, $n \geq 0$ is a sequence of step-sizes that satisfy

$$\sum_n a(n) = \infty, \ \sum_n a^2(n) < \infty. \tag{23.6}$$

The parameter θ_n is d-dimensional and has components (say) $\theta_n(1), \ldots, \theta_n(d)$. Thus, $\theta_n = (\theta_n(1), \ldots, \theta_n(d))^{\mathrm{T}}$. The algorithm (23.5) is called the TD learning (TD(0)) algorithm for the discounted cost case and has been analyzed for its convergence in a more general setting (TD(λ) with $\lambda \in [0, 1]$) in [8]. The average cost version of this algorithm has also been analyzed in [9, 21].

It can be shown (see [8]) that the TD(0) algorithm (23.5) converges according to $\theta_n \to \theta^*$ with probability one, as $n \to \infty$, where

$$\theta^* \triangleq (\theta_1^*, \ldots, \theta_d^*)^{\mathrm{T}} = (\Phi^\top D^\mu(\gamma P^\mu - I)\Phi)^{-1}\Phi^\top D^\mu c^\mu.$$

In Section 23.3, we describe our feature adaptation scheme.

23.3 THE FEATURE ADAPTATION SCHEME

We let all entries of the feature matrix Φ to take values between 0 and 1. Recall that we approximate $V^\mu(i) \approx \phi(i)^{\mathrm{T}}\theta$. Thus, $V^\mu = (V^\mu(1), \ldots, V^\mu(|S|))^{\mathrm{T}}$ can be approximated as

$$V^\mu \approx \Phi\theta = \begin{bmatrix} \sum_{j=1}^d \phi_j(1)\theta_j \\ \sum_{j=1}^d \phi_j(2)\theta_j \\ \cdot \\ \cdot \\ \cdot \\ \sum_{j=1}^d \phi_j(|S|)\theta_j \end{bmatrix}.$$

Alternatively, note that

$$V^\mu \approx \sum_{j=1}^d \phi_j\theta_j.$$

We run the algorithm using a nested loop sequence with the outer loop being run for a total of R iterations. The feature matrix is updated in the outer loop procedure. In between two successive updates of the outer loop, that is, for a fixed feature matrix update, the inner loop comprising of TD recursions is run for a given (large) number M of iterates to obtain a final parameter corresponding to the aforementioned feature matrix (update). We call each such M-iterate run of TD for a given feature matrix an episode. Thus, our scheme is run for a total of R episodes. The feature adaptation scheme is described in detail below:

23.3.1 The Feature Adaptation Scheme

- *Step 0 (Initialize): Select a* $|S| \times d$ *feature matrix* Φ^0 *with columns* $\phi_j^0 = (\phi_j^0(s), s \in S)^{\mathrm{T}}$, $j = 1, \ldots, d$. *Let* Φ^0 *have all its entries between 0 and 1 and let* *it satisfy Assumption 23.2. Set $r := 0$ and $n := 0$, respectively. Set $V_0^{\mu,0} = 0$ to be the initial estimate of the value function. Let $\bar{V}_0^{\mu,0} = 0$ be the initial estimate of the normalized value function. Also set M and R to be given large integers. Set θ_0^0 as the initial parameter value.*

- *Step 1: Let Φ^r denote the feature matrix during the rth step of the algorithm. The $(n + 1)$st update of TD in the rth step of the algorithm is given as follows:*

$$\theta_{n+1}^r = \theta_n^r + a(n)\delta_n^r \phi^r(X_n), \tag{23.7}$$

where $a(n)$, $n \geq 0$ satisfy (23.6) and δ_n^r, $n \geq 0$ are defined according to

$$\delta_n^r = (c(X_n, \mu(X_n)) + \gamma\phi^r(X_{n+1})^{\mathrm{T}}\theta_n^r - \phi^r(X_n)^{\mathrm{T}}\theta_n^r).$$

Set $n := n + 1$. If $n = M - 1$, set $V_M^{\mu,r} = \Phi^r \theta_M^r$, and go to Step 2; else repeat Step 1.

- *Step 2: Let $\theta_M^r \triangleq (\theta_{M,1}^r, \ldots, \theta_{M,d}^r)^{\mathrm{T}}$. Find $\theta_{M,k}^r$, $\theta_{M,l}^r$, $k, l \in \{1, \ldots, d\}$, $k \neq l$ such that*

$$\theta_{M,k}^r \leq \theta_{M,l}^r \leq \theta_{M,j}^r \;\forall j \in \{1, \ldots, d, j \neq k, j \neq l\}.$$

(Any tie is resolved according to some prescribed rule.) Obtain a new feature matrix Φ^{r+1} as follows: First normalize $V_M^{\mu,r}$ (by first letting any negative component to be zero and then dividing every component by one with the largest value) to obtain $\bar{V}_M^{\mu,r}$. Thus, all entries of $\bar{V}_M^{\mu,r}$ lie between 0 and 1. Set $\phi_k^{r+1} := \bar{V}_M^{\mu,r}$. Next obtain ϕ_l^{r+1} by picking all its entries independently according to the uniform distribution on $[0, 1]$. Retain the remaining columns in the Φ^{r+1} matrix from the matrix Φ^r itself, that is, set $\phi_i^{r+1} := \phi_i^r$ for all $i \in \{1, \ldots, d\}$ with $i \neq k, l$. Set Φ^{r+1} to be the matrix with columns ϕ_j^{r+1}, $j = 1, \ldots, d$. Set $r := r + 1$. If $r < R$, go to Step 1; else go to Step 3.

- *Step 3 (Terminate): Output θ_M^R as the final parameter value and $V_M^{\mu,R} = \Phi^R \theta_M^R$ as the corresponding value function estimate.*

We summarize the notation used with regards to the value function below (as this will also be used in the convergence analysis):

- V^{μ}: exact value function under policy μ,
- $V_n^{\mu,r} = \Phi^r \theta_n^r$: estimate of the value function at the rth iterate of algorithm after the nth run of Step 1 ($M \geq n \geq 0$),
- $V_M^{\mu,r} = \Phi^r \theta_M^r$: estimate of the value function after completion of Step 1 during the rth iterate of algorithm (i.e., with $n = M$),

- $V_*^{\mu,r} = \Phi^r \theta_*^r$: estimate of the value function from Step 1 of algorithm at the rth iterate if Step 1 was run until convergence of TD, that is, where $\theta_*^r = \lim_{M \to \infty} \theta_M^r$ with probability one,
- $\bar{V}_M^{\mu,r}$: normalized estimate of value function after completion of Step 1 during the rth iterate of algorithm,
- $\check{V}_r^{\mu}$: projection of V^{μ} to S_r.

Note above that we obtain the normalized value function $\bar{V}_M^{\mu,r}$ at each iterate r of the algorithm only after termination of Step 1 on that iterate. Convergence of the TD scheme for a given feature matrix Φ^r, as $M \to \infty$, has been shown in the literature [4, 8] under the weighted Euclidean norm $\| \cdot \|_{D^{\mu}}$ defined according to

$$\| x \|_{D^{\mu}} = \sqrt{x^{\top} D^{\mu} x},$$

for any $x \in \mathcal{R}^{|S|}$. Let S_r, $r = 1, 2, \ldots$ denote the subspace

$$S_r = \{\Phi^r \theta \mid \theta \in \mathcal{R}^d\}.$$

Let $\check{V}_r^{\mu}$ represent the best approximation to the value function V^{μ} within the subspace S_r. This is obtained by simply projecting V^{μ} to S_r. Note that TD (or for that matter any other algorithm that operates only within the subspace S_r) does not in general converge to $\check{V}_r^{\mu}$. In fact the parameters θ_n^r, $n \geq 0$ obtained from TD, see Step 1 of the feature adaptation scheme, converge to θ_*^r almost surely where

$$\theta_*^r = (\Phi^{r\top} D^{\mu} (\gamma P^{\mu} - I) \Phi^r)^{-1} \Phi^{r\top} D^{\mu} c^{\mu}.$$

This would correspond to a value function estimate of $V_*^{\mu,r} = \Phi^r \theta_*^r$ which is not the same as $\check{V}_r^{\mu}$ even though both $V_*^{\mu,r}$ and $\check{V}_r^{\mu}$ are vectors in the subspace S_r. Since we run TD for a fixed number M of instants during each visit of the algorithm to Step 1, the estimate of the value function as given by TD after the rth cycle (of M iterates) is $V_M^{\mu,r} = \Phi^r \theta_M^r$. By choosing M sufficiently large, we can bring down the difference between $V_M^{\mu,r}$ and $V_*^{\mu,r}$.

The key idea behind our adaptation scheme is the following: Since we replace the "worst basis vector" by the current best estimate of the value function and the "next-to-worst basis vector" by uniformly generated random numbers, we ensure that (a) the best linear combination of the basis functions in the current subspace is retained while (b) in addition, the scheme does a random exploration to find a potentially better subspace than the current one. As our experiments demonstrate, it indeed seems to be the case that our adaptive scheme results in significant performance improvement.

Remark 23.1 *One possible variation is to pick new basis vectors from a prespecified overcomplete basis. These vectors could be sparsely represented. Nevertheless, the running estimate of the value function need not have a sparse representation. Thus, the algorithm stores all but one sparse vectors, thus saving on storage.*

Remark 23.2 *Another possible tweak is to perform a Gram–Schmidt step on each new independent basis vector picked relative to the rest and replace it by the unit normal vector thus obtained. While this is aesthetically more pleasing and will avoid certain "low probability" pathologies, it has a computational overhead. We did not use this tweak in our experiments as we obtained good results regardless.*

Remark 23.3 *A good error estimate for fixed point of a projected linear contraction relative to fixed point of the contraction itself that holds with "high probability" under certain dimensionality requirements on the range of the projection is given in [22].*

In Section 23.4, we give a partial analysis of the convergence behavior of our feature adaptation scheme.

23.4 CONVERGENCE ANALYSIS

We first introduce some notation. Let U denote the unit sphere in $\mathcal{R}^{|S|}$. Since we normalize the basis vectors, they can be identified with points in U. Let $V^{\perp}$ denote the subspace of $\mathcal{R}^{|S|}$ orthogonal to V^{μ} (recall that V^{μ} is the true value function). Let $B \subset U$ denote the ϵ-neighborhood of $V^{\perp} \cap U$ in U for a small $\epsilon > 0$. Let D denote the ϵ-neighborhood of V^{μ}. Let Π_r denote the projection to Range(Φ^r). Then $\check{V}_r^{\mu} = \Pi_r(V^{\mu})$. Let $\Psi : \mathcal{R}^{|S|} \times \mathcal{R}^{|S|} \rightarrow \mathcal{R}^{|S|}$ denote the map

$$\Psi(x, y) := \frac{x - \langle x, y/\|y\|\rangle y/\|y\|}{\|x - \langle x, y/\|y\|\rangle y/\|y\|\|},$$

that is, the component of x orthogonal to y, normalized to have unit norm. Let

$$\beta_r = \frac{\| V^{\mu} - V_*^{\mu,r} \|}{\| V^{\mu} - \check{V}_r^{\mu} \|}.$$

We have the following lemma.

Lemma 23.1

(a) $\beta_r \in \left[1, \dfrac{1}{1-\gamma}\right] \; \forall r \geq 0.$

(b) For β_r close to one (for any $r \geq 0$), $V_*^{\mu,r} \approx \check{V}_r^{\mu}$.

Proof: (a) It is easy to see that the vectors $V^{\mu} - V_*^{\mu,r}$, $V^{\mu} - \check{V}_r^{\mu}$, and $\check{V}_r^{\mu} - V_*^{\mu,r}$ form a right angle triangle with $V^{\mu} - V_*^{\mu,r}$ as the hypotenuse. Further, $V^{\mu} - \check{V}_r^{\mu}$ and $\check{V}_r^{\mu} - V_*^{\mu,r}$ form a right angle with $\check{V}_r^{\mu} - V_*^{\mu,r}$ being a vector within the subspace S_r and $V^{\mu} - \check{V}_r^{\mu}$ being orthogonal to S_r. Thus,

$$\| V^{\mu} - V_*^{\mu,r} \| \geq \| V^{\mu} - \check{V}_r^{\mu} \|,$$

implying that $\beta_r \geq 1\ \forall r \geq 0$. Further, from Lemma 6 of [8], we also obtain that

$$\| V^{\mu} - V_*^{\mu,r} \| \leq \frac{1}{(1-\gamma)} \| V^{\mu} - \check{V}_r^{\mu} \| .$$

Thus, we also have $\beta_r \leq \dfrac{1}{1-\gamma}\ \forall r \geq 0$.

(b) By the Pythagorean theorem, we have

$$\| V^{\mu} - V_*^{\mu,r} \|^2 = \| V^{\mu} - \check{V}_r^{\mu} \|^2 + \| \check{V}_r^{\mu} - V_*^{\mu,r} \|^2 . \tag{23.8}$$

From (23.8), we obtain

$$\beta_r^2 = 1 + \frac{\| \check{V}_r^{\mu} - V_*^{\mu,r} \|^2}{\| V^{\mu} - \check{V}_r^{\mu} \|^2},$$

or that

$$\| \check{V}_r^{\mu} - V_*^{\mu,r} \| = \sqrt{\beta_r^2 - 1}\, \| V^{\mu} - \check{V}_r^{\mu} \|$$

$$\leq 2K\sqrt{\beta_r^2 - 1},$$

where K is an upper bound on $\| V^{\mu} \|$. Thus if β_r is close to one, we have that $V_*^{\mu,r} \approx \check{V}_r^{\mu}$. ■

Now let $\nu > 0$ be such that $\| \check{V}_r^{\mu} - V_*^{\mu,r} \| < \nu$. Let $\alpha > 0$ be such that if $V \notin D$ and $z \notin B$ is a unit vector, then for $\Pi' :=$ projection onto Range(Φ^r) and $\Pi'' :=$ projection onto span(Range(Φ^r) $\cup \{z\}$), we have

$$\|V - \Pi''(V)\| \leq \|V - \Pi'(V)\| - \alpha.$$

Next, we will consider the actual error $\| V_M^{\mu,r} - V^{\mu} \|$ and show that it decreases on the average as r is increased when M is large enough. We have the following result.

Theorem 23.1 *$\exists \alpha_0 > 0$ and $M_0 > 0$ such that for $\alpha > \alpha_0$, and $M \geq M_0$,*

$$E[\| V_M^{\mu,r+1} - V^{\mu} \|] < E[\| V_M^{\mu,r} - V^{\mu} \|].$$

Proof: Note that because of the finite (M) step termination of the TD update in between two successive updates of the feature matrix, the quantity $V_M^{\mu,r}$ is random for any given $r \geq 0$. The dynamics of $\{V_M^{\mu,r}, r \geq 0\}$ can be described according to

$$V_M^{\mu,r+1} = V_*^{\mu,r+1} + \eta((V_M^{\mu,r}, \xi_{r+1}), \zeta_{r+1}),$$

where (1) the i.i.d. variables $\{\xi_r\}$ denote the new basis vectors being inducted, (2) ζ_{r+1} captures the random inputs of the algorithm with the fixed basis, between the rth and $(r+1)$th basis change, and (3) $\eta_{r+1} := \eta((V_M^{\mu,r}, \xi_{r+1}), \zeta_{r+1}) = V_M^{\mu,r+1} - V_*^{\mu,r+1}$ is the error due to inexact convergence (because of the finite run of TD).

Let $\chi_r := V_*^{\mu,r} - \check{V}_r^{\mu}$. Further, as before, let K be an upper bound on $\| V^{\mu} \|$. Note that by Corollary 14, Chapter 4 of [23], $\exists M_0$ such that for $M \geq M_0$, η_{r+1} can be bounded by a given $\kappa > 0$ with probability $1 - \delta$, $\delta > 0$ prescribed. Let

$$A_r = \{\Psi(\xi_{r+1}, V_M^{\mu,r}) \in B\},$$
$$C_r = \{\|\eta_{r+1}\| \geq \kappa\}.$$

Then for $\epsilon_1 > P(A_r)$, we have

$$\begin{aligned}
E[\|V_M^{\mu,r+1} - V^{\mu}\|] &\leq E[\|V_*^{\mu,r+1} - V^{\mu}\|] + E[\|\eta_{r+1}\|],\\
&\leq E[\|\check{V}_{r+1}^{\mu} - V^{\mu}\|] + E[\|\chi_{r+1}\|] + E[\|\eta_{r+1}\|],\\
&\leq E[\|\check{V}_{r+1}^{\mu} - V^{\mu}\| I_{A_r}] + E[\|\check{V}_{r+1}^{\mu} - V^{\mu}\| I_{A_r^c}]\\
&\quad + E[\|\chi_{r+1}\|] + E[\|\eta_{r+1}\|],\\
&\leq E[\|\check{V}_r^{\mu} - V^{\mu}\|] - \alpha + 2\epsilon_1 K + E[\|\chi_{r+1}\|] + E[\|\eta_{r+1}\|],\\
&\leq E[\|V_M^{\mu,r} - V^{\mu}\|] - \alpha + 2\epsilon_1 K + E[\|\chi_r\|] + E[\|\eta_r\|]\\
&\quad + E[\|\chi_{r+1}\|] + E[\|\eta_{r+1}\|],\\
&\leq E[\|V_M^{\mu,r} - V^{\mu}\|] - \alpha + 2((\epsilon_1 + \delta)K + (1-\delta)\kappa + \nu).
\end{aligned}$$

Let $\alpha_0 := 2((\epsilon_1 + \delta)K + (1-\delta)\kappa + \nu)$. Then for $\alpha > \alpha_0$,

$$E[\|V_M^{\mu,r+1} - V^{\mu}\|] < E[\|V_M^{\mu,r} - V^{\mu}\|].$$

Further, $\alpha > \alpha_0$ will hold if δ, ν, κ, ϵ_1 are sufficiently small and $V_M^{\mu,r} \notin D$. ∎

From Theorem 23.1, the average norm difference between the true value function and the approximate value function to which TD(0) converges for a given set of bases decreases as the set of bases is updated after each update of the basis adaptation scheme.

23.5 APPLICATION TO TRAFFIC SIGNAL CONTROL

We consider the problem of traffic signal control as an application setting. The aim in general for this problem is to find an optimal policy for switching signals at traffic junctions in a road network so as to maximize flows and minimize delays. The signals that can be switched to green simultaneously form a sign configuration. Q-learning with linear function approximation has been applied for this control problem in [20]. We, however, consider the problem of adaptively tuning the feature matrix when actions are chosen according to a given policy. The policy that we use is obtained using the Q-learning algorithm from [20], that is, after convergence of the Q-learning algorithm. We apply the feature adaptation scheme described in Section 23.3.

Consider a road network with $m \geq 1$ junctions and N signaled lanes. We let the state be the vector of queue lengths on individual lanes and of the elapsed times since the signal turned red on each of those lanes that have a traffic signal (at the various junctions). We assume that there is a central controller that receives the state information from the various lanes and based on the given policy, conveys information to the individual junctions regarding which traffic lights to switch green during a cycle. We assume no delays in the transfer of information from (to) the controller to (from) the various lanes. (Accounting for bounded delays would add another asymptotically vanishing error term—see Chapter 7 of [23].) The state at instant n is the tuple

$$s_n = (q_1(n), \ldots, q_N(n), t_1(n), \ldots, t_N(n)),$$

where $q_1(n), \ldots, q_N(n)$ are the queue lengths on each of the N lanes at instant n. Similarly, $t_1(n), \ldots, t_N(n)$ are the elapsed times on each of the above lanes. The actions a_n correspond to the sign configurations, that is, feasible combination of traffic lights to switch at each of the m junctions in the road network. Thus, $a_n = (a_1(n), \ldots, a_m(n))$, where $a_i(n)$ is the sign configuration at junction i in the time slot n. We only allow those sign configurations to be in the action set that are feasible and ignore all other (infeasible) sign configurations. This helps keep the computational complexity manageable.

As with [20], we allow lanes on the main road to have a higher priority than those on the side roads. This is accomplished through the form of the cost function as explained below. Let I_p denote the set of lanes that are on the main road. Then, the form of the cost function $c(s_n, a_n)$ is given by

$$\begin{aligned} c(s_n, a_n) = r_1 * &\left(\sum\nolimits_{i \in I_p} r_2 * q_i(n) + \sum\nolimits_{i \notin I_p} s_2 * q_i(n)\right) \\ &+ s_1 * \left(\sum\nolimits_{i \in I_p} r_2 * t_i(n) + \sum\nolimits_{i \notin I_p} s_2 * t_i(n)\right). \end{aligned}$$

Here $r_i, s_i \geq 0$ are certain weights that satisfy $r_i + s_i = 1, i = 1, 2$. Further, we let $r_2 > s_2$. In our experiments, we let $r_1 = s_1 = 0.5$. Thus, we assign equal weightage to both queue lengths and elapsed times. Further, we let $r_2 = 0.6$ and $s_2 = 0.4$, respectively. Thus, queue lengths and elapsed times for lanes on the main road are weighted more than those on the side roads. We let the discount factor $\gamma = 0.9$ in the experiments.

We consider two different road traffic networks: (a) a network with a single traffic signal junction and (b) a corridor with two traffic signal junctions. The two networks are shown in Figures 23.1 and 23.2, respectively. We implemented these network settings and our algorithm on the green light district (GLD) open source software for road traffic simulations [24].

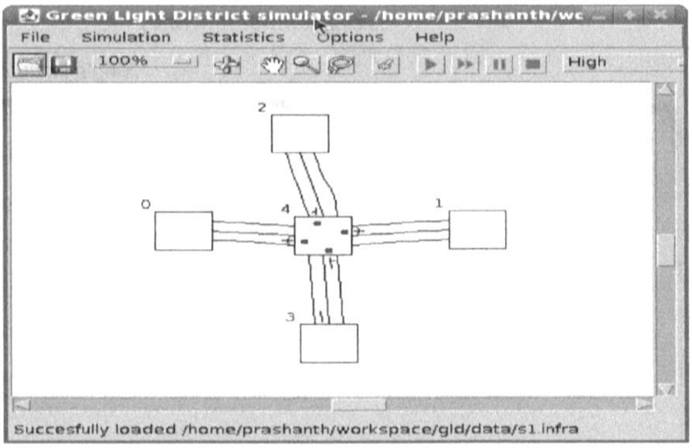

FIGURE 23.1 A single-junction road network.

We study the performance of our feature adaptation scheme using estimates of $E[\| V_M^{\mu,r} \|]$. Recall that from Theorem 23.1, the estimates $E[\| V_M^{\mu,r} - V^{\mu} \|]$ diminish with r. The value function V^{μ}, however, is not available, so we use the fact that by the foregoing, $E[\| V_M^{\mu,r} \|]$ will tend to increase. For estimating $E[\| V_M^{\mu,r} \|]$, we use the sample averages of the estimates of $\| V_M^{\mu,r} \|$. As mentioned previously, we let $V_n^r = \Phi^r \theta_n^r$ denote the nth estimate of the value function when the feature matrix is Φ^r. We obtain the aforementioned sample averages by running the recursion

$$Z_{n+1}^r = (1-a)Z_n^r + a \| V_n^r \|,$$

where for any given $r \in \{0, 1, \ldots, R-1\}$, the above recursion is run for M iterations, that is, with $n \in \{0, 1, \ldots, M-1\}$. Next, the value of r is updated and the above procedure is repeated. Here, a is a small step-size that we select to be 0.001. By abuse of notation, we denote Z_n^r as Z_m in these figures where m denotes the number of cycles or time instants (in absolute terms), that is, $m \in \{0, 1, \ldots, RM-1\}$, when Z_m is updated.

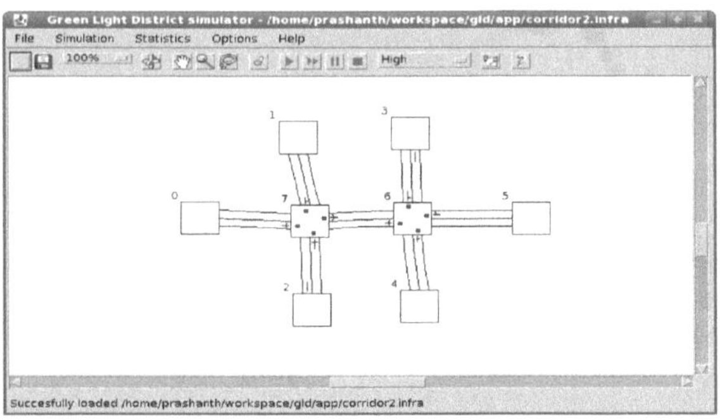

FIGURE 23.2 A corridor network with two junctions.

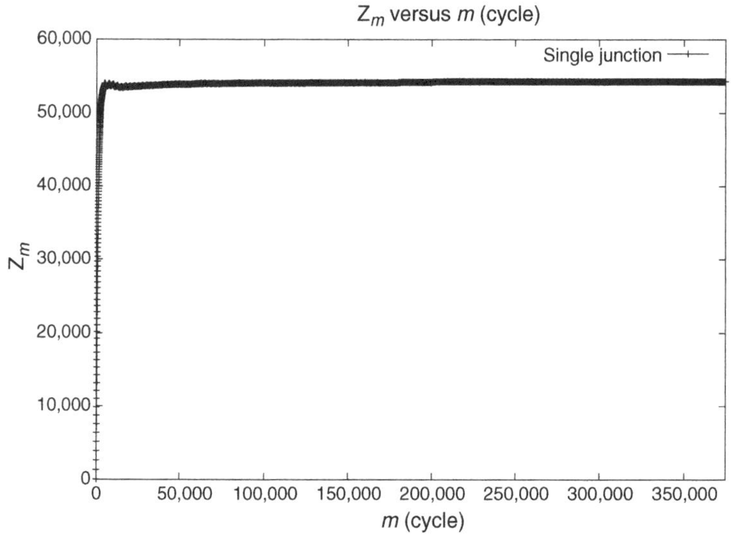

FIGURE 23.3 Plot of Z_m versus m in the case of single-junction road network.

We call each group of M cycles when the feature matrix Φ^r is held fixed for some r, an episode. We conducted our experiments for the cases of single-junction and two-junction-corridor networks for a total of 150 episodes in each where each episode comprised of 2500 cycles. Thus, $M = 2500$ and $R = 150$ in our experiments.

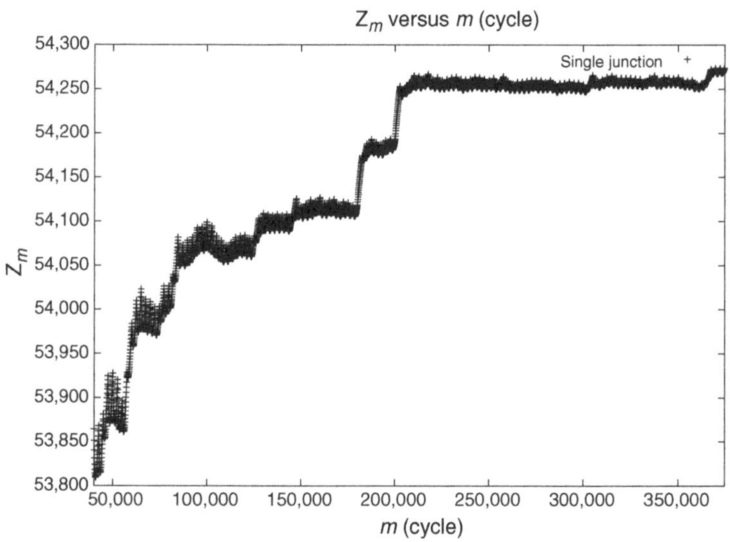

FIGURE 23.4 Plot of Z_m versus m after 40,000 cycles in the case of single-junction road network.

TABLE 23.1 Performance Improvement with Feature Adaptation for the Single-Junction Road Network

# Cycle (m)	Z_m	$Z_m - Z_{M-1}$, ($m \geq M-1$)
2,499	51,042.23	
74,999	54,003.00	2,960.76
149,999	54,116.59	3,074.36
224,999	54,260.28	3,218.05
299,999	54,255.38	3,213.15
374,999	54,274.72	3,232.49

For the single-junction case, we show in Figure 23.3 the plot of Z_m as a function of m (the number of cycles). We observe that there is a significant improvement after the first episode, which results in a steep jump in the Z_m value. In subsequent (r) iterations when the Φ^r matrix is updated, the performance improvement continues, though in smaller steps. The improvement in performance from feature adaptation can be seen more clearly in Figure 23.4, when the values of Z_m are plotted for $m \geq 40{,}000$. The value of Z_m as well as the difference $Z_m - Z_{M-1}$, that is, improvement in "Z_m performance" in relation to its value obtained after the completion of the first episode (i.e., with the originally selected feature matrix) is shown at the end of the 30th, 60th, 90th, 120th, and 150th episodes respectively, in Table 23.1. As expected, the values of Z_m are seen to consistently increase here.

Next, for the case of the two-junction corridor road network, we show a similar plot of Z_m as a function of m in Figure 23.5. Further, in Figure 23.6, we show the

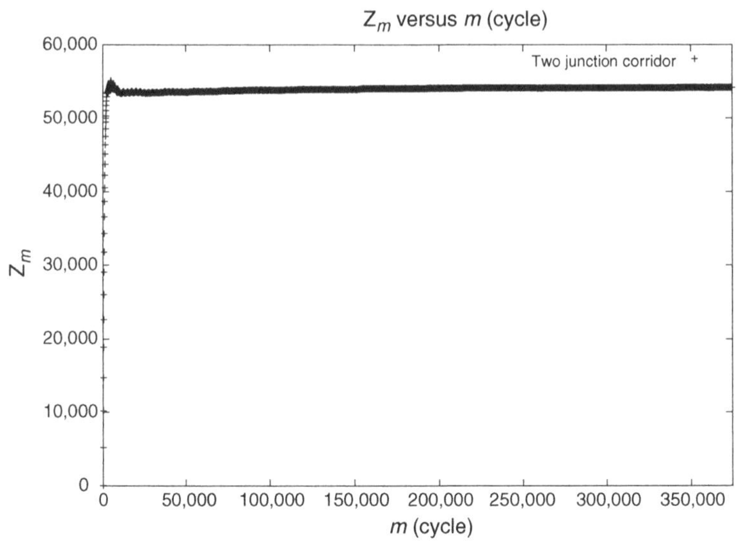

FIGURE 23.5 Plot of Z_m versus m in the case of two-junction corridor network.

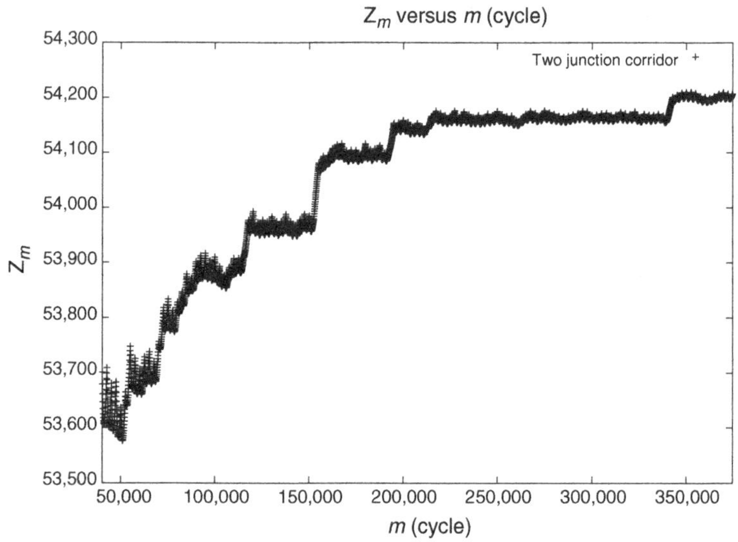

FIGURE 23.6 Plot of Z_m versus m after 40,000 cycles in the case of two-junction corridor network.

TABLE 23.2 Performance Improvement with Feature Adaptation for the Two-Junction Corridor Road Network

# Cycle (m)	Z_m	$Z_m - Z_{M-1}$ ($m \geq M - 1$)
2,499	53,480.65	
74,999	53,834.04	353.39
149,999	53,985.54	504.89
224,999	54,166.69	686.04
299,999	54,167.99	687.34
374,999	54,207.78	727.13

same plot for cycles 40,000 onwards to show performance improvement resulting from feature adaptation. Similar observations as for the single-junction case hold in the case of the two-junction corridor as well.

Finally, as before, we show in Table 23.2, the values of Z_m as well as of the difference $Z_m - Z_{M-1}$ at the end of the 30th, 60th, 90th, 120th, and 150th episodes, respectively. The values of Z_m are again seen to consistently increase here as well.

23.6 CONCLUSIONS

We presented in this chapter a novel online feature adaptation algorithm. We observed significant performance improvements on two different settings for a problem of traffic signal control. We considered the problem of prediction here and applied our feature adaptation scheme in conjunction with the TD learning algorithm. As future work, we may consider the application of our algorithm together with other schemes such as least squares temporal difference (LSTD) learning [25, 26] and least squares policy evaluation (LSPE) [27, 28]. Moreover, we may apply a similar scheme for the problem of control, for instance, in conjunction with the actor–critic algorithms in [29] and [21].

ACKNOWLEDGMENTS

Borkar was supported in his research through a J.C. Bose Fellowship from the Department of Science and Technology, Government of India. The work of Bhatnagar and Prashanth was supported by the Automation Systems Technology (ASTec) Center, a program of the Deparment of Information Technology, Government of India.

REFERENCES

1. D.P. Bertsekas. *Dynamic Programming and Optimal Control*, 3rd edition, Vol. 1. Athena Scientific, Belmont, MA, 2005.
2. D.P. Bertsekas. *Dynamic Programming and Optimal Control*, 3rd edition, Vol. 2. Athena Scientific, Belmont, MA, 2007.
3. M.L. Puterman. *Markov Decision Processes: Discrete Stochastic Dynamic Programming*. John Wiley, New York, 1994.
4. D.P. Bertsekas and J.N. Tsitsiklis. *Neuro-Dynamic Programming*. Athena Scientific, Belmont, MA, 1996.
5. D.P. Bertsekas. Approximate dynamic programming, (Online) Chapter 6 of *Dynamic Programming and Optimal Control*, 3rd edition, Vol. 2. Available at: http://web.mit.edu/dimitrib/www/dpchapter.html, 2011.
6. R.S. Sutton and A. Barto. *Reinforcement Learning: An Introduction*. MIT Press, Cambridge, MA, 1998.
7. R.S. Sutton. Learning to predict by the method of temporal differences. *Machine Learning*, 3:9–44, 1988.
8. J.N. Tsitsiklis and B. Van Roy. An analysis of temporal difference learning with function approximation. *IEEE Transactions on Automatic Control*, 42(5):674–690, 1997.
9. J. Tsitsikis and B. Van Roy. Average cost temporal-difference learning. *Automatica*, 35:1799–1808, 1999.
10. I. Menache, S. Mannor, and N. Shimkin. Basis function adaptation in temporal difference reinforcement learning. *Annals of Operations Research*, 134:215–238, 2005.
11. H. Yu and D.P. Bertsekas. Basis function adaptation methods for cost approximation in MDP. In *Proceedings of the IEEE International Symposium on Adaptive Dynamic Programming and Reinforcement Learning*, Nashville, TN, USA, March 30–April 2, 2009.
12. P.W. Keller, S. Mannor, and D. Precup. Automatic basis function construction for approximate dynamic programming and reinforcement learning. In *Proceedings of the 23rd International Conference on Machine Learning*, Pittsburgh, PA, June 25–29, 2006.
13. D.P. Bertsekas and H. Yu. Projected equation methods for approximate solution of large linear systems. *Journal of Computational and Applied Mathematics*, 227:27–50, 2009.

14. S. Mahadevan and B. Liu. Basis construction from power series expansions of value functions. In *Proceedings of Advances in Neural Information Processing Systems*. Vancouver, B.C., Canada, 2010.
15. Y. Sun, F. Gomez, M. Ring, and J. Schmidhuber. Incremental basis construction from temporal difference error. In *Proceedings of the Twenty Eighth International Conference on Machine Learning*. Bellevue, WA, USA, 2011.
16. D. Di Castro and S. Mannor. Adaptive bases for reinforcement learning. In J.L. Balcźar, F. Bonchi, A. Gionis, and M. Sebag, editors. *Machine Learning and Knowledge Discovery in Databases, Proceedings of ECML PKDD 2010*. Barcelona, Spain, September 20–24, 2010. Part I, Lecture Notes in Computer Science Vol. 6321, pp. 312–327, 2010.
17. R. Parr, C. Painter-Wakefield, L. Li, and M. Littman. Analyzing feature generation for value-function approximation. In *Proceedings of the 24th International Conference on Machine Learning*. Corvallis, OR, June 20–24, 2007.
18. D. Huang, W. Chen, P. Mehta, S. Meyn, and A. Surana. Feature selection for neuro-dynamic programming. In F.L. Lewis and D. Liu, editors. *Reinforcement Learning and Approximate Dynamic Programming for Feedback Control*, Chap. 24. IEEE Press Computational Intelligence Series, 2011.
19. J.S. Baras and V.S. Borkar. A learning algorithm for Markov decision processes with adaptive state aggregation. In *Proceedings of the 39th IEEE Conference on Decision and Control*, IEEE Press, Piscataway, NJ, Vol. 4, pp. 3351–3356. Sydney, Australia, Dec. 12–15, 2000.
20. L.A. Prashanth and S. Bhatnagar. Reinforcement learning with function approximation for traffic signal control. *IEEE Transactions on Intelligent Transportation Systems*, 12(2):412–421, 2011.
21. S. Bhatnagar, R.S. Sutton, M. Ghavamzadeh, and M. Lee. Natural actor-critic algorithms. *Automatica*, 45:2471–2482, 2009.
22. K. Barman and V.S. Borkar. A note on linear function approximation using random projections. *Systems and Control Letters*, 57(9):784–786, 2008.
23. V.S. Borkar. *Stochastic Approximation: A Dynamical Systems Viewpoint*. (Jointly published by) Cambridge University Press, Cambridge, U.K., and Hindustan Book Agency, New Delhi, India, 2008.
24. M. Wiering, J. Vreeken, J. van Veenen, and A. Koopman. Simulation and optimization of traffic in a city. *Proceedings of IEEE Intelligent Vehicles Symposium*, IEEE Press, Piscataway, NJ, pp. 453–458, Parma, Italy, June 14–17, 2004.
25. S.J. Bradtke and A.G. Barto. Linear least-squares algorithms for temporal difference learning. *Machine Learning*, 22:33–57, 1996.
26. J.A. Boyan. Least-squares temporal difference learning. In *Proceedings of the Sixteenth International Conference on Machine Learning*, pp. 49–56. Morgan Kaufmann, San Francisco, CA, 1999.
27. A. Nedic and D.P. Bertsekas. Least-squares policy evaluation algorithms with linear function approximation. *Journal of Discrete Event Systems*, 13:79–110, 2003.
28. D.P. Bertsekas, V.S. Borkar, and A. Nedic. Improved temporal difference methods with linear function approximation. In A. Barto, W. Powell, and J. Si, editors. *Handbook of Learning and Approximate Dynamic Programming*. IEEE Press, 2004, pp. 231–255.
29. V.R. Konda and J.N. Tsitsiklis. On actor–critic algorithms. *SIAM Journal on Control and Optimization*, 42(4):1143–1166, 2003.

CHAPTER 24

Feature Selection for Neuro-Dynamic Programming

DAYU HUANG,[1] W. CHEN,[1] P. MEHTA,[1] S. MEYN,[2] and A. SURANA[3]

[1]Coordinated Science Laboratory, Department of Electrical and Computer Engineering, University of Illinois at Urbana-Champaign, Urbana, IL, USA
[2]Department of Electrical and Computer Engineering, University of Florida, Gainesville, FL, USA
[3]United Technologies Research Center, East Hartford, CT, USA

ABSTRACT

Neuro-dynamic programming encompasses techniques from both reinforcement learning and approximate dynamic programming. *Feature selection* refers to the choice of basis that defines the function class that is required in the application of these techniques.

This chapter reviews two popular approaches to neuro-dynamic programming, TD- and Q-Learning. The main goal of the chapter is to demonstrate how insight from idealized models can be used as a guide for feature selection for these algorithms. Several approaches are surveyed, including fluid and diffusion models, and the application of idealized models arising from mean-field game approximations. The theory is illustrated with several examples.

24.1 INTRODUCTION

If you have taken a course that mentioned a Riccati equation, then you have already been exposed to approximate dynamic programming: *No physical system is linear!* The model is assumed to be linear, and the cost function is assumed quadratic, so that a closed-form expression for the optimal feedback law can be computed.

In many cases, a linear approximation is not easily justified. In particular, in applications from operations research or chemical engineering, the state space is usually

Reinforcement Learning and Approximate Dynamic Programming for Feedback Control, First Edition. Edited by Frank L. Lewis and Derong Liu.

constrained. To obtain an effective feedback law for control, we must find alternative approaches to approximation.

Techniques from approximate dynamic programming (ADP) and reinforcement learning (RL) are designed to obtain such approximations [1, 2]. In particular, in TD-Learning and related approaches, there is no attempt to approximate the system or cost function. Instead, the solution to a dynamic programming equation is approximated directly, within a prescribed finite-dimensional function class. A key determinant of the success of these techniques is the selection of this function class, also known as the *basis*.

The goal of this chapter is to describe techniques to construct a basis for approximation. All of these techniques return to *approximate modeling*: We first construct a highly idealized approximate model, and then compute or approximate the solution to a dynamic programming equation for this idealization. These approximations lead naturally to a basis for approximation in the original stochastic system.

Several specific approaches are surveyed. The simplest and most easily applied is through consideration of a fluid model in which variability is disregarded. Diffusion models, and models obtained from the solution of *mean field games* are also considered. The main message is that ADP and RL techniques provide a general approach to applying insight gained from naive models of real-life systems.

The remainder of the chapter is organized as follows. Section 24.2 contains a brief review of optimal control theory for both stochastic and deterministic models. Here we demonstrate the similarity between the HJB equation seen in a first-year graduate control course, and the average-cost equations appearing in average-cost optimal control for both finite-state MDPs, and controlled diffusions. We argue that the continuous-time dynamic programming equations are frequently more tractable than their deterministic counterparts. Using Taylor-series approximations, we also explain why the solution to the dynamic programming equations for a deterministic or diffusion model might provide a good approximation for the MDP model. Section 24.3 contains a brief survey of techniques from *neuro-dynamic programming*—the term is used here to cover both reinforcement learning and approximate dynamic programming.

TD-, Q-Learning, SARSA, and approximate linear-programming techniques are all designed to obtain an approximate solution to a dynamic programming equation, within a prescribed finite-dimensional function class. The question of how to select an appropriate function class is discussed based on the idealized models introduced in the first part of the chapter. The application of "fluid models" (or ODE models) is explored in greater depth in Section 24.4, followed by diffusion approximations in Section 24.5. There are many other approaches for approximations. We briefly discuss the value of *mean-field games* in Section 24.6. Conclusions and directions for future work are contained in Section 24.7.

24.2 OPTIMALITY EQUATIONS

Here we briefly review optimality equations for dynamic control systems—We consider both stochastic and deterministic models. The state space is denoted X, and the space of inputs is denoted U. It is assumed that a cost function $c\colon \mathsf{X}\times\mathsf{U}\to\mathbb{R}_+$ is given, and the goal is to minimize either average or discounted cost.

Throughout the chapter we assume full state measurements are available. The input at time t is assumed to be a function of present and past state values.

To unify the presentation, we introduce a *generator* for each class of models. This is standard practice in the theory of control and performance evaluation for diffusions, but less common in other settings.

We begin with the deterministic model in continuous time, and recall the HJB equations for the associated optimal control problems. The total cost criterion (24.3) is highlighted because this is what is typically seen in control courses. Its most natural cousin in the stochastic setting is *average-cost* optimal control. Theory to explain the solidarity between solutions of the various optimality equations is contained in [3–5], and the references therein.

24.2.1 Deterministic Model

We have a state process $\boldsymbol{x}$ evolving on $\mathsf{X}=\mathbb{R}^\ell$. Given the initial condition $x(0)$, and the input $\boldsymbol{u}$ evolving on $\mathsf{U}=\mathbb{R}^{\ell_u}$, the output is assumed to be uniquely specified by the ODE,

$$\dot{x}=f(x,u). \tag{24.1}$$

This is the deterministic, nonlinear state space model.

The controlled generator $\mathcal{D}^{\mathrm{F}}$ maps functions to functions: If $h\colon \mathsf{X}\to\mathbb{R}$ is differentiable, then the generator defines a new function on $\mathsf{X}\times\mathsf{U}$,

$$\mathcal{D}^{\mathrm{F}}_u h\,(x)=\frac{\mathrm{d}}{\mathrm{d}t}h(x(t))\quad \text{with } t=0,\ x(0)=x, \text{ and } u(0)=u.$$

By the chain rule of calculus, we have

$$\mathcal{D}^{\mathrm{F}}_u h\,(x)=f(x,u)\cdot\nabla h\,(x). \tag{24.2}$$

The generator appears in dynamic programming equations for the deterministic model. Consider the total-cost criterion

$$J^*(x)=\inf_{U}\int_0^\infty c(x(t),u(t))\,\mathrm{d}t\,,\qquad x(0)=x. \tag{24.3}$$

Under general conditions, the Hamilton–Jacobi–Bellman (HJB) equation holds,

$$\min_u\big\{c(x,u)+\mathcal{D}^{\mathrm{F}}_u J^*\,(x)\big\}=0, \tag{24.4}$$

and an optimal policy is defined as a minimizer: $u^*(t)$ is a minimizer of (24.4) when $x^*(t)=x$.

24.2.2 Diffusion Model

The state space and action space remain the same, but we now introduce noise

$$\mathrm{d}X = f(X, U)\,\mathrm{d}t + b(X, U)\,\mathrm{d}N, \tag{24.5}$$

where $\boldsymbol{N}$ is an ℓ_n-dimensional Brownian motion, and $b(x, u)$ is an $\ell \times \ell_n$ matrix for each x and u. We use upper case letters to emphasize that $(\boldsymbol{X}, \boldsymbol{U}, \boldsymbol{N})$ are stochastic processes.

The controlled generator is again defined as a derivative

$$\mathcal{D}_u^{\mathrm{D}} h\,(x) = \frac{\mathrm{d}}{\mathrm{d}t}\mathsf{E}[h(X(t))] \quad \text{with } t = 0,\ X(0) = x, \text{ and } U(0) = u, \tag{24.6}$$

whenever the derivative exists; this will require smoothness and growth conditions on h. Under these conditions, the generator can be expressed as a second-order differential operator

$$\mathcal{D}_u^{\mathrm{D}} h\,(x) = f(x, u)\cdot\nabla h\,(x) + \tfrac{1}{2}\mathrm{trace}\,\big(b(x, u)^{\mathrm{T}}\nabla^2 h\,(x) b(x, u)\big). \tag{24.7}$$

A common control criterion is average cost. Its minimum over all policies is denoted as

$$\eta^* = \inf_{U}\Big\{\limsup_{T\to\infty}\frac{1}{T}\int_0^T c(X(t), U(t))\,\mathrm{d}t\Big\}. \tag{24.8}$$

Subject to conditions on the model and the cost function, the average cost η^* is a deterministic quantity, independent of the initial condition. Under further assumptions, an optimal policy achieving this minimal average cost can be obtained by solving the average cost optimality equation (ACOE):

$$\min_u\big\{c(x, u) + \mathcal{D}_u^{\mathrm{D}} h^*\,(x)\big\} = \eta^*. \tag{24.9}$$

The minimizer again defines an optimal policy.

Observe the close similarity between (24.9) and the HJB equation (24.4).

The ACOE is a fixed point equation in the *relative value function* h^*, and the optimal cost η^*. Its solution is not unique, because we can always add a constant to h^* to obtain another solution. Fix any point $x^\bullet \in \mathsf{X}$, and normalize h^* so that $h^*(x^\bullet) = \eta^*$. Then, the ACOE (24.13) is a fixed point equation in h^* alone:

$$\min_u\big\{c(x, u) + \mathcal{D}_u^{\mathrm{D}} h^*\,(x)\big\} = h^*(x^\bullet). \tag{24.10}$$

24.2.3 Models in Discrete Time

Controlled models in discrete time will be expressed in the recursive form

$$X(t+1) - X(t) = f(X(t), U(t), N(t+1)). \tag{24.11}$$

where $\boldsymbol{N}$ is i.i.d.. This is known as a *Markov Decision Process* (MDP).

The controlled generator is defined as in the previous cases, except that now we consider differences rather than derivatives,

$$\begin{aligned}\mathcal{D}_u h\,(x) &= \mathsf{E}[h(X(t+1)) - h(X(t)) \mid X(t) = x,\ U(t) = u] \\ &= \mathsf{E}[h(x + f(x, u, N))] - h(x).\end{aligned} \tag{24.12}$$

The minimal average cost is defined as in the diffusion model (24.8), but as a discrete-time average. The ACOE is expressed as in the case of diffusions

$$\min_u \{c(x, u) + \mathcal{D}_u h^*\,(x)\} = \eta^*, \tag{24.13}$$

and h^* is again called the *relative value function*.

In this chapter, most of our attention is focused on the MDP model (24.11). A running example is the controlled queue

$$X(t+1) = X(t) - U(t) + A(t+1), \tag{24.14}$$

in which $\boldsymbol{X}$, $\boldsymbol{U}$, and $\boldsymbol{A}$ are nonnegative scalar-valued stochastic processes. The state $X(t) \in \mathsf{X} = \mathbb{R}_+$ is interpreted as the queue-length, $U(t) \in \mathbb{R}_+$ the quantity of service, and $A(t)$ the number of new "customers" arriving at time t. It is assumed that $\boldsymbol{A}$ is i.i.d., so that (24.14) defines an MDP model. When $U(t)$ is constrained to the binary set $\{0, 1\}$, this is a special case of the controlled random walk (CRW) queue [5].

24.2.4 Approximations

The various models lead to very similar dynamic programming equations. The models in continuous time are frequently attractive because we can apply calculus for computation or approximation.

One way to approximate the MDP model with a fluid or diffusion model is through Taylor series. The *mean drift* for (24.11) is defined by

$$\overline{f}(x, u) = \mathsf{E}[X(t+1) - X(t) \mid X(t) = x,\ U(t) = u], \tag{24.15}$$

and the associated fluid model,

$$\frac{\mathrm{d}}{\mathrm{d}t} x(t) = \overline{f}(x(t), u(t)), \qquad x(0) \in \mathsf{X}. \tag{24.16}$$

A first-order Taylor series approximation can be used to obtain the following approximation, subject to conditions on h,

$$\begin{aligned}\mathcal{D}_u h\,(x) &= \mathsf{E}[h(X(t+1)) - h(X(t)) \mid X(t) = x,\ U(t) = u] \\ &\approx \overline{f}(x,u) \cdot \nabla h\,(x).\end{aligned}$$

The right-hand side defines the generator for the fluid model (24.16).

The diffusion model is a refinement of the fluid model to take into account volatility. It is motivated similarly, using a *second-order* Taylor series expansion.

To bring in randomness to the fluid model, it is useful to first introduce a fluid model in discrete-time. For any $t \geq 0$, the random variable denoted $\Delta(t+1) = f(X(t), U(t), W(t+1)) - \overline{f}(X(t), U(t))$ has zero mean. The evolution of $\boldsymbol{X}$ can be expressed as a discrete time nonlinear system, plus "white noise,"

$$X(t+1) = X(t) + \overline{f}(X(t), U(t)) + \Delta(t+1), \qquad t \geq 0. \tag{24.17}$$

The fluid model in discrete time is obtained by ignoring the noise

$$x(t+1) = x(t) + \overline{f}(x(t), u(t)). \tag{24.18}$$

This is the discrete time counterpart of (24.16). Observe that it can be justified exactly as in the continuous time analysis: If h is a smooth function of x, then we may approximate the generator using a first order-Taylor series approximation as follows: $\mathcal{D}_u h\,(x) :=$

$$\begin{aligned}&\mathsf{E}[h(X(t+1)) - h(X(t)) | X(t) = x, U(t) = u] \\ &\quad\approx h(x + \overline{f}(x,u)) - h(x) + \mathsf{E}[\nabla h(x + \overline{f}(x,u)) \cdot \Delta(1) \mid X(0) = x, U(0) = u] \\ &\quad= h(x + \overline{f}(x,u)) - h(x).\end{aligned} \tag{24.19}$$

The right-hand side is the generator applied to h, for the *discrete-time* fluid model (24.18).

Consideration of the model (24.18) may give better approximation for value functions in some cases, but we lose the simplicity of differential equations that characterize value functions in the continuous time model (24.16).

The representation (24.17) also motivates the diffusion model. Denote the conditional covariance matrix by

$$\Sigma_f(x,u) = \mathsf{E}[\Delta(t+1)\Delta(t+1)^{\mathsf{T}} \mid X(t) = x,\ U(t) = u]\,,$$

and let b denote an $\ell \times \ell$ "square-root," $b(x,u)b(x,u)^{\mathsf{T}} = \Sigma_f(x,u)$ for each x, u. The diffusion model is of the form (24.5)

$$\mathrm{d}X(t) = \overline{f}(X(t), U(t))\,\mathrm{d}t + b(X(t), U(t))\,\mathrm{d}N(t), \tag{24.20}$$

where the process $\boldsymbol{N}$ is a standard Brownian motion on $\mathbb{R}^\ell$.

To justify its form we consider a second order Taylor series approximation. If $h\colon \mathsf{X} \to \mathbb{R}$ is a C^2 function, and at time t we have $X(t) = x$, $U(t) = u$, then the standard second order Taylor series about $x^+ = x + \overline{f}(x, u)$ gives

$$h(X(t+1)) - h(X(t)) \approx \nabla h\,(x^+) \cdot \overline{f}(x, u) + \tfrac{1}{2}\Delta(t+1)^{\mathrm{T}}\nabla^2 h\,(x^+)\Delta(t+1).$$

Suppose we can justify a further approximation, obtained by replacing x^+ with x in this equation. Then, on taking expectations of each side, conditioned on $X(t) = x$, $U(t) = u$, we approximate the generator for $\boldsymbol{X}$ by the second-order ordinary differential operator,

$$\mathcal{D}_u^{\mathrm{D}} h\,(x) := \nabla h\,(x) \cdot \overline{f}(x, u) + \tfrac{1}{2}\mathrm{trace}\left(\Sigma_f(x, u)\nabla^2 h\,(x)\right). \tag{24.21}$$

This is precisely the differential generator for (24.20) (see (24.7)).

For the controlled queue (24.14), the associated fluid and diffusion models take similar forms:

$$\dot{q}(t) = -u(t) + \alpha \quad \text{Fluid model,} \tag{24.22}$$

$$\mathrm{d}Q(t) = (-U(t) + \alpha)\,\mathrm{d}t + \mathrm{d}I(t) + \sigma_A\,\mathrm{d}N(t) \quad \text{Diffusion model,} \tag{24.23}$$

where α denotes the mean of $A(t)$, and σ_A^2 its variance.

We relax integer constraints in these models: $\boldsymbol{x}$ and $\boldsymbol{X}$ evolve on $\mathbb{R}_+$, and are assumed to be continuous functions of time. Consequently, we must also relax integer constraints on the inputs. If $U(t)$ is constrained to $\{0, 1\}$ in (24.14), then $u(t)$ and $U(t)$ are constrained to the closed interval $[0, 1]$ in (24.22) and (24.23).

The diffusion model (24.23) is expressed in the more abstract form to emphasize that this is a stochastic differential equation. The *idleness process* $\boldsymbol{I}$ is interpreted as a projection of $\boldsymbol{Q}$ to $\mathbb{R}_+$, so that,

$$\int_0^\infty \mathbb{I}\{Q(t) > 0\}\,\mathrm{d}I(t) = 0. \tag{24.24}$$

Moreover, it is a nondecreasing function of t, with $I(0) = 0$.

If in addition we have $U(t) \equiv 1$, then $\boldsymbol{Q}$ is called a *Reflected Brownian Motion* (RBM). On an interval of time for which $dI(t) = 0$, this corresponds to draining the queue at maximal rate.

In applications found in operations research, it is usually taken for granted that $c(x, u) \equiv x$ (or $\sum x_i$ in network models). Theorem 24.4.1 combined with (24.61) demonstrate that the dynamic programming equations for the CRW queue, and the corresponding diffusion and fluid models, are each quadratic, and very similar in form. Similar results hold for network models under general conditions [5].

24.3 NEURO-DYNAMIC ALGORITHMS

The algorithms considered in this chapter are designed to construct approximations to dynamic programming equations. There are several approaches, based on a few different approximation criteria. We begin with some further discussion on the MDP model in discrete time.

24.3.1 MDP Model

The controlled stochastic model (24.11) is the general MDP model considered in this paper. The controlled transition law is denoted as

$$P_u(x, A) := \mathsf{P}\{x + f(x, u, W(1)) \in A\}, \ \ A \in \mathcal{B}(\mathsf{X}).$$

Hence, the controlled generator defined in (24.7) is simply $P_u - I$. The state space X will be taken to be $\mathbb{R}^\ell$, or a subset. The set of inputs (or *action space*) U is taken to be a subset of $\mathbb{R}^{\ell_u}$. The constraints are sometimes state dependent: We let $\mathsf{U}(x)$ denote the set of feasible inputs u for $U(t)$ when $X(t)$ takes the value $x \in \mathsf{X}$.

A cost function $c\colon \mathsf{X} \times \mathsf{U} \to \mathbb{R}_+$ is given, and our goal is to find an optimal control based on this cost function. We focus on the average cost problem, with associated average cost optimality equation (ACOE) (24.13).

Without loss of generality, we restrict to inputs that are defined by a stationary policy. This means that the input is defined by state feedback, $U(t) = \phi(X(t))$, $t \geq 0$, where $\phi\colon \mathsf{X} \to \mathsf{U}$. Sometimes we allow randomization,

$$U(t) = \phi(X(t), Z(t)), \qquad t \geq 0, \tag{24.25}$$

where $\boldsymbol{Z}$ is i.i.d., and independent of $\boldsymbol{N}$ in (24.11). This amounts to a deterministic stationary policy for the extended state process $(\boldsymbol{X}, \boldsymbol{Z})$. If $\boldsymbol{X}$ is controlled using a stationary policy ϕ, then it evolves as a time-homogeneous Markov chain, with transition law denoted P_ϕ. For a deterministic policy, $P_\phi(x, dy) = P_{\phi(x)}(x, dy)$ is the resulting transition law for the chain. If the policy is randomized, then $P_\phi(x, dy) = \int P_{\phi(x,z)}(x, dy)\, \mu_Z(dz)$, where μ_Z is the marginal distribution for $\boldsymbol{Z}$. We let c_ϕ denote the resulting cost function on X alone: $c_\phi(x) = c(x, \phi(x))$ for deterministic policies, and $c_\phi(x) = \int c(x, \phi(x, z))\, \mu_Z(dz)$ for the general randomized stationary policy. The *average cost* is denoted η_ϕ.

Key to the theory of average-cost optimal control is *Poisson's equation*, which is a degenerate version of the ACOE: If $\boldsymbol{X}$ is a Markov chain with transition kernel P, and generator $\mathcal{D} = P - I$, then Poisson's equation is the fixed point equation,

$$c + \mathcal{D}h = \eta, \tag{24.26}$$

where η is the steady-state mean of c. To make sense of "steady state," and to ensure the existence of a solution, it is necessary to impose both irreducibility conditions and stability conditions on $\boldsymbol{X}$ [5, 6].

Two common approaches to computation of a solution to the ACOE are policy iteration and value iteration. The first approach is usually avoided because of higher computational cost, but it is in fact the most convenient approach for integration with the learning techniques described in this section.

In the policy iteration algorithm (PIA), a sequence of deterministic stationary policies are obtained, with increasingly improved performance, in the sense that the corresponding average costs are decreasing. The algorithm is initialized with a policy ϕ^0 and then the following operations are performed in the kth stage of the algorithm:

- given the policy ϕ^k, find the solution h^k to Poisson's equation $P_{\phi^k} h^k = h^k - c_k + \eta_k$, where $c_k(x) = c(x, \phi^k(x))$, and η_k is the average cost,
- update the policy via

$$\phi^{k+1}(x) \in \arg\min_u \{c(x, u) + P_u h^k(x)\}, \quad x \in \mathsf{X}. \tag{24.27}$$

Complexity in the PIA may come from one or two sources:

(i) the solution to Poisson's equation can be cast as a matrix inversion problem, whose dimension coincides with the *number of elements in* X,

(ii) the update step (24.27) may require considerable computation if U or X is large, especially if the minimization cannot be expressed as a convex program.

TD-Learning is designed to address (i), and both SARSA and Q-Learning are methods designed to address (ii).

24.3.2 TD-Learning

TD-Learning is a technique for approximating value functions of MDPs within a parameterized class. Letting h denote a value function, and $\{h^r\}$ a family of approximations, the goal is to find r^* that solves

$$\min_r \|h - h^r\|^2. \tag{24.28}$$

If $\{h^r\}$ is a linearly parameterized family of functions, and the norm is defined by an "inner product", then this is a least-squares problem. It is simplest to begin with an abstract least squares problem that mirrors TD-Learning, following [5, 7, 8].

Least Squares Let $\mathcal{X}$ denote a vector space of functions, let $R\colon \mathcal{X} \to \mathcal{X}$, and for $c \in \mathcal{X}$ given, denote $h = Rc$. We are also given a d-dimensional subspace $\mathcal{L} \subset \mathcal{X}$, with basis $\{\psi_i : 1 \le i \le d\} \subset \mathcal{X}$. We let $h^r = \sum r_i \psi_i$, a typical element of $\mathcal{L}$, and seek the best r to approximate h.

For approximation, we assume that there is an inner-product on $\mathcal{X}$ that defines a norm. Our goal then is to solve the quadratic optimization problem,

$$\min_r \|h^r - h\|^2 = \min_r \langle h^r - h, h^r - h \rangle. \tag{24.29}$$

That is, we seek the projection of h onto $\mathcal{L}$. This is the least-squares problem that has an explicit solution: Let M denote the $d \times d$ matrix with entries $M_{i,j} = \langle \psi_i, \psi_j \rangle$, and b the vector in $\mathbb{R}^d$ defined by $b_i = \langle \psi_i, h \rangle$. If M is invertible, so that $\mathcal{L}$ is d-dimensional, then the unique minimizer in (24.29) is

$$r^* = M^{-1} b. \tag{24.30}$$

The problem with this representation is that b depends on $h = Rc$, the function we are attempting to approximate. This issue is solved by introducing the *adjoint* $R^\dagger$, defined so that

$$b_i = \langle \psi_i, Rc \rangle = \langle R^\dagger \psi_i, c \rangle. \tag{24.31}$$

The adjoint has an elegant form in the specific MDP setting, provided the norm is chosen consistently with the Markov model.

Markovian Setting. The simulation algorithm used to compute the optimal parameter r^* given in (24.30) is called *least-squares TD-Learning* (LSTD). To obtain the adjoint which leads to the LSTD algorithm, we must first define the vector space $\mathcal{X}$, and we must then define R in this setting.

Let P denote the transition kernel for the Markov chain, and assume that there is an invariant probability distribution π. In the context of this chapter, P will be obtained by fixing a stationary policy ϕ, so that $P = P_\phi$. We let $\mathcal{X} = L_2(\pi)$, the usual Hilbert space of square-integrable functions, with inner product $\langle f, g \rangle = \mathsf{E}[f(X(0))g(X(0))]$, $f, g \in L_2(\pi)$, where the expectation is in steady-state, so that $X(0) \sim \pi$. It is assumed that a cost function $c \in L_2(\pi)$ is given.

For the class of Markov chains considered in this chapter, and more generally in [6], it is known that the solution to Poisson's equation can be expressed $h = Rc$, where R is the *fundamental kernel*. Under the assumptions we have imposed, the function h exists (as an a.e. finite function on X). Stronger conditions are required to guarantee that $h \in L_2(\pi)$.

Suppose that there is a fixed state $\vartheta \in \mathsf{X}$ satisfying $\pi(\vartheta) > 0$. Let τ_ϑ denote the first return time to this state, $\tau_\vartheta = \min\{t \geq 1 : X(t) = \vartheta\}$. In this case, the fundamental kernel can be expressed as the *total relative cost until reaching* ϑ:

$$Rc\,(x) = \mathsf{E}\Big[\sum_{t=0}^{\tau_\vartheta - 1} [c(X(t)) - \eta] \,\Big|\, X(0) = x\Big].$$

The existence of a so-called atom ϑ is not at all necessary, but this assumption saves us from introducing additional technical machinery [6].

Given this representation, it is not difficult to arrive at a representation for the adjoint of R: for any $f \in L_2(\pi)$ with mean η we have

$$R^{\dagger} f(x) = \mathsf{E}\Big[\sum_{\tau_{\vartheta}^{-} < t \leq 0} [f(X(t)) - \eta] \Big| X(0) = x \Big], \qquad (24.32)$$

provided the expectation exists and is finite-valued.

LSTD Algorithm. Let $\psi\colon \mathsf{X} \to \mathbb{R}^d$ denote the vector of basis functions, and $h^r = r^{\mathrm{T}}\psi$, exactly as in the abstract least-squares problem. LSTD takes the mean-square error criterion

$$\|h - h^r\|^2 = \mathsf{E}_{\pi}[(h(X(0)) - h^r(X(0)))^2] := \int (h(x) - h^r(x))^2\, \pi(\mathrm{d}x).$$

Hence the optimal parameter satisfies the fixed point equation,

$$\mathsf{E}_{\pi}[(h(X(0)) - h^r(X(0)))\psi_i(X(0))] = 0 \qquad 1 \leq i \leq d. \qquad (24.33)$$

This is the least squares solution described previously: in this Markovian setting, the optimal parameter (24.30) can be expressed as follows:

$$\begin{aligned} r^* = M^{-1} b \quad \text{with} \quad M &= \mathsf{E}_{\pi}[\psi(X(0))\psi(X(0))^{\mathrm{T}}] \\ b &= \mathsf{E}_{\pi}[\psi(X(0)) h(X(0))]. \end{aligned} \qquad (24.34)$$

The algorithm is defined in the following steps, to recursively estimate M, b, and the steady-state mean of c. For each $t \geq 0$,

(i) the steady-state average cost η_{ϕ} is estimated via

$$\eta_{\phi}(t+1) = \eta_{\phi}(t) + \frac{1}{t+1}\big(c(X(t+1) - \eta_{\phi}(t)\big),$$

(ii) the *eligibility vector* is updated as follows:

$$\psi^{\vartheta}(t+1) = \psi^{\vartheta}(t) + \mathbb{I}\{X(t) \neq \vartheta\}\psi(X(t)), \quad \text{with } \psi^{\vartheta}(0) = \psi(X(0)),$$

(iii) estimates of the matrix M and vector b are updated as follows:

$$M(t+1) = M(t) + \frac{1}{t+1}\big(\psi(X(t))\psi^{\mathrm{T}}(X(t)) - M(t)\big), \qquad (24.35)$$

$$b(t+1) = b(t) + \frac{1}{t+1}\big([c(X(t)) - \eta_\phi(t)]\psi^{\vartheta}(t) - b(t)\big), \qquad (24.36)$$

(iv) the estimate of r^* is $r_t = M_t^{-1}b_t$.

Observe that several of these steps are simply formulations of standard Monte-Carlo. In particular, M_t is a running average of $\psi(X(t))\psi^{\mathrm{T}}(X(t))$, and b_t approximates the running average of $[c(X(t)) - \eta]\psi^{\vartheta}(t)$. The algorithm is consistent, $r_t \to r^*$ as $t \to \infty$, provided ψ and h are square integrable, M is invertible, and X is stationary. The stationarity assumption can be relaxed under further assumptions.

Consideration of the *algorithm in steady-state* explains the introduction of the eligibility vectors. Suppose that $\{X(t), \psi^{\vartheta}(t) : -\infty < t < \infty\}$ is a stationary realization of the coupled Markov model and LSTD algorithm, in which $\eta(t) \equiv \eta$ is the mean of c under π. By the law of large numbers for stationary processes we obtain

$$\lim_{t\to\infty} b(t) = \lim_{T\to\infty} \frac{1}{T}\sum_{i=1}^{T}[\psi^{\vartheta}(t)(c(X(t)) - \eta)] = \mathsf{E}[\psi^{\vartheta}(0)(c(X(0)) - \eta)].$$

By the recursion that defines the eligibility vectors, the right-hand side is expressed

$$\mathsf{E}[\psi^{\vartheta}(0)(c(X(0)) - \eta)] = \mathsf{E}\Big[\sum_{\tau_{\vartheta}^{-} < t \le 0} \psi(X(t))[c(X(0)) - \eta] \,\Big|\, X(0) = x\Big],$$

which is precisely the representation of b given in (24.31, 24.32).

24.3.3 SARSA

Recall that to apply policy iteration, a potential difficulty is the policy update formula (24.27): the computation of $P_u h$ itself may be difficult if the state space is large.

Consider the following approach to avoid integration: Let H denote the function of two variables,

$$H(x, u) = c(x, u) + P_u h\,(x). \qquad (24.37)$$

If we can estimate H directly, then the policy update can be obtained by minimizing $H(x, u)$ over u, for each state x.

To place the analysis within the prior setting, assume that a policy ϕ has been specified, and that h is the solution to Poisson's equation for P_ϕ, with cost c_ϕ. Then

H and h satisfy the following equation,

$$H_\phi(x) := H(x, \phi(x)) = c_\phi(x) + P_\phi h\,(x) = h(x) + \eta. \tag{24.38}$$

Substituting this back into the definition of H (replacing h by $H_\phi - \eta$), we obtain the fixed point equation,

$$\eta + H = c + P_u H_\phi. \tag{24.39}$$

We conclude that H solves Poisson's equation for the state-control process:

Proposition 24.1 Suppose that X is controlled using a randomized stationary policy ϕ, and that h is the solution to Poisson's equation using c_ϕ. Then,

(i) the state-control process $\Phi(t) = (X(t), U(t))$ is also Markov,
(ii) the function H is a solution to Poisson's equation for the Markov chain $\Phi(t)$, and cost function $c(x, u)$.

Proof: Part (i) is obvious: The process $\mathbf{\Phi}$ evolves according to a controlled stochastic model of the recursive form (24.11).

To see (ii) we reinterpret (24.39):

$$\eta + H(x, u) = c(x, u) + \mathsf{E}[H(\Phi(t+1)) \mid \Phi(t) = (x, u)],$$

which is indeed Poisson's equation for $\mathbf{\Phi}$. ■

A natural parameterization is of the form $H^r = c + r^{\mathsf{T}}\psi$, where $\psi \colon \mathsf{X} \times \mathsf{U} \to \mathbb{R}^d$. Given a basis $\{\psi_i^0 : 1 \le i \le d\}$ intended for application in TD-Learning, we might choose, $\psi_i(x, u) = P_u \psi_i^0\,(x)$, $x \in \mathsf{X}$, $u \in \mathsf{U}$. If the integration is difficult to compute, then we might seek an approximation using a fluid or diffusion model. Note that $P_u = I + \mathcal{D}_u$. So, for a fluid approximation with generator $\mathcal{D}^{\mathrm{F}}$, we might justify the approximation $P_u \approx I + \mathcal{D}^{\mathrm{F}}$, and define a basis using $\psi_i(x, u) = \psi_i^0\,(x) + \mathcal{D}_u^{\mathrm{F}} \psi_i^0\,(x)$, $x \in \mathsf{X},\ u \in \mathsf{U}$.

24.3.4 Q-Learning

The Q-Learning algorithm is designed to skip the policy improvement step entirely. Rather than attempt to minimize the mean-square error criterion (24.28), for a given approximation we consider the error in the optimality equations.

Q-Learning is more difficult than TD-Learning because the ACOE cannot be interpreted as a *linear* fixed-point equation, and hence least-squares concepts are not directly applicable. The successful algorithms do not attempt to approximate h^* directly, but instead consider the so-called Bellman error. If h^r is an approximation to

h^*, then the Bellman error measures the error in the ACOE equation (24.10):

$$\underline{\mathcal{E}}^r(x) := -h^r(x^\bullet) + \min_u \{c(x,u) + \mathcal{D}_u h^r(x)\}. \tag{24.40}$$

If $\mathcal{E}^r(x)$ is a zero, then the ACOE is solved, where $\eta^* = h^r(x^\bullet)$.

The successful approaches to this problem consider not h^*, but the function of two variables,

$$H^*(x,u) = c(x,u) + P_u h^*(x). \tag{24.41}$$

Watkins introduced a *reinforcement learning* technique for *computation of* H^* in his thesis, and a complete convergence proof appeared later in [9]. An elementary proof based on an associated "fluid limit model" is contained in [10]. Unfortunately, this approach depends critically on a finite state space, a finite action space, and a *complete parameterization*, that includes all possible Markov models whose state space has a given cardinality. Consequently, complexity grows with the size of the state space.

Progress has been more positive for special classes of models. For deterministic linear systems with quadratic cost (the LQR problem), a variant of Q-Learning combined with an adaptive version of policy iteration is known to be convergent—see [11] for an analysis in discrete time, and [12] for a similar approach in continuous time. A complete solution to the deterministic problem can be found in [13]. There is also a complete theory for optimal stopping problems [8, 14].

A general approach to the construction of an algorithm proceeds as follows. First, note that just as in the construction of SARSA, we can write the ACOE as a fixed point equation in H^*: on denoting $\underline{H}^*(x) = \min H^*(x,u)$, $x \in \mathsf{X}$, the ACOE (24.13) gives

$$\underline{H}^*(x) = h^*(x) + \eta^*.$$

Substituting $h^* = \underline{H}^* - \eta$ into the definition of H^* then gives

$$H^*(x,u) = c(x,u) + P_u \underline{H}^*(x) - \eta^*. \tag{24.42}$$

Based on this expression, the Bellman error (24.40) must be extended for approximation of H^* among functions on $\mathsf{X} \times \mathsf{U}$. If H^r is an estimate of H^*, we denote $\underline{H}^r(x) := \min_u H^r(x,u)$, $x \in \mathsf{X}$, and define the Bellman error as the function of two variables,

$$\mathcal{E}^r(x,u) := -H^r(x^\bullet, u^\bullet) + \{c(x,u) + P_u \underline{H}^r(x) - H^r(x,u)\}, \tag{24.43}$$

where the pair $(x^\bullet, u^\bullet) \in \mathsf{X} \times \mathsf{U}$ is again arbitrary. Once again, if $\mathcal{E}^r(x,u)$ is zero, then the ACOE is solved, where $\eta^* = H^r(x^\bullet, u^\bullet)$.

One version of Q-Learning can be obtained as steepest descent. For simplicity consider a linear parameterization,

$$H^r(x,u) = c(x,u) + \sum_{i=1}^{d} r_i \psi_i(x,u)\,, \quad r \in \mathbb{R}^d, \tag{24.44}$$

where $\{\psi_i : 1 \le i \le d\} \subset \mathcal{X}$, and $\mathcal{X}$ now consists of real-valued functions on $\mathsf{X} \times \mathsf{U}$. Let $\|\cdot\|$ denote any norm on $\mathcal{X}$, and consider the steepest descent algorithm,

$$\frac{\mathrm{d}}{\mathrm{d}t} r(t) = -\nabla \|\mathcal{E}^r\|^2 \Big|_{r=r(t)}. \tag{24.45}$$

An RL or ADP algorithm can be constructed by mimicking this ODE. For the norm, we choose an ergodic norm as in TD-Learning. A major difference is that in this case we choose a randomized stationary policy, designed to explore the state-action space $\mathsf{X} \times \mathsf{U}$. For any function $g \in \mathcal{X}$ we define,

$$\|g\|^2 = \mathsf{E}_\pi[g(X,U)^2],$$

where $X \sim \pi$, the steady-state distribution obtained with the given policy ϕ, and U is a randomized function of X: given $X = x$, we have $U \sim \phi(x,z)\mu_Z(dz)$ (see definition below (24.25)). A descent direction is expressed,

$$-\nabla \|\mathcal{E}^r\|^2 = -2\mathsf{E}[\mathcal{E}^r(X,U)\nabla \mathcal{E}^r(X,U)], \tag{24.46}$$

where, in general, "∇" denotes a subgradient.

From the definition of the Bellman error (24.43) we have

$$\nabla \mathcal{E}^r(x,u) = -\nabla H^r(x,u) + \nabla\big\{P_u \underline{H}^r(x)\big\}.$$

In the finite state space case this can be expressed,

$$\nabla \mathcal{E}^r(x,u) = -\nabla H^r(x,u) + \sum_{x' \in \mathsf{X}} P_u(x,x') \nabla \underline{H}^r(x').$$

A similar expression can be obtained in general, subject to integrability constraints. For each x', the function $\underline{H}^r(x')$ is concave as a function of r. A subgradient is

$$\nabla \underline{H}^r(x') = \psi(x', u^*_{x',r}) \in \mathbb{R}^d\,,$$

where $u^*_{x',r}$ solves $\min_u H^r(x',u)$. The gradient of H^r is obviously

$$\nabla H^r(x,u) = \psi(x,u) \in \mathbb{R}^d\,.$$

So, a subgradient of the Bellman error is given by

$$\nabla\mathcal{E}^r(x,u) = -\psi(x,u) + \mathsf{E}\big[\psi(X(t+1), U_r^*(t+1)) \mid X(t)=x,\ U(t)=u\big], \tag{24.47}$$

where $U_r^*(t+1) = u^*_{x',r}$ when $X(t+1) = x'$.

In conclusion, the descent direction (24.46) can be expressed,

$$-\nabla\|\mathcal{E}^r\|^2 = -2\mathsf{E}[\mathcal{E}_t^r \nabla\mathcal{E}_t^r], \tag{24.48}$$

$$\begin{aligned}\mathcal{E}_t^r &= c(X(t), U(t)) + \mathsf{E}\big[\underline{H}^r(X(t+1)) \mid X(t),\ U(t)\big] - H^r(X(t), U(t)) \\ &\quad - H^r(x^\bullet, u^\bullet), \\ \nabla\mathcal{E}_t^r &= -\psi(X(t), U(t)) + \mathsf{E}\big[\psi(X(t+1), U_r^*(t+1)) \mid X(t),\ U(t)\big].\end{aligned}$$

The representation (24.48) is amenable to the construction of a simulation-based algorithm for approximate dynamic programming. For each t, having obtained values $(x,u) = (X(t), U(t))$, we obtain *two values of the next state*, denoted $X(t+1)$ and $\widetilde{X}(t+1)$. These random variables are each distributed according to $P_\phi(x,\ \cdot)$, but conditionally independent, given $X(t)=x$. That is, we are taking two draws from P_ϕ. We can thus express

$$\mathsf{E}[\mathcal{E}_t^r \nabla\mathcal{E}_t^r] = \mathsf{E}[B_{t+1}(r)\Phi_{t+1}(r)], \quad \textit{where,}$$

$$B_{t+1}(r) = -H^r(x^\bullet, u^\bullet) + \big\{c(X(t), U(t)) + \underline{H}^r(\widetilde{X}(t+1)) - H^r(X(t), U(t))\big\},$$

$$\Phi_{t+1}(r) = -\psi(X(t), U(t)) + \psi(X(t+1), U_r^*(t+1)).$$

Given these representations, a stochastic approximation of the ODE (24.45) is given by the recursion,

$$r_{t+1} = r_t - a_t B_{t+1}(r_t)\Phi_{t+1}(r_t), \qquad \text{where } \{a_t\} \text{ is the stepsize.}$$

24.3.5 Architecture

These techniques require a basis for approximation of a value function, or some generalization. Another approach not considered in this chapter is based on linear programming techniques: it is well known that dynamic-programming equations can be expressed as linear programs, and this leads to approaches to approximate dynamic programming [15]. These methods require a basis, exactly as in Q- or TD-Learning.

The remainder of this chapter concerns the question of basis construction. There are many ways to approach this: Taylor or Fourier series, linearization, approximations using various asymptotics (such as congestion—"heavy traffic" in networks, or large-state in general models [5]). All of the approaches considered in the following sections rely on approximate modeling.

24.4 FLUID MODELS

24.4.1 The CRW Queue

The ACOE for this model is given by (24.13), with generator $\mathcal{D}_u h^*(x) = \mathsf{E}[h^*(x - u + A(1)) - h(x)]$. The input u is constrained to $\{0, 1\}$.

With the cost function $c(x, u) \equiv x$, it is not hard to guess that the optimal policy for the stochastic model is nonidling: $U^*(t) = \mathbb{I}\{X^*(t) \geq 1\}$. Hence, the ACOE coincides with *Poisson equation*,

$$\mathcal{D}h^*(x) = -x + \eta^*, \qquad x \in \mathbb{Z}_+, \tag{24.49}$$

where $\mathcal{D}$ without the subscript "u" denotes the conditional expectation (24.12) under the nonidling policy. We can compute h^* in this case, and we will see that it is a quadratic function of x.

Consider first the fluid model approximation. Under the optimal policy that sets $u(t) = 1$ when $q(t) > 0$, the resulting state trajectory is given by $q(t) = \max(q(0) - (1 - \alpha)t, 0)$. We denote $T^* = x/(1 - \alpha)$, which is the first time that $q(t)$ reaches zero.

The value function J^* defined in (24.3) can be interpreted as the area under a right triangle with height $x = q(0)$, and width T^*. Consequently,

$$J^*(x) = \tfrac{1}{2} x T^* = \tfrac{1}{2} \frac{x^2}{1 - \alpha}.$$

It is easy to see that the HJB equation (24.4) is satisfied,

$$\min_u \{c(x, u) + \mathcal{D}_u^{\mathrm{F}} J^*(x)\} = \min_u \{x + (-u + \alpha) \cdot \nabla J^*(x)\} = 0.$$

Theorem 24.4.1 is taken from [5, Theorem 3.0.1]. The identity (24.51) establishes that h^*, the solution to Poisson's equation for the CRW model, is a perturbation of the fluid value function J^*. The formula (24.50) for the steady-state mean of $\boldsymbol{Q}$ is a version of the *Pollaczek–Khintchine formula*.

Theorem 24.1 *Consider the CRW queueing model* (24.14) *satisfying* $\alpha := \mathsf{E}[A(t)] < 1$, $\sigma_A^2 := \mathsf{E}[(A(t) - \alpha)^2] < \infty$. *Then,*

(i) $\boldsymbol{Q}$ *is positive recurrent: With steady-state mean,*

$$\eta := \mathsf{E}_\pi[Q(0)] = \tfrac{1}{2} \frac{\sigma^2}{1 - \alpha}, \tag{24.50}$$

where $\sigma^2 = \sigma_A^2 + \alpha(1 - \alpha)$.

(ii) A solution to Poisson's equation (24.49) *is*

$$h^*(x) = J^*(x) + \tfrac{1}{2}\Big(\frac{(1-\alpha)^2 - \alpha^2}{1-\alpha}\Big)x\,, \qquad x \in \mathbb{Z}_+. \tag{24.51}$$

24.4.2 Speed-Scaling Model

Dynamic speed scaling is an increasingly common approach to power management in computer system design. The goal is to control the processing speed so as to optimally balance energy and delay costs: Reducing (increasing) the speed in times when the workload is small (large). Speed scaling has traditionally been used in processor design, however in recent years speed scaling approaches have been applied in a variety of other contexts, including disks and wireless transmission.

The primary model is described in discrete time as the controlled queue (24.14). The cost function is chosen to balance the cost of delay with the energy cost associated with the processing speed. A special case is quadratic,

$$c(x, u) = x + \beta u^2\,, \tag{24.52}$$

with $\beta > 0$, where u^2 models the power required as a function of the speed u.

Applying the definition of $\overline{f}$ in (24.15), the fluid model corresponding to the speed scaling model is again given by (24.22). To obtain the total-cost value function for the fluid model we modify the cost function so that it vanishes when $x = 0$ and $u = \alpha$ (its equilibrium value). A simple modification is the following:

$$c^{\mathrm{F}}(x, u) = x + \beta[u - \alpha]_+^2, \tag{24.53}$$

where $[u - \alpha]_+ = \max(0, u - \alpha)$. With this perturbation of the cost, the total cost J^* defined in (24.3), and the optimal policy ϕ^{F} (the minimizer of (24.4)) are computable in closed form. In particular, $J^*(x) = 1/3(2x)^{3/2}$, when $\beta = 1/2$.

Rather than modify the cost function, another approach is to modify the objective function in the control problem for the fluid model. Consider

$$K^*(x) = \inf_{\boldsymbol{u}} \int_0^{T_0} c(x(t), u(t))\, \mathrm{d}t\,,$$

where T_0 is the first hitting time to the origin, and $x(0) = x \in \mathbb{R}_+$. This function is also computable in closed form. When $\beta = 1/2$ we obtain

$$K^*(x) = \alpha x + \tfrac{1}{3}[(2x + \alpha^2)^{3/2} - \alpha^3]. \tag{24.54}$$

This solves the TCOE (24.4) using the cost function $c(x, u) = x + 1/2u^2$.

We can also obtain an expression for the total *relative cost*: for any $\eta > 0$, and any policy for the fluid model, denote $T_\eta = \min\{t : c(x(t), u(t)) \le \eta\}$. Let K^*_η denote the minimal total *relative* cost,

$$K^*_\eta(x) = \inf \int_0^{T_\eta} \big(c(x(t), u(t)) - \eta\big)\, dt. \tag{24.55}$$

As in the definition of J^* and K^*, the infimum is over all policies. For $x > \eta$, the value function K^*_η solves the dynamic programming equation,

$$\min_u \big\{c(x, u) + \mathcal{D}^{\mathrm{D}}_u K^*_\eta(x)\big\} = \eta.$$

An approximate solution is given by,

$$h_\eta(x) := K^*(x) - 2\eta\big(\sqrt{2x + q^2} - \sqrt{q^2}\big), \tag{24.56}$$

where $q > 0$ is a constant. It can be shown that $\sup_x |h_\eta(x) - K^*_\eta(x)| < \infty$, regardless of the values of the nonnegative scalars η and q.

Numerical Results In the results that follow, the arrival process $\boldsymbol{A}$ is taken to be a scaled geometric distribution on $\{0, 1, \dots\}$ with parameter p_A. The mean and variance of $A(t)$ are given by, respectively,

$$\alpha = \frac{p_A}{1 - p_A}, \qquad \sigma_A^2 = \frac{p_A}{(1 - p_A)^2}. \tag{24.57}$$

The solution to the ACOE (24.13) was computed for this model, and with the cost function $c(x, u) = x + 1/2u^2$, using value iteration. This required approximation: the input was restricted to the non-negative integers, and the state space X was restricted to a finite set of the form $\{0, 1, \dots, N\}$, by truncating arrivals.

Shown in Figure 24.1 is a comparison of the optimal policy, computed numerically using value iteration, and the (c, K^*)-myopic policy defined in analogy with the optimal policy,

$$\phi^{K^*}(x) = \arg\min_{0 \le u \le x} \{c(x, u) + P_u K^*(x)\}. \tag{24.58}$$

Shown in Figure 24.2 is a comparison of the fluid value function K^*, the relative value function h^*, and the output of the LSTD algorithm using the basis $\psi_1(x) \equiv x$, and $\psi_2(x) \equiv K^*(x)$ (defined in (24.54)). The policy (24.58) was used in the application of the TD-Learning algorithm.

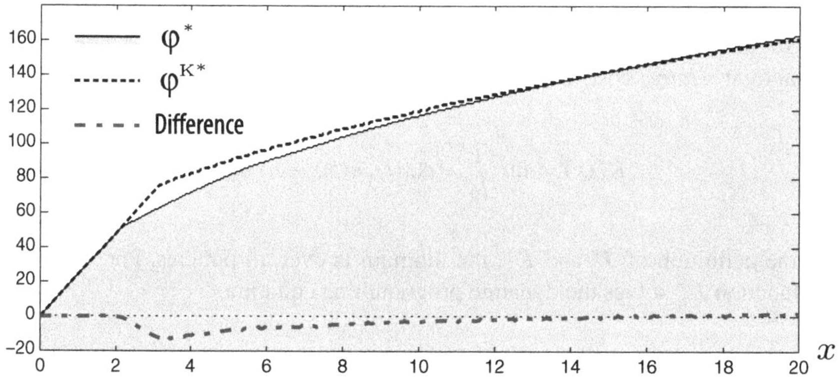

FIGURE 24.1 The optimal policy compared to the (c, K^*)-myopic policy for the quadratic cost function $c(x, u) = x + 1/2u^2$.

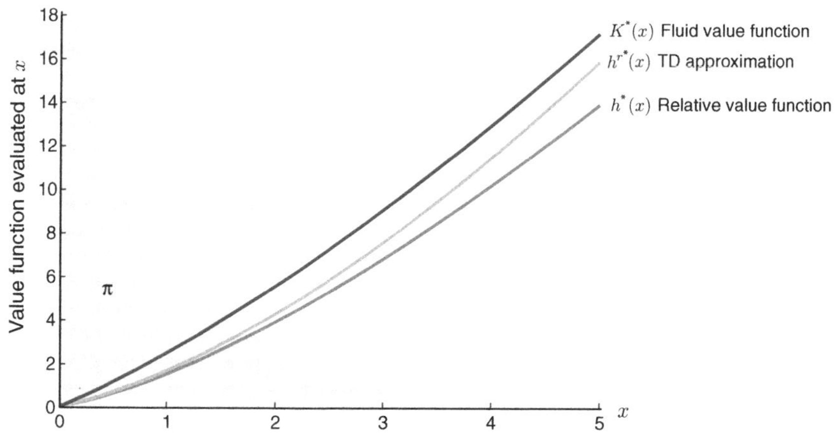

FIGURE 24.2 Value functions for the dynamic speed scale model with quadratic cost. The x-axis denotes the state value, and the three plots compare the final approximation h^{r^*} obtained using LSTD-Learning with the fluid value function, and the relative value function h^* (computed using value iteration). The histogram is an estimate of the steady-state distribution π obtained using the optimal policy.

24.5 DIFFUSION MODELS

In the two examples that follow we will see that the fluid value function is a very good approximation to the relative value functions for both the CRW model and its diffusion approximation. This solidarity holds in part because $\nabla^2 J^*$ is not large. In other network applications, the diffusion model might result in a significant refinement of the fluid model [5]. We omit details here due to lack of space.

24.5.1 The CRW Queue

The generator (24.6) for the RBM model (24.23) is

$$\mathcal{D}^{\mathrm{D}}_u h\,(x) = -(u-\alpha)h'(x) + \tfrac{1}{2}\sigma_A^2 h''\,(x).$$

For the nonidling process, the queue evolves as a *reflected Brownian motion* on $\mathbb{R}_+$. The differential generator is given with $u \equiv 1$,

$$\mathcal{D}^{\mathrm{D}} h\,(x) = -\delta h'(x) + \tfrac{1}{2}\sigma_A^2 h''\,(x), \tag{24.59}$$

where $\delta = 1-\alpha$. The interpretation of the generator requires *Itô's formula.*

Itô's formula provides a representation of $h(Q(t))$ for any twice continuously differentiable (C^2) function $h\colon \mathbb{R} \to \mathbb{R}$,

$$\begin{aligned} h(Q(t)) = h(x) &+ \int_0^t [-\delta h'(Q(s)) + \tfrac{1}{2}\sigma_A^2 h''(Q(s))]\,\mathrm{d}s \\ &+ \int_0^t h'(Q(s))\,\mathrm{d}N(s) + \int_0^t h'(0)\,\mathrm{d}I(s). \end{aligned} \tag{24.60}$$

The idleness process can only increase when $\boldsymbol{Q}$ is zero, giving

$$\int_0^t h'(Q(s))\,\mathrm{d}I(s) = \int_0^t h'(0)\,\mathrm{d}I(s) = h'(0)I(t).$$

Suppose that the function h solves Poisson's equation, $c + \mathcal{D}^{\mathrm{D}} h = \eta$. Then, Itô's formula (24.60) implies the representation,

$$\int_0^t [c(Q(s)) - \eta]\,\mathrm{d}s = h(x) - h(Q(t)) + h'(0)I(t) + \int_0^t h'(Q(t))\,\mathrm{d}N(t).$$

Subject to growth conditions on h, we have

$$\lim_{t\to\infty} \frac{1}{t}\int_0^t h'(Q(t))\,\mathrm{d}N(t) = \lim_{t\to\infty} \frac{1}{t} h(Q(t)) = 0.$$

Consequently, we can conclude that the steady-state mean of c is equal to η, provided $h'(0) = 0$.

A function satisfying this condition is the fluid value function, $J^*(x) := 1/2\delta^{-1}x^2$. Applying the generator (24.59) we obtain

$$x + \mathcal{D}^{\mathrm{D}} J^*\,(x) = \tfrac{1}{2}\sigma_A^2[\delta^{-1}]. \tag{24.61}$$

Hence J^* solves Poisson's equation provided $\delta > 0$. That is, $\alpha < 1$, which is the necessary and sufficient condition for stability. Under this condition, the average cost is in fact given by $\eta^* = 1/2(1-\alpha)^1\sigma_A^2$, a formula similar to (24.50), and $h^* = J^*$ solves the ACOE (24.9).

24.5.2 Speed-Scaling Model

The diffusion model is again given by (24.23), but with $U(t) \in \mathbb{R}_+$, rather than constrained to a bounded interval.

Looking back at the analysis of the fluid model in Section 24.4, we see that h_η is in the domain of the generator, in the sense that $d/d2h_\eta(0) = 0$, provided we choose $q = \eta/\alpha$ (see (24.56)). The function h_η is strictly convex, so we have $h_\eta(x) > 0$, for all $x > 0$, when this derivative condition holds.

Let $\underline{\mathcal{E}}^\eta$ denote the Bellman error for the approximation h^η, defined via (24.40), with the differential generator:

$$\underline{\mathcal{E}}^\eta(x) := \min_u\{c(x,u) + \mathcal{D}_u^{\mathrm{D}} h_\eta(x)\}.$$

The function h_η approximately solves the ACOE for the diffusion model, in the sense that the Bellman error is bounded: $\sup_x |\underline{\mathcal{E}}^\eta(x)| < \infty$ for any η.

24.6 MEAN FIELD GAMES

Q-Learning is a natural candidate for applications in distributed control and games (see [16, 17] for closely related messages). We illustrate this with results obtained for the *large-population cost-coupled LQG problem* introduced in [18].

The model consists of n nonhomogeneous autonomous agents. The ith-agent is modeled as a deterministic, scalar linear system, in *continuous time*,

$$\frac{\mathrm{d}}{\mathrm{d}t}x_i = a_i x_i + b_i u_i, \tag{24.62}$$

where x_i and u_i denote the state and the control of the ith-agent, respectively. The agents are coupled through their respective quadratic cost functions: For each $1 \le i \le n$, $c_i(x_i, u_i) = (x_i - z)^2 + u_i^2$, where z is the mean, $z = n^{-1}(x_1 + \cdots + x_n)$. The optimal control problem of [18] is the *discounted cost*, linear-quadratic control problem. The HJB equation is modified as follows:

$$\min_u\{c(x,u) + \mathcal{D}_u^{\mathrm{F}} J^*(x)\} = \gamma J^*, \tag{24.63}$$

where $\gamma > 0$ is the discount rate. As in the definition (24.42), the "Q-function" will remain the function appearing in the brackets in (24.63). Because the cost is assumed to be quadratic, $c(x,u) = \frac{1}{2}x^{\mathrm{T}}Mx + \frac{1}{2}u^{\mathrm{T}}Ru$, we take the parameterization,

$$H^r(x,u) = c(x,u) + x^{\mathrm{T}}E^r x + x^{\mathrm{T}}F^r u, \tag{24.64}$$

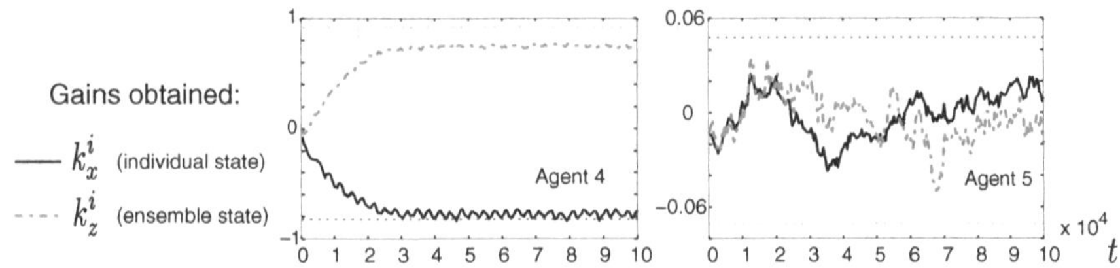

FIGURE 24.3 Sample paths of estimates of (k_x^i, k_z^i) for $i = 4$ and 5. The dashed lines show the asymptotically optimal values obtained in Ref. [18].

where $\{E^r, F^r\}$ depend linearly on r. In this notation, the corresponding policy obtained by minimizing H^r over u is given by,

$$\phi^r(x) = -R^{-1}(F^r)^{\mathrm{T}}x. \tag{24.65}$$

Hence $u = \phi^r(x)$ is linear state-feedback for any r.

An approximate model is proposed in [18] (see their Equations (4.6)–(4.9)): Each agent solves the optimal control problem with a two-dimensional approximate state, given by $(x_i, z)^{\mathrm{T}}$ for the ith-agent.

The conclusions of this prior work suggest the following architecture for Q-Learning. The Q-function for the LQ problem is defined according to the matrix parametrization (24.64). Each agent has three parameters (E^r) that are coefficients of the basis functions $\{x_i^2, z^2, x_i z\}$, and two parameters (F^r) that are coefficients of the basis functions $\{x_i u_i, z u_i\}$.

The following numerical results are based on an example with five agents described by (24.62). Each of the five inputs u_i were taken to be sinusoidal, with irrationally related frequencies. Five applications of the Q-Learning algorithm were run in parallel, one for each agent. For details see [13].

Figure 24.3 depicts the evolution of estimates of the two components of the local optimal gain (24.65) for two of the five agents ($i = 4$ and 5), expressed $u_i = -k_x^i x_i - k_z^i z$. Also shown are the gains introduced in [18] that were found to be asymptotically optimal for large n. The limiting values of the estimates of (k_x^i, k_z^i) were close to those predicted in [18] in all cases. The first plot shows typical behavior of the algorithm.

The sequence of gains shown in the second plot appear more volatile, and less consistent. However, note that the vertical scale is bounded by ± 0.1, so that the gains are nearly zero.

24.7 CONCLUSIONS

We have shown through theory and examples that insight from a naive model can be extremely valuable in the creation of an architecture for reinforcement learning or approximate dynamic programming. We have focused on fluid and diffusion value functions to approximate the stochastic value function, but other approximations may be useful, depending on the application. When the approximate model cannot be solved exactly, then further approximation can be applied, as we have seen in Sections 24.4 and 24.5.

In this chapter we have largely ignored the subject of computing error bounds, and applying these bounds for performance approximation. We refer the reader to [4, 5, 7, 8] for some results in this direction.

One open problem in this area concerns the formulation of Q-Learning algorithm for reinforcement learning, in a parameterized setting, in which the system dynamics are not given *a priori*. One possible approach to parameterized reinforcement learning might emerge via the approximate LP approaches.

ACKNOWLEDGMENTS

Financial support from UTRC, AFOSR Grant FA9550-09-1-0190, ITMANET DARPA RK 2006-07284, and NSF Grant CPS 0931416 is gratefully acknowledged. Any opinions, findings, and conclusions or recommendations expressed in this material are those of the authors and do not necessarily reflect the views of the National Science Foundation.

REFERENCES

1. D.P. Bertsekas and J.N. Tsitsiklis. *Neuro-Dynamic Programming*. Atena Scientific, Cambridge, MA, 1996.
2. C. Szepesvári. *Algorithms for Reinforcement Learning*. Synthesis Lectures on Artificial Intelligence and Machine Learning. Morgan & Claypool Publishers, 2010.
3. S.G. Henderson, S.P. Meyn, and V.B. Tadić. Performance evaluation and policy selection in multiclass networks. *Discrete Event Dynamic Systems: Theory and Applications*, 13(1-2):149–189, 2003. Special issue on learning, optimization and decision making (invited).
4. W. Chen, D. Huang, A.A. Kulkarni, J. Unnikrishnan, Q. Zhu, P. Mehta, S. Meyn, and A. Wierman. Approximate dynamic programming using fluid and diffusion approximations with applications to power management. In *Proceedings of the 48th IEEE Conference on Decision and Control*, pp. 3575–3580, 2009.
5. S.P. Meyn. *Control Techniques for Complex Networks*. Cambridge University Press, Cambridge, 2007.
6. S.P. Meyn and R.L. Tweedie. *Markov chains and stochastic stability*. Cambridge University Press, Cambridge, 2nd edition, 2009. Published in the Cambridge Mathematical Library, 1993 edition online.
7. J.N. Tsitsiklis and B. Van Roy. An analysis of temporal-difference learning with function approximation. *IEEE Transaction on Automatic Control*, 42(5):674–690, 1997.
8. J. Tsitsiklis and B. van Roy. Optimal stopping of Markov processes: Hilbert space theory, approximation algorithms, and an application to pricing high-dimensional financial derivatives. *IEEE Transaction on Automatic Control*, 44(10):1840 –1851, 1999.
9. C.J. C.H. Watkins and P. Dayan. *Q*-Learning. *Machine Learning*, 8(3–4):279–292, 1992.

10. V.S. Borkar. *Stochastic Approximation: A Dynamical Systems Viewpoint.* Hindustan Book Agency and Cambridge University Press (jointly), Delhi, India and Cambridge, UK, 2008.

11. S.J. Bradtke, B.E. Ydstie, and A.G. Barto. Adaptive linear quadratic control using policy iteration. In *Proceedings of the 1994 American Control Conference*, Vol. 3, pp. 3475–3479, 1994.

12. D. Vrabie, O. Pastravanu, M. Abu-Khalaf, and F.L. Lewis. Adaptive optimal control for continuous-time linear systems based on policy iteration. *Automatica*, 45(2):477–484, 2009.

13. P.G. Mehta and S.P. Meyn. Q-Learning and Pontryagin's minimum principle. In *Proceedings of the 48th IEEE Conference on Decision and Control; held jointly with the 2009 28th Chinese Control Conference*, pp. 3598–3605, December 2009.

14. H. Yu and D.P. Bertsekas. Q-Learning algorithms for optimal stopping based on least squares. In *Proceedings of European Control Conference (ECC)*, July 2007.

15. D.P. de Farias and B. Van Roy. The linear programming approach to approximate dynamic programming. *Operations Research*, 51(6):850–865, 2003.

16. R. Cogill, M. Rotkowitz, B. Van Roy, and S Lall. An approximate dynamic programming approach to decentralized control of stochastic systems. In *Control of Uncertain Systems: Modelling, Approximation, and Design*, pp. 243–256. Springer, 2006.

17. S.P. Meyn and G. Mathew. Shannon meets Bellman: feature based Markovian models for detection and optimization. In *Proceedings of 47th IEEE CDC*, pp. 5558–5564, 2008.

18. M. Huang, P.E. Caines, and R.P. Malhame. Large-population cost-coupled LQG problems with nonuniform agents: individual-mass behavior and decentralized ε-Nash equilibria. *IEEE Transaction Automat. Control*, 52(9):1560–1571, 2007.

CHAPTER 25

Approximate Dynamic Programming for Optimizing Oil Production

ZHENG WEN,[1] LOUIS J. DURLOFSKY,[2] BENJAMIN VAN ROY,[1,3] and KHALID AZIZ[2]

[1]Department of Electrical Engineering, Stanford University, Stanford, CA, USA
[2]Department of Energy Resources Engineering, Stanford University, Stanford, CA, USA
[3]Department of Management Science and Engineering, Stanford University, Stanford, CA, USA

ABSTRACT

In this chapter, a new ADP algorithm integrating (1) systematic basis function construction, (2) a linear programming (LP) approach in dynamic programming (DP), (3) adaptive basis function selection, and (4) bootstrapping is developed and applied to oil production problems. The procedure requires the solution of a large-scale dynamic system, which is accomplished using a subsurface flow simulator, for function evaluations. Optimization results are presented for cases involving single-phase primary oil production and water injection. In the first case, the global optimum can be computed, and the ADP results are shown to essentially achieve this optimum. Clear improvement, relative to various baseline strategies, is similarly observed in the second case. Components of the new algorithm, or even the entire procedure, may also be applicable in other domains.

25.1 INTRODUCTION

Recently, there has been increasing interest in applying advanced computational optimization techniques to improve the performance of petroleum reservoir operations. Many optimization algorithms have been applied to maximize reservoir performance. Most of these optimization algorithms can be classified into two categories: gradient-based/direct-search (e.g., [1–3]) and global stochastic search (e.g., [4–6]). Both classes

Reinforcement Learning and Approximate Dynamic Programming for Feedback Control, First Edition. Edited by Frank L. Lewis and Derong Liu.

of optimization algorithms face limitations: gradient-based and direct-search algorithms settle for local optima, while global stochastic search algorithms such as genetic algorithms typically require many function evaluations for convergence, and even then there is no assurance that the global optimum has been found.

As we will see in Section 25.2, the petroleum reservoir production optimization problem can be formulated as a dynamic optimal control problem. Thus, in principle, the globally optimal control (global optimum) can be computed via dynamic programming (DP) [7]. However, most reservoir production problems of practical interest are large-scale, nonlinear dynamic optimization problems. Thus, except for a few special cases, if we directly apply DP to such problems, the computational requirements become prohibitive due to the curse of dimensionality. In particular, time and memory requirements typically grow exponentially with the number of state variables.

One classical way to overcome the curse of dimensionality in DP is to apply approximate dynamic programming (ADP), which aims to address this computational burden by efficiently approximating the value function (see [8–10] for more on ADP). The result is an approximate value function, which is typically represented by a linear combination of a set of predefined basis functions. ADP algorithms provide methods for computing the coefficients associated with these basis functions. ADP has been successfully applied across a broad range of domains such as asset pricing [11–13], transportation logistics [14], revenue management [15–17], portfolio management [18], and even to games such as backgammon [19].

In this chapter, we propose a new ADP algorithm based on the linear programming (LP) approach and apply it to two reservoir production problems. We also compare the performance of ADP with that achieved using various baseline strategies. As we will show, ADP performs well relative to the baseline results for the problems considered.

Our contribution in this work is twofold: from the perspective of reservoir production problems, in addition to proposing a new approach that achieves promising performance, our ADP algorithm offers some advantages relative to alternative optimization techniques. For example, ADP can readily handle general nonlinear constraints in reservoir production, such as enforcing a maximum production rate for water (which is often produced along with oil), which may not be straightforward to handle using alternative algorithms. In addition, the ADP algorithm we propose is essentially deterministic, yet it aims to approximate a global optimum. This distinguishes it from stochastic global search algorithms that have been applied to oil production optimization.

Our proposed ADP algorithm combines four key ingredients: (1) systematic basis function construction based on proper orthogonal decomposition (POD), (2) coefficient computation based on smoothed reduced linear programming (SRLP), (3) selection of basis functions via an adaptive procedure, and (4) performance enhancement through bootstrapping. To our knowledge, both POD-based basis function construction and our adaptive basis function selection method are new. In addition, this work represents the first attempt to incorporate all four of these techniques together and apply them to large-scale dynamic optimization problems. The simulation results indicate that the overall approach is promising, and we expect that the ideas developed here are also applicable in other domains.

The remainder of this chapter is organized as follows. In Section 25.2, we introduce the reservoir production optimization problem. The basic DP and ADP procedures, especially the LP approach, are reviewed in Section 25.3. Then, in Section 25.4, we propose our ADP algorithm and discuss its computational demands. Simulation results for linear and nonlinear examples are presented in Section 25.5. We conclude this contribution in Section 25.6.

The basic formulation appearing in this chapter was presented previously (in a Society of Petroleum Engineers conference paper [20]) and the current description closely follows the earlier presentation.[1] The examples presented in Section 25.5 are, however, different from (and more realistic than) those in the earlier work.

25.2 PETROLEUM RESERVOIR PRODUCTION OPTIMIZATION PROBLEM

In this section, we introduce the petroleum reservoir production optimization problem and formulate it as an optimal control problem. Our treatment here is brief—interested readers can refer to Refs. [1–4, 21, 22] for detailed explanations of petroleum reservoir simulation and production optimization.

Reservoir fluids, such as oil and water, reside within the pore space of porous rock formations. Oil can be produced via primary production, in which reservoir fluid is removed but no fluid is injected, as well as via water injection. In the latter case, injected water maintains reservoir pressure and sweeps oil to production wells. To model subsurface flow systems such as oil reservoirs, we must specify a geological model (which defines rock properties and the flow domain), fluid and rock–fluid properties, and well locations and controls, among other parameters. The key rock property that impacts flow is permeability, which relates flow rate to pressure gradient (permeability can thus be viewed as a flow conductivity). Flow dynamics are described by one or more (depending on the number of components being tracked) partial differential equations that combine statements of mass conservation with constitutive relations.

In practical applications, the governing equations are discretized on a grid and a numerical solution technique is applied (finite-volume procedures are typically used in reservoir simulation). Following spatial discretization, these equations can be represented as a system of ordinary differential equations (ODEs):

$$\mathrm{d}\mathbf{x}(t)/\mathrm{d}t = F(\mathbf{x}(t), \mathbf{u}(t)), \tag{25.1}$$

subject to an initial condition $\mathbf{x}(0) = \mathbf{x}_0$ and instantaneous constraints $S(\mathbf{x}(t), \mathbf{u}(t)) \leq 0$. Here $\mathbf{x}(t)$ and $\mathbf{u}(t)$ denote the reservoir states and the control action at time t, respectively. Typically, $\mathbf{x}$ includes the grid-block pressure and saturation values (saturation

[1] SPE holds the copyright of the problem formulation and algorithm description portions of this chapter (Sections 25.2–25.4), which are reproduced with the permission of the copyright owner. Further reproduction is prohibited without permission.

is the volume fraction of a particular phase in the grid block) and $\mathbf{u}$ encodes the well settings (well pressures in the reservoir, referred to as bottomhole pressures or BHPs, and/or well flow rates) of the injection and production wells. The constraints $S(\mathbf{x}(t), \mathbf{u}(t)) \leq 0$ restrict the state and control action at time t. Constraints in reservoir production problems typically include upper/lower bounds on BHPs and upper/lower bounds on flow rates of fluids (e.g., maximum water rate and minimum oil rate).

We also assume that at time t, a payoff accumulates at a rate given by an instantaneous payoff function $L(\mathbf{x}(t), \mathbf{u}(t))$ that depends on the state $\mathbf{x}(t)$ and control action $\mathbf{u}(t)$. A widely used instantaneous payoff function is

$$\begin{aligned} L(\mathbf{x}(t), \mathbf{u}(t)) &= \text{revenue from producing oil} \\ &\quad - (\text{cost for producing water} + \text{cost for injecting water})\,. \end{aligned} \tag{25.2}$$

In our optimizations, we aim to maximize the net present value (NPV) of the cumulative payoff $\int_0^{\mathrm{T}} e^{-\alpha t} L(\mathbf{x}(t), \mathbf{u}(t))\mathrm{d}t$ by determining the optimal well settings (BHPs in this case) over a time horizon $[0, T]$, where $\alpha \geq 0$ is the continuous-time discount rate that captures the time value of the payoff.

In most oil production optimization problems, the termination time T is large enough so that the difference between the NPV of the cumulative profit

$$\int_0^{\mathrm{T}} e^{-\alpha t} L(\mathbf{x}(t), \mathbf{u}(t))\mathrm{d}t,$$

and its infinite horizon approximation

$$\int_0^{\infty} e^{-\alpha t} L(\mathbf{x}(t), \mathbf{u}(t))\mathrm{d}t,$$

is sufficiently small. Moreover, it is well known that there exists a stationary policy $\mathbf{u}^*(t) = \mu^* (\mathbf{x}(t))$ that maximizes the infinite horizon discounted objective. As we will see later, the existence of a stationary globally optimal policy μ^* simplifies our ADP algorithms. For this reason, we will focus on solving the infinite horizon dynamic optimization problem:

$$\begin{aligned} \max_{\{u(t), t\geq 0\}} \quad & \int_{t=0}^{\infty} e^{-\alpha t} L(\mathbf{x}(t), \mathbf{u}(t))\mathrm{d}t, \\ \text{s.t.} \quad & \mathrm{d}\mathbf{x}(t)/\mathrm{d}t = F(\mathbf{x}(t), \mathbf{u}(t)), \\ & \mathbf{x}(0) = \mathbf{x}_0 \text{ and } \quad S(\mathbf{x}(t), \mathbf{u}(t)) \leq \mathbf{0}. \end{aligned} \tag{25.3}$$

The optimization problem (25.3) for practical reservoir production problems is very challenging for the following reasons. First, for two-phase or multiphase flow cases, F is a highly nonlinear function, so the ODE (25.1) is nonlinear. Second, the dimension of the state space is very high. Specifically, $\mathbf{x} \in \Re^{N_b N_p}$, where N_b is the number of grid blocks used to discretize the system and N_p is the number of components. For instance, for a case with $40 \times 40 \times 5$ grid blocks and two components (oil and water), the dimension of the state space is 16,000.

The performance (NPV of the cumulative payoff) of any control strategy (policy) μ is evaluated through numerical solution of the governing equations (i.e., by performing a reservoir simulation). As described in Section 25.4.2, reservoir simulation is also used for constraint sampling. Since most of the computation of our proposed ADP algorithm is consumed by reservoir simulations, it is appropriate to measure the computational demands of the algorithm in terms of the number of reservoir simulations that must be performed. This issue is discussed in detail in Section 25.4.5.

25.3 REVIEW OF DYNAMIC PROGRAMMING AND APPROXIMATE DYNAMIC PROGRAMMING

In this section, we briefly review DP and ADP, especially the LP approach. DP offers a class of optimization algorithms that decompose a dynamic optimization problem into a sequence of simpler subproblems. At each time t, an optimal control action $\mathbf{u}^*(t)$ is computed by solving one subproblem. These subproblems are coordinated across time through a value function denoted by J^*. The value function captures the future impact of the current action; the algorithm must balance immediate payoff against future possibilities. In our model, the value function is defined by:

$$\begin{aligned} J^*(\mathbf{x}_0) = \max_{u(t), t\geq 0} & \int_{t=0}^{\infty} e^{-\alpha t} L(\mathbf{x}(t), \mathbf{u}(t)) \mathrm{d}t, \\ \text{s.t.} \quad & \mathrm{d}\mathbf{x}(t)/\mathrm{d}t = F(\mathbf{x}(t), \mathbf{u}(t)), \\ & \mathbf{x}(0) = \mathbf{x}_0 \text{ and } S(\mathbf{x}(t), \mathbf{u}(t)) \leq 0. \end{aligned} \tag{25.4}$$

Note that $J^*(\mathbf{x}_0)$ is the maximum NPV starting from state $\mathbf{x}_0$ at time 0. Further, given the time homogeneity of our model, the value $J^*(\mathbf{x}(t))$ at any time t and for any state $\mathbf{x}(t)$ is the maximum NPV of payoffs that can be accumulated starting at that time and state.

Given the value function J^* and current state $\mathbf{x}(t)$, an optimal control action at time t can be selected by solving the following optimization problem [7]:

$$\begin{aligned} \max_{\mathbf{u}} \quad & L(\mathbf{x}(t), \mathbf{u}) + F(\mathbf{x}(t), \mathbf{u})^{\mathrm{T}} \nabla J^*(\mathbf{x}(t)), \\ \text{s.t.} \quad & S(\mathbf{x}(t), \mathbf{u}) \leq 0, \end{aligned} \tag{25.5}$$

where ∇J^* is the gradient of J^*. Note that this problem decouples the choice of control action at time t from that at all other times. It is in this sense that DP decomposes the dynamic optimization problem into simpler subproblems through use of the value function. As we will see later, for reservoir production problems, the subproblem (25.5) can be solved efficiently.

In light of the relative ease of generating optimal control actions given the value function, the challenge in optimizing reservoir production using DP (or ADP) is in the computation of the value function. As discussed in Ref. [18], the optimal value function for a continuous-time optimal control problem can, in principle, be computed by solving the Hamilton–Jacobi–Bellman (HJB) equation:

$$\max_{\mathbf{u}:\, S(\mathbf{x},\mathbf{u})\leq \mathbf{0}} \left\{ L(\mathbf{x},\mathbf{u}) + F(\mathbf{x},\mathbf{u})^{\mathrm{T}} \nabla J(\mathbf{x}) - \alpha J(\mathbf{x}) \right\} = 0, \qquad (25.6)$$

which is a partial differential equation.

An alternative though closely related characterization of the value function is an infinite dimensional linear program (LP):

$$\begin{aligned} \min_{J\in C^2} \quad & \int J(\mathbf{x})\pi(d\mathbf{x}), \qquad (25.7)\\ \text{s.t.} \quad & L(\mathbf{x},\mathbf{u}) + F(\mathbf{x},\mathbf{u})^{\mathrm{T}} \nabla J(\mathbf{x}) - \alpha J(\mathbf{x}) \leq 0 \quad \forall \mathbf{x}, \mathbf{u} \text{ s.t. } S(\mathbf{x},\mathbf{u}) \leq 0, \end{aligned}$$

where π is a probability measure chosen with exhaustive support such that the integral of J^* is finite. This LP poses an infinite number of decision variables and an infinite number of constraints since there is one decision variable $J(\mathbf{x})$ per state $\mathbf{x}$ and there is one constraint $S(\mathbf{x}, \mathbf{u}) \leq 0$ per state-action pair $(\mathbf{x}, \mathbf{u})$.

From Theorem 1.1.3 of Ref. [18], we have the following theorem:

Theorem 25.1 *If $J^* \in C^2$, then J^* uniquely attains the optimum in (25.7).*

For a given reservoir production problem, it is not clear whether or not the value function J^* satisfies the technical condition $J^* \in C^2$. However, as will be described in Section 25.4, the approximate value function $\tilde{J}$ in our proposed ADP algorithm does satisfy $\tilde{J} \in C^2$. As a result, the technical condition $J^* \in C^2$ does not impose any additional restriction for our proposed ADP algorithm.

Similar to other large-scale optimization problems, in most petroleum reservoir production problems of practical interest, solving the HJB equation (25.6) or the infinite dimensional LP (25.7) exactly is impossible, as this would require computing and storing J^* for each state in a continuous state space. Even if we were to discretize the state space by quantizing each state variable into multiple discrete values, the number of discrete states would grow exponentially with the dimension of the state space D ($D = N_p N_b$), again making storage and computation impractical.

In a few special cases, the value function assumes a special structure which facilitates efficient computation and storage. This is the case, for example, with single-phase reservoir models with no constraints on control actions and an instantaneous payoff function that is quadratic in state and control action. In this context, the value function is itself quadratic and the problem reduces to one of linear-quadratic (LQ) control. Such problems are well known to be tractable. For more realistic problems, involving for example two-phase flow, the reservoir dynamics are nonlinear and computing the value function exactly is infeasible.

One way to overcome the curse of dimensionality is to approximate the value function J^*. Deriving a near-optimal control strategy $\tilde{\mu}$ through appropriate value function approximation is known as approximate dynamic programming (ADP). As is common in ADP, we will approximate the value function of our dynamic optimization model in terms of a linear combination of a selected set of basis functions:

$$J^*(\mathbf{x}) \approx \tilde{J}(\mathbf{x}) = \sum_{k=0}^{K} r_k \phi_k(\mathbf{x}), \tag{25.8}$$

where $\phi_k(\cdot)$, $k = 0, \ldots, K$ is a set of basis functions, and the r_k's are their respective coefficients. As we will see, the main advantage of the linear parameterization in the approximation structure (25.8) is that various "approximate" versions of the LP (25.7) (e.g., ALP (25.9), RLP (25.10), and SRLP (25.11)) are themselves LP problems and hence can be solved efficiently if they have finite constraints and decision variables. Replacing the exact value function J^* with the approximation $\tilde{J}$ when solving each subproblem (25.5) results in a control strategy $\tilde{\mathbf{u}} = \tilde{\mu}(\mathbf{x})$. We expect that if $\tilde{J}$ is "close" to J^*, the use of controls $\tilde{\mu}$ will provide performance that is "close" to the global optimum.

With the approximation structure (25.8), we only need to specify how to construct basis function ϕ_k's and compute their coefficients (weights) r_k's. In Section 25.4, we will describe how we implement our ADP algorithm and discuss some techniques to further improve its performance.

25.4 APPROXIMATE DYNAMIC PROGRAMMING ALGORITHM FOR RESERVOIR PRODUCTION OPTIMIZATION

In this section, we develop a new ADP algorithm based on the LP approach to compute a near-optimal control for the petroleum reservoir production problem presented in Section 25.2. Our ADP algorithm proceeds as follows. We first construct a pool of candidate basis functions $\phi_k(\cdot)$, where $k = 0, 1, \ldots, \tilde{K}$. Next, we apply adaptive basis function selection to effectively choose basis functions from the candidate basis function pool. Third, we compute the coefficients (weights) r_k of the selected basis functions. We then derive near-optimal control actions $\tilde{\mathbf{u}}$ using the approximate value function $\tilde{J}$ by solving subproblem (25.5). Finally, we use bootstrapping to further enhance the performance of the ADP algorithm.

In the rest of this section, we will discuss our implementation of each of these key steps. The overall procedure is illustrated in Figure 25.1.

25.4.1 Basis Function Construction

In this section, we describe the construction of the candidate basis functions $\phi_k(\cdot)$. Although there are studies focusing on constructing basis functions automatically (e.g., [23]), in most applications, to achieve a more accurate approximation of the value function, problem-dependent approaches are used. For oil production problems, we propose an approach to construct the candidate basis functions using POD. The candidate basis function pool is built using these basis functions as well as some other prespecified basis functions.

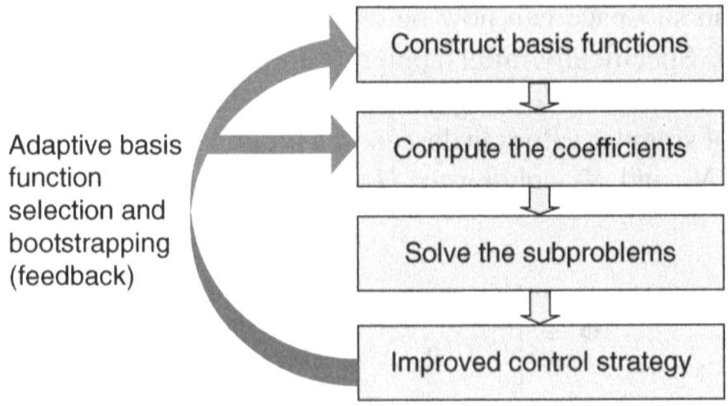

FIGURE 25.1 Flowchart of proposed ADP algorithm for oil production optimization.

In our approach, the value function J^* is the NPV of the future total profit under the optimal control strategy. Hence, the basis functions should encode information about aspects of the states that strongly impact future profit. We have observed that the following two categories of functions are correlated to future profit, which has motivated us to choose basis functions accordingly:

1. Functions reflecting the "global status" of the reservoir, such as the average oil and water saturations and the average pressure. As we have discussed earlier, a systematic way to capture the global status is through POD of global snapshots. We will discuss how to construct basis functions based on POD later.
2. Functions reflecting the status of the near-well regions, such as the average oil saturation near production wells, average pressure near injection wells, and so on.

In addition to the values described earlier, it is also useful to include higher-order polynomial functions of these values, as discussed below.

POD identifies a low-dimensional subspace that captures most of the system variability. It is used in model-order reduction for petroleum reservoir simulation (e.g., [22]) and other application areas, so it seems reasonable to apply POD for the construction of ADP basis functions. Our use of POD for basis construction proceeds as follows.

We first run the reservoir simulator under a prescribed baseline strategy μ_0 and record "snapshots" of the states. That is, we use the reservoir simulator to solve $\mathrm{d}\mathbf{x}/\mathrm{d}t = F(\mathbf{x}, \mu_0(\mathbf{x}))$, and sample the state trajectory $\mathbf{x}(t)$ at $t = t_0, t_1, t_2, \ldots, t_{L-1}$. We define $\mathbf{x}_p(t)$ and $\mathbf{x}_S(t)$ to be vectors containing the pressure and saturation components of $\mathbf{x}(t)$, respectively (for an oil–water problem). The sample means are denoted by $\bar{\mathbf{x}}_p$ and $\bar{\mathbf{x}}_s$.

The centered snapshot matrices for pressure and saturation are

$$X_p = \left[\mathbf{x}_p(t_0) - \bar{\mathbf{x}}_p, \mathbf{x}_p(t_1) - \bar{\mathbf{x}}_p, \cdots, \mathbf{x}_p(t_{L-1}) - \bar{\mathbf{x}}_p\right],$$
$$X_S = \left[\mathbf{x}_S(t_0) - \bar{\mathbf{x}}_S, \mathbf{x}_S(t_1) - \bar{\mathbf{x}}_S, \cdots, \mathbf{x}_S(t_{L-1}) - \bar{\mathbf{x}}_S\right],$$

respectively. The subspace can now be characterized using singular value decomposition (SVD). Specifically, after applying SVD, we decompose X_p and X_S as $X_p = U_p \Sigma_p V_p^{\mathrm{T}}$ and $X_S = U_S \Sigma_S V_S^{\mathrm{T}}$, respectively, where Σ_p and Σ_S are diagonal matrices made up of singular values in descending order. Let $U_p(1:N_p)$ and $U_S(1:N_S)$ denote the first N_p and N_S columns of U_p and U_S, respectively. We define the projection matrix

$$\mathbf{\Phi} = \begin{bmatrix} U_p(1:N_p) & \mathbf{0} \\ \mathbf{0} & U_S(1:N_S) \end{bmatrix}.$$

Note that span($\mathbf{\Phi}$) is a subspace that captures most of the variation in pressure and saturation.

Now, we construct basis functions that are polynomials defined on span($\mathbf{\Phi}$). Specifically, letting φ_i denote the ith column of $\mathbf{\Phi}$, the basis functions take the form $\phi_k(\mathbf{x}) = M_k \prod_{i=1}^{I} \left(\varphi_{k_i}^{\mathrm{T}} \mathbf{x}\right)^{m_i}$, where m_i are nonnegative integers, M_k is a normalization constant, and I is the number of terms in the basis function ϕ_k. We say ϕ_k is a one-term polynomial if $I = 1$.

This approach is based on the assumption that the reservoir dynamics under the baseline strategy μ_0 are sufficiently similar to those under the optimal strategy μ^*. This is not always the case. One way to address this issue involves bootstrapping, as we will describe later.

It is evident that we can potentially construct many (polynomial) functions reflecting either the status of the entire reservoir or the status of near-well regions. Thus, there are many candidate basis functions. However, the use of too many basis functions can result in onerous computational requirements or performance loss due to overfitting. Thus, in practice, only a subset of all of these candidate basis functions are used to approximate the value function. We will later propose an adaptive basis function selection scheme, which chooses an effective subset of basis functions.

Furthermore, no matter how we construct and choose basis functions, we always include a constant basis function $\phi_0(\mathbf{x}) = 1$ to achieve better approximation results and numerical stability. We also normalize all basis functions such that $|\phi_k(\mathbf{x}_0)| = 1$ for $k = 0, \ldots, K$, where $\mathbf{x}_0$ is the initial state.

Finally, we observe that any candidate basis function ϕ_k constructed as described above satisfies the technical condition $\phi_k \in C^2$. Since the approximate value function $\tilde{J}$ is a linear combination of a selected set of candidate basis functions, $\tilde{J} \in C^2$.

25.4.2 Computation of Coefficients

As discussed earlier, for a given set of basis functions, we need to compute coefficients to approximate the value function. A number of ADP algorithms have been proposed for this purpose. Examples include approximate value iteration, approximate policy iteration, temporal-difference learning, Bellman error minimization, and approximate linear programming (ALP) (see Refs. [8–10]). In this chapter, we use smoothed reduced linear programming (SRLP) with L_1 regularization. This is a variant of ALP proposed in Ref. [24]. Compared with other ADP algorithms, the main advantage of this procedure (in our context) is that it provides "optimal" solutions in much less time than alternative approaches. In the rest of this section, we describe the SRLP algorithm with L_1 regularization. An important aspect of this algorithm—constraint sampling—is discussed in detail in Ref. [20].

Theoretically, the value function can be obtained by solving the infinite dimensional LP (25.7). However, this is not feasible because there are an infinite number of decision variables and an infinite number of constraints. This technical difficulty can be addressed by constraining the function to the span of the basis functions. In particular, we consider solving

$$\min_{\mathbf{r}} \quad \int \sum_{k=0}^{K} r_k \phi_k(\mathbf{x}) \pi(d\mathbf{x}), \tag{25.9}$$

$$\text{s.t.} \quad L(\mathbf{x}, \mathbf{u}) + F(\mathbf{x}, \mathbf{u})^{\mathrm{T}} \sum_{k=0}^{K} r_k \nabla \phi_k(\mathbf{x}) - \alpha \sum_{k=0}^{K} r_k \phi_k(\mathbf{x}) \leq 0 \quad \forall (\mathbf{x}, \mathbf{u}) \text{ s.t. } S(\mathbf{x}, \mathbf{u}) \leq 0,$$

which is known as the approximate linear program (ALP). Notice that we have removed the technical condition $\sum_{k=0}^{K} r_k \phi_k(\cdot) \in C^2$ from the constraints of (25.9) since it always holds for the basis functions considered. However, there are still an infinite number of constraints in the ALP (25.9), so it cannot be solved exactly. We overcome this issue through constraint sampling; that is, only $\mathcal{M}$, a finite set of sampled states and control actions, is used to constrain variables. This results in what is known as the reduced linear program (RLP). Let $\mathcal{M} = \left\{(\mathbf{x}^{(1)}, \mathbf{u}^{(1)}), \cdots, (\mathbf{x}^{(M)}, \mathbf{u}^{(M)})\right\}$, where M is the cardinality of $\mathcal{M}$. The RLP takes the form

$$\min_{\mathbf{r}} \quad \frac{1}{M} \sum_{m=1}^{M} \sum_{k=0}^{K} r_k \phi_k \left(\mathbf{x}^{(m)}\right), \tag{25.10}$$

$$\text{s.t.} \quad L\left(\mathbf{x}^{(m)}, \mathbf{u}^{(m)}\right) + F\left(\mathbf{x}^{(m)}, \mathbf{u}^{(m)}\right)^{\mathrm{T}} \sum_{k=0}^{K} r_k \nabla \phi_k \left(\mathbf{x}^{(m)}\right) - \alpha \sum_{k=0}^{K} r_k \phi_k \left(\mathbf{x}^{(m)}\right) \leq 0,$$
$$\forall m = 1, 2, \cdots, M.$$

The way in which constraints are sampled significantly impacts the performance of the resulting approximation, and our approach to constraint sampling is nontrivial. Roughly speaking, we sample the state/action pairs based on a randomized baseline strategy and then bootstrap this process. Interested readers can refer to the appendix of Ref. [20] for the constraint sampling algorithm. In practice, a rule of thumb is to choose the number of sampled states M to be 40–60 times the size of K, the number of basis functions. It has been observed in some studies [24] that RLP (25.10) might be in some sense over-constrained. We have also observed that the L_1 regularization of $\mathbf{r}$ can improve performance. These observations motivate the SRLP [24] with L_1 regularization:

$$\min_{\mathbf{r},\mathbf{s}} \quad \frac{1}{M}\sum_{m=1}^{M}\sum_{k=0}^{K} r_k\phi_k\left(\mathbf{x}^{(m)}\right) + \epsilon||\mathbf{r}||_1, \tag{25.11}$$

$$\text{s.t.} \quad L\left(\mathbf{x}^{(m)},\mathbf{u}^{(m)}\right) + F\left(\mathbf{x}^{(m)},\mathbf{u}^{(m)}\right)^{\mathrm{T}}\sum_{k=0}^{K} r_k\nabla\phi_k\left(\mathbf{x}^{(m)}\right) - \alpha\sum_{k=0}^{K} r_k\phi_k\left(\mathbf{x}^{(m)}\right) \le s^{(m)},$$

$$\forall m = 1, 2, \ldots, M,$$

$$s^{(m)} \ge 0 \quad \forall m = 1, 2, \ldots, M \quad \text{and} \quad \sum_{m=1}^{M} s^{(m)} \le \theta\sum_{m=1}^{M}\left|L\left(\mathbf{x}^{(m)},\mathbf{u}^{(m)}\right)\right|,$$

where $\mathbf{s} = \left[s^{(1)}, s^{(2)}, \cdots, s^{(M)}\right]$. There are two parameters to be specified in (25.11), θ and ϵ. The variable $\theta \ge 0$ characterizes the extent to which the constraints of the RLP are relaxed, and $\epsilon \ge 0$ characterizes a penalty on the magnitude of $\mathbf{r}$, in the sense of the L_1 norm. We have observed that the selection of θ and ϵ influences $\mathbf{r}_{\text{srlp}}$, the solution of (25.11). In practice, θ and ϵ are either prespecified, or chosen based on line search.

25.4.3 Solving Subproblems

Once the approximate value function $\tilde{J}$ is available, control actions $\tilde{\mathbf{u}}$ can be obtained by solving (25.5), with J^* replaced by $\tilde{J}$. A useful observation relevant to oil production problems can significantly simplify this step. Consider the reservoir dynamics $d\mathbf{x}/dt = F(\mathbf{x},\mathbf{u})$ and constraints $S(\mathbf{x},\mathbf{u}) \le 0$. In most cases, although $F(\mathbf{x},\mathbf{u})$ and $S(\mathbf{x},\mathbf{u})$ are nonlinear functions of the state $\mathbf{x}$, F is affine in the control action $\mathbf{u}$ for fixed $\mathbf{x}$. In addition, for common constraints used in practice, such as maximum/minimum flow rates, S is also affine in the control action $\mathbf{u}$ for fixed $\mathbf{x}$.[2]

This means that

$$F(\mathbf{x},\mathbf{u}) = F_1(\mathbf{x}) + F_2(\mathbf{x})\mathbf{u},$$
$$S(\mathbf{x},\mathbf{u}) = S_1(\mathbf{x}) + S_2(\mathbf{x})\mathbf{u},$$

thus, the subproblem at time t is

$$\max_{\mathbf{u}} \quad L(\mathbf{x}(t),\mathbf{u}) + \left[\nabla\tilde{J}(\mathbf{x}(t))\right]^{\mathrm{T}} F_2(\mathbf{x}(t))\mathbf{u}, \tag{25.12}$$
$$\text{s.t.} \quad S_1(\mathbf{x}(t)) + S_2(\mathbf{x}(t))\mathbf{u} \le 0.$$

[2]For interested readers, this is the case because the so-called well transmissibility does not depend on BHP or flow rate.

This problem (25.12) is easy to solve if the instantaneous payoff function L is concave in $\mathbf{u}$. For example, with the instantaneous payoff function L defined in (25.2), the subproblem at each time reduces to a low-dimensional linear program.

25.4.4 Adaptive Basis Function Selection and Bootstrapping

The ADP algorithm described thus far is "open-loop" in the sense that once a new control $\tilde{\mathbf{u}}$ is available, it does not get used to further improve the result. In this section, we introduce two advanced techniques, adaptive basis function selection and bootstrapping, to "close the loop" and enhance the performance of the ADP algorithm.

25.4.4.1 Adaptive Basis Function Selection As discussed earlier, in practical implementations, we need to select a subset of basis functions from a large set of candidates. One way to effectively choose basis functions is through use of adaptive basis function selection. Specifically, we initialize the set of basis functions $\mathcal{F}$ as a set containing only the constant basis function $\phi_0(\cdot) = 1$ (or a known set of "good" basis functions such as those selected in the previous bootstrapping round). At each iteration, we choose a new basis function and add it to $\mathcal{F}$. Our rule in selecting the new basis function is that, compared to other candidate basis functions, including the new basis function in $\mathcal{F}$ results in the largest increase in the NPV. We continue to add basis functions until the addition of any new basis function leads to a decrease in NPV. In other words, adaptive basis function selection is a greedy algorithm for adding new basis functions. Refer to Ref. [20] for a detailed description of this algorithm.

At each step of the adaptive basis function selection procedure, the algorithm iterates over all of the candidate basis functions to choose the optimal basis function to add to $\mathcal{F}$. This approach can be time consuming. However, similar to the constraint sampling algorithm, parallel computation can significantly accelerate this algorithm.

25.4.4.2 Bootstrapping Another technique to further improve the performance of our ADP algorithm is bootstrapping. Specifically, to construct basis functions based on POD and sample the constraints for SRLP (25.11), we need to sample states and state/action pairs based on a known strategy μ. Theory [25] suggests that, to obtain the best results, we should use an optimal control policy when sampling. Since an optimal policy is not available (the objective of the ADP algorithm is to find a policy close to optimal), we sample using a baseline strategy μ_0, which inevitably leads to some performance loss.

Bootstrapping is proposed to address this issue. Specifically, we start from a baseline strategy μ_0, then apply the above described ADP algorithm to compute an approximate value function $\tilde{J}_1$, which generates a new strategy μ_1. Then, we sample states and state/action pairs based on μ_1, and apply the ADP algorithm again to obtain a new approximate value function $\tilde{J}_2$ and policy μ_2. We repeat this procedure until it no longer increases NPV. A detailed description of the bootstrapping algorithm is provided in Ref. [20].

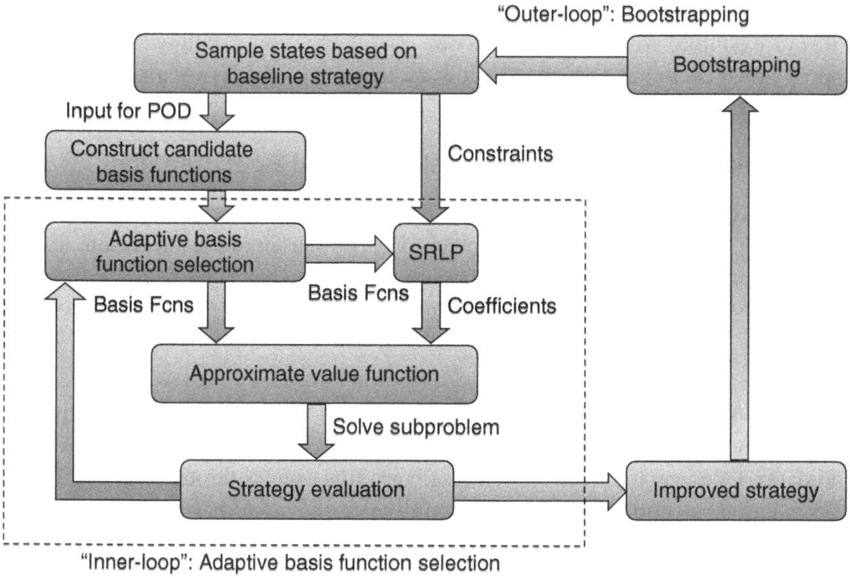

FIGURE 25.2 Architecture of the overall ADP algorithm.

Figure 25.2 illustrates the architecture of our ADP algorithm with the incorporation of adaptive basis function selection and bootstrapping.

25.4.5 Computational Requirements

We now briefly discuss the computational requirements of our proposed ADP algorithm. As is the case for other production optimization algorithms, most of the computation in ADP is consumed by the reservoir simulations. Thus, it is appropriate to assess the computational demands for the proposed ADP algorithm in terms of the required number of simulations.

As shown in Figure 25.2, the ADP algorithm uses the reservoir simulator to sample states and evaluate the strategy $\tilde{\mu}$. As discussed earlier, in general, SRLP (25.11) requires 40–60K state samples, where K is the number of basis functions, and each state sample requires one simulation.

Adaptive basis function selection requires a large number of additional simulations. For this approach, assume there are $\tilde{K}$ candidate basis functions and we are allowed to use up to $K_{\max}$ basis functions. One obvious upper bound on the number of required simulations is $60K_{\max} + \tilde{K}K_{\max}$. However, this bound is far from tight for the following reasons: first, as is observed in Case 2 in the following section, for large $K_{\max}$ the number of basis functions ultimately used (K) tends to be much less than $K_{\max}$. Second, the ADP algorithm does not need to re-evaluate the strategy if the computed coefficient of the new basis function is zero, which happens frequently due to the L_1 regularization in SRLP (25.11). Thus, the number of required simulations in practice is much less than the theoretical bound above.

For the ADP algorithm with bootstrapping, assuming the algorithm applies the bootstrapping B times, the required number of simulations will be B times the above results. In practice, it has been observed that the performance tends to degrade after several iterations of bootstrapping (see Ref. [26]; this is also observed in the examples here). Hence, $B \leq 5$ in most cases.

It is evident that ADP runtime scales with the number of (candidate) basis functions. In general, the use of more candidate basis functions in adaptive basis function selection will improve algorithm performance though it requires more computation. Thus, there is a trade-off between performance and computational effort. It would be of interest to further investigate this trade-off in future work. As we have discussed earlier, it is important to note that both state sampling and adaptive basis function selection can be implemented in parallel. Thus, by using multiple processors, the elapsed time for the ADP algorithm need not be excessive.

One way to design the candidate basis function pool is to consider only one-term polynomials. In this case, the number of candidate basis functions is $\tilde{K} = O(N_\mathrm{p} + N_\mathrm{S} + N_\mathrm{w})$, where N_w is the number of wells and N_p and N_S are determined during POD. However, in some cases, the "cross terms" (i.e., functions with more than one term) are useful in approximating the value function. Of course, adding all such cross terms will result in an intractably large candidate basis function pool. One way to address this issue is to add only the cross terms between the first few columns of $\mathbf{\Phi}$ (which correspond to the largest singular values) to the candidate basis function pool.

25.5 SIMULATION RESULTS

In this section, we apply the above proposed ADP algorithm to two example cases and compare our results against various baseline strategies. The first example is a single-phase primary production case, and the second example is a two-phase oil–water case with general (nonlinear) constraints.[3] Both examples involve three-dimensional reservoir models. In our implementation, we use Stanford's general purpose research simulator (GPRS) [27, 28] for all reservoir simulations and CPLEX, a commercial LP solver, to solve the SRLP (25.11). Other components of the ADP algorithm are implemented in MATLAB.

We have also applied the proposed ADP algorithm to three different examples involving two-dimensional reservoir models in Ref. [20]. In that work, we compared the performance of the ADP algorithm with that of the gradient-based method described in Ref. [2]. Interested readers can refer to Ref. [20] for the detailed results.

Case 1: Primary Production with Penalty Terms

In this first example, we apply the proposed ADP algorithm to a single-phase primary production case. In this example, the reservoir contains only oil and no fluid is injected

[3]Mathematically, these constraints are nonlinear inequalities involving $\mathbf{x}$ and $\mathbf{u}$. However, as discussed in Section 25.4.3, for a given state $\mathbf{x}$, they are affine inequalities of $\mathbf{u}$.

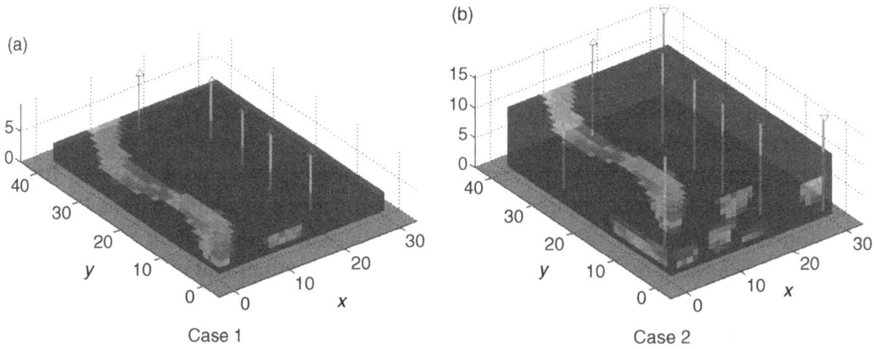

FIGURE 25.3 Geological models (x-direction permeability, where blue indicates low and red indicates high permeability) and well configurations for simulation examples. Production wells are indicated by green lines with upward-pointing triangles and injection wells by green lines with downward-pointing triangles. (*See the color version of this figure in color plates section.*)

into the reservoir. Thus, the state vector $\mathbf{x}$ only includes the pressure of each grid block. The reservoir dynamics are described by a linear flow equation in this case. The geological model is the top three layers of the Stanford VI synthetic reservoir model [29]. The model contains $30 \times 40 \times 3$ grid blocks and has five production wells, each of which is controlled by a bounded BHP. The geological model is shown in Figure 25.3a (the x-direction permeability field is the quantity depicted). We simulate for 300 days with 300 control periods. The instantaneous profit function is

$$L(\mathbf{x}, \mathbf{u}) = \text{revenue from producing oil} + \lambda \sum_{i=1}^{5} \log(\mathbf{u}_i - \text{LB}_i),$$

where $\mathbf{u}_i$ is the BHP for producer i and LB_i is its lower bound. The logarithm term penalizes BHPs that are close to the lower bound. Including this term (in this form) enables the analytical determination of the global optimum (i.e., globally optimal control and the associated NPV of the cumulative profit) using DP. Key parameters for Case 1 are summarized in Table 25.1. Note that STB stands for stock tank barrel (1 STB = 0.1590 m^3) and 14.50 psi = 1 bar (STB and psi are standard units in petroleum engineering).

We apply the ADP algorithm with both adaptive basis function selection and bootstrapping. Note in this case the projection matrix from POD is $\mathbf{\Phi} = U_p(1 : N_p)$, and

TABLE 25.1 Parameters for Case 1

BHP range for Producer 1	800–5000 psi	BHP range for Producer 2	800–5000 psi
BHP range for Producer 3	1200–5000 psi	BHP range for Producer 4	1000–5000 psi
BHP range for Producer 5	500–5000 psi	Oil price	\$40/STB
Logarithm penalty coefficient λ	1×10^4	Discount rate α	1×10^{-2}

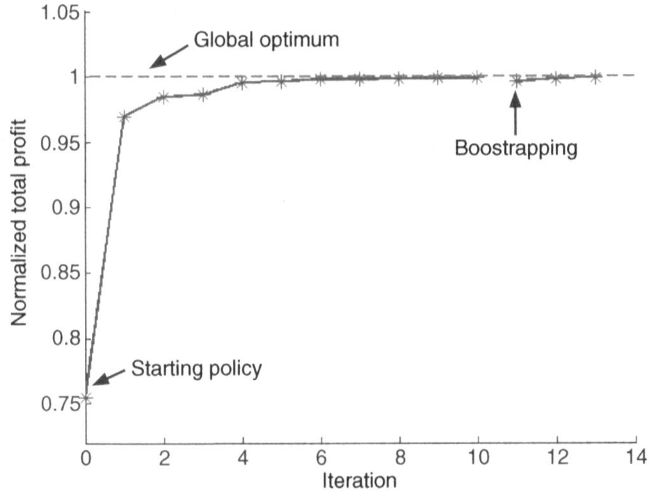

FIGURE 25.4 Performance of ADP for Case 1, where the normalization is with respect to the global optimum.

we choose $N_p = 12$. The candidate basis function pool includes (1) a constant basis function $\phi_0 = 1$, (2) average pressure $\bar{p}$ of the entire reservoir, (3) average pressures in near-producer regions, and (4) all the first-order polynomials defined on span($\mathbf{\Phi}$). There are a total of $\tilde{K} = 19$ candidate basis functions, and we set $K_{\max} = \tilde{K} = 19$.

The progress of the ADP algorithm is presented in Figure 25.4. We start from an (arbitrary) constant control strategy $\mu_0(\mathbf{x}) = LB + 10\mathbf{e}$, where $\text{LB} = [\text{LB}_1, \cdots, \text{LB}_5]^{\text{T}} \in \Re^5$ is the vector encoding the lower bounds of the producer BHPs and $\mathbf{e}$ is the vector of all ones. With this starting strategy, POD characterizes a subspace of dimension 12 that captures more than 99.9999% of the variation in pressure (in the sense of singular values). For the constraint sampling, we use randomized μ_0 to sample 800 states, with randomization magnitude $\eta = 100$ psi. Then, we apply the adaptive basis function selection algorithm, with the initial basis function set $\mathcal{F} = \{\phi_0(\cdot) = 1\}$. To compute the coefficients based on SRLP (25.11), we choose $\epsilon = 0.1$ and choose θ based on a line search from 0 to 0.2. As is evident in Figure 25.4, after adaptively adding 10 basis functions, addition of any new basis functions in $\mathcal{F}$ leads to a decrease in NPV. Thus, the ADP algorithm terminates adaptive basis function selection after the 11th iteration and records the associated suboptimal control strategy μ_1. The result from μ_1 is 99.88% of the global optimum.

The ADP algorithm then bootstraps once using control strategy μ_1, that is, it reconstructs $\mathbf{\Phi}$ based on μ_1 and then performs the constraint sampling based on randomized μ_1, followed by another round of adaptive basis function selection. As in the first round, we use $\eta = 100$ psi, $\epsilon = 0.1$ and choose θ based on a line search from 0 to 0.2. The difference here is that the adaptive basis function selection starts from the 11 basis functions selected in the first bootstrapping round. We observe that, after adding two new basis functions, neither adding a new basis function nor bootstrapping further increases NPV. Thus, the ADP algorithm terminates, with a result that is 99.93% of the global optimum. We note that the ADP algorithm was also applied with different starting policies and achieved similar results.

In addition to the global optimum, another interesting control strategy is the myopic policy (the policy that aims at maximizing the instantaneous profit at each time). In this case, the myopic policy achieves 93.51% of the global optimum. Thus, we see that the ADP result is not only very close to the global optimum for this case, it is also significantly better than the result using the myopic policy.

Although this case is quite simple, it represents one of the few cases where the global optimum can be obtained analytically. It is significant that, for a case where the global optimum is known, ADP does indeed provide a result that is very close to the global optimum.

Case 2: Waterflood with General Constraints

We now consider a two-phase (oil–water) model with general constraints. The geological model, which contains $30 \times 40 \times 10$ grid blocks, is again a portion of the Stanford VI model. Flow is driven by four injection wells and five production wells, as indicated in Figure 25.3b. We impose the following constraints at each time t: (1) maximum field water injection rate, (2) minimum field oil production rate, (3) maximum field liquid (water and oil) production rate, (4) maximum watercut (fraction of water in the produced fluid) for each production well, and (5) upper and lower bounds on BHP of each well.

In this case, the reservoir dynamics are described by nonlinear equations, and most of the constraints listed above are also nonlinear. The instantaneous profit function is specified by (25.2). We simulate for 365 days with 10 control periods. The constraints and parameters are given in Table 25.2.

For oil–water problems with the constraints specified above, it can be difficult to find a feasible control strategy without applying an optimization technique. In particular, the myopic policy is infeasible for this case. However, for the subproblems (25.12), feasible control can be found as follows. We randomly sample parameters $\mathbf{g} \in \Re^9$ (each component of $\mathbf{g}$ is sampled i.i.d. from a uniform distribution over $[-1, 1]$) and replace the unknown affine function $L(\mathbf{x}(t), \mathbf{u}) + t[\nabla J^*(\mathbf{x}(t))]^{\mathrm{T}} F_2(\mathbf{x}(t))\mathbf{u}$ by $\mathbf{g}^{\mathrm{T}}\mathbf{u}$.

TABLE 25.2 Parameters for Case 2

BHP range for producers	2500–6000 psi	BHP range for injectors	4500–9000 psi
Maximum field water injection rate	4×10^4 STB/day	Minimum field oil production rate	1×10^4 STB/day
Maximum field liquid production rate	5×10^4 STB/day	Maximum watercut	0.7
Oil price	\$50/STB	Water production cost	\$10/STB
Water injection cost	\$5/STB	Discount rate α	1×10^{-2}

At time t with state $\mathbf{x}(t)$, we then solve the following LP to obtain a feasible control at time t:

$$\begin{aligned} \max_{\mathbf{u}} \quad & \mathbf{g}^{\mathrm{T}}\mathbf{u} \\ \text{s.t.} \quad & S_1(\mathbf{x}(t)) + S_2(\mathbf{x}(t))\mathbf{u} \leq 0. \end{aligned} \tag{25.13}$$

If (25.13) is infeasible at time t, we restart this procedure from time 0. We use this approach to generate 100 feasible control strategies, and choose the best one as the baseline for Case 2.

Both adaptive basis function selection and bootstrapping are used for this case. We start the ADP algorithm by using the above-derived baseline strategy to construct the candidate basis function pool through POD. Specifically, we partition the projection matrix as $\mathbf{\Phi} = [\mathbf{\Phi}_1, \mathbf{\Phi}_2]$, where $\mathbf{\Phi}_1$ includes the first six left singular vectors. The candidate basis function pool includes (1) a constant basis function $\phi_0 = 1$, (2) one-term polynomials of average reservoir pressure and saturation and oil–water ratio of the entire reservoir up to third order, (3) one-term polynomials of the average oil saturations in near-producer regions up to third order, (4) all the polynomials defined on span($\mathbf{\Phi}_1$) up to third order, and (5) all the other one-term polynomials defined on span($\mathbf{\Phi}$) up to third order. There are thus a total of $\tilde{K} = 340$ candidate basis functions and we choose $K_{\max} = 50$. Then, we use the randomized baseline strategy (with the randomization magnitude $\eta = 100$ psi) to sample $M = 2000$ constraints for the SRLP (25.11). Finally, we start adaptive basis function selection with the initial basis function set $\mathcal{F} = \{\phi_0(\cdot) = 1\}$. For a given choice of basis functions, the coefficients are computed by SRLP (25.11), and we set $\epsilon = 5 \times 10^{-2}$ and choose θ by a line search from 0 to 0.05. We observe that, after adding seven basis functions, inclusion of any new basis functions decreases the NPV of the cumulative profit. Compared with the baseline, without any bootstrapping, the ADP algorithm achieves a 16.03% improvement.

The ADP algorithm then bootstraps once using the obtained strategy. We reconstruct the candidate basis function pool and resample the constraints. Then, we start the basis function selection from the eight basis functions chosen in the first bootstrapping round. We observe that, after adding three new basis functions, NPV is not increased further by adding new basis functions or bootstrapping. Thus, the ADP algorithm terminates, at which point it has achieved an 18.67% improvement compared with the baseline strategy. Optimization results for Case 2 are plotted in Figure 25.5.

Figure 25.6 presents simulation results together with the associated constraints over time for the near-optimal control obtained by the ADP algorithm. It is evident that the maximum water injection rate and maximum liquid production rate are tight at the beginning of the simulation, while the minimum oil production rate and watercut of Producer 4 are tight at the end of the simulation. In addition, for most of the wells, bounds on the BHPs are tight in all or some simulation time steps. The constraints that are nontight over the entire simulation could be relaxed. In any event, this example demonstrates that our ADP algorithm does satisfy all of the constraints throughout the simulation.

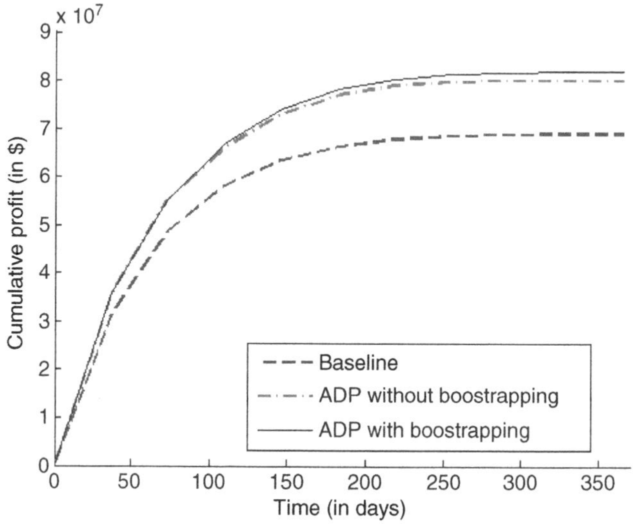

FIGURE 25.5 Comparison of ADP (both with and without bootstrapping) with baseline strategy for Case 2.

25.6 CONCLUDING REMARKS

In this chapter, we developed an LP-based ADP algorithm for oil production optimization. We discussed the general algorithm, described how to efficiently construct and select basis functions, and how to compute their coefficients based on SRLP. Results were presented for primary production and water injection problems. Compared with the baseline strategies, the ADP algorithm was shown to perform well.

It is worth noting that the ADP algorithm proposed in this chapter should also be applicable for problems other than oil production. In fact, except for constructing some prespecified candidate basis functions (such as average oil saturation) and the separation of pressure and saturation in POD, the other steps in the algorithm are not specific to the oil production problem. The main characteristics of oil production problems that we exploited in our new ADP algorithm are (1) the availability of a reservoir simulator, which enables model-order reduction and control strategy evaluation and (2) the tractability of the subproblem (25.5) (since it is convex). These two characteristics together ensure an efficient evaluation of the approximate value function $\tilde{J}$, and hence render the adaptive basis function selection and bootstrapping computationally tractable. For optimal control problems in other fields that share these characteristics, we expect that some of the techniques proposed in this chapter, such as basis function construction using model-order reduction and the incorporation of adaptive basis function selection and bootstrapping, or even this ADP algorithm in its entirety, could be applied after suitable modifications.

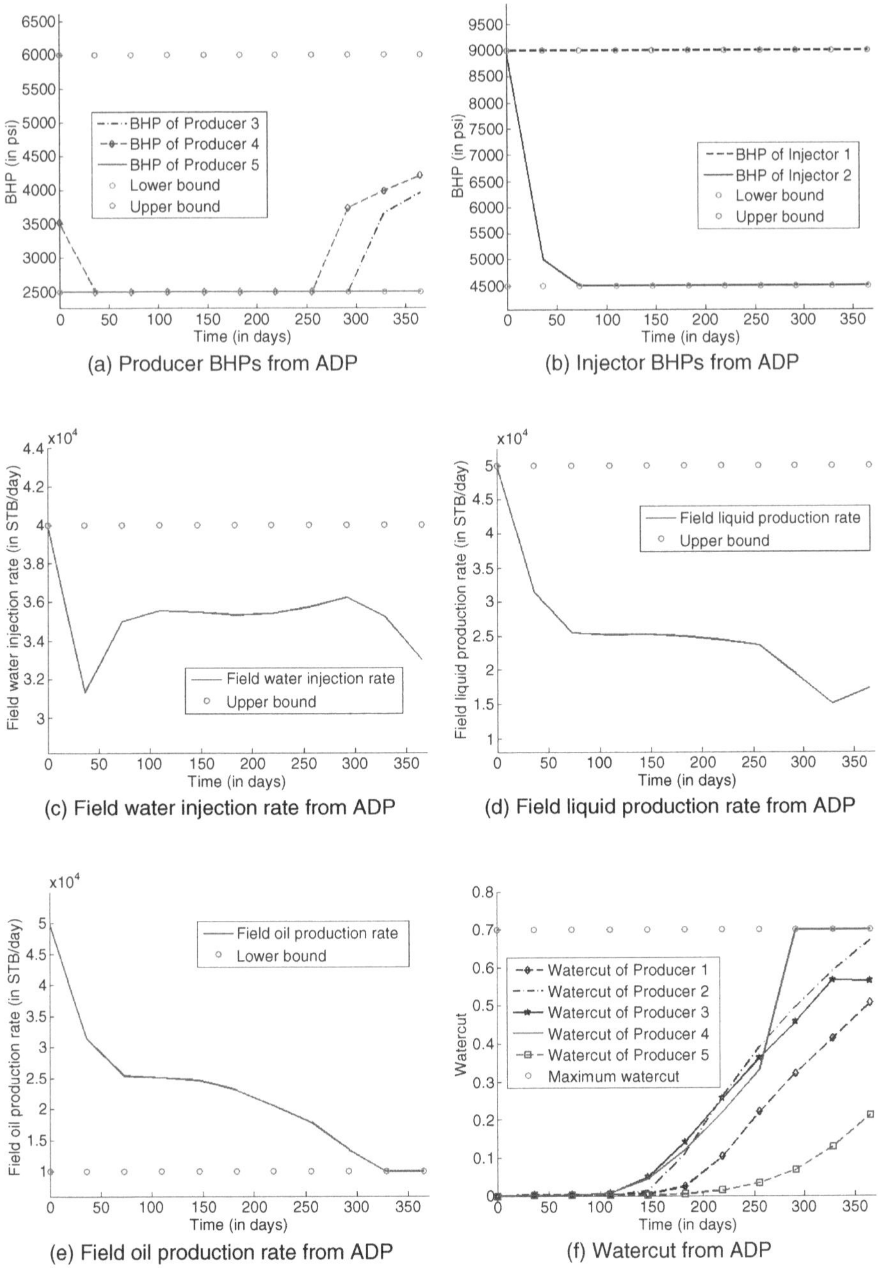

FIGURE 25.6 Simulation results for the near-optimal control achieved by ADP for Case 2.

ACKNOWLEDGMENTS

We are grateful to the industry sponsors of the Stanford Smart Fields Consortium for partial funding of this work. We also thank David Echeverría Ciaurri and Huanquan Pan for useful discussions and suggestions.

REFERENCES

1. D.R. Brouwer and J.D. Jansen. Dynamic optimization of waterflooding with smart wells using optimal control theory. *SPE Journal*, 9(4):391–402, 2004.
2. P. Sarma, L.J. Durlofsky, K. Aziz, and W.H. Chen. Efficient real-time reservoir management using adjoint-based optimal control and model updating. *Computational Geosciences*, 10:3–36, 2006.
3. D. Echeverría Ciaurri, O.J. Isebor, and L.J. Durlofsky. Application of derivative-free methodologies for generally constrained oil production optimization problems. *International Journal of Mathematical Modelling and Numerical Optimization*, 2:134–161, 2011.
4. T.J. Harding, N.J. Radcliffe, and P.R. King. Optimization of production strategies using stochastic search methods. *SPE paper 35518 Presented at the 1996 European 3-D Reservoir Modeling Conference*. Stavanger, Norway.
5. A.S. Cullick, D. Heath, K. Narayanan, J. April, and J. Kelly. Optimizing multiple-field scheduling and production strategy with reduced risk. *SPE paper 84239 Presented at the 2003 SPE Annual Technical Conference and Exhibition*. Denver, Colorado.
6. V. Artus, L.J. Durlofsky, J. Onwunalu, and K. Aziz. Optimization of nonconventional wells under uncertainty using statistical proxies. *Computational Geosciences*, 10:389–404, 2006.
7. D.P. Bertsekas. *Dynamic Programming and Optimal Control*, Volume 1. Athena Scientific, 1995.
8. D.P. Bertsekas and J.N. Tsitsiklis. *Neuro-Dynamic Programming*. Athena Scientific, 1996.
9. B. Van Roy. Neuro-dynamic programming: overview and recent trends. In E. Feinberg and A. Shwartz, editors. *Handbook of Markov Decision Processes: Methods and Applications*. Kluwer, 2001.
10. W.B. Powell. *Approximate Dynamic Programming*. John Wiley and Sons, 2007.
11. J.N. Tsitsiklis and B. Van Roy. Optimal stopping of Markov processes: Hilbert space theory, approximation algorithms, and an application to pricing high-dimensional financial derivatives. *IEEE Transactions on Automatic Control*, 44(10):1840–1851, 1999.
12. F.A. Longstaff and E.S. Schwartz. Valuing American options by simulation: a simple least-square approach. *Review of Financial Studies*, 14(1):113–147, 2001.
13. J.N. Tsitsiklis and B. Van Roy. Regression methods for pricing complex American-style options. *IEEE Transactions on Neural Networks*, 12(4):694–703, 2001.
14. H.P. Simão, A. George, W.B. Powell, T. Gifford, J. Nienow, and J. Day. Approximate dynamic programming captures fleet operations for Schneider National. *Interfaces*, 40(5):342–352, 2010.
15. D. Zhang and D. Adelman. Dynamic bid-prices in revenue management. *Operations Research*, 55(4):647–661, 2007.

16. V.F. Farias and B. Van Roy. An approximate dynamic programming approach to network revenue management. Available at: www.stanford.edu/~bvr/psfiles/adp-rm.pdf. Working paper.
17. D. Zhang and D. Adelman. An approximate dynamic programming approach to network revenue management with customer choice. *Transportation Science*, 43:381–394, 2009.
18. J. Han and B. Van Roy. Control of diffusions via linear programming. In G. Infanger, editor. *Stochastic Programming: The State of the Art, in Honor of George B. Danzig*. Springer, 2011, pp. 329–354.
19. G.J. Tesauro. Temporal difference learning and TD–Gammon. *Communications of the ACM*, 38:58–68, 1995.
20. Z. Wen, L.J. Durlofsky, B. Van Roy, and K. Aziz. Use of approximate dynamic programming for production optimization. *SPE paper 141677 Presented at the 2011 SPE Reservoir Simulation Symposium*. The Woodlands, Texas.
21. K. Aziz and A. Settari. *Petroleum Reservoir Simulation*. Applied Science Publishers, 1979.
22. M.A. Cardoso and L.J. Durlofsky. Use of reduced-order modeling procedures for production optimization. *SPE Journal*, 15(2):426–435, 2010.
23. P.W. Keller, S. Mannor, and D. Precup. Automatic basis function construction for approximate dynamic programming and reinforcement learning. In *Proceedings of the 23rd International Conference on Machine Learning*, 2006.
24. V.V. Desai, V.F. Farias, and C.C. Moallemi. The smoothed approximate linear program. In *Advances in Neural Information Processing Systems 22*. MIT Press, 2009.
25. D.P. de Farias and B. Van Roy. On constraint sampling in linear programming approach to approximate dynamic programming. *Mathematics of Operations Research*, 29(3):462–478, 2004.
26. V.F. Farias and B. Van Roy. Tetris: a study of randomized constraint sampling. In G. Calafiore and F. Dabbene, editors. *Probabilistic and Randomized Methods for Design Under Uncertainty*. Springer-Verlag, 2006.
27. H. Cao. Development of techniques for general purpose simulators. Ph.D. thesis, Stanford University, 2002.
28. Y. Jiang. Techniques for modeling complex reservoirs and advanced wells. Ph.D. thesis, Stanford University, 2007.
29. S. Castro. A probabilistic approach to jointly integrate 3D/4D seismic, production data and geological information for building reservoir models. Ph.D. thesis, Stanford University, 2007.

CHAPTER 26

A Learning Strategy for Source Tracking in Unstructured Environments

TITUS APPEL,[1] RAFAEL FIERRO,[1] BRANDON ROHRER,[2] RON LUMIA,[3] and JOHN WOOD[3]

[1]MARHES Lab, Electrical & Computer Engineering, University of New Mexico, Albuquerque, NM, USA
[2]Sandia National Laboratories, Albuquerque, NM, USA
[3]Mechanical Engineering, University of New Mexico, Albuquerque, NM, USA

ABSTRACT

Reinforcement learning is difficult to use on real robots. Noise and delays can prohibit a system from learning to perform its task correctly or learn the task at all. Control delay—the delay from a commanded action to when the commanded action is actually reached—can cause problems when using Q-Learning. In this chapter, we give a brief overview of Q-Learning and its applications to robotics. Then, we formulate a source tracking problem where a robot is supposed to learn to navigate toward a light source. Next, the light-following problem is implemented in simulation and hardware using ROS, the Robot Operating System. In the simulation and hardware experiments, control delays and noise were encountered which affected the agent's learning process. To overcome the effect of the control delays, we present an approach that modifies the Q-Learning algorithm by using the sensed actions to update the learning rule. This modification improves both the learning and the resulting paths of the light-following robot greatly.

26.1 INTRODUCTION

Ideally, mobile robots should be able to perform navigation tasks autonomously without human intervention. Because of noise and a dynamically changing environment,

Reinforcement Learning and Approximate Dynamic Programming for Feedback Control, First Edition. Edited by Frank L. Lewis and Derong Liu.

this becomes a very difficult task. The navigation tasks are normally implemented with a type of feedback controller that processes the data from the sensors and makes decisions to complete the robot's navigation task. These controllers need to be properly tuned, using gains, to ensure that the controllers are stable and behave optimally with respect to the most direct route to reach a goal. Because the sensors and actuators vary among different robots, these parameters will be different. Additionally, over time, the sensors and actuators of robots can change due to wear and tear. Consequently, the navigation controller parameters need to be manually fine-tuned to ensure optimal performance. For example, robot brush motors may experience wear and their response may change. As a result, the controller may have to be changed to accommodate the change in motor response. The manual tuning of parameters on each robot can be very costly and time consuming when the number of robots increases from a couple of robots to thousands.

Learning, specifically reinforcement learning (RL), can be used to overcome the problem of having to manually fine-tune and update navigation controllers. The robot can learn through reinforcement to perform various navigation problems, thus eliminating the need for the time-consuming task of fine-tuning navigation controllers. It can learn navigation behaviors based on positive and negative rewards from the RL algorithm [1]. In this work, Q-Learning (QL), an RL algorithm, was used to have the robot learn to navigate toward a light source. This problem was implemented both in simulation and on real hardware. The organization of this chapter is as follows: Section 26.2 will give a brief introduction to Q-Learning and its implementation, Section 26.3 will explain how the problem was organized, Sections 26.4 and 26.5 will discuss the results of both the simulation and the implementation in hardware. Then the experimental conclusions and future work will be discussed.

26.2 REINFORCEMENT LEARNING

The field of machine learning is broken up into three categories: supervised, unsupervised, and reinforcement learning. In a few words, reinforcement learning is when an agent learns a behavior by trial-and-error interactions with a dynamic environment and receives a delayed reward. Typically, the RL problem is posed as follows: given a discrete set of states, $\{S\}$, a discrete set of actions, $\{A\}$, and reward signal, R, find the optimal policy, π, that maximizes the future reward. The optimal policy is a function that maps states to actions where the agent completes the task at hand correctly, determined by an expert observer. For example in the light-following task, optimal means that the robot takes the most direct route to navigate to the light source. This is unlike supervised learning where the correct action/output is given based on the states/inputs. In RL, the agent only knows the previous and current states and the reward of how good the previous action was. From this information, the agent tries to learn the mapping of states to actions that maximizes the future reward [1, 2]. In this model, the agent takes action, a_t, at time, t, based on its state, s_t. Then the action produces a new state, s_{t+1}, from the environment, and a reward, r_{t+1}, is observed. According to its RL algorithm, the agent then updates its policy, π, from $s_t,\ s_{t+1},\ and\ r_{t+1}$. This is shown in

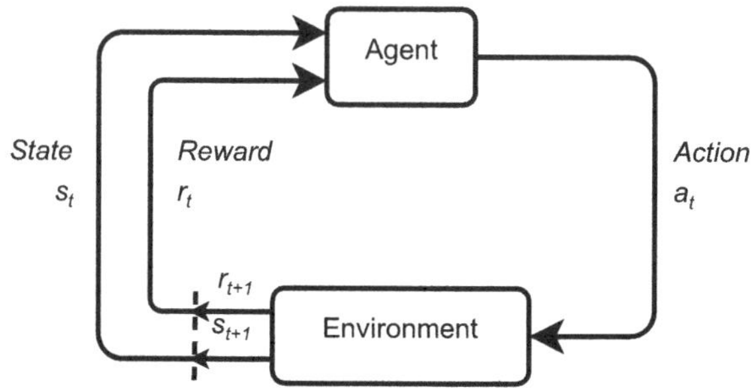

FIGURE 26.1 Reinforcement learning description.

Figure 26.1. To converge toward the optimal control policy quickly, the trade-off between exploration and exploitation needs to be understood. Exploration is when the agent selects random actions while learning, regardless of the current state, to visit every state–action pair. The agent might select an action that it thinks is suboptimal to find out that it might be good. Exploration allows the agent to explore the entire state-action value space to avoid the problem of learning local maxima. For example, in the light-following case, if the agent does not visit a state-action pair while learning, it may learn to take a suboptimal action for a given state, like turning left when the light is to the right of the agent. This would result in the robot taking more time to navigate to the light goal.

In contrast to exploration, exploitation is the use of the agent's knowledge in selecting actions. The agent selects the action that it thinks will produce the maximum reward given the current state [3]. Care needs to be given when choosing the trade-off between exploration and exploitation. Using mostly exploration or exploitation can cause the agent to take a long time to learn the task or not even learn the task at all. Usually when the agent is first learning, exploration is used, and toward the end of the learning process, exploitation is used. The gradual transition from exploration to exploitation is decided by the algorithm developer. One approach to transition between the two strategies is the Boltzmann distribution, which will be described in the next section. For the light source navigation task, reinforcement learning methods were used because they allow the state-action mapping to be learned through reinforcement in real-time.

26.2.1 Q-Learning

Q-Learning is a temporal difference reinforcement learning algorithm developed by Watkins [4]. Temporal difference learning methods use changes in predictions over consecutive time steps in order refine the prediction of a quantity. They continuously update their estimates based on previously learned estimates, also known as bootstrapping. This is done until a terminal state is reached. Temporal difference methods use a combination of Monte Carlo methods and dynamic programming methods [1]. Like Monte Carlo methods, temporal difference methods do not need a model of the

```
Initialize Q(s,a)  arbitrarily
while repetitions count < number repetitions  do
   state = current state
   repeat
      Action = GetAction(State)
      TakeAction, Observe Reward and Resulting State
      Update Q-Tablewith Update Equation
      State = Resulting State
   until State is in Goal State
end while
```

FIGURE 26.2 Q-Learning algorithm.

environment's dynamics. Like dynamic programming, temporal difference methods use bootstrapping.

Q-Learning is an off-policy, temporal difference method; off-policy means that it can separate exploration from control. Q-Learning learns a state-action value function, Q, that directly approximates the optimal state-action value function, Q^*, independent of the policy being followed. Q^* is optimal in the sense that it is the best state-action value function for this problem as deemed by the expert user. This function informs the agent of the action with the greatest reward when the agent is in the current state. The QL states are the inputs to the system and the actions are the outputs from the system. It has been proven that Q converges to Q^* with probability of 1 if all state-action pairs are continuously updated [5]. The QL value function, $Q(s_t, a_t)$, is updated after the agent takes action, a_t, using Equation (26.1) and the general QL algorithm is shown in Figure 26.2.

$$Q(s_t, a_t) \leftarrow Q(s_t, a_t) + \alpha \left[r_{t+1} + \gamma \max_a Q(s_{t+1}, a) - Q(s_t, a_t) \right]. \tag{26.1}$$

Q-Learning uses a finite set of discrete states, $s_i \in S$, where $0 \leq i < N_A$, and actions, $a_i \in A$, , where $0 \leq i < N_S$. N_A and N_S are the number of discrete actions and states, respectively. The value function, referred to as the Q-Table in the rest of the chapter, is in tabular form. $Q(s_t, a_t)$ is the value of Q at state, s_t, and action, a_t, at time, t [1]. The Q-Table is stored in the agent's memory, which is typically either a computer's hard disk or RAM. Memory can be a concern if the task to learn has many state and action pairs resulting in the system running out of memory. However, it is hard to learn tasks with many state-action pairs because the time it takes to learn these tasks is very long.

In the QL algorithm, action, a_t, is chosen based on the policy, π, and the current state, s_t. The policy, π, is chosen based on the exploitation/exploration trade-off discussed earlier. The selection of the policy is important to make sure that the system converges to the optimal solution. There are many approaches to choose a policy and three will be discussed here. The first approach, the greedy policy, is a policy that exploits the value function. Basically, the agent chooses the action based on the maximum Q value over all possible actions for a given state, or $a_t = \max_a Q(s_t, a)$. This policy is usually used toward the end of the training and once the agent is trained. The second approach is a nongreedy policy where the agent chooses random actions at every time step, or $a_t = random(a \in A)$, where the actions have a uniform probability. Using this approach is usually good to apply at the beginning of training. Both of these policies can be used as stated; however, they can theoretically be used for the entire learning process and the Q values should converge to their optimal values regardless of the policy.

One of the more popular policies is the Boltzmann distribution [6] which is used to move from exploration to exploitation. The Boltzmann distribution is shown in Equation (26.2). Equations (26.3) and (26.4) show the limits of the Boltzmann distribution with respect to T.

$$\Pr(a_i) = \frac{e^{Q(s,a_i)/T}}{\sum_{b\in A} e^{Q(s,b)/T}}, \quad a_i \in A, \tag{26.2}$$

$$\lim_{T\to\infty} \Pr(a_i) = \frac{1}{N_A}, \tag{26.3}$$

$$\lim_{T\to 0} \Pr(a_i) = \begin{cases} 1 & \text{if } a_i = \max_a Q(s_t, a), \\ 0 & \text{otherwise.} \end{cases} \tag{26.4}$$

The temperature parameter, T, in the Boltzmann distribution controls the randomness of selecting an action. When T is large, the probability of choosing an action is approximately equal to the probability of choosing any other action, favoring exploration. When T is small, the probability of an action increases with its Q value, favoring exploitation. In practice, the temperature parameter is decreased over time as $T = C\alpha^t$ where C is the temperature at time, $t = 0$, and α is the rate of decay. Using the Boltzmann distribution along with the decay function allows the agent to gradually transition from exploration of the state-action space in the beginning of the learning stage to exploitation of its acquired knowledge towards the end, thereby speeding up the learning procedure. Some examples of how the Boltzmann distribution affects the Q values are shown in Table 26.1. As the parameter T increases, the probability to choose any action is approximately equal. Once the temperature decreases, the probability favors the highest Q value. To use the Boltzmann distribution, first a random number is generated. Then the probability of each action corresponding to the current state, s_t, is calculated using Equation (26.2). Next, the cumulative distribution is generated from the calculated probabilities. The generated

TABLE 26.1 The Probabilities of Taking Actions with Different Temperatures and Q Values.

Q-value	0.2	0.5	1.0
$T = 0.1$	3.33e-04	6.69e-03	9.92e-01
$T = 0.5$	0.128	0.234	0.637
$T = 1$	0.218	0.295	0.486
$T = 5$	0.309	0.328	0.362

random number is compared to the cumulative sum and the action whose probability range contains the random number is chosen as the action to use.

The following description will explain the Q-Learning algorithm in detail. Figure 26.3 will be used in this explanation. First, the Q-Table is initialized. It can be initialized to a constant value, such as zero, or to some random numbers. Then the current state, s_t is observed and an action, a_t, is chosen using any of the policies discussed above. Figure 26.3 shows the use of the Boltzmann distribution. After the action is selected, it is then executed and the agent is delayed for a period of time, t_d. After the delay, the next state, s_{t+1}, and reward, r_{t+1}, are observed. Then the Q value, $Q(s_t, a_t)$, is updated according to the update equation, Equation (26.1). The update equation uses the next state, s_{t+1}, and finds the maximum Q value over all actions for that state. It then calculates the expected discounted reward, which is the reward added to the maximum Q value multiplied by the discount factor, γ. The expected discounted reward tells the agent how good the action that was taken actually was. If it was good, $Q(s_t, a_t)$, will increase, and if it was bad, $Q(s_t, a_t)$ will decrease. The discount factor, γ, adjusts how much future rewards affect the update. The expected total discounted reward can be written as:

$$\begin{aligned} R_t &= r_t + \gamma \max_a Q(s_{t+1}, a), \\ &= r_t + \gamma \left[r_{t+1} + \max_a Q(s_{t+2}, a) \right], \\ &= r_t + \gamma r_{t+1} + \gamma^2 r_{t+2} + \cdots + \gamma^n r_{t+n} + \cdots . \end{aligned} \tag{26.5}$$

The discount factor is on a range of [0, 1]. When it is small, the agent will only weigh current rewards. When γ is large, future rewards are weighted more heavily. Next the expected discounted reward is multiplied by the learning rate, α. The learning rate is a number from [0, 1] and controls how much of the new information is used to update the old information. When $\alpha = 0$, nothing is learned, and when $\alpha = 1$, full weight is given to the new information. The learning rate does not have to be constant while the agent is learning. The learning rate can start at 1 and decrease to 0 so that the agent uses the new information more heavily in the beginning of training and uses less in the end. The learning rate can even be different for each state. If the agent visits a state more frequently, that state's learning rate can decrease faster than others.

After the update step, the next state is set as the current state. Then this algorithm is repeated until the agent enters the terminating state. Then the process is repeated for a number of times or until the agent has learned the task. After the agent learns the task, the task can be accomplished by the agent by using the state-action mapping that was learned. After the task is learned, the agent normally uses a greedy approach by exploiting the learned mapping and does not update the Q-Table. However, the agent can be designed to continuously learn changes in the environment that may occur after the initial training. Q-Learning is a simple, popular reinforcement learning algorithm that trains an agent to learn a task from discrete states and actions and a reward signal.

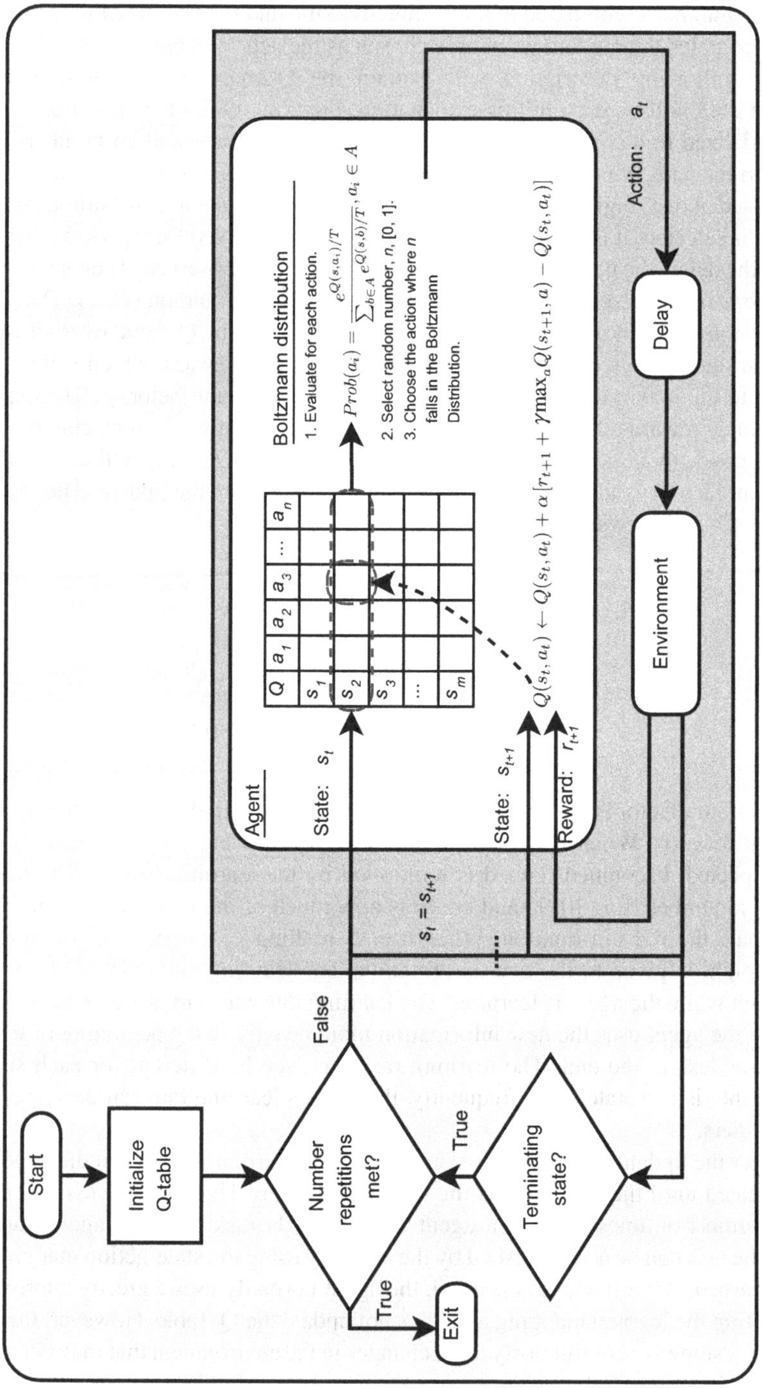

FIGURE 26.3 Q-Learning flowchart.

26.2.2 Q-Learning and Robotics

Q-Learning has been used in several robot hardware experiments on a wide variety of platforms, including RoboCub players [7, 8], Lego Mindstorms NXT [9], and a self-steering automobile [10]. As is commonly reported, implementation on physical hardware posed some significant challenges [9, 11]. Even though navigation and tracking tasks were overwhelmingly popular among robot Q-Learners (as opposed to more complex tasks, such as grasping and manipulation), the robots and their physical environments provided unavoidably noisy learning problems. Such problems can in principle be solved with Q-Learning, but in practice the learning time required can exceed the mean time between failure for most robot actuators.

One way that the problems were kept tractable was to limit the number of sensory channels (inputs) and actions (outputs) available to the agent. In some cases, these were kept coarsely discretized, to minimize the number of states that must be visited [7, 9, 12]. Where inputs and outputs were continuous, function approximators were used [8, 10, 11]. These were multilayer perceptrons that extrapolated values from previously visited state-action combinations to cover the entire state-action space. Another effective strategy was to pair the Q-Learner with either a higher level mediating control law [7] or with lower level sensory and action primitives and heuristics [9, 12]. Using these strategies, these Q-Learners were able to achieve a broad set of learning objectives across a rich sampling of available robot platforms, illustrating the generality of Q-Learning in robotic applications.

26.3 LIGHT-FOLLOWING ROBOT

In order for the robot to learn to navigate towards a light source, the problem needs to be formulated with more details. In the following sections, the robot model, defined actions, environmental states, reward function, and the policy used will be described in more detail.

The Robot Model

A nonholonomic mobile robot is used in the experiment. The nonholonomic constraint means that the robot cannot move side-to-side without first rotating or moving forward. The discrete time model is shown in Figure 26.4. However, this model is not used in the Q-Learning algorithm because it is model free. Figure 26.4 shows the position of the light sensors in the four corners of the robot. There are four light sensors that measure the intensity of the light on the robot. The light sensors measure human perceptible light on the range of 1–1000 lux. The sensors are positioned in a rectangle pattern on the top of the robot. They are mounted at a distance of ± 0.2286 m in the X direction and ± 0.127 m in the Y direction from the center of the robot. The sensors are labeled front-right (FR), front-left (FL), rear-right (RR), and rear-left (RL). The sensor readings are not corrupted by noise in the simulation but are in the real experiment.

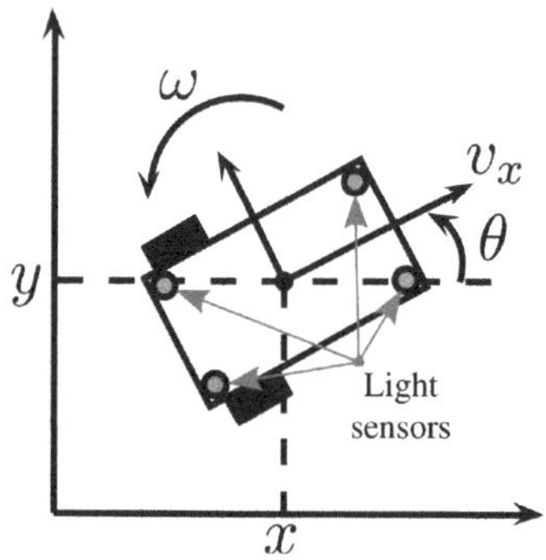

$$\begin{aligned} x_{t+1} &= x_t + v_x \cos(\theta_t)dt, \\ y_{t+1} &= y_t + v_x \sin(\theta_t)dt, \quad (26.6) \\ \theta_{t+1} &= \theta_t + \omega dt. \end{aligned}$$

FIGURE 26.4 The robot kinematic model.

Robot Actions

Since Q-Learning requires discrete actions, three actions were chosen: forward left, forward straight, and forward right. Only forward actions were chosen because it limits the number of state-actions pairs and the learning time. These actions with their associated linear and angular velocities are listed in Table 26.2. The forward velocity is maintained constant in order that the navigation algorithm can be applied to both differential drive and car-like robots. The selected angular velocities are the maximum achievable angular velocities corresponding to the selected linear velocity satisfying $v_x = \omega R$, where R is the minimum radius of curvature of the car-like robot used in the experiment.

Light Sensor States

From the mounted light sensors, the direction of the light is calculated. The magnitudes of the light readings are separated into their X and Y components and then added to compute the direction of the strongest intensity of the light source. Then the direction is discretized into eight states as seen in Figure 26.5. In the figure, if the direction falls within the specified boundaries, then the direction is discretized according to the corresponding area. Therefore, the total number of states is eight and the number of actions is three, resulting in 24 state-action pairs. The small number of states permits the robot to learn quickly.

TABLE 26.2 Chosen Robot Action Velocities

Action	v_x (m/s)	ω (rad/s)
Forward left	0.3	0.54
Forward straight	0.3	0.0
Forward right	0.3	−0.54

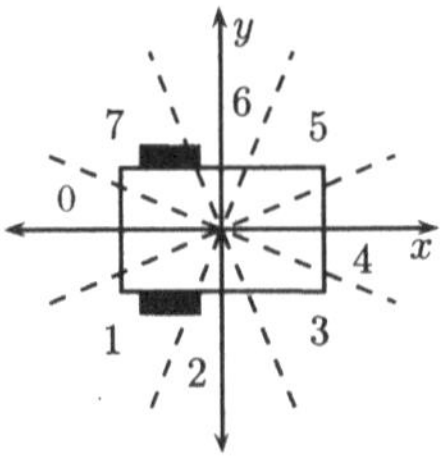

0. Rear center
1. Rear right
2. Center right
3. Front right
4. Front center
5. Front left
6. Center left
7. Rear left

FIGURE 26.5 Discrete light direction states.

Reward Function

The reward function uses the current and next light direction states to calculate the reward. Equations (26.6–26.8) show how the reward function is calculated.

$$s'_t = s_t - 4, \tag{26.6}$$

$$s'_{t+1} = s_{t+1} - 4, \tag{26.7}$$

$$r_{t+1} = \begin{cases} 1 & \text{if } s'_{t+1} = s'_t = 0 \text{ or } |s'_t| - |s'_{t+1}| > 0 \\ 0 & \text{if } s'_{t+1} = s'_t \neq 0 \neq -4 \\ -1 & \text{otherwise.} \end{cases} \tag{26.8}$$

First the current and next states are subtracted by 4. This results in the front center state equal to zero and the rear center state equal to -4 allowing for an easier calculation in the next step. Next the reward is given based on the conditions in Equation (26.8). These conditions give a reward that encourages the robot to navigate to the light source. The reward gives positive reinforcement when the heading error to the light source is reduced and when the light is continuously in front of the robot. Negative reinforcement is given when the heading error to the light source is increased or the light source is continuously behind the robot. A reward of 1 is given when the current and next states are the front center state, meaning that the light source is in front of the robot. This reward is also given to the robot when its direction state gets closer to the front center state, resulting in the robot's heading error decreasing. A reward of zero is observed by the robot when the light direction state does not change and the light is not directly in front of or behind the robot. A reward of -1 is given to the robot when the heading error increases or when the light source is behind the robot. This reward function produces a light following behavior as seen in the following simulation and experiment.

Test Environment

The test environment consisted of a room, safety boundary, starting radius, light source, robot, and light terminating position. Figure 26.6 shows all of these in detail. First of all, the light source was positioned in the center of the test room at $(0, 0)$.

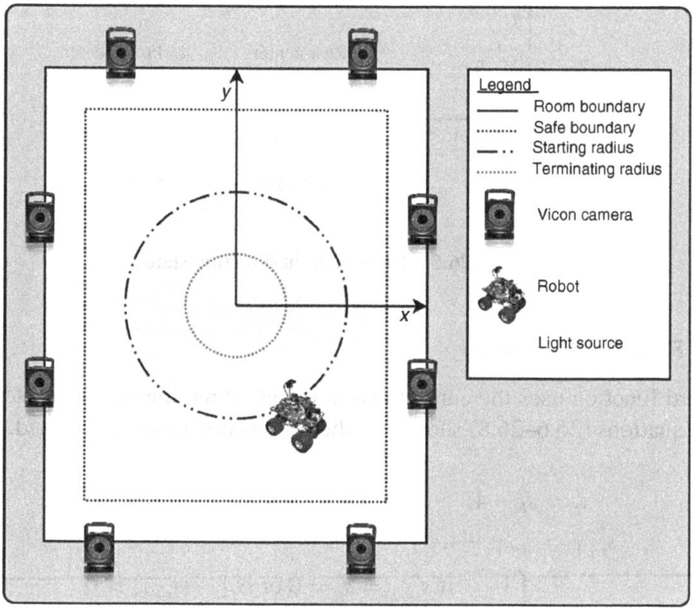

FIGURE 26.6 Learning environment.

At the beginning of each learning repetition, the robot started at a specified radius from the center of the room at a random angle from the positive x-axis of the room. This allowed the robot to start in different locations of the room. At the starting location for each repetition, the robot was also set at a random heading toward the light to generalize the learning. During the learning repetitions, it was likely that the robot would navigate and collide with one of the walls of the room. Therefore, a safety boundary was created to terminate the learning repetition when a collision was eminent. The learning repetition was also terminated after the robot passed through the terminating radius. The terminating radius was a boundary that represented the robot finishing its navigation task towards the light. After the learning repetition was terminated, the robot was moved to a new starting location and heading towards the light and the next learning repetition was started. Both of the terminating states needed global frame positioning information. This was provided through Stage, a robot simulator. In the hardware experiment, the Vicon motion capture system [13] was used. These are described in the Sections 26.4 and 26.5.

26.4 SIMULATION RESULTS

The formulated Q-Learning navigation problem was then simulated and implemented on a real robot. Both the simulation and hardware experiments were implemented in ROS, the Robot Operating System, developed by Willow Garage [14]. ROS is an open-source meta-operating system for use on robots. It provides the hardware abstraction, low-level device control, message passing between processes, and package management that is necessary in robotics. It also provides tools and libraries for building, writing, obtaining, and running code on multiple computers. ROS has simulation capabilities in Stage and Gazebo [15] while having robot hardware and sensor device drivers to make the transition from simulation to real hardware fairly seamless. ROS provides a standard interface for various types of data. For example, there is a standard message for laser scans, allowing different lasers to be used with the same application code. Also laser scan data is the same in the simulation environment. ROS allows algorithms to be tested and then implemented with a few code changes.

The light-following learning algorithm was implemented in the Stage simulation environment. Stage provides a 2D simulated environment with a map, autonomous ground vehicles, and some sensors. In this simulation, the simulated light and the light sensors had to be artificially simulated because Stage did not have light sensors and light objects. The intensity was modeled as a point light source where the intensity of light decreases at a rate of the reciprocal of the distance squared. The intensity of the light at a position, $(x, \ y)$, assuming the light is at $(0, \ 0)$ is shown in Equation (26.9).

$$I(x, y) = \frac{1}{(x^2 + y^2)^2}. \tag{26.9}$$

The map chosen was a 5 by 5 m^2, which allowed for ample room during the learning phase. During the experiment, a sample time of 2 Hz was used to update the Q-Table. This presented the robot with enough change in heading and position to make it possible to learn. The terminating state of the simulation was that the robot was 0.5 m away from the center of the light source or if the robot drove out of the safety boundary. Stage provides global frame position information to check whether one of the terminating states was reached. Once the termination state was reached, the robot was driven to a position at a fixed radius of 1.25 m from the light source at a random angle from the x-axis and a random heading to the light source. For the policy, a Boltzmann distribution was used. The temperature parameter, T, was decreased over time according to $T = C\alpha^t$. In the simulation, C was chosen to be 5 and α was chosen to be 0.65. This allowed for the robot to utilize exploration at the beginning and exploitation towards the end of the learning process. A learning rate of $\alpha = 0.3$ and a discount factor of $\gamma = 0.5$ was chosen in simulation. When starting the learning phase, the Q-Table was initialized with random values from -0.1 to 0.1.

When testing with the simulator, it was noticed that the robot can reach the commanded velocities instantaneously. To make the dynamics of the simulated and real robots similar, a velocity ramp was introduced to the robot in the simulator. This caused some problems when learning. Because a significant control delay was introduced, the robot would update its Q-Table with the wrong data. The action would be commanded; however, when the resulting state and reward were obtained, the action was not yet reached by the robot, resulting in the Q-Table being updated with incorrect data.

To compensate for the control delay, two approaches were used. In these approaches, the Q-Learning algorithm was slightly modified. In the first approach, the

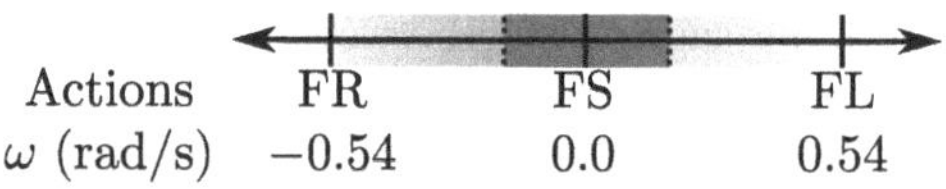

FIGURE 26.7 The sensed angular velocity boundaries.

Q-Table is not updated until after the commanded action is reached. After the velocity action is selected and sent to the robot, the actual velocities are continuously read until they are within a certain threshold of the commanded velocity. Then the state, s_t, is read and the velocities of the selected action are held until the delay period is finished. After the delay period, the Q-Table is updated and the algorithm repeats as normal. By waiting for the velocities of the action to be reached, the robot can use the correct information to learn the task. This approach resulted in the robot learning the correct Q-Table; however, the trajectories were suboptimal because the robot had to wait for the action to be reached which resulted in oscillations.

The second approach was to use the sensed action after each delay interval to update the Q-Table without waiting for the robot to reach the commanded action. In this approach, the selected action is sent to the robot. Then the delay time is waited and the action is sensed. The sensed action is used to update the Q-Table instead of the commanded action. The action is sensed by quantizing the sensed velocity into an appropriate action. This is accomplished by sensing the velocities with the wheel encoders and dividing the angular velocity space of ± 0.54 rad/s by the number of actions, 3. Then the sensed action is selected by comparing the sensed angular velocity with the angular velocity boundaries. The boundaries are shown in Figure 26.7 where each color represents a different action. Finally, the Q-Table is updated with the sensed quantized action, current reward and state. This modification to the Q-Learning algorithm limits the effects of the control delay on learning and provides optimal paths when completing the light following task. This approach to minimize the effects of the control delay is used in the simulation and experiment.

In the simulation, the robot performed 31 repetitions of the task starting at a new position and orientation to the light source each time. In the beginning of the training, the robot both succeeded and failed in navigating towards the light source. During the final repetitions of the simulation, the robot succeeded in navigating towards the light source every time. A couple of the paths of the robot are shown in Figures 26.8 and 26.9. The light is shown as the yellow circle in the center of the map and the starting and ending positions are shown as red circles. It can be seen that the robot did not learn the task completely in Figure 26.8.

Figure 26.9 shows the final path of the robot. The path in this figure is almost the direct path to the light source from the starting location.

At the end of the simulation, the Q-Table was visually checked to see if the values were intuitively correct. For example, if the light was to the right of the robot, one would expect the robot to turn right. Therefore, the action that is expected to be executed for a given state should have the highest Q-value among all the action Q-values for that state. After the simulation results were obtained, the algorithm was validated on hardware through a series of experiments.

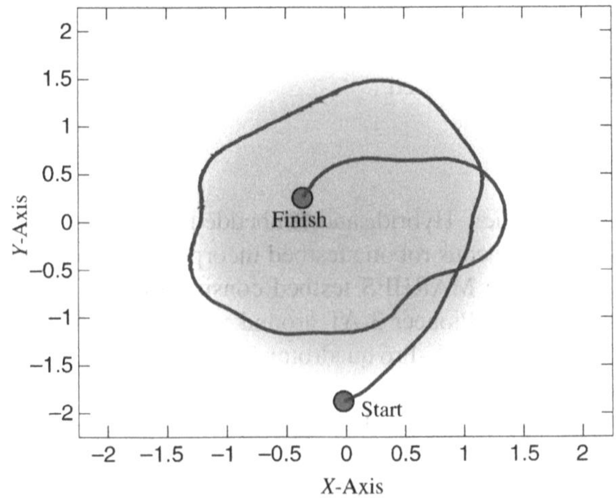

FIGURE 26.8 Simulated robot's path on the 10th learning trial. The path is not the shortest to the goal, because the agent has not learned all of the state-action pairs.

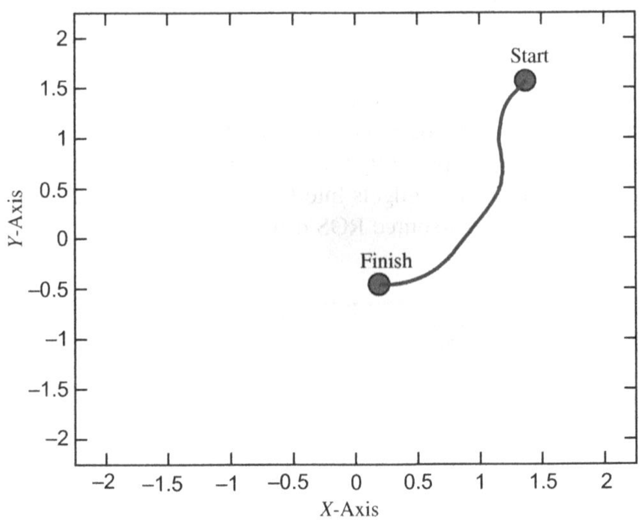

FIGURE 26.9 Simulated robot's path on the 31st learning trial. The path is shorter than the path in Figure 26.8 because the agent has learned all of the state-action pairs.

26.5 EXPERIMENTAL RESULTS

Implementing the learning algorithm in hardware produces more problems than in simulation. Noise was encountered in the light sensors and delays were experienced in the robot following the commanded velocities, both of which debilitated the learning process. However, results were obtained by filtering the light sensor signals and using the learning algorithm from the simulations to avoid the effects of control delays in learning. In this section, the hardware used, problems with the hardware implementation, and results will be discussed.

26.5.1 Hardware

The Multi-Agent, Robotics, Hybrid, and Embedded Systems (MARHES) laboratory testbed is a heterogeneous robotic testbed incorporating both ground and aerial autonomous vehicles. The MARHES testbed consists of 10 TXT-1 off-road vehicles [16], 5 MobileRobots Pioneer 3-AT ground vehicles, 3 AscTec Hummingbird quadrotors, and a Dragonflyer X-Pro quadrotor. These vehicles are used in the design and testing of controllers, multirobot coordination algorithms, sensor networks, and learning algorithms. The testbed hardware that was used will be described in detail in this section.

Figure 26.10 shows the hardware setup used in this experiment. First, the TXT-1 robot was utilized for the experiment. The TXT-1 is a car-like off-road robot developed by the MARHES laboratory for research in learning, smart cars, and cooperative control. This robot uses the Tamiya TXT-1 chassis and is capable of moving at up to 1.5 m/s and can operate indoors or outdoors. It also has a payload of capable sensors and an on-board computer with wireless capabilities. The light sensors used are a part of the Phidgets Interface Kit [17]. The kit allows for eight analog inputs, eight digital inputs, and eight digital outputs to be connected and read or written to via the USB port. Only four of the analog inputs were used with the light sensors. The light sensors measure human perceptible light on the range of 1 lux (moonlight) to 1000 lux (TV Studio Lighting). The Phidgets Interface Kit was connected to the on-board computer of the TXT-1. Open-source ROS drivers were used to interface with both

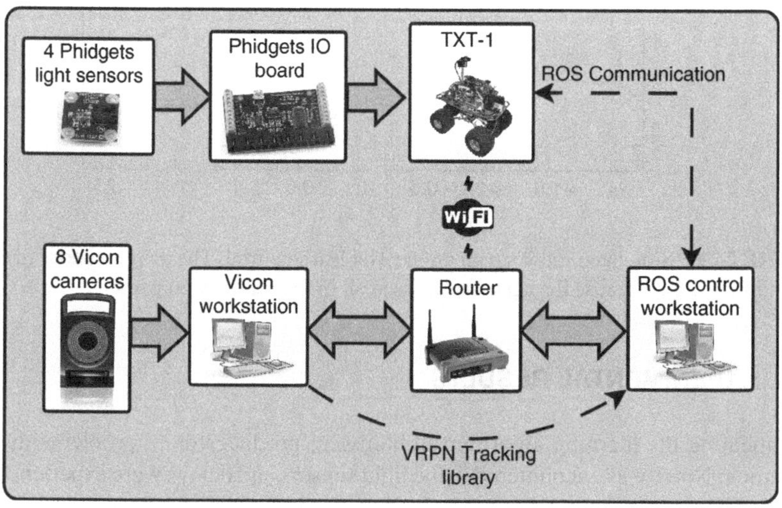

FIGURE 26.10 The experimental hardware setup diagram.

FIGURE 26.11 The robot and sensors.

the TXT-1 robot and the Phidgets interface kit. To get an accurate position of the robot to stay within the safety boundary of the map, the TXT-1's odometry was not used due to wheel slippage and drift. Instead the Vicon motion capture system was used. The Vicon motion capture system uses multiple cameras with reflective spheres to track rigid bodies. To use the system, the reflective spheres were positioned on it in a unique pattern to get accurate positioning information to stop the robot if it breaches the safety boundary. Then a ROS driver was developed using the VRPN library [18] to use the global position data.[1] The light sensors and interface kit were also mounted on the robot with the spheres for the Vicon cameras. The mounting is shown in Figure 26.11. Finally, a 100 W light was mounted in the center of the room on the ceiling. Because of ROS's ability to abstract hardware drivers, most of the same code used for simulation was used in the experiment. The only change was the switching of the simulated light sensor nodes to the hardware light sensors and addition of a Vicon positioning node.

26.5.2 Problems in Hardware Implementation

Noise in the light sensor readings affected the learning in the experiment greatly. For example, the calculated direction of the light source would oscillate around the actual direction by about $\pm 45^\circ$. When learning, the learner would receive the wrong reward a majority of the time, causing it to unlearn the task. Therefore each light sensor reading was filtered with the exponential moving average filter shown in Equation (26.10). I_t'

[1]The VRPN Library was used for this experiment, which was developed by the CISMM project at the University of North Carolina at Chapel Hill, supported by NIH/NCRR and NIH/NIBIB award #2P41EB002025.

is the measured light intensity and I_t the filtered light intensity at the current time. Because the light sensors were sampled every 16 ms, a value of $\alpha = 0.9$ was able to be used to produce an accurate calculated direction without delay from the filter.

$$I_t = \alpha I_{t-1} + (1 - \alpha) I'_t, \tag{26.10}$$

Another problem was the delay in the robot following the velocities of the commanded actions. This would have caused problems with learning; however, the effects of the control delay were minimized by using the Q-Learning modification used in simulation of using the sensed actions for updating the Q-Table.

26.5.3 Results

Despite the problems encountered during the hardware implementation of the Q-Learning light-following algorithm, results were still obtained. Table 26.3 shows the resulting Q-Table after the agent learned for 30 repetitions in hardware. Once learned, the highest value in the row corresponding to the current state is the action that the robot thinks it should take for the state. All the actions with the highest Q value for each state logically correspond to the correct action that the robot should take. Therefore, the table shows that the robot has learned to navigate to the light in hardware. The results of this Q-Table are shown in Figure 26.12. This figure shows the path of the robot navigating toward the light. The resulting path is optimal in the sense that it moves directly toward the light. During learning, it was noticed that the quantization error of the sensed actions resulted in the some of the actions having close Q values. For example, this occurred mostly with the Front Center light state. Table 26.3 shows that the Q values for this state are close compared to the rest of the table. In this state with the quantization error, it was easy for the right and left actions to produce good rewards also. Therefore, care should be taken when quantizing the sensed actions. Despite the noise and delays that were encountered in hardware experiments, the algorithm performed reasonably well.

TABLE 26.3 Q-Table After a Learning on Hardware.

State/action	Front right	Front center	Front left
Rear center	0.9494	−0.6495	−0.5848
Rear right	0.7950	−0.0449	−0.2863
Center right	0.7120	0.1365	−0.3091
Front right	0.4466	0.1255	−0.0286
Front center	0.8803	1.4476	1.1834
Front left	0.3014	0.3008	0.8993
Center left	−0.5585	−0.0546	1.1205
Rear left	−0.3775	0.0077	0.8443

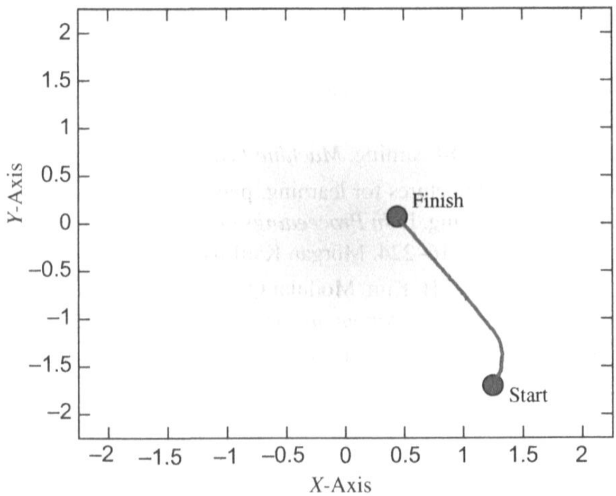

FIGURE 26.12 The resulting path of the robot using hardware.

26.6 CONCLUSIONS AND FUTURE WORK

In conclusion, the Q-Learning algorithm produced encouraging results in simulation and hardware. Using this approach, the robot navigated towards the light, following close to the most direct path. The hardware results were affected by noise in the sensors and delays in the robot response. More work will be done to learn with system delays. Another approach that may be taken is to learn from the raw sensor inputs and use a neural network to learn the actions [19, 20]. A further possibility for improving performance in the noisy system described here is to use a biologically-inspired reinforcement learner [21]. Biological systems deal with inherently noisy sensors and environments as a matter of course and have adapted sophisticated mechanisms for handling the uncertainty.

ACKNOWLEDGMENTS

This work was sponsored by the DOE University Research Program in Robotics (URPR), Grant #DE-FG52-04NA25590, awarded to the UNM Manufacturing Engineering Program, and by NSF Grant ECCS #1027775.

REFERENCES

1. R. Sutton and A. Barto. *Reinforcement Learning: An Introduction*. MIT Press, 1998.
2. L.P. Kaelbling, M. Littman, and A. Moore. Reinforcement learning: a survey. *Journal of Artificial Intelligence Research*, 4:237–285, 1996.

3. P. Dayan and C. Watkins. Reinforcement learning. In *Encyclopedia of Cognitive Science*. MacMillan Press, London, England, 2001.
4. C.J.C.H. Watkins. Learning from delayed rewards. In *International Symposium on Physical Design*, 1989.
5. P. Dayan. Technical note: Q-Learning. *Machine Learning*, 292(3):279–292, 1992.
6. R. Sutton. Integrated architectures for learning, planning, and reacting based on approximating dynamic programming. In *In Proceedings of the Seventh International Conference on Machine Learning*, pp. 216–224. Morgan Kaufmann, 1990.
7. K.-H. Park, Y.-J. Kim, and J.-H. Kim. Modular Q-Learning based multi-agent cooperation for robot soccer. *Robotics and Autonomous Systems*, 35(2):109–122, 2001.
8. M. Riedmiller, T. Gabel, R. Hafner, and S. Lange. Reinforcement learning for robot soccer. *Autonomous Robots*, 27(1):55–73, 2009.
9. C. Kroustis and M. Casey. *Combining heuristics and Q-Learning in an adaptive light seeking robot.* Technical Report CS-08-01, University of Surrey, Department of Computing, 2008.
10. M. Riedmiller, M. Montemerlo, and H. Dahlkamp. Learning to drive a real car in 20 minutes. In *FBIT*, pp. 645–650, 2007.
11. C. Gaskett. Q-Learning for robot control. Ph.D. thesis. The Australian National University, 2002.
12. C. Chen, H.-X. Li, and D. Dong. Hybrid control for robot navigation—a hierarchical Q-Learning algorithm. *Robotics Automation Magazine IEEE*, 15(2):37–47, 2008.
13. V. Tracker. 2011. Available at http://www.vicon.com/products/vicontracker.html.
14. M. Quigley, B. Gerkey, K. Conley, J. Faust, T. Foote, J. Leibs, E. Berger, R. Wheeler, and N. Andrew. Ros: an open-source robot operating system. In *International Conference on Robotics and Automation*, 2009.
15. The Player Project. November 2010. Available at http://playerstage.sourceforge.net/.
16. Titus Appel. The development of a multivehicle testbed with applications in Q-Learning. Master's thesis, University of New Mexico, 2011.
17. Phidgets Inc. - Unique and Easy to use USB Interfaces. Available at: http://www.phidgets.com/index.php, 2010.
18. R. Taylor II, T. Hudson, A. Seeger, H. Weber, J. Juliano, and A. Helser. Vrpn: A device-independent, network-transparent vr peripheral system. In *Proceedings of the ACM Symposium on Virtual Reality Software & Technology 2001, VRST 2001*, November 15–17, 2001.
19. V. Ganapathy, C. Y. Soh, and W.L.D. Lui. *Utilization of Webots and the Khepera II as a Platform for Neural Q-Learning Controllers*, pp. 783–788. IEEE, 2009.
20. B.-Q. Huang, G.-Y. Cao, and M. Guo. Reinforcement learning neural network to the problem of autonomous mobile robot obstacle avoidance. In *Proceedings of 2005 International Conference on Machine Learning and Cybernetics*, pp. 85–89, 2005.
21. B. Rohrer. A developmental agent for learning features, environment models, and general robotics tasks. In *Joint IEEE International Conference on Development and Learning and on Epigenetic Robotics, Frankfurt, Germany*, August 24–27, 2011.

INDEX

Reinforcement Learning and Approximate Dynamic Programming for Feedback Control, First Edition. Edited by Frank L. Lewis and Derong Liu.

IEEE Press Series on
COMPUTATIONAL INTELLIGENCE

Series Editor, **David B. Fogel**

The IEEE Press Series on Computational Intelligence includes books on neural, fuzzy, and evolutionary computation, and related technologies, of interest to the engineering and scientific communities. Computational intelligence focuses on emulating aspects of biological systems to construct software and/or hardware that learns and adapts. Such systems include neural networks, our use of language to convey complex ideas, and the evolutionary process of variation and selection. The series highlights the most-recent and groundbreaking research and development in these areas, as well as the important hybridization of concepts and applications across these areas. The audiences for books in the series include undergraduate and graduate students, practitioners, and researchers in computational intelligence.

Computational Intelligence: The Experts Speak. Edited by David B. Fogel and Charles J. Robinson. 2003. 978-0-471-27454-4

Handbook of Learning and Appropriate Dynamic Programming. Edited by Jennie Si, Andrew G. Barto, Warren B. Powell, and Donald Wunsch II. 2004. 978-0471-66054-X

Computationally Intelligent Hybrid Systems. Edited by Seppo J. Ovaska. 2005. 978-0471-47668-4

Evolutionary Computation: Toward a New Philosophy of Machine Intelligence, Third Edition. David B. Fogel. 2006. 978-0471-66951-7

Emergent Information Technologies and Enabling Policies for Counter-Terrorism. Edited by Robert L. Popp and John Yen. 2006. 978-0471-77615-4

Introduction to Evolvable Hardware: A Practical Guide for Designing Self-Adaptive Systems. Garrison W. Greenwood and Andrew M. Tyrrell. 2007. 978-0471-71977-9

Computational Intelligence in Bioinformatics. Edited by Gary B. Fogel, David W. Corne, and Yi Pan. 2008. 978-0470-10526-9

Computational Intelligence and Feature Selection: Rough and Fuzzy Approaches. Richard Jensen and Qiang Shen. 2008. 978-0470-22975-0

Clustering. Rui Xu and Donald C. Wunsch II. 2009. 978-0470-27680-8

Biometrics: Theory, Methods, and Applications. Edited by: N.V. Boulgouris, Konstantinos N. Plataniotis, and Evangelia Micheli-Tzanakou. 2009. 978-0470-24782-2

Evolving Intelligent Systems: Methodology and Applications. Edited by Plamen Angelov, Dimitar P. Filev, and Nikola Kasabov. 2010. 978-0470-28719-4

Perceptual Computing: Aiding People in Making Subjective Judgments. Jerry Mendel and Dongrui Lui. 2010. 978-0470-47876-9

Reinforcement Learning and Approximate Dynamic Programming for Feedback Control. Edited by Frank L. Lewis and Derong Liu. 2012. 978-1118-10420-0

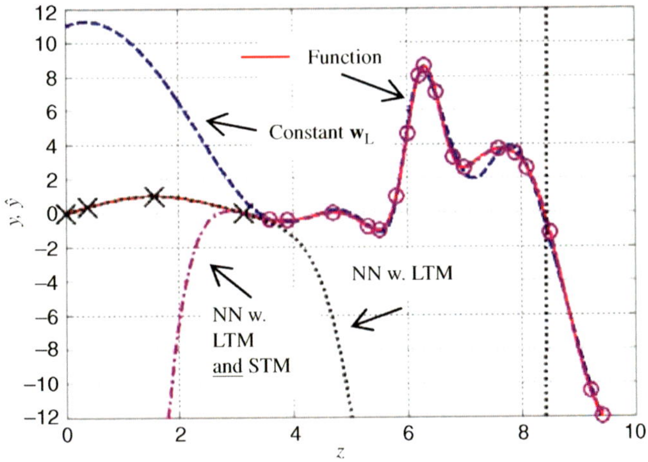

FIGURE 8.1 Function to be approximated incrementally and performance of existing algorithms (taken from [8]).

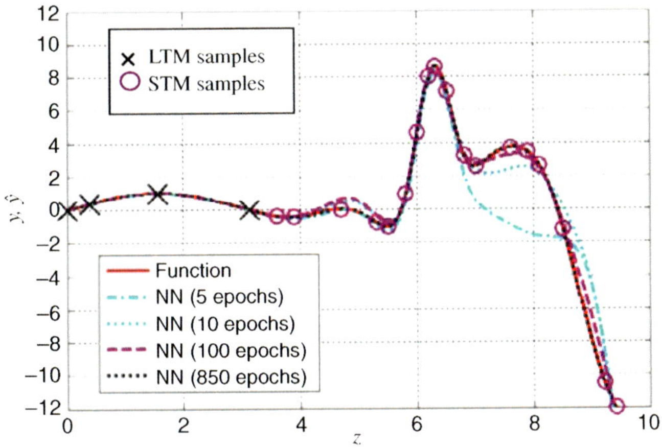

FIGURE 8.2 CPROP preserves LTM accurately at every epoch, until it properly learns the STM at approximately 850 epochs (taken from [8]).

Reinforcement Learning and Approximate Dynamic Programming for Feedback Control, First Edition. Edited by Frank L. Lewis and Derong Liu.

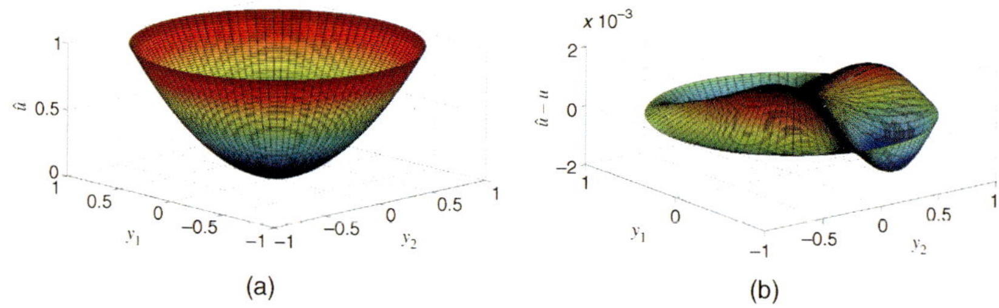

FIGURE 8.3 Neural network solution (a) and error surface (b) when $j = 1$.

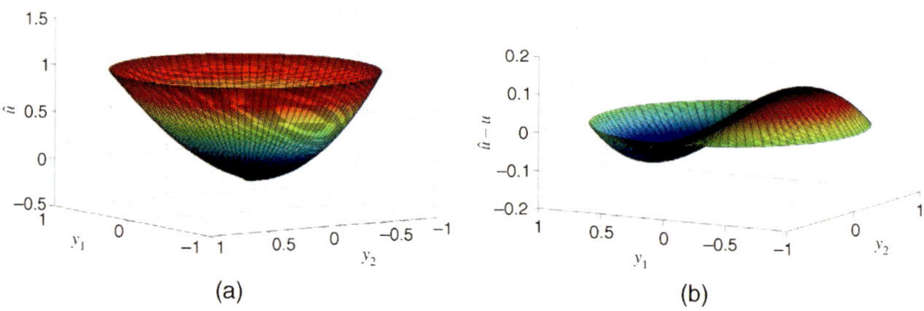

FIGURE 8.4 Neural network solution (a) and error surface (b) when $j = 2$.

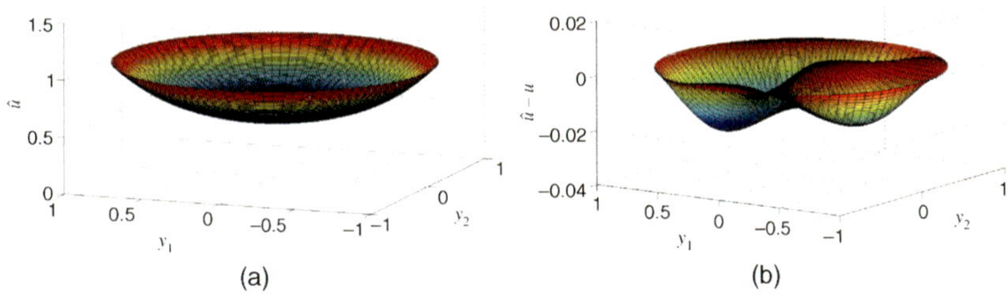

FIGURE 8.5 Neural network solution (a) and error surface (b) when $j = 3$.

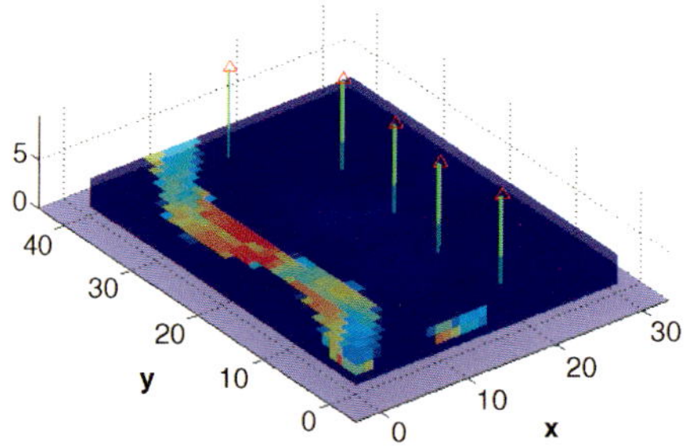

(a) Case 1

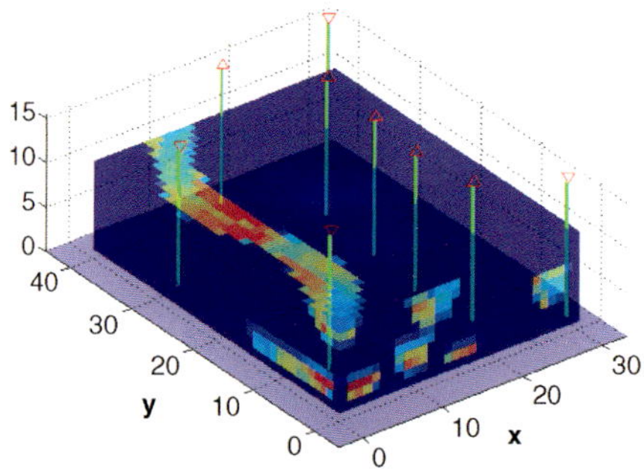

(b) Case 2

FIGURE 25.3 Geological models (x-direction permeability, where blue indicates low and red indicates high permeability) and well configurations for simulation examples. Production wells are indicated by green lines with upward-pointing triangles and injection wells by green lines with downward-pointing triangles.

Printed in the USA/Agawam, MA
September 16, 2026

910661.031